ISBN 978-1-5276-5047-3
PIBN 10877741

This book is a reproduction of an important historical work. Forgotten Books uses state-of-the-art technology to digitally reconstruct the work, preserving the original format whilst repairing imperfections present in the aged copy. In rare cases, an imperfection in the original, such as a blemish or missing page, may be replicated in our edition. We do, however, repair the vast majority of imperfections successfully; any imperfections that remain are intentionally left to preserve the state of such historical works.

BULLETIN

OF THE

UNITED STATES FISH COMMISSION.

VOL. XXII,

FOR

1902.

GEORGE M. BOWERS, Commissioner.

WASHINGTON:
GOVERNMENT PRINTING OFFICE,
1904.

CONTENTS.

ILLUSTRATIONS.

IV

ILLUSTRATIONS

VIII   BULLETIN OF THE UNITED STATES FISH COMMISSION.

PLATE 1.

VIEWS OF THE HERRING INDUSTRY OF YARMOUTH, ENGLAND.

OBSERVATIONS ON THE HERRING FISHERIES OF ENGLAND, SCOTLAND, AND HOLLAND.

By HUGH M. SMITH.

The herring (*Clupea harengus*) has justly been called the "king of fishes." Although its importance is now relatively less than it was several centuries ago, it is to-day a leading fish in the United States, Canada, Newfoundland, England, Scotland, Ireland, Holland, France, Norway, Sweden, and Russia. A species very similar to that of the Atlantic Ocean is found in the North Pacific Ocean, and is caught in large quantities in Japan and Alaska. In point of number of individual fish taken for market, no species exceeds the herring. The annual value of the herring fisheries is about $25,250,000, representing 1,500,000,000 pounds of fish.

In 1900 the writer visited the principal herring-fishing centers of England, Scotland, and Holland. The following notes, based on the observations then made and the information there collected, are presented chiefly because of the large consumption of European herrings in the United States and because of the desirability of applying the foreign fishing and preserving methods to the herring industry on our east and west coasts. No attempt is made herein to furnish a complete account of the herring fisheries of the countries mentioned.

The writer was very courteously assisted in his inquiries by the following persons, to whom special acknowledgments are due: Mr. Charles E. Fryer, London, one of the three Government inspectors having jurisdiction over the fisheries of England and Wales; Mr. W. C. Robertson, Edinburgh, secretary of the fishery board for Scotland; Mr. J. R. Nutman, of Great Yarmouth, England; Mr. James Ingram, Government fishery officer, Aberdeen, Scotland; Mr. E. A. Man, United States consular agent at Schiedam, Holland, and Messrs. C. Van der Burg & Son, Vlaardingen, Holland.

Although the capture of herring is already one of the leading fisheries of the United States, the writer believes that the industry may be increased and the trade made more profitable by the adoption of foreign methods with a view (1) to supply from local fisheries the very large quantities of pickled herring now imported from Europe and Canada, and (2) to open a large trade with southern Europe and other regions.

The following letter from Mr. F. F. Dimick, secretary of the Boston Fish Bureau, dated April 7, 1900, is pertinent to this subject:

The herring imported from Norway, Holland, and Scotland are of a different quality from the herring found on our coast. They are fatter, and great care is taken of the fish when caught and in packing them. The herring caught on our coast of the same size are not so fat. Our fishermen generally find a good demand for their herring at from $1.50 to $3.50 per barrel fresh for bait for the cod and haddock fisherman. The packers generally receive from $3 to $4.50 per barrel for United States shore herring. and there is generally not enough to supply the demand. The foreign herrings are consumed principally by foreigners, and sell at from $8 to $14 per barrel.

The quantities of pickled herring imported into the United States in 1900 from the countries stated were as follows:

Countries.	Smoked.		Pickled.	
	Pounds.	Value.	Pounds.	Value.
United Kingdom	299,322	$12,043	8,960,272	$375,546
Holland	69,123	3,681	12,191,397	674,685

THE ENGLISH AND SCOTCH HERRING FISHERIES.

GENERAL NOTES.

Grimsby and Yarmouth, the important English fishing centers on the shore of the North Sea, are extensively engaged in the herring fishery. At Grimsby the herring fishery is overshadowed by the beam-trawl fishery for bottom fish, but at Yarmouth the herring fishery predominates. The great herring markets of Scotland are Aberdeen, Fraserborough, and Peterhead, also on the North Sea. While these notes are based primarily on visits to Yarmouth and Aberdeen, supplementary information was gathered in London, Edinburgh, and other places. Although the same general methods are adopted in the herring industry in different parts of each country, it must be remembered that the accompanying notes are especially applicable to the particular places mentioned.

The principal fishing season at Yarmouth is from some time in October to the last of December. A few fish are caught earlier in special nets; these are small, well-flavored fish known as "longshore herring," and are for local consumption. There is also a spring fishery, involving most of the month of March and lasting five or six weeks. The fish then caught are small and poor, and are used for bait in the line fisheries.

The Yarmouth herring fishing is carried on with sailing vessels called luggers, and also by steamers. The use of the latter is increasing, 50 to 70 being operated in 1900. The crew of each kind of vessel consists of 11 men. Each vessel carries 200 gill nets, 30 yards long, 20 yards (or 260 meshes) deep, and costing £2 apiece. The number of meshes to a yard is from 28 to 31, the average size of mesh, bar measure, being equal to a shilling. The nets are cotton, machine-made, and, with proper care, may last seven years. They are first tanned with a solution of hot "kutch" or catechu from Burma, which is said to be better for this purpose than tan bark. After drying they are thoroughly soaked in linseed oil, and again dried by spreading on the ground. Oiling is done only once, but soaking in the astringent solution is repeated from time to time.

The entire complement of nets is shot at one time, the nets being tied together. A duplicate set of nets is held in reserve. The water is thick for 15 or 20 miles off Yarmouth, and fishing may be done at any hour of the day or night, but the best times are about sunrise and sunset. The herring do not remain long in the nets unless storms prevent hauling. Some vessels run fish in fresh, others dry-salt their catch at sea. and store it in the hold in compartments. A vessel may leave port, set nets, make a catch, and be back the same day, or it may be out two weeks.

There has never been a failure of the Yarmouth herring fishery; although fish are less abundant some years, they have never been so scarce as to make fishing unprofitable. Some of the Yarmouth fishermen think that the herring frequenting that part of the English coast constitute a distinct body which spawn and remain off that coast, and do not come down from the North. Herring are taken much earlier in Scotland and northern England than here.

When a vessel arrives in port the fish are lifted out of the hold in baskets and spread on deck, where they are counted into baskets by hand, 100 fish to a basket. These baskets are then passed over the rail to the dock and emptied into large, peculiarly shaped baskets holding 500 fish, arranged on the dock in lines or tiers of 20 baskets each. The fish are heaped in 10 piles over the edges of adjoining baskets to facilitate counting. A line of the large baskets constitutes a last, which is the unit of measure in the herring trade. A last represents about 1¼ tons of herring, or, theoretically, 10,000 fish; but, as a matter of fact, 13,200 fish of any size, as 132 fish are called 100 in counting.

Herring are sold at public auction by lasts. The buyer puts his card or tag on the first basket of the tier, and his drayman comes shortly afterwards and takes the fish to the pickling-house or smoke-house. Sometimes, at the height of the fishery, 1,000 lasts (or 3,000,000 pounds) are landed and sold daily in Yarmouth, and the wharves present scenes of great activity and excitement.

The Scotch herring fishery is rather uncertain. In 1900 it was poor on the Scotch coasts, except about the Shetland Islands, where there was a phenomenally large run. A number of years ago, after expensive curing establishments had been built in those islands, the fish disappeared, fishing had to be abandoned, and the packers lost all they had invested.

Different races of herring are recognized as frequenting different parts of the Scotch seaboard. Thus, according to Mr. W. C. Robertson, of the fishery board, the best herring are those taken near Barra and Loch Fyne, on the west coast. These are fine, fat fish, which have brought as much as £6 per barrel.

The different kinds of cured herring to which reference may be made are ordinary pickled fish, kippered herring, bloaters, and red herring.

BLOATERS.

A favorite form of preserved herring for local consumption is the bloater. In the United States this term has come to mean a large, lightly smoked herring, but in Great Britain a herring of any size may be a bloater, which may be defined as a round herring, lightly salted and lightly smoked, and intended for immediate consumption.

The extensive trade in Yarmouth bloaters which formerly existed with London and other cities away from the coast has to a great extent died out, owing to the fact that the smoking is now done at the place of consumption. The fish bear the rail shipment better before smoking than after, so that the bloater trade now consists largely in shipping lightly salted fish to cities where there are smoke-houses. Bloaters remain in good condition for two or three days, but are regarded as being best when smoked and eaten the day after being caught.

The essential steps in the preparation of bloater herring are as follows: Immediately after being caught the fish are dry-salted from 12 to 24 hours if fat, or only 6 hours if lean. They are then smoked for 4 to 16 hours and are ready for consumption.

Yarmouth bloaters bring a good price; sometimes as much as 17s. 6d. is received for 100 fish.

RED HERRING.

A special grade of salted and smoked herring is known to the English and Scotch trade under this name. The fish are destined chiefly for the Italian, Grecian, and general Mediterranean trade, but some are sold in London and other parts of Great Britain. Some of the herring dealers handle only red herring; but, as a rule, the preparation of red herrings is incidental to the packing of other grades.

The fish which are destined to be made into red herrings are often those which have been kept at sea for several days to a fortnight, and hence have become too hard, from prolonged salting, to be made into bloaters, kippers, or regular pickled herring. If they have been salted too long on the vessels they are spitted on sticks and softened by steeping them in fresh water. The special peculiarities of red herring are that they are round, are rather heavily salted, and are smoked for a long time to give them a good rich color.

When intended for export to very warm countries red herring are salted for 30 to 48 hours in strong brine and are then smoked for a fortnight or three weeks. For temperate or cold countries the fish are kept for a shorter time in pickle and are smoked 10 or 11 days. Hard-wood sawdust and hard-wood sticks are considered necessary in producing the smudge and heat required to give to these fish their peculiar flavor.

Red herring for the Mediterranean trade are packed with their heads against the barrel and their tails at the center, in dry-ware barrels holding 500 to 600 fish, half barrels holding 300 to 350 fish, and kegs or third barrels holding 180 to 200 fish. The average gross prices for these packages in recent years have been 10s., 5s. to 6s., and 3½s. to 4s., respectively. The expenses on a barrel for freight and commissions are about 3s., the fish being sent by rail to Liverpool and thence by water to the Mediterranean. They are sometimes packed in tin cases when destined for especially warm countries, and for the London market they are packed in flat boxes holding 50 to 60 fish.

KIPPERED HERRING.

Among the various kinds of prepared herring none ranks higher than kippered herring. The essential characteristic of kippered herring (and of all kippered fish) is that, before being salted and smoked, they are split and eviscerated. Fish intended to be made into kippers should be very fresh when received from the vessels. At Yarmouth large fish are preferred for this method, while at Aberdeen small, fat fish are preferred.

As soon as received they are split down the back from tail to head, eviscerated, and then salted in strong brine of Liverpool salt for 15 to 60 minutes, according to fatness. They are then spread on square sticks by means of hooks, and smoked over a hot fire of hard-wood shavings for 6 to 8 hours (Aberdeen) or 10 to 16 hours (Yarmouth), requiring constant attention. The color imparted to the skin is either golden or light, to suit the markets. After cooling they are packed in boxes holding 4 to 6 dozen, and are sold at good prices throughout the United Kingdom. They have longer life than bloaters, will easily keep for three to five days, and in cool weather a fortnight, but they should be eaten as soon as possible.

WHITE-CURED HERRING.

Under this name are officially recognized the herring brine-salted and packed in barrels and half barrels. Such fish are more extensively prepared than all others combined, and give to the English and Scotch herring trades the importance they have attained.

Various grades of herring are recognized by the salt-herring trade. These grades are sharply distinguished and are usually indicated on the outside of the barrel by a brand. Branding is more generally practiced in Scotland than elsewhere.

The grades of salt herring in England are "mattie," "mat full," "full," and "spent" or "shotten." "Matties" are the smallest herring, 8¾ to 9¼ inches long, with undeveloped reproductive organs; "mat fulls" are fish 9¼ to 10¼ inches long, with the ovaries and spermaries left in; "fulls" are fish 10¼ inches or more in length, with roe or milt; and "spent" fish are at least 10¼ inches long, with eggs or milt discharged.

The grades as recognized in Scotland are "mattie," "mat full," "full," "large full," and "spent," and several other minor grades. The word "mattie" originally meant a maiden herring, "mattie" being the terminology of the east coast and "matje" of the west coast. "Matje" still retains the original meaning, the herring so designated being caught in May and June; all such fish when salted are now sent to Russia. "Mattie," however, represents a small herring, full of either roe or milt, or even spent. The official requirements of the herring of the various grades are as follows: "Mattie," not less than 9 inches long; "mat full," not less than 9¼ inches long, with roe or milt well developed and clearly seen at throat; "full," not less than 10¼ inches long, with roe or milt; "large full," not less than 11¼ inches long, with roe or milt; "spent," not less than 10¼ inches long, without roe or milt.

The lengths of salted herring specified under the different grades apply to the fish after shrinking, and are measured from the end of the snout to the end of the compressed tail fin. Special measuring sticks or gages are employed by the fishery officers.

The continental markets require fish that are gilled and gutted but not split. Herring are gutted through the gill cavity, the heart, liver, and reproductive organs being left in situ, but the greater part of the stomach and intestines being removed.

Gutting is done as soon as the fish are landed, by a crew of three women, two of whom do the gutting while the third first "rouses" the fish (i. e., stirs them by hand in "rousing" tubs of water to remove dirt, blood, etc.) and then packs them in barrels with the proper amount of salt. In gutting, a small knife is inserted through the isthmus and, with the forefinger or thumb, draws out the viscera. The roes and spermaries are always left in, as they are considered food delicacies and, in addition, give the fish a fat or plump appearance. Sometimes the roes are so large that in packing they rupture the abdominal wall. Although excellent fish, they can not, in this condition, be sold as the best grade.

For rousing, Liverpool salt is used; but for packing, coarser Spanish salt is employed, about 100 pounds of salt being required for each barrel of fish.

In packing herring, it is customary to pack 7 barrels with a ton of fish (2,100 pounds), there being 300 pounds in a barrel. Each barrel contains from 850 to 1,100 fish, according to size. In packing, each herring is carefully arranged in a definite position, with the abdomen upward and with the head against the side of the barrel, the fish in a given layer or tier being parallel. The fish in the next layer are arranged in the same way, but their long axis is at right angles to that of the fish in the adjoining layer. The barrels are filled with alternate layers of fish and salt, and then headed. In packing, the fish are compressed vertically and their bellies are flattened, giving them the appearance of being larger and rounder. Laterally compressed fish are not in demand.

During the process of curing, the fish shrink considerably and the barrels have to be refilled. In Scotland the law requires that the barrels rest on their side and be refilled after 11 days. In England, where there is no law, about 8 days are allowed to elapse. A bunghole is bored 13 inches above the bottom, the barrel is placed on end, the head is removed, and the pickle is allowed to run off; then the hole is closed, 2 to 5 tiers of fish of the same catch are placed on top, and the barrel is closed, placed on its side, and the original pickle is returned through the bunghole. No new pickle is introduced, and under no circumstances are the fish washed in water. After branding, the barrel is ready for market. A well-cured and well-packed barrel, after the lapse of 10 full days, should contain no more undissolved salt than would fill a cylindrical tub 9 by 9 inches.

The prices of salt herring vary greatly, depending on the supply. The average price of the best grades is usually about 30s., but it may drop to 20s. or rise to 40s. In 1899 the prices in the German cities of Stettin, Konigsburg, and Danzig, and also in Russia, were the best ever known, "matties" bringing 24 to 34 marks per barrel, "mat fulls" 32 to 36 marks, and "fulls" 36 or more marks. From these gross prices, expenses amounting to about 4½ marks per barrel were deducted. In 1896–97, when there was a large catch in Scotland, the average prices of salt herring in Germany were 13 or 14 marks for "matties," 16 or 17 marks for "mat fulls," and 22 or 23 marks for "fulls."

The authorities and fishermen of Scotland fully appreciate the importance of plainly designating on the barrels the quality of salted herring, and the fishery board has formulated a very complete system of regulations governing branding. In view of the benefits which have accrued to the Scotch herring fishery from the operation of the branding regulations, and because of the importance with which the present writer regards branding as applied to the United States herring trade, the following detailed references to the subject are made.

The official branding of barrels of salted herring is not compulsory, and only about half the packers resort to branding, but it is generally regarded as facilitating the sale of fish. A good judge of herring would be able, from personal inspection, to buy just as good fish without the brand as with it; but in distant markets the brand carries a guaranty. The fee charged by the government for affixing the official brand, certifying to the quality of the fish, is 4d. (10 cents) per barrel. During the years 1898 and 1899 the fees from this source aggregated £11,500, or about $57,500.

The following are the regulations now in force governing the official branding of "white-cured" herring in Scotland. They are presented *in extenso* because of their thoroughness and the model they afford:

FISHERY BOARD FOR SCOTLAND.

Regulations for examining barrels and half barrels intended to be filled, and branding and stenciling barrels and half barrels filled with cured white herrings, for the guidance of fishery officers and the fish-curing trade.

I. Capacity and mode of construction of barrels and half barrels filled or intended to be filled with cured white herrings.

1. *Capacity:* (1) Every barrel shall be capable of containing 26⅔ gallons imperial measure, being equal to 32 gallons English wine measure. (2) Every half barrel shall be capable of containing 13⅓ gallons imperial measure, being equal to 16 gallons English wine measure.

2. *Tightness:* Every barrel and half barrel shall be perfectly tight.

3. *Staves and ends:*

(a) Thickness: The staves and ends of every barrel and half barrel shall, when completed, be not less than one-half part of an inch, and not more than three-fourths parts of an inch, in thickness throughout.

(b) Breadth: (i) The staves of every barrel and half barrel shall not exceed 6 inches in breadth at the bulge. (ii) The head end of every barrel and half barrel must contain not less than three pieces and the bottom end not less than two pieces.

(c) Quality, etc., of staves: The staves of every barrel and half barrel shall be well seasoned and well fired, so as to bring them to a proper round. The staves shall not be cracked, broken, or patched, and there shall not be a double croze. The chime shall not be less than 1 inch in length.

(d) Fitting of ends in crozes: The ends of every barrel and half barrel shall fit properly in the crozes, and shall not be turned inside out, nor bent outwards nor inwards so as to affect the sufficiency of the barrel or half barrel.

4. *Hooping:* (1) Every barrel or half barrel shall be hooped in one of the three following ways, viz: (a) Entirely with wooden hoops: (b) entirely with iron hoops: or (c) partly with wooden hoops and partly with iron hoops.

(a) *Entirely with wooden hoops:*

Every barrel or half barrel hooped entirely with wooden hoops shall be hooped in either of the two following ways, viz: (i) Every barrel and half barrel shall be full-bound at the bottom end and have at least three good hoops on the upper quarter, and every barrel shall have four good hoops and every half barrel three good hoops on the head end; the distance between the nearest hoops on opposite sides of the bulge of every barrel shall not exceed 11 inches after the hoops have been properly driven: the distance for half barrels shall be in like proportion. Or, (ii) every barrel and half barrel shall be quarter-hooped, the barrels with four good hoops on each end and three good hoops on each quarter, and the half barrels with three good hoops on each end and three good hoops on each quarter.

(b) *Entirely with iron hoops:*

(1) Every barrel hooped entirely with iron hoops shall be hooped in either of the two following ways, viz: (i) Every barrel shall be hooped with at least four hoops, one of those to be on each end of the barrel and not to be less than 2 inches wide, of wire gage No. 16, and the other two to be on the quarters of the barrel and not less than 1½ inches wide, of wire gage No. 17, the four hoops to be placed at proper relative distances on the barrel. Or, (ii) every barrel shall be hooped with six hoops, one of these to be on each end of the barrel and not to be less than 1⅝ inches wide, of wire gage No. 16, one to be on each of the quarters and not to be less than 1 inch wide, of wire gage No. 18, and one to be on each side of the bulge and not to be less than 1½ inches wide, of wire gage No. 17, the six hoops to be placed at proper relative distances on the barrel.

(2) Every half barrel hooped entirely with iron hoops shall be hooped with at least four hoops, one of these to be on each end of the half barrel and not to be less than 1⅝ inches wide, of

wire gage No. 17, and the other two to be on the quarters of the half barrel and not less than 1¼ inches wide, of wire gage No. 18, the four hoops to be placed at proper relative distances on the half barrel.

(c) *Partly with wooden hoops and partly with iron hoops:*

(1) Every barrel hooped partly with wooden hoops and partly with iron hoops shall have either (a) the hoop of the head end alone, or (b) the hoops of both ends, made of iron at least 2 inches wide, of wire gage No. 16. (a) If the hoop of the head end alone be of iron, the remaining portion of the barrel shall be bound with wooden hoops in either of the two following ways, viz, the bottom end full bound, with at least three good hoops on the upper quarter, or quarter hooped, with three good hoops on each quarter and four good hoops on the bottom end. (b) If the hoops of both ends be of iron, each of the two quarters shall be bound with at least three good wooden hoops.

(2) Every half barrel hooped partly with wooden hoops and partly with iron hoops shall have either (a) the hoop of the head end alone, or (b) the hoops of both ends, made of iron at least 1¼ inches wide, of wire gage No. 17. (a) If the hoop of the head end alone be of iron, the remaining portion of the half barrel shall be bound with wooden hoops in either of the two following ways, viz, the bottom end full bound with at least three good hoops on the upper quarter, or quarter hooped, with three good hoops on each quarter and three good hoops on the bottom end. (b) If the hoops of both ends be of iron, each of the two quarters shall be bound with at least three good wooden hoops.

II. Marks which curers are required to put, or are prohibited from putting, on barrels and half barrels filled, or meant to be filled, with cured white herrings.

(a) On the outside of the bottom of every barrel and half barrel, at the time when they are given by the curer to the packer to be packed with herrings, there shall be legibly written or marked with red keel or black lead a description of the herrings to be packed, the date of their cure, and the number of the packer; and neither chalk nor any other substance shall be allowed to pass as a substitute for red keel or black lead, and no barrel or half barrel unmarked as here prescribed shall be examined for branding.

(b) When any barrel or half barrel has been emptied of the herrings it contained, the old marks on the bottom shall be obliterated, and the barrel or half barrel, at the time it is given to a packer to be again packed with herrings, legibly marked anew, in red keel or black lead, with the description of herrings it is intended to pack therein, the date of the cure, and the number of the packer.

(c) The curer's name and the name of the port or place of cure shall be branded on the side of all barrels or half barrels presented for the crown brand, and in addition the name of the district may be added thus: to Sandhaven may be added Fraserburgh, and to Boddam Peterhead, and above these impressions there shall be legibly scrieved a description of the herrings contained in the barrels or half barrels, and the date of their cure—the month of cure to be expressed by the first letter thereof, except in the cases of January, April, May, and June, which shall be designated by JA, AP, MA, and JE, respectively; the following being given as examples of scrieving: 12th July 1895, La. Full, L 12 J/95; Full, F 12 J/95; Mat. Full, MF 12 J/95; Spent, S 12 J 95; Mattie, E 12 J 95. On crown-branded barrels of herrings the year need not be branded, as that is given in the scrieving and also in the crown brand, which should be placed in close proximity to the curer's name and the name of the port or place of cure.

(d) No descriptive mark or marks shall be placed on the ends of barrels or half barrels of crown-branded herrings under penalty of removal of the crown brand without return of fees.

III. Heading up of barrels and half barrels after filling them up with cured white herrings.

After filling up, according as the barrel or half barrel has been opened at the head end or the bottom end, it shall be flagged round the head or bottom, made perfectly tight to contain the pickle, and pickled at the bunghole. The bunghole shall be bored within 1¼ inches of the foremost hoop of the left end; and both chime and quarter hoops of each end of every barrel or half barrel shall be properly nailed.

IV. Quality, method of cure, packing, etc., of white herrings necessary to secure the official brand.

Quality: The herrings shall be of good quality.

Gutting, curing, and packing: They shall be gutted with a knife, and cured, and packed in barrels or half barrels within twenty-four hours after being caught.

They shall be well cured and regularly salted, and all fish broken or torn in the belly shall be excluded.

They shall be carefully laid in barrels or half barrels, each tier being completed with head herrings, and the herrings in each successive tier being arranged transversely to those in the next tier underneath, and drawn closely together, care being taken that the heads of the herrings are kept close to the sides of the barrel or half barrel until it is completely filled.

None of them shall be laid in bulk after being cured in barrels or half barrels.

They shall, if intended for the La. Full. Full, or Mat. Full brands, be pined in salt for not less than ten free days; and, if intended for the Spent or Mattie brands, they shall be similarly pined for not less than eight free days, these periods to be exclusive of the day of cure and the day of filling up for branding; and this requirement shall apply to the herrings used in filling up as well as to those in the original packing.

Filling up:

(*a*) The surplus pickle shall be run well off through the bunghole, and the seastick herrings then left in the barrel or half barrel be pressed down by the cooper steadily and uniformly, by daunt or otherwise (use of daunt preferable), thus testing the firmness of the original packing, and whether the surplus pickle has been sufficiently poured off or not. Pickle shall alone be used for the purpose of washing herrings offered for the crown brand.

(*b*) The space left in the head end of the barrel or half barrel shall then be tightly packed with herrings carefully laid, regularly and lightly salted, the barrel or half barrel being firmly filled with herrings round the sides, as well as in the center. The herrings shall be pressed firmly to the sides of the barrel or half barrel with both hands, each tier being completed with head herrings, and the herrings in each successive tier being arranged transversely to those in the next tier underneath, and the weight of the hands being pressed on each tier when finished, care being taken that the heads of the herrings in every tier are kept close to the sides of the barrel or half barrel until it is completely filled.

(*c*) No herrings which have lost their original pickle shall be used in filling up.

V. Conditions on which cured white herrings which have lost their original pickle may secure the official brand.

No herrings which have lost the original pickle shall be accorded the crown brand unless they have been repacked, washed in pickle, and presented separately for inspection, when if found worthy in every other respect they shall, in addition to the crown brand, receive the "Repack" iron across the St. Andrew's cross on the shoulders of the crown, so that it can not be removed without effacing the crown brand. If barrels or half barrels of repacked herrings, instead of being offered separately, are found mixed up with any parcel of bung-packed herrings presented for the brand, the whole parcel shall be rejected.

VI. Reassortment of rejected herrings for the official brand.

When herrings once presented for branding have been rejected by an officer for bad quality, bad cure, bad gutting, or for being mixed with overday's fish (see penalty for presenting overday's fish, etc., on the back of request note, and at the end of these regulations), they can not be reassorted and presented again for branding.

Herrings rejected for bad selection, or for too many undersized herrings for the standard of the iron applied for, may be reselected and presented anew, but they must be pickled with original pickle, when they may be crown-branded, if found otherwise satisfactory, with the "Repack" iron added. The daunt must be used with all repacked herrings.

Early-caught herrings: Herrings caught on the north and east coast of Scotland and on the coasts of Orkney and Shetland before 12th July shall not be crown-branded with the "Mattie" iron, while those caught on the coasts mentioned from 12th to 19th July, both days inclusive, shall have the long gut taken out before being eligible for the "Mattie" brand.

Winter-caught herrings: Winter herrings may, from the 1st November of the ~~one~~
1st April of the succeeding year, be crown-branded, with the word "Winter" branded ~~right~~
the St. Andrew's cross on the shoulders of the crown, so that it can not be removed without
the crown brand.

VII. *Examination of barrels and half barrels in respect of their capacity and mode of constr*

(a) Barrels and half barrels intended to be filled with cured-white herrings: Officer
examine at least four in every hundred barrels or half barrels intended to be filled with he
the capacity of one being tested (if necessary) by liquid measure, and the capacity of the r
ing three by diagonal rod 23 inches long for barrels and 18¼ inches long for half barrels, mea
from the croze of the bottom end to the croze of the head end; the examination to be made
time or times suitable for the officers themselves.

(b) Barrels and half barrels filled with cured white herrings: Officers shall examine all barr
and half barrels filled with herrings, and (if necessary) shall empty the herrings out of at le
one barrel or half barrel in every hundred and test its capacity by liquid measure and test th
capacity of at least three others by callipers.

VIII. *Examination for branding and stenciling barrels and half barrels in respect of the quality, method of cure, packing, etc., of the white herrings they contain.*

(a) The barrels or half barrels presented for branding shall be laid out so that the bottom ends
come at once under the eye of the branding officers.

(b) The curer, or his authorized manager at the place of cure, having delivered to the officer the
proper account of cure of the herrings presented for branding, and a request note containing the
number of barrels and half barrels to be presented, the officer shall see, first, that the number of
barrels and half barrels is correctly stated in the request note; second, that the request note is
signed by the curer or his authorized manager (as the case may be); and third, that the branding
conditions attached thereto are likewise signed by the curer or his authorized manager.

N. B. It shall be understood that no manager can be recognized as an authorized manager
except under authority obtained from the board upon application previously made by the curer.

(c) Brand fees (at the rate of 4*d.* per barrel and 2*d.* per half barrel) corresponding to the
correct number of barrels and half barrels in the request note shall be deposited with the officer
before branding, subject to the condition that if the parcel be not branded the amount of brand
fees so deposited shall be returned to the curer; or, if only a portion of the parcel be rejected, the
brand fees corresponding thereto shall be returned.

(d) The curer's declaration shall then be taken and signed by him.

(e) The minimum number of barrels to be examined per hundred shall be seven. Two barrels
per hundred or smaller parcel shall be examined down through the original, the remaining five
down to the lower quarter hoop of either end.

Officers are not restricted to this scale, but if need be shall open as many more barrels or half
barrels as they may deem requisite to satisfy them that the herrings are fit for branding, for
which they will be held responsible to the board; but they shall understand that in no case whatever shall fewer barrels or half barrels than what is prescribed in the above scale be opened for
examination previous to branding.

(f) The barrels or half barrels selected for examination shall, as a general rule, be opened at
the bottom and head end alternately—that is to say, No. 1 shall be opened at the head end, No. 2
shall be opened at the bottom end, and so on until the whole examination is concluded. The
herrings in all barrels or half barrels opened shall be searched down to the lower quarter hoop
of either end, two barrels per hundred or under as in note (e), and as much farther as may be
deemed necessary.

But where an officer, from any cause, sees reason to examine a larger proportion of barrels or
half barrels at the one end than at the other, he shall be at liberty to substitute the examination
of such larger proportion for the above alternate examination, only observing that not less than
the full proportions per hundred or smaller parcel which are laid down for the minimum scale of
examination shall be examined in all, and that as many more shall be examined as he may see fit.

(g) In examining a parcel the work of different packers shall be selected, as well as herrings
of different dates of cure.

(h) All objectionable herrings shall be removed from the barrels examined before affixing the crown brand.

(i) When, on the first examination of herrings for branding, they are found of bad quality, badly cured, or badly gutted, refusal of the brand shall be final and absolute. When, however, this refusal has been entirely owing to the barrels or half barrels being too slackly packed with herrings, or the filling up badly selected, the case shall be treated exceptionally and shall be remedied by filling up or reselection only of the filling up, and the herrings may thereafter be branded, if in accordance with the following conditions:

(I) They shall be presented for renewed inspection only to the officer who previously rejected them, who shall satisfy himself by full examination that they are, apart from slack packing or bad selection of the filling up, worthy of the brand.

(II) The filling up shall have been properly completed; but, failing this, the herrings shall be finally rejected and no further examination permitted.

(III) The officer shall retain the fees until the herrings are branded or finally rejected; in the latter case returning the fees.

(IV) He shall state upon the request note the particulars of the first refusal; and if the herrings be afterwards branded, the date of branding.

When, however, a parcel is small, and upon the first inspection the deficiency in filling up is seen to be so very trifling that it can be supplied at once in the presence of the officer without difficulty or detention, the above conditions shall not apply, but the filling up may be done upon the spot and the branding proceeded with immediately afterwards, the officer being careful to satisfy himself previous to the branding that the herrings are in all other respects entitled to be branded.

(k) The officers shall see that the barrels opened are filled up and headed with proper care.

(l) The officers shall put a double crown on the bilge of the barrel examined and toward the end examined.

(m) Oversalting shall be determined by the measure known as the cog; and the quantity of salt left in any barrel emptied of fish must not exceed this measure.

IX. *Branding and stenciling barrels and half barrels in respect of the quality, method of cure, packing, etc., of the white herrings they contain.*

Every barrel or half barrel containing white herrings presented to one of the officers for examination shall, if the capacity and mode of construction of the barrel or half barrel, and the quality, cure, selection, packing, etc., of the herrings are, in his opinion, such as to satisfy the requirements of these regulations, (1) have branded in his presence, by means of a hot iron, on the bilge, in close proximity to the curer's name and the name of the port or place of cure, a crown surrounding the word "Scotland," a description of the herrings, viz: La. Full, Full, Mat. Full, Spent, or Mattie (as the case may be), the initial letters of the examining officer's name and the year; and (2) have stenciled in his presence, on the head end, a crown surrounding the same word, description, and letters as those branded on the bilge, with the words "Fishery Board, Crown Brand," stenciled below.

X. *Requirements of the different brands.*

In addition to what are contained in the foregoing regulations, the requirements in respect of the different brands shall be as follows:

Crown "La. Full" Brand.

Barrels or half barrels of herrings for this brand shall contain large full herrings of not less than 11¼ inches in extreme length, as measured by the fishery officer's gage, made for the purpose. Rejections under this brand shall be:

(1) On original packing for spent, torn, broken herrings, or herrings of bad or indifferent quality if more than fifteen; or, on filling up, if more than six.

(2) On original packing if the undersized amount to more than fifteen; or, on filling up, to more than six. And the parcel shall also be rejected if it should appear that the larger herrings suitable for this brand have been previously taken out.

Crown "Full" Brand.

Barrels or half barrels of herrings for this brand shall contain full herrings of not less than 10¼ inches in extreme length, as measured by the fishery officer's gage, made for the purpose. Rejections under this brand shall be:

(1) On original packing for spent, torn, or broken herrings, or herrings of bad or indifferent quality, if more than eighteen; or, on filling up, if more than nine.

(2) On original packing if the undersized amount to more than eighteen; or, on filling up, to more than nine. And the parcel shall also be rejected if it should appear that the larger herrings, suitable for this brand, have been previously taken out.

Crown "Mat. Full" Brand.

Barrels or half barrels of herrings for this brand shall contain full herrings well developed—the roe or milt being clearly seen at neck or throat without pressure—of not less than 9¼ inches in extreme length, as measured by the fishery officer's gage, made for the purpose. Rejections under this brand shall be:

(1) On original packing for spent, torn, or broken herrings, or herrings of bad or indifferent quality, if more than twenty-one; or, on filling up, if more than nine.

(2) On original packing if the undersized amount to more than twenty-one; or, on filling up, to more than nine. And the parcel shall also be rejected if it should appear that the larger herrings suitable for this brand have been previously taken out.

Crown "Spent" Brand.

Barrels or half barrels of herrings for this brand shall contain spent herrings of not less than 10¼ inches in extreme length, as measured by the fishery officer's gage, made for the purpose. Rejections under this brand shall be:

(1) On original packing for torn or broken herrings, or herrings of bad or indifferent quality, if more than eighteen; or, on filling up, if more than nine.

(2) On original packing if undersized amount to more than eighteen; or, on filling up, to more than nine. And the parcel shall also be rejected if the larger herrings have been previously taken out.

Crown "Mattie" Brand.

Barrels or half barrels of herrings for this brand shall contain herrings not eligible for any of the foregoing brands, and of not less than 9 inches in extreme length, as measured by the fishery officer's gage, made for the purpose, but shall not contain headless herrings. Rejections under this brand shall be:

(1) On original packing for torn or broken herrings, or herrings of bad or indifferent quality, if more than thirty; or, on filling up, if more than twelve.

(2) On original packing if undersized amount to more than thirty; or, on filling up, to more than twelve.

"Repack" Brand.

For exportation out of Europe:

1. The herrings for this brand shall have been pined in salt for not less than ten days, exclusive of the day of catch and the day of beginning to repack for branding.

2. They shall be emptied out of each barrel or half barrel in which they were originally cured, and they shall be washed clean.

3. They shall have the crown gut, if adhering to them, removed.

4. They shall be repacked into the barrels or half barrels from which they were emptied and into as many additional barrels or half barrels as may be necessary.

5. They shall be salted sufficiently, and be pickled with strong pickle made of clean salt.

6. Every barrel or half-barrel shall be full bound at the head end as well as at the bottom end, and shall have at each end an iron hoop of 1 inch in breadth.

"Lozenge" Brand.

1. This brand shall be applied to herrings previously branded which have been repacked in the manner required for the "repack" brand, and the lozenge shall be stamped immediately under and close to the crown brand already upon the barrels or half barrels.

2. Upon the additional barrels or half barrels derived from the repacking the "repack" brand shall be affixed, subjoining thereto the lozenge brand as above.

> *Note.*—By the strict letter of the act the curer or proprietor of the herrings ought to give twenty-four hours' notice, in writing, of his intention to repack for this brand, but, of course, where the officer can accomplish his examination of the herrings sooner he should endeavor to do so and accommodate the curer as far as he can.

Under any crown brand, if the officer is satisfied with the cure, quality, etc., of the herrings, but considers them generally too flatly packed, in addition to the crown brand he shall cause the lozenge brand to be affixed to cover the St. Andrew's cross on the top of the crown.

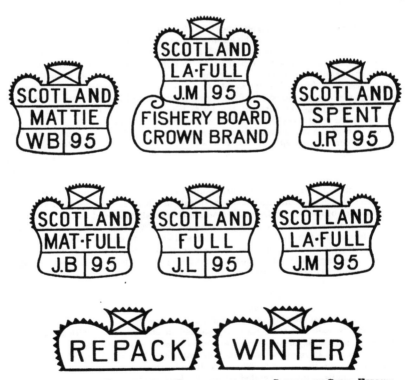

REPRESENTATIONS OF BRANDS USED IN SCOTLAND IN MARKING BARRELS OF CURED HERRING.
About one-half actual size.

THE DUTCH HERRING FISHERY.

Vlaardingen, situated on the Maas, a few miles below Rotterdam, is the center of the Dutch herring trade. There the herring boats fit out, there they land their catch, and there are the houses in which the fish are prepared for shipment.

The herring fishery is conducted by steam and sail vessels, which use tanned cotton gill nets 360 meshes deep and 720 meshes long, the size of mesh being 2-inch stretch. From 80 to 150 nets are carried by each vessel, this outfit usually costing from 5,000 to 7,000 guilders. The nets are set about 6 feet below the surface, being held in position by leads and by corks (5 inches long, 3 inches wide, and 2 inches thick) at intervals of a foot, the cork line being attached to the top of the net by numerous gangings 8 inches long. At times, when the herring are close inshore, some surface fishing is done.

The dressing and salting of the fish immediately after the nets are hauled on the vessels are considered of great importance by the Dutch herring fishermen, and no doubt contribute largely to the quality of the cured fish.

The Dutch method of cleaning herring is similar to the Scotch. Provided with a short knife, attached to the fourth and fifth fingers of the right hand by a string tied to the handle, the fishermen take the herring in the left hand, with the belly up and the head forward, and thrust the knife crosswise directly through the gill cavities, entering the left side and emerging from the right. The edge of the knife being turned upward or outward, the knife is pulled directly through the tissues, cutting and tearing away the gills, branchiostegals, heart, œsophagus, stomach, and often a part of the intestines; the pectoral fins, with the skin and muscle at their base, also come away with the same movement. There is apparently little effort made to remove anything except the gills and pectorals, the other organs coming away incidentally. The men become very expert in cutting, and some of them can handle 1,200 fish per hour (20 per minute).

The removal of the gills and heart results in opening the large blood vessels, and free bleeding ensues; this leaves the flesh pale or white, in contrast with the dark reddish color of the Scotch herring, in which the blood is allowed to clot. It sometimes happens in the Dutch fishery that when there is a large catch the blood has clotted in the last fish handled. The chief and only genuine benefits of cutting as now practiced are the bleeding and the opening of the abdominal cavity to the brine.

Some herring examined by me as brought in by the fishing vessels at Vlaardingen contained pyloric cæca and part of the intestines, as well as the liver and reproductive organs. The intestine, with or without the cæca, often hangs outside the wound made with the dressing knife; it is called the "zeele" (soul), and is frequently eaten by the packers of salt fish, being regarded a choice morsel.

The packing of herring is done while the vessels are still at sea. The fish are first rolled in salt and then carefully packed in straight rows, with backs down. The fish in a given layer are at right angles to those of adjoining layers. One barrel of St. Ubes salt is required to pack four barrels of herring at sea, the salt being disposed between the layers of fish. The barrels are headed up and stored in the hold until the fishing trip is ended or all the barrels filled. On reaching port the catch is unloaded and sold at auction.

The buyer almost always repacks the fish in order to sort them by size and grades of quality, no sorting being attempted on the vessels. Sometimes purchasers or agents prefer the sea-packed, unsorted fish, but as a rule the dealers or jobbers wish to know how many fish are in a barrel and what their quality is. Some shrinkage ensues; this is usually made up with fish of the same lot before the sale, the refilling being done either on shore or on the vessel.

After coming into the hands of the packer the herring are emptied into large vats or tanks, from which they are repacked according to the prevailing practice. The original brine (called "blood brine" or "blood pickle") is considered much better than any newly-made brine, and is always saved and poured back on the fish after repacking. The fish are placed in the barrels in the same manner as at sea, and fresh salt is added in the proportion of 1 barrel to 8 barrels of fish. The shrinkage in repacking is about 8 per cent—that is, 100 barrels of sea-packed herring will make about 92 barrels of fish ready for market.

The Dutch herring barrels contain about 125 kilograms of fish, and most of the catch is marketed in such barrels. Smaller receptacles—$\frac{1}{2}$, $\frac{1}{4}$, $\frac{1}{8}$, $\frac{1}{16}$, and $\frac{1}{32}$ barrels— are also used, but are not nearly so much in demand as they were a few years ago. The barrel staves are oak and are imported from New York, Baltimore, and Newport News in the form of rough pieces, which are cut into proper shape at Vlaardingen, where several thousand coopers are regularly employed. The hoops are made from willow trees grown on the dykes. White, clean barrels are required for the American trade; dark, dirty barrels are accepted by the continental countries. The Scotch herring barrel is regarded as a very good, strong barrel, and is imported by the Dutch packers. The preferred arrangement of the hoops is to have four between the bung and each end as well as four at each end, so that when the barrel is rolled its weight rests on the hoops. The bung is large and central. Some barrels have a single iron hoop at the top.

In Holland there is no official regulation of packing or branding, but the packers have a standard which is generally observed, as it is to their interest to have the fish properly packed and labeled. The different grades of herring recognized are similar to the Scotch, and are based on the spawning condition of the fish. The ripe or full fish are branded "VOL" (=full); the matties (maatjis) are branded "M"; the spent herring (ijlen) are branded "IJ" or "IJLE." Of each of these there are several qualities designated No. 1, No. 2, etc., and there are several other grades. The barrels are usually marked with a stencil.

The Dutch herring trade is not restricted to the fish caught and packed by the Dutch fishermen. Considerable quantities of salt Scotch herring are received at Vlaardingen, which, after repacking, are sold as Dutch herring. Furthermore, the Dutch sell some of their own herrings in Scotch barrels in the continental countries, where the Scotch pack is well known.

SUGGESTIONS IN REGARD TO THE AMERICAN HERRING FISHERY.

The experience of the European herring-packers has resulted in a prepared product, which meets with ready sale throughout the world at better prices than are received for other cured herring. If American herring-curers wish to supply the home markets and to establish a profitable trade with other countries, they must take cognizance of the demands of those markets and make their fish conform thereto. That there is an opportunity for a large increase in both the domestic and foreign trade in American herring there can be no doubt; and the following suggestions are made to this end:

1. While the demand for fresh herring for bait, for smoking, and for canning takes a large part of the catch on certain parts of the New England coast, there are localities where the salting of herring could be made very profitable. Even in the canning and smoking districts it may prove more remunerative to the weir fishermen to salt their large-sized herring. It seems probable that the excellent herring of the Pacific coast can be salted to great advantage and ought to find a ready market.

2. Care must be exercised in all steps of the curing and packing processes. Only plump fish in the best condition should be salted, and only sound fish should be packed. Herring of different grades should not be packed in the same barrel.

3. Different standard qualities of salt herring should be recognized and conscientiously adhered to. The organization of the United States Government would probably not warrant the Federal authorities in exercising jurisdiction in the matter of inspecting and branding fish. While this jurisdiction could doubtless be acquired (as has been done in the case of meats intended for export), it can not be regarded as essential. Each State is competent to superintend the inspection and branding of its own fish, to adopt special regulations and brand marks, and take such other measures as will tend to promote the salt-herring trade. This systematic branding under State authority is regarded as one of the most essential factors in the development of the salt-herring trade.

4. The establishment of a large trade with southern Europe, the Philippines, Australia, and elsewhere in salted river herrings or alewives is entirely feasible. These fish, which are in excellent condition when they ascend the Eastern rivers in untold millions each season, should, if properly prepared, sell almost as well as the sea herring. An especially good opportunity for promoting the alewife fishery appears to exist in the Middle and South Atlantic States, where the catch is only imperfectly utilized and where labor is abundant and cheap. The river herrings might be prepared as white-cured herring and also as red herrings.

HERRING VESSELS AND HERRING-PACKING ESTABLISHMENTS AT VLAARDINGEN, HOLLAND.

JAPANESE OYSTER-CULTURE.

By BASHFORD DEAN,

Adjunct Professor in Zoology, Columbia University.

European oyster-culture, especially as practiced in France and Holland, is generally regarded as the most refined development of this ancient art; and from our knowledge of this—if we admit that those cultural methods are the most perfect which produce the greatest number of oysters in a given area—we can reasonably conclude that some of the devices at least of the European culturist will ultimately come to be adopted in our own system. In view, accordingly, of the prospective value of foreign methods in the development of American oyster-culture, the United States Commission of Fish and Fisheries has already published (in its bulletins[a] for 1890 and 1891) reports upon the practical workings of the best forms of European oyster parks. From the character of the methods there in use, we can, I believe, conclude positively that similar establishments could be operated successfully at suitable points—e. g., in Chesapeake Bay or Long Island waters, as soon, that is, as the demand for oysters will warrant the use of what will prove at first a more expensive system.

While these European methods are applicable on our Atlantic coast, it still remains to be determined whether they include the best that could be employed along the Pacific, should artificial oyster-culture be here attempted. For in these waters different conditions have produced oysters which differ widely from those of the North Atlantic. The Pacific culturist may therefore feel a more lively interest in the oysters of Japan, for not merely are they closer akin to his own, in structure and in habitat, and therefore more readily acclimatable, but they are larger, better shaped, and certainly of greater value, commercially speaking, than the local product, *Ostrea californica*. Moreover, the Japanese oysters have long been cultivated, and with great success. Indeed, by some experts the Japanese methods have been commended as the simplest and most practicable of all forms of "artificial" oyster culture, and thus of possible interest in somewhat broader lines.

Unfortunately there is no literature accessible dealing in detail with the culture or living conditions of this western Pacific oyster, and it is with the aim of filling this gap that the present report was prepared. Its material was collected by the writer during a stay in Japan in 1900–1901. He there acted under special instructions from Commissioner George M. Bowers, and in aid of his inquiries was designated as a biologist of the United States Commission of Fish and Fisheries.

[a] Report on the Present Methods of Oyster-culture in France. Bulletin U. S. Commission of Fish and Fisheries, 1890, pages 363-398, plates LXVIII-LXXVIII.
Report on the European Methods of Oyster-culture. Op. cit. for 1891. pages 357-406, plates LXXV-LXXXVIII. (Italy, Spain, Portugal, Germany, Holland, Belgium, and England.)

17

For the rest, Japanese oyster-culture proved to be worthy of careful study, not only for its merits, but because of the suggestions it affords for cultural experiments. One may frankly doubt whether it can *at once* be employed profitably—for example, at many points on the Oregon coast—in view of the expense for labor which it entails, but I believe that there is a reasonable chance that it could be made profitable if employed in a favorable locality. In any event, so far as the Pacific coast is concerned, the Japanese methods are the most practicable, and experiments with them could be made readily and at little expense, and would soon demonstrate whether artificial oyster-culture can here be employed commercially.

The Japanese industry is largely seated along the north shore of the Inland Sea near Hiroshima, in the gulf-like Sea of Aki, famous for its oysters. From what period, indeed, this oyster-culture has been carried on is not known accurately, but from its present condition it is evidently the product of centuries. As early as 1708 records show that concessions at Osaka were granted to an oyster company or to oyster companies of Aki for storage of their output pending the final marketing.[a]

Regarding the origin of the oyster-culture in this region I may here quote a paragraph from a tract on fishery matters published by Hiroshima-Ken.

In ancient times certain shellfish, *Tapes*, were gathered in great numbers on the flats of Aki; and while awaiting their shipment to market the fisher people came to keep them in shallow-water inclosures, the fences of which they formed of bamboo stalks. The discovery was then made that the brushy fences became incrusted with young oysters, and thus it soon became evident that under certain conditions and at certain places it would be more profitable to plant bamboo and to cultivate oysters than to continue the Tapes industry. This was the first instance, it is said, that bamboo collectors, or "shibi," were employed in oyster-culture.

The first detailed report upon the oyster industry of Aki was prepared for the Japanese government by Prof. Kakichi Mitsukuri, the head of the department of zoology of the Imperial University of Tokyo. It was published in 1894 (Tokyo) by the department of agriculture and commerce, a royal octavo of about 50 pages, containing many figures and several plates.[b] Unfortunately for the foreign reader it has not yet been translated into a European language. My own knowledge of it is due to my friend (who has also kindly drawn for me the text figures here reproduced), Mr. Naohidé Yatsu, Rigakushi, a pupil of Dr. Mitsukuri. I have availed myself freely of its substance, and if there is anything of value in the present paper it owes its merit to my Japanese colleague. He has, moreover, given me generous assistance in connection with my visit to Aki, advising me as to ways and means and providing me with personal and official letters to the local authorities.

In Hiroshima I was courteously received by his honor Governor Senshi Egi, of the prefecture of Aki, and to him and to his staff I am indebted for favors extended me in many ways. To Mr. Shinobu Suda, official engineer of the oyster properties, and to Mr. Masugi Shiraishi, a government expert in oyster matters, I am under especial obligation for information regarding details in cultural methods; and finally to Mr. Y. S. Murai, for many personal courtesies.

[a] A probable trace of this early privilege is seen at the present day, for the people of Aki are granted the best places in the river mouths of Osaka.

[b] All the present text figures, except fig. 27, are reproduced with unimportant changes from the Japanese report. Plates 4 and 5 are from photographs taken by an artist in Hiroshima, and are copied by Dr. Mitsukuri; Plates 3 and 6 are original.

THE JAPANESE OYSTER—ITS KINDS AND NATIVE CONDITIONS.

Three kinds of Japanese oysters are to be considered in this connection. First, a small one, probably a dwarfed salt-water variety of *Ostrea cucullata* Born., is abundant along the southern and eastern coasts of Japan—by far the most abundant of its kind. This is a bay oyster, occurring in shallow water of specific gravity of about 1.020 to 1.026, where it forms an almost barnacle-like incrustation upon the tidal rocks. It is collected in great numbers for local consumption; fisher people open them on the spot, not detaching the shell from the rock, and market them by bulk. In actual size this oyster is rarely larger than one's finger nail; but its flavor makes amends for its size. It is plumbeous in color. The shell measures rarely more than 2 inches in length; it is deeply crenulate, Gryphæa-like, at the margin. In size and flavor it suggests very closely the California oyster.

The second form, *Ostrea cucullata* (*cf.* pl. 6, fig. D), is the important one from the culturists' standpoint. Although not large, it averages the size of a "Blue Point," or of an English "native." The oyster itself is cream-colored, its shell delicately nacreous, well shaped, thin, deep, and with a series of imbricating, horn-like outgrowths, which suggest the shell of the European oyster, *O. edulis*. This species occurs abundantly throughout the Inland Sea, in the small bays along the northeast coast of the main island and at certain points in the Hokkaido (Yezo). It thrives best in the bays well tempered by fresh water, of specific gravity of 1.017 to 1.023. Its young are more abundant in the shallow and fresher water. The best that are marketed grow at a depth of a fathom or two below low-water mark; it is practically absent in water deeper than 8 fathoms. It is this species which will be considered through the remainder of the present report in connection with cultural methods.

The third form, *Ostrea gigas* Thunb., is of large size, specimens weighing with shell 4 or 5 pounds being not infrequent. It rarely occurs in water less than 2 fathoms deep and is most abundant in about 10 fathoms. The specimens which I examined were taken by divers in water of about 35 feet. It is a typical sea oyster, occurring in water of specific gravity of about 1.026. As far as I have been able to ascertain, its value is purely local, no region producing sufficient numbers to warrant a definite fishery. A large bank of oysters occurs in the Hokkaido, off the northeast coast, not far from the town of Akkeshi. The oysters here are said to be of extraordinary size, but during my visit to the Hokkaido I was not able to ascertain whether they represent this third species or whether they are large examples of *O. cucullata*. The latter species certainly occurs in the neighborhood.

The oyster-producing region of Japan is *par excellence* the Inland Sea, and it is here that the culturists have carried on their industry with greatest success. This body of water can indeed be looked upon as one of the most important natural preserves of fish and shellfish in the world. It can be compared to a deep marine lake, but it is sufficiently open to the sea to insure favorable conditions of density and of renewal of its waters, while its occupants are free from the dangers of an open gulf. From the oyster-culturists' standpoint the Inland Sea is remarkable in that its connection with the ocean is established both at its ends and near its middle point. Thus at the extreme east it opens to the ocean through the Straits of Naruto, as well as at the mouth of the Izuminada. At the west, 240 miles away, it opens again, this time to the Japan Sea, through the Straits of Shimonoseki, and to the south again to the Pacific through Bungo Channel. In this middle region the large island Shikoku approaches closely the mainland, and the Inland Sea is broken up by a maze of islands extending from Shodoshima on the east to Iwai on the west, a stretch of 130 miles; and it is here that the most favorable conditions exist for the

FIG. 1.—Map of the region of Japanese oyster-culture on the north shore of the Inland Sea near Hiroshima (Sea of Aki). Oyster parks are indicated in black areas. Particular reference in the present report to Kaida Bay, Kusatsu, and Nihojima. The distance from Kaida to the island Itsukushima is 12 miles.

growth of shellfish. Everywhere are bays and harbors, and the density of the shallow water is favorably tempered by the incoming streams of Aki, Bingo, Bitchu and Bizen on the north and Iyo and Sanuki on the south. Add to these advantages that there is a favorable tide fall of from 10 to 15 feet and an abundance of sandy and gravelly bottom, thus enabling the culturist to operate his submerged farms conveniently. Throughout this entire region oyster-culture is carried on more or less generally, but the most important seats of the industry are at Okayama in the east and near Hiroshima (prefecture of Aki) in the west. In the former locality a small nearly-inclosed bay, which suggests that of Arcachon in France, proves very productive and supplies no little part of the market of Kobe and Osaka. Here also are canning factories. Near Hiroshima, however, the industry is conducted on a somewhat larger scale, although on the same general lines.

It is in the latter region, as already noted, that Professor Mitsukuri secured the material for his report upon Japanese oyster-culture.

FIG. 2.—Fisherwoman opening oysters. The drawing shows the block and opener, oyster basket and trays.

OYSTER-CULTURAL METHODS IN THE NEIGHBORHOOD OF HIROSHIMA.

In addition to its natural advantages the region of Aki is especially favorable for oyster-culture, since close by is Hiroshima, a city of nearly 100,000, to furnish a ready market for its product and to provide the necessary capital and labor for the growth of the industry. The culture is carried on in a gulf-like area, known as the Sea of Aki, a dozen miles in width, whose mouth is protected from southern storms by the islands of Itsukushima (Miyajima), Nomijima, and Etajima (fig. 1). The most favorable points for the cultural work are along the northern coast on either side of the Otagawa, which flows through the city and tempers the salinity of the neighboring water. On the east of Hiroshima are the establishments of Nihojima, and somewhat further up the bay, Kaida; to the west, the grounds of and near Kusatsu. The entire extent of these most favorable points is about 10 miles. This cultural area, it may be noted, corresponds closely to that of Tarente in the south of Italy, of Arcachon in France, and of the best part of Long Island Sound.

It is an interesting fact that the culturists in Aki have at certain points developed independently branches of the industry which are strikingly similar to those employed, for example, in France. We thus find that a clear distinction is made between the regions in which young oysters—"spat"—can be obtained and those having the best

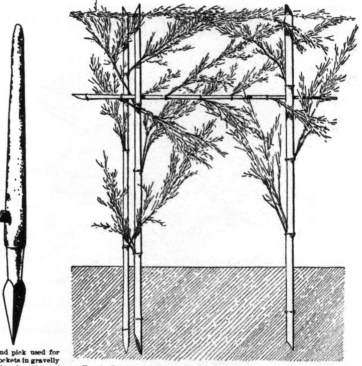

FIG. 3.—Hand pick used for making sockets in gravelly bottom for insertion of shibi.

FIG. 4.—Bamboo collectors, shibi, arranged so as to form a boundary hedge of an oyster farm.

conditions for different stages of growth; also regions in which the final touches are given in preparing the oyster for market. It is convenient, therefore, to describe the Japanese cultural methods from the standpoint of locality. Thus at Nihojima, where the water is freshened by the entrance of the Otagawa, the "production" of young oysters is an especial feature of the industry. At Kaida Bay there is a region favorable for a combination of production and growth (élevage), and at Kusatsu, and further along in the direction of Miyajima, are the best conditions for élevage.

In the description of the methods employed in these three localities, it will be best to consider them in the order from the simplest and the most complex, (1) Kaida, (2) Kusatsu, (3) Nihojima, for at Kaida Bay all stages of the industry are represented in the same oyster park; at Kusatsu the methods become more complicated, and finally, in the region of Nihojima, specialization in the cultural devices has reached a point surpassed in but few European localities.

FIG. 5.—Arrangement of branched collectors as a close-set fence to form one of the lines of an oyster park.

CULTURAL METHODS OF KAIDA BAY.

In this well-protected region (cf. the accompanying map of the sea of Aki, fig. 1) there is a large area of shallow water, and at low tide great flats are exposed. Here it has been ascertained that the conditions of water density are favorable rather for the growth of young oysters than for the production of "spat," but at certain points production is carried on with marked success. The greatest disadvantage of the region is the lack of space in which the oysters can be kept covered by water during all times of the month. Probably it is for this reason that the growth of the oyster comes to be checked, since it is well known that they rarely increase in size after the end of the second

FIG. 6.—Arrangement of branched collectors in close-set hedge, common in most types of Japanese oyster farms. Vertical projection.

FIG. 7.—Shibi of different rows, new and old, in boundary hedge, showing how they are implanted to give mutual support.

year. At this age, then, they are marketed, their small size distinguishing them from the oysters of Aki cultivated in other localities. The bay of Kaida is, however, so fertile in its class of production that it takes a high place among Japanese oyster-

grounds and its concessions are keenly sought.   At low tide it bristles with closely
set oyster farms and from a distance reminds one, save in color, of a region of
European vineyards.   Each farm is a simple inclosure formed by "shibi" or bamboo
stalks, with or without interlacing branches.   (*Cf.* figs. 4 to 9.)   Bamboo in this, as
in many other arts and trades of the Japanese, possesses many advantages.   It is
durable even in salt water (good material lasting three years or thereabouts); it is

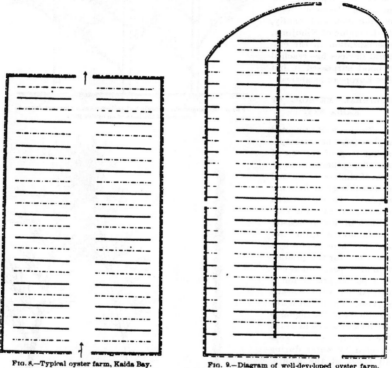

FIG. 8.—Typical oyster farm, Kaida Bay.   FIG. 9.—Diagram of well-developed oyster farm.
The black lines in figs. 8 and 9 represent newly arranged bamboo collectors, the dotted lines
the collectors of the second year.   Direction of current is indicated by arrows.

light and strong, gives an interlacing series of branches and leaves which in texture
serve admirably for the attachment of spat, and which give, moreover, a great extent
of attachable surface.   In addition the bamboo stalks can be readily put in place and
removed; they are easy to obtain in any locality, and their cheapness is not one of
their smallest virtues.

In the present farms shibi are planted in position every spring at a time which the

PLATE 3.

culturists have determined by experience to be most favorable for the set of spat, usually about the middle of April. Their arrangement and mode of replacement are as follows: Shibi are brought to the oyster-grounds in skiffs and deposited on the flats as they become exposed by the receding tide. The culturist will have had the boundaries of his concession staked out, and he has but to construct his fences of shibi as quickly as possible to take advantage of the hours of low tide. As a time-saving device, he has already had the ends of the shibi sharpened so that they can be thrust deeply (one-quarter or one-fifth their entire length) into the soft bottom. Should the bottom prove hard, however, holes are first made for the shibi by means

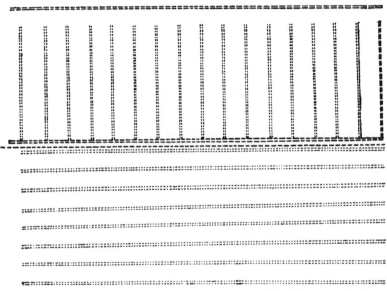

Fig. 10   Diagram of an oyster farm in which are combined the longitudinal and transverse modes of arranging the bamboo collectors. Kusatsu.

of an iron-shod pick, shown in fig. 3. This the workman sometimes presses down with his foot, hence the lateral support near the head of the implement.

According to the usual type of oyster farms in Kaida, the main boundaries are planted nearly parallel to the tide marks. Similar rows of shibi are next thrown out in the direction of the middle line of the park. (Figs. 8, 9.) They do not, however, meet, but leave an open median space passing through like an aisle. Thus on either side of the main aisle of the oyster park there are rows of transverse alleyways, each about 6 feet wide, which terminate often blindly at the main fence of the park. Details in arrangement are given in figs. 4, 5, 6, and 7, fig. 6 showing a horizontal projection of the shibi of fig. 5. The fences and partitions of shibi stand about waist high.

Such an inclosure in Kaida Bay is practically the whole stock in trade of the culturist, for in it spat is taken and the oysters grow attached to the bamboo until the time when they are sent to market. From the permanent character of the park, therefore, the culturist has been led to ingenious arrangements by which the shape of the park can be retained and the oysters grown, young and old, side by side. To accomplish this each fence is usually formed of a double row of shibi, as indicated in

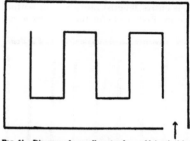

figs. 8, 9, of which one row is of the second year and the other a new one. The latter is often added as the older row is removed. The fences in this locality are made of the more delicate species of bamboo, "hachiku." The stalks are trimmed in lengths of about 5 feet and are inserted a foot deep in the bottom. The distance between the double rows of shibi is intentionally narrow, so that their opposing branches can interlace. (Fig. 7.)

The foregoing description refers to the commoner type of park in Kaida. Occasionally some are seen of a more complicated pattern, as in fig. 9, and,

Fig. 11.—Diagram of a small oyster farm of labyrinthine pattern. Ondo.

rarely, some in which the double rows of shibi are placed parallel instead of transverse to the main axis of the park, as in fig. 10. An excellent example of this type is pictured in plate 3, from a photograph which was taken at the end of a season, the shibi showing well-grown oysters (from park at Tanna). Rarer still is a form in which a labyrinth-like arrangement of the hedges of shibi prevails (fig. 11), or even a concentric pattern (fig. 12). In all these forms, however, the arrangement is such that many eddies will be formed about and within the rows of the shibi, since these eddies have been found conducive to the attachment of the young oyster.

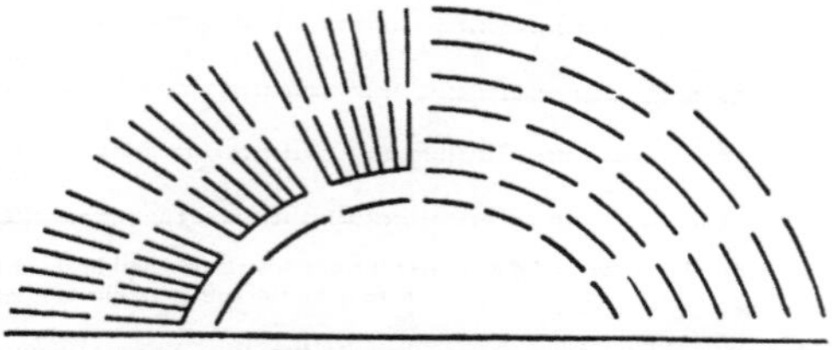

In the foregoing types of parks the visitor notes that the height and strength of the shibi, their simple or branching character, together with the closeness in their arrangement, vary somewhat widely in different localities. Occasionally an

Fig. 12.- Diagram of an oyster farm in which the collectors are arranged in circular and concentric order. Middle clear space used as a living ground. Itsukushima.

arrangement which alternates old and new shibi in the same row is found to be adopted advantageously.

At the close of the cultural process, that is, at the end of the second year, it remains only to remove the marketable oysters from the shibi. This is done during the favorable tides, the culturist using a pick-like instrument (fig. 13) with one hand and seizing the shibi with the other. For the protection of this hand a curious heavy but open mitten is used, figured in fig. 17. After the oysters are detached they are taken from the ground by means of the rake shown in fig. 15, placed in baskets, fig. 18, and carried thence in the usual oyster boat often to the mouth of some adjacent river, where they are thrown out and raked over. By the latter process, "drinking" the oysters in fresher water, they increase in size and become cleaner, a process, by the way, quite similar to that employed in France, in England, and often in America.

Fig. 13.—Oyster hook used for dislodging well-grown oysters from the shibi.

At Kanawa, an important cultural ground, a similar method to that of Kaida Bay is employed. The cultural area is not large but it is very productive, and here they have found it profitable to plant shibi in close rows at right angles to the coast line, as shown in fig. 19.

OYSTER-CULTURAL METHODS OF KUSATSU.

The oyster-grounds of Kusatsu are the best of those situated west of Hiroshima, but all of these, and of this entire region, are essentially the same, as far as cultural

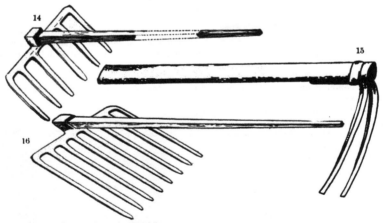

Fig. 14.—Oyster rake, *gofuzt-guwa*, used for "cultivating" the oysters, i. e., stirring them about roughly as they lie on the living grounds, so as to break off the delicate shell margin.
Fig. 15.—Oyster rake, *nihon-zuml*, used to gather oysters fallen from the shibi, or to select oysters to be detached.
Fig. 16.—Oyster rake, *yutsugo*, used to collect marketable oysters from the living ground.

methods are concerned. They extend along the western coast 7 or more miles from Hiroshima, at points indicated on the map (fig. 1). In these localities oysters are cultivated at greater depths than in other waters of Aki, for it has here been found that under the deeper conditions the shellfish continue to increase in size after the second year, unlike those at Kaida. Therefore, at Kusatsu, to the end of growing a larger oyster, the culturist divides his parks into three classes: those of shallow water (largely for spat collecting), early rearing, and deep water (for late rearing).

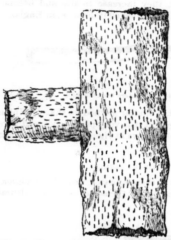

The cultural concessions accordingly have come to be arranged with their long axes at right angles to the shore, thus providing a range of water passing from shallow water to deep, each lot measuring, in round numbers, a thousand feet by fifty. In shallower waters, tempered by fresh streams from the direction of Hiroshima, are the best conditions for spat-collecting. Here the specific gravity is very nearly the same as in Kaida Bay—in summer about 1.018; in winter 1.020. The density of the water rises gradually and attains about 1.025 in the deepest zone. Accordingly, the shallowest region in each park usually becomes laid out in a zone of collectors, or shibi-ba, and resembles somewhat Kaida Bay. In the next and deeper ground are the rearranged and oyster-bearing shibi (of the shallower zone), toya-ba, and in the deepest water are the typical oyster beds, or miire-ba. Of course such an arrangement is sometimes modified, since practice demonstrates that the local conditions of an oyster park, *i. g.*, water currents, are apt to warrant widely different treatment.

FIG. 17.—Mitten of heavy sack cloth, open at thumb and finger tip, used to hold the oyster-bearing shibi while separating the oysters.

(*a*) In the shallow zone an arrangement of shibi in lines parallel to the shore is the common one. Between the rows are intervals of about 4 feet, the park in this case resembling one of the common type of Kusatsu (fig. 8). Often, however, the shibi are shorn of their branches and planted like canes (3 to 5 feet in length) in close-set rows. Such, for example, are the shibi photographed in plate 6, *A*, *B*, and *C*. The first of these, *A*, has had the oysters attached to it for about a month; the second, *B*, for about 6 months, and the third, *C*, for about 18 months. It is at the last period that the masses of oysters come to separate somewhat readily from the bamboo. In these parks the arrangement and treatment of the shibi varies greatly in accordance with local conditions. In rapid currents, which distinguish

FIG. 18. Basket, *taragama kago*, for collecting and storing marketable oysters.

the region of Kusatsu, short and branchless shibi are commonest, these, too, made of the strong and short-jointed species of bamboo, "madake." Their arrangement is frequently in clumps, or toya, as indicated in figs. 20 to 23. Of these the toya

PLATE 4.

OYSTER PARK NEAR NIHOJIMA. TYPICAL LIVING GROUND, OR IKE-BA, WITH HEDGE OF SHIBI.
Maximum tide fall of about 15 feet.

of figs. 20 and 21 are specialized for shallow water, and that of fig. 22 to a rapid current. In such clumps branching shibi frequently occur, and in this event the tips of the stalks are more apt to diverge than in Kaida (contrast figs. 23 and 24),

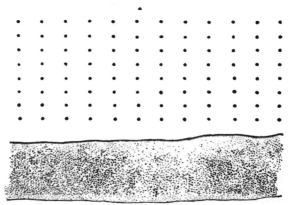

Fig. 19. Diagram of oyster farm in which shibi are planted in rows parallel and at right angles to shore line.

another adaptation to more rapid current. In general the bases of the component shibi are implanted about a foot. Thus made the toya remain in position for about two years.

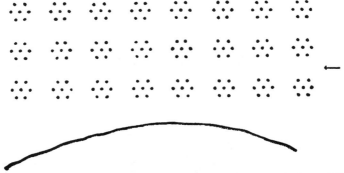

Fig. 20.—Arrangement of collectors in clumps, dotted parallel to the shore, leaving straight shoreward alleyways between. Clumps 3 or 4 feet apart, arranged for sluggish current, the direction of the latter denoted by an arrow.

(b) In the next deeper zone, typical toya-ba, the grouping of shibi becomes more massive, and when at low tide this region is exposed (10 to 15 feet is the tide fall) one sees them in long rows which suggest diminutive haycocks. An excellent idea of such toya-ba is had in plate 5. Closer inspection shows that each toya is in nearly every case a complex of shibi of many ages. Transplanted shibi, with oysters of one or two years' growth, usually form the nucleus of the cluster, and around them are planted concentric rows, one, two, or three, of branched shibi, old and new, to the end that all ultimately bind or mat themselves into a living, springy, cone-shaped mass, well suited to resist currents or storms. (Fig. 25 and Pl. 5.) Such

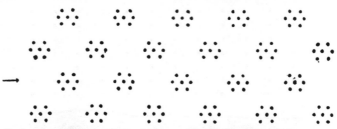

Fig. 21.—Arrangement similar to fig. 20, but in oblique order, adapted to a somewhat more rapid current.

toya are made, or remade, toward the end of each spawning season, *i. e.*, during the end of August or early part of September. They are then pulled to pieces (in this work the rake shown in fig. 14 is used), and from each bamboo there are shaken and broken off the oysters which are least securely attached. During this process the shibi found to be still useful are put aside to form the nucleus of new toya;

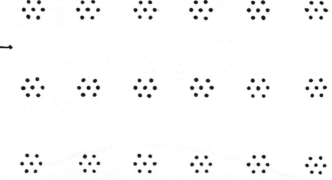

Fig. 22.—Arrangement similar to fig. 20, but adapted to rapid current, rows of collectors separated by intervals of 8 or 10 feet.

the gleanings, twigs, detached oysters, and all, are now raked up and carried to a third locality, the living grounds (ike-ba). Here in swifter current the débris washes away and the oysters remain, the shells usually becoming clean in the meanwhile.

(c) The living ground, or ike-ba, of which a good example is pictured in plate 4, is generally located in a zone of deeper water. It has a clean, gravelly bottom, which in most cases becomes exposed only at the lowest tides. Here the oysters are brought together which have become detached from shibi in various parts of the farm, and thus they are retained in classified beds until they have attained their second year's growth. During this time the culturist has but to keep them well spread out and to see that the beds are kept clean. Always at lowest tides, and sometimes as often as fortnightly, the laborers give the oysters a vigorous raking (using for this purpose the short-toothed rake shown in fig. 14), which scatters them about the bed, removes foreign bodies, and, best of all, gives the shells a better shape and a firmer rim, for in this treatment the delicate, cuticular outgrowths of the shell are removed and a more symmetrical growth results. Of especial importance is the process of raking in cases where the oysters are sent directly from the living ground to market; for it has been found (here as elsewhere) that those oysters

Fig. 23.—Bamboo collectors arranged after the fashion common in Kusatsu. The shibi stand about 3 feet above the bottom and their tips diverge: the clumps are set 4 or 5 feet apart.

whose shell rims are strong and accurately fitted together fare better during transportation and in the market. Those with delicate and brittle margins soon suffer injury and lose their fluid through leakage and evaporation.

In some farms, on the other hand, it is maintained that oysters of different sizes should be mixed on the same living ground. For it is claimed that the young oysters grow better side by side with the older ones, and even that if the more perfectly grown oysters of different grades can be mixed together during the process of raking, the better will be the general output.

(d) The final stage in the culture of Kusatsu takes place in maturing grounds, or miire-ba. Here the larger oysters of the second year's growth are laid down, transplanted from the one or more ike-ba of each establishment. Usually they are in the deepest water cultivated, i. e., in water a few feet in depth at lowest tides up to water of 3 or 4 fathoms. I was told that in one miire-ba oysters were cultivated successfully in water of 50 feet. Each farm in Kusatsu has its separate miire-ba, and these adjoin one another, forming a continuous zone in deeper water, each miire-ba designated by a number. In the shallowest portions the oysters are usually protected against displacement and invasion of mud by means of low fences arranged with wing-like expansions, as shown in ground plan in fig. 26, and in detail in fig. 27. This device has been developed largely in view of the storms of the typhoon season. The effect of the maturing ground is to give the oysters greater size and

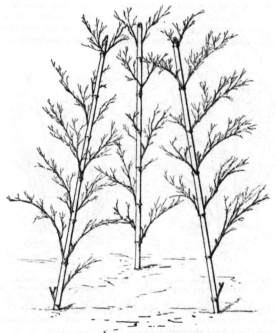

FIG. 21.—Collectors arranged in a way common in Kaida Bay. The shibi stand about 4 feet from the bottom and their tips converge.

weight, and to give the meat a whiter color. The finished product resembles closely a well-grown oyster of Long Island Sound. Marketing takes place after the oysters have remained from six months to a year in the deeper ground, making their total age about three years.

Each oyster farm has its separate houses, situated usually on the adjacent shore, and the details in handling and packing the oysters appear to be closely similar to those of Continental Europe—baskets, blocks, rakes, arrangement, and

PLATE 5.

OYSTER PARK NEAR NIHOJIMA. GENERAL VIEW, SHOWING NEWLY ARRANGED TOYA.

The fall of tide is here at a maximum 15 feet.

storage. I may note in passing the curious "knife," a combination of knife, mallet, and lever, with which oysters are opened with surprising rapidity. The *modus operandi* is shown in fig. 2, page 21.

OYSTER-CULTURE AT NIHOJIMA.

In this locality (cf. fig. 1), finally, oyster-culture has been developed on more special lines than anywhere else in the East. As at Kusatsu, the industry embraces three distinct branches—(*a*) spat collecting; (*b*) rearing the young, and (*c*) maturing and fattening the oysters for market. But, unlike at Kusatsu, these three branches of the industry are carried on, not on the same shore reach, but at points widely

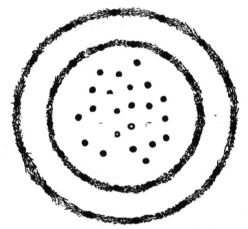

Fig. 25.—Ground plan of a mound toya of collectors. Shibi with well-grown oysters are indicated by the black spots within the two circles of branching shibi.

separate. In other words certain special tracts are taken advantage of in collecting spat; others are specially arranged for early rearing; others, in turn, for maturing. In these regards oyster-culture at Nihojima resembles closely that of the coast of Brittany or of Holland. The details of the management of the farms are in essentials, however, like those previously described.

(*a*) *Spat collecting.*—The spat is collected in very shallow water, less than a fathom deep at the usual tides, tempered considerably by incoming streams. The specific gravity is said to rarely exceed 1.017. In such a region shibi are put in position, usually in very close order, at the beginning of each spawning season, say from the middle of April to the middle of May. After a period of about three months the entire mass of shibi will be uprooted and transplanted, sometimes a mile or more, to a locality better fitted for rearing the young oysters. This transportation, I was told, is the most difficult part of the work of the culturist of Nihojima, for the minute oysters are, as everywhere, peculiarly liable to injury;

careless handling will destroy great numbers of the delicate shells; hasty packing of
the shibi on the scows, whereby the branches are allowed to rub together, is another
palpable source of loss; drying, direct sunlight, changes of temperature, are all
deadly, and, above all, severe thunderstorms—the latter, according to my informant,
causing death by *fear*. I suspect, however, in the last regard, that a fresh, cold
shower bath is more apt to be a moving cause, although I was assured that the scows
are covered with the ever-present Japanese oiled paper to guard against such a
contingency.

(b) *Rearing the young oysters.*—The shibi, covered with young spat, are now
arranged in toya-ba, like those of Kusatsu, but closer in arrangement and usually
of many varieties. Here they remain from one to two and a half years. In the

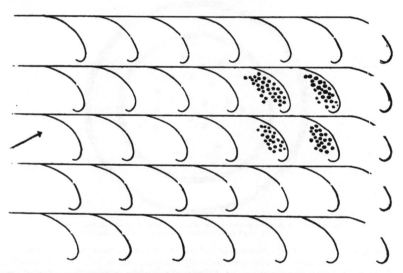

Fig. 26.—Diagram of living ground of two to three year old oysters. The spots represent oysters and the dark
lines are rows of shibi placed so as to provide shelters against currents or storms. The arrow denotes direction of strongest current.

case of the older and rearranged toya a long, mound-like type is commoner than
the circular ones described above. As a rule the toya are covered with water save
at lowest tides.

(c) *Later rearing: Maturing.*—The living inclosures in deeper water, ike-ba,
correspond to those of Kusatsu. They contain the oysters which have been separated or are readily separated from the brush of the toya. A similar process prevails
of raking the oysters roughly, and I was shown some shells of the older oysters
from this region which were of very regular shape. At favorable tides, furthermore, the oyster beds are cleaned and the boring whelks—*Purpura clavigera* Kaster

and *Rapana bezoar* L.—are removed. Starfish are not troublesome. Further transplanting takes place, usually at the end of the second season, and in still deeper water the oyster attains finally its marketable size.

REGULATION OF OYSTER-CULTURE BY THE JAPANESE GOVERNMENT.

All cultivable grounds, whether for oysters, other shellfish, or seaweed, are—in Aki, at least—the property of the prefecture and can be neither sold nor subrented. All cultivable tracts are surveyed, the lots tending to decrease in size as the estimated value of the property increases. The mode of laying out concessions can perhaps be best understood by reference to plate 7, copied from a Government

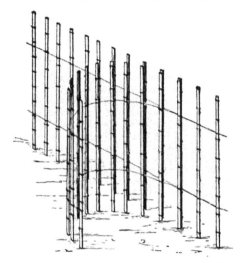

FIG. 27. - Detail of the living ground shown in ground plan in preceding figure. Bamboo rods 3 to 4 feet high.

chart. The farms are rented auction fashion to the highest bidder, and the tenants have the privilege of renewing their leases indefinitely at the original rentals, a privilege, however, which can not be used speculatively. When at the termination of a lease the property passes again into the hands of the prefecture it is at once advertised and rerented. In addition to the yearly rental the property is subject to a small local tax upon the total area of each farm, and to a charge of 1 per cent of the rental to cover the expenses of administering the oyster-cultural bureau of the prefecture. This series of taxes impresses the stranger as formidable, until he learns that it does not represent an accumulation of taxes, but rather an itemized statement as to the apportionment of public funds made thus in accordance with local custom. Rental of concessions from the prefecture, it should further be said, brings with it certain privileges in marketing the oysters in Osaka.

THE QUESTION OF INTRODUCING AND CULTIVATING THE JAPANESE OYSTER IN THE UNITED STATES.

To what degree the Japanese oyster would flourish on our coast can be determined, obviously, only by experiment; and if experiments are to be made, the following suggestions seem to me of some practical value:

On the Pacific coast, on grounds which have been found especially favorable for the reproduction and growth of our Western species, *O. californica*, consignments of Japanese oysters may be planted—in the north with oysters from the Hokkaido beds, preferably from the region of Mororan or from Akkeshi Bay, and in the south from the region of Hiroshima. In this way similar conditions of temperature will be obtained. To fulfill a second favorable condition an effort should be made to secure oysters from water of nearly the same specific gravity as in the chosen American localities. There would be, I fancy, little difficulty in the matter either of securing oysters from equivalent localities or of transporting them. Through the Imperial Bureau of Fisheries, under the able headship of Dr. K. Kishinouyé, one could promise, *à priori*, prompt and efficient aid in getting in touch with the Japanese oyster-culturists whose establishments are known to be favorably located. And with fast freight service from Yokohama the oysters could be transported with a minimum of loss, as similar exportation (*e. g.*, of American oysters to England) demonstrates. In a case of this kind, however, extra precautions would not be out of place. Large oysters should first be selected, and, preferably, treated with the raking process on ike-ba for a few weeks. By this means the shell margins will be thickened, and thus the oysters will lose a minimum amount of fluid during transportation. Care should also be taken to pack the oysters each with the more concave valve downward and to mark the cases so that during shipment they shall be kept in the right position. Other details—if not indeed the above—can safely be left to the skilful Japanese work people.

A further suggestion is that the shipments be made during the months of February and March, to the end that the coldest season on our coast would be avoided, and thus the oysters would have the chance of becoming in good condition and somewhat acclimated by the following winter. Moreover, at this season of the year it has been found that the oysters have laid up the maximum supply of nutriment against the breeding time, a supply which could be used as reserve nutriment during transportation.

No experiment, however, could be regarded as a fair one, I believe, which did not deal with an adequate number of the imported oysters. If but a few oysters—a score or two—are laid down in each experimental locality they may be lost through accidents which would not befall a larger number. For in oyster banks there is certainly strength in numbers, an œcological feature in oyster-culture which governmental regulations in France, Holland, and Germany keenly recognize. By this numerical factor it appears a true paradox that one thousand planted oysters have *more* than ten times the chances of survival than a hundred. As oysters are cheap, it seems to me better economy therefore to make the more convincing experiment.

To what degree is it practicable to introduce the methods of Japanese oyster-culture into the United States? This, I take it, is a question which can be answered

BAMBOO OYSTER COLLECTORS, OR SHIBI, AFTER HAVING BEEN IN USE ABOUT ONE MONTH, *A*; SIX MONTHS, *B*; EIGHTEEN MONTHS, *C*.

Detached oysters are shown at *D*. Figures about one-third natural size.

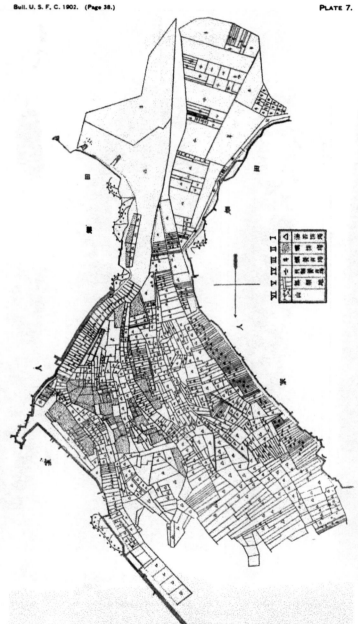

MAP OF THE OYSTER AND SEAWEED CONCESSIONS IN ONE OF THE ESTUARIES OF NIHOJIMA, TO SHOW HOW COMPLETELY THE CULTURAL AREA IS DEVELOPED.

I, shidi bearing purple seaweed; II, shidi bearing oysters; III, grounds for rearing oysters; IV, cultural area for other mollusks, Tapes, Area, etc.

THE HABITS AND CULTURE OF THE BLACK BASS.[a]

By DWIGHT LYDELL.

This paper sets forth the experiences of nine seasons, beginning with 1894, during which I have had charge of the black-bass work of the Michigan Fish Commission. The work was begun at Cascade, Mich., and after four seasons was transferred to Mill Creek, where it is now carried on. The methods of pond culture finally adopted are based on a knowledge of the breeding habits of the fish under natural conditions. The account has reference to the small-mouthed bass, unless the large-mouthed is specified.

In studying the habits of the bass, it is necessary to distinguish the males from the females; ordinarily this is not possible except by dissection, but at spawning time the female is distinguishable, even at a distance of 10 or 20 feet, on account of her distension with eggs, and this makes it possible to determine the part taken by each sex in nest-building and the rearing of young.

The nests of the black bass are built by the male fish working alone. The small-mouthed bass prefers a bottom of mixed sand and gravel, in which the stone ranges from about the size of a pea to that of one's fist. As the spawning season approaches the males are seen moving about in water of 2 or 3 feet depth seeking a suitable resting-place. Each male tests the bottom in several places by rooting into it with his snout and fanning away the overlying mud or sand with his tail. If he does not find gravel after going down 3 or 4 inches, he seeks another place. Having found a suitable place, he cleans the sand and mud from the gravel by sweeping it with his tail. He then turns over the stones with his snout and continues sweeping until the gravel over a circular spot some 2 feet in diameter is clean. The sand is swept toward the edge of the nest and there forms a few inches high, leaving the center of the nest concave like a saucer. The nest is usually located near a log or large rock so as to be shielded from one side. If the bank is sheer and the water deep enough, the nest may be built directly against the bank. If possible, it is placed so that the fish can reach deep water quickly at any time.

During nest-building no females are in sight, but when the nest is done—and this takes from 4 to 48 hours—the male goes out into deep water and soon returns with a female. Then for a time—it may be for several hours—the male exerts himself to get the female into the nest and to bring her into that state of excitement in which she will lay her eggs. If she lies quiet, he turns on his side and passes beneath her in such a way as to stroke her belly in passing. If she delays too long, he urges her ahead by biting her on the head or near the vent. If she attempts to escape, he heads her off and turns her back toward the nest. If, after all, she will not stay in the nest, he drives her roughly away and brings another female.

a Read at the thirty-first annual meeting of the American Fisheries Society.

Some 15 to 30 minutes before the female is ready to enter the nest and spawn, her excitement is made evident by a change of color. Ordinarily she appears to be of a uniform dark olive or brown above, changing to a light green below. The only markings readily seen are four stripes on each cheek; but in reality the sides of the fish are mottled with still darker spots on the dark-olive background. The spots are arranged so as to form irregular, vertical bands like those on the perch, but these are not usually visible. Now, as the excitement of the female increases the background becomes paler and finally changes to a light-green or yellowish hue, so that the spots and bands stand out in strong relief. The whole surface of the fish becomes thus strongly mottled. This is a visible sign that the female will soon spawn. The male undergoes a similar but less pronounced change of color.

Soon after this the female enters the nest and the male continues to circle about her, glide beneath her, and to bite her gently on the head and sides. At times he seizes her vent in his mouth and shakes it. When this has continued for a time spawning takes place. The two fish turn so as to lie partly on their sides with their vents together and undergo a convulsive fluttering movement lasting 3 to 5 seconds. During this time the eggs and milt are extruded. The circling movements are then resumed to be interrupted after a few seconds by spawning. This alternate circling and spawning continue for about 10 minutes. The male then drives the female away, biting her and showing great ferocity. She does not return.

The male, and the male only, now continues to guard the nest, fanning sediment from the eggs and repelling enemies. At 66° F. the eggs hatch in 5 days and the young fish swarm up from the bottom in 12 to 13 days from the time of hatching.

Henshall in his "More About the Black Bass," published in 1898, quotes, with approval, Arnold's observations to the effect that the nests are built and then guarded by the female. The Manual of Fish Culture, published in 1897 by the United States Fish Commission, speaks of the nests as being built by the mated fish, sometimes working together, sometimes separately. These seem to be the latest published observations, and are not at all in accord with my observations in Michigan.

Shortly after the young small-mouthed bass rise from the nest they scatter out over a space 4 or 5 rods across—not in a definite school with all the fish moving together, but as a loose swarm, moving independently or in small groups. This makes it impossible to seine the young fry, as upon the approach of the seine, instead of keeping together, they at once scatter and escape the seine. The fry may be at the surface or on the bottom in weeds or clear water and are attended by the male until they are 1¼ inches long. The swarm then gradually disperses and the young fry, which were previously black, take on the color of the old fish.

The breeding habits of the large-mouthed black bass are similar to those of the small-mouthed, but differ in some respects, which are of importance in pond culture.

1. The nests of the large-mouthed are not made on gravel, but by preference on the roots of water plants. These are cleaned of mud over a circular area, and on them the eggs are laid. As the eggs of the large-mouthed bass are smaller and more adhesive than those of the small-mouthed they are apt, when laid on gravel, to become lodged between the stones and to stick together in masses, and are then likely to be smothered. When laid on fibrous roots of water plants this does not occur.

2. The young large-mouthed bass remain together in a compact school very much smaller than that of the small-mouthed, and the fry usually move all in the same direction. This makes it easy to seine the large-mouthed fry when wanted.

BLACK BASS POND, SHOWING FRY RETAINERS IN USE—BED FRAMES REMOVED.

FRY RETAINER, TO BE PLACED AROUND BED JUST
BEFORE THE FRY RISE.

BASS BED, TO BE PLACED ON POND IN EARLY
SPRING AND FILLED WITH GRAVEL.

PLATE 8

CULTURE OF BLACK BASS.

Ponds and stock fish.—After some experimenting all our ponds, both for stock fish and fry, are built on the model of a natural pond. There is a central deeper portion or kettle about 6 feet deep, and around the shore a shallow area where the water is about 2 feet deep. The bottom is the natural sand, and water plants are allowed to grow up in the ponds. All ponds are supplied with brook water, and silt from this furnishes a rich soil for the aquatic plants. The water of these ponds contains *Daphnia, Bosmina, Corix,* and other small aquatic forms in great numbers. These furnish food for the bass fry. The ponds run in size from 120 feet by 190 feet to 100 feet by 100 feet.

At first we were unable to feed the stock fish on liver, but after a time we found that by cutting the liver into strips about the size and shape of a large angleworm and by throwing the strips into the water with the motion that one uses in skipping stones they wriggle like a worm in sinking and are then readily taken. The liver must be fresh. If bass are fed on liver alone they do not come out of winter quarters in good condition. Of eleven nests made by bass thus fed only three produced fry. Although eggs were laid in all, they seemed to lack vitality, owing to the poor condition of the parent fish, and in eight of the nests the eggs died.

In order to bring the fish through the winter in good condition it is necessary to begin feeding minnows in September and to continue this until the fish go into winter quarters. The bass eat minnows until they go into winter quarters, after which they take no food until spring. The minnows are left in the ponds over winter, so that the bass, when they come out of winter quarters, find a plentiful supply, which lasts them until the spawning season. At this time the minnows are seined from the pond, as their presence interferes with the spawning. Before this, however, some of the minnows have spawned, and their fry later serve the young bass as food. Bass fed in this way come out of winter quarters in fine condition and their eggs are found to be hardy.

Artificial fertilization.—During the first two or three seasons of our work numerous attempts were made at artificial fertilization, but only twice with success. On one occasion the female was seined from the nest after she had begun to spawn. She could then be readily stripped. The male was cut open and the eggs were fertilized with the crushed testes. About 75 per cent of the eggs hatched on a wire tray in running water, the eggs being fanned clean every day with a feather.

In the second case the fish were seined while spawning, and it was found that in the case of one female pressure on the abdomen caused a reddish papilla to protrude from the vent. This had the appearance of a membrane closing the vent. It was pinched off, and the female was then stripped readily and the eggs fertilized and hatched.

Pond culture.—Having abandoned artificial fertilization, our attention was turned to pond culture, and this we have carried on for about six years. Our earlier ponds not furnishing natural spawning-grounds, we constructed alongside each of the large ponds six smaller ponds for use as spawning-ponds, each about 16 by 24 feet, 16 inches deep, with gravel bottom, and connected to the central pond by a 4-foot channel.

The fish entered these and spawned. In one case we had eight nests in a single pond of this sort. Where so many nests were made, usually but one or two of them came to any good, the others being destroyed by the fighting of the males. Ordinarily but one or two nests were built in each spawning pond. The male first to enter and begin the construction of a nest generally regarded the whole pond as his property and held it against those that tried to enter after him. On one occasion the male thus holding the pond was attacked by 10 or 20 other males at one time and after a long struggle was killed and his nest destroyed.

The attempt to use small spawning-ponds was then abandoned and all the ponds were made of good size and with a central kettle and shallow shore area, as already described. The problem now was to prevent the fighting of the male fish and the consequent destruction of nest and eggs. I finally hit upon remedies for what seemed to be the two chief causes of this fighting. I had noticed that in the natural water the nests of the small-mouthed bass were frequently built against a stone or log, so as to be shielded on one side. When so built the nests might be quite close together, as near as 4 feet, and the fish did not fight, because they did not see one another when on the nest. On the other hand, if a bass nest was built where it was not shielded the bass on that nest would prevent any other bass from building within 25 or 30 feet of him. It occurred to me to try to construct artificial nests and shield them so that the fish on the nests could not see one another, placing the nests so near together as to fully utilize the pond area.

In the spring, before the spawning season opened, the ponds were drawn down so as to expose the shallow terrace along the shore. The terrace was then cleaned to a depth of about 2 inches of sediment and vegetation which had accumulated since the previous summer. Rectangular nest frames of inch board were made 2 feet square and without bottoms. On two adjacent sides these frames were 4 inches high, while on the other sides they were 16 inches high. They were set where there would be about 2 feet of water when the pond was filled, and so placed that the corner formed by the junction of the two lower sides pointed to the center of the pond, while the opposite corner, formed by the higher sides, pointed toward shore. The frames were set directly on the bottom, not in excavations, and each was filled with gravel containing sand suitable for nest-building. A board was laid diagonally across the two higher sides and a heavy stone laid on this to keep the frame in place. The two higher sides form a shield on two sides of the nest, while the board across the top affords shade. The frames were set in two rows about the pond, parallel to the shore line.

The rows were about 6 feet apart and the nests in each row about 25 feet apart, alternating with those in the other row. There was thus about one nest to each 100 square feet of suitable bottom, or in each area of 10 by 10 feet. When the bass were on the nests no one was able to see any other and the fighting from this cause was practically eliminated. The number of rows of nests may be increased to three or four, or more where the area of shallow water is wide enough.

The bass selected these nests in preference to any other spawning-ground. They cleaned up the gravel and behaved in the nests in every particular as they would on natural spawning-grounds. The first time we tried these shielded nests not a single bass made a nest outside of them, though there was plenty of good gravel bottom available for this purpose.

As to the second cause of fighting: In 1900, when these nests were first tried, from 475 stock fish we obtained 315,000 fry and 750 fingerlings. In the season of 1891 the output was very much less and there was considerable fighting among the fish. This remained unexplained till the ponds were drawn down after the spawning season, when it appeared that although the fish had been sorted, the number of male fish was considerably in excess of the number of females, and these excess males, banding together, went about breaking up the nests of their more fortunate brothers. It is now the practice when setting the nests to seine out the stock fish and sort them, putting about 40 males to 60 females, thus removing the second source of fighting.

During the present season from 493 adult fish we had produced 430,000 fry up to May 26, and we believe that we can do as well every year.

Up to the present year there have been two sources of loss incident to the water supply. The supply is a spring-fed brook, which runs over an open country before it reaches us. The water in this brook becomes quite warm on a hot, sunny day and cools off at night. The temperature thus falls at night sometimes as much as 13° F. and becomes as low as 46° F. This is disastrous, since when the temperature gets below 50° F. the adult fish desert the nests and the eggs or young fry are killed by the sediment. By watching the temperature of the water and, when it approaches 50° F., shutting off the supply until the water warms up, this difficulty is obviated. Since the ponds are well stocked with water plants the fish do not suffer from lack of oxygen when the water is shut off. Indeed, if the water did not leak out of the ponds, I doubt if it would be necessary to introduce any running water into them during the breeding season.

The second difficulty with the water supply is from sediment brought down by the brook after heavy rains. This sometimes accumulates over the nests so thick as to smother the eggs and drive away the parent fish. By shutting off the water supply whenever the water is much roiled this trouble is avoided.

The water supply, however, must be kept fairly constant. If the level lowers more than about 6 inches the fish leave their nests and the eggs die. For the purpose of maintaining a constant water level it would probably be best to have the ponds made with clay bottoms. The difficulties arising from roily water of variable temperature are, however, local, and would probably not be usually encountered.

Handling the fry after they rise from the nest.—The small-mouthed fry have the habit of scattering into a large swarm when they leave the nest and it is consequently difficult to seine them when wanted. It is therefore desirable, just before the fry rise from the bottom, to set over each nest a cylindrical screen of cheesecloth supported on a frame of band iron, first removing the wooden nest frame. The screen keeps the fry together. They thrive and grow within it and may be left there until one desires to ship them. The old fish stays outside and watches the screen. When this supply is gone other crustacea may be taken from the pond with a tow net and placed inside the screen. The fry are removed from these screens directly to the shipping cans, as wanted.

Raising the fingerlings.—The water in one of the ponds is lowered, the old fish seined out of the kettle and transferred to another pond; the pond is then refilled, and the fry, now about one-half to three-fourths of an inch long, are put in. The water in the pond is thick with *Daphnia* and other crustacea, and these do not get out when the water is drawn off. The fry feed on them and the supply is usually sufficient; but if it gives out, a fresh supply may be gathered from another pond and placed in the nursery pond. As the young bass grow they eat not only the *Daphnia* but young *Corixa*, and doubtless other aquatic animals.

In 1901, fry one-half to three-fourths inch long were introduced into the nursery pond on July 12; on August 5 they were seined out and shipped, and were then 2 or 3 inches long. They had had none but the natural food. In three months these fish, under the same conditions, are 4 to 6 inches long.

I have spoken so far of the small-mouthed bass, and it remains to say something of the large-mouthed, with which my experience is more limited. It is less necessary to resort to pond culture with them since, owing to the habit of the fry of keeping in a close swarm, they may be readily seined from their natural waters shortly after they have left the nests. In cultivating them in ponds I use the shielded nests already described, but make the bottom of some fiber, preferably Spanish moss bedded in cement, as has been suggested by Mr. Stranahan. This imitates the natural nest bottom and gives better results in our locality than the gravel nest. I do not place screens about the nests, since the young fry are so small that it is difficult to hold them with a screen, and since they may be readily taken with a seine when wanted. I allow the large-mouthed fry to leave the nests with the parent fish and seine them when wanted.

Finally, I will sum up what seem to me to be important points in pond culture of small-mouthed black bass, the ponds being constructed, as is usual, on the model of a natural pond with a central kettle and shallow shore region, well grown up with water plants, and supplied with lake or brook water:

1. Fish should be so fed (with minnows) as to be in good condition in the spring.
2. They should be sorted into the ponds in the spring in about the proportion of 4 males to 6 females.
3. Shielded nests should be used, arranged as already described—about 1 to each 100 square feet of shallow water.
4. The gravel in the nests should be carefully selected; it should contain sand and plenty of small stones.
5. Water on the nesting-grounds should be kept constantly at a level between 18 inches and 2 feet.
6. The water temperature should be kept constantly between 66° and 75° F. (in our locality).
7. Roily water should be, as far as possible, kept out of the ponds during the spawning season.
8. Fish should not be disturbed until the eggs are hatched.
9. The nests of the small-mouthed bass should be screened just before the fry rise from the bottom.
10. The water should contain an abundance of natural food for the fry.

The processes described are perhaps susceptible of improvements, viz:

1. Special nursery ponds might be provided for rearing fingerlings.
2. It is perhaps desirable to have the nest frames shielded on three sides instead of two sides, and made with a bottom; then when the fry rise from the nest, close the fourth side of the nest frame by sliding a screen into it. In this way it would not be necessary to remove the nest frame and put a screen over it, but the frame could be left in place and the open side closed with a screen.
3. If the ponds were made with clay bottoms, the water supply could be entirely shut off during the breeding season, if necessary.

Contributions from the Biological Laboratory of the U. S. Fish Commission.
Woods Hole, Massachusetts.

HEARING AND ALLIED SENSES IN FISHES.

By G. H. PARKER,

Assistant Professor of Zoology, Harvard University.

It is a well-known fact that many fishes are extremely sensitive to disturbances in the water such as are caused by splashing with an oar, stamping in a boat, or striking the side of an aquarium. When, for instance, the opaque wall of a fish tank containing young king-fish, sea-robins, or killi-fish is struck a vigorous blow with the fist, the fishes usually respond by giving a short, quick leap, and, if such blows are frequently repeated, surface fishes are often driven to the bottom and kept there. Notwithstanding the sensitiveness indicated by such reactions, most of these fishes appear to be unaffected by loud talking or other like noises originating in the air. Fishermen are familiar with these peculiarities and often take them into account in the practice of their art.

Such facts as these are also usually accepted as evidence that fishes can hear (as an example, compare the statements made by W. C. Harris in Dean Sage's "Salmon and Trout," 1902, p. 311), but a simple experiment will show, I believe, that this assumption is not necessarily correct. If one end of a wooden rod is vigorously tapped while the other is beneath the level of the water a disturbance is produced that will call forth an obvious response from most fishes of moderate sensitiveness. Such a disturbance will likewise affect a human being, for if one holds the head beneath the water the vibrations from the rod can be easily heard, and if the hand be placed in the water near the rod they can be distinctly felt.

Since, as Müller (1848, p. 1229) long ago pointed out, we can feel as well as hear these vibrations, it follows that such evidence as that already given can not be accepted as conclusive proof that fishes hear, for it is conceivable that their responses may be entirely through their sense of touch, i. e., dependent on their skins. Moreover, fishes possess a special system of tegmentary sense organs, the lateral-line organs, which are completely absent from us, and it may be that these are in some way the recipient organs for the disturbances already described. When, therefore, a fish responds to water vibrations of the kind mentioned, we are not justified in concluding that it hears, for it may respond through the skin or the lateral-line organs and not through the ears.

45

It may be reasonably asked at this point, What constitutes hearing? Everyone will agree, I believe, that the sensation we get through the skin from a vibrating rod in water should not be called hearing, and what is true for us should hold for the lower vertebrates. Hearing in these animals may therefore be defined as that sensory activity resulting from a stimulation of the ear by material vibrations. This is in essential accord with the definition given by Kreidl (1895, p. 461) to the effect that hearing is that sensation which is mediated by the nerve that is homologous with the auditory nerve of man. When, therefore, a fish responds to sound vibrations the question at once arises whether the stimulus is received by the skin, the lateral-line organs, or the ear. And until this question can be answered, at least so far as the ear is concerned, the query whether fishes hear or not must remain open. In dealing with this general subject I shall take up, first of all, the question whether fishes respond to sound vibrations through the ears.

THE EARS.

Introductory.—The internal ears of fishes were described as early as 1610 by Casserius, and were studied in some detail in the following century by Geoffroy, Scarpa, Comparetti, and Hunter. The attitude taken by many of these early workers on the question of the ability of fishes to hear or not is well illustrated by a quotation from Hunter (1782, p. 383), who at the conclusion of his paper on the organs of hearing in fishes made the following statement:

> As it is evident that fish possess the organ of hearing, it becomes unnecessary to make or relate any experiment made with live fish which only tends to prove this fact; but I will mention one experiment to shew that sound affects them much and is one of their guards, as it is in other animals. In the year 1762, when I was in Portugal, I observed in a nobleman's garden, near Lisbon, a small fish-pond full of different kinds of fish. Its bottom was level with the ground and was made by forming a bank all round. There was a shrubbery close to it. Whilst I was lying on the bank, observing the fish swimming about, I desired a gentleman, who was with me, to take a loaded gun and go behind the shrubs and fire it. The reason for going behind the shrubs was that there might not be the least reflection of light. The instant the report was made the fish appeared to be all of one mind, for they vanished instantaneously into the mud at the bottom, raising, as it were, a cloud of mud. In about five minutes after they began to appear, till the whole came forth again.

This passage shows very clearly that in the opinion of Hunter the internal ears of fishes, like those of the higher vertebrates, are organs of hearing. Without further experimental evidence this view was accepted by Müller (1848, p. 1238) in his well-known chapters on the physiology of the senses, and by many other eminent authorities, such as Owen (1866, pp. 342 and 346), Günther (1880, p. 116), and Romanes (1892, p. 250). To these investigators the presence of the internal ears seemed, as it did to Hunter, sufficient ground for concluding that these animals could hear.

Within recent years, however, this opinion has been called in question, or even denied. Some of the grounds for this change of view may be stated as follows: Bateson (1890, p. 251), in some investigations on the sense organs and perception of fishes, observed that the report from the blasting of rocks caused congers to draw back a few inches, flat-fishes (like the sole, plaice, and turbot) to bury themselves, and pouting to scatter momentarily in all directions; other fishes seemed to take no notice of the report. When the side of a tank containing pollock or soles was struck with a heavy stick, the fishes behaved as they did toward the report of the blasting. Pollock did not respond, however, to the sound made by rubbing a wet finger on the glass window of an aquarium or to the noise made by striking a piece of glass under water with a stone, provided the means of producing the noise was not seen by the fishes. Bateson concluded that, while it may be regarded as clear that fishes perceive the sound of sudden shocks and concussions when these are severe, they do not seem to hear the sounds of bodies moving in the water but not seen by them.

Without knowledge of Bateson's observations, Kreidl (1895) carried out a series of experiments with the view of testing the powers of hearing in the gold-fish, *Carassius auratus*. This species was chosen because of the ease with which it could be kept in the laboratory and, further, because it is one of those fishes that have long been reputed to come at the sound of a bell. After an extended series of experiments, Kreidl (1895, p. 458) concluded that normal gold-fish never respond to sounds produced either in the air or in the water, though they do react to the shock of a sudden blow given to the cover of the aquarium. Individuals rendered abnormally sensitive by strychnine gave no response to the sound of a tuning-fork or a vibrating-rod even when these were in contact with the water, though the fishes responded at once to such slight shocks as tapping the aquarium, etc., or even clapping the hands vigorously in the air.

To test whether these responses were dependent upon the auditory nerves, Kreidl removed these nerves and the attached ear-sacs from a number of individuals, and, after poisoning them with strychnine, subjected them to stimulation by sound. In all cases they were found to respond precisely as the poisoned animals with ears did. Kreidl, therefore, concluded that gold-fishes do not hear by the so-called ear, but that they react to sound-waves by means of an especially developed cutaneous sense, or, to put it in other words, the gold-fish *feels* sound but does not *hear* it (Kreidl, 1896, p. 581).

After having reached this conclusion, Kreidl was led to take up a specific case of the response of fishes to the sound of a bell, and an opportunity for doing this was found at the Benedictine monastery in Krems, Austria. Here the trout of a particular basin were said to come for food on the ringing of a bell. Kreidl (1896, p. 583), however, found that they would assemble at sight of a person and without the ringing of the bell. If they were not then fed, they soon dispersed and no amount of bell-ringing would induce them to return. If, however, a pebble or a small piece of bread was thrown into the water they immediately swam vigorously toward the spot where the disturbance had occurred. Moreover, if a person approached the basin without being seen and rang the bell vigorously no response was observed. From these facts Kreidl (1896, p. 584) concluded that the assembling of the fishes was brought about through sight and the cutaneous sense, and not through hearing, and that the conclusion reached with the gold-fish might be extended to other kinds of fishes.

Kreidl's conclusions were supported by the observations of Lee (1898), who studied the reactions of several species of fishes to such sounds as the human voice, the clapping of hands, and the striking of stones together in air and under water. In all of his experiments Lee (1898, p. 137) obtained no evidence whatever of the existence of a sense of hearing, as the term is usually employed, although he found that the fishes were exceedingly sensitive to gross shocks, such as the jarring of their tank or concussions upon its walls. Lee, moreover, called attention to the fact that the papilla acustica basilaris, which is the special organ of hearing in the internal ears of the higher vertebrates, did not occur in the fishes. From the observations and experiments of Bateson and of Kreidl, and from his own work, Lee (1898, p. 138) believed that the conclusion was justified beyond doubt that fishes do not possess the power of hearing, in the sense in which the term is ordinarily used, and that the sole function of the ear in fishes is equilibration.

The generalization to be drawn from the work just summarized, viz, that fishes do not hear, though they may respond to sound-waves by the skin, has seemed to me not wholly in accord with certain well-known facts in the natural history of these animals. Among these facts may be mentioned the undoubted ability on the part of some fishes to make sounds. If a fish has this power it might naturally be supposed to hear the sounds it makes. Lee (1898, p. 137) has called attention to the small number of sound-producing fishes as evidence against the view that fishes in general hear. But the fact that there are such fishes has always appealed to me in quite the reverse way and should, in my opinion, serve to indicate the species most worthy of attention in any investigation of the sense of hearing. It must be admitted, however, that fishes may possibly produce sounds that they themselves can not hear, but that other animals may hear and take warning from. Thus when small swell-fish, *Chilomycterus schœpfi*, are thrown into a tank containing hungry scup, *Stenotomus chrysops*, they are immediately set upon by the latter. In defense the swell-fishes inflate themselves with sea water till their tegmentary spines stand out rigidly, and at the same time they make a peculiar sound by gritting the two front teeth of the lower jaw against the inner surface of those of the upper jaw. It is not known that this sound is heard by the swell-fish, though it may be. All that one can say with certainty is that the sound seems to be directed against the foe, for it is made, so far as I know, only when the swell-fish is molested. Granting, however, that the swell-fish does not hear its own sound, one would still be rash to conclude that this was an argument against the hearing of fishes, for the vast majority of animals toward which the sound is directed are fishes themselves, and these presumably hear the sounds.

Another good instance of the production of sound by a fish is found in the squeteague or weak-fish, *Cynoscion regalis*. The grunting noise made by this fish is, however, produced only by the males, and this specialization is very difficult to understand unless one assumes an ability on the part of one or other sex to hear. Since the sounds made by both the swell-fish and the squeteague are in no sense shocks or concussions but resemble more closely, in rate of vibration and in intensity, such sounds as might be obtained from the ordinary action of an instrument like a tuning-fork of low pitch, it seems to me that they afford evidence in favor of the sense of hearing rather than the reverse.

A second reason for questioning the generalization advocated by Kreidl, and by Lee, is the character of the observations upon which it is based. Both authors state that no positive evidence in favor of hearing could be obtained. But it must be borne in mind that in many animals known to possess a sense of hearing the auditory reflexes are perhaps the least conspicuous of any connected with the more important sense organs, and that consequently the most careful scrutiny of the movements of fishes must be made before one can with certainty declare that hearing is absent. A perusal of the papers already summarized led me to the conclusion

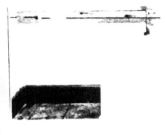

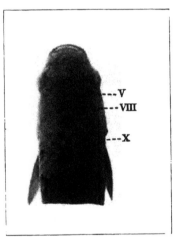

1. Side view of aquarium, showing sounding apparatus at right-hand end and suspended glass cage in which the fish were confi

2. End view of aquarium, showing sounding apparatus.

3. Dorsal view of brain of *Fundulus heteroclitus*, dissected to show positions of the roots of the fifth and seventh nerves (V), the r the ninth and tenth nerves (X), and the internal ear as indicated by its otolith. Enlarged about twice.

4. Dorsal view of head of *Fundulus heteroclitus*, to show the region where the following nerves were cut: The fifth and seventh (V eighth (VIII), and the lateral-line nerve (X). Enlarged about twice.

5. Side view of *Fundulus heteroclitus*, showing the region where the lateral-line nerve was cut (X). Slightly enlarged.

that something more might be attained in this direction, and I therefore resolved to give particular attention to the reactions of a few fishes with the view of ascertaining whether or not they showed any evidence of hearing.

At the outset I thought it best to experiment on some common sound-producing species, and for this purpose I did some preliminary work on the swell-fish (*Chilomycterus schœpfi*), the squeteague (*Cynoscion regalis*), and the sea robin (*Prionotus carolinus*). To all of these, practical objections were found, and I was at last obliged to abandon them for fishes that produce no sounds. Among these, three species were found to be especially sensitive to slight vibrations—the king-fish (*Menticirrhus saxatilis*), and the two common species of killi-fish (*Fundulus majalis* and *F. heteroclitus*). Because of the great abundance of *F. heteroclitus*, the ease with which it could be operated upon, and its great hardiness, I chose it for study, and the observations recorded on the following pages, unless otherwise stated, refer to this species.

The ears in Fundulus heteroclitus.—When a tank containing a number of *Fundulus heteroclitus* is struck with the open hand so that the fish can not see the movement of striking, they respond to the vibrations by springing suddenly an inch or so through the water. The question to be considered is whether these vibrations stimulate the fishes through the skin, the lateral-line organs, the ears, or some combination of these. If it could be shown that the ears were not stimulated by the vibrations, it seems to me that we would have evidence pointing to the conclusion that the fishes did not hear. If on the other hand it could be demonstrated that the vibrations did stimulate the ears, the evidence would be conclusive that the animals possessed the sense of hearing. To test these points considerable experimentation was necessary.

Much of the work that has been carried out heretofore has been done with sound generated in air but intended to affect fishes in water. That this method is extremely inefficient I found by trying the following experiment. If a dinner bell is rung in the air by a person standing breast-deep in water, it will, of course, be heard easily by a second person standing in a similar way a yard or two off. If, however, the second person puts his head under the water during the ringing of the bell the sound seems to cease almost entirely and is not again heard clearly by the diver till he emerges. In like manner a bell rung or hit with a stone under water is heard, at best, very faintly by a person standing in the water unless his head is under the surface. In other words, the plane separating air and water is, under ordinary circumstances, an almost impenetrable one for most sounds, whether they are generated on one side or the other of it, and many of the negative results obtained by previous investigators on the sense of hearing in fishes may have been due not so much to the absence of hearing in the animals experimented upon as to their inaccessibility to the sound, or at least to sound of an intensity sufficient to stimulate. This difficulty has been recognized by Kreidl, and in devising apparatus I have profited by his experience and used sound-producing appliances that were in direct contact with the water containing the fishes.

The chief piece of apparatus that I used consisted of an ordinary marine aquarium (pl. 9, fig. 1) with a slate base, two heavy glass sides, and originally two slate ends, one of which, however, I replaced by a piece of deal board free from knots, to serve as a sounding-board. The inside dimensions of the aquarium were as follows: depth, 40 cm. (16 in.); breadth, 37 cm. (15 in.); and length, 87 cm. (35 in.). To the middle of one edge of the sounding-board a stout beam of wood was attached so that it stood out horizontally about 1 meter (40 in.) in the plane of that end (fig. 2). From the free end of the beam a bass-viol string was stretched to the opposite side of the sounding-board. This string could be tightened by a bolt and nut at the free end of the beam, and it was made to pass over a bridge placed near the middle of the sounding-board. The length of the string from the attached end on the sounding-board to the bridge was 25 cm. (10 in.), and from the bridge to the attachment near the free end of the beam 1.15 meters (45 in.). Thus the end of the aquarium might be regarded as something like a large one-stringed bass viol resting sidewise, with the sounding-board for a body and the beam for a neck.

When the string was tightened and plucked or bowed a good tone was obtained, which was transmitted directly through the sounding-board to the water within the aquarium. On keying the string up to a good clear tone, I found by writing off its vibrations on a revolving cylinder that it produced on an average 40 per second, and I retained this pitch by frequent adjustment for the experiments that I subsequently performed. I was led to adopt this low tone since most of the noises that I have heard fishes make were in the nature of low-pitched grunts.

Each time the string was plucked the note began with maximum intensity and then gradually died away. It was, consequently, impossible to get any very significant record of the intensity, but I endeavored to use the apparatus in a uniform way by drawing the string out a constant distance from its position of rest each time I plucked it. The distance usually employed was about 1.5 cm. (0.6 in.). The amount of weight required, when hung at the middle of the longer segment of the string, to depress it 1.5 cm. was found to be about 2.15 kilograms (4.75 pounds), so that each time the string was liberated on being plucked in the usual way, it moved forward with an initial force equal to the pull of 2.15 kilograms, a rough measure of the maximum intensity of the sound produced.

The fishes to be experimented upon were not allowed to swim unrestricted in the aquarium, but they were placed in a small cage (fig. 1) suspended from a cord attached at its ends to the walls of the room. Thus the support for the fish cage was entirely independent of the walls of the aquarium and any vibration that reached the fishes must have done so almost entirely through the water. The cage could be moved in a horizontal direction back and forth on the cord, and thus the fish could be placed at any desired distance from the sounding-board up to 75 cm. (30 in.). The inside measurements of the cage were as follows: Height, 10 cm. (4 in.); length, 20 cm. (8 in.); and breadth, about 10 cm. (4 in.). The bottom of the cage was wood, padded on the inside with cotton wool covered with cloth to provide a deadened surface on which the fishes might rest. The top and three sides were glass; the fourth side was made of coarse netting to retain the fish but to interfere as little as possible with the entrance of sound, and this side was always directed toward the sounding-board. As the fishes averaged about 7 cm. (2¾ in.) in length, the cage gave them ample room for moving about.

My plan was to introduce fishes in various conditions into the cage, and, after they had become accustomed to their surroundings, to subject them to stimulation by sound and observe their reactions. I found it desirable to experiment with three classes of fishes; first, normal ones for a basis of comparison; secondly, fishes from which the ears had been removed; finally, fishes in which the general integument had been rendered insensitive, but in which the ears were intact. The methods of obtaining fishes in these conditions and the responses that they showed will be described for each class of fishes.

Normal fishes.—When a normal fish is first liberated in the cage it swims vigorously about for a few moments, after which it may, sooner or later, come to rest on or near the bottom. The animals are extremely quick-sighted, and, if after they have come to rest the observer makes any sudden movements near the aquarium, they are very likely to begin active swimming anew. It is, therefore, extremely necessary to work in such a way that all movements, and particularly quick ones, are made out of sight of the fish. When the fish is resting on the bottom of the cage, two sets of motions will usually be observed: first, the respiratory movements of the operculum; and secondly, the alternate vibratory movements of the pectoral fins. The opercular movements, as might be expected, always continue, but the movements of the pectoral fins, which seem to be connected also with the respiratory function, often cease entirely.

When a fish has become quiet, except for the respiratory movements, the vibrations from the string may call forth any of four kinds of responses. The first of these is the vibratory movement of the pectoral fins, either a few slight beats, if the fins were previously at rest, or an increased rate or extent of swing if they were previously in motion. The vibration of the string at the intensity ordinarily employed almost invariably called forth this reaction; thus, in ten observations taken from each of ten fishes at a distance of about 25 cm. (10 in.) from the sounding-board there were 96 pectoral-fin responses and 4 failures. Since this response is so readily observed, it has afforded one of the most satisfactory criteria of stimulation.

The second form of response is a change in the rate of the respiratory movements. In a quiescent fish measuring 8 cm. (3.2 in.) in length the respiratory rate was 114 per minute. On stimulating by sound this rate rose suddenly to 138 per minute for some ten or a dozen movements and then fell rapidly to about the former number. This is probably a very usual form of response, perhaps quite as much so as the movement of the pectoral fins, but the shortness of its duration and its inconspicuousness make it less satisfactory as an indication of stimulation than that afforded by the pectoral fins. If the sound from the string is of considerable intensity, the third form of response may appear, a slight motion of the caudal fin, beginning usually at the dorsal edge and proceeding as a wave ventrally. Finally, with strong stimulation, the fish may make a short but quick spring forward.

All these reactions have been obtained from fishes even at 75 cm. (30 inches) from the sounding-board, although the springing movements are more frequently observed when the animals are not so far from the source of sound. One very interesting fact about these reactions is that they can not be repeated rapidly for even a short period. A fish that responds to the first stimulus by a spring, may react to the second or to the third only by moving the pectoral fins, and to the fourth in no observable way. It is only when a considerable period of rest intervenes that the reactions may be repeatedly obtained; and I have found that the minimum period of rest is not far from one minute, though, even then, reactions may sometimes fail to appear.

Earless fishes.—The removal of the ears from a fish is a serious operation, but it is one which, after a little practice, may be accomplished with success and from which the fishes generally recover. These animals are easily etherized by putting them in sea water containing enough ether to give it a strong odor. On being transferred to pure sea water they quickly recover, and an individual that I etherized six times in the course of one afternoon finally recovered without showing any ill effects. The first method I used in operating on the etherized fishes was to open the cranium in the region of the ears and, after cutting the auditory nerves, to remove those nerves and the attached ear-sacs. These parts were easily identified from the fact that the auditory nerve emerges from the medulla almost exactly ventral to the cleft between that organ and the optic lobe, and the ear-sac, which is only partly surrounded by cartilage, lies in the cranial cavity only slightly peripheral to the point where the nerve leaves the medulla (pl. 9, fig. 3). After the operation the fishes were returned to pure sea water and, notwithstanding the exposure of the brain, a considerable number recovered and survived. One of these I kept for more than six weeks, and, though its swimming was characteristically irregular, it was alert and active and, except for a brief intervening period, it fed normally.

From the operation just described about one fish in ten recovered. This proportion was greatly increased by a second form of operation in which the auditory nerve was cut without opening the cranium (fig. 4). After a little practice I found that this could be done with great certainty and about eight out of ten fishes usually recovered. All fishes that had been operated on were kept at least twenty-four hours before they were subjected to experimentation.

Fishes in which the auditory nerves have been cut have very characteristic reactions. When resting or when swimming slowly they behave for the most part as normal fishes do, and, in fact, are often undistinguishable from individuals upon which no operation has been performed. When, however, they are stimulated to rapid locomotion, they swim either in irregular spirals, the same individual revolving sometimes to the right and sometimes to the left, or they turn over and over in irregular circles without accomplishing much real progression. This loss of orientation on attempting rapid locomotion has for some time been recognized as indicative of one of the chief functions of the ears in fishes—i. e., equilibration. It is probable that in resting or in swimming slowly the fish depends upon the eye for orientation, but in quick movements the ears come more into play, and hence after their loss quick movements are accompanied with lack of orientation. The forced movements thus observed may be taken with perfect certainty, so far as my experience goes, as evidence of the successful outcome of an attempt to cut the auditory nerves, for in the few cases where these movements failed to appear, subsequent dissection showed that the nerves had not been cut, and in all instances where the movements were observed and the animals afterwards dissected, the nerves were found severed.

A second feature of interest that generally characterized fishes with severed auditory nerves was the color that they finally assumed. Under ordinary circumstances the color of this species is a light greenish-gray. When etherized the fishes become very dark, with a mottling of blue-green on the sides and belly. After recovery from cutting the auditory nerves, the dark coloration disappears and the fish assumes a tint even paler than that of a normal individual. This tint is retained throughout life. Etherizing probably influences the chromatophores of the skin directly, but cutting the auditory nerves introduces changes that are probably dependent upon the nervous control of the chromatophores.

When the earless fishes were tested in the sounding apparatus, they yielded very interesting results. Unlike the gold-fishes experimented on by Kreidl (1895), they differed markedly from normal fish. In an extended series of observations on over 20 fishes I never once observed with certainty the springing reflex as a result of sounding the bass-viol string. The fishes were usually very active, and I was never able to ascertain with certainty whether they showed a change in the respiratory rate on stimulation. The pectoral-fin movements, however, were observed with much certainty. On 10 earless fishes I succeeded in getting 10 observations each to sound stimuli at about 25 cm. (10 inches) from the sounding-board. The total result was that in 82 observations there were no reactions and in the remaining 18 the reactions were at best slight ones. As the fishes often moved the pectoral fins without apparent cause, some of the 18 reactions may have been accidental coincidences, but others were so precise and typical that I am convinced they were due to stimulation. Earless fishes, therefore, differ from normal ones in that their pectoral-fin responses to vibrations from the bass-viol string are enormously reduced, though not entirely obliterated.

Fishes with insensitive skins.—For reasons already given it is imperative, before drawing conclusions from the condition of earless fishes, to examine the evidence afforded by those whose general surface has been rendered insensitive. In this way it is possible to ascertain what part the integument plays in the reception of sound vibrations. I had hoped that the integument of *Fundulus heteroclitus* could be rendered insensitive by immersing the fish for a short time in a solution of cocaine, but all attempts in this direction proved failures, since the drug acted much more vigorously as a poison than as an anæsthetic, and I was finally obliged to abandon this method altogether and resort to nerve-cutting.

The following operation performed on etherized fishes insures an almost complete insensibility of the surface. The fifth and seventh cranial nerves can be cut just posterior to the eyeball (pl. 9, figs. 3 and 4), the lateral-line branch of the tenth nerve can next be cut at the posterior edge of the pectoral girdle (fig. 5), and finally the spinal cord can be severed at the fourth or fifth vertebra. Severe as this operation is, almost all fishes recover from it and respire and feed normally, though they seldom live beyond two weeks after the operation.

Fishes that have recovered from this operation show certain well-marked characteristics. The integument, particularly that of the dorsal surface, is unusually dark, as a result of the expanded condition of the chromatophores. The fish's mouth is gaping and motionless in consequence of the motor portion of the fifth nerve having been cut. This condition, however, does not interfere with respiration or with the sucking in of pieces of food, an act which the fish performs with avidity. Since in cutting the fifth and seventh nerves, the three small nerves to the muscles of the eyeballs, the third, fourth, and sixth, must also be cut, the eyes are motionless and usually protrude somewhat. Finally, as a result of cutting the spinal cord, the whole trunk of the animal is, as a rule, passive and is drawn after the head, the swimming being performed by the pectoral fins. Since the greater part of the cord is intact, a more or less vigorous stimulus applied to the trunk is followed by movements in the dorsal, anal, and caudal fins, or even by a locomotor response of the whole trunk, but such movements are made only after special stimulation, and the trunk is ordinarily carried passively, like a paralyzed appendage. As a result of having so little of the normal locomotor apparatus intact, the fishes often swim ventral side up, for the action of the pectoral fins is not always sufficient to overcome the physical effects of the specific gravity of the fish's body.

Fishes that have recovered from the operation just described have intact the ears, the central nervous organs from the anterior end of the brain caudad to the fourth or fifth vertebra, and the sensory and motor apparatus for the region of the gills and the pectoral fins. Excepting in these two rather restricted regions, the whole integument is insensitive, at least so far as its capacity to originate impulses to movements in the gills or pectoral fins is concerned. Such fish, therefore, are in a condition to receive stimuli through the ears and to respond by respiratory or pectoral-fin movements.

The reactions that these fishes showed to the sound apparatus were surprisingly clear and decisive. From the nature of the operation one would not expect them to be able to give the sudden spring that the normal fishes often showed, and, as a matter of fact, such responses were never observed. Were the skin of the trunk sensitive, it is conceivable that the caudal-fin reaction might occur, for the cord, though severed from the rest of the central nervous organs, was in itself intact. Caudal-fin reactions were, however, also never observed. The respiratory reactions and the pectoral-fin responses occurred with great regularity. When the bass-viol string was made to vibrate, the respiratory rate increased for a very brief period. In a fish 7 cm. (2¾ inches) long the rate previous to stimulation was 120 per minute; immediately after stimulation it was 156. The reactions of the pectoral fin were also well marked. In ten observations on each of ten animals at a distance of about 25 cm. (10 inches) from the sounding-board the pectoral-fin responses occurred 94 times in the total hundred. This is in close agreement with the normal fishes and in strong contrast with the earless ones. So far, then, as reactions to the vibrating chord are concerned, these fishes show the essential characteristics of normal individuals.

Discussion of the results of the experiments.—It is clear from the experiments described in the preceding sections that fishes whose ears were rendered functionless, but whose skins were normally sensitive, reacted only slightly to the stimulus from the sound-producing apparatus, whereas those with insensitive skins but functional ears responded to this stimulus, as far as their conditions would permit, almost exactly as normal fishes did. It might be assumed that the failure to respond on the part of earless fishes was due not to the loss of the ear, but to the shock of the operation they had undergone. This, however, does not seem to be the case, for, after the fishes had recovered from the immediate effects of the operation, they were active, fed well, and sometimes lived many weeks. Moreover, if the operation were as severe as is implied in the above assumption, one might expect some indications of this in fishes in which only one auditory nerve had been cut. As a matter of fact, immediately after this operation fishes with only one ear intact did swim irregularly, but in from six to eight hours this tendency disappeared entirely, and the fish in its quickness, precision, and normality of response became, so far as my observations went, absolutely indistinguishable from a normal individual. Further, fishes with the fifth, seventh, and lateral-line nerves and spinal cord cut have without doubt suffered a more severe shock than those that have had only the eighth nerve cut, and yet the pectoral-fin reactions of the former were essentially normal. It therefore seems to me that the great reduction in the number of pectoral-fin reactions of earless fishes is due to the loss of the ear as a sense organ and not to secondary complications accompanying the operation.

Although some of the observations recorded on the preceding pages make it certain that in these fishes the ears are stimulated by disturbances such as those set up in the water by the sounding apparatus, it may still fairly be asked whether these disturbances are in the nature of sounds. When the bass-viol string attached to the aquarium was plucked, a series of sound waves of diminishing intensities was delivered to the water. To ascertain something of the nature of this sound I immersed my head in the water of the aquarium and had an assistant pluck the string in the usual way. The sound that I thus heard was, so far as I could judge, of nearly the same pitch as that which the string gave to the air and of only slightly greater intensity. This sound certainly reached the fishes.

The sounding apparatus, however, did more than give rise to this sound. When the string was plucked two things besides the production of sound certainly happened: First, the whole aquarium, including its supporting table, trembled slightly, and, probably as a consequence of this, ripples started from the ends and sides of the aquarium and proceeded toward the center. These ripples, though chiefly surface effects, indicated a wave motion that penetrated the water to some extent, and that was doubtless the cause of the very slight swaying movement of the fish cage occasionally noticed after the string had been vigorously plucked. Moreover, a distinct tremor could be felt in the water when the hand was held 5 to 8 cm. (2 to 3 inches) from the sounding-board and the string was plucked. The question naturally arose whether the fishes did not respond to the movement of the aquarium as a whole or to the wave movement indicated by the ripples rather than to the true sound waves.

To answer this question, at least so far as the ripple movement was concerned, I was led to study the reaction time of the fishes. Unfortunately circumstances prevented me from reducing this to a very accurate process; but, by listening to the beat of a chronometer and at the same time watching the fish, I am confident that the fin reactions occurred in less than 0.2 second after the string had been plucked. The sound waves and ripples mentioned above traveled from the sounding-board toward the fish at very different rates. The sound waves must have passed over the 25 cm. of water between the sounding-board and the fish almost instantly. The surface ripple traveled much less rapidly and its rate could be easily measured. This proved to be a meter (39¾ inches) in 4.8 seconds; hence, to traverse 25 cm. (10 inches) the ripple required about 1.2 seconds. Since the fishes responded in less than 0.2 second, they must have reacted to something other than the disturbance indicated by the ripples.

Having eliminated the ripples as the initial stimulus for the fishes, it remained to be shown whether this stimulus was the movement of the whole aquarium or the sound waves proper. I succeeded in doing this by substituting an electric tuning-fork for the bass-viol string. The tuning-fork was placed so that its base was within about a millimeter (¹⁄₂₅ inch) of the sounding-board. The iron frame holding the fork rested on supports made of rubber bottle-stoppers. These flexible supports allowed the fork to be moved enough to bring its base into contact with the sounding-board without moving the supports over the surface on which they rested. As this could be done without any initial jar, it was possible to communicate to the water in the aquarium a sound of uniform intensity and pitch without moving the aquarium as a whole and also without producing any ripple. The fork, moreover, produced a tone much purer than that obtained from the string. It had a pitch of 128 vibrations per second.

Earless fishes, when subjected to sound waves from the tuning-fork, showed nothing that I could identify as a reaction. Normal fishes and fishes with normal ears but insensitive skins very usually reacted by pectoral-fin movements. The occasional failure to respond was attributed by me to the faintness of the vibrations, for the most intense sound obtained from the fork was much less than that produced by the bass-viol string. That the fishes, however, always did react, even to this relatively faint tone, was pointed out to me by my friend Dr. F. S. Lee, who while watching one of the experiments thought he detected an increase in the respiratory rate even when no pectoral-fin reaction occurred. Subsequent study showed this to be entirely correct, for, irrespective of pectoral-fin responses, at each sounding from the tuning-fork an increase of the respiratory rate did take place for a very short period. There is, then, no question but that these fishes respond to sound waves, and, since this response is through the ear, I conclude that *Fundulus heteroclitus* may be said to hear. Since I never succeeded in getting reactions of any kind to the tuning-fork from earless fishes with skins and lateral lines intact, I have no reason for believing that these parts are stimulated by true sound waves, and I attribute the responses that earless fishes occasionally showed to the vibrating bass-viol string not to the action of its sound waves on the skin or the lateral-line organs, but, as will be shown later, to the influence of the accompanying movement of the whole aquarium and its contained water on these parts.

Although the experiments already described remove every reasonable doubt from my mind as to the ability of these fishes to hear, the objection may still be raised that the conditions under which they were carried out were so artificial that they may be said to have almost no bearing on the ordinary habits of *Fundulus*, and it must be admitted that the relatively small volume of water in the aquarium and the character of its walls as reflecting ·surfaces for sound, may possibly have introduced factors to which the fishes, in their natural surroundings, were not accustomed. To ascertain how much weight should be given to this objection the following experiment was tried. The sounding apparatus, consisting of the sounding-board and the bass-viol string, was taken from the aquarium and set up in the open water of the outer pool at the Fish Commission wharf. The fish cage was hung at a distance of 50 centimeters (20 inches) from the sounding-board and toward the center of the pool, which is about 100 feet wide. The sound, therefore, was as unrestricted as that which naturally reaches these fishes. On experimenting with normal fishes, fishes without ears, and those with insensitive skins, results were obtained essentially like those observed in the aquarium, and I therefore concluded that the restriction of the water in the aquarium played no essential part in the results obtained from that apparatus. There is, thus, good reason to believe that *Fundulus heteroclitus* not only hears, but that for it hearing is a normal process.

Having determined that hearing was one of the normal functions of the ears in *Fundulus*, I had hoped to be able to ascertain by experiment the particular part of the ear, if such there be, that was concerned with this sense. The internal ear in *Fundulus heteroclitus* is like that in most teleosts. It consists of the usual three semicircular canals and a large sacculus, at whose posterior end a well-developed lagena is present. The sacculus is a thin-walled chamber, vertically flattened and containing a thin, flat otolith of considerable size. Sometimes this otolith is represented by two pieces—a small one at the anterior end of the sacculus and a much larger one occupying the more central part of this chamber. The lagena, which is well separated from the sacculus, also contains an otolith. On the median face of the sacculus is an extensive macula acustica sacculi, formed by the termination of the major part of the eighth nerve. There is also a well-developed papilla acustica lagenæ, as well as the usual three cristæ acusticæ ampullarum. I am unable to state whether other sensory patches, such as the macula acustica neglecta, occur here or not.

Having made a preliminary study of the anatomy of the internal ear, I had hoped to be able to cut in different individuals different branches of the eighth nerve, and, by further experimentation on fishes thus prepared, to determine the functions of the several sense organs of the internal ear. After numerous unsuccessful attempts I was at last obliged to abandon this plan because of the small size of the branches and their somewhat intricate relations, and I am, therefore, in possession of no observations that show which part or parts of the internal ear are concerned in hearing.

THE LATERAL-LINE ORGANS AND THE SKIN.

Introductory.—The lateral-line canals were regarded by most of the earlier investigators as glands for the production of the mucus so characteristic of the skins of fishes. About the middle of the last century Leydig (1850, p. 171) discovered the numerous sense organs contained in these canals, and declared that the whole system represented a sensory apparatus peculiar to fishes. Subsequently Leydig (1868, p. 2) expressed the opinion that these organs implied the possession of a sixth sense, one in addition to the five usually attributed to vertebrates, though he admitted that this sense was probably closely related to touch. Two years later the lateral-line organs were investigated by Schulze (1870), who demonstrated that true lateral-line organs were found only in the water-inhabiting vertebrates. From a study of their structure Schulze (1870, p. 86) was led to the belief that they were stimulated by the mass movement of the water, as when a current passes over the surface of a fish or when the fish swims through the water. He further believed that they were stimulated by sound waves whose length was greater than that of waves to which the ear was adapted. In this respect they were organs somewhat intermediate in character between those of touch and of hearing. These opinions were opposed by Merkel (1880, p. 54), who pointed out the inaccessibility of the organs to moving water in many cases, and who regarded them merely as organs of touch. The opposite extreme was taken by P. and F. Sarasin (1887–1890, p. 54), who designated them accessory ears, a view suggested some years previously by Emery (1880, p. 48).

The opinions thus far given were based for the most part on an interpretation of the anatomy of the lateral-line organs, and not upon any positive experimental evidence as to the function of these parts. Fuchs (1895, p. 467) seems to have been the first to attempt work in this direction. His experiments were made chiefly on the torpedo, a fish in which, in addition to the lateral line proper, two other sets or organs, the vesicles of Savi and the ampullæ of Lorenzini, may be regarded as parts of the lateral-line system. In an active torpedo Fuchs cut the nerves connected with these two special sets of organs without, however, being able to detect any significant change in the subsequent movements of the fish. He then exposed the nerve innervating the vesicles of Savi, and having placed it in connection with the appropriate electrical apparatus, he found that on pressing lightly upon the vesicles a negative variation in the current from the nerve could be detected. As this negative variation is evidence of the momentarily active condition of the nerve, it follows that pressure differences may be assumed to be a means of stimulating the vesicles of Savi. No such results were obtained from the nerves distributed to the ampullæ of Lorenzini, but the nerves from the unmodified lateral-line organs in *Raja clavata* and *R. asterias* showed negative variations when their terminal organs were subjected to pressure. Dilute acids and changes of temperature were not stimuli for any of the terminal organs tested, and Fuchs (1895, p. 474) concluded that pressure was the normal stimulus in the skate for the lateral-line organs, and in the torpedo for the vesicles of Savi, but not for the ampullæ of Lorenzini.

Apparently without knowledge of the work done by Fuchs, Richard (1896, p. 131) performed some experiments on the gold-fish. These consisted in the removal of the scales from the lateral line and the destruction of the sense organs under these scales by cauterizing with heat, silver nitrate, or potassic hydrate. After this operation some of the fishes were unable to keep below the surface of the water, and though they soon died, Richard (1896, p. 133) believed that he had evidence enough to show that the lateral-line organs were connected with the production of gas in the hydrostatic apparatus.

Richard's conclusions were called in question by Bonnier (1896, p. 917), who pointed out the severity of the operations employed by the former and intimated that the results were more probably dependent upon the excessive amount of tissue removed than upon the destruction of the lateral line. Bonnier (1896, p. 918) further recorded experiments of his own in which the lateral-line organs were destroyed by electro-cautery. Fishes thus operated upon showed two characteristics—they could easily be approached by the hand and even seized, and they failed to orient themselves in reference to disturbances caused by bodies thrown into the water. Bonnier concluded from his experiments that the lateral line, in addition to other functions, had to do with the orientation of fishes in reference to centers in the water from which shock-like vibrations might proceed.

Lee (1898, p. 139), whose experimental methods were much the same as those used by Bonnier, obtained some significant results, particularly with the toad-fish, *Batrachus tau*. When the pectoral and pelvic fins of this fish were removed, so that the animal might be said to be without its usual mechanical support, and the lateral-line organs were destroyed by thermo-cautery, the animal would lie quietly for some time, either on its side or back, and acted as though it had lost its "sense of equilibration." That its condition was not due to excessive injury was seen from the fact that a finless fish in which an equal amount of skin had been cauterized, but in which the lateral-line organs were intact, showed no lack of equilibration, and in its general behavior closely resembled a normal fish. Moreover, stimulation of the central end of the lateral-line nerve resulted in perfectly coordinated fin movements, and Lee (1898, p. 144) therefore concluded that the organs of the lateral line are equilibrating organs. How these are stimulated Lee does not attempt to decide, though he suggests (1898, p. 143) that pressure changes in the surrounding medium may be the means of stimulation.

From this brief historical résumé it must be evident that there is still very little unity of opinion as to the functions of the lateral-line organs.

The lateral line in Fundulus heteroclitus.—The lateral-line system in *F. heteroclitus* presents a condition typical for teleosts. Its sense organs are contained in canals that open by pores on the surface of the skin. A lateral-line canal as indicated by its pores (pl. 9, fig. 5) extends along the side of the trunk from near the tail forward to the head. Here the arrangement of the pores (figs. 4, 5) gives evidence of a mandibular, a suborbital, a supraorbital, and an occipital branch. By cutting the fifth and the seventh nerves behind the eye (fig. 4), and the lateral-line nerve near the pectoral girdle (fig. 5), the innervation of this whole system, except a small tract above the gills, can be rendered inoperative; the sense organs in the small tract can be easily excised. Fishes that have undergone this operation recover almost invariably and in a very short time; the integument of their heads is insensitive owing to the necessity of cutting the fifth as well as the seventh nerves; but that of their trunks, which is of course innervated from spinal sources, retains its normal sensitiveness, except so far as the lateral-line organs are concerned. In seeking for evidence as to the function of the lateral-line organs, I compared carefully the reactions of normal fishes with those in which the nerves of the lateral-line organs had been cut.

When a normal fish is liberated in an aquarium, it swims at once to the bottom. Here it may move about excitedly for some minutes, after which it usually begins to make upward excursions. At first it will swim only part way to the top, returning each time quickly to the bottom. Eventually it may make several quick excursions to the upper surface of the water, and ultimately may remain there playing about close to the top. If now any disturbance is made the fish will again swim at once to the bottom, and only after some time will it return to the top, in the same cautious way as before. Almost any disturbance seems to drive the fish to the bottom—a flash of light on the water, a quick but noiseless movement of the observer, or an unseen blow on the aquarium, conditions all of which suggest that the movements of the fish are of a protective nature.

To one form of disturbance the fishes were particularly sensitive, and this was the slight movement of the whole aquarium that occurred whenever the bass-viol string was plucked. This movement could be produced without the accompanying sound by giving a slight vibratory motion to the beam attached to the sounding-board on the aquarium (fig. 2). It was remarkable how accurately the fishes responded to this stimulus. If the fish was playing at the top of the water, the slightest movement of the aquarium as a whole would cause it to descend immediately to the bottom; if it was on its upward course, it could be checked and made to descend at any point; and if it was near the bottom, it could be kept there as long as the movement continued. In all of the several hundred trials of this kind that I made, I never found a normal fish that would remain high in the water or swim upward while such movements were being imparted to the aquarium. Whenever the fish was above the bottom, the response was an instantaneous downward course.

With fishes in which the nerves to the lateral-line organs had been cut, the reactions were totally different. Such fishes, when left to themselves in an aquarium, were scarcely distinguishable from normal ones. As with the toad-fishes observed by Lee (1898, p. 140), the loss of the lateral-line organs seemed to interfere in no essential respects with the movements of the animals; they were active and quick, returned at once to a normal position when displaced, and oriented with accuracy, so far as I could see, in that they at once swam away from such centers of disturbance as come from dropping a stone in the water. In this last particular they were very unlike the fishes reported on by Bonnier (1896, p. 918). In one important point they differed absolutely from the normal fishes; they would swim upward and remain near the top during even a considerable agitation of the whole aquarium, though they would dart downward at any sudden movement on the part of the observer. Hence these fishes must have lost their capacity to be stimulated by the mass movement of the water, and since this defect was observed only after the lateral-line organs had been rendered inoperative, I concluded that the normal stimulus for these organs was a very slight mass movement of the surrounding water. Since such movements always accompanied the sound produced by the bass-viol string, it follows that the disturbances set up by this string must have acted as a stimulus for the lateral-line organs as well as for the ears, and it is therefore not surprising that earless fishes sometimes reacted when the string was plucked.

If the lateral-line organs are stimulated by a slight mass movement of the water, it occurred to me that I ought to be able to separate, in a mixed school of fishes, those with lateral-line organs intact from those in which the nerves to the organs had been cut, by simply imparting a slight mass movement to the water. Under such conditions the normal fishes ought to swim to the bottom, leaving the defective ones above; but on trying the experiment I found that the fishes were so accustomed to form a school that when the normal ones started for the bottom the others did the same, and I was entirely unable by this means to separate the normal from the defective individuals. But I finally succeeded in doing this by modifying the experiment, in that I used only two individuals, one normal and one defective, and agitated the aquarium only when they were widely separated. The result was very decisive in that the normal one invariably took the initiative in descending, and in fact was often not followed by its defective companion.

Having found the conditions under which the lateral-line organs were stimulated, it is natural to inquire as to the exact nature of the stimulus. Ordinarily the fishes were induced to react by making the whole aquarium swing at about ten vibrations per second; but a like reaction was obtained from the normal fishes when a single swing, or what was as near as possible a single swing, was given to the aquarium. The stimulus therefore is not necessarily of a vibratory kind, but consists in a slight movement of the body of water as a whole. It might be supposed that since the fish was suspended in the water, the motion of the aquarium as a whole could have no influence on it. But it must be remembered that the fish was somewhat heavier than the water, and that each time the aquarium was moved the fish, from its inertia, must have lagged a little behind or, once set in motion, moved a little ahead, and it is this slight difference in the rate of movement of the fish and of the adjacent water that, in my opinion, induces stimulation. I am not prepared to say how this affects the sense organs in the lateral-line canals; but it is not impossible, as Schulze (1870, p. 85) suggested, that slight currents are thereby set up that move and thus stimulate the bristle cells of the lateral-line organs.

The extreme sensitiveness of animals to slight motions of this kind has already been pointed out by Whitman (1899, pp. 287 and 302) in the leech and salamander, and I suspect that the sensitiveness of the blind fish, as observed by Eigenmann and quoted by Whitman (1899, p. 303), may also be in the nature of a lateral-line response.

Having reached the conclusion that the downward swimming of the fishes could be brought about by stimulating the lateral-line organs through slight mass-movements of the water, I next attempted to ascertain the relative importance of different parts of the lateral-line system in this reaction. I prepared one set of the fishes in which the lateral-line nerves were cut close to the pectoral girdles, thus rendering ineffective the lateral-line organs of the trunk while those of the head were left intact. These individuals responded in all respects, so far as I could see, as normal fishes did, and I therefore concluded that the lateral line proper was not an essential part of this system of sense organs.

In the second set of fishes I cut the fifth and seventh nerves of both sides, thus preventing the lateral-line organs of the head from acting. These animals always descended when the aquarium was shaken, but with noticeably less precision than in the cases of normal individuals. It therefore seemed probable to me that the portion of the lateral-line system on the head was more effective than that on the trunk, but as this experiment involved cutting the general cutaneous nerves of the head as well as the lateral-line nerves, the experiment is not wholly conclusive.

Finally, in a third set of fishes, I cut the lateral-line nerves and the fifth and seventh nerves of the right sides only, leaving the left sides intact. These fishes, though a little sluggish, reacted in an essentially normal way. From these three sets of experiments I conclude that the lateral-line organs may be considerably reduced without seriously impairing the action of the system as a whole, though the portion of the system on the head is less easily dispensed with than that on the trunk.

The skin in Fundulus heteroclitus.—While I was experimenting on fishes in which the lateral-line organs had been rendered inoperative I was at times puzzled by getting reactions that seemed contradictory to the general conclusion that such fishes were not stimulated by a slight movement of the whole mass of water. Occasionally on making the whole aquarium move slightly a fish without lateral-line organs would swim rapidly to the bottom. On watching for instances of this kind I soon found that they occurred only when the fishes were close to the top of the water, and in fact were within the range of wave action. When the whole aquarium was moved, even only slightly, the upper surface of the water was thrown into small waves. These waves, as could be seen by the motion of small suspended particles, extended only a few centimeters below the surface of the water, but they established a region into which the fishes without lateral-line organs would not ascend, and from which, if overtaken by the waves there, they immediately escaped by swimming downward. As fishes without ears as well as without lateral-line organs were stimulated by these surface waves, I concluded that in this instance the motion of the water must affect the general cutaneous nerves (touch).

If the motion of surface waves is a stimulus for the general cutaneous nerves, it would seem probable that currents in the water would also affect these nerves and that the ability of a fish to head up a stream might depend rather on the stimulation of its skin than, as Schulze has implied, on the stimulation of its lateral-line organs. *Fundulus* is in a marked degree rheotactic, i. e., it swims vigorously against a current, and I therefore resolved to test this fish to ascertain whether its rheotaxis depended on its lateral-line organs or not. Six specimens, in which the nerves to the lateral-line organs had been cut, were placed one after another at the open end of a large glass tube through which a moderately strong current of sea water was flowing. All swam energetically up the tube, and, so far as this reaction was concerned, they were in no observable respect to be distinguished from normal fishes. Their rheotaxis certainly did not depend upon their lateral-line organs, but was undoubtedly the result of cutaneous stimulation. Unfortunately I was unable so to operate on other individuals that I could obtain active specimens whose cutaneous nerves were severed but whose lateral-line systems were intact, and hence the only conclusion I can draw is that the general cutaneous nerves are stimulated by wave and current action and that this is sufficient to account for rheotaxis, but I can not state whether or not the lateral-line organs are also stimulated by these means.

Conclusions concerning the lateral-line organs and the skin.—The observations on *Fundulus* recorded in the preceding pages give no support to the view of P. and F. Sarasin that the lateral-line organs are to be regarded as accessory ears, for individuals in which the eighth nerves had been cut and in which the lateral-line organs were intact did not respond to the sound-waves from a tuning-fork to which fishes with ears reacted with certainty. I have also seen no reason to suppose that the lateral-line organs are especially connected with the production of gas in the air-bladder, as suggested by Richard, or that they are particularly concerned with equilibration, as advocated by Lee. Since they are stimulated by slight disturbances in the water that do not affect the general cutaneous sense organs, I can not agree with Merkel in classing them as tactile organs. Their appropriate stimulus is a slight mass-movement of the water, which may or may not be vibratory, and which induces the fish to swim into deeper regions. This form of stimulus is of precisely the kind that was attributed to these organs by Schulze (1870), but I have not been able to confirm Schulze's further opinion that current and surface wave movements stimulate these parts. Such stimuli certainly do affect the general cutaneous sense organs, but whether or not they influence the lateral-line organs I am unable to say.

GENERAL REACTIONS OF OTHER FISHES.

Although my studies were made almost exclusively on *Fundulus heteroclitus*, I tested, as opportunities offered, other species of common fishes. These were placed without being operated upon in the aquarium with the bass-viol string as a means of producing sound. Because of the mixed character of the stimulus produced by this apparatus and also because of the fact that the fishes were not operated upon in any way, the results are significant in only one or two instances.

Young mackerel, while swimming in the aquarium, always moved downward when the string was vibrated. The same was found true of adult mackerel, but whether this reaction was an ear or a lateral-line response was not determined.

Menhaden, after they became somewhat accustomed to the aquarium, gave a sudden leap each time the string vibrated, but showed no tendency to descend. In this instance, too, no clew was obtained as to the organs stimulated.

Three specimens of smooth dog-fish, each about 18 inches long, were tested. When these fish were resting quietly on the stone bottom of the aquarium, the vibration of the string would cause them to move their pectoral and pelvic fins, or even begin swimming, but when they rested on some 3 inches of cotton wool covered with a cloth to afford a deadened surface on the bottom of the aquarium, no reaction of any kind was ever obtained. Apparently the ears, lateral-line organs, and skins of these fishes are not open to any of the stimuli produced by the vibrations of the bass-viol string and transmitted through the water, and they thus differ markedly from the other fishes examined.

These few notes serve to show that different species respond very differently to the same forms of stimuli and emphasize the importance of refraining from generalizations on the functions of the lateral-line organs and the ears in fishes before a considerable number of species have been fully examined.

SUMMARY.

1. Normal *Fundulus heteroclitus* reacts to the sound waves from a tuning-fork of 128 vibrations per second by movements of the pectoral fins and by an increase in the respiratory rate. It probably also responds to sound waves by caudal-fin movements and by general locomotor movements.

2. Individuals in which the eighth (auditory) nerves have been cut do not respond to sound waves from the tuning-fork.

3. The absence of responses to sound waves in individuals with severed eighth nerves is not due to the shock of the operation or to other secondary causes, but to the loss of the ear as a sense organ.

4. *Fundulus heteroclitus* therefore possesses the sense of hearing.

5. The ears in this species are also organs of prime importance in equilibration.

6. Normal *Fundulus heteroclitus* swims downward from the top of the water and remains near the bottom when the aquarium in which it is contained is given a slight noiseless motion.

7. Individuals in which the nerves to the lateral-line organs have been cut will swim upward or remain at the top while the aquarium is being gently and noiselessly moved.

8. The lateral-line organs in this species are probably stimulated by a slight mass movement of the water against them. They are not stimulated by sound waves such as stimulate the ears.

9. Individuals in which the nerves to the lateral-line organs have been cut swim downward and thus escape from regions of surface wave action. They also orient perfectly in swimming against a current. Since surface waves and current action stimulate fishes in which the nerves to the lateral-line organs and to the ears have been cut, these motions must stimulate the general cutaneous nerves (touch).

10. The vibrations from a bass-viol string when transmitted to water stimulate the ears and the lateral-line organs of *Fundulus*. They also stimulate mackerel and menhaden, but not the smooth dog-fish, which responds only when in contact with solid portions of an aquarium subjected to vibrations.

The work recorded on the preceding pages was done at the biological laboratory of the United States Fish Commission at Woods Hole, Mass., and I take this opportunity of expressing my indebtedness to the Director, Dr. Hugh M. Smith, and to his assistants for much help rendered me. I am also under obligations to Prof. W. C. Sabine, of Harvard University, for advice and assistance in connection with the sound-producing apparatus, and to Prof. F. S. Lee, of Columbia University, for friendly criticism and many suggestions.

LIST OF REFERENCES.

BATESON, W.
  1890. The sense-organs and perceptions of fishes: with remarks on the supply of bait. Journal
          of the Marine Biological Association of the United Kingdom. New series, vol. 1, pp.
          225–256, pl. xx.
BONNIER, P.
  1896. Sur le sens latéral. Comptes rendus des séances et mémoires de la société de biologie.
          Série 10, Tome 3, pp. 917–919.
EMERY, C.
  1880. Le Specie del Genere Fierasfer nel Golfo di Napoli, e Regioni limitrofe. Fauna und Flora
          des Golfes von Neapel. 2 Monographie, Leipzig. 76 pp., Tav. I–IX.
FUCHS, S.
  1895. Ueber die Function der unter der Haut liegenden Canalsysteme bei den Selachiern.
          Archiv für die gesammte Physiologie. Bd. 59, pp. 454–478. Taf. VI.
GÜNTHER, A. C. L. G.
  1880. An introduction to the study of fishes. Edinburgh. XVI · 720 pp.
HUNTER, J.
  1782. Account of the organ of hearing in fish. Philosophical transactions of the Royal Society
          of London. Vol. 72, pp. 379–383.
KREIDL, H.
  1895. Ueber die Perception der Schallwellen bei den Fischen. Archiv für die gesammte
          Physiologie. Bd. 61, pp. 450–464.
  1896. Ein weiterer Versuch über das angebliche Hören eines Glockenzeichens durch die
          Fische. Archiv für die gesammte Physiologie. Bd. 63, pp. 581–586.
LEE, F. S.
  1898. The functions of the ear and the lateral line in fishes. American Journal of Physiology.
          Vol. 1, pp. 128–144.
LEYDIG, F.
  1850. Ueber die Schleimkanäle der Knochenfische. Archiv für Anatomie, Physiologie und wis-
          senschaftliche Medicin. Jahrgang 1850, pp. 170–181. Taf. IV, figs. 1–3.
  1868. Ueber Organe eines sechsten Sinnes. Dresden. 108 pp., Taf. I–V.
MERKEL, F.
  1880. Ueber die Endigungen der sensiblen Nerven in der Haut der Wirbelthiere. Rostock. 214
          pp. Taf. I–XV.
MÜLLER, J.
  1848. The physiology of the senses, voice, and muscular motions, with the mental faculties.
          Translated by W. Baly. London. XVII + pp. 849 to 1419 +32 + 22 pp.
OWEN, R.
  1866. On the anatomy of vertebrates. Vol. I. London. XLII+650 pp.
RICHARD, J.
  1896. Sur les functions de la ligne latérale du Cyprin doré. Comptes rendus des séances et
          mémoires de la société de biologie. Série 10, Tome 3, pp. 131–183.
ROMANES, G. J.
  1892. Animal intelligence. International Scientific Series. vol. 44. New York. XIV+520 pp.
SAGE, D., C. H. TOWNSEND, H. M. SMITH, and W. C. HARRIS.
  1902. Salmon and Trout. New York. X+417 pp.
SARASIN, P., und F. SARASIN.
  1887–1890. Zur Entwicklungsgeschichte und Anatomie der ceylonesischen Blindwühle Ichthyo-
          phis glutinosus, L. Ergebnisse naturfwissenschaftlicher Forschungen auf Ceylon.
          Bd. 2. 263 pp. Taf. I–III.
SCHULZE, F. E.
  1870. Ueber die Sinnesorgane der Seitenlinie bei Fischen und Amphibien. Archiv für mikro-
          skopische Anatomie. Bd. 6, pp. 62–88, Taf. IV–VI.
WHITMAN, C. O.
  1899. Animal Behavior. Biological Lectures from the Marine Biological Laboratory, Woods
          Holl, Mass., 1898, pp. 285–338.

NATURAL HISTORY OF THE QUINNAT SALMON.

A REPORT OF INVESTIGATIONS IN THE SACRAMENTO RIVER, 1896-1901.

By CLOUDSLEY RUTTER,

Naturalist, United States Fish Commission Steamer Albatross.

TABLE OF CONTENTS.

Daily observations of the migrations of the young salmon in the main river were made at two stations, about 325 miles apart, from January to May, 1899. The hatchery experiments were carried on during two seasons, 1897 and 1898, at Battle Creek hatchery. Twenty-five thousand eggs were hatched at the Hopkins Seaside Laboratory at Pacific Grove during the winter of 1898–99, and the young were used in experiments testing the effect of sea water on alevins and fry. Experiments in planting alevins and fry were made at Olema, Marin County, and at Sisson, Siskiyou County, in 1897 and 1898. The investigations in 1900 and 1901 consisted of observations on adult salmon at Pacific Grove and at various points on the Sacramento, especially at Black Diamond, Rio Vista, and Mill Creek and Battle Creek fisheries.

Observations of the habits of the young were made first by watching them in the water. This, of course, could be done only in the small streams, but was very successful in the work the first year at Olema. Specimens were secured by means of the ordinary Baird seine; the one most used in this investigation was 50 feet long and 7 feet deep; smaller seines, 20 and 15 feet long, were used in small streams; in some of the work in San Pablo Bay a net 150 feet long was used. Traps constructed to suit particular purposes were employed where occasion required.

The work was carried on under the direction of the United States Fish Commission, with the cooperation of the California Fish Commission, through Mr. N. B. Scofield, during the first two years. Mr. F. M. Chamberlain, of the United States Fish Commission steamer *Albatross*, assisted in the work from May, 1898, to April, 1899. Mr. A. B. Alexander, also of the steamer *Albatross*, began the work at Olema in 1897. Much of the success of the investigation is due to the interest and counsel of Mr. J. P. Babcock, of the California Fish Commission; and Prof. Charles H. Gilbert, of Stanford University, has aided much in planning the work and in affording facilities for studying the collections.

The author is under obligations to the agents of the Southern Pacific Company at the shipping-points along the Sacramento River, to the various fish-dealers in Sacramento, and to the salmon-packing associations at Benicia and Black Diamond for statements of the catch of salmon at various places along the river and bays and for other courtesies; also to the officials of the United States Weather Bureau at San Francisco and Red Bluff and to the agent of the Southern Pacific Company at Sacramento Bridge for river statistics. The directors of the Hopkins Seaside Laboratory at Pacific Grove gave the free use of laboratory facilities for carrying on the experiments at that place.

GENERAL RESULTS OF THE INVESTIGATIONS.

A great many points of more or less interest have been considered in this series of investigations, of which the following have the most practical value and deserve special mention:

1. The original object of the investigation has been carried out in determining that young salmon should be released from the hatcheries soon after the yolk has been entirely absorbed and that they should not be released in the headwaters late in the spring.

2. A method has been found for removing and fertilizing the eggs left in the fish after artificial spawning, thus increasing the take of eggs from a given number of fishes by about one-fifth.

3. The site for a new hatchery, Mill Creek station, has been discovered.

Other points of interest determined are:

a. The spermatozoa of the milt are active for only 3 to 5 minutes after the milt is mixed with water.

b. Ova become incapable of fertilization after 5 minutes' immersion in water, and good results can not be obtained after 1 minute.

c. Ova can be exposed to air for half an hour provided they are kept moist by the ovarian fluid.

d. Ova may be fertilized while immersed in the ovarian fluid, or in the slime from the skin, or in unclotted blood.

e. Ova are not affected by immersion in normal salt solution for half an hour, and are capable of fertilization thereafter.

f. Between the ages of 6 and 16 days, when the water temperature is about 50°, the embryo is especially sensitive and liable to injury. During this period the eggs should be handled with the greatest care.

g. Fungus is not a great pest at Battle Creek hatchery, probably owing to the considerable amount of silt carried in the water and deposited on the mat of filaments.

h. Alevins have many enemies in the streams; fry but few.

i. The fry begin feeding and commence their downstream migration as soon as the yolk is absorbed and they are able to swim.

j. The fry drift downstream tail first, traveling mostly at night and averaging about 10 miles a day. They are 4 or 5 months old when they reach the ocean.

k. A few of the later winter fry, about 10,000 to the mile in the Upper Sacramento, remain in the headwaters all summer, which is deleterious on account of slow growth.

l. The food of young salmon at all places and seasons is insects, larval or adult.

m. Salmon spend from 2 to 4 years in the ocean.

n. They usually return to the river through which they reached the ocean, because during their ocean life they do not get far away from its mouth.

o. The later fall salmon ascend the Sacramento River at the rate of 4 or 5 miles a day, being about 65 days reaching Tehama from Rio Vista. The spring salmon, without doubt, travel faster.

p. Salmon do not eat after leaving the ocean, and the stomach shrivels up to about a tenth of its normal size.

q. Salmon lose from 15 to 20 per cent of their weight in migrating, and from 10 to 15 per cent more in spawning.

r. The sexes can not be distinguished in salt water, but they differ greatly in fresh water. The males develop the long hooked jaw, the large canine teeth, the deep slab-sided body, and the color usually becomes more or less reddish. The females do not change in appearance except as is due to the loss of flesh, the development of the ova, and to the change in color from silvery to olive.

s. The males vary more in size than the females and are of two forms, adult and grilse; the grilse resemble the females, but are much smaller.

t. The percentage of fertilization in natural propagation is high, probably about 85 per cent.

u. The injuries received in fresh water are mostly due to exertions in spawning the last few ova.

v. It is well known that all Pacific salmons die immediately after spawning once, and this investigation simply bears out the fact.

THE SACRAMENTO RIVER AS A SALMON STREAM.

The Sacramento is a large river, navigable for boats as far as Red Bluff, which is 225 miles by rail from San Francisco. It is quite crooked, and the distance by water from Red Bluff to the Golden Gate is about 375 miles. The river rises in several small lakes in the mountains about 20 miles west of Sisson, in Siskiyou County, California, and for nearly half its length flows through a narrow canyon. The upper portion is a typical mountain stream, with innumerable pools and rapids and gravel beds, forming ideal spawning-places for the salmon, though it has not been visited by many of them during the past few years. Near the lower end of the canyon it receives Pit River from the east.

Pit River is a much larger stream than the Sacramento above their union. Its lower portion lies in a canyon and except in size is similar to the Upper Sacramento. About 75 miles above its mouth are the Pit River Falls, which, until a fishway was blasted out, were impassable for salmon. The upper portion of Pit River lies on a plateau and during the summer is a very unimportant stream.

The salmon that pass Pit River Falls spawn in Fall River, which enters Pit River a few miles above the falls. When seen in August, 1898, Fall River flowed several times as much water as Upper Pit River, though it is only 12 or 15 miles in a direct line from its mouth to its source. It is about 100 feet wide and 3 to 4 feet deep, flowing through a level plain and taking its rise in several large springs.

Hat Creek, draining Mount Lassen on the north, empties into Pit River a few miles below the falls. It is a considerable stream, but its ascent is difficult for salmon on account of very steep rapids.

McCloud River, draining Mount Shasta on the south, empties into Pit River near its mouth. It is two or three times the size of the Sacramento River above the mouth of Pit River and is an important salmon stream. Baird hatchery is located near its mouth.

The most important salmon stream of the basin, excepting the main river as noted below, is Battle Creek, which drains Mount Lassen on the west and empties into the Sacramento between Redding and Red Bluff. Battle Creek hatchery is located at the mouth of this stream.

A few miles below the mouth of Pit River, and just above Redding, the Sacramento emerges from the canyon through which it runs from its source and widens into a broad, shallow stream, though the current continues swift. Below Redding, for perhaps 100 or 150 miles as the river winds, it continues broad and shallow, with many short riffles and usually a gravel bank along one side. In ordinary years when the river is in its normal low-water condition the principal spawning-beds of the fall salmon are in this portion of the main river, notably in the vicinity of Red Bluff and Tehama. In November, 1900, the river was examined carefully between the mouth of Battle Creek and Tehama. Few salmon were seen until within a few miles of Red Bluff, but from that point on every riffle was covered with spawning-beds and dead salmon were everywhere abundant in their vicinity. Seventy-five dead fishes were counted at one time in the lower 100 yards of Mill Creek and in the river within 50 yards of its mouth.

A few miles above Red Bluff the river cuts through a range of hills, and for 2 or 3 miles consists of a series of rapids, the longest of which is known as Iron Canyon. After passing Iron Canyon the river again assumes the character found at and below Redding. Farther downstream the channel becomes deeper, gravel banks disappear, sand banks become less frequent, and rapids are wanting. Such is the character from Colusa to Sacramento. Below Sacramento it runs through a level country and for most of the distance is affected by tides. There are many sloughs, some connecting it with the San Joaquin. The Sacramento and San Joaquin rivers join as they empty into Suisun Bay.

The water of the upper part of Sacramento River and the upper tributaries is quite clear, and continues so until the mouth of Feather River is reached, from which point to the mouth it is very muddy. It is in the muddy water between the mouth of Feather River and Vallejo that the salmon for the markets are taken.

The only species of salmon regularly frequenting the Sacramento River is the quinnat. The dog salmon is found occasionally, four specimens having been seen during this investigation. Mr. Chamberlain reports finding single specimens of the blueback and humpback at Baird in 1899. The humpback has also been reported by others. The only record of the silver salmon from the Sacramento River is that given by Jordan & Jouy (Proc. U. S. National Museum 1881).

THE GENERAL LIFE-HISTORY OF THE PACIFIC SALMONS.

The salmon under consideration in this report is the Sacramento or quinnat salmon (*Oncorhynchus tschawytscha*), also known as Columbia River salmon, king salmon, and chinook salmon, and is the largest and most important of the five species of Pacific salmons. The others, in the order of their importance, are (1) blueback (*O. nerka*), also called red salmon, redfish, and sockeye; (2) silver salmon (*O. kisutch*); (3) humpback salmon (*O. gorbuscha*), and (4) dog salmon (*O. keta*).

The Pacific salmons, as above named, are different from the Atlantic salmon, which is related to the steelhead or salmon trout of the Pacific coast. One of the important characters that separate the Pacific salmons from the steelhead and its relatives, the Atlantic salmon and the rainbow and cut-throat trouts, is the larger number of rays in the anal fin—the unpaired fin on the under side of the tail. The steelhead and its allies have fewer than 13 rays (usually 11) in this fin, while the Pacific salmons have more than 13, the number for the quinnat being 16.[a]

An equally fundamental, though physiological, difference lies in the fact that the Atlantic salmon and steelhead trout spawn several times while the Pacific salmon, of whatever species, dies as soon as it spawns once. This is a very striking difference and its importance can hardly be overestimated. A further difference lies in the habits of the young. The young of the Pacific salmon seek the ocean as soon as they are able to swim; their migration is accelerated by high water and retarded by low water, and they do not return to fresh water till mature. On the contrary, young trout do not seek the ocean for several months after they are able to swim, low water is an incentive to migration, and they run back and forth between fresh and salt water seeking food.

The quinnat salmon is found throughout the Pacific coast from Monterey Bay northward, but is less abundant north of Puget Sound. At spawning time it frequents the larger streams, especially those with estuaries. The blueback salmon is the most abundant of the Pacific salmons, and is most numerous in Alaska. Its favorite spawning streams are those tributary to lakes. The silver salmon prefers

---

[a] For other anatomical characters, see Jordan & Evermann, Fishes of North and Middle America, Bulletin 47, United States National Museum.

the larger streams, though not necessarily those emptying through estuaries. The other two species are of little importance. They spawn in any kind of stream, frequently in mere brooks that empty directly into the ocean.

The Pacific salmon, of whatever species, passes most of its life in the ocean, and upon arriving at maturity ascends the rivers to spawn. Sometimes enormous numbers pass upstream together; stories of their being "thick enough to walk across on" are often told, but I have never seen them quite so numerous as that.

The upstream migration occurs sometime during the warmer half of the year, the earlier fish going farther upstream. Spawning occurs late in the summer or during the fall, and as soon as it is completed the salmon dies.

Most of the eggs deposited are eaten by other fishes, or are killed by being covered with sand and gravel. Those not destroyed hatch in from seven to ten weeks, according to the temperature of the water. In the cold waters of Alaska they are four or five months in hatching. It requires about six weeks more for the yolk-sac to be absorbed, when the fry are able to swim and are ready for their seaward migration. Most of the alevins, however, are devoured by other fishes before they are able to swim. It is to prevent this great mortality among eggs and alevins that artificial propagation has been employed.

The young salmon start downstream as soon as they are able to swim, and reach brackish water when three to five months old, according to the distance they have to go. Those from the vicinity of Battle Creek hatchery reach Benicia in about seven weeks. It is not known when they reach the ocean, but probably soon after. The variation in the time of spawning and hatching makes the period of migration very long. The fry from the summer run begin passing Battle Creek in September, and from that time until April following there is a continuous stream of young salmon, about 1¼ inches long, passing that point.

Although a large majority go downstream as soon as they can swim, many, especially those hatched in the spring, remain in the pools in the headwaters all summer and fall. There were estimated to be from 700 to 1,000 in each of several pools in the vicinity of Sims during the summer of 1898, and there were probably as many as 10,000 to the mile in that portion of the river. These remained in the headwaters until the first of December, when rains caused a rise in the river.

The food of young salmon in fresh water, at all times, places, and ages, consists of insects, either larval or adult.

SALMON EGGS AND MILT.

The following notes embody observations and experiments made at Battle Creek hatchery in 1897 and 1898. Although dealing largely with artificial propagation, they are not intended to give even a general account of the methods of fish-culture as applied to the Pacific salmon. For such an account reference is made to the Manual of Fish-culture issued by the United States Fish Commission.[a]

For facilities placed at my disposal I am under obligations to Mr. G. H. Lambson, superintendent of the station. Special acknowledgments are due to Mr. William Shebley, superintendent of Sisson hatchery, and to Mr. Robert Radcliff, of Baird, who had charge of the spawn-taking operations in 1897 and 1898, respectively, and heartily cooperated in the experimental work.

---

[a] Report United States Fish Commission 1897, pages 1-340.

PLATE 11.

MENTO RIVER NEAR SIMS, "POOL B," REFERRED TO ON PAGE 102.

SACRAMENTO RIVER IN THE VICINITY OF PRINCETON.

PLATE II.

MIXED RIVER BLUE BOG, "FOWL'S", REFERRED TO ON PAGE 98.

SANDSTONE ESKER IN THE VICINITY OF MONCTON.

EXPERIMENTS ON THE VITALITY OF SPERMATOZOA.

Vitality in water. Experiment No. 1.—Milt was mixed with water until the mixture became of a milky appearance, which was the condition desired for fertilizing the ova at the spawning platform. Small quantities of the mixture were taken from time to time and examined under a compound microscope. After several repetitions of the experiment it was found that the spermatozoa remained active in the water from 3 to 5 minutes, the length of time varying slightly in each experiment. Each time the spermatozoa were placed on the slide they were very active for about 30 seconds, after which time nearly all became attached by the tail to the slide or cover. The head continued to move for a few seconds longer, but all motion ceased after 65 seconds from the time the milt was placed on the slide.

Experiment No. 2.—Water was mixed with milt as in the above experiment and a small quantity used to fertilize eggs every half minute for 8½ minutes, with the following results, the eggs being freshly spawned in each case:

Time milt had been in the water.	Percentage of fertilization.	Time milt had been in the water.	Percentage of fertilization.
0.25 minute	98	4.5 minutes	7
0.5 minute	98	5 minutes	0
1 minute	98	5.5 minutes	0
1.5 minutes	4	6 minutes	0
2 minutes	8	6.5 minutes	0
2.5 minutes	2	7 minutes	2
3 minutes	8	7.5 minutes	0
3.5 minutes	1	8 minutes	0
4 minutes	0	8.5 minutes	0

The experiment was tried three times, but the results were practically the same. At one time a number of eggs were immersed in water taken from the top of the can in which the spawn was taken from the spawning platform to the hatchery. This water was white from the superfluous milt which had been spawned from 3 to 10 minutes. None of the eggs so treated were fertilized.

From these experiments it will be seen that the milt and eggs should be thoroughly mixed while in the spawning pan and within 30 seconds from the time the milt is mixed with water.

Activity in normal salt solution.[a]—Milt was mixed with normal salt solution until the liquid was distinctly whitish, and a portion of it was at intervals poured over freshly spawned eggs. After a short time the eggs were washed with fresh water. The per cent fertilized in each case is given in the following table:

Time solution had been spermatized.	Percentage of fertilization.	
	First attempt.	Second attempt.
1 minute	74	98
2 minutes	83	
4 minutes	55	
5 minutes		79
6 minutes	70	
8 minutes	34	15

[a] Normal salt solution, 0.75 per cent common salt in water.

The results of this experiment are too varying to be of any practical value. It seems probable that both spermatozoa and ova remain passive in the salt solution and that fertilization takes place only after the addition of fresh water in washing. Where a considerable quantity of fresh water was added and the ova mixed well with it before it was poured off the percentage of fertilization was high. When the water was poured off immediately and without mixing the ova well the fertilization was incomplete. Normal salt solution apparently preserves the vitality of the spermatozoa longer than fresh water.

Vitality in air.—It was found that milt kept in an open, large-mouth bottle for 24 hours fertilized 74 per cent of the eggs it was mixed with. Milt that had been so exposed 48 hours did not fertilize any eggs, nor did that kept in a tightly corked vial for 24 hours.

"Watery" milt.—Milt when taken from the fish varies greatly in consistency. That from some fishes is very thin and is known as watery milt. Experiment proves that it fertilizes eggs as well as any and that no larger quantity is needed.

Amount of milt required in artificial fertilization.—In taking spawn it was the custom at Battle Creek to express the eggs from one female into a pan containing about a pint of water and add enough milt to make the water distinctly whitish. The amount of milt necessary for this varies, depending on the amount of abdominal fluid mixed with it, but is never less than 3 or 4 fluid ounces. This method gives good results and should be followed when there is an abundance of males, which is always true at Battle Creek after the first few days of the season. A smaller amount of milt, however, will suffice. Ninety-six per cent of the eggs from one female were fertilized by a tablespoonful of milt; 85 per cent were fertilized by a teaspoonful; 35 and 57 per cent were fertilized by spawning fishes in the creek and letting the milt float over the eggs, which had been caught on a screen.

Of course, all that is necessary is to bring a very minute quantity of milt in contact with each egg. A single drop of milt if thoroughly disseminated through the water would be sufficient to fertilize all the eggs from one female. In the experiments above noted the milt could not have been thoroughly mixed until after it had become inactive. It is not advisable to use less than a fifth of an ounce of milt to fertilize 1,000 eggs. More water is necessary where a small amount of milt is used in order to facilitate thorough mixing.

EXPERIMENTS WITH OVA.

How to test fertilization.—The quickest and surest way to determine whether ova have been fertilized is to put them into a dilute (5 to 10 per cent) acetic acid. This can be made from commercial acetic by adding from two to five parts of water. A few minutes after the ova have been placed in the acid the embryos turn white, while the yolk remains clear. The embryo can be distinguished in this manner within 15 hours after the ovum has been fertilized. In making this test for the first time it is best to make a comparative test with unfertilized ova that have been kept in water during the same period.

Short exposure to water detrimental.—A quantity of eggs were spawned into a pan of water and some were removed and spermatized every half minute for several minutes, and for various periods up to several hours. The milt was, of course, taken fresh each time.

The results were as follows:

Time eggs had been in water.	Percentage of fertilization.	Time eggs had been in water.	Percentage of fertilization.
0.25 minute	98	4 minutes	4
0.50 minute	96	4.5 minutes	5
1 minute	95	5 minutes	4
1.5 minutes	68	5.5 minutes	1
2 minutes	57	6 minutes	0
2.5 minutes	33	6.5 minutes	0
3 minutes	17	7 minutes	1
3.5 minutes	17	7+	0

The susceptibility of ova to fertilization decreases rapidly after they are placed in fresh water, and the milt can not be added too quickly. Fifteen seconds is as long as they should be in the water before the milt is added, and it is preferable to add the milt at the same time that the ova are spawned.

On one occasion the eggs remaining in the body cavity after artificial spawning were removed by cutting the fish open, which mixed them with much blood. They were washed as quickly at possible and then spermatized. Only 11 per cent of the eggs were fertilized. In another experiment 85 per cent were fertilized in the blood without washing it off.

Effect of exposure to air.—A number of eggs were spermatized after having been exposed to the air, temperature 76°, for various periods with the following results, the ordinary method giving fertilization of 99 per cent:

Time ova were exposed to air.	Percentage of fertilization.
12 minutes	100
30 minutes	99
50 minutes	78

Apparently there is no injury to eggs by an exposure to the air for half an hour. It must be noted that although the eggs were in an open pan they were practically immersed in the liquid from the body cavity.

Fish slime not deleterious.—It is sometimes said that the slime on the fishes is fatal to the spermatozoa, and at some stations much care is taken to wipe the fish dry before spawning. To test the truth of the supposition, a pan of eggs were covered with slime scraped from several fishes, and then spermatized without the use of water. Out of 174 eggs examined only two were unfertilized, which is as good as is obtained by the ordinary process.

Fertilization in body fluid.—At another time, 392 eggs were spermatized "dry," and the milt entirely washed off with normal salt solution before water was added. All but six, or 98.5 per cent, were fertilized. This proves that water is not necessary to excite the activity of the spermatozoa, and that fertilization may be effected in the abdominal fluid alone.

Immersion in normal salt solution.—A quantity of eggs were immersed in normal salt solution and at the end of certain periods were taken out and fertilized in the usual manner. The results as shown in the table, while as good as could be desired, are scarcely better than are obtained by the ordinary method, which gives a fertilization of 99 per cent. The value of the experiment lies in the fact that it gives us a method of washing bloody eggs without preventing their fertilization, as will be noted below.   By making a chemical analysis of the fluid of the body cavity, a liquid could probably be prepared that would be entirely passive and in which the ova could be kept for days.   This, however, is unnecessary, as a saltness of three-fourths of 1 per cent gives a liquid sufficiently passive for washing out bloody eggs.

Time eggs were in salt solution.	Percentage of fertilization.
2 minutes	99
4 minutes	100
6 minutes	100
8 minutes	99
15 minutes	97
25 minutes	86

To make artificial spawning complete.—Even the best spawn-takers can not get all the eggs from the fish.   Often the fish is not entirely ripe; but whatever the condition may be many of the eggs are entangled in the folds of the ovary and viscera and are not spawned.   Under natural conditions the ovary shrivels up and does not obstruct the outward passage of the eggs.   The number of eggs remaining in the fish after the artificial spawning varies from 200 to 1,500, depending upon the size and condition of the fish.   I have found an average of 900 eggs remaining in 55 fishes after artificial spawning.   The spawning was done by experienced men and could not well be improved upon.

The average number of eggs taken from a fish in ordinary spawning is 5,000.   This was the average during the season of 1897.   By removing the remaining 900 eggs the yield can be increased about 18 per cent.   They can all be removed only by slitting the abdomen from the pectoral fins backward, but this allows a large quantity of blood to mix with them.   It is possible to fertilize 85 per cent of these eggs in the blood, but in this case the unfertilized 15 per cent have to be picked out of the hatching baskets, which would be a considerable expense if the plan were followed.   But the blood can be removed from the eggs without any detriment to fertilization by washing them in normal salt solution (one ounce of common salt to one gallon of water).   They can then be fertilized in the ordinary manner.

. This method has been used at Battle Creek hatchery since 1900 with satisfactory results, the loss with the "remnant" eggs being but little greater than with the ordinary take.   By care in handling the loss need not be any greater.   By such means the take of eggs can be increased from 10 to 20 per cent without increasing the cost appreciably.[a]

An aid to spawn-taking.—It was found that fishes were much more easily and rapidly spawned after cutting the body walls across the opening of the oviduct.   Unless the cut was made considerably in advance of the vent no perceptible amount of blood issued.   A greater percentage of the ova were spawned than if the gash had not been made, and no eggs were broken in spawning, which is an important point.   The shells or "shucks" from eggs broken in spawning are a great nuisance in the hatching basket, being difficult to pick out and forming a basis for the growth

---

[a] From his study of the physiology of the Sacramento salmon in 1902, Prof. C. W. Greene, of the University of Missouri, has determined the amount of salt in the ovarian fluid to be 0.94 per cent, which, therefore, is the density of the solution that is normal for salmon ova, and should be used in washing the blood from eggs, rather than 0.75 per cent as used in the experiments here noted.   A solution of 0.94 per cent can be made by adding 1¼ ounces of pure dry salt to 1 gallon of water.

of fungus if they are not removed. It was found by counting the number of broken eggs in several lots spawned in the ordinary manner that they averaged nearly 1 per cent of the entire take.

In the ordinary method of expressing the eggs they leave the oviduct under considerable pressure and strike the spawning pan with as much force as if they had fallen several feet. This manner of spawning, as already seen, breaks nearly 1 per cent of the eggs, and there may be many among those not broken that are injured; this may account for the heavy loss of the first day in the hatching-house.

Dry process of fertilization.—The method of fertilization used at Battle Creek in 1897 and 1898 was to spawn the eggs into a pan containing a little less than a pint of water, spermatizing them at the same time. They were then allowed to stand about 2½ minutes, when they were poured into a large bucket and gradually washed by adding fresh water. Basket No. 6 of the table given in the notes below on the critical period experiment was treated in this manner. The eggs of basket No. 5 were fertilized without any water, but otherwise were treated the same as No. 6. There was a difference of only 0.3 per cent in fertilization. There was a difference of 0.2 per cent between baskets No. 6 and No. 7, and they were from the same fishes and treated in the same way, so far as fertilization was concerned.

The method used at Battle Creek seems the better, as the eggs can be mixed with the milt more easily. A half minute, or just long enough to mix the eggs thoroughly, is an abundance of time for them to remain in the spawning-pan.

Killing the female before spawning.—It has been claimed by some fish-culturists that killing the female before spawning causes deformed fry. Basket No. 1 (see table on p. 79) contained eggs from fishes killed by a blow on the head. There were not even so many deformities in it as in others. This method of procedure is not recommended, however, as green fishes would sometimes be killed, and their eggs therefore lost.

Quality of bloody eggs.—Occasionally a female has been injured before spawning, and the eggs when pressed from the body were mixed with blood. Eggs from three such fishes were kept separate; 7.7 per cent of the eggs died within five days; of the remainder, 2 per cent (3 out of 154) were unfertilized. The fertilization was about as good as the average, and a small amount of blood seems not to be detrimental to fertilization. Several females were opened after spawning, and the eggs remaining were removed. The eggs were mixed with a great deal of blood, and only 85 per cent could be fertilized, so that a large amount of blood is detrimental to fertilization, probably because clots of blood prevent thorough mixing with the milt rather than from any injurious effect upon the ova or spermatozoa.

Foamy eggs.—Often the ovarian liquid becomes foamy as the eggs are spawned. It was not known whether such eggs were fertilizable. In the foamy eggs experimented with 99 per cent were fertilized.

Granular eggs.—The eggs from a certain small salmon, owing to the arrangement or superabundance of oil globules, had a peculiar granular appearance. Fertilization by ordinary process was 99 per cent; apparently healthy when 26 days old.

Eggs dead when spawned.—Occasionally eggs at the time of spawning have a dull, yellowish appearance, and are evidently not healthy. They are always thrown away. Eggs of this kind from one fish were kept. Seventeen days after spawning 30 per cent had died, and of the remainder 23 per cent were unfertilized. They were not kept for further observations. Such eggs were not found in 1898.

Eggs from dead fish.—On two occasions a ripe female was removed from the water, and after it had been dead 2 hours a few eggs were spawned and fertilized in the usual manner.   A few eggs were spawned from time to time until the fish had been dead 34 hours.   The following table gives the results:

Number of hours fish had been dead.	Percentage of fertilization.		Percentage of eggs that died within 10 days.
	First fish.	Second fish.	
2 hours	99		
4 hours	98		
6 hours	92	94	
8 hours	92	89	
10 hours		27	10
12 hours		62	31
16 hours		21	68
24 hours	0		100
34 hours		0	100

The first four lots of eggs of 2, 4, 6, and 8 hours were kept 26 days and were apparently entirely healthy at the end of that time.

At another time eggs were taken from two fishes that had died in the water. One had been dead 1 hour, the other over 6 hours. Of the first, over 97 per cent hatched and were healthy fry; of the other, 85 per cent.

From the above it is evidently safe to take the eggs from fishes that have been dead less than 5 hours, and fairly good results can be obtained up to 8 hours.

Spotted eggs.—Sometimes a considerable number of eggs, a few weeks after fertilization, have a small, irregular white spot about the size of the head of a pin in the yolk near the surface. This does not mean that the egg is about to die. Fifteen such eggs were put into a separate basket, and all hatched as perfectly healthy fry excepting one, which died in breaking through the shell. The spot did not appear on the yolk-sac.

Yellow eggs.—When eggs are nearly ready to hatch a yellowish fluid sometimes collects around the embryo. This does not affect them very seriously, as most of them hatch into healthy alevins.

The critical period for eggs.—It is a well-known fact that at a certain stage in development, from about the sixth to the sixteenth day, eggs are much more liable to injury than at other stages. When first taken they can be handled with comparative roughness with impunity. At Battle Creek in 1897 the spawning platform was about half a mile from the hatchery. The eggs were hauled this distance over a road that lacked much of being smooth, yet the loss traceable to such handling was slight. Of nearly 700,000 eyed eggs sent from Battle Creek to Olema at one time, less than 300 were killed in shipping. They were hauled about 10 miles in a heavy wagon, were on the train some 15 hours, and out of the water 48 hours. At the time of shipment the eggs were 43 days old. But at an earlier date, when 6 to 16 days old, such treatment would have killed every egg.

For purposes of comparison 60,000 eggs from several fishes, fertilized in the ordinary manner, were mixed in a can at the spawning platform, and at the hatchery were equally divided between two baskets. The eggs of one of the baskets were picked over daily regardless of results in order to remove the dead or addled eggs. The eggs in the other were picked over in the same manner on the first, third, twenty-second, twenty-fourth, and forty-first days, and occasionally after that date. The former of these experiments was called No. 7, the latter No. 6.   As a further test another basket, No. 1, of 30,000 eggs was picked daily. The loss of the baskets picked daily was from three to seven times greater than that of baskets not so treated.

The following table indicates the comparative loss in baskets No. 6 and No. 7, and shows that eggs are very sensitive between the sixth and sixteenth days, and that they should be disturbed as little as possible during that period. Basket No. 6 lost but 785 eggs from the fourth to the twenty-second day, while basket No. 7, which was picked daily, lost over 8,000. After the twenty-second day the loss in this basket, No. 7, was only 613, while it was 1,369 in No. 6, the one not picked daily. The loss on the forty-third day was the result of being shipped from Battle Creek to Olema, in which case the loss in No. 6 was nearly eight times that of No. 7. It is evident, therefore, that daily picking takes out nearly all the weak eggs, but it is also strikingly evident that it takes out a very great many that are not weak.

Table showing loss of eggs in baskets No. 6 and No. 7, taken November 15, 1897.

Age, days.	Temperature of water.	Loss.		Age, days.	Temperature of water.	Loss.	
		No. 6.	No. 7.			No. 6.	No. 7.
	°F.				°F.		
1	49	303	306	35	47		16
2	49		54	43		665	84
3	52	56	114	46			18
4	52		60	47			5
5	53		131	48			2
6	51		208	49		64	3
7	50		a 267	50			5
8	51		a 225	51			3
9	50		a 420	52			4
10	49		b 1,240	53			9
11	47		b 1,000	54			11
12	48		c 797	56			2
13	48		1,200	57			2
14	49		1,582	58		62	
15	52		489	60			14
16	51		331	61		23	
17	48		132	62			20
18	46		47	63		25	
19	46		100	64			16
20	48		54	65			17
21	51		28	68		21	13
22	50	785	31	69		15	9
23	50		18	72		32	24
24	48	430	24	74		11	15
25	50		31	75		5	14
26	51		28	77		5	9
27	51		24	78		4	
28	49		12	79		3	
29	49		44	80		2	
30	47		28	83			8
31	46		28	84		2	5
32	45		23				
33	46		16	Total		2,497	9,404
34	44		14				

a Many killed while picking (went over basket but once).   b Not so many killed while picking.
c Died almost as fast as one person could pick them out.

The table below gives a summary of the loss in four baskets; two, No. 1 and No. 7, picked daily, and two, No. 5 and No. 6, picked so as to avoid the critical period:

	Picked daily.		No. 5.	No. 6.
	No. 1.	No. 7.		
Loss of eggs	16,889	9,404	3,292	2,497
Loss of alevins	7	14	21	a 112
Deformed fry	35	4	38	56
Total loss	16,931	9,422	3,351	2,665
Percentage of loss	56	32	11	9
Percentage unfertilized	0.3	0.6	0.7	0.4

a The large loss of alevins of No. 6 was caused by accidental smothering.

There were 30,000 eggs in each basket at first. The eggs of basket No. 1 were from fish that were killed before spawning. Those of No. 5 were fertilized according to the dry process. Those of No. 6 and No. 7 were fertilized by the ordinary process used at Battle Creek. Nos. 1 and 7 were picked daily. Nos. 5 and 6 were not disturbed during the critical period. The number unfertilized among the dead eggs was not determined before the twenty-eighth day; therefore the last item of the table is only relative. To get the true percentage of unfertilized eggs it would probably be about right to double that given. (It can be determined whether addled eggs are fertilized by putting them in about 8 per cent acetic acid. The yolk of the egg becomes clear, and the embryo, if there be one, turns white. A strong solution of common salt will clear addled eggs, but it also disintegrates very young embryos.)

Even after the critical period has passed, the most careful handling kills some fertilized eggs. Several tests show that from 10 to 20 per cent of the loss after the critical period is in fertilized eggs that have been killed in handling. They should therefore be disturbed as little as possible.

Fungus in the hatchery.—Fungus is a considerable pest in a hatchery, but the loss of eggs at Battle Creek traceable to this cause is very small. Numerous experiments were made in order to determine if the fungus would attack and destroy living eggs. Only on one occasion have I found a live egg attacked by fungus. This one had a few filaments of fungus growing on one side, and the egg had begun to die where the fungus was attached. Whether the egg had started to die before the fungus attacked it or whether it was attacked first I can not say, but all other observations and experiments indicate that the fungus attacks only the dead eggs.

Fungus grows rapidly on dead eggs, and the filaments extend in all directions and entwine the adjacent eggs in a thick mat. This interrupts the circulation of the water and often smothers the eggs so matted. When eggs are smothered the embryo turns white before the yolk becomes addled, so that death from that cause can be distinguished.

At Battle Creek, in 1897 and 1898, the baskets of eggs were gone over on the second and third days after spawning and all of the dead eggs picked out. They were not disturbed again until after the critical period, or about the twentieth day. This method was followed even where baskets (size, 23 by 15½ by 6 inches) contained 40,000 eggs each. The number of eggs that died after the third day was small, and at the "breaking out," that is, the first picking after the critical period at about the twentieth day, the few dead eggs were found scattered here and there through the baskets. Each dead egg was covered with fungus, the filaments of which had entangled the live eggs lying in its immediate neighborhood, holding them together in a bunch. It was seldom that more than fifteen eggs were held together in such a bunch, and the dead eggs never exceeded three or four.

Such treatment is not recommended for other stations, as the difference in the character of the water supply makes it necessary to carry on separate investigations for each station in order to determine methods of treatment.

The reason Battle Creek is so free from fungus is that the water contains a considerable quantity of silt or dirt, and if the eggs are left undisturbed a couple of days they become covered with a fine sediment. This collects on the fungus, which acts as a kind of filter, making of it a black muddy mass and impeding its growth. Clay or other dirt free from organic matter is often mixed in the water to destroy a growth of fungus on fry.

While carrying on the experiments at Pacific Grove in January, 1899, when they were 38 days old the fry were attacked by a very serious growth of slime, sometimes called gill-fungus by fish-culturists. This slime was composed mostly of a microscopic unicellular animal with a silicious shell, belonging to the order *Flagellatae*. Some other microscopic animals and unicellular plants, such as diatoms, were present. The slime collected on the gills of the fish and killed about two-thirds of them. They were treated with a 25 per cent mixture of sea water, which was very effective. Those which had been removed to a mixture of sea water before the appearance of the slime were not affected at all.

It must not be supposed from the statements given above that hatching troughs and baskets must never be touched during the critical period, nor that fungus is the only disease to which salmon ova and the alevins are liable. As has been stated above, I am not giving a general method of fish-culture, but an account of certain investigations. If the deposits of sediment on the eggs and troughs show traces of decaying organic matter, especially if there is a growth of slime on the sides of the trough, everything must be cleaned immediately. There is no doubt that microscopic plants and animals, such as bacteria and those mentioned above as having injured the fry at Pacific Grove, are very injurious to the eggs, alevins, and fry and must be scrupulously guarded against.

THE ALEVIN.

UNDER NATURAL CONDITIONS.

The eggs that are not destroyed in one way or another when deposited by the spawning fishes lie among the rocks, where they lodge and hatch in from 6 to 9 weeks, the time depending on the temperature of the water. The alevins also remain among the rocks at the bottom for a few weeks, and their movements, slight though they are, expose them greatly to such fishes as the sculpin and trout. During this time the yolk supplies them with what nourishment they need, but about four weeks after hatching the quantity of yolk has become so small that it is not absorbed rapidly enough by the blood to meet the needs of growth. At this time also the alevin is able to swim a little, and it frequently leaves the bottom to snap at some floating object. Its movements are necessarily awkward on account of the unabsorbed yolk, and it therefore attracts predaceous fishes. This is the most critical period in the life of the salmon after hatching. The yolk-sac disappears entirely at the age of 5 or 6 weeks, when the young are known as fry. This is the age at which they begin feeding.

OBSERVATIONS ON ALEVINS ARTIFICIALLY REARED.

General account.—In December, 1896, 855,000 eyed eggs were shipped from Battle Creek hatchery to Bear Valley hatchery in Marin County. Here they were hatched early in February, 1897, under the care of Mr. Frank Shebley, of the California Fish Commission. After the yolk-sac was absorbed, which was about 35 days later, they were fed for a few days on curds of milk, and then, in the second week of March, were turned into Paper-mill Creek and its tributaries, Nicasio,

Olema, and Hatchery creeks. The fry were strong and healthy, and were turned into the streams in the best of condition. The young salmon were watched day after day, and systematic observations were made of their movements, habits, enemies, and rate of growth. The work was first begun by the United States Fish Commission and carried on until the middle of May. After a break here of three weeks the California State Commission carried it to completion.

In the winter of 1897–98 eggs were again shipped from Battle Creek to Bear Valley. This time the number was increased to 2,000,000, necessitating the planting of the alevins as soon as they began to hatch. All were planted before the yolk-sac was absorbed. This, when taken in connection with the previous year's work, gave an opportunity to study the comparative effectiveness of planting alevins and fry. The alevins were transported from the hatchery to the planting-grounds in 20-gallon cans. It was possible to carry 40,000 alevins in such a can for two hours at a temperature of 40° without loss, though 20,000 or 30,000 to the can was the usual number carried. They were carried by wagon or rail as the case required. A wagon was found to be preferable, the jolting being an advantage, as the splashing in the cans kept the water well aerated. If the road was smooth or if carried by rail, the water had to be aerated frequently, and it was necessary to put fewer in the can. Although alevins appear to be very delicate, they stand transportation much better than fry, and a much larger number can be safely carried in each can.

Paper-mill Creek and its tributaries, where the young salmon were planted, were never visited by the quinnat salmon. This was one reason that they were selected for the experiment, as any young of that species that we might find would be known to have been planted there. The streams are rich in aquatic insect life, affording an abundance of food for the salmon fry. Trout and sculpins (*Cottus*) are the only predaceous fishes. The streams do not flow directly into the ocean but through several miles of brackish tidewater into Tomales Bay, and the transition from fresh to salt water is very gradual, removing the danger of the fry being rushed too quickly from river to ocean water. It was thought that if the fry could thrive in these streams and pass successfully into salt water it would be of advantage to utilize coast hatcheries and plant in the smaller streams, where the young salmon would not be subjected to their supposed enemies during the long journey from the Upper Sacramento to the sea. The thing feared in the experiment was that the streams would prove too short and that the young salmon would arrive at salt water before they were ready to conform to the conditions of life they would have to encounter there.

Observations indicate that fry can be as safely planted in Paper-mill Creek and its tributaries as in the Sacramento River, and they reach the ocean six weeks earlier. If it is true, as the experiments made at the Clackamas hatchery in Oregon indicate, that most of the salmon return to fresh water to spawn after being in the ocean two years, a difference of one or two months in the time of reaching the ocean is worth considering. If the full growth is attained in 24 to 36 months, the average gain in weight is from 12 to 16 ounces a month. As the gain is necessarily slight at first, it must be much more than a pound a month later. Any extension of time for living in salt water is an increase of the rapid-growing period, as the early period of slight increase in weight must be passed through in any case. This argument holds good only on the supposition that the individual would leave the ocean in a particular month. But the great variation in the time in which the Sacramento salmon leaves tne ocean makes it quite certain that the time is determined by other influences than the season. If so, it is doubtful whether those planted in Paper-mill Creek would have any advantage over those planted in the Sacramento River.

Planting alevins.—Alevins on being liberated in swift water swim frantically and scatter in all directions as they are swept downsteam. Most of them seek the bottom and crowd into crevices between the pebbles or get into quiet places under or behind large bowlders. Others find their way into still water along the edge of the stream, where they remain exposed to view. In moderately swift water some find a lodging-place on the bottom or near the shore before they have been carried a hundred feet downstream, and it has to be very swift water that will carry them a hundred yards. For several hours after being planted in swift water many of them keep moving about. Often the place where they first lodge is too much exposed to the current, and they are repeatedly swept downstream, lodging here and there for a few moments, until they finally reach a quiet place where they can stay. After a few hours this moving about ceases and they remain quiet, retaining their places for at least several days. In one instance 60,000 alevins were liberated on a very swift riffle in Sacramento River 200 yards above a quiet pool. The riffle was shallow, at no place over a foot deep, but so swift as to make it almost impossible for a person to stand. The alevins all found shelter before they were carried a third of the distance to the pool. On visiting the riffle a day later none could be found much over a hundred yards below the place of planting and none was found in the pool below, which was seined thoroughly. All had found sheltered places and had ceased to move down with the current. When alevins are planted only a few yards above a pool, even in moderately quiet water, large numbers will drift into it, where they remain if they are not eaten by trout or other fishes.

When alevins are liberated in a pool or pond they at first scatter out near the surface, but soon settle to the bottom, where they keep up a constant wriggling of the tail and pectoral fins. Within a day or so they collect in bunches, appearing as brilliant salmon-colored blotches over the bottom of the pond. The constant motion of the individuals stirs up the silt until it is washed away from them and each bunch rests on the solid bottom. Alevins in a hatchery, by this constant motion, keep the hatching-trough free from sediment.

During a freshet at the California State hatchery at Eel River a thick sediment of sand was washed into the hatching-troughs and came so fast that the alevins, just hatched, were unable to keep it from settling. It covered the bottom of the troughs to a depth of 2 inches, becoming hard and compact. The alevins, instead of being covered, were found above the cement-like deposit, and none of them had been lost. This interesting incident demonstrates their ability to keep from being covered by sediment during a freshet, whether they be in pools, ponds, or troughs.

Alevins in the pond at the Bear Valley hatchery began swimming about in schools before the yolk-sac was entirely absorbed. The presence of predaceous fishes might have caused them to do otherwise.

Enemies.—When alevins are planted on riffles they are inclined to congregate in eddies and sheltered places behind bowlders. In these places several thousand of them may be found huddled together in a bunch plainly exposed to view. They are not very shy at this age and do not appear to try so much to get out of sight as to get out of swift water. The brilliant salmon color of the yolk makes them very conspicuous, especially when they collect in bunches. One such bunch in Olema Creek, in which there were four or five thousand, could be plainly seen from a point 50 yards distant. In such instances they are very much exposed to the ravages of ducks and geese—both tame and wild—herons, cranes, and other wading birds, and even of hogs, to say nothing of the fish in the stream. Alevins are very tempting morsels, and there is scarcely an animal that will not eat them when given a chance. However, most of the alevins can get into crevices between and under the pebbles and bowlders, where they are much safer from the attacks of the fish that may be in the stream than they would be if they were in the pools or quiet places.

Observations were also made on planting alevins in Sullaway Creek near Sisson. This is a very favorable stream in which to release young salmon, so far as predaceous fishes are concerned. The only fishes of the stream are rainbow trout, sculpins, and quinnat salmon parrs remaining from the season before. The plants were made on the riffles, but as these were all rather short, many of the alevins drifted into the pools. They were liberated in the morning, and in the afternoon the pools were seined with a small-meshed net. The fish caught were examined to find to what extent they had eaten the alevins. The result is shown by the following table, which gives the length of each fish examined and the contents of its stomach:

Species examined.	Size, in inches.	Number alevins eaten.	Other material in stomach.	Species examined.	Size, in inches.	Number alevins eaten.	Other material in stomach.
Trout	6	17	1 small pebble.	Sculpins	3.5	3	
	5.5	9	6 insect larvæ.		3.5	4	
	5	7	1 caddis larva.		3.3	2	
	4.5	6			3.3	2	3 insect larvæ.
	2.5	0	Insect larvæ.		3.3	3	
	2.5	0	Do.		3	2	
Salmon parrs.	3.8	1			3	2	1 insect larva.
	3.5	1			2.8	2	
	3.5	1			2.8	2	
	3.5	2			2.8	1	
	3.5	2			2.7	1	
	3.5	2	1 insect larva; 1 water bug.		2.4	0	
					2.4	1	
	3.4	1			2.4	1	
	3.4	1			2	0	
	3.4	2			2	0	
	3	1	2 winged insects.	Species and size unknown		3	

In the case of the largest trout, 6 inches long, 7 of the 17 alevins were in its throat and mouth. It had evidently gorged itself to the limit.

In all cases where salmon parrs had eaten two alevins, the tail of the second remained sticking out of the mouth, their stomachs being large enough to accommodate only one. The sculpins also had gorged themselves in the same manner. All of the fish caught were examined and only three had not eaten alevins, being too small. Three alevins had been disgorged by some of the fish. Evidently alevins are a favorite food for trout, sculpins, and salmon parrs; and when they remain exposed to such enemies from 2 to 4 weeks, it is a wonder that any escape.

Alevins planted in the Marin County streams in 1898 met even a worse fate. Here the trout are more numerous and larger. The sculpins are also larger and more abundant. There were no salmon parrs to feed on alevins, but there were myriads of sticklebacks, which, though unable to swallow an alevin, killed many by nibbling at the yolk. The only other fish in these streams was the roach (Rutilus symmetricus), which as far as could be learned did not feed on the alevins. Four is a moderate estimate of the average number of alevins that a trout will eat in a day, at which rate each trout would destroy about 150 before the absorption of the yolk-sac; and 1,000 trout would destroy 150,000 alevins.   The lesson is obvious.

Just here it may be well to state that in 1897, although only 150,000 fry—not alevins—were planted in Olema Creek, large numbers of them were yet to be found in June following, and quite a number in August.   In 1898, 850,000 alevins were planted in Olema Creek, and in June following there was a smaller number left in the stream than was found in August the year before.   There are two ways to account for this.   One is that the alevins were washed out to sea before they began swimming; but it is more probable that they were eaten by trout and sculpins.

In the spring of 1898, 7,000,000 salmon were planted at the hatchery on Battle Creek about two weeks before the yolk-sac was absorbed.   Although trout are not numerous there, the stream swarms with sculpins (*Cottus gulosus*), salmon fry remaining from the season before, Sacramento pike (*Ptychocheilus grandis*), black pike (*Orthodon microlepidotus*), hitch (*Lavinia exilicauda*), split-tail (*Pogonichthys macrolepidotus*), and suckers (*Catostomus occidentalis*).   All of these, though they do not feed exclusively on animal matter, take salmon eggs and alevins when they can get them.   The Sacramento pike is very destructive to young fish.   The split-tail is the most numerous species, and lives on salmon eggs during the spawning season.

Each day while the hatchery was in operation the bad or addled eggs picked from the hatching baskets were thrown into the stream.   Usually they were thrown into a small brook near its entrance to the creek.   In a very short time after emptying a can of eggs the split-tails always began to appear, running in from the creek.   In a few minutes the water would be alive with them, almost a solid mass tumbling one over the other, splashing the water and crowding each other in their frantic efforts to get the eggs, until some were forced into the mud at the edge, while others were lifted upward till their backs or bellies were out of water, or one might get into a vertical position with its head or tail out of water.   Frequently one would gorge itself till throat and mouth were so full that the passage of the water over the gills was shut off and it suffocated.   It usually required about 5 minutes to consume 5 gallons of eggs.

Alevins are almost as helpless as eggs and fully as palatable, and there can be little doubt of their fate when planted in such an environment.

Results of observations.—The egg and alevin stages are the periods in the life of the salmon when the care of the fish-culturist is most needed.   The art of taking and caring for the spawn has been so perfected that the loss in hatching need not be over 10 per cent, and is often less.   The loss of alevins, if they are retained in the hatching-troughs or nursery ponds, need not be over 2 per cent.   If the young are planted during the alevin stage, the loss is very great.   If large numbers of alevins are released in unsuitable places, where the bottom is comparatively free from stones, and where such predaceous fishes as the split-tail and trout abound, the loss may even be greater than if the parent salmon had been allowed to take their natural course in spawning.

Young salmon should never be planted until the yolk-sac has entirely disappeared and their swimming power has fully developed, even though they have to be fed a few days.   There is no advantage in holding them after this time.

THE FRY.

NOTES ON YOUNGER FRY.

Planting fry from the hatchery.—Fry are transported from the hatchery to the streams in the same manner as already described for alevins, but it is not practicable to carry over 10,000 in a can, even for a short distance. They require more care than alevins, it being necessary to aerate the water constantly.

When fry are liberated in running water, they immediately head upstream and try to stem the current. Owing to their being more or less faint from confinement in the can the current nearly always carries them downstream a short distance, but they soon find their way into the more quiet water along the edge of the stream, in the eddies or quiet pools, or among the stones at the bottom. Some even move a few yards above the place of planting before they come to rest. On gaining quieter water they rest themselves, moving only enough to keep from drifting downstream. When in such position they begin feeding on any particles of food that float within their vision, often snapping viciously at insects half as large as themselves.

In a small stream there is no marked tendency of the fry to form schools, each appearing to act independently; but in a larger stream, and especially in the large pools, they often swim about in schools. It appears, too, from our observations in the Sacramento, that they run in schools after gaining the main river in their migration to the sea.

After planting, the fry soon begin to drift downstream from one resting-place to another. This movement in small streams is not in schools. If many are planted at one place the movement downstream is quite rapid, and within 24 hours they will be scattered evenly along the stream for over a mile below the place of planting. The movement, though marked in the daytime, is more general at night. In one instance a screen was placed across a small stream a quarter of a mile below where 50,000 fry were released. Although but few reached the screen that day, the following morning apparently every one had reached it. Other observations have shown the same thing. Muddy water hastens the movement downstream, as does also high water, which is usually muddy.

In Hatchery Creek, in Marin County, 150,000 fry 10 weeks old were released. They gradually scattered downstream, floating tail first. In four or five days they were about evenly distributed along the creek for 1½ miles below the hatchery. At the end of this time a net with a 10-inch circular mouth was placed in the current in the daytime with mouth upstream. In one hour 30 or 40 fry were caught. This illustrates well the decided movement downstream after planting.

When released in a large pool or pond the fry collect in schools immediately and travel toward the inlet.

In 1898, 150,000 alevins were placed in a pond at the Bear Valley hatchery. These remained in the pond without being fed until four weeks after the absorption of the yolk-sac. As it had but 600 square feet of surface and was only 2 or 3 feet deep, there were obviously too many in the pond to do well without being fed. As would be expected, they grew but little, though few, if any, died. At the end of four weeks all were very nearly of one size—1.4 inches long. Those of the same age in the creek a mile below the pond varied from 1.5 to 1.9 inches; specimens from Olema Creek only two weeks older were from 2 to 2.4 inches long.

At any time during the four weeks that the fry were so crowded in the pond they could have gone out, as the overflow trough was unobstructed. Very few if any of them did so, however. Indeed, it was difficult to get them to go out at all, very few escaping till nearly all the water was drawn off. As soon as they came near enough to the overflow to feel the course of the current they would dart back into the pool again. It has often been noticed that fry have an aversion to going over a waterfall or swift rapid. The observations at Sims during the summer of 1898 indicate the same thing. On account of this, fry should not be planted above falls or swift rapids, especially in small streams, as it is desirable that they should move downstream as soon as possible.

Observations of a particular fry.—Fry were observed daily from September 18 to October 3, 1900, in a pool between a rock and the shore in Battle Creek. The pool was about 18 inches across, 4 feet long, and 2 or 3 feet deep. There was but one fry until the 25th, when another appeared. It is probable that only two individuals were seen during the observations, though we can not be sure that such was the case.

When first seen the fry swam near the surface, but after a few days it remained a few inches below. It stayed most of the time in the rather strong current, and was continually snapping at minute floating objects. When swimming near the surface it made from two to ten strikes a minute. Observations could not be made so easily after it began swimming deeper. It was seen to make at least 150 strikes, but each time whatever was caught was immediately ejected. Apparently it had to make a great many efforts before finding anything edible.

One of the fry was seen to leave the pool and resume its migration. It had been in the lower portion of the pool all day, and as evening approached allowed itself to be carried down into the shallow and swift water of the outlet, always keeping its head upstream. Several times it was carried halfway through the outlet, but darted back into the pool. Once it got entirely through the outlet and into the deep water below the rock and then darted back, but finally it was carried out into the main current, tail first, and was lost sight of.

Enemies.—As already stated, 855,000 young salmon were planted in the streams of Marin County, Cal., in 1897, after having been kept in the hatchery until the yolk-sac was absorbed and they had begun to feed. In order to determine to what extent they were preyed upon by the other fishes of the stream, large numbers of trout and a few sculpins were caught and examined, being the only fishes in the stream that could be suspected of eating salmon fry. Beginning at the time the plants were made and continuing for three weeks, 30 or 40 trout, ranging from 6 to 10 inches in length, were daily caught and examined. In not one instance had a salmon been eaten. The only fish eaten by them was the small minnow (*Rutilus symmetricus*), and no more than 10 of these were found in about 700 trout examined. Of the sculpins (*Cottus gulosus*), only 25 of size large enough to eat a salmon fry were caught. None of these had eaten fish of any kind.

In 1898, after the young salmon planted that year had absorbed the yolk-sac, a number of trout were examined. None were found to have eaten salmon fry.

On one occasion a small pool 8 feet across and about 18 inches deep was seined. Over 100 young salmon were caught, averaging 2.1 inches in length. Along with them about a dozen trout from 6 to 8 inches long were taken. It would seem that if ever trout ate young salmon it would be here. These trout were examined, and it was found they had eaten only caddis larvæ and periwinkles. On the Upper Sacramento River I have examined many trout taken while the stream was full of the small salmon fry, but have never found that they had eaten young salmon. The same is the case with the sculpin, and these are the only fishes to be feared in the Upper Sacramento. Farther down stream many of the smaller Sacramento pike have been examined, but none of them were guilty of eating young salmon.

In the spring of 1899, while observing the migration of the salmon fry on the lower Sacramento River by means of a fyke-net trap, we occasionally caught cat-fish along with the young salmon. In every case it was found that the cat-fish had eaten salmon fry. Their capacity for young salmon was greater even than that of the trout for alevins. Several cat-fish 9 inches in length were found with over 60 salmon fry in their stomachs, and one of this same size had eaten 86 of the fry which averaged a little over 1¼ inches long. To determine whether the cat-fish captured the fry only while in the bag of the net we caught nearly 50 with hook and line. The stomach of none of them contained a young salmon. Thus it is evident that the cat-fish likes salmon fry and would catch them regularly if it could. It is too sluggish a fish, however, to catch salmon fry under ordinary circumstances.

The only other fish at all likely to prey upon the young salmon in fresh water is the striped bass (*Roccus lineatus*), which is found in the lower river and in large numbers in the brackish water of Suisun Bay. It is also found in San Pablo and in San Francisco bays. I have no information on the subject, except that the striped bass preys to a large extent on the carp in the sloughs of the lower rivers and in the salt or brackish water feeds almost exclusively on small crabs. It is significant, however, that both striped bass and salmon are increasing in numbers in California waters, the former enormously, and it can not, therefore, be very detrimental to salmon. Young pike, suckers, and split-tails are abundant in the waters inhabited by the bass, and all are sluggish in comparison with the salmon. It would seem that young salmon would be the last fish upon which they would prey. A young salmon is very active and strong and much more shy than even a trout of same size; after it has begun to swim about and feed it is perfectly able to take care of itself, and the number killed by enemies in the Sacramento is very small.

MIGRATION OF FRY.

In Olema Creek.—The first year at Olema 150,000 fry, and the second year 850,000 alevins, were released in Olema Creek. The stream was seined about a month after the fry were planted in 1897, and in 1898 about a month after the time when the fry should have begun swimming. Very few young salmon were taken in either year, and the results show that over 95 per cent had left the stream within the month.

Battle Creek station.—The observations in Battle Creek were made while we were engaged with the hatchery experiments during October and November, 1898. In obtaining data concerning the young salmon we used a 50-foot seine such as was employed in nearly all of the investigations; but the most important device for this work was a trap which caught the young salmon as they were going downstream. The trap was made by sewing a piece of fine-meshed webbing across the mouth of the bag of a 30-foot seine and fixing a funnel to extend back into the bag from the middle of the webbing. It was set in a strong current just below the upper rack at the Battle Creek fishery, with the wings extending obliquely upstream, their ends being about 10 feet apart. The fish were deflected by the wings to the middle portion of the net, and found their way through the funnel into the bag. No effective means could be devised to prevent the funnel from becoming choked with leaves or other trash, which often happened within an hour or two after the net was set. There were many adult salmon below the rack, and they often tore the net with their teeth and frequently got fast in the funnel. Part of the time the net was set during the day, more often during the night. Sometimes it was set for only an hour or two during the night. The following is a record of the catch, showing the date, the time of day, and the number and size of the fry taken:

Record of salmon fry taken in trap at Battle Creek fishery, Oct. 7 to Nov. 30, 1898.

Date.	Time.	No.	1.4	1.5	1.5+	1.6	1.7	1.8	4.0	4.3	4.6	4.7	4.8	5.+	5.2
Oct 7	Day	0													
8	Night	5		5											
9	...do	1		1											
11	...do	3		2						1			1		
12	...do	17	3	8		5						1			
13	Day	0													
	Night	7		5					2						
14	Day	0													
	Night	10	2	6							2				
16	Day	0													
	Night	8	2	3						1	1			1	
17	Day	0													
	Night	6		2									4		
21	...do	11		8		2					1			1	
24	Day	0													
	Night	2		2											
25	5-9 p.m.	4		1		1			1		1				
	Night	6		2		4									
Nov 12	...do	6		2		4									
13	Day	0													
	Night	26	2	10		8	5		1						
14	...do	27		26						1					
17	8-9 p.m.	17		17											
21	3-4 a.m.	83	12	37		6	27	1							
22	4-5 a.m.	6		6											
24	12-1 a.m.	11													
25	...do	5	2	6		8	5								
26	1-2 a.m.	15													
27	2-3 a.m.	0													
28	...do	0													
30	1-2 a.m.	49		49											
	8-9 a.m.	24		24											
	Total		23	100	144	53	1	1							

NOTE.—The numbers in the column headed 1.5+ and 5.+ indicate the number taken that were about 1.5 inches long, or about 5 inches long, as the case may be, but were released without measuring.

From the preceding record it will be seen that all of the fry (not including the parrs) were practically of the same size, 1.5 inches long. Of the 322 fry examined, only two were over 1.6 inches long, one being 1.7, the other 1.8. The 1.5-inch specimens had just absorbed the yolk-sac. Indeed, there was often a small amount of yolk remaining in the body, although the sac had disappeared. The size of these specimens shows that they begin their downstream migration as soon as they begin swimming, or at the age of six weeks; their continuing the same size during the two months shows that practically all start downstream at the same age. If part of them had held back for two or three weeks, this would have been indicated by a greater variation in size.

The record also shows that ordinarily the young salmon travel at night. The trap was so set that they could not have avoided it had they traveled during the day. That they can be caught during the day is proved by their being taken in the open tow net set in Hatchery Creek, as noted above, under "Planting fry from the hatchery." Fry were seen quite often lying in a pool near the shore during the day, and were seen to rise to small insects that lighted on the water. They probably feed more during the day, which makes their migration slower, or stops it altogether.

On November 30 there was a rise in the creek and the water was muddy. The catch from 8 to 9 a. m. was larger than the average night catch at other times, showing that high and muddy water induces salmon fry to travel during the day. This fact is borne out also by the work at Walnut Grove, an account of which is given below.

The great variation of the catch when the net was set for an hour or two during the night indicates that they travel in schools.

A trap similar to that used in 1898 was set in Battle Creek in 1900 from September 13 to October 4. Salmon fry 1.5 inches long were taken, two or three at a time, from September 18 on. The downstream migration, therefore, begins at least as early as the middle of September.

Adult salmon can be found in some part of the river throughout the year, and the spawning season is therefore very long. It is probable that there are salmon spawning at some place in the river or its tributaries in every month of the year. They are spawning in considerable numbers from July till January, inclusive. With such an extensive spawning period, it is obviously difficult to separate the young according to size, and say that those of a certain size belong to the spring or fall run of a certain year. A variation in rate of growth, noted elsewhere, adds to the difficulty. However, in the following table of measurements of specimens taken with the seine at Battle Creek fishery during October and November, 1898, three sizes may be distinguished, which doubtless represent three runs of adults. Those from 1.4 to 2.2 inches in length were from the summer run of 1898; the 3.7 to 4.7 inch specimens from the fall of 1897 (and they doubtless were among the last to hatch); and the 6.2-inch specimens an earlier run, probably the summer run of 1897.

Measurements of young salmon taken with the seine, Battle Creek, October 18 to December 1, 1898.

Size.	No.	Size.	No.
1.4 inches	1	3.9 inches	..
1.5 inches	13	4 inches	1
1.6 inches	1	4.1 inches	..
1.7 inches	1	4.2 inches	..
1.8 inches	1	4.3 inches	2
1.9 inches	..	4.4 inches	1
2 inches	..	4.5 inches	2
2.1 inches	2	4.6 inches	1
2.2 inches	1	4.7 inches	2
2.3 to 3.6 inches	None.	4.8 to 6.1 inches	None.
3.7 inches	1	6.2 inches	2
3.8 inches	..		

Balls Ferry station.—An observation station equipped with a trap similar to that used at Battle Creek fishery was established on the river at Balls Ferry, about 3 miles above the mouth of Battle Creek. Observations were made by Mr. Chamberlain, beginning January 6 and closing April 25, 1899. The following table gives the data obtained at this station.

Record of salmon fry taken in trap at Balls Ferry, January 6 to April 25, 1899.

Date	Catch A.M.	Catch M.	Catch P.M.	1.4	1.5	1.6	1.7	1.8	1.9	2.0	2.1	3.1	Average size
Jan. 6	24			4	18	2							1.40
7	37	2	0										
8	14	11											
9	33		1										
10	39												
13	25	3	10	7	24	5	1	1					1.51
14	9												
21		35	15	9	25	14	1			1	1		1.53
22	19	44	29										
23	250	22	15										
24	158	13	2										
25	104	1	6										
26	39	6	4										
27	56	12	6										
28	42	13	11										
29	87	15	5										
30	153	18	3	30	80	30	2						1.51
31	173	5	3										
Feb. 1	192	5	2										
2	322	15	4										
3	326	5	2										
4	241	6	0								1		
5	230	0											
6	131	1	2										
7	70												
9	100	7											
10	117												
11	213												
12	157												
13	217	0	2										
14	237		1										
15	143	2	1										
16	163	4	0										
17	100	2	0										
18	137		10										
19	100	9		25	41	4	2						1.48
20	59												
23	77	0	36										
24	128	82	24										
25	227	70	14										
26	137	61											
27	45	61	5										
28	108	45	15										
Mar. 1	70	41	20										
2	58	12	55										
3	95	28	14										
4	35	5	25										
5	60	14	5										
6	27	14	7										
7	86	29	20	29	78	23	3	1	1				1.51
8	109	30	20										
9	121	31	24										
10	189	28											
11	65	15	7										
12	114	10											
13	45	13	0										
14	76	2	0										
15	254			34	101	88	13						1.52
21		1	5										
22	8												
31			2	1									
Apr. 1			3	1									
2	4	0	1	8	2	6	7	1	1				1.58
3	1	0	1										
4	1	1	0										
5	1	0	0										
6	3	0	5										
7	8	0											
9	5												
10		2											
11	7	1	0										
12	5	0	0										
13	0	0	0										
14	2	0	0										
15	1		0										
16	1		0										
17	0		0										
18	2		0										
19	2		0										
20	2		0										
21	1		0										
22	0		0										
23	0		0										
24													
25	0		0										

The preceding table needs but little explanation. It indicates that the greater part of the young salmon from the fall run passed Balls Ferry between the middle of January and the middle of March. Practically all had passed by March 20. Measurements taken January 6, 13, 21, and 30, February 19, March 7, 15, and 31, and April 1, 2, 3, and 4 show that the average size of those taken on the dates specified, during a period of $3\frac{1}{2}$ months, varied but one-tenth of an inch. The average of all measurements is 1.53 inches. No satisfactory estimate of the number passing could be made, except that there were probably many millions. This record also proves that salmon fry begin migrating as soon as they are able to swim, and that practically all start downstream at that age, otherwise the later ones would have been larger.

It was also ascertained that a large migration was not coincident with remarkably high water. It is probable that when the fry once enter the main river their migration is not impeded by low water; but it seems probable, from observations noted in another place (see "Summer residents") that many of the late fry that hatch in the headwaters are detained there during the summer by low water.

Walnut Grove station.—Our knowledge of migrations through the lower part of the river was gained from the general investigation of 1898, and especially from observations made by means of a trap established at Walnut Grove from January to May, 1899. This trap was constructed especially for the work, but was hardly more efficient than traps made from seines and used at Battle Creek and Balls Ferry. It consisted of a bag with a short funnel hung to a 4-foot hoop, with wings 20 feet long. At Walnut Grove the Sacramento makes a sharp bend, changing its direction from southeast to southwest. At this bend Georgeanna Slough breaks off and continues the southeasterly direction of the river above. It thus gets a large amount of water, probably half as much as the river below, and is in the direct path of the migrating fry. The trap was set about 150 yards from the head of the slough, which at that place is about 75 feet wide and 15 to 20 feet deep. The banks are abrupt and covered with bushes. One end of the trap was fastened by a long rope to a tree on the bank, the other to a buoy anchored about the middle of the stream. It was sometimes set in other positions in the slough or in the river, but without results of particular value. During a sudden rise in the river it could not be set on account of the great amount of trash in the water.

The following gives the record of the catch. In the column headed "A. M." is given the number of fry found in the trap at 8 a. m., and in the "P. M." column the number caught between noon and 5 p. m.

Record of salmon fry taken in the trap at Walnut Grove, January 7 to May 8, 1899.

Date.	Catch.			Size, inches.																										Average size.
	A. M.	M.	P. M.	1.4	1.5	1.6	1.7	1.8	1.9	2	2.1	2.2	2.3	2.4	2.5	2.6	2.7	2.8	2.9	3	3.1	3.2	3.3	3.4	3.5	3.9	4.1	
Jan. 7-13			0																										
20			22	1	7	11		1		1																			1.57
21			8	1	1	5	1																						1.58
23		7	7	1	5	5	3																						1.57
25			25																										
26			28																										
27			14																										
28			12	4	5	3																							1.58
29			18																										
30			28	1	6	12	8								1														1.65
Feb. 1			6																										
2			9																										
3			13																										
4			3																										
5			4																										
6																													

Record of salmon fry taken in the trap at Walnut Grove, January 7 to May 8, 1899—Continued.

Date.	Catch.			Size, inches.																								Average size.
	A.M.	M	P.M.	1.4	1.5	1.6	1.7	1.8	1.9	2	2.1	2.2	2.3	2.4	2.5	2.6	2.7	2.8	2.9	3	3.1	3.2	3.3	3.4	3.5	3.9	4.1	
Feb. 7	.....	3																										.....
10	.....																											1.79
11	.....		0				4	2	3																			
12	1	5	2																									.....
13	5	...	3																									.....
14	5	3	0																									.....
15																												
16	0	0	4	1		5	4	3	3																			1.81
17	6	4	1																									
18	2	0	5			2	3	1	1																			1.81
19	1		4																									
20	3		0																									1.82
21	2	2	5	1	2	1	1	1																				2.00
22	2	2	0		1		1	1																				
23	1	5	1				1	1	1																			2.11
24	1	3					1	1																				
25	1																											
Mar. 1	.....	8	5	1	3	3		1	2		2	2	1															1.85
2	5	2	13	1	6	5	3		1																			1.75
3	17	14	30	5	14	17	9	3	1	2	3	1	1	1														1.81
4	89	19	124	1	12	27	24	8	1																			1.83
5	44	29	4		6	16	13	5		1																		1.74
6	.....	.....	45																									1.79
7	81				2	7	4	2																				1.79
8	246	300	138	1	3	2	10	8	1					1														1.75
9	36	240	168																									
10	39		51																									1.70
11	175	225	43	2	3	12	6																					1.68
12	183	.....	43	1	6	10	4																					
13	101		82																									
14	12	64	87																									
15	17	36	33																									
16	80	21	25																									
17	19																											1.77
18		4	12																									1.70
19	16	316	376	1	3	8	3	1																				
20	278	153	283																									
21		64	68																									
22		2																										
28																												
30		11		1	4	2	2	1		2			1															1.70
31		10		1	1	2	1	1		2		1																1.85
Apr. 1	.....	8																										1.94
2	.....	12		1	1	3	5		1		1																	1.87
3	.....	10		3	3	1	1		3	1		1																1.85
4	.....	14		3	3	1	4	1	1			1																1.91
5	.....	25	1	2	3	3	4	4		1																		1.99
6	.....	18			2	2	4	5	4		1																	2.03
7	.....	13			1	1	5	5																				2.07
8	.....	17		1	1	4	7	7	1		1																	2.08
9	.....	48			4	10	5	7	5	3	2	2	1	2	1													2.07
10	.....	24			3	1	5	4	4	2	1	2	1	2	1													2.19
11	.....	14					4	4	4	2																		2.11
12	.....	19			3		4	4	2		1	2																2.27
13	.....	15		1		3	1	4		1	2		1		1													2.33
14	.....	8																										2.34
15	.....	8						1	5		1	3	2	2	2	2		1										2.48
16	.....	12				1		1	1		1	1	1															2.46
17	.....	8					1		1		1		3	2		1												2.50
18	.....	8									1			2		1												2.48
20	.....	5																										2.50
21	.....	12								2		1	2		6		3		1	3	3	1						2.60
22	.....	25			1			1	2		3		1	1		1												2.63
23	.....	10													1				1	1				1	1			3.08
24	.....	5		1		1								1				1										2.69
25	.....	3											1		2													3.00
26	.....	2														1		2		1	1	1						3.00
27	.....	5																										3.00
28	.....	7											2						3		1	1						2.74
30	.....	5							1									1	3							1		3.00
May 1	.....	4																										2.60
3	.....	4		1		1											1		1	1								2.60
4	.....	5														1				1	1	1						2.78
5	.....	3																1			1	1	1					3.17
6	.....	3																										3.00
7	.....	4								1						1							1					2.77
8	.....	3					1		1											1				1				

From the data above given it will be seen that—

1. From the middle of January to the middle of May there were salmon fry in various numbers passing Walnut Grove.

2. The height of migration was from March 4 to about the 24th, about 20 days.

3. On March 8 and 20 there were two large runs of fry.

4. Practically all had passed by the 22d of April.

5. The average size of those taken during January was 1.6 inches, during February 1.8 inches, during March 1.7 inches. From the 30th of March till May 7 the size gradually increased from 1.7 inches to 3 inches.

6. No fry were taken during the first 9 days in January.

7. Young salmon traveled as much during the day as during the night.

Comparing the information for Walnut Grove with that for Balls Ferry, as shown in the accompanying diagrams, plate 12, it appears that—

(*a*) The Balls Ferry run of February 2 reached Walnut Grove March 8 and was 34 days making the distance. The fry increased in size 0.3 inch.

(*b*) The Balls Ferry run of February 14 reached Walnut Grove March 20, and was 34 days making the distance. The fry increased in size 0.3 inch.

(*c*) The runs that passed Balls Ferry February 25, and later, were caught by high water the latter part of March, which probably carried them down faster. The runs were not noticed at Walnut Grove; the net could not be worked during the beginning of high water, March 23 to 29, during which time they may have passed.

(*d*) The fry taken at Walnut Grove after April 1 had grown more than 0.3 inch since starting downstream, and were therefore the stragglers from the regular migration. Those taken during May had probably been three months on the way.

It is evident, therefore, that the fry of the regular migration require about 34 days to pass from Balls Ferry to Walnut Grove.

The distance between the two stations is about 350 miles. An object floating as fast as the current would make the distance in about 9 days. It requires 8 days for a rise in the river to travel from Red Bluff to Sacramento. If the fry traveled only at night, and simply kept with the current, they would make the distance in 18 days. There is no doubt that in migrating the fry drift downstream tail first, keeping the head upstream for ease in breathing as well as for convenience in catching food floating in the water. In this way they would drift much more slowly than the current. At Battle Creek hatchery fry have been observed traveling with the current, and always with the head upstream unless frightened.

The later and larger specimens found had simply been longer on the way. The larger they became the more slowly they drifted, as they swam against the current more strongly. Those taken at Walnut Grove in January were but 1.6 inches long, being brought down by the high water in January, the short time they had traveled being indicated by their smaller size.

The failure to catch any fry during the first 9 days in January indicates that the fry from the summer run had all passed and that those from the fall run had not yet reached Walnut Grove. Without doubt there were a few passing at that time, for there were some passing Battle Creek as late as December 6, but they were so few that none were taken in the trap. It is possible that there are a few passing down the river all summer, though we have been unable to find any after June.

Observations at Benicia.—February 21 and 24 and March 3, 1899, five specimens 1.8 to 1.9 inches long were taken in Carquinez Straits at Benicia. The average size at Walnut Grove after February 10 was 1.8 inches and the size of the Benicia specimens indicates a short passage between the two places, probably not over a week. This would make the time from Battle Creek hatchery to brackish water 6 weeks.

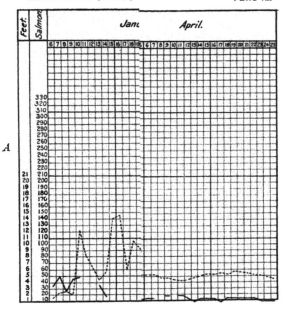

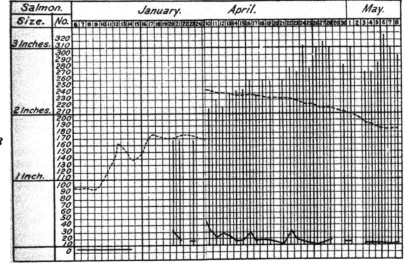

Heavy lines indicate number of fry; dotted lines ⟨⟩ represent the exact average length of the fry taken.

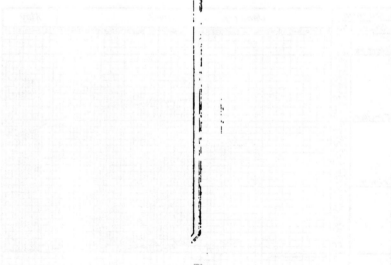

General investigation during 1898.—The following notes give the data obtained by the general investigation in 1898.

Table showing greatest number of fry taken at one haul of the seine.

Station.	Apr. 28 to May 4.	May 18 to 30.	July 8 to 27.	Station.	Apr. 28 to May 4.	May 18 to 30.	July 8 to 27.
Redding	50	66	8	Princeton	15	0	
River at mouth of Battle Creek	60		2	Colusa	6	0	
Red Bluff	150	1	0	Grimes	4	0	
Six miles below Red Bluff		6	0	Wilson's farm	4	0	
Tehama	100	31	0	Twenty miles below Grimes	0	0	
River at mouth of Thomas Creek		1	0	Knights Landing	2	0	
River at mouth of Deer Creek		8	0	Mouth of Feather River	0	0	
Chico bridge		24	0	Sacramento	10	6	0
Jacinto		10	0	Ryde	0	0	
Butte City		10	0	Rio Vista	2	0	
				Collinsville	0	0	
				Benicia	2	0	
				San Pablo Bay	4	0	

Table showing size of young salmon taken in the Sacramento River in May, 1898.

Dates and localities.	Size in inches.																											Average size.
	1.4	1.5	1.6	1.7	1.8	1.9	2.0	2.1	2.2	2.3	2.4	2.5	2.6	2.7	2.8	2.9	3.0	3.1	3.2	3.3	3.4	3.5	3.6	3.7	3.8	3.9	4.5	
Sisson, May 15			2	2	5	2	2	1	2		4	2	1	3	3		1						2			1	1	2.4
Sims, May 17	2	3			2	1	2	5	1	2	3	1		2		3	2		3	1	2							2.6
Hazel Creek, May 17	2	5	3	1				4	2																			2.0
Redding, May 4 and 18		1	1	3	3	4	2	2	3	1	3	2		2	1	2	1	1			1							2.4
River at Battle Creek, April 30	1	1	3	4	5	3	4	4	2	1	2	2		1	2	1		1										2.1
Red Bluff, April 28, May 20		2	1	1	4	4	4	1	1	1	1	2		1	2													2.1
Tehama, May 5 and 21	1	3	4	1	1	1	2	1	1	1	2		1				1		1									2.1
River at Thomas Creek, May 22					1			4	1	2		1																2.5
River at Deer Creek, May 22	1			1		1	1	2								1												2.2
Chico Bridge, May 23			1	1			1	1	1	1		1																2.7
Jacinto, May 24			1	1		1		1				1			1													2.4
Butte City, May 25			1		1		1																					2.1
Princeton, May 25							2	3	1																			2.6
Colusa, May 26					1		1																					2.4
Grimes, May 27							1																					2.3
Wilson farm, May 27				1		1		1																				2.3
Knights, May 29				1				3	2																			2.3
Sacramento, Apr 23	2	1	1		1	3	2																					1.9
Rio Vista, May 11 and 30						2	2																					2.3
Benicia, May 13 and 23			1	1		1		1	.	1	1			1														2.6
San Pablo Bay, May 18, June 17						2	2	1	1			1				1												2.7

From the tables given it will be seen that—

1. Young salmon were abundant in the river the first of May, at least between Redding and Tehama. As a few were taken at Sacramento April 23, it is probable that they were distributed throughout the river.

2. The last of May they were nowhere so abundant as they had been at Red Bluff and Tehama three weeks previously.

3. A few were found throughout the river May 18 to 30.

4. The fry found in the river in May had an average size of about 2.2 inches.

5. In July there were no fry found except a few at Redding and Battle Creek.

It is known from the work at Balls Ferry in 1899 that practically all the fry leave the river before March 20. It is evident, therefore, that the fry found between Redding and Tehama in May, 1898, were the stragglers left from the regular migration. Their size, 2.2 inches, precludes their belonging to the regular run. They had collected in the pools where we did our seining, as they do in the headwaters during the summer (see notes on "Summer residents"), which made them appear to be more abundant than they probably were. The rise in the river, which occurred from May 15 to 25, and the accompanying muddy water caused them to pass downstream. This is indicated by finding fewer in the pools in the latter part of May (when migrating they would not be collected in the pools) and by finding none at all in July.

The conditions in 1898 were exceptional on account of the early occurrence of low water—from the middle of March till the middle of May—though doubtless there are always a few stragglers from the regular migrations. These decrease in number and increase in size (slightly) till the rains of the following winter, when all leave the river.

In the mountain streams the young salmon prefer the pools, where they are often abundant. Nearly 500 were taken at one haul of the seine in a pool at the head of Box Canyon, near Sisson, in August, 1897, and it was not at all uncommon to catch over a hundred at a time in many of the pools of the headwaters. The rapids have been fished a number of times, but young salmon were scarcely ever caught unless the water was at least 2 feet deep.

Below Redding more salmon were found in the water with moderate current, gravelly bottom, and a depth of over 2½ feet, but none was found in absolutely still water, and none over a soft mud bottom. Not much seining was done over rocky bottom, on account of the strong current and the injury to the seine by its picking up cobblestones. A few salmon were caught by putting enough floats on the seine to keep it at the surface and then hauling in water 15 feet deep.

The following table indicates the various characters of stream in which young salmon were found in the main river, with the number taken in one haul of the seine under the various conditions.

Table showing number of young salmon taken in various stream conditions.

Localities.	Month.	Current.			Bottom.					2½-feet depth.	
		Slight.	Medium.	Strong.	Mud.	Clay.	Sand.	Gravel.	Rocks.	Under.	Over.
Redding	May			116			116				116
Mouth of Battle Creek	do		60	60				60	60	60	60
Red Bluff	do		36	178		36		178			214
Tehama	do		31	100			31	100	100		31
Vina	do		13				13				13
Chico	do		24	54			32	46		18	60
Jacinto	do		22	11½			10	13	1	13	11
Butte City	do	1		8½			16	7		8	15
Ten miles below Princeton	do			4½			40				4½
Colusa	do		17					17			17
Grimes	do		8				8				8
Wilsons	do		4			4					4
Knights	do		5			2		2			5
Sacramento	April	10					10			10	
Rio Vista	May		4								
Total		11	224	580		46	283	355	161	289	608

Movements in estuary and bay.—Much seining was done both years at Olema in trying to learn something of the movements of young salmon in brackish water. None was found in 1898. A few were caught near the mouth of Paper-mill Creek in 1897, and one was taken 2 or 3 miles from the mouth of the creek, across the head of the bay. The net was stretched across the mouth of the creek for 15 minutes during the flood-tide, and two salmon fry were taken, indicating that they run back and forth with the tide.

The fishermen at Marshall, on Tomales Bay, about 20 miles from the mouth of Paper-mill Creek, reported having taken young salmon in considerable numbers the last of April, 1897, about 50 days after they were liberated in the streams near Olema. At that time the salmon were about 100 days old and were large enough to be taken in the seines used by the fishermen. They caught as many as 15 or 20 at a haul for about a week, and caught them occasionally till the middle of June. I think the report reliable, as the salmon was a new fish for the bay, and would attract much attention. This indicates that the fry may reach the ocean at the age of three months. The water at Marshall is pure sea water.

Specimens have been taken in brackish water in Suisun and San Pablo bays, but not enough to determine their movements. A few about 10 weeks old have been taken at Benicia in water that was about 20 per cent sea water.

Effect of sea water on alevins and fry.—To determine the effect of sea water on alevins and fry, 25,000 eyed eggs were taken from Battle Creek hatchery to the Hopkins Seaside Laboratory at Pacific Grove. The eggs were received at the laboratory December 10, 1898, and most of them hatched on the 17th, which date was taken as the basis for determining their age in the various experiments. Those not being experimented with were cared for as alevins and fry ordinarily are.

The first experiments were made by putting a few alevins directly from fresh water into battery jars filled with various mixtures of fresh and sea water. In the later experiments glass tanks 2 and 3 feet long were used, and the water was kept running. The experiments were begun when the alevins were 6 days old. It was found that at this age they could live indefinitely in water that was 25 per cent sea water. Those about 40 days old could live in 50 per cent sea water, and at 50 days 75 per cent. Those 60 days old could live in 95 per cent, though there was considerable loss. Ninety-five per cent was as nearly pure sea water as could be obtained, the laboratory pump being broken and the tank partly filled with fresh water. The loss was much less when the density alternated between a high and low percentage, which indicates the value of the change of density in the estuaries with the rise and fall of the tides.

When the younger alevins were placed in 50 per cent sea water or stronger, the yolk was solidified, becoming much like soft rubber. The blood was driven from the body, making it appear bleached, and the adipose membrane at each edge of the tail adjacent to the caudal fin turned white. The circulation was retarded and the fish became sluggish. The only noticeable effects on the older alevins were sluggish movements and an inability to keep a horizontal position. Sometimes death was immediately preceded by violent and spasmodic swimming in any and all directions. The same actions were noticed in minnows placed in a strong mixture of sea water.

The following table gives a record of the experiments. The percentage of sea water is reckoned by taking pure sea water as a standard of 100, a mixture of equal parts of fresh and sea water being 50 per cent. The percentage of dead alevins at any one time is not based on the original number with which the experiment began, but on the number left in the vessel the previous day, e. g., if we start with 40 alevins and 10 die the first day, that is 25 per cent; if 15 die the second day, that is 50 per cent.

Table showing effect of salt water on alevins.

Day of experiment.	Age.	Percentage died.						
		Exp. 1. Fresh water.	Exp. 2. 25 per cent sea water.	Exp. 3. 50 per cent sea water.	Exp. 4. 75 per cent sea water.	Exp. 5. Sea water.	Exp. 6. Few changed from No. 2 to 50 per cent.	Exp. 7. Few changed from No. 6 to 75 per cent.
1	6	0	0	0	0	0		
2	7	0	0		20	100	0	
3	8	0	2	0	100		0	0
4	9	0	0	100			0	100
5	10	0	0				100	
6	11	0	0					
7	12	0	0					
8	13	0	0					
9	14	0	0					
10	15	0	7					
11	16	0	0					
12	17	0	3					
13	18	0	0					
14	19	0	0					
15	20	0	0					
16	21	2	17					
17	22	0	0					
18	23	0	7					
19	24	0	0					
20	25	0	0					
21	26	0	0					
22	27	0	8					
23	28	0	0					
24	29	2	0					
25	30	0	0					
26	31	0	0					
27	32	0	0					
28	33	2	0					
29	34	0	0					
30	35	0	0					
31	36	Discontinued.						

Day of experiment.	Age.	Percentage died.				Experiment 12.		
		Density raised gradually to—			Placed directly into sea water.			
		25 per cent. Exp. 8.	50 per cent. Exp. 9.	75 per cent. Exp. 10.	Exp. 11.	Age.	Percentage sea water.	Percentage died.
1	12	0	0	0	0	19	50	0
2	13	2	0	85	100	20	50	0
3	14	0	7	50		21	50	50
4	15	5	8	100		22	50	7
5	16	0	7			23	50	0
6	17	20	80			24	50	33
		50 per cent.						
7	18	11	10			25	50	43
8	19	9	100			26	50	0
9	20	0				27	50	50
10	21	10				28	75	0
11	22	15				29	75	50
12	23	65				30	75	100
13	24	8						
14	25	63						
15	26	50						
		75 per cent.						
16	27	25						
17	28	33						
18	29	100						

Table showing effect of salt water on alevins and fry.

Day of experiment.	Experiment 13.			Experiment 15.			Experiment 16.		
	Age	Percentage sea water.	Percentage dead.	Age.	Percentage sea water.	Percentage dead.	Age.	Percentage sea water.	Percentage dead.
1	26	10	0	33	0	0	33	0	0
2	27	15	0	34	10	2	34	5	1
3	28	25	1	35	35	0	35	15	5
4	29	35	1	36	40	3	36	20	5
5	30	50	7	37	50	5	37	25	5
6	31	65	84	38	55	7	38	30	1
7	32	75	100	39	60	7	39	35	2
8				40	65	34	40	40	4
9				41	65	58	41	45	3
10				42	70	83	42	50	2
11				43	80	66	43	52	6
12				44	90	100	44	70	10
13							45	70	20
14							46	80	41
15							47	65	23
16							48	60	2
17							49	60	8
18							50	60	7
19							51	65	0
20							52	85	0
21							53	70	0
22							54	65	0
23							55	83	2
24							56	50	0
25							57	93	0
26							58	93	0
27							59	95	x
28							60	95	0
29							61	Discontinued.	

Experiment 1 was made to check the others, showing that the loss was not the result of being confined in battery jars.

Experiment 2 shows that 25 per cent sea water has but little deleterious effect upon alevins over 5 days old.

Experiments 3 to 7 show that alevins of 6 to 10 days of age can not live in sea water of 50 per cent or over, either when put directly from fresh water into the mixture or when the density is gradually raised.

Experiments 8 to 11 show that alevins 12 days old can live longer in 50 and 75 per cent sea water than those only 6 days old. Some of the 12-day alevins lived 7 days in 50 per cent sea water, while the 6-day ones lived but 3 days. The older alevins lived one day longer in 75 per cent, but died the first day in pure sea water, as did the 6-day individuals.

Experiment 12 shows that when 19 days old they live longer yet in 50 per cent. Two lived 2 days in 75 per cent after having been in 50 per cent for 9 days. The one that lived till the twelfth day was a week older.

Experiment 13 shows that a gradual rise from a density of 10 per cent sea water when 26 days old to 75 per cent when 32 days old was fatal. In a similar way a gradual rise from fresh water when 33 days old to a density of 90 per cent when 44 days old was fatal, as is shown in experiment 15.

Experiment 16 indicates that alevins 50 to 60 days old can bear a high density of sea water, but that they can withstand it better if the density, instead of increasing regularly, alternates between high and low. It does not show the exact age at which the fry can live in sea water, and it is doubtful whether this can be determined accurately in aquaria. Whether it can or not, the time was not at my disposal to carry the observations further.

As a whole, the experiments show three important points. First, the fry can not live in sea water until several weeks after the yolk-sac is absorbed; second, when able to live in sea water they can not go directly from fresh water to sea water, but must pass gradually; third, they are greatly aided by an alternation of densities such as is obtained by passing through an estuary. For these reasons it would not be well to plant fry in a stream that does not reach the ocean through an estuary.

Change of color during migration.—The color of young salmon depends much on the character of the water in which they live. Those in small, cold streams are much more dusky and have the parr marks strongly developed. They become lighter in color upon entering the main river. Those in brackish and salt water are bright silvery on the sides, with the back sea green. Parrs 4 to 6 inches in length, found in Battle Creek and similar places, have the sides bright silvery, the back olive brown, with the upper end of the parr marks making regular shadings along the back. Specimens 2.6 inches in length from Rodeo, San Pablo Bay, May 18, have distinct parr marks; 2.7 inches, from Benicia, May 13, have lost them. Sometimes the caudal fin is reddish; sometimes there are yellowish stripes on the ventral and anal fins, especially with fry about 1.5 inches long.

Summary of observations on migration.—The fry begin their downstream migration as soon as they are able to swim. In the clear water they travel more at night; in muddy water, as much or more during the day. Much of the time they float downstream tail first, and in the larger streams they travel more or less in schools. In the larger streams their downstream movement is not dependent upon the height of the water, but upon age. From October to April, inclusive, over 99 per cent that pass the vicinity of Battle Creek are of the same size, 1.5 inches long. They pass down the river at the rate of about 10 miles a day, and are about 6 weeks reaching brackish water, being 3 months old at that time. They are probably 4 or 5 months old when they reach the ocean. The ebb and flow of the tide in the estuary, causing an alternation in the density of the water, is apparently beneficial.

SUMMER RESIDENTS IN THE RIVERS.[a]

General account.—In the upper portion of the Sacramento River there yet remained, after the winter and spring migration in 1898, a large number of young salmon. In the vicinity of Sims we found from 700 to 1,000 in the various pools. We found them common in the McCloud at Baird in September, and in Fall River in August. These summer residents, as they may be called, are confined to the headwaters—the clear streams with rocky bottoms. They do not stay much of the time in the very swift current or riffles, but remain in the more quiet pools. Here they feed on aquatic insects and take the angler's fly the same as trout.

Most of the data concerning the summer residents was obtained from investigations near Sims, in Hazel Creek, and the river below its mouth. Hazel Creek is a small mountain stream, with many pools and gravelly riffles, and is a favorite spawning stream both for salmon and trout. The two lower pools, which are about a quarter of a mile from the mouth, were seined several times during the summer and fall, and it was from this work that we learned much that we know of the habits of the fry remaining in the streams during the summer. In the table below one of these pools is called the upper and the other the lower.

The Sacramento River in the vicinity of Sims is about 40 or 50 feet wide, and during the summer has an average depth of about 3 feet. It is very swift except

---

[a] These notes on the "summer residents" are given largely as a matter of record. While the conclusions drawn in some cases are scarcely warranted, yet the available data point toward them, and too many of the estimates closely approximate each other to be the result of mere chance.

in the pools, which were the only places that could be seined. Seven of these pools were seined frequently, and for convenience in keeping notes we numbered them A, B, C, D, E, F, and G, beginning with the upper.

In July and August all specimens taken in Hazel Creek and in the river near by were marked by cutting off the adipose fin with a pair of small curved scissors. This enabled us to know when we were taking specimens that had been taken before.

Upstream movement.—The following is a record of the seining in Hazel Creek. The data for each seine haul consists of the date, the pool where it was made, the number of fry previously marked in this pool, number of days since the last were marked, total number taken, and number of marked fishes taken.

Record of seine hauls in Hazel Creek.

Date.	Pool.	Previously marked.	Days since last marking.	Total catch.	Catch of marked fish.	Date.	Pool.	Previously marked.	Days since last marking.	Total catch.	Catch of marked fish.
May 17.	Lower			14		Sept. 18.	Lower	104	32	52	39
				21			Upper	107	32	40	36
July 9.	....do			55		Oct. 18.	Lower	104	62	125	37
Aug. 17.	....do	48	39	49	12		Upper	107	62	92	48
	....do	48	39	45	5	Nov. 18.	Lower	104	93	93	22
	....do	48	39	21	2		Upper	107	93	113	54
	Upper	0		78	1	Dec. 18.	Lower	104	123	4	1
	....do			53	1		Upper	107	123	0	0
	....do			37	0						

It will be seen from the table that in seining the lower pool in August, 19 young salmon were obtained that had been marked in July. As there had been but 48 marked in this pool in July, it indicates that at least 40 per cent of the fishes that were in this pool July 9 remained until August 17.

In the upper pool we found two July-marked fishes in August, where none had been marked in July. As the lower pool is about 100 yards from the upper, this indicates that at least 4 per cent of the fishes in the lower pool had ascended the stream that distance.

There were 104 specimens marked in the lower pool during July and August; 39, or 38 per cent of these, were found there in September; 36, or 35 per cent of the 107 marked in the upper pool in August, were found in September. Only one haul of the seine was made in each place in September.

In August four-sevenths of the marked fishes found in the lower pool were taken in the first haul of the seine. (When more than one haul was made the marked fishes taken were held till the seining was over, in order that they might not be counted twice.) Assuming that the same proportion was taken in the one haul in September, we would reason that there were 68 marked salmon in the lower pool that month; 68 would be 65 per cent of the number marked—that is, 65 per cent of the fishes in Hazel Creek on August 17 remained until September 18. This approximates the estimates made for the pools in the river below the mouth of Hazel Creek. (See notes below.)

In a similar way, 36 per cent of the 104 marked in the lower pool were found in one haul in October, and 45 per cent of the 107 marked in the upper pool. Thus there was a loss of 2 per cent over the previous month in the lower pool, and a gain of 10 per cent in the upper, indicating an upward movement.

But the upward movement is indicated better by the simple statement of numbers, as given in the table. In September there were 39 marked fishes taken in the lower pool to 36 in the upper; in October the ratio was 37 to 48, and in November it was 22 to 54. It is difficult to see how this can mean anything else than that the young salmon in Hazel Creek continued to work their way upstream during September, October, and November.

This table shows an increase in the number of salmon in October. The average of two hauls in May was 17; one in July gave 55; six in August gave an average of 49; two in September averaged 46; two in October 110; two in November 103. The simple fact of there being more fishes found on later dates would not indicate an upward movement; they might have come downstream. The larger percentage of marked fishes in the upper pool, however, would indicate an upstream movement.

Several places above the upper pool were fished in October—one place within 15 yards of it—but no marked salmon were found. This would indicate a lack of upward movement; but, all the data being considered, it is evident that there was at least a slight upward movement in October and November.

In December the young salmon had all disappeared from the upper pool, and only 4 were found in the lower. One of these was a specimen marked in July or August. That is, 1 out of 200 remained after the December rise. It is evident from this that practically all the young salmon left the creek between November 18 and December 18. There was a heavy rain in the vicinity on November 28 and 29 (precipitation over 2 inches), and there can be little doubt that the salmon all left at that time.

Migration during summer.—Pool B (see plate 11) of the river is separated from Pool A (at the mouth of Hazel Creek) by a rapid about 150 yards long with a fall of about 4 feet. It is over 6 feet deep, with large angular rocks along one edge affording excellent hiding-places for young salmon. It required three men to seine this pool well, one to throw out the seine from a large rock at the upper end, and two to pull it in. When there were but two of us, one would pay out the seine from a riffle above the pool; the other would wade out as far as possible in the lower end of the pool and pull the seine down with a rope. When the seine was stretched through the pool it was pulled ashore. Obviously such work was not very satisfactory.

The following table gives a record of the catch at each haul of the seine in Pool B in July and August. Four hauls were made in this pool in May, about 50 young salmon being taken in one haul. The specimens caught in May represented two sizes, such as were found in Hazel Creek on the same date, but there were very few of the smaller size. These smallest were probably from a few late-spawning individuals.

The table gives for each haul of the seine in July and August, (1) the date, (2) number of young salmon caught, (3) number previously marked and released in the pool in July, (4) time since the July-marked fishes were released, (5) number of July-marked fishes caught, (6) number previously marked in August, (7) time since the August-marked fishes were released, (8) number of August-marked fishes taken.

These data may be used to estimate the number of young salmon in the pool by making the following proportion for any particular haul of the seine: The number of marked fishes taken is to the total number of marked fishes known to be in the pool (having just been released), as the total number taken is to *the total number in the pool.* The results, of course, are variable, and it is only by a number of trials that we can get near the probable truth. The value of the estimates is not enhanced by there being no marked fishes taken at certain hauls. In such cases, however, there were but few of either kind. The estimate made from each seine haul is given in the table.

The freshly marked specimens could be distinguished from those marked a month previously by the latter having the scar healed.

Table of data obtained from seining Pool B.

Date.	Number of young salmon caught.	Number previously marked in July.	Time since July marking.	Number of July-marked fishes taken.	Number previously marked in August.	Time since last August-marked fishes released.	Number of August-marked fishes taken.	Estimated number of young salmon in pool.
July 9	177							
	23	167	1 to 2 hours	0				
	25	190	3 hours	6				702
Aug. 14	22	209	36 days	5				435
	20	209	do	9		1 to 4 hours	1	435
Aug. 15	17	209	37 days	4	34	1 day	1	578
	9	209	do	0	46	30 minutes	0	
	20	209	do	6	55	do	1	1,595
	9	209	do	1	77	do	0	
Aug. 16	30	209	38 days	9	83	1 day	6	415
	56	209	do	13	98	4 hours	9	421
	14	209	do	2	132	30 minutes	3	616
Aug. 18	3	209	40 days	0	141	2 days	1	423
Average								685

The chief value of this table is in the estimates given in the last column. The estimate of the number of young salmon in the pool ranges from 415 to 1,595. Half of the estimates come within 107 of the average, which is 685. This average is probably not far from the actual number.

Pool C is quite similar to Pool B, and was seined in much the same way. When two worked it, one had to hang the net on the rocks on one side, then swim across to the other side, when both pulled the seine off the rocks and hauled it inshore at the lower end of the pool.

The following table gives a record for Pool C similar to that given for Pool B. In the only haul made in July, two of the four fishes taken had been marked. They had evidently come down from the pool above. Likewise in August four August-marked fishes were taken in the first haul, though none had yet been marked in this pool. In making the estimates of the number in the pool these four are considered, being added to the "number previously marked in August."

Table of seine hauls in Pool C.

Date.	Number of young salmon taken.	Number previously marked in July.	Time since July marking.	Number of July-marked fishes caught.	Number previously marked in August.	Time since last August marking.	Number of August-marked fishes caught.	Estimated number in pool in August.
July 9	4	0		2				
August 15	92	2	36 days	3			4	1,131
	26	2	do	2	83	2 hours	2	1,908
	30	2	do	2	105	30 minutes	2	288
	2	2	do	0	140	do	1	
August 16	32	2	37 days	0	141	1 day	5	1,028
	10	2	do	0	168	30 minutes	5	344
August 18	32	2	39 days	1	173	2 days	10	566
Average								861

The data obtained from Pool C gives a larger estimate for the total number in the pool than that for Pool B (C, 861; B, 685). In the seven hauls of the seine in this pool in August there were 8 July-marked fishes secured. In the ten hauls in Pool B 49 were secured. From these two statements we determine that 18 per cent of the July-marked fishes of the two pools were in the lower pool in August. $(8 \div 7 \times 10 = 11, 49 + 11 = 60, 11 \div 60 = .18.)$ Only 1 per cent was released there, leaving 17 per cent to migrate. Some of these may have drifted over while faint from being confined in the net, but we think not many. We never saw any do so, though we often watched them for that purpose. It is safe to say that most of them went over voluntarily. We would expect this, as the connection between the two pools is quite deep, though swift. It is even remarkable that no more than 17 per cent passed to the lower pool. It is conceded that the estimates in the tables above are liable to considerable errors. There are always some unknown quantities in the equations, yet the results appear trustworthy.

Pool D is below the lower railroad bridge, and was quite unimportant. It was seined but once, a large rock at the lower end of the pool making seining impracticable. Six young salmon were taken. It was one of these six that was taken in the first haul in Pool E. The two pools are continuous.

Pool E is a portion of the river about 75 yards long, immediately below and not separated from D, ending above a long riffle. It was the only pool that could be entirely covered by the seine. The bottom is mostly covered with cobblestones, and there are large rocks along one shore. These afforded hiding-places for the young salmon while the seine was being drawn. It was seined many times in August.

The following table gives the record of the catch at each haul of the seine made in the pool, giving (1) the date, (2) the number caught, (3) the number previously marked and released in the pool, (4) the number of marked fishes taken, (5) the estimate of the number of young salmon in the pool, and (6) the variation of this estimate from the average estimate. On the 16th and 17th of August the seining was carried on continuously, the time required for making a haul and counting and marking the parrs being from 20 to 30 minutes. The marked specimen taken in the first haul was one of the six from Pool D.

Record of seine hauls in Pool E.

Date.	Number of young salmon caught.	Number previously marked.	Number marked fish caught.	Estimated number of young salmon in pool.	Variation of estimate from average estimate.
August 16, p. m.	66	1	1	......	......
	148	68	13	751	271
	146	197	30	982	30
August 17, a. m.	83	312	38	682	340
	47	357	19	883	139
	19	385	4	1,804	+782
	64	400	25	1,024	+2
	149	438	62	1,055	+33
	71	525	40	932	90
	35	556	15	1,297	+275
August 17, p. m.	19	576	9	1,216	+194
	3	586	3	586	436
Total for 24 hours	850	586	250	a 1,022	......
August 18	0		0		
September 18	178	......	97	a 1,075	
October 18	20	......	5	......	

a Average.

The purpose of the work in this pool was to determine the number of young salmon that might be found in a pool. The estimates are made in the same way as in the case of Pools B and C. The third haul, as noted in the table, may be taken as an example: 30, the number of marked fishes taken, is to 197, the number marked previous to this haul, as 146, the total number taken, *is to 992, the total number in the pool.* The estimates vary from 586 to 1,829, but several of them are not far from a thousand. The average of the estimates is 1,022, which is probably near the truth.

September 18 we seined the pool again, catching 178 young salmon, 97 of which, a little over half, had been marked. By looking at the table above it can be seen that a little over half of the estimated number in the pool were marked in August—586 out of 1,022. If an estimate is made of the number in the pool in September, by assuming that all the marked fishes remained in the pool, the result will be the following proportion: 97, the number of marked fishes taken in September, is to 586, the number marked during August, as 178, the total catch in September, *is to 1,075, the total number in the pool*, which is remarkably close to 1,022.

If there had been much of a migration between August 18 and September 18 we would have taken a larger proportion of unmarked fishes. If there were no migration we would expect to get marked fishes in proportion to the total catch as 586 to 1,022, which was the proportion of marked fishes in the pool in August. The record for September, 97 marked out of a total of 178, is very nearly that ratio, indicating that there was little or no migration between August 18 and September 18.

August 17 we fished two pools, F and G, about a quarter and a half mile, respectively, below where any fish had been marked, catching 13 and 115 salmon. We also caught 57 in Pool G in September. Neither at this nor at any other time have we found marked salmon below the pool in which they were released, except in the case of Pool C, above referred to. The one marked salmon taken in the first haul in Pool E was released in the upper end of that pool.

It is especially worthy of note that none of the 591 fishes taken in Pool E had been marked in the pools above the previous month. If any had left the upper pools, they had not stopped in the vicinity.

Summary on number and movements.—The estimate of the number of young salmon in August in Pool B is 685, in C 861, and in E 1,022. Pool E is much longer than the others, and might very well have more fishes than either. From these estimates it is probable that there were about 10,000 young salmon to the mile in the Upper Sacramento during the summer of 1898, or between a half and three-quarters of a million in all the headwaters of that stream.

There is little migration of the young salmon between May and December. Where pools are separated by shallow riffles, no evidences of migration could be found. If connected by deep water, it was found that about 17 per cent of those in the upper passed to the lower. In Hazel Creek there was an upward migration of 4 per cent during July, and a larger—about 12 per cent—during September. There was no diminution in number either in Hazel Creek or the river up to November—even an apparent increase. There was a slight migration, however, during the whole period, indicated by the disappearance of the larger marked specimens.

This residence in the headwaters during the summer is probably due to low water. It has been noticed many times, both in the streams and in the hatcheries, that young salmon dread going over a fall. When the river is very low, as it is every summer, the rapids become almost like waterfalls, thus preventing downstream migration. A slight rise obliterates the fall and at the same time makes it difficult to find food; hence the decided migration in December. The abundance of food appears to be of some importance when we notice that there was a scarcity of food in September and also a slight increase in migration that month, although the water was the lowest of the season.

Young salmon were reported abundant in the pools near Sims the 1st of May, 1899, and they doubtless remained during the summer, as was found during 1898.

Growth in fresh water.—It has been shown above ("Migration, general investigations of 1898") that the size of young salmon found in May was the same for all parts of the river. This was true also in July for the portions of the river in which they were found.

The following shows the measurements of young salmon taken in the Sacramento in July, 1898. Bold-faced type indicates where the average sizes fall. For Dunsmuir, Redding, and Battle Creek the numbers given indicate the total number of fish taken. None were taken below the latter point. All were taken between July 9 and 13.

Table showing size of young salmon taken during July, 1898.

Size.	Number of specimens.					Size.	Number of specimens.				
	Duns- muir.	Sims.	Hazel Creek.	Red- ding.	Battle Creek.		Duns- muir.	Sims.	Hazel Creek.	Red ding.	Battle Creek.
2.1 inches			1			3 inches	1	1	1		
2.2 inches						3.1 inches	1			**0**	
2.3 inches			1		1	3.2 inches			1		
2.4 inches					1	3.3 inches				1	
2.5 inches						3.5 inches		1		1	2
2.6 inches	1	3		1		3.6 inches	1		1		
2.7 inches			1			3.7 inches	1		1		
2.8 inches	**1**		**0**			3.9 inches	1				
2.9 inches	1	1		1	**0**	4.3 inches	1				

Note that the average sizes in the above table are from 2.8 inches in Hazel Creek to 3.1 inches at Redding. In comparing this table with that for the month of May (see above), it will be seen that the average size increased from 2.4 inches in May to 3 inches in July, an increase of 0.6 inch in two months. This is also the amount of increase if only the smallest specimens in Hazel Creek are considered. They increased from 1.5 inches in May to 2.1 inches in July. This is an increase of 0.3 inch per month for fishes averaging under 3 inches in length. Each table shows that there was a greater variation at the upper stations.

The growth of 0.3 inch per month is also shown by the following table of measurements of specimens taken at Sisson in May and August. The average sizes are indicated by heavy-faced type. In computing the average for May, the four largest fish are not counted, as they evidently belonged to the summer run of adults instead of the fall. Measurements of specimens taken in the river at the mouth of the creek in August are given for comparison with those from the creek. The average size in May was 2.2 inches; in August, a little over three months later, it was 3.3 inches, the increase in size of those remaining in the creek being one-third inch per month. The growth was probably a little greater than that, the larger specimens migrating.

Table showing increase in size of young salmon at Sisson.

Size.	Number of specimens.			Size.	Number of specimens.		
	Sullaway Creek.		River, Aug. 19.		Sullaway Creek.		River, Aug. 19.
	May 15.	Aug. 19 and 25.			May 15.	Aug. 19 and 25.	
1.6 inches	2			2.9 inches	1	2	1
1.7 inches	2			3 inches			
1.8 inches	5			3.1 inches		2	4
1.9 inches	2			3.2 inches		1	4
2 inches	2			3.3 inches		1	5
2.1 inches	1			3.4 inches		2	2
2.2 inches	2			3.5 inches		1	1
2.3 inches	4			3.6 inches		2	3
2.4 inches	2			3.7 inches	2		
2.5 inches	1			3.8 inches		3	1
2.6 inches	3		1	3.9 inches	1	1	
2.7 inches	3		2	4.3 inches		1	
2.8 inches		3	3	4.5 inches	1		

The above shows the amount of variation in the young salmon of approximately the same age. All were released from the Sisson hatchery. The oldest were hatched

December 23, 1897, and the youngest January 23, 1898. The largest were 2.9 inches long, the smallest 1.6, a difference of 1.3 inches, which can not be accounted for by the one month difference in age. There was still a difference of 1 inch in August, when they were 3.8 and 2.8 inches long.

The following is a table of measurements of specimens taken in Hazel Creek during various months. They are thought to be representative, though the specimens were selected. We picked out extremes and what we thought to be average sizes.

Table of measurements of specimens from Hazel Creek.

Size	Number measured.				Size	Number measured.			
	May.	July.	Sept.	Dec.		May.	July.	Sept.	Dec.
1.5 inches	2				3.0 inches	1	1		
1.6 inches	5				3.1 inches				1
1.7 inches	3				3.2 inches				1
1.8 inches	1				3.7 inches		1	2	1
2.1 inches		1			3.8 inches				
2.3 inches		1			3.9 inches				1
2.6 inches		1			4.2 inches		1		1
2.7 inches	4	1			4.3 inches			2	
2.8 inches	2		1						

The above measurements indicate two ages in May, but the youngest were doubtless from a few, probably a single pair of fishes, that spawned much later than usual. The oldest were from the regular fall run of adults. The difference between these two sizes in May was 0.9 inch. The two sizes are not discernible after May, those shown in the table being due to the selections of specimens, which is not the case for May, however. The growth is indicated by the increase in size of the smaller specimens. The smallest specimens were: In May 1.5 inches, July 2.1 inches, September 2.8 inches, December 3.1 inches, the intervening period in each case being 2, 2, and 3 months, and the increase being 0.6, 0.7, and 0.3 inch, respectively, or 0.30, 0.35, and 0.10 inch per month. The total growth in 7 months, as shown by the smallest specimens, was only 1.6 inches, and for the last 3 months 0.1 per month.

Pool A is at the mouth of Hazel Creek. It is a semicircular pool of quiet water at one side of, but not at all separated from, the main channel. It is over 6 feet deep, and the seine had to be hauled by means of ropes. As the seine was stretched across the mouth of the pool and hauled in at the upper end, with the ends close to the banks, there was but little chance for the fish to escape. The pool was fished monthly, beginning with August.

The following table gives the number and size of the young salmon taken in Pool A during the season, and also indicates how many were marked fishes and had therefore remained since August. As the measurements were made on live fishes they could not be made accurately enough to be given in tenths. There was one 2-inch fish taken in October, but it was not counted, as it evidently belonged to a different run. We marked and returned to the pool all of the fishes taken in August. None were marked in any other month. At another haul in September, not recorded in the table, 82 specimens were taken, 32 being marked, which was a larger proportion by 7 per cent than in the haul recorded in the table for that month. No marked fishes were taken in December. The record of one haul in pool E for September is given for comparison with one made in pool A on the same date. It shows that specimens from pool E were smaller than those from pool A, which was the deeper pool.

Table giving size of young salmon taken in Pool A, Sacramento River near Sims, with a comparative record from Pool E.

Date.	3	3¼	3½	3¾	4	4¼	4½	4¾	5	5¼	5½	5¾	6	Average size.	Total number.
POOL A.															
August 18	3	11	25	11	11	6	3	4	3		1	1	2	3.91	81
September 18:															
Marked		2	3	10	4	3		2						3.86	24
Unmarked	1	2	11	15	12	7	2	1						3.84	51
Total	1	4	14	25	16	10	2	3						3.85	75
October 18:															
Marked				2	7	8	4		1					4.20	22
Unmarked	1	3	1	7	12	14	15	3	2	1				4.19	59
Total	1	3	1	9	19	22	19	3	2	1				4.19	81
November 18:															
Marked				1	6	4	3	1						4.20	15
Unmarked		4	9	9	20	27	34	8	2					4.19	113
Total		4	9	10	26	31	37	9	2					4.20	128
December 18:															
Marked															
Unmarked	1	2	2			1	1							3.61	7
Total	1	2	2			1	1							3.61	7
POOL E.															
September 18:															
Marked			23	37	21	13	3							3.56	97
Unmarked	4	25	30	21	1									3.47	81
Total	4	48	67	42	14	3								3.53	178

The main value of this table is in showing the size of the young salmon in different months. The average size in August was 3.91 inches; that of the August-marked specimens taken in September, 3.86 inches, a very slight decrease. It will be noted later that there was a scarcity of food in September, which would account for a slight migration. In October the average size of the marked specimens increased to 4.20 inches, but remained the same in November. In the one month, September 18 to October 18, there was an increase of 0.34 inch, but the total increase for the three months, August 18 to November 18, was but 0.29 inch, or 0.10 inch per month. This small increase for the total period indicates either that the growth was very slow or that the migration during that time almost compensated for the growth. During September, probably owing to a scarcity of food, the migration was a little greater, and as a result the fishes were smaller September 18 than they were a month before. There was a decrease of 0.6 inch in December, when nearly all had left the pool.

From measurements made on specimens taken at Olema, we have the following table for age and size of young salmon remaining in fresh water:

Age.	Smallest.	Largest.	Average.	Increase per month, average.
Three months	1.7	2.6	2.2	0.4
Five months	2.8	3.3	3.0	.4
Twelve months			5.5	.36
Seventeen months			7.0	.35

The growth in fresh water is, therefore, very slow, and in artificial propagation every effort should be made to prevent their remaining in the river over summer. The growth in salt water is much more rapid. The salmon should reach the ocean when about 3 inches long, and grow to be 36 inches in twenty-four months, which would be an increase of about 1.4 inches per month.

Gastric parasites.—Of 209 fresh-water specimens examined in the investigation of food of young salmon, 31 had parasites in the stomach. The parasites were of two or three kinds, one elongated, the others short and grain-like. They have not been studied, except to note the date and size of the fish. It is evident that residence in fresh water is conducive to the growth of parasites in the stomachs of young salmon.

Month.	Number examined	Number with parasites.	Percentage with parasites	Size.	Number examined	Number with parasites.	Percentage with parasites.
January	9	1	11	1.4 to 2 inches	61	3	5
February	10	0	0	2.1 to 3 inches	57	3	5
March	10	0	0	3.1 to 4 inches	53	10	19
April	15	0	0	4.1 to 5 inches	30	12	40
May	30	1	5	5.1 to 6.3 inches	8	3	38
July	21	1	5				
August	20	3	15	Total	209	31	15
September	18	3	17				
October	30	8	23				
November	15	3	20				
December	11	8	73				
Total	209	31	15				

Diseased parrs.—Only two diseased young salmon from the streams have been met with. One was found dead, covered with fungus, near Sims in 1898; the other, 5 inches long, was taken in the trap at Battle Creek, September 28, 1900. The upper lobe of the caudal fin was wanting, and the remainder, with the caudal peduncle, was covered with fungus.

Mature male parrs.—In October, 1897, several mature males, between 4 and 5 inches long, were taken at Sisson. In January, 1898, two males, 5.5 inches long and known to be only a year old, were taken above the Bear Valley dam near Olema; one was mature. In August, 1898, a 4-inch mature male was taken at Sisson. Four of the 6 young salmon taken at Fall River Mills in August, 1898, were males, all with the genital organs mature. Mature male parrs were frequently taken at Battle Creek fishery in October and November, 1898. The sex of a number of parrs, 4 to 6 inches long, from the general collection was determined; 15 were mature males, 2 immature males, and 12 were females. These mature male parrs can usually be distinguished by their more dusky color and by the slightly distended abdomen. Examined under the microscope, the milt is apparently the same as that from adults. A few eggs from a female of ordinary size were fertilized by milt from a 4.7-inch male. The fertilization was complete, all of the eggs hatched, and the alevins were of normal appearance.

No explanation of this early maturing of males can be made, and nothing is known of their future history. They feed the same as other young salmon and apparently are not attracted by mature females as the adult males are. It may be that they return from the ocean as the stunted form known as grilse. It is probable that several months' residence in fresh water causes the generative organs to mature both in the adults and in the young males.

Temperature notes.—The following table shows the number of young salmon taken at one haul of the seine in water of various temperatures. It indicates but little, except that young salmon may reside during the summer in water having a temperature of 64 degrees. The 25 taken in Thomas Creek, with a temperature of 68 degrees, were landlocked in a shallow pool.

Table of seine-haul and water-temperature records.

Place.	Date.	53°	54°	55°	56°	57°	58°	59°	60°	61°	62°	63°	64°	65°	66°	67°	68°
Dansmuir	July 8										6						
Hazel Creek	May 17		11														
Do	July 9									55							
River at Sims	May 17			50													
Do	July 9											177					
Do	Aug. 18												118				
Do	Sept. 18						178										
Do	Oct. 18	81															
Redding	May 4							50									
Do	May 18							66									
Do	July 11													8			
Do	Aug. 13																0
Battle Creek	Apr. 30				(3)				(3)								
Do	July 13													2			
Redbluff	Apr. 28												130				
Below Redbluff	July 14 to 31													0	0	0	
Mouth of Thomas Creek	May 22																25
Vina	May 22								8								
Jacinto	May 23						28										
Grimes	May 27														6		
Rio Vista	May 11														2		
Benicia	May 13							2									
San Pablo Bay	June 17																

Conclusion drawn from study of summer residents.—It seems evident from these observations that the later fry that hatch in the headwaters, or are planted there after the spring freshets have passed, are liable to remain till the rains of the following winter. This means a slow growth for at least 6 months, or about a fifth of their growing period. It means the precocious maturing of the males, which may be responsible for the great number of dwarfs known as grilse; and it means that 15 per cent will become infested with gastric parasites. For these reasons it is imperative that the fry from our hatcheries should not be released above Redding after the spring freshets, though they may be released in the headwaters earlier without any detriment, and they certainly should not be held after this time merely for the purpose of feeding. Superintendent Shebley of Sisson Hatchery states, as this paper is going to press, that there are not nearly so many young salmon remaining in the Sacramento River near Sisson during the summer since he has quit holding the fry in the hatchery during the spring for feeding as there were when he did so hold them. There is no advantage in holding fry in the hatcheries for feeding.

FOOD OF YOUNG SALMON.

General study of food in fresh water.—The young salmon feed principally upon floating or drifting insects, either immature or adults. When feeding they often take a station below a stick or rock and catch their food as it floats down on either side. They eagerly catch small insects and larvæ if thrown into the water. Fry 1.5 inches long have been observed to rise to a small fly that alighted on the water. They hardly ever eat encased caddis larvæ, although that is the main food of the trout.

The following is a tabular statement of the stomach contents of 225 young salmon, being based on an examination of about five specimens from each locality each month in the year in which any were taken. The record for each fish examined consists of the station, date, size of fish, and number of specimens of each kind of food or other material found in the stomach. Four forms of insects are recognized in the table, viz, larvæ; pupæ, including nymphs; flying insects; and "terrestrial insects," including adult wingless insects and spiders.

Table of stomach contents of young salmon, Sacramento Basin.

+ indicates presence of certain objects, the number of which was not determined. The totals in last two columns indicate number of fishes in whose stomachs parasites or indigestible material was found.

Station.	Specimen number.	Size of salmon, inches	Larvae.	Pupae.	Flying insects.	Terrestrial insects.	Crustacea.	Gastropods.	Worms.	Unidentifiable.	No food.	Parasites.	Wood, seeds, etc.
Sullaway Creek, May 15, 1898	1	1.8	6	2									
	2	3.0	38	16			1						
	3	3.4		7	8	1	1						
	4	3.4		7		1						−	
	5	4.4		4									
Total		16.6	44	37	2	1	1					3	
Average		3.3	9	7									
Sullaway Creek, Aug. 25, 1898	6	3.0	4	20									+
	7	3.1	4	5			1						
	8	3.1	8	5	14								+
	9	3.3	3	2	1			1	1				
	10	3.5	27	1	1			1	1				
Total		16.0	46	33	16		1	2	1				2
Average		3.2	9	7	3								
Sisson, Aug. 19, 1898	11	3.1	11	6									+
	12	3.1	1										+
	13	3.3	20	4	3							+	+
	14	3.5		5	3	1							
	15	3.7		2	1								
Total		16.7	39	17	7	1					1		3
Average		3.2	8	3	1								
Dunsmuir, July 8, 1898	16	2.5	3	12									
	17	3.0	13	4	1								
	18	3.0		5	2								
	19	3.6	2	11									
Total		12.1	18	32	3								
Average		3.0	5	8	1								
Sims, May 17, 1898	20	1.6	13	20									
	21	2.6	7	24	1	1							
	22	3.7	1	18									
	23	3.8		1	2								
Total		11.7	21	63	3	1							
Average		2.9	5	16	1								
Sims, July 9, 1898	24	2.6	17	11	1								
	25	2.6	39	41	3								
	26	3.5	5	5									
	27	3.9	15	9	1								
	28	4.3	40	13									
Total		16.9	116	79	5								1
Average		3.4	23	16	1								
Sims, Aug. 17, 1898	29	3.3	18	16	1								
	30	3.4	7	2	2								
	31	3.5	5	5									
	32	3.6	19	2	4								
	33	3.8	3	50	7								
Total		17.6	54	75	14								
Average		3.5	11	25	3								
Sims, Sept. 18, 1898	34	3.7	3	2	1								+
	35	3.8	5	10	7								+
	36	4.5											
Total		12.0	8	12	8						1		2
Average		4.0	3	4	3								
Sims, Oct. 18, 1898	37	3.1	3	11	3								−
	38	3.6	2	7	1								
	39	4.1		3									
	40	4.1	1	5	2								−
	41	4.6	7	6	1								
Total		19.5	13	45	7								2
Average		3.9	3	9	1								

Table of stomach contents of young salmon, Sacramento Basin—Continued.

Station.	Specimen number.	Size of salmon, inches.	Larvæ.	Pupæ.	Flying insects.	Terrestrial insects.	Crustacea.	Gastropods.	Worms.	Unidentifiable.	No food.	Parasites.	Wood, seeds, etc.
Sims, Nov. 18, 1898..............	42	4.3	6	19	2	2							
	43	1.5	7	2								+	
	44	1.6		4									
	45	3.5	21	37	1								
Sims, Dec. 18, 1898	46	3.5	16	23								+	
	47	3.9	16	18								+	
	48	4.3	13	34								+	
	49	4.7	20	26	3							+	
Total...............		23.0	93	148	4							5	
Average...............		3.3	13	21	1								
	60	2.0	2	3									
Hazel Creek, July 9, 1898	61	2.6	3	8									
	62	3.7		3		1							
	63	4.3		3		1							
Total...............		12.6	5	17		2							
Average...............		3.2	1	4									
	64	2.8	6	5	2	1							
	65	3.0	4	2	5								
Hazel Creek, Sept. 18, 1898	66	4.1			3								
	67	4.4	1	4	2							+	
	68	4.9	2	2	6	1						+	
Total...............		19.2	13	13	20	3						2	
Average...............		3.8	3	3	4								
	69	2.6	6	4	2	1							
	70	2.8	5	1	2	8							
Hazel Creek, Oct. 18, 1898	71	3.6	9	17	2								
	72	3.9	1	1	2								
	73	4.3		2									
Total...............		17.2	21	25	7	4							
Average...............		3.4	4	5	1	1							
	74	2.9		3	2	1							
Hazel Creek, Nov. 18, 1898	75	3.9	3	9									
	76	4.1	7	3									
	77	4.7		3									
Total...............		15.6	10	18	2	1							
Average...............		3.9	3	4									
	78	3.1	1	7	1							+	
Hazel Creek, Dec. 18, 1898	79	3.2		1							+	+	+
	80	3.9	1	1	1							+	
	81	4.2										+	
Total...............		14.4	2	9	2						1	3	1
Average...............		3.6	1	2	1								
Fall River at Dana, Aug. 28, 1898...	50	3.8	45	5									
	51	5.2		48	5							+	
Fall River at mouth, Aug. 29, 1898.	52	5.2	1	32	18	1							
	53	5.3		10	26								
	54	5.6		18	13								+
Total...............		21.3	1	108	62						1		2
Average...............		5.3		27	16								
	55	3.3											
	56	3.7		3	1								
McCloud River at Baird, Sept. 16, 1898.	57	3.8											
	58	4.4										+	
	59	4.8										+	
Total...............		20.0		3	1						4		
Average...............		4.0											
	82	1.7	4	5	1								
	83	1.8	1	9									
Redding, May 4, 1898.............	84	2.4		12	2								
	85	3.5											
	86	3.2		4									
	87	3.7		15	1								
Total...............		15.1	5	45	4						1		
Average...............		2.6	1	8	1								

Table of stomach contents of young salmon, Sacramento Basin—Continued.

Station.	Specimen number.	Size of salmon, inches.	Larvæ.	Pupæ.	Flying insects.	Terrestrial insects.	Crustacea.	Gastropods.	Worms.	Unidentifiable.	No food.	Parasites.	Wood, seeds, etc.
	88	2.6	1	6	3					1			
Redding, July 11, 1898	89	3.0	2									+	
	90	3.3	4		1								
	91	3.5	1										
Total		12.4	8	6	4					1	1	1	
Average		3.1	2	2	1								
	92	1.4	7	2									
	93	1.5	2	2									
Balls Ferry, Jan. 21, 1899	94	1.5	2	6									
	95	1.6	1	4									
	96	1.6	2	1		1							
Total		7.6	14	15	1								
Average		2.5	3	3									
	97	1.5		1									
	98	1.5		1									
Balls Ferry, Feb. 19, 1899.	99	1.5		1									
	100	1.7	2										
	101	1.7										+	
Total		7.9	2	3							1		
Average		1.6		1									
	102	1.5	1	4									
	103	1.5										+	
Balls Ferry, Mar. 15, 1899.	104	1.5		1									
	105	1.6		2									
	106	1.6	1	2									
Total		7.7	2	9							1		
Average		1.5		2									
	147	2.9	1	15	2		1	1					
	148	3.1	10	12	1	1	1						
Battle Creek at Longs, Sept. 14, 1898	149	3.5	10	24	1								
	150	4.0	15		1								
	151	4.0	11	12									
Total		17.5	47	63	6	1	1	1				1	
Average		3.5	9	13	1								
	107	2.3		4	1								
	108	2.4	2	3	1								
Balls Ferry, July 13, 1898	109	3.5		1	1								
	110	3.5	1	48	1								
Total		11.7	3	56	3								
Average		2.9	1	14	1								
	111	1.5		5									
	112	2.1	1	6	13								
Battle Creek, Apr. 30, 1898.	113	2.5	2	8	1	1							
	114	3.0	12	12	3								
	115	3.5										+	
Total		12.6	15	31	17	1					1		
Average		2.5	3	6	3								
	136	1.5	1		1							+	
	138	1.6		1									
	139	1.7	2	3	2								
	140	4.2	2	3									
Battle Creek, Nov. 4 to 13, 1898	141	4.4									+		+
	142	4.6		2							+		+
	143	4.8									+		
	144	5.4		2	3	1							
	145	6.2		2								+	
	146	6.3											
Total		40.7	5	13	6	2					3	3	
Average		4.1		1	1								

Table of stomach contents of young salmon, Sacramento Basin—Continued.

Station.	Specimen number.	Size of salmon, inches.	Larvæ.	Pupæ.	Flying insects.	Terrestrial insects.	Crustacea.	Gastropoda.	Worms.	Unidentifiable.	No food.	Parasites.	Wood, seeds, etc.
	116	1.4		1						+			
	117	1.8								+	(a)		
	118	1.4									(a)		
	119	1.5		1									
	120	1.5								+			
	121	1.5		1									
	122	1.5											
	123	1.5									(b)		
	124	1.6											
	125	4.1	1	1									+
Battle Creek, Oct. 9 to 25, 1896.	126	4.1	2	12	2								+
	127	4.3		10									+
	128	4.3		8									+
	129	4.3		2									
	130	4.6		3	1								
	131	4.6		3	4								
	132	4.7	5	8	15	13							
	133	4.7		2						+			+
	134	4.8		1						+		+	+
	135	5.2								+			
Total		53.0	9	61	34	13					5	1	6	1
Average		3.1	1	3	2	1								
	152	1.7											
	153	1.8		26									
Red Bluff, May 21, 1896.	154	2.1		8	1								
	155	2.5	4	4						+		1	
	156	2.7		29	1								
Total		10.8	5	67	2						1	1	
Average		2.2	1	13									
	157	1.5	18	2									
	158	2.2			1					+			
Tehama, May 5, 1896	159	2.4	9	3									
	160	3.1		4									
	161	3.5	1	3									
Total		12.7	28	12	1	2			1				
Average		2.5	6	2									
	162	1.8								1		1	
	163	1.9								1			
Chico, May 23, 1896.	164	2.6	3	4									
	165	3.1	7	2									
	166	3.6	9	5	1								
Total		13.0	19	11	1				1	2			
Average		2.6	4	2									
Jacinto, May 24, 1896.	167	2.1	3	2									
	168	2.3		3									
Total		4.4	3	5									
Average		2.2	2	2									
	169	1.8	16	3		1							
Butte City, May 25, 1896	170	2.1	4	20									
	171	2.5	8	2							1		
Total		6.4	28	25		1					1		
Average		2.1	9	8									
	172	2.1	30		1							1	
Colusa, May 27, 1896	173	2.4								+	1		
	174	2.7	8							+			
Total		7.0	38		1						2	1	
Average		2.4	13										

a Yolk not yet all absorbed.   b Salmon egg, waste from spawning platform.

Table of stomach contents of young salmon, Sacramento Basin—Continued.

Station.	Specimen number.	Size of salmon, inches.	Larvæ.	Pupæ.	Flying insects.	Terrestrial insects.	Crustacea.	Gastropods.	Worms.	Unidentifiable.	No food.	Parasites.	Wood, seeds, etc.
Grimes, May 27, 1888	175 176	2.3 2.4	2	6 1	1 1								
Total Average		4.7 2.3	2 1	7 4	2 1								
Knights Landing, May 28, 1888.	177 178 179 180 181	1.9 2.4 2.4 2.5 2.6			1 3 6	1 1						+	
Total Average		11.8 2.4		10 2	2						2		
Sacramento, Apr. 23, 1889.	182 183 184 185 186	1.4 2.0 2.0 2.0 2.0	2 3 1 1	3 4 2	1	1							
Total Average		9.4 1.9	7 1	9 2	1	1							
Walnut Grove, Jan. 29, 1889	187 188 189 190	1.8 1.8 1.9 2.0	1 2 1 6	1			75 75 50						+
Total Average		7.5 1.9	10 3	1			200 50					1	
Walnut Grove, Feb. 19, 1889	191 192 193 194 195	1.5 1.6 1.7 1.8 1.8	3 1 2	2 2	2							+	1
Total Average		8.4 1.7	6 1	2 1	4 1						1		1
Walnut Grove, Mar. 21, 1889	196 197 198 199 200	1.7 1.7 1.8 1.9 2.1	7 14 50	1	2 2 1 3								1
Total Average		9.2 1.8	71 14	1	8 2								1
Walnut Grove, Apr. 23, 1889	201 202 203 204 205	2.3 2.5 2.7 2.7 3.0	1	3 2 2 2	1 1	1				+		1	
Total Average		13.2 2.6	1	9 2	2	1						1	
Rio Vista, May 11 and 30, 1888	206 207 208 209 210	2.0 2.1 2.2 2.2 2.9		1 2 1	1 6 10 1							1	+	+
Total Average		11.4 2.3		4 1	18 4						1		
General summary of fresh-water stations, 135 specimens.													
Total Average		2.96	788 4.0	1,008 5.4	376 1.9	37 .2	208 1.0	3 +	2 +	25	12	24	9

This study shows that young salmon in fresh water feed exclusively on insects, and that immature aquatic insects form by far the larger portion of their food. The general summary of the table shows that approximately half of the food of the specimens studied consisted of pupæ (or nymphs, which were not distinguished from pupæ), one-third of larvæ, and one-sixth of adult winged insects.

There was an increase of flying insects in the food of specimens taken in Sulla way Creek in August, and an increase in amount of food in specimens from Sims during July and August. It was during September. when apparently there was a scarcity of food, that the larger young salmon disappeared from Sims. There was an increase in flying insects in food of specimens from Hazel Creek in September; a scarcity of food and a noticeable lack of larvæ in specimens from Battle Creek in October and November, and a smaller amount of food in specimens from the lower stations. (See summary for May, page 117.)

Two specimens from above Bear Valley Dam, near Olema, taken January 18, 1898, had stomachs gorged with larvæ and pupæ, one having about 50 of the former and 25 of the latter, but no indications of adult insects.

Three specimens, Nos. 117, 118, and 137, of food table, were taken in October and November before the yolk was yet absorbed. One had nothing in its stomach; another had some food, but it was unrecognizable; the third had eaten one larva and two adult insects, besides some other food that was unrecognizable. This indicates that they begin feeding even before the yolk is all absorbed.

The food data, if arranged according to size of fish, would give the following average amounts per fish. This table shows that pupæ and nymphs are the favorite food for all sizes. Those from 1.4 to 2 inches in length feed very little upon adult insects; the largest size feed very little upon larvæ.

Size.	Number examined	Average number in stomach.		
		Larvæ.	Pupæ.	Adult insects.
1.4 to 2 inches	59	4	3	(a)
2.1 3 inches	53	4	6	3
3.1 4 inches	46	6	8	2
4.1 5 inches	32	3	4	1
5.1 6.3 inches	8	1	14	8

a Indicates an average of less than one

The following table brings together a statement of the average amount of food found in the stomachs of the young salmon from various stations for the month of May, the only month in which we secured young salmon from many of the lower stations. The table indicates that the important food of the young salmon throughout the basin in May was larvæ and pupæ, of which there was an average of 4.4 of the former and 6 5 of the latter per fish. The fish examined averaged 2.5 inches. It also shows there was a slightly smaller amount of food in specimens from the lower portion of the river. They were not starving, however, and there is no evidence that the passage down the river is detrimental on account of the lack of food. " indicates an average of less than one. The numbers in the columns headed "No food" and "Parasites" indicate the number of fishes examined that had empty stomachs or parasites, as the case may be. The totals are taken from the complete table of food, but only for the month of May, and are not the sums of the averages given in this table

Table showing the average amount of food in stomachs of young salmon from the various stations in May.

Station	No. of speci-mens.	Size.	Larvæ.	Pupæ.	Winged insects.	Terres-trial insects.	Crusta-cea.	Gastro-pods.	Uniden-tifiable.	No food.	Para-sites.
Sullaway Creek	5	3.3	9	7	(a)	(a)	(a)	(a)			3
Sims	4	2.9	5	16	1	(a)					
Redding	6	2.6	1	8	1				(a)		
Battle Creek fishery	5	2.5	3	6	3	(a)			(a)		1
Red Bluff	5	2.2	1	13	(a)				(a)	1	1
Tehama	5	2.5	6	2	(a)	(a)			(a)	2	
Jacinto	5	2.6	4	2	(a)				(a)		
Chico Bridge	2	2.2	2	2							
Butte City	3	2.1	9	2	2	(a)					
Colusa	3	2.4	9		(a)				(a)		
Grimes	2	2.3	1	4	1						
Knights	5	2.4		2	(a)					2	
Sacramento	5	1.9	1	2	(a)	(a)			(a)		
Rio Vista	5	2.3		1	4				(a)	1	
Total	60		22	89	20	6	2	1	7	6	5
Average		2.5	4.4	6.5	.1	.1	(a)	(a)	.1	.1	.1

a Indicates an average of less than one.

Food in brackish water.—Relatively few specimens of young salmon have been obtained from brackish water, and the following table gives a list of the food found in nearly all that were caught:

Station.	No.	Size.	Amphi-pods.	Cope-pods.	Fish.	Adult insects.	Seeds.	Para-sites.
Benicia, Feb. 21 to Mar. 3, 1880.	26	1.8				25		
	26	1.9						+
	28	1.9	1			30		
	29	1.9		10		10		
	210	1.9				1		
	211	2.2		150		15		
	212	2.7	2	100				
Total		14.3	3	350		81		1
Average		2.0		50		12		
Benicia, May 13-30, 1880.	213	2.1	2			5		
	214	2.3	1			13		
	215	2.4				31		
	216	2.7	1	30		1	1	
	217	3.3						
Total		11.9	5	30		50	1	
Average		2.4	1	40		10		
San Pablo Bay, May 17-21, 1880.	218	1.6						
	219	2.1	2					
	220	2.4				35		
	221	2.4				5		+
	222	2.5				8		
	223	2.8	3					+
	224	3.0	2	8				
	225	3.4			1	5		
Total		20.2	7	8	1	53		2
Average		2.5	1					

Summary for 20 brackish-water specimens.

Total			15	388	1	104		3
Average		2.3	1	20		9.2		

The chief food of the few brackish and salt water specimens studied were adult insects. Only 5 of the 20 specimens had fed to much extent on copepods, and only 1 had eaten a fish; the species of the fish could not be determined, though it was evidently a smelt (*Osmerus*). No aquatic insects were found, such as were found in specimens from the lower river, which indicates that the fish had been in brackish water at least long enough to digest all fresh-water food.

Classification of the insect food of young salmon.—The following is a classified list of the insects found in the stomachs of young salmon from the Sacramento River, collected during this investigation, as reported by Miss Bertha Chapman. It is evident that in many cases the fish confined themselves largely to food collected at the surface of the stream, as is the case with fish taken at Rio Vista in May, or those taken at Fall River Mills in August; others sought the immature forms living under water, as can be seen from the majority of cases in the list. But in no case can this distinction in feeding habits be definitely made. They seem to have fed indifferently on water and surface forms. These surface forms are almost invariably insects living about streams, and which might therefore easily have fallen into the water from overhanging plants. Much of the stomach contents had been so far digested that it was not possible to identify the insects. Other insects have been partially determined by single wings or particles of the body; but it seemed not so important to carry the classification to species as to determine the types of insects forming the food of young fish. The results of the study are given in the following table:

Classification of insect food for young salmon.

Place.	Date.	Number of fish.	Libellula.	Ephemeridæ.	Plecoptera.	Orthoptera.	Acridæ.	Corisidæ.	Nabidæ.	Notonectidæ.	Capsidæ.	Jassidæ.	Membracidæ.	Psyllidæ.	Aphidæ.	Trechoptera.	Lepidoptera.	Diptera.	Tipulidæ.	Blepharoceridæ.	Culicidæ.	Chironomidæ.	Ceratopogon.	Simulidæ.
Balls Ferry	January	9																15					† 2
Walnut Grove	do	30													† 1			‡			†		†1
Balls Ferry	February	†2		3	†									‡ 1				†					‡
Walnut Grove	do	†29			‡													†					‡
Sims	March			1	2													1					
Balls Ferry	do	†23		1	†1									‡				2					‡
Walnut Grove	do	†30		1	†									‡							† 2		
Benicia (brackish)	do	4										13									‡ 2		
Battle Creek	April	5		2			1	2													p3		†1
Sacramento	do	†18																15			‡		†1
Olema	do	†28		‡	2													†1			p2		
Sisson	May	5			†	1								p1				†1					
Sims	do	4	6	†1	†1			3	‡				2	1	†3	†4		†1					
Redding	do	6			‡p									1									
Red Bluff	do	5				2															p2		†2
Tehama	do	5			2		2							2			†3						†2
Chico	do	5																	‡p				†1
Jacinto	do	2																			‡p		1
Butte City	do	3																†1					†1
Colusa	do	3								1											p1		†1
Grimes	do	3										1											
Knights	do	5							‡	2													
Rio Vista	do	5							‡	3		‡											
Benicia (brackish)	do	5							‡	2		‡							+				†
Pinole (brackish)	do	7					2		‡	2													
Point Richmond (brackish)	June	1					2																
Dunsmuir	July	†6			†1	†								1	†								3
Sims	do	5			†										†			†			×		†1
Hazel Creek	do	†6			‡										†1								2
Redding	do	4	4											1							p2		†1
Battle Creek	do	1			†									3									1
Sisson	August	5			†1													‡1					†4
Dana	do	1		1														†2	p1			†	
Fall River Mills	do	4		†	†1													†2					
Sims	do	5		†	†1													‡					†1
Sullaway Creek	September	†12		†	†1		2			2								‡					
Sims	do	†9		2	3										11		‡p		3	†1		×
Battle Creek	do	†12		†1	†1										2	p1			6	1		×1
Sims	October	5		†1	†1						×	×				6		†1			†p		×1
Hazel Creek	do	5		†	†													‡1					×1
Battle Creek	do	20		†	†	1	1							2	2				1		‡p		×1
Sims	November	1			†									2	2						p2	†4	†3
Battle Creek	do	10			†					2	2										p2	†4	3
Sims	December	5			†												†1						†
Hazel Creek	do	†14		†	†						2		‡				†1						

Classification of insect food for young salmon—Continued.

Place.	Date.	Number of fish.	Leptidæ.	Empidæ.	Phoridæ.	Muscidæ.	Dexiinæ.	Anthomyiinæ.	Muscinæ.	Coleoptera.	Staphylinidæ.	Hymenoptera.	Ichneumonidæ.	Braconidæ.	Formicidæ.	Arachnida.	Pseudoscorpions.	Crustaceans.	Parasites.	Woody particles.	Unclassified.	Fish.	Eggs.
Walnut Grove	January	39			*l*					:					2	+							
Balls Ferry	February	*2		2				3															
Walnut Grove	do	*29						1								+							+
Benicia	do	3								:											+		
Balls Ferry	March	*23		2																			
Walnut Grove	do	*30						2	1	:				1							1		+
Benicia (brackish)	do	4		*l*				1		1	2			1							*l*		
Battle Creek	April	5	+			:		2		:				1							+*l*		
Sacramento	do	*18						4															
Olema	do	*28																					
Sisson	May	5																					
Sims	do	4		*l*				:		*l2*				1	1	:					3		
Redding	do	6								*l1*				:									
Red Bluff	do	5								+				2									
Tehama	do	5								*l3*				4									
Jacinto	do	2							1														
Rio Vista	do	5	*p2*									1	1										
Benicia (brackish)	do	5		:	2	1				1	3				1	1	2		*l1*				
Pinole (brackish)	do	7		2						:					:								
Dunsmuir	do	*6												2									
Hazel Creek	do	*6												3									
Redding	do	4								1				1									
Battle Creek	do	1						2															
Sisson	August	5		2					1					1									
Dana	do	1							1														
Fall River Mills	do	4	2	1		5	*l* 2			:				4									
Sims	do	5	5	1						:											+*l*		
Sullaway Creek	September	*12			3					:				2	3	1							
Sims	do	*9								:			1										
Baird	do	5								:			1										
Battle Creek	do	*12	*p3*		3			:		*l* 1			2	3	3						+		
Sims	October	5											2		2						:		
Hazel Creek	do	5												2	2								
Battle Creek	do	20						*l*	1					1									
Battle Creek	November	10						*l2*		1				2									
Sims	December	5								1													

+ Means more than 10.   · Means present, but number not known.   *p* Means pupa.
: Means more than 5.   *l* Means larva.   *o* Means adult.

The following list of common names will be of assistance to those who are not familiar with the scientific terms used in the above table:

Odonata:
  Libellulidæ . . . Dragon-flies.
  Ephemerida . . . . . May-flies.
  Plecoptera . . . . Stone-flies.
Orthoptera:
  Acrididæ . . . . Grasshoppers.
Hemiptera . . . . . . Bugs, lice, aphids, etc.
  Corisidæ . . . . Water-boatmen.
  Notonectidæ . . . Back-swimmers.
  Capsidæ . . . . . . Leaf-bugs.
  Jassidæ . . . . Leaf-hoppers.
  Membracidæ . . . Tree-hoppers.
  Psyllidæ . . . . . Jumping plant-lice.
  Aphididæ . . . . Plant-lice or aphids.
Trichoptera . . . . Caddis flies or caddice worms.
Lepidoptera . . . . . Moths, skippers, butterflies.
Coleoptera . . . . . . Beetles.
  Staphylinidæ . . . Rove beetles.

Diptera:
  Tipulidæ . . . . Crane-flies.
  Blepharoceridæ . . . . Net-winged midges.
  Culicidæ . . . . Mosquitoes.
  Chironomidæ . . . . . . Midges.
    Ceratopogon . . . Punkies.
  Simuliidæ . . . . Black flies.
  Leptidæ . . . . Snipe flies.
  Empididæ . . . . Dance flies.
  Phoridæ . . . . Humpbacked flies.
  Muscidæ . . . . Muscids.
    Dexiinæ . . . . Nimble-flies.
    Anthomyiinæ . Anthomyiids.
    Muscinæ . . . . Common flies.
Hymenoptera . . . . . . Ants, bees, wasps, etc.
  Ichneumonidæ . . Ichneumon flies.
  Braconidæ . . . . Braconids.
  Formicidæ . . . . Ants.

The following is a list of insects having the early stages passed in the water, selected from the set worked over:

Ephemerida.
Plecoptera.
Corisidæ (entire life in water).
Notonectidæ (entire life in water).

Trichoptera.
Tipulidæ.
Blepharoceridæ.
Culicidæ.

Chironomidæ.
Simuliidæ.
Leptidæ.
Coleopterous larvæ.

THE PERIOD OF GROWTH.

Of the ocean life of the salmon very little is known, although it comprises two-thirds of their existence. They have been taken in Tomales Bay when four months old, about 20 miles from the mouth of the stream in which they had been planted, and a few of about the same age have been taken in San Pablo Bay. These were on their way to the ocean and were already in nearly pure sea water. Specimens 8 to 15 inches long are sometimes taken by anglers in San Francisco Bay. In the estuary and along the beach at Karluk, Kadiak Island, Alaska, the writer has seen schools of several hundred 8-inch red salmon that had come inshore with the adults. These were feeding and were not dwarfs on their way to spawning-grounds.

Salmon are sometimes taken in the ocean near San Francisco in paranzella nets and are also captured in large numbers with the troll in Monterey Bay, where they appear in February and are found until the middle of August.

Something of the ocean life of the salmon might be learned by making a study of the food found in the stomachs of individuals taken when they first enter Monterey Bay. At such times a portion of the food taken on their offshore feeding grounds should yet be in identifiable condition. The presence of deep-water fishes or crustaceans would indicate a deep-water life for the salmon.

During its life in the sea the salmon is, of course, not entirely free from enemies, and something might be learned of them by studying the scars left by the injuries they inflict where they fail to kill. Injuries and deformities received before entering fresh water were of frequent occurrence among spawning fishes at Battle Creek in 1897, but no particular attention was given them. The males often had the snout twisted or split, or even cut off. Very often there were one or more scars extending obliquely backward and downward on the side above the anal fin. Sometimes two or three were parallel, as if they were scratches made by teeth of some other fish while the salmon was smaller. These scars were more often present on the females. The dwarf females were always injured in some way. Very few injured fishes were observed in 1898. Two had lost the ventral fins, one had lost the lower two-thirds of the opercle, two had deformed backbones. Only one fish, a female, had the oblique and parallel scars mentioned; they were slightly curved and in two series of seven each. This subject is worthy of further consideration.

A very characteristic scar, and one that always attracts attention, is that left by the sucking disk of the lamprey. The lamprey has no lower jaw, its mouth being circular and of the size of the end of its head. The gullet ends in the middle of the mouth and is bounded on the upper and lower sides by hooked teeth. There are other smaller hooked teeth above and below these on the disk, and on each side of the disk there are about four short cross rows of teeth, and the whole circumference of the disk is beset with small teeth. When the lamprey attaches itself to another fish the outer row of small teeth leaves a scar somewhat resembling the milling on a coin, which has led imaginative persons to see in the whole scar a resemblance to a brand made by a heated coin. The illustration on plate 13 is made from the photograph of a lamprey scar on the opercle of a blueback salmon found at Karluk, Alaska.

In 1896 5,000 marked fry were released in a tributary of the Columbia River; about 400 of these were taken in 1898, and a few more each in 1899 and 1900. This indicates that most salmon remain in the ocean two years, though a few remain three or four years, as will be seen from the following chapter.

PLATE 13.

ADULT AND GRILSE FORMS OF MALE SALMON, WITH GENITAL ORGANS MATURE.
The upper is the adult, the lower the grilse. The measure shown is 24 inches long.

LAMPREY SCAR ON OPERCLE OF SALMON.

DIED FROM GILL PARASITES, LAST OF
SUMMER RUN, SEPTEMBER, 1900.

THE ADULT SALMON.

MIGRATION.

Do salmon return to their native streams?—There is a widespread belief that when a salmon returns to fresh water to breed it seeks the stream in which it was hatched, though there is very little evidence that such is true. Various fishermen claim that they can distinguish the salmon of particular streams by their general appearance, which is incredible. The employees of the Alaska Packers' Association state that the red salmon taken at Uganuk are always smaller than those taken at Karluk, both places on the north coast of Kadiak Island, Alaska; that 13 of the former are required to make a case of canned salmon, while only 11 of the latter are necessary. This seems to indicate that the salmon of the two localities are distinct, but the larger salmon may go to Karluk, not because they have been hatched in Karluk Lake, but because they are larger.

In 1897 855,000 quinnat salmon fry were released in Paper-mill Creek and its tributaries draining into Tomales Bay, California, and 2,000,000 alevins were released in the same streams in 1898. (See "Observations on alevins artificially reared.") In 1900 a few salmon were seen in Paper-mill Creek, and in 1901 they were abundant. In one haul of the seine in the tide-water portion of Paper-mill Creek, covering a section about 150 feet long, 7 quinnat salmon were taken November 16, 1901. It is well known that quinnat salmon did not breed in Paper-mill Creek or its tributaries previous to 1897, for which reason these streams were selected for the experiment. Mr. Thomas Irwin reports that he saw two large salmon in Paper-mill Creek about 1890, but with these exceptions he never saw any fishes in the stream that might be taken for quinnats until 1900. He lives on the banks of the creek and knows the stream thoroughly. His statement agrees with that of other persons. Paper-mill Creek is not suitable for quinnat salmon, being entirely too small, but it is frequented by dog salmon and steelheads.

But there is no conclusive evidence that the fishes which were found in Paper-mill Creek in 1900 and 1901 were the same individuals released there three or four years previously. They may have been merely stray fishes, and their being found there at that time only a coincidence; or their coming into Tomales Bay may have been caused by there being an extra large number of salmon in the ocean, which might very well be, owing to the large output of young from the hatcheries; or those found in Paper-mill Creek in 1900 and 1901 may have been some of those released there, in which case it is very probable that they had never reached the ocean at all, but remained in Tomales Bay. Paper-mill Creek would then be their only stream.

It is incredible that the salmon remember their native stream during their two or three years of ocean life and that they consciously seek it when they desire to return to fresh water. Probably most of them do return to the stream from which they entered the ocean, not because it is their native stream, but because they do not get far away from its mouth, and when ready to return to fresh water it is the first to attract them.

The two runs of salmon.—Adult salmon may be found in the Sacramento River at almost any time of the year. There are, however, two more or less distinct runs, the first of which passes up the river during April, May, and June, and the latter during August, September, and October. The former is known as the spring run, the latter as the fall run.

The salmon of the spring run ascend the river to the headwaters, such as the Upper Sacramento, McCloud River, and Hat Creek, and some of the earlier ones even pass Pit River Falls and ascend Fall River to its source. They are not found in Pit River above the mouth of Fall River. By the time they reach this portion of the stream, the Upper Pit River is very low and the water impure, and the salmon all turn into Fall River. The salmon of this, the spring run, spawn mainly in August.

The fall salmon do not ascend the river as far as the spring run, but turn into the lower tributaries or spawn in the main river. They reach their spawning-grounds during the latter half of October, November, and the first half of December, and spawn soon after. The main river is very low at that time of the year, and the portion between Tehama and Redding is an important spawning-ground. (See chart of spawning-grounds, plate 17.)

As a matter of fact there is no definite distinction between the spring and fall runs; that is, there is no time during the summer when there are no salmon running. First there are a few very early salmon that begin running up the river in February, and the number increases until May when it decreases till July; then it increases till the first of September when it again decreases, there being a very few each month until the next spring run.

The spawning seasons merge in the same way. The earliest salmon go farthest upstream, and as the season advances they stop at lower points. The localities and dates of the spawning of the earlier salmon have not been determined, except that Superintendent Lambson, of Baird, reports having seen a pair of spawning salmon in the McCloud at the hatchery on the 20th of April, 1902, which is the earliest record known. By the 1st of October spawning fishes are found as far downstream as Redding, and as far as Tehama by the first of November.

Details of migration.—When the salmon enter San Francisco Bay they come in against the ebb tide, stem the current till the tide changes, and then run out against the flood tide, losing much of the distance gained during the ebb. How it is that they do not lose altogether as much as they gain will be understood from the following explanation: The tide runs up the bay and river as a broad, low wave, on the upper side of which is flood tide and on the lower side ebb tide. When the crest of a wave—that is, slack high water—is at Isleton, the trough, or slack low water, is about at the Golden Gate. This wave is about three hours reaching Benicia and four in reaching Collinsville. The farther up the bay and river it reaches the smaller the wave becomes, the shorter the flood, and (as the flood and ebb combined must equal 12 hours) the longer the ebb.

The following diagram will illustrate the movements of a salmon in passing through the bays: *a*, *b*, and *c* represent the tide wave at successive points as it passes up the bay, ←— indicates ebb tide, and —→ flood tide. Suppose that a

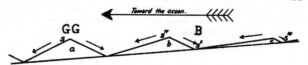

salmon enters the Golden Gate, *GG*, at the beginning of ebb tide, which would be the most favorable time. His position on the wave will be at *s*. If he is able to travel up the bay as fast as the wave he will keep his position near the crest, that is at *s*. But he can hardly do that, especially as the current would be very slight, and in the broad bay hardly strong enough for his guidance. Suppose that by the time he reaches Benicia, *B*, he has fallen behind the wave until he has the position at *s'*. It is then slack low water, and he can make no headway. Soon the next wave reaches him and he is in flood tide. He will therefore swim back against the current. As the wave is going up the bay and he is going down, he soon gets past the crest and finds himself in the ebb tide at *s''*. He then turns and stems the ebb tide, and as the wave is going in the same direction he is, he goes much beyond Benicia, *B*, before he again falls back to slack low water at *s'''*, and gets into the flood of the next tide wave.

There is no way of tracing the passage of the salmon through the bays, but from records made at Vallejo, Benicia, and Collinsville it seems to require about a week to reach the mouth of the river after they enter the Golden Gate.

Plate 14 indicates the catch of fish at various places from Vallejo to Sacramento for a certain period, and is intended to show the passage of two schools between the two places. Each division of the diagram indicates the relative amount of salmon taken at the ten places during one day, the unit being the average daily catch at the given place. The vertical spaces indicate tenths of the average daily catch. By a careful study of the diagram the following points will be noted:

On Monday, April 25, there were few fish taken anywhere, the catch being less than the average at all points. This is the more marked because the Monday catch is on an average 25 per cent greater than that of other days, on account of there being no fishing on Sunday. On Tuesday there was a big catch at Vallejo (3.0 times the average), and a slight increase at Benicia (1.1), Dutton (1.2), Black Diamond (1.2), and Collinsville (0.9). There was little or no increase at other points.

On Wednesday, the second day of the run, the catch at Vallejo had fallen off, and by Thursday the run had entirely passed that point. The points on Suisun Bay and along the river as far as Isleton were gained on the second and third days, and the run reached Courtland on Friday, the fourth day. There was no fishing at some of the upper stations on Saturday, that is, Friday night, the law prohibiting fishing from sunrise Saturday to sunset Sunday, and the record for the fifth day is incomplete. This run was two days in passing Vallejo, and four days in going from Vallejo to Courtland.

On Friday, April 29, another run began passing Vallejo, the catch being over three times the average, and the next day it had increased to over six times the average. On Monday the Vallejo catch decreased to 1.9, on Tuesday to 1.7, and on Wednesday to 0.4, the run being five days in passing that place. This new run was not noticed at the other points on Friday, but on Saturday, the second day, it had reached all points up to Collinsville at the mouth of the river. By Monday, the fourth day of the run, it had reached all points from which we have records, the greatest increase being at the stations farther up the river. During the remainder of this week the catch continued to fall off at the lower stations, but continued very large at Sacramento. By Wednesday, the sixth day, it had passed Rio Vista, and Walnut Grove by the seventh. On Friday there was still a big catch at Sacramento (5.9) and at Courtland (3.8). The record is imperfect for Saturday as usual, but apparently the run had passed all stations. To summarize: This run was five days in passing Vallejo. The foremost were four days going from Vallejo to Sacramento, and the run was five days passing Sacramento.

The spring run passes upstream quite rapidly, reaching their spawning-grounds on the McCloud River in about six weeks after entering the river at Collinsville.

The fall run moves more slowly. They are about two months reaching their spawning-grounds, which are not so far upstream. The flood and ebb tides are more nearly equal, owing to the smaller amount of water coming from the rivers, making the passage of the salmon through the bay a little longer. The nets of the fishermen also offer a greater obstruction during the low water and in this way hold the salmon back. In 1900 salmon were taken in abundance in Suisun Bay and in the river as far up as Rio Vista by the middle of August, but were not taken at Sacramento until after the first of September. The low water doubtless made the movement slow, and the taking of from 2,000 to 10,000 daily out of a slow run would account for their nonappearance at Sacramento.

Upon reaching the shoals in the middle portion of the river they cease their migration, having already found good spawning-grounds. In 1898, 1899, and 1900 the water was normally low and a large proportion of the salmon found spawning-places in the main river. The early high water and frequent fall rains in 1897 sent them into the tributaries.

The latter part of September, 1901, 150 salmon were weighed and branded with serial numbers and released in the river near Rio Vista. Three of these were taken at the hatcheries the latter part of November, just at the close of the season. The following is a particular account of these three specimens:

No. 8, a female, was branded September 20, when it weighed 13,930 grams. It was taken again at Mill Creek fishery November 23, when it weighed 10,180 grams, having been 64 days on the road and having lost 26 per cent of its weight.

No. 91, also a female, was branded September 24, when it weighed 8,470 grams. It was retaken at Mill Creek November 20, when it weighed 7,160 grams, its time in passing up the river being 56 days and its loss in weight being 15 per cent. This specimen was returned to the creek after being weighed November 20. It was found dead on the racks 8 days later, when it had spawned all but 20 of its ova. Its weight had decreased 1,860 grams.

No. 43, a male, was branded September 20, when it weighed 10,080 grams. It was taken at Battle Creek November 25, when it weighed 6,275 grams, making its time from Rio Vista 66 days and its loss in weight 25 per cent.

This important experiment proves that the fall salmon travel very slowly, at a rate of 4 or 5 miles a day, and require about two months to reach the spawning-grounds from the mouth of the river.

The salmon of the spring run arrive at their spawning-grounds from two to six weeks or even longer before they are ready to spawn. This time they spend lying quietly in the pools. The fall salmon are more nearly ripe when they reach their spawning-grounds. Indeed, it is probable that many of them cease to ascend the streams only when they are ready to spawn.

Downstream movement.—Under ordinary conditions there is probably little or no downstream movement, yet when the salmon meet with such obstructions as the racks at the fish-culture stations, there is a tendency to go back downstream. At Battle Creek fishery more salmon are taken at the lower end of the pool than at the upper, indicating that they go as far downstream as possible under the circumstances. The large number of fishes in good condition that get caught on the rack

PLATE 14.

Sixth Day.   Seventh Day.   Eighth Day.

7.0

6.0

5.0

4.0

3.0

2.0
1.9
1.8
1.7
1.6
1.5
1.4
1.3
1.2
Line of Average 1.1
catch for Season. 1.0
.9
.8
.7
.6
.5
.4
.3
.2
.1

V. B. D. Bk. C. R. I. W. Ct. S.   V. B. D. Bk. C. R. I. W. Ct. S.   V. B. D. Bk. C. R. I. W. Ct. S.   V. B. D. Bk. C. R. I. W. Ct. S.

Monday,
April 25.

Thursday,
May 5.

Friday,
May 6.

Saturday,
May 7.

indicates an attempt to go downstream. They may frequently be seen coming downstream toward the rack, though I have never seen any try to get through it. When they get close enough to the rack to feel the force of the swift current, they always try to turn back. Eventually they become so weak that they are unable to keep from being carried onto the rack, where many of them perish.

There are also a few fishes that drop downstream as they become exhausted from long residence in fresh water; rarely from spawning. Such were found almost daily on the upper rack at Battle Creek fishery in 1900. Very few of them had spawned, though they were almost completely exhausted and hardly ever lived over a day after coming near the rack. Such specimens usually lie in the less swift water some 10 or 15 yards above the rack, where little effort is required to maintain a position. As they become weaker and the current carries them back toward the rack, they swim back and forth across the creek, their bodies set obliquely to the current, and their tails frequently almost touch the rack. A very little of such exertion soon exhausts them, and frequently they go but a few feet before being carried against the rack, where they die in a few minutes.

Relation between weather and migration.—It is popularly supposed that the movement of salmon in the rivers is largely determined by weather conditions. Almost any fisherman can tell of a notable run of salmon that accompanied or followed a south wind. Observations made during two years at Battle Creek fishery show that there is no relation whatever between the direction of the wind and the movements of salmon. A strong wind of any direction, however, does apparently cause them to move upstream when they have been lying in a pool for some time. The most notable movement at Battle Creek fishery in 1898 was coincident with a strong north wind. A rain or a slight rise in the water usually causes them to run upstream, but not always. There is apparently no relation whatever between weather conditions and ripening of fish.

CHANGES IN SALMON AFTER ENTERING FRESH WATER.

The alimentary canal.—It is not uncommon for fishes of the salmon family to fast during the breeding season. Such is the case with the Atlantic salmon, the various white-fishes of the Great Lakes, and probably with other species, and it is well known that adult Pacific salmons do not eat while in fresh water. The Sacramento salmon will often snap at bright floating objects and can frequently be taken with the spoon while on their spawning-grounds or while passing up the river. Seventy-five specimens were taken in this way at Jelly Ferry during October and November, 1900. They have never been known to take food, though indigestible material, such as leaves, is sometimes found in the stomach. A 13-inch, mature, sea-run male salmon was taken at Battle Creek fishery in October, 1898. The stomach was contracted the same as in the ordinary adults, but contained two small bits of chitinous substance looking somewhat like portions of the thorax of a grasshopper, but may have been portions of a crustacean.

As they do not eat after leaving salt water, the digestive organs immediately begin to shrivel up. In most of the specimens of the fall run that reach the head of Suisun Bay the stomach and caeca have already contracted and their walls have become firm. Only rarely are they thin and flaccid, as if food had recently been digested. The longer the time since leaving salt water, the more the digestive organs become contracted.

The figures on pages 126–128 illustrate the successive changes in the alimentary canal as observed in specimens from Monterey Bay, head of Suisun Bay, Sacramento River at Sacramento, and at Battle Creek fishery.

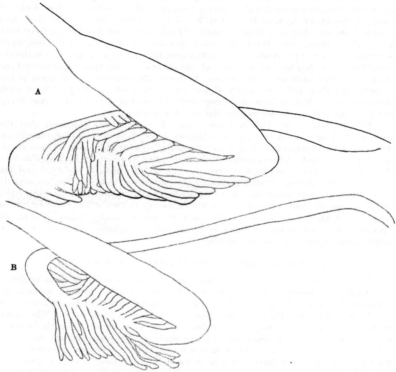

Stomach, pyloric appendages, and part of intestine of two female salmon taken in Monterey Bay, Cal., drawn to same scale. *A*, July 6, 1900, stomach containing food. *B*, July 10, 1900, stomach empty.

The skin.—The most immediate change noticeable in the salmon after leaving the ocean is the great increase in the amount of slime that exudes from the skin upon removal from the water. This point is of physiological interest, but has not yet been studied. By the time the fish reaches the spawning-grounds the skin in most cases has thickened considerably, and frequently the scales are entirely embedded and invisible. In the upper figure of plate 13 the scales can be seen only where the skin has been worn off.

Loss in weight.—Many weights and measurements were made in 1900 for the purpose of determining the loss in weight sustained by the salmon during their residence in fresh water, but our scales proved somewhat inaccurate, and the data can not be used in detail. The loss was shown to be very large, about 35 per cent.

Weights were again taken in 1901 with accurate scales. The results are shown in the tables on pages 128 and 129, which give:

First, the length in centimeters of the specimens weighed, the measurements being made from the nostril to the last joint in the spinal column. The nostril was selected as a point of measurement rather than the tip of the snout because the snout becomes lengthened in breeding males.

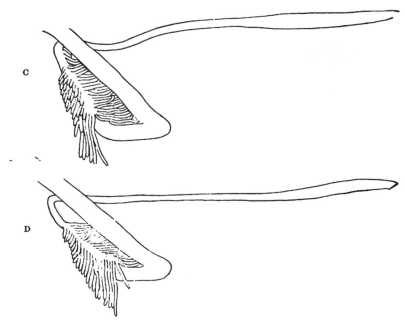

Stomach, pyloric appendages, and intestine of two female salmon drawn to same scale. *C.* Head of Suisun Bay, July 31, 1901, after a fast of a week or two. *D.* Sacramento, August 29, 1901, after a still longer fast.

Second, the average weights of specimens of various lengths delivered at the cannery at Black Diamond September 5 to 11.

Third, similar weights and averages for specimens taken upon their arrival at the spawning-grounds, but before they had begun spawning; also the percentage of loss in weight in these specimens, the average weight of Black Diamond specimens of the same length being taken as a basis in each case. The males were weighed at Battle Creek October 29, the females at Mill Creek November 15.

Fourth, similar weights, averages, and losses of spent fishes, taken either before or immediately after death. Specimens found between October 10 and November 30.

In determining the percentage of loss in the weight of males no account is taken of the loss of milt in spawning, which is very slight, the average total weight of the spent fishes or those just arrived at Battle Creek being compared with the average total weight of Black Diamond specimens of the same length.

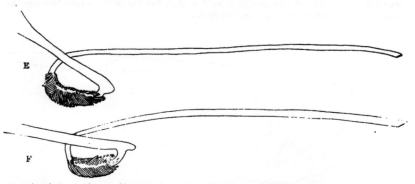

Stomach, pyloric appendages, and intestine of two salmon from upper Sacramento River drawn to scale. E, D. Creek, October 20, 1900, a spent female. F, Battle Creek, September 15, 1900, a male dying on spawning-grounds.

Table of average weights of male salmon.

Length, centimeters.	At Black Diamond.		On arrival at Battle Creek.			Spent fishes.		
	Number of specimens.	Weight.	Number of specimens.	Weight.	Loss, per cent.	Number of specimens.	Weight.	Loss, per cent.
44	1	2,200	2	1,905	13			
45	1	2,650	2	1,905	24			
46	1	2,580	2	2,375	8			
47	1	2,690	4	2,198	18			
48	3	3,057	5	2,430	20			
49	3	3,167	3	2,700	15	1	2,860	9
50	1	3,060				1	2,770	9
51	2	3,720	2	2,928	21			
52	1	3,910				1	1,630	61
53	1	4,105	2	3,308	17			
56	2	5,295				2	4,025	24
60	2	6,280	1	5,495				
61	2	6,455	1	5,510	15			
63	4	7,236				2	5,583	23
64	3	7,378				1	5,980	20
65	3	8,040	1	6,970	13			
75	2	12,390	1	8,870	28	1	8,585	30
76	4	12,855	1	10,950	15			
78	5	14,192				2	10,768	18
79	3	16,797				1	10,015	40
80	1	15,360	2	13,707	11			
81	1	13,415				1	10,770	20
82	2	16,135				1	11,250	33
83	1	15,670	2	14,175	10			
84	2	17,010	1	15,060	12			
89	1	18,490	1	17,665	10			
Average					16			26

The percentage of loss in ripe females is determined by comparing the total average weight with the average weight of specimens of the same length weighed at Black Diamond, the latter being taken as a basis. The loss percentage as stated for the spent fishes does not include the loss of ova. The percentage is determined by comparing the weight of spent fishes plus the weight of the extruded ova with the weight of Black Diamond fishes of the same length. The weight of the extruded ova is determined by finding the difference in weight between the ovaries of spent fishes and those of fishes just arrived at the spawning-grounds.

Table of average weights of female salmon.

Length, centimeters.	Black Diamond.		Ripe, unspawned.				Spent fishes			
	Number of specimens.	Average weight.	Number of specimens.	Average weight. Total	Ovary.	Loss, per cent.	Number of specimens.	Average weight. Total.	Ovary.	Loss, per cent.
65	1	7,725					1	5,680	67	13
66	17	8,249	2	7,048	1,595	15	1	4,680	80	24
67	15	8,134					1			
68	9	8,681	2	7,760	2,018	11	1			
69	6	8,949	1	7,720	1,921	14	2	4,988	140	25
70	6	8,981	1	8,340	2,057					
71	4						1	6,340	113	15
72	7	10,227					2	6,553	80	16
73		10,090					2	6,388	89	15
74	1	10,700	2	9,085	2,192	15	2	7,450	92	11
76	3	11,508	1	9,930	2,014	14	1	6,090	90	31
77	3	12,350	1	11,090	2,775	6				
78	2	12,220	2	11,985		2				
80	2	14,763	3	11,983	2,372	19	1	10,990	159	11
82	1	14,405					1	8,730	104	24
Average						12				19

The averages of the loss percentages are: For males upon their arrival at the spawning-grounds, 16 per cent; for males at time of death, including loss of milt, 26 per cent; for females upon arrival at the spawning-grounds, 12 per cent; for females at the time of death, not including loss of ova, 19 per cent. The difference between the loss as determined in 1900 and 1901 is accounted for by there having been many more grilse weighed at the spawning-grounds the former year. (See "Two forms of males," page 130.)

Under the heading "Details of migration" (page 124) will be found an account of three salmon that were released in the Sacramento near Rio Vista, after being weighed and branded, and subsequently taken at the Government fisheries. One had lost 15 per cent of its weight, another 25 per cent, and the other 26 per cent during the migration.

One important point to be considered in this study of the loss in weight during migration is the deterioration in the value of the flesh as a food. The loss of 12 or 16 or 25 per cent is entirely in nutriment. If even a very fat beef were starved two months, or until it had lost 16 per cent of its total weight, no one would care to eat of its flesh. But such is the condition of the fall salmon upon their arrival at the upper portion of the river. They have eaten nothing for over two months, and nutriment to the extent of about 16 per cent of their weight has been absorbed, almost wholly from the flesh.

It is evident, therefore, that the fall salmon taken at the upstream points have but little value as food, and their capture should be prohibited.

COMPARISON OF THE SEXES.

Relative changes in fresh water.—Before entering the fresh water the two sexes are identical in appearance. With the fall run the sex can often be distinguished by the external appearance in specimens taken at the head of Suisun Bay, and it can always be distinguished in normal specimens by the time of their arrival at the spawning-grounds. Soon after entering fresh water several cartilaginous teeth appear in the front of the jaws of the males, and at the same time the jaws begin to grow longer. By the time the males reach the spawning-grounds the jaws are much prolonged and hooked, and the teeth have grown to be large and solid canines; the body becomes deep and slab-sided, and the color usually more or less reddish.

The principal change in the females lies in the diminution in the muscular tissue of the back and sides and in the distension of the abdominal walls on account of the development of the ova. Their color is usually more or less olive. After spawning the female is as thin as the male, but the jaws are not prolonged.

The following illustrations indicate the changes better than descriptions:

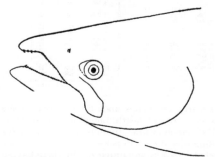

Head of male salmon taken at Sacramento, September 5, 1900, showing the beginning of the jaw prolongation and canine teeth while yet cartilaginous.

Head of female salmon taken at Sacramento September 5, 1900. This is also the head of the male in salt water.

Two forms of males.—The illustrations on page 131 show the two forms of male salmon, known as adult and grilse, that are found in the headwaters. Those here shown were nearly of the same length, though it is very rare to find as small a specimen as the upper that has the adult form, and the lower was a rather large grilse: (See also plate 13, photographs of these same specimens.)

The differences are obvious. The grilse simply fails to develop the characteristics of the breeding male, viz, the prolonged and hooked jaws, the large, hooked teeth, the deep, slab-sided body, and red color, and retains its salt-water appearance except in the loss of flesh. Grilse weigh from a half pound up, and intergrade with the adults both in weight and appearance; specimens with a length of 90 centimeters (35.5 inches) are occasionally found. I have seen two, which, from their olive color, could be distinguished as sea-run fishes, that were only 13 inches long.

At Battle Creek fishery in 1900 the males were nearly all grilse, though there were almost as many of the adult males as there were females. The great preponderance of the grilse over the adult males and females is due to their being too small to be taken by the nets of lawful mesh.

The cause or causes that lead to the production of the grilse form are not known. Mention is made in another place in this report of the sexual maturing of the male parrs that remain in fresh water during their first summer. It it possible that this stunts them and causes the production of grilse. If grilse were simply young individuals that followed the adults into fresh water, we would expect to find females among them. The two forms can not be distinguished except in breeding fishes.

The dwarfed form of the female is practically unknown. Among many thousand specimens handled at Battle Creek fishery only one dwarfed specimen was found. This was 16 inches in total length and weighed 2 pounds. Its ova were mature.

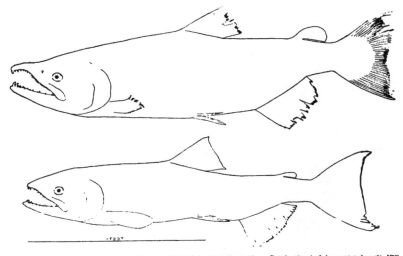

Two spent male salmon, adult and grilse forms, found dead on the rack at Battle Creek fishery October 28, 1900. Length from hinder edge of eye to base of tail, larger specimen 500 mm., smaller 455 mm.; weight, larger specimen, 3,100 grams, smaller 1,290 grams.

Comparative statement of weight and length of the snout in adult and grilse salmon of the same body length.

Length from hinder edge of eye to base of caudal fin (centimeters).	Weight in grams			Length of snout from hinder edge of eye (millimeters).		Length from hinder edge of eye to base of caudal fin (centimeters).	Weight in grams			Length of snout from hinder edge of eye (millimeters).	
	Adult.	Grilse.	Excess of adult.	Adult	Grilse.		Adult.	Grilse.	Excess of adult.	Adult.	Grilse.
40	1,125	1,025	100	68		49	2,475	2,375	100	77	70
41	1,650	1,100	650	75	55	49	2,700	1,425	1,275	79	67
43	1,725	1,575	150	66	66	50	2,975	1,525	1,225	92	71
43	2,500	1,900	600	66	65	51	2,775	1,875	900	90	75
45	2,350	1,075	1,275	72	65	54	3,700	2,425	1,275	80	71
46	1,875	1,675	200	72	62	57	3,950	2,775	1,175	95	82
46	2,450	1,525	925	87	65	57	4,000	2,400	1,600	100	72
47	1,855	1,800	75	73	64	77	11,600	7,050	4,550	137	128
47	2,825	1,725	1,100	80	60	79	10,800	8,550	2,250	145	102
47	2,175	1,300	875	71	64						

Hermaphrodites.—I am indebted to Mr. J. P. Babcock, of the California Fish Commission, and to Mr. Chamberlain for two hermaphrodite salmon. Mr. Chamberlain's specimen was discovered by the spawning crew at Battle Creek hatchery in December, 1900; the other specimen was discovered by Mr. F. A. Coles while cleaning fish for the cannery at Black Diamond, and was taken in that vicinity in May, 1901. The accompanying illustration represents the Black Diamond specimen.

The genital organs of the two specimens are essentially the same in structure. There is but one pair, as in ordinary individuals, but each organ is developed partly as testis and partly as ovary. One organ in the Black Diamond specimen has about 3 inches of the anterior portion well developed as a testis, and nearly mature.

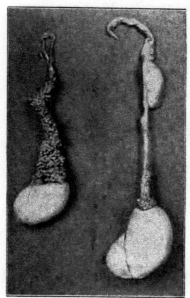

Genital organs of hermaphrodite salmon.

Immediately behind this are a few ova that are about as large as ova ordinarily are in salmon taken in this part of the river. The next 4 or 5 inches of the organ consists merely of the supporting membrane and seminal duct, with half a dozen ova developed in one place. Then follows a portion about 2 inches long developed as testis. The usual seminal duct leads posteriorly. The other organ of this specimen also has the anterior portion developed as testis, while all of the posterior portion is ovarian. While the ova are of normal size, the locality being considered, their number is not over one-fourth as great as would be produced by a similar portion of a normal ovary.

The Battle Creek specimen is similarly developed. Some of the ova are attached, while others are free, as if the fish had been only partly ripe. All are variable in size, but none of normal appearance are as large as the average ova taken at Battle Creek. Many of the ova evidently were dead at the time the fish was taken, and some of these were abnormally large. I understand that some of the free ova were spermatized with milt from the testis portion of the same organs, but none of them lived. It is not known whether they were fertilized.

The genital organs only of these specimens were sent me, and I do not know the condition of the cloacal openings.

Relative number of males and females.—In measuring and weighing salmon at Black Diamond, on August 20 and 21, 1900, it was noticed that the females were more numerous than the males. To determine whether this was merely a peculiarity of the catch of those two days, Mr. F. V. Hubler, the weigher for the Black Diamond cannery, was employed to make notes of the relative number of males and females during the season, with the following result.

Table showing relative number of male and female salmon taken at mouth of river.

Date.	Males.	Females.	Date.	Males.	Females.
Aug. 21	74	154	Sept. 1	108	214
Aug. 22	87	113	Sept. 3	130	163
Aug. 25	4		Sept. 4	101	135
Aug. 26	15	22	Sept. 5	124	163
Aug. 29	73	219	Sept. 6	137	195
Aug. 29	135	201	Sept. 7	161	146
Aug. 30	121	212	Sept. 8	136	148
Aug. 31	76	121	Sept. 10	241	316
			Total	1,780	2,345
			Per cent	41	59

During October and November, 1900, Mr. Arthur Sergison caught 76 male and 32 female salmon on spoon hooks at Jelly Ferry.

The following is a statement of the relative number of males and females taken in the seine at Battle Creek fishery in 1900:

Table showing relative number of salmon taken at Battle Creek fishery.

Date.	Males.	Females.	Date.	Males.	Females.
Oct. 12	9	8	Oct. 23	238	22
	86	14		184	18
Oct. 14	187	20		73	4
Oct. 20	45	10	Nov. 1 (river)	12	6
	22	4	Nov. 19	5	2
	22	6	Total	1,107	156
	175	34	Per cent	88	12
Oct. 23	111	11			

Relative weight.—The average weight of all salmon taken at Black Diamond cannery from August 20 to September 10, 1900, varied daily from 21.3 pounds to 23.8 pounds. As the largest specimens taken were invariably males, it is probable that the average weight of the males was greater than that of the females. The reverse was true in the upper river, as will be seen from the following statement of weights of salmon taken by Mr. Sergison at Jelly Ferry:

Table showing relative weight of male and female salmon taken at Jelly Ferry.

Date	Weight, pounds.		Date.	Weight, pounds.		Date.	Weight, pounds.	
	Males.	Females		Males.	Females.		Males.	Females.
Sept. 28	4.5	11.5	Oct. 7	10.0		Oct. 24	10.0	
Oct. 1	9.0	8.0	Oct. 8		24.0		10.0	
	31.0	11.5	Oct. 9	3.0	8.0		37.0	
Oct. 2	3.0	13.0		4.0		Oct. 25	4.0	
	6.0	8.0		12.0			5.5	
	7.0	10.0	Oct. 10		14.0		10.0	
	8.0	13.5	Oct. 12	3.0	7.0	Oct. 26	4.5	
	11.0			4.0	12.0	Oct. 27	8.0	
	12.0			5.0	15.0		4.0	
	14.0			8.0	32.0	Oct. 28	4.0	24.0
Oct. 3	1.5	9.0	Oct. 13	6.0	18.0		10.0	
	3.0	12.0	Oct. 14	4.0	22.0	Oct. 29	14.0	
	3.0	14.0			24.0		14.5	
	3.0	24.0	Oct. 16	3.0		Oct. 31	38.0	
	3.5	26.0		4.0		Nov. 1	4.5	
	3.5	34.0		5.5			12.5	
	4.0			6.0		Nov. 3	8.0	
	4.0			7.0			10.5	
	4.0			6.0	11.5	Nov. 6		18.0
	4.5		Oct. 17	5.5		Nov. 10	6.0	10.5
	5.0		Oct. 18	6.0			34.0	13.0
	6.0			8.0		Nov. 12	11.5	
	8.0			34.0		Nov. 13	10.0	12.0
	9.0					Nov. 15		24.0
	12.0		Oct. 22	3.5	32.5	Nov. 20	21.0	
	34.0		Oct. 23	14.0				
Oct. 7	4.5	12.0		20.0		Average.	9.1	16.8
	7.0		Oct. 24	4.0				

The average weight of the Battle Creek specimens could not be determined, owing to the selection of the larger males for spawning, but it was certainly less than that given in the Jelly Ferry record.

It will thus be seen that throughout the fall season of 1900 there was a greater proportion of female salmon taken by fishermen in Suisun Bay and the lower river. On the other hand, the small males, being too small to be taken in the regulation net of the market fishermen, were greatly predominant in the headwaters. The evidence here given does not indicate that more of one sex is produced than of the other.

This point should be considered in making laws governing salmon fishing. The small males are not desirable for propagation, either natural or artificial, and on account of their great number they are a nuisance at the Government fisheries. They are simply so much valuable food wasted. The present law prohibits the use of nets that would catch them, and it should be amended. As there are no small females, the small-mesh net would not affect the supply of breeding females. A law prohibiting the taking of small fishes is of value only when the small fishes are growing fishes. But the small salmon that come in from the ocean are not growing fishes. None of the salmon ever return to salt water. Their sole value lies in adding so many pounds to the market supply or in reproducing their kind on the spawning-beds. A large fish is worth more on the markets than a small fish; but so are large cattle worth more on the market than small cattle, yet a stock-raiser would never think of selling his fine cattle and keeping only the runts to breed from. It would be better for the salmon as a species, and therefore better for the salmon industry, if the present minimum net-mesh were made the maximum. A small-meshed net does not catch so many large fishes, which would allow the larger individuals to reach the spawning-grounds. The salmon will certainly deteriorate in size if the medium and larger sizes are taken for the markets and only the smaller with a few of the medium allowed to breed.

NATURAL PROPAGATION.

Spawning habits.—Salmon in spawning usually take a position at the upper end of a riffle where the current is strong and where there are gravel and cobblestones among which the eggs may lodge. After selecting the place the female extrudes a few eggs and then moves away. The male immediately takes her exact position, or perhaps a point one or two feet downstream from it, and extrudes a small quantity of milt. In about five minutes the process is repeated, the female always taking the position first occupied. This they continue day and night for over a week, usually nearly two weeks. I have observed salmon spawning at night, but have never been able to watch one pair until spawning was completed. Branded salmon No. 91, previously referred to, was only eight days in spawning, although some eggs had been extruded before it was taken. Two weeks is the spawning time usually assigned by persons living in the vicinity of salmon streams, which is probably about right.

On account of the difficulty in seeing eggs under water, it has been impossible to determine the rate at which ova are deposited. The motions of the fish show just when ova are being extruded, but observation at a distance of 5 feet, with the aid of a field glass, has failed to disclose the eggs.

The female at irregular intervals turns over on her side and digs her tail into the gravel. If the gravel is fine there is often a considerable hillock thrown up, leaving a hole 6 or 8 inches deep and 2 feet across. This digging is probably not for the purpose of covering the eggs, nor to make a space for them to lie in, but by the violent exercise to loosen the eggs from the ovaries. If the purpose were to cover the eggs it would be repeated every time any were deposited. Gravel does not drift so far as the eggs, and if such were the purpose it would not be accomplished. Besides, it is almost impossible to cover eggs with gravel; the eggs, being almost as light as the water, slide away from the gravel. More than that, a covering of over an inch of even fine gravel kills them. The hillock, by forming an eddy at the bottom of the stream, prevents many eggs from floating away and being devoured by other fishes, but such are liable to be covered too deeply and killed in that way. Some of the fine sediment, however, may settle on the eggs and tend to make them invisible to egg-eating fishes. The "nest" can hardly be made as a place for the eggs to lie in, for the current always carries them below it.

The presence of the other sex is not necessary to excite either to spawning efforts. I have seen the female spawning alone at Battle Creek fishery, and other persons have reported similar observations from other places. In September, 1900, I saw a male spawning alone near Sims, the female having been killed by a sportsman in order to get trout bait. Like observations have been reported by other persons.

Percentage of fertilization.—As one pair of salmon deposits an average of 6,000 eggs the increase would be enormous unless there was great loss at some period. It is usually supposed that the greater part of this loss is due to a lack of fertilization of the ova. The great care necessary to secure perfect fertilization artificially has led fish-culturists to suppose that the percentage of fertilization under natural conditions must necessarily be very low. In artificial fertilization the ova and milt are mixed together in a vessel, insuring a coating of milt or spermatized water over each ovum. In natural spawning the ova are caught in the eddies among the rocks, either near the nest or within a few yards below it. A few seconds after the ova are spawned a small quantity of milt is disseminated in the current to be carried against them. It seems very unlikely that a large percentage could be fertilized under such conditions. The following experiments throw some light on the question:

To determine the percentage of fertilization under as nearly natural conditions as possible a box was built 4 feet wide, 14 feet long, and 15 inches deep, and a strong current of water turned through it. About 5 inches of gravel was put in the upper three-fifths. A pair of salmon were placed in the box October 28, 1897. A female not quite ripe was selected, in order to allow a few days to become accustomed to the place. Pickets nailed to the side prevented the fishes from jumping out. By November 2 they seemed to be at home in the box, and their actions indicated that they were ready to spawn. A few eggs were deposited the next day. On the 4th the male died, having become almost entirely covered with fungus in the one week. Another was put in immediately, but the spawning was interrupted, as it required a day or two to get used to the place. The female died November 12, having deposited but few eggs. No cause of death could be ascertained. Of the 512 eggs deposited, 343 were killed while being deposited. Of the remaining 169, 129 or 76 per cent were fertilized. In the second attempt 82 per cent were fertilized.

In 1898 a pair of salmon were put in a ravine with simply a rack to prevent their going downstream. No eggs were deposited.

So far as the number of eggs killed is concerned this experiment is not a fair test. The level floor of the box made few eddies, and the eggs were washed into the corners and killed. The percentage of fertilization would certainly be no greater

than under entirely natural conditions. We would expect the death of the male
and the introduction of a new one to cause some eggs to be left unfertilized.

November 18, 1897, I dug up five or six nests in comparatively still water where
fishes had been seen spawning for a month. The sand and gravel were thrown into a
screen and carefully sifted, but no eggs were found; but on stirring the gravel and
cobblestones in a strong current and setting the screen below so as to catch floating
objects, 13 eggs were secured; 11 were alive and all fertilized, 4 about 3 days old and
the others about 28 days. I could see no indication of fertilization in the dead eggs.

In November, 1898, in order to obtain eggs naturally spawned, I placed a screen
obliquely in the water 2 or 3 feet below where salmon were spawning. The screen
was covered with small cobblestones, that the eggs might lodge among them and
be protected from spawn-eating fishes. The first trial was unsuccessful. The second
secured 48 eggs; 30 were dead when found. All of the live eggs were fertilized, the
others could not be tested.

The experiment with the screen is not a fair test for the number killed, as the
screen caught much gravel which pressed the eggs against the wires, and without
doubt killed many more than would have been killed under natural conditions. The
1897 experiment in securing eggs from the stream, when 15 per cent were dead, was
a fair test, but the number of eggs was too small to warrant definite conclusions.
In both experiments all live eggs were fertilized.

In November, 1900, 39 eggs were secured from natural spawning-beds in Battle
Creek, 25 of them evidently fertilized, and 1 certainly not fertilized. The condition
of the others could not be satisfactorily determined, as they were killed in securing.

At another time a fish was artificially spawned in the creek on natural spawning
beds, a screen being placed below to catch the eggs. A male was held in the same
position immediately afterwards and milt expressed. The test was not quite fair,
for although there was probably a larger quantity of milt than is discharged at once
naturally, yet there was also a larger number of eggs. The eggs being caught by the
screen and thereby remaining closer to where the male was stripped was of no
advantage, as the life of milt in water is ample to allow it to come to rest. If they
had been farther away it would have given time for the milt to become more thoroughly
disseminated through the water. The eddies caused by our standing in the water and
holding the fish prevented the eggs and milt from floating off in a natural manner.
In two trials, 35 per cent were fertilized in the first, and 50 per cent in the second.

These various experiments indicate a high percentage in natural fertilization, probably over 80 per cent. It is significant that all live eggs found that had been spawned
naturally, excepting one, were fertilized. The fertilization of dead eggs could not be
determined, though they were no more liable to be unfertilized than live ones.

Mortality among ova.—These experiments also point to a high but indefinite
mortality from being covered by the gravel. The greatest loss, however, is probably
due to spawn-eating fishes. In the middle portion of the river, as at Battle Creek,
when salmon are spawning, great numbers of other fishes, mostly the split-tail
(*Pogonichthys*), gather about to feed on the spawn. Fifty or more split-tails may
often be seen lying a few feet downstream from a spawning salmon, and although
each fish may eat but few eggs, all together they probably destroy a large percentage of the eggs spawned in the middle portion of the river.

Trout have been taken near the spawning platform at Battle Creek station with
the stomach and throat gorged with eggs, the waste from artificial spawning, and 3

ova were found in the stomach of a trout taken in Mill Creek December 1, 1901, several days after any artificial spawning had been done at the station. As there were several salmon spawning in the creek at that time, there is little doubt that the eggs were secured from natural spawning-beds. Trout are adapted to catching floating objects and are doubtless very destructive to salmon spawn where the salmon breed naturally.

Other fishes, such as the hitch (*Lavinia*), hardhead (*Mylopharodon*), and sucker (*Catostomus*), have not been found to eat salmon spawn, though they probably do; the black-fish (*Orthodon*) has not even been found near the spawning-beds. A large Sacramento pike (*Ptychocheilus*) that had secured spawn from natural spawning-beds was taken in the river near the mouth of Battle Creek in 1900.

Natural versus artificial propagation.—Probably the most important problem yet remaining unsolved in connection with the natural history of the salmon is the efficiency of natural propagation. If we could segregate a certain number of fishes in a small stream, then put a fine screen across below where they are to spawn, and later catch all the alevins and fry produced, we could solve the problem. But a small stream, such as could be experimented with, is liable not to have an average number of fishes to prey upon the spawn and alevins, and the conditions would not be entirely natural. The following statement represents approximately the comparative value of natural and artificial propagation:

Items.	Percentage of loss in propagation.	
	Natural	Artificial
Not spawned	1	a 1
Unfertilized	15	2
Killed before hatching	b 70	8
Alevins killed	b 13	c 2
Total loss	99	13

a 10 to 30 per cent if un-spawned eggs are not removed by abdominal section
b No definite data
c At least 50 per cent if alevins are planted

From the foregoing it will be seen that the heavy loss in artificial propagation has been in not spawning all the eggs and in planting alevins, both of which can be remedied, as is elsewhere shown in this report. The total loss in artificial propagation should not be 15 per cent.

There is a much greater loss when alevins are planted artificially than when they hatch out naturally.

(*a*) From a given number of ova, as those produced by one fish, which is the basis of the percentage, there are more alevins to be destroyed in the case of artificial propagation. In natural propagation they have been already largely destroyed before they become alevins, and there are not 87 per cent left to be destroyed, as in the case of artificial propagation.

(*b*) But even with a given number of alevins the percentage is greater in artificial planting. It is not possible to scatter them as well in artificial planting as in natural propagation. No amount of care will prevent their collecting in bunches, which has not been seen in natural propagation.

Something of the value of artificial propagation can be learned from an experiment tried at Clackamas hatchery, Oregon. In March, 1896, 5,000 salmon fry 2.5 inches long were marked by cutting off the adipose fin. The eggs from which the fry were hatched were spawned at Baird hatchery in September, 1895. Mr. Hubbard, superintendent of Clackamas hatchery, who tried the experiment, reported that 375 of the marked fishes were taken in 1898. The smallest weighed 10 pounds, the largest 57 pounds, and the average was 27.7 pounds. Besides these, 5 were taken in the Sacramento River in 1898. A few more were taken both in the Columbia and in the Sacramento in 1899, and also in 1900. The 1900 specimens, however, may have been of those marked in the Sacramento in 1898. From those 5,000 fry 2.5 inches long, costing less than a dollar to produce, fish weighing over 5 tons were taken. That means that for every female fish stripped at the hatchery the fishermen should catch about 5 tons three years later. About 400 of the 5,000 marked fishes were reported taken. We have no means of knowing how many came back to fresh water and escaped the nets, or how many were caught but not noticed.

INJURIES AND DISEASES.

General effects of spawning.—Notwithstanding their long journey from the ocean, the salmon reach their spawning-grounds in good condition. They are not nearly so fat as when they left the ocean, but all their bruises are received after arrival at the spawning-grounds. This fact has already been noted by Evermann (Bulletin United States Fish Commission 1896, p. 191).

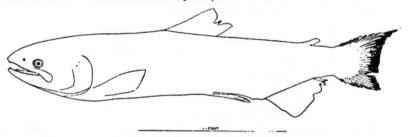

Spawned-out female. Battle Creek, October 20, 1900.

As spawning progresses the abdominal walls of the female contract and she becomes as thin as the male. Her caudal fin is worn off to a mere stub. All fins of both sexes become more or less frayed, the skin wears off the sides of the tail and the prominent portions of the body, such as the edges of the jaws and bases of the fins. Fungus nearly always attacks the gills and the various bruised places and frequently destroys one or both eyes.

It has been supposed that the exertions of spawning completely exhaust the female and that she dies immediately upon its completion. It would seem rather strange if there were just enough energy to spawn all the ova, and that with the extrusion of the last one the fish should die at once. Observations indicate that the female has considerable energy left after spawning all of the ova, and that she continues on the spawning-beds for some time thereafter. The injuries are received only after most of the ova have been spawned. She probably does not know when the ova have all been extruded, and her instinct compels her, when once spawning has begun, to continue the spawning efforts as long as energy lasts. The complete extrusion of the ova, since it is not noticed, is merely incidental.

In 1900, 14 spent females were taken alive on natural spawning-beds; 7 of them had extruded all ova, in one specimen there was 1 ovum yet unspawned, in two

PLATE 15.

SPENT AND DISEASED SALMON.

A. Female that had spawned all but about 500 ova showing that the injuries are received while spawning the last few ova, or after all have been spawned.

B. Male, apparently exhausted from long residence in fresh water, but not from being on spawning beds; typical condition of late summer-run males, Battle Creek, September 15, 1900.

C. Female, with ova but half developed, Battle Creek September 15, 1900; died from long residence in fresh water.

D, E, Two males, grilse and adult, showing extreme cases of fungous growth, October 22, 1900.

there were 2 each, and in the others there were 3, 58, 92, and 107, respectively. If they died immediately after spawning the last ovum we would not have found such a large proportion of live specimens completely spawned out; and if spawning completely exhausted them we would not have found them on riffles but in the more quiet water. In one instance when only one female was taken she was entirely spawned out and had been seen spawning just before the seine was hauled. Of course, it is possible that the last ova were spawned just before hauling the seine, but in any case the fish was far from being in a dying condition. (See plate 16 and fig. A, plate 15.)

Spent salmon.—A few sample field notes on the condition of spent salmon found dead at Battle Creek in 1900 are here given. Several hundred similar specimens were examined during the season.

The following notes refer to spent females:

September 30. Nearly spawned out. Numerous parasites (copepods) and a small patch of fungus on each gill. Top of head without fungus, but with skin worn off. No fungus on body. Fins and skin in good condition.

October 10. But two eggs left in body cavity. Gills about one-fourth covered with fungus; several gill parasites. Skin worn off caudal fin and the rays about half worn off.

October 15. All but 10 eggs spawned. Died in shallow water. Caudal fin entirely worn off, but fish otherwise in good condition. But little fungus. Gills but slightly injured.

October 26. Entirely spawned out, except 2 eggs yet attached to ovary. Half of caudal fin worn off evenly; other fins in good condition. Gills one-third destroyed. Small patches of fungus in various places on body. One eye blinded.

November 1. Specimens Nos. 2 and 3 from the river were of the same length. No. 2 had spawned all but 92 eggs, and No. 3 all but 442. No. 3 weighed 900 grams more than No. 2. The skin was entirely worn off the caudal fin of No. 2, and the rays half worn off. Caudal fin of No. 3 in good condition.

The following notes refer to spent males:

November 5. Badly scarred, one eye blinded, skin worn off edges of fins and jaws. Another specimen, not badly scarred, blind in both eyes, skin worn off snout and edges of fins.

November 6. Skin and flesh worn off in several places behind dorsal and on tail nearly to backbone; skin worn off edges of fins, jaws, and the whole snout; both eyes blinded; gill filaments half destroyed by fungus and parasites.

November 10. One eye blinded; much scarred; little fungus. Another specimen, blind in both eyes; skin worn off jaws and edges of fins; skin dead all over tail and caudal fin; nearly every gill filament with one or more parasitic copepods, and many sloughed off for one-third their length.

Diseases of intestine.—The intestine of the spawning salmon is frequently inhabited by tapeworms, which sometimes completely fill it and extend into the cæca, but I have never found them in the stomach. They were much more abundant in 1898 than in 1900. In addition to the tapeworms the intestine, especially posteriorly, is filled with a viscid, greenish yellow fluid. No examination of this has been made, but it is probably formed by the disintegration of the lining of the intestine, a catarrhal desquamation such as has been found in the Scottish salmon.

Fungus.—Fungus as related to salmon deserves special investigation. Nearly all the salmon that reach the vicinity of Battle Creek fishery during September and October become affected with fungus, which grows in velvety patches on various parts of the body. The points most commonly affected are the top of the head, the gills, fins, and eyes. Of 31 specimens noted on the racks at Battle Creek fishery during October 16, 17, and 18, 1900, 5 were blind in both eyes and 14 others blind in one eye, as a result of fungus.

The following extracts from my field notebook indicate the rapidity of growth of fungus, and show a condition rather worse than the average, though by no means exceptional. The two descriptions refer to the same specimen; first, on September 30, when it was caught and tagged and returned to the creek; and second, on October 4, when it was found dead against the rack:

September 30.—Male, ripe, weight 2,800 grams. A notch in left pectoral, a slit in dorsal; caudal with a few small dead spots, one worn through; 3 parasites on left gill, 7 on right; whole top of head and upper edge of pectoral fins covered with fungus; skin partly worn off sides of tail.

October 4.—Fungus covering following portions of fish: whole top and sides of head to below eyes, lower jaw, back in front of dorsal, edges and bases of pectoral fins, upper side of ventrals, a spot behind right pectoral and one on back before adipose fin, half of adipose fin, spot behind left pectoral, left side below dorsal nearly to lateral line and half way to adipose fin, base of anal on left side, and belly behind ventrals. Left gill with seven streaks of two or three dead filaments each; a little fungus on each dead portion; a small patch of fungus at tip of gill matting together the filaments of all the arches. Right gill with a patch of fungus at tip matting together filaments from all the arches, and another anteriorly on the inner arches. Skin of tail and most of caudal fin dead; some of caudal rays gone.

There were worse cases of fungus than the one here described, but this shows what can grow in four days. Another specimen that was in good condition when tagged November 1 was "half covered with fungus" when seen last on the 8th.

The pest almost disappeared in December. Figures D and E of plate 15 show the extent to which salmon are sometimes affected with fungus.

Gill parasites.—Another common pest of the salmon in fresh water is a parasitic copepod which attaches itself to the gill filaments. There are usually not very many on one fish, but sometimes the gills are almost destroyed by them. Plate 15 shows an extreme example. The gills sometimes decay without being affected with fungus or parasites, as was found in a specimen at Battle Creek fishery, October 7, 1900, in which one-third of the gill filaments were dead. (See also plate 13.)

Diseased ova.—In all of the females found dying during September and October, 1900, the ova were more or less diseased. Sometimes there were only a few addled and misshapen ova crowded into the interstices of the healthy ova, but sometimes almost all were addled. Occasionally there were a few abnormally large ova, half an inch or more in diameter, and in a specimen taken October 12, 1900, nearly all were in this condition. In another taken about the same time half were of this character, while the normal ova had been spawned. In another a third of the ova were addled, and the others had absorbed water and were turgid. See fig. C, plate 15.

Length of life of fall salmon after reaching spawning-grounds.—September 30, 1900, numbered metallic tags were attached to 3 male salmon, which were then released in the pool between the racks at Battle Creek fishery; 1 of these was found dead October 5, having survived 5 days. October 22, 36 others were tagged and released in the pool; 27 of these were seen at various times, some of them quite frequently, up to November 1, and 5 of them were found dead within that time, the maximum time being 10 days. On October 25 36 were tagged and released in the creek below the racks; 8 of these were found dead on the racks up to November 10, a period of 16 days. Four were tagged and released in the mouth of the creek, about 2 miles below the fishery, on November 4. One of these was seen on the 5th and again on the 8th, when it was almost dead, a period of 4 days. November 9, 39 were tagged and released in the river below the mouth of Battle Creek; 3 were seen at the fishery on the 16th, 7 days afterwards.

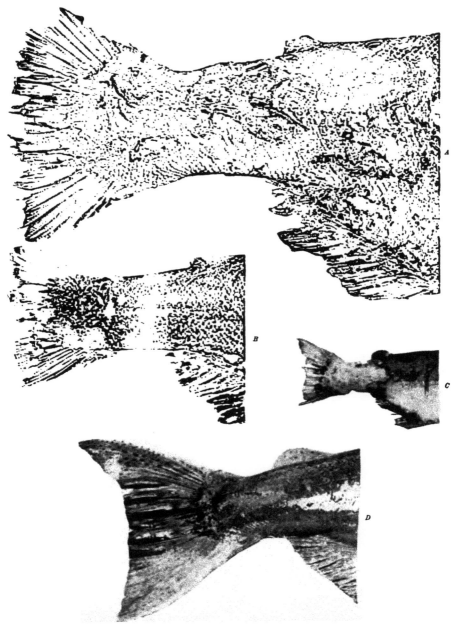

TAILS OF FEMALE SALMON FROM SPAWNING BEDS.

Altogether 12 tagged fishes were seen after dying, and the average time that they lived after tagging was 11 days. The longest time was 16 days; some had probably been in the creek a few days when tagged, though the freshest were selected. Two weeks is a very fair estimate of the length of life after reaching the spawning-grounds. Branded specimen No. 91, a female, lived but 8 days after reaching the spawning-grounds.

DEATH.

The salmon of the genus *Oncorhynchus* apparently has no instinct whatever to return to salt water after spawning. Worn-out specimens are sometimes seen drifting down stream and have been found as far down as Sacramento, though it is by no means certain that such have been on spawning-beds. In such cases they are simply too weak to stem the current and, according to a Sacramento fisherman, "not fit to look at." Dead salmon rarely float, though the current sometimes washes them along the bottom a short distance I have seen dead salmon lie for several days in rapids and have seen them in all stages of decay in strong currents. Of the 200 or more dead salmon that were marked and thrown over the upper rack at Battle Creek fishery in 1900, only 2 were carried to the lower rack, which was a half mile further down stream. In small streams the water is often greatly contaminated by the dead fish, and the stench is a great nuisance to people living in the vicinity.

The great variation in size of spawning salmon, together with the occasional presence of certain scars, such as a broken nose, has led many people to doubt whether they all die after spawning once. The variation in size amounts to nothing as an argument, when we know that with about 60 marked fishes known to be of the same age, taken in the Columbia River in 1898, the variation in size was from 10 to 57 pounds. The broken nose could be received at many other times than when spawning.

It is sometimes thought that if a spawned-out salmon would float down stream to salt water it would revive, but such is not the case. Humpback and dog salmon often spawn in small creeks and brooks that empty directly into the ocean, yet they die like other species. They have been seen dying and dead in brackish water. The investigation of the blueback salmon or redfish in Idaho in 1895 (see Bulletin United States Fish Commission 1896, p. 192), when a net was placed across the mouth of a small stream containing about a thousand salmon, proved that that species has no tendency to return to salt water after spawning. Lake Karluk, Kadiak Island, Alaska, is but about 20 miles from the ocean and is a great spawning-place for the blueback salmon. The outlet is shallow near the mouth, and if the salmon ever went back the Indians would be sure to see them, but they do not. In June, 1897, the shore of the lake for miles was lined with the bones of the salmon that had died six to eight months previously.

The fact that all salmon of the genus *Oncorhynchus* die very shortly after spawning once can not be questioned.

PLATE 17.

Bull.

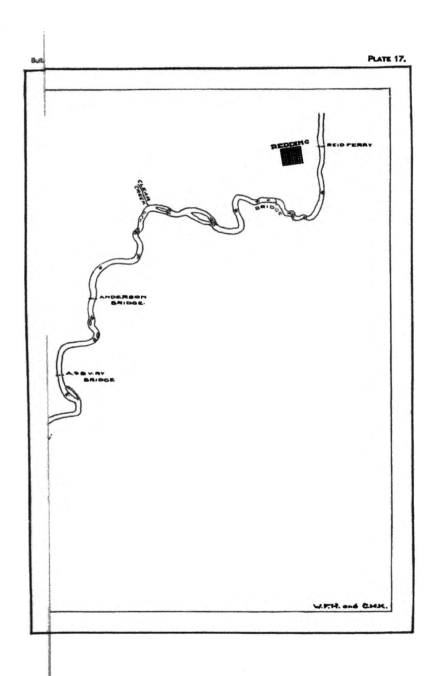

REDDING   REID FERRY

CLEAR CREEK

BRIDGE

ANDERSON BRIDGE.

A. & B. V. RY BRIDGE

W.F.H. and C.H.K.

i

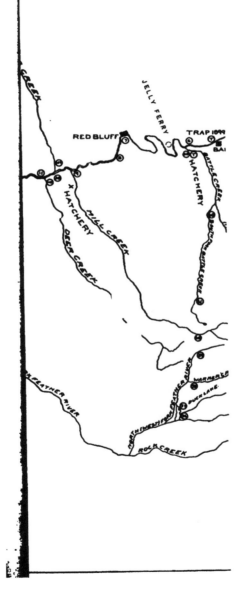

NOTES

ON

FISHES FROM STREAMS AND LAKES OF NORTHEASTERN CALIFORNIA NOT TRIBUTARY TO THE SACRAMENTO BASIN.

By CLOUDSLEY RUTTER,

Naturalist, United States Fish Commission Steamer Albatross.

NOTES ON FISHES FROM STREAMS AND LAKES OF NORTHEASTERN CALIFORNIA NOT TRIBUTARY TO THE SACRAMENTO BASIN.

By CLOUDSLEY RUTTER,

Naturalist, United States Fish Commission Steamer Albatross.

The fishes forming the basis of the following report were collected in 1898 and 1899 while studying the distribution of the fishes of the Sacramento Basin. The collection was studied at Leland Stanford Junior University, where special facilities for study and comparison were afforded by the ichthyological museum.

The localities from which the collection was obtained represent four basins, now distinct, though at one time probably tributary to Lake Lahonton. These basins are Grasshopper Lake, Eagle Lake, Honey Lake, and Truckee River.

Grasshopper Lake is an alkaline pond, with no outlet, at the southern end of Grasshopper Plains, in Lassen County, and contains no fishes. A species of *Agosia* was found to be abundant in a spring emptying into the lake.

Eagle Lake also has no outlet, the lowest point in the surrounding watershed being over 50 feet above the surface of the lake. Its water is slightly alkaline, though very clear, and near the shore supports thick aquatic vegetation. Only two species of fishes were obtained, the Eagle Lake white-fish (*Rutilus olivaceus*) and a sucker (*Chasmistes chamberlaini*) here described as new. A trout is known to inhabit the lake, but none was obtained.

Two streams were fished in Honey Lake Basin—Willow Creek and Susan River. The former rises at the lowest point in the Eagle Lake watershed, and its upper part is a rough mountain stream. It was fished about 15 miles north of Susanville, where it passes through an extensive meadow. Susan River rises on the eastern slope of Lassen Butte, and above Susanville is a mountain torrent. Its lower part lies in the plains adjacent to Honey Lake and is dry during part of the year.

Collections were made in three streams of Truckee Basin—Little Truckee River, Sage Hen Creek, and Prosser Creek. The former, a considerable stream, 15 to 30 feet wide, with very rocky bottom, drains Webber and Independence Lake, and was fished a short distance below the outlet of the latter. Sage Hen Creek is tributary to Little Truckee River. It is but a small stream, flowing through a narrow wooded valley. Prosser Creek is tributary to Truckee River and drains the table-land north of Truckee. It was fished near Prosser Bar, where it is a meadow stream from 6 to 10 feet wide and 6 inches to 6 feet deep.

The fish fauna of these waters is very limited. Three species are described as new. The collection consists of 9 native and 2 introduced species, distributed as follows: One species, *Agosia robusta*, is common to three of the four basins and is probably to be found in the other basin, Eagle Lake. Another, *Rutilus olivaceus*, is found in all the basins except that of Grasshopper Plains. *Chasmistes chamberlaini*, of Eagle Lake, is not found elsewhere. Honey Lake and Truckee River basins have 5 native species in common—*Pantosteus lahonton, Catostomus tahoensis, Rutilus olivaceus, Agosia robusta,* and *Cottus beldingii.* *Salmo irideus* is found in both basins, but it has been introduced into Truckee Basin, and possibly into Honey Lake Basin. *Leuciscus egregius* is also known from Honey Lake Basin, and *Coregonus williamsoni* and *Salmo henshawi* from Truckee Basin. *Salvelinus fontinalis* has been introduced into the latter basin.

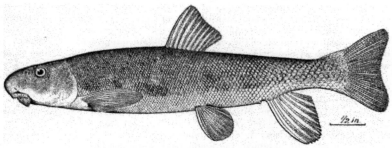

PANTOSTEUS LAHONTON Rutter, new species.

Pantosteus lahonton Rutter, new species.

Head 4.5 in length, depth 5.5; eye 6 in head; D. 10 or 11; A. 7; scales 17–81 to 96–12, 47 to 50 before dorsal. Body terete, caudal peduncle but little compressed; interorbital slightly convex, or flat, width of bone 2.8 in head; eye posterior, 3 in snout, 2.5 in interorbital space, 1.5 in distance between eye and upper end of gill-opening; snout equal to half of head, broadly rounded both vertically and horizontally, projecting beyond the large mouth; 4 rows of papillæ on upper lip, 4 rows across symphysis of lower lip, 10 papillæ in an oblique row from corner of mouth to inner corner of lobe of lower lip; isthmus broader than interorbital, equal to distance between pupils; fontanelle present, but less than half width of pupil in a 6-inch specimen; dorsal inserted from 49 to 52 hundredths of body from tip of snout; ventrals inserted under ninth ray of dorsal, halfway between tip of snout and tip of middle caudal rays; caudal 1.5 in head, deeply emarginate, not forked; pectoral 1.3 in head; height of dorsal about 1.4 in head, the base equal to head, margin slightly emarginate; ventrals 1.7 to 1.8 in head. Very dark, almost black above, abruptly paler below, lower fins slightly dusky. Maximum length, about 6 inches.

Closely related to *Pantosteus generosus,* but with the following differences, determined by comparison with specimens of that species of the same size from Provo, Utah. The Provo specimens have the dorsal 10 or 11 instead of 9 or 10, as described by Jordan & Evermann.

Species.	Interorbital space.	Scales in lateral line.	Scales before dorsal.	Distance of ventrals from snout equals distance from their insertion—	Caudal fin contained in head.	Width of lower jaw cartilage in head.	Rows of papillæ across symphysis of lower lip.	Papillæ on lower lip in oblique row from corner of mouth.
P. lahonton	Flat	81 to 96	45 to 50	To tip of middle caudal rays.	1.2	3.5 to 4, margin convex.	4	10
P. generosus	Convex.	77 to 87	41 to 45	To middle of middle caudal rays.	1	3 to 3.7, margin nearly straight.	2	7

Found in abundance in Susan River, and also in Little Truckee River and Prosser Creek. Types (No. 50587, U. S. Nat. Mus.) from Susan River, collected by Rutter and Chamberlain.

Catostomus tahoensis Gill & Jordan.

Head 4.4 in length, 5 to tip of middle caudal rays; depth 4.8 in length; width of head through opercles equal to its depth; eye 5.5 in head, 2.5 in snout, 2.7 in interorbital space, 1.7 in distance from eye to upper end of gill-opening (by eye is meant the orbital opening, not the eye-ball nor socket); interorbital (bone) 2.3 in top of head; width of isthmus 3.8 in head, a little less than distance between eye and gill-opening, equal to width of opercle, and also equal to distance between corners of mouth. (Measurements made on a 7.3 inch specimen.) D. 11; A. 7; scales 17–89 to 105–16. Body rather slender, profile steep; snout blunt; mouth large, with full lips, covered with rather coarse papillæ which do not become much smaller toward margin of lips; upper lip with about six rows of much-crowded papillæ; lower lip with two rows across symphysis, and about 8 papillæ in a longitudinal row through lobes; posterior margin of lower lip reaching vertical through posterior nostril. Orbital rim but little developed, middle ridge of skull broad, the interorbital space rather high and rounded. Insertion of dorsal in middle of body, its length 1.4 in its height. Insertion of ventrals under fifth ray of dorsal. Length of caudal 1 to 1.1 in head, rather deeply forked, middle ray 1.6 in longest. Anal reaching past base of caudal, its height equaling length of caudal; length of ventrals equals height of dorsal; pectoral a little shorter than caudal. Least depth of caudal peduncle 2.6 in head. Lateral line complete, straight. Peritoneum black. Color nearly black above, slightly mottled with pale yellowish below.

Taken in Willow Creek, Susan River, Little Truckee River, and Prosser Creek. Description based on specimens from Susan River.

Chasmistes chamberlaini Rutter, new species.

One young example, 1.7 inches long, and a dried head 3.3 inches long, which can not be referred to any hitherto described species, were obtained at Eagle Lake. Eye 7 in head, 3 in snout, 2.8 in interorbital bone, 2 in distance from eye to upper end of gill-opening. Premaxillary spines forming a prominent hump, maxillary inclined about 40 , falling far short of anterior nostril, its length from free end to tip of snout just equal to snout in front of nostril, 3.2 in head; lower jaw 3.5 in head. Interorbital (bone) 2 in head, considerably arched transversely; a low, sharp longitudinal ridge along middle suture, showing even in the young example. Nasal spines very prominent; fontanelle closed, covered by a thin bone. Mucus canals prominent, but probably intensified in dried specimen. Lips thin, two rows of papillæ on upper; lower incised to base, lobes small, with scattered papillæ. (The above data from the dried head.)

Cross series of scales 93; D. 10; A. 7. Origin of dorsal in middle of body, ventrals inserted under sixth or seventh ray of dorsal. Pectoral broad, reaching two-thirds distance to ventrals. Ventrals scarcely reaching vent, the outer two rays longest. Anal low, when depressed reaching halfway to caudal. Caudal peduncle long and slender.

Has smaller scales than any other species of the genus. The dorsal and anal are the same as in *C. copi*, but it differs from that species in the broad interorbital and the papillose lips, in addition to the small scales. The sharp ridge on interorbital also seems to be a distinctive character.

Named for Mr. F. M. Chamberlain, of the U. S. Fish Commission steamer *Albatross*.

Type (No. 50588 U. S. Nat. Mus.). Collected in Eagle Lake by Rutter and Chamberlain.

Leuciscus egregius (Girard).

The specimens here noted are not quite so deep as specimens from Winnemucca, Nev., but otherwise can not be distinguished. They have two red stripes along side, with a darker stripe between. Lower part of cheek yellowish, with some yellow along edge of belly. Scales in lateral line 55 to 63. D. 8 or 9; A. 8 or 9. Common in Willow Creek and Susan River.

Rutilus olivaceus (Cope). *Eagle Lake White-fish.*

This species was met with in Eagle Lake and Willow Creek, where it attains a length of 8 inches. Head 3.3 to 3.7 in body; depth 3.7 to 4.5; eye 4.4 to 5 in head; insertion of dorsal 0.53 to 0.57 of body from snout. Scales 15–58 to 64–8; D. 8; A. 8; teeth 5–4 or 5–5. Body elongate, little compressed, little elevated, regularly curved from occiput to dorsal, highest over tip of pectoral. Head long; mouth oblique; jaws even, the lower forming a distinct though very obtuse angle with lower profile. Premaxillary on level with lower half of pupil. Top of head slightly concave. Lateral line but little decurved. Tip of depressed dorsal over front of anal. Caudal peduncle long, but little tapering, its length from anal equal to head behind front of eye, its thickness over end of anal equal to snout. This species differs from *Rutilus bicolor* in the finer scales and in having the same number of rays in the anal that it has in the dorsal, *R. bicolor* having one fewer in the anal.

Agosia robusta Rutter, new species.

Body heavy, highest above insertion of pectorals; the ventral outline curved almost as much as the dorsal. Head 3.8 to 4 in body; snout blunt, but little overlapping the premaxillary and never extending beyond it; mouth oblique, barbels usually absent, present on 10 to 50 per cent of specimens from any one locality. Fins small; D. 8; A. 7; pectoral about equal to head behind nostril, variable; caudal moderately forked, middle rays two-thirds length of longest; rudimentary caudal rays forming prominent keels along upper and lower edges of tail; margin of anal slightly rounded, the anterior rays not all produced, not extending beyond posterior rays when fin is depressed. Lateral line nearly always incomplete, but with scattered pores frequently extending to base of caudal; scales 56 to 77, varying about 12 in any one locality. Usually two dusky lateral stripes, the upper extending from snout to caudal, the lower branching off from the upper behind the head and ending along base of anal; cheek abruptly silvery below lateral stripe; tinged with orange about lower jaw, upper end of gill-opening, and at base of lower fins.

Type (No. 50589 U. S. Nat. Mus.). Collected in Prosser Creek by Rutter and Atkinson.

Taken in Spring Creek, Willow Creek, Susan River, Little Truckee River, and Prosser Creek.

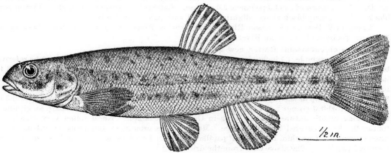

AGOSIA ROBUSTA Rutter, new species.

Coregonus williamsoni Girard.

Abundant in streams tributary to Truckee River. A "native white-fish," probably this species, is reported from Bigler (Tahoe) and Donner lakes in the California Fish Commission Report for 1883–84.

Salmo henshawi (Gill & Jordan). *Lake Tahoe Trout.*

Occurs in only the Truckee Basin, and taken in Little Truckee River, Sage Hen Creek, and Prosser Creek. The black spots of sides much larger and fewer than in *Salmo irideus*.

Salmo irideus Gibbons. *Rainbow Trout.*

Readily distinguished from the above by the very small and numerous black spots, as well as by the absence of the red blotch on inner edge of mandibles. Introduced into Truckee Basin, and possibly also into Honey Lake Basin. It was observed in Susan River and Prosser Creek.

Salvelinus fontinalis (Mitchill). *Brook Trout.*

This species has been introduced into Prosser Creek, where specimens were taken.

Cottus beldingii Eigenmann & Eigenmann. *Blob.*

Palatine teeth wanting; no prickles on skin; lateral line broken under posterior rays of dorsal, sometimes a few pores on caudal peduncle, usually none. Top of head covered with minute pimples. Dorsal spines 6 to 8, dorsal rays 15 to 19, anal rays 11 to 14.

Found in Susan River, Little Truckee River, Sage Hen Creek, and Prosser Creek.

BREEDING HABITS OF THE YELLOW CAT-FISH.

By HUGH M. SMITH and L. G. HARRON.

BREEDING HABITS OF THE YELLOW CAT-FISH.

By HUGH M. SMITH and L. G. HARRON

In view of the paucity of information in regard to the spawning habits of cat-fishes, and owing to the possible inauguration of cat-fish culture in response to a wide-spread demand, we think it worth while to present these observations on one of the most important members of the family. The notes may be taken in conjunction with Dr. Albert C. Eycleshymer's "Observations on the breeding habits of *Ameiurus nebulosus*," published in the *American Naturalist* for November, 1901.

On July 3, 1902, it was observed that among a lot of yellow cat-fish (*Ameiurus nebulosus*) from the Potomac River near Washington, which had been in the Fish Commission aquarium since May 19, 1902, two had paired and exhibited breeding tendencies. They had withdrawn to one end of the aquarium tank and maintained themselves there, the male driving away any others which approached. The other fish were thereupon removed and the two in question left unmolested. They were kept under daily observation, and their behavior furnished the principal data on which this paper is based. In the latter part of July another pair of fish in the same lot showed an inclination to spawn and afforded additional information, as did also a lot of eggs of the same species found in a pool in the Fish Commission grounds on June 16; these eggs, which were about ready to hatch, were removed to an aquarium, where two-thirds hatched the same night, the others being dead the next morning.

Nest-making.—The aquarium in which the fish were held was 5 feet long and 16 inches wide on the bottom and 18 inches high, the posterior wall inclining obliquely backward so that at the surface of the water the tank was 2 feet 4 inches wide. The front was of solid glass, and the sides, bottom, and back were of slate. The bottom was covered with gravel and a little sand to the depth of 1½ or 2 inches.

The nest-making, as modified by the artificial conditions of the aquarium, consisted in removing all the stones and sand from one end and keeping the slate bottom scrupulously clean from all foreign objects, even the smallest particles of food, sediment, etc. In moving the pebbles, which were mostly from one-half to three-fourths of an inch in diameter, the fish took a vertical or slightly oblique position and sucked a pebble into the mouth, usually beyond the lips and out of sight, then swam toward the other end of the tank and dropped it by an expulsive or blowing effort. Sometimes the gravels were carried only a few inches and sometimes the entire length of the aquarium. Usually the fish swam horizontally near the bottom when carrying a stone, but sometimes turned obliquely upward and dropped it from near the surface. Both fish participated in this operation. The removal of finer sediment was effected by a quick lateral movement of the body which caused a whirl that lifted and floated the particles beyond the limits of the nest.

The pair of fish more particularly under consideration, during the first night they were in the aquarium, removed all the gravel from over a space nearly 2 feet long and 1¼ feet wide, upward of a gallon of stones being transferred as described. After the second pair of fish had cleared a similar space, a pint or more of gravel was scattered on the nest: the fish immediately began to remove the stones, and in a few minutes had completely freed the nest from gravel. The gravel—regarded by bass and other fishes as desirable material for the bottom of nests - may be removed by the cat-fish for two reasons: (1) To have a clean place for the eggs and young, so that they may be better guarded and agitated as hereafter described; (2) to provide a smooth place on which to rest and against which to rub the abdomen.

Upward of twenty years ago the yellow cat-fish was much more abundant in the Potomac River than at present. The marshes in Piscataway Creek were a favorite place for the fish to spawn, and large quantities were there taken each season in spring and summer, mostly by colored people living near the river. The fish at that time of year were found in shallow water occupying depressions in the muddy bottom, with most or all of their bodies concealed in an excavation extending laterally from the rounded depression. The fishermen easily made large catches by wading and thrusting their hands into the depressions. An old colored man whom we knew used to refer to a marsh as his "meat market," and would often bring ashore a sackful of yellow cat-fish caught in this way. We are inclined to believe that these fish were brooding, but we have had no opportunity of late years to examine them critically.

Behavior of adult fish before spawning.—Two days intervened between the beginning of the nest-making and the laying of the eggs. As soon as the nest was made ready, the fish became very quiet. During most of the time they rested on the bottom, with practically no body or fin movement, except at intervals. The fish lay close together, often parallel, with their abdomens just clear of the bottom, their weight being borne on the anal and ventral fins. At frequent intervals the female compressed her distended abdomen against the smooth slate bottom with a quivering or convulsive movement, the male often accompanying or following the female in this action, which is obviously for the purpose of loosening the eggs.

The second pair remained on the nest from July 18 to September 30, when they were removed to make room for other species, as it was evident no eggs would be laid. During this time they behaved in the same way as the other pair and their failure to spawn can not be positively accounted for, though such an outcome has been the rule among fish retained in the Fish Commission aquarium. The enlargement of the abdomen and ripening of the eggs go on to a point when spawning seems imminent; the actions of the fish suggest the arrival of breeding time; but no eggs are laid. After a few weeks the enlargement of the abdomen subsides, and dissection has sometimes shown a liquefaction of the egg mass. It has been suggested that the presence of alum in the circulating water has an injurious astringent action on the mucous membrane of the vent, and it is a significant fact that the change from an open to a closed circulation, with consequent elimination of the alum filter, was soon followed by the spawning of the cat-fish first mentioned—an unprecedented occurrence at the Fish Commission aquarium. The second pair of fish had been in the alum-filtered water for a few days, some time before the spawning season.

Number, character, and incubation of eggs.—On July 5, between 10 and 11 a. m., the eggs were deposited in four separate agglutinated masses on the clean slate bottom. Unfortunately, the fish were not under observation at this time, although they were watched for about fifteen minutes after the extrusion of the first two lots of eggs, when it was supposed the spawning had been completed. The masses of eggs were of nearly uniform size, about 4 inches long, 2½ inches wide, and half an inch thick. The newly laid eggs are one-eighth of an inch in diameter, nearly transparent, and of a pale yellow color. The number of eggs deposited was estimated at 2,000. The incubatory period was 5 days in a mean water temperature of 77° F., the lowest temperature being 75° and the highest 80°. About 12 hours intervened between the hatching of the first and last eggs. Active movement was observed in the embryos 40 hours after the eggs were laid. Fully 99 per cent of the eggs hatched into normal fry, a few weak and deformed fry and a few unfertilized or dead eggs being noticed.

Growth of young.—When the fry first emerged from the egg they were about one-quarter of an inch in length, and of a yellowish, transparent color. By the second day the skin of the back had begun to darken, and by the end of the fourth day the entire upper parts were uniformly bluish black and the under side had become whitish. On the third day the barbels at the angles of the mouth and the pectoral and dorsal spines were clearly visible through the glass front of the aquarium.

Until 6 days old they remained on the bottom in densely packed, wriggling masses, the largest lot in the nest and several smaller lots in other parts of the aquarium. On the sixth day they began to rise vertically a few inches above the bottom, at first falling back at once, but gradually remaining longer above the bottom. By the end of the seventh day they were swimming actively, and practically all collected in a school just beneath the surface, where they remained for two days. They then began to scatter, and subsequently did not school.

The relatively large yolk-sac had nearly disappeared by the sixth day, when they began to eat finely ground beef liver, and they were feeding ravenously by the eighth day. Between feeding times, they passed much of the time on the bottom of the aquarium in search of food, which they ate in an almost vertical position, head downward; they also browsed on the sunny side of the aquarium, where there was a short growth of algæ. The early growth was rapid, but not uniform; on the eleventh day their length varied from one-half to three-fourths of an inch. At the age of 2 months the average length was 2 inches; but after that time the growth was very slight, and in January, 1903, six months after hatching, the length of the survivors was only 2¼ to 2¼ inches. The slow growth was undoubtedly due to the fact that the fry were retained in small troughs where the conditions were unnatural.

Care of eggs and young.—During the entire hatching period both parents were incessant in their efforts to prevent the smothering of the eggs, to keep them clean, and to guard against intruders. The eggs were kept constantly agitated and aerated by a gentle fanning motion of the lower fins, and foreign particles, either on the bottom of the nest or floating near the eggs, were removed in the mouth or by the fins. The most striking act in the care of the eggs was the sucking of the egg masses into the mouth and the blowing of them out, this being repeated several times with each cluster before another lot was treated.

The male was particularly active in watching for intruders, and savagely attacked the hands of the attendant who brought food, and also rushed at sticks or other objects introduced into the aquarium. Practically the entire work of defense was assumed by the male, although the female occasionally participated.

During the time the fry were on the bottom the attentions of the parents were unrelaxed, and, in fact, were increased, for the tendency of the different lots to become scattered had to be corrected, and the dense packing of the young in the corners seemed to occasion much concern. The masses of fry were constantly stirred as the eggs had been by a flirt of the fins, which often sent dozens of them 3 or 4 inches upward, to fall back on the pile.

The very young fry were also taken into the mouths of the parents and blown out, especially those which became separated from the main lot and were found in the sand and sediment. The old fish would take in a mouthful of fry and foreign particles, retain them for a moment, and expel them with some force. After the young began to swim and became scattered, the parents continued to suck them in and mouth them, and, as subsequently developed, did not always blow them out.

An interesting habit of the parents, more especially the male, observed during the first few days after hatching was the mixing and stirring of the masses of young by means of the barbels. With their chin on the bottom, the old fish approached the corners where the fry were banked, and with the barbels all directed forward and flexed where they touched the bottom, thoroughly agitated the mass of fry, bringing the deepest individuals to the surface. This act was usually repeated several times in quick succession. The care of the young may be said to have ceased when they began to swim freely, although both parents continued to show solicitude when the attendant approached the aquarium from the rear.

When 12 days old, about 1,500 of the fry were removed from the aquarium to relieve crowding, and placed in a hatching-trough such as is employed for salmon and trout. For some unknown cause, about 1,000 of these died during the first three days. The others survived with little or no loss, and are still on hand.

The fry which were left with their parents continued healthy, but their number steadily decreased. There being no way for them to escape, and a closely woven wire screen preventing inroads from the exterior, it was suspected that the old fish were eating their young, though they were liberally fed at suitable intervals. They were kept under close observation during the day, and were seen to be fond of mouthing the fry, more especially the weaker ones --a habit which at this stage seemed unnecessary. They were frequently seen to follow leisurely a fry, suck it in their mouth, retain it for a while, and then expel it, sometimes only to capture it again. There was no active pursuit of the fry, and the tendency seemed to be to spit them out. In one or two instances, however, it appeared that fry taken into the mouth were not liberated, the feeding instinct becoming paramount to the parental instinct. After all the fry which had been left with their parents had disappeared--in about 6 weeks after hatching --18 fry from the trough were placed in the aquarium one evening, and only 2 of these had survived on the following morning.

During the entire period covered by these observations liver and beef were fed regularly to the brood fishes, and at no time did their appetites fail. There was apparently no interference with deglutition, or closure of the œsophagus, such as has been observed in some other cat-fishes, as half-inch cubes of meat were readily ingested during the entire time the fish were under observation.

External sexual characters of adults. --Besides the fullness of abdomen which the mass of eggs gives to the female, there was in both pairs of fish under consideration another external feature by which the sexes could be distinguished. This was the shape of the snout and interorbital region, which in the males were noticeably flatter and broader than in the females. The males in both these cases were about 12 inches long and were an inch longer than their partners.

THE DESTRUCTION OF TROUT FRY BY HYDRA.

By A. E. BEARDSLEY,

Professor of Biology, Colorado State Normal School.

THE DESTRUCTION OF TROUT FRY BY HYDRA.

By A. E. BEARDSLEY,

Professor of Biology, Colorado State Normal School.

The following observations were made during an investigation at the United States Fish Commission hatchery, Leadville, Colo., in August, 1902. On August 4, some eggs of the black-spotted trout in a number of the hatching-troughs were just hatching, while in others the young fish were several days old. Each trough was separated by screens into three divisions. The first division—that into which the water enters from the supply pipes—contained no eggs, these having all been removed several days before on account of the great mortality of the young fishes hatched in this division of the troughs. In the second, or middle, division, the newly hatched fry were dying in considerable numbers, some before leaving the egg trays. In the third division of these troughs, as well as in the troughs not directly fed from the supply pipes, the death rate was merely nominal.

These facts clearly indicated that the cause of the mortality was directly connected with the water supply, which was found to be derived chiefly from two sources. The main supply pipes were fed from Rock Creek, and an auxiliary set of pipes led from a spring near the hatchery. Connected with the main pipes was a branch leading from the third or lowest of the Evergreen lakes. This was closed at the time, only a small quantity of water coming through leaks around the gate at the head of the pipe.

The water from the main pipes was clear, containing very little sediment, and with a temperature of 48° F.; that from the spring was very clear and pure, without sediment, its temperature being 43° F. There was very little sediment in the hatching-troughs. In this, however, microscopical examination disclosed the presence of great numbers of a very transparent hydra, which had been discovered by the attendants at the hatchery a few days before, when the sun's rays, just before sunset, had fallen obliquely into one of the troughs. In the dim light of the hatchery this hydra was quite invisible, but by placing a large mirror outside of the building so as to throw a beam of sunlight through the window, with a hand mirror reflecting this beam so as to throw it into the trough, the hydras could be plainly seen as slender, whitish threads, 1 to 2 centimeters in length and 0.15 to 0.30 millimeter in diameter, fixed by one end to the bottom or to the side of the trough, and bearing a crown of 5 or 6 long tentacles around the mouth at the free end. The

157

hydras were found quite equally distributed through the first division of all troughs supplied directly from the main pipes. A careful count of the number on several square inches in different troughs gave an average of 131 hydras per square inch (20+ per square centimeter). Comparatively few were found in the middle division of the troughs, most of them having fixed themselves before reaching the first screen. Very little animal life other than hydra was found in the sediment of the troughs.

Since no other cause for the mortality of the young fishes could be discovered, and as the hydras were exceedingly abundant and are well known to be armed with great numbers of dart cells or nettling cells which secrete a fluid that quickly causes paralysis in small crustaceans and other minute forms of animal life, it appeared that the injury was probably due to the hydras. In so far as the writer is aware, no injury to fishes by hydra has heretofore been known. The following experiment was therefore instituted to determine what injury, if any, was to be attributed to this cause:

Five beakers, each of 250 cubic centimeters capacity, were filled with water from the supply pipes; in each of the first four of these was placed the sediment from 21 square centimeters of the bottom of the first division of one of the hatching-troughs, containing about 430 hydras; the fifth beaker was intended as a control, and contained water only. Five trout newly hatched and apparently in good health were taken from the hatching-trays and placed in each beaker. Nos. 1 and 2 were filled with water from the spring and were placed in running water, so that the temperature was nearly constant; Nos. 3, 4, and 5 were filled from the main supply pipes, No. 4 having been kept over night in the office, and all three were set on a shelf in the hatching-room. At the end of the experiment, Nos. 1 and 2 were at nearly the same temperature as at the beginning, while Nos. 3, 4, and 5 had acquired the temperature of the hatching-room.

The following table shows the result of this experiment:

	1.	2.	3.	4.	5.
Temperature at beginning of experiment.	43° F.	43° F.	48° F.	58° F.	48° F.
Hour of beginning	9.13 a. m	9.16 a. m	9.20 a. m	9.22 a. m	9.23 a. m.
Result at 9.45 a. m	1 dead	1 dead	2 nearly dead.	3 dead *a*	0 dead.
10 a. m	4 dead	1 dead	3 dead	4 dead	0 dead.
10.15 a. m	4 dead	4 dead	4 dead	4 dead	0 dead.
10.30 a. m	5 dead	5 dead	5 dead	5 dead	0 dead.
Temperature at end of experiment.	44° F.	44° F.	55° F.	55° F.	55° F.

a One of these had burst the yolk-sac in its struggles.

In this experiment 25 per cent of the trout were killed by hydras in less than 30 minutes, 60 per cent in 45 minutes, 80 per cent in 60 minutes, and 100 per cent in 75 minutes; those trout which were least active in the beginning of the experiment were the ones that survived longest, probably because they came in contact with a smaller number of stinging cells of the hydra. With the aid of a lens, the hydras could be seen with their mouths closely applied to the surface of the fish, particularly on the yolk-sac; in some cases more than a dozen hydras were seen attached to a single fish. Soon after the fishes were placed in the beakers most of them were seen to struggle violently, one of them bursting its yolk-sac in its struggles and dying immediately; these struggles recurred at intervals, but with diminishing intensity, until death supervened. The five trout in the beaker without hydras were kept in the beaker until the next day and were then found to be all alive and in good health.

No other cause of injury having been discovered after the most careful search, the destructive effects of the hydras upon the fishes in the foregoing experiment were taken as conclusive evidence that these were the cause of the unusual mortality of the trout fry. This fact being demonstrated, a careful examination of all the sources of water supply to the hatchery was made. The lower of the three lakes was first visited. This lake is quite shallow, being about 12 feet (4 meters) in depth in the deepest part; along the borders there is considerable aquatic vegetation, consisting of sedges and cat-tails; here the hydras were found in immense numbers, clinging to the submerged stems and leaves as well as to the green filamentous alga which was growing abundantly on the bowlders which are scattered over the bottom. The other lakes, and Rock Creek for a distance of about half a mile above the head of the supply pipe, as well as the spring, were examined in turn, but although very careful search was made, no hydras were found in either of these waters.

The temperature of the water in each of the three lakes at 1 foot below the surface was taken August 8, about 2 p. m., and was found to be as follows: Upper Evergreen Lake, 60° F.; Middle Evergreen Lake, 64 ; Lower Evergreen Lake, 65°.

The leaks about the head of the pipe leading from the lower lake were immediately stopped and no water from the lake is now entering the hatchery.

The natural causes which control the development of the different species of hydra, favoring or retarding it, are as yet but little understood. At one period hydras may be very abundant at a given point, and soon afterwards entirely disappear without any apparent cause. They have been found in the vicinity of Greeley, Colo., during all months of the year, sometimes in great abundance; sometimes, however, a whole year has passed without a single one being seen, although searched for most diligently. They occur in lakes, ponds, and marshes, usually in clear water. Warm water (60° to 80° F.) appears to favor their rapid multiplication, since they are usually most abundant in summer and early autumn; cold water does not apparently injure them, however, as the writer has frequently taken vigorous individuals in the winter, through holes in the ice. Hydras reproduce at certain times by eggs, which settle to the bottom and probably remain dormant through the winter, but the usual and most rapid mode of multiplication is by budding. Little buds arise from the side of the parent, soon acquire a mouth and tentacles like the parent, and after a time break loose and lead an independent existence. In the lake most of the hydras examined were bearing from two to six buds, showing that the conditions there were favorable to their rapid multiplication. In the hatchery troughs, on the contrary, very few were found bearing buds; and these were probably recent arrivals. The conditions within the hatchery do not, therefore, appear favorable for their increase, and it only remains to rid the troughs of them in the most practicable manner.

As the hydra is very tenacious of life and may even be cut into several pieces without serious injury, each piece developing the lost parts and becoming, in a few days, a complete hydra, it is not probable that it can be destroyed in the troughs without injury to the fish eggs or young fry. By removing all eggs and fry, briskly scrubbing the bottom and sides of the trough with a stiff brush so as to cause the hydras to loosen their hold, then quickly flushing the trough into the waste-pipe, most of them can be removed.

When the lower divisions of the trough contain hatching eggs and fry that can not readily be removed without injury, the first or upper division (in which nearly all the hydras will have fixed themselves) may be cleaned by shutting off the supply pipes, placing a temporary partition between the upper and lower divisions, and, after a brisk scrubbing, quickly siphoning off the water and floating hydras from the upper division of the trough.

As the water now entering the hatchery is taken from Rock Creek and from the spring, both of which are free from hydras, it is probable that loss from this cause will cease as soon as all the hydras now in the hatching troughs and supply pipes can be removed.

It was impossible to find characters other than those of color and size by which to differentiate this hydra specifically from the well-known *H. fusca* Linnæus. It differs from that common form only in being of larger size and in the entire absence of coloration. Among the large number of individuals observed, both in the troughs of the hatchery and at the lake, not one showed a trace of fuscous coloration. These differences appear to be constant, and I propose the name *pallida* for the new species, in allusion to its lack of color. It may be described as follows:

Hydra pallida Beardsley, new species.

Characters.—Body cylindrical, 1 to 2 cm. in length and .15 to .30 mm. in diameter; tentacles 5 or 6, when fully extended two or three times as long as the body; color, milk-white in reflected light, whitish and translucent in transmitted light.

Differs from typical *Hydra fusca* in being somewhat larger in average size and in the entire absence of fuscous coloration.

Type locality.—United States fish-cultural station, Leadville, Colorado.

DESCRIPTIONS

OF

NEW GENERA AND SPECIES OF FISHES FROM THE HAWAIIAN ISLANDS.

By DAVID STARR JORDAN AND BARTON WARREN EVERMANN.

DESCRIPTIONS OF NEW GENERA AND SPECIES OF FISHES FROM THE HAWAIIAN ISLANDS.

By DAVID STARR JORDAN AND BARTON WARREN EVERMANN.

During the investigations of the aquatic resources of the Hawaiian Islands carried on by us in 1901 under the direction of the Hon. George M. Bowers, United States Commissioner of Fish and Fisheries, very large collections of the fishes and other animals occurring in the waters of those islands were made.

A detailed report, covering the entire aquatic fauna of that group of islands, is now in preparation, which, it is hoped, will be ready for publication within the year.

Among the fishes collected are many species which appear to be new. Descriptions of 57 of these are given in the present paper. Illustrations of these new species, together with more extended notes regarding their abundance, distribution, habits, and commercial value, will be given in the general report to follow.

The types of all the new species have been deposited in the United States National Museum, and, when possible, one or more cotypes have been donated to each of the following museums and institutions: Museum of Leland Stanford Junior University (L. S. Jr. Univ. Mus.), U. S. National Museum (U. S. N. M.), Reserve series of the U. S. Fish Commission (U. S. F. C.), Museum of Comparative Zoology at Cambridge, Mass. (M. C. Z.), American Museum of Natural History, New York City (Am. Mus. Nat. Hist.), Academy of Natural Sciences of Philadelphia (Ac. Nat. Sci. Phila.), University of Indiana (Mus. Ind. Univ.), Field Columbian Museum, Chicago (Field Col. Mus.), California Academy of Sciences, San Francisco (Cal. Ac. Sci.), and the Bernice Pauahi Bishop Museum at Honolulu (Bishop Mus.). When possible, we have given in this paper the numbers which the types and cotypes bear on the records of the various museums to which they have been assigned.

The majority of specimens here described were obtained by us in the market or directly from the fishermen at Honolulu, Oahu Island. Others were obtained in the market or from the fishermen at Hilo, island of Hawaii; others at Kailua, island of Hawaii; others on the reef at Waikiki, near Honolulu, and one at Heeia, Oahu Island.

Family CARCHARIIDÆ. The Sand Sharks.

1. Carcharias phorcys Jordan & Evermann, new species.

Head 4.8 in length; depth 6.5; width of head 1.75 in its length; depth of head 1.8; snout about 2.2 in head; interorbital space 2.2; space between tip of snout and front of mouth 2.5; width of mouth 2.5; eye 6 in interorbital space; internasal space 1.8; least depth of caudal peduncle a little over 4.8; caudal 3.5 in body; pectoral 5.75.

Body elongate, rather robust, the tail compressed; head elongate, somewhat narrow and depressed; snout long and narrowly pointed when viewed above, the tip rounded; eyes small, their

163

posterior margins about midway between tip of snout and first gill-opening; nictitating membrane well developed; mouth large, very convex, the anterior margin of mandible below front rim of orbit; teeth in upper jaw narrow, with broad bases, compressed, serrate, not notched, and with 4 or 5 basal cusps behind; teeth in mandible rather long, pointed, and not serrate, the edges smooth; nostrils without flap, inferior, and nearer eye than tip of snout; interorbital space broad and convex, the upper profile of the head rising gradually in a nearly straight line to back of head; gill-openings of moderate length, the posterior over the base of the pectoral; peritoneum white or pale; body very finely roughened when stroked forward; height of first dorsal less than depth of body, its origin a little nearer tip of snout than origin of second dorsal; origin of second dorsal nearer origin of first dorsal than tip of caudal, the fin small and about over the anal so that the origins of the two fins are opposite; caudal long, with a notch at its tip, deep, the lower lobe 2.25 in the length of the fin; pectoral with margin of fin slightly concave; ventrals small, their origins a little nearer base of lower caudal lobe than origin of pectoral; back convexly ridged, broader between the dorsals; base of caudal with a pit above and below.

Color in alcohol, pale brown, the lower parts pale or whitish with a brown streak the color of the back along side from gill-opening to origin of ventral; tips of dorsals, edge of caudal, and tip of pectoral blackish.

This description is based upon the type, No. 50612, U. S. N. M. (field No. 03747), a specimen 27.5 inches long, obtained by us at Honolulu. The collection contains 4 other examples, all from Honolulu, which we take as cotypes. They are: No. 12715, L. S. Jr. Univ. Mus. (field No. 03745); No. 12715, L. S. Jr. Univ. Mus. (field No. 03746); No. —, M. C. Z. (field No. 03748); and No. 1685, Bishop Mus. (field No. 03749).

We have examined 2 other specimens, each about 29 inches long, obtained by Dr. Jenkins in 1889, and one fœtus obtained by us at Honolulu.

Family OPHICHTHYIDÆ.　The Snake Eels.

2. Microdonophis fowleri Jordan & Evermann, new species.

Head about 5 in trunk measured from tip of snout to vent; tail shorter than head and trunk by the length of the former; eye nearly 1.6 in snout or 1.5 in interorbital space; snout 6 in head; interorbital space about 6.75; mouth 2.75; pectoral a little over 4.25 in head.

Body elongate, cylindrical, the tail tapering gradually to a conical horny point; head cylindrical and pointed; snout moderately long and pointed, slightly flattened above, projecting over and beyond the mandible; eye elongate, small, anterior and superior, about midway in length of mouth; mouth rather large; lips somewhat fringed; teeth large and canine-like in front of jaws, and on vomer in a single row; tongue small, adnate to floor of mouth; anterior nostrils in short tubes near tip of snout, the posterior with broad flaps on the lips and opening downward; interorbital space concave, each supraocular ridge slightly elevated; peritoneum silvery; skin perfectly smooth; head with mucous pores, a series of which encircle head above and about midway in its length; lateral line well developed, pores about 140; origin of dorsal slightly in advance of gill-opening or base of pectoral; pectoral small, the rays just above the middle the longest, fin rounded; dorsal fin long and low, its height about equal to length of snout; anal similar to dorsal, its height a trifle less.

General color, when fresh, white, rendered somewhat shaded on upper portions by very minute points (seen only with a good lens) of gray; back and upper surface with numerous round brown spots and about 17 indistinct transverse dark brown crossbands which do not extend over the dorsal; the interspaces between the spots on the head yellow, the pectoral bright lemon yellow; end of tail for about 1 inch from point bright yellow; spots on margin of dorsal brown, with yellow borders; a band of yellow runs from under one eye backward, upward, across the top of head, and down under the other eye; the transverse series of pores which encircles the head above and about midway in its length, with black margins, and also a similar series over head along the margin of mouth, and then up, back of eye, over head; pores of lateral line without black margins.

This species is based upon a single specimen, the type, No. 50613, U. S. N. M. (field No. 03431), an example 23 inches long obtained by us in the market at Honolulu, July 21.

Family MURÆNIDÆ. The Morays.

3. Muræna kailuæ Jordan & Evermann, new species.

Head 7 in total length; depth 11.5; eye 14 in head; snout 6; interorbital 12; gape 2.75.

Body short, stout, and moderately compressed; distance from tip of snout to vent less than that from vent to tip of tail by a distance equal to two-thirds length of head; head very small and pointed; snout long, quadrate, the jaws equal, the lower curved so that the mouth does not completely close; lips thin, the teeth showing; each side of upper jaw with a single series of unequal, sharpish canine-like teeth, inside of which is a single depressible fang-like tooth near middle of side; front of median line with 2 long, sharp, fang-like, depressible teeth; shaft of vomer with a single series of short, movable teeth; each side of lower jaw with a single series of unequal, sharp canines, those in front largest; eye small, midway between angle of mouth and tip of snout; anterior nostrils each in a pointed filament whose length is about half that of eye, situated at tip of snout just above lip; posterior nostrils each with a long filament, equal to snout in length, and situated just above anterior edge of eye; interorbital space very narrow and flat; gill-opening small, nearly circular; dorsal fin very low anteriorly, increasing much in height on tail; anal low.

Ground color in life, dark brown, with fine yellow and blackish spots and reticulating lines, the yellow predominating on anterior part of body; end of tail dark purplish brown; edge of dorsal and anal dull dark red, with short pale bands bordered with darker and with small pale spots interspersed; ground color of cheek and throat yellow, with pale spots bordered with black; jaw orange red, with pale black-edged bars; tips of jaws bright coral red; tips of nostril filaments bright red.

Color in alcohol, body with a ground color of light grayish brown, marked with fine whitish lines or specks, and profusely covered with numerous small, round, white spots, each ocellated with black; among these are scattered larger black spots and blotches; white spots smallest on back and largest on belly, where some are as large as eye; a broad, dark brown bar over nape, extending on side to level of eye; top of head and snout with fine white spots; side of snout with a well-defined vertical dark bar about midway between eye and tip; a short white line downward to mouth from front of eye, and a similar longer one downward and backward from posterior lower angle of eye; lower jaw crossed by 3 V-shaped white bars opening forward and bordered by darker; tip of jaw with 2 oblique white bars separated by a narrow brown line; last V-shaped white bar extending across angle of mouth and forming a large white area at base of upper jaw, behind which the angle of the mouth is dark brown; inside of mouth mottled brown and white; nasal filaments mottled with brown and white; throat light brown, with large white spots, some of which unite to form oblong spots or lines; gill-opening not surrounded by dark; anal fin dark brown, crossed by about 28 short white bars; posterior portion of tail crossed by about 12 distinct but somewhat irregular vertical white bars, which extend upon dorsal and anal fins; tip of tail brownish black, with 1 or 2 whitish specks. Only one specimen known.

Type, No. 50614, U. S. N. M. (field No. 03709), a specimen 19 inches long, obtained August 9, 1901, by Messrs. Goldsborough and Sindo at Kailua, Hawaii.

4. Gymnothorax vinolentus Jordan & Evermann, new species.

Head 7.2 in total length, 3.6 in distance from tip of snout to vent; depth 14.5 in total length; eye 14 in head; snout 6.4; gape 2; interorbital 8.6; vent a little nearer tip of snout than tip of tail.

Body long, but stout and not greatly compressed; tail moderately stout and compressed; head much swollen above; snout long and slender, the anterior profile ascending somewhat abruptly from interorbital region; mouth large, extending beyond eye a distance equal to eye and snout; lower jaw projecting, strongly curved, so that the mouth does not completely close; eye small, over anterior half of gape; interorbital narrow, about half greater than diameter of orbit; anterior nostril in a tube whose length is 1.6 times eye, situated near tip of snout; posterior nostril slightly anterior to vertical at front of orbit, oval, surrounded by a narrow, raised, flattened flap whose diameter is two-thirds that of orbit; lips rather thin, not covering the teeth; gill-opening small, its length less than diameter of orbit. Teeth in a single series on each side of upper jaw, the posterior ones short, sharp, and close-set; the anterior ones, about 12 in number, slender, sharp canines of unequal length; inside of these is a series of 5 or 6 long, slender, depressible canines; median line of roof of mouth with 2 long, sharp, depressible canines in front, and a third somewhat farther back; vomer with a single series of short, blunt teeth; lower jaw with a single series of rather close-set, short, backwardly directed canines, somewhat compressed, inside of which anteriorly are 3 or 4 much longer depressible canines on each side. Origin of dorsal midway between gill-opening and angle of mouth; height of dorsal 2 in distance from tip of snout to posterior edge of orbit; anal much lower than dorsal.

Color in alcohol, rich purplish brown or wine-color, almost uniform over entire body and head; side of head with about 7 shallow longitudinal grooves which are darker than ground-color; under side of lower jaw yellowish white, blotched with brown; throat blotched with yellowish white and brown; gill-opening rather paler than surrounding parts; body everywhere with numerous, but very obscure, dark points, posteriorly with numerous narrow vertical dark lines appearing as shallow grooves in the skin; dorsal and anal fins uniform dark brown, not white-edged; tip of tail not white.

The only specimen of this species which we have is the type, No. 50615, U. S. N. M. (field No. 03726), 29 inches long, obtained by Messrs. Goldsborough and Sindo, at Kailua, Hawaii.

5. Gymnothorax steindachneri Jordan & Evermann, new species.

Head 7.3 in length; depth 9.5; eye 9.5 in head; snout 5; interorbital 7.2; gape 2; distance from from tip of snout to vent less than distance from vent to tip of tail by more than half length of head.

Body moderately long and slender, much compressed; head small; snout small and pointed, the anterior dorsal profile concave above the eyes; the nape and sides of head much swollen; gape long, extending far behind eye; lower jaw shorter than the upper, curved so that the mouth does not quite completely close; lips moderately thick, entirely covering the teeth in the closed mouth; eye small, about midway between tip of snout and angle of mouth; teeth on sides of upper jaw in a single series, rather close-set, short, compressed, triangular canines, those in front scarcely enlarged; vomer with a single row of bluntly rounded teeth; each side of lower jaw with a single series of rather strong, backwardly directed canines, the anterior ones somewhat enlarged, those on tip of jaw movable; anterior nostril in a long tube, its length about half diameter of eye, situated near tip of snout just above lip; posterior nostril without tube, just above anterior edge of eye; pores on sides of jaws inconspicuous. Origin of dorsal fin about midway between gill-opening and angle of mouth, its height about equal to length of snout; anal similar to soft dorsal, but much lower; tail moderately slender and pointed; a series of inconspicuous pores along middle of side; gill-opening a long oval slit exceeding diameter of orbit.

Color in alcohol, pale brown or whitish, sprinkled with ragged or dendritic brown spots formed more or less into irregular vertical blotches or crossbands; margins of fins narrowly creamy white or yellowish, that of the anal much wider; corner of mouth and space about gill-opening deep blackish-brown; about 5 longitudinal blackish-brown grooves on lower side of head; under side of lower jaw with 2 blackish longitudinal lines which meet at an acute angle under chin; throat and belly creamy white, with few scattered brownish markings; sides and top of head whitish, with small, sparingly scattered, irregular brownish spots most numerous around and between the eyes.

This species is related to *G. kidako* (Schlegel), from which it differs much in coloration, the present species being much paler and less reticulated, the angle of the mouth with more black, the gill-opening being surrounded by a broad black area (nearly or quite absent in *kidako*), and in having the white border to the dorsal fin much more distinct.

This species is known only from Laysan (whence Dr. Steindachner had 2 examples) and from Honolulu, where the *Albatross* obtained 1 specimen in 1891 and the Fish Commission 3 examples in 1901.

The specimens from Laysan which Dr. Steindachner identified with *Muræna flavomarginata* Rüppell, and of which he gives a good figure, evidently belong to this species. As suspected by Dr. Steindachner, the species is quite different from *G. flavomarginatus*, of which species we have examined several specimens from Pedang, on the west coast of Sumatra. The present species is therefore known from the 2 examples which Dr. Steindachner had from Laysan, one specimen obtained by the *Albatross* at Honolulu in 1891, and 3 specimens secured by us at Honolulu in 1901.

Field No.	Length.	Locality.	Final disposition of specimen.
	Inches.		
03775	24	Honolulu	Type, No. 50616, U. S. N. M.
01904	14	....do	Cotype, No. 7447, L. S. Jr. Univ. Mus.
04905	8	....do	Cotype, No. 2097, U. S. F. C.
1318	17	Honolulu (Albatross)	

Muræna flavomarginata var., Steindachner, Denks. Ak. Wiss. Wien, LXX, 1900, 514, pl. VI, fig. 3 (Laysan); not of Rüppell.

6. Gymnothorax goldsboroughi Jordan & Evermann, new species.

Head nearly 3 in trunk (exclusive of head and tail), or 9 in total length; head and trunk about 1.5 in tail; eye 1.75 in snout, 1.2 in interorbital space; snout 5 in head; interorbital space 7.5; mouth 2.

Body rather compressed, the tail gradually tapering narrowly behind; head compressed, swollen above; snout pointed, the tip blunt and the sides compressed; eye rather small, a trifle nearer tip of snout than corner of mouth; mouth large, snout slightly projecting beyond mandible; lips rather fleshy and concealing the teeth when the mouth is closed; teeth in a single series in jaws, anteriorly large and canine-like, and the vomer with a single large, depressible fang; anterior nostrils at tip of snout in small tubes; posterior nostrils directly above eye in front; interorbital space more or less flattened like top of snout; gill-opening about equal to eye; skin smooth; head with a number of mucous pores; origin of dorsal a little nearer corner of mouth than gill-opening; caudal small.

Color in alcohol, brown, covered all over body except anal fin with round or roundish white spots, those on anterior part of body small, very small and numerous on head, becoming larger on trunk, and finally increasing very much in size on tail where they are scattered and rather far apart; reticulations around the light spots blackish brown upon posterior part of dorsal fin, same color as base of anal; margins of anal and dorsal fins whitish; gill-opening and anus bordered with blackish brown. General color of body in life, brown, rather pale olivaceous anteriorly, and covered all over with small white spots which are close-set and small on head where the dark color forms a network; spots sparse and irregular on posterior parts, and also much larger; vent and gill-opening dusky; dorsal colored like the body, with a broad white edge, growing broader behind; anal dark brown, unspotted, and with a broad pale border.

This species is known only from the type, No. 50617, U. S. N. M. (field No. 03392), a specimen 21 inches long, obtained by us at Honolulu.

7. Gymnothorax hilonis Jordan & Evermann, new species.

Head 8.2 in length; depth 16; eye 7 in head; snout 6; interorbital 6; gape 2.4; distance from tip of snout to vent 1.2 in distance from vent to tip of tail.

Body rather short, moderately compressed, the tail more compressed and bluntly pointed; head short, the nape swollen; interorbital space broad; a distinct median groove from near the tip of snout to origin of dorsal; angle of mouth posterior to eye a distance equal to eye's diameter; lower jaw but slightly curved, shorter than the upper; front of upper jaw with 3 short, bluntly pointed, movable teeth; side of upper jaw with a single series of short, pointed canines directed backward; shaft of vomer with short, blunt teeth; lower jaw on each side with a single series of rather long, pointed canines, longest in front and curved backward; anterior nostril in a long tube, about 2 in eye, near tip of snout just above lip; posterior nostril small, round, without tube, situated just above anterior part of eye; gill-opening small, its direction obliquely forward toward nape; a series of 4 pores on each side of upper jaw; similar pores on lower jaw. Origin of dorsal fin on nape midway between gill-opening and middle of eye; dorsal fin well developed, its greatest height somewhat exceeding length of snout; anal similar to dorsal, but lower.

Color in alcohol, rich, velvety black above, paler below where it is marbled and reticulated with narrow white lines; series of pores on side of upper jaw and those on tip of lower, white; cheek with a few irregular white spots; gill-opening whitish; side of body anteriorly with some small white specks and irregular whitish markings; lower jaw with larger, oblong, white cross-lines; dorsal fin rich brownish black, the edge posteriorly with a narrow, irregular, white border, sometimes interrupted by black; anal brown, with a narrow white edge from which extend narrow intrusions of white, some reaching base of fin; end of tail with a few small white spots, the tip narrowly white.

The only known example of this species is the type, No. 50618, U. S. N. M. (field No. 04902), a specimen 9.5 inches long, obtained by us at Hilo, Hawaii.

8. Echidna zonophæa Jordan & Evermann, new species.

Head 3 in trunk, or 6.5 in total; tail longer than head and trunk by a little more than the snout; eye 2 in snout, 1.5 in interorbital space; snout 6; interorbital space 7.75; mouth 2.8.

Body compressed, the tail tapering rather narrowly posteriorly; head deep and compressed, pointed in front; snout rather long and pointed, the tip obtusely rounded and projecting considerably beyond the mandible; eye rather small, midway between tip of mandible and corner of mouth; mandible shutting completely, arched below so that only the anterior teeth touch the front of the jaw above, though the thick fleshy lips conceal them all; teeth molar, those in front of jaws pointed; anterior nostrils in short tubes, the posterior pair above the eye with a slightly elevated margin; interorbital space convex; top of head more or less swollen or convex in profile; gill-opening 1.67 in eye; skin smooth; head with a few pores; origin of dorsal beginning at last fourth of space between corner of mouth and gill-opening; caudal small.

Color in alcohol, grayish white, the body and tail crossed by about 25 broad rich brown bands, extending upon the dorsal and anal fins; dark bands anteriorly broadest above and not meeting across belly, their width about equal to the distance from tip of snout to middle of eye; first brown band through eye, second across nape, the fourth across gill-opening; gray bands of ground color anteriorly broad, and widening much upon belly; posteriorly the gray bands are narrower and better defined, especially on the fins, their width scarcely greater than half that of the brown bands; tip of tail very narrowly white; body anteriorly, especially within the gray bands, profusely covered with numerous small, roundish, black specks, less numerous and more scattered posteriorly; no black spots on head; angle of mouth black, with a small white blotch immediately in front on lower jaw, continued across under jaw as a broad whitish band; side of head with about 4 or 5 narrow blackish lines between mouth and gill-opening; region of gill-opening marbled with dark brown and whitish, the opening dark.

One example (No. 03545) had much yellow on the head and between the brown zones.

This species is known from the type and 3 cotypes, all obtained by us at Honolulu.

Field No.	Length.	Locality.	Final disposition of specimen.
	Inches.		
03899	21	Honolulu	Type, No. 50821, U. S. N. M.
03361	16	....do	Cotype, No. 2698, U. S. F. C.
03545	17	....do	Cotype, No. 7108, L. S. Jr. Univ. Mus.
03900	15	....do	Cotype, No. 3965, Field Col. Mus.

Family MYCTOPHIDÆ.   The Lantern-fishes.

9. Rhinoscopelus oceanicus Jordan & Evermann, new species.

Head 3.5 in length; depth 4.1; eye 2.5 in head; snout very short, about 6; interorbital 3.5; D. about 12; A. about 18; scales 2-35-3.

Body strongly compressed, particularly posteriorly, where it tapers into the long, slender caudal peduncle; head exceeding depth of body; mouth large, somewhat oblique, the jaws equal, the maxillary reaching beyond the orbit, its posterior end club-shaped; eye large; anterior profile rather evenly convex from tip of snout to nape; teeth difficult to make out, but a single row of minute ones can be seen on the edge of each jaw, the exterior granular or short, villiform stripe, if it exists, being invisible even with the aid of a good lens; teeth on vomer and edges of palatines more distinct than those on jaws, and forming a broader line as if there were 2 or 3 rows; no granular patches visible on disk of palatine bone; an elevated acute mesial line separating one nasal prominence from the other; interorbital space convex, rounded; preopercle nearly vertical, sloping slightly backward from above downward; scales large, undulated and very irregularly and sparingly toothed or crenate, and having about 3 basal furrows; scales of lateral line conspicuous and more persistent; 7 photophores along base of anal, 5 along lower edge of caudal peduncle, 2 at base of caudal, 1 on middle of side above last anal photophore, 4 on each side of belly between ventrals and origin of anal fin, 5 between base of ventral and gill-opening, 1 on side above base of ventral, a row of 3 upward and backward from front of anal, 1 above and 1 below base of pectoral, and 1 on lower anterior portion of opercle; origin of dorsal somewhat behind base of ventrals, the posterior rays, together with those of anal, divided to the base; no spine at base of caudal.

Color in alcohol, uniform brownish, the scales, especially on middle of side, metallic steel blue; top of head brownish; side of head bluish; photophores black with silvery center; fins dusky whitish.

This species was recorded by Fowler from "near the Sandwich Islands," as Rhinoscopelus coruscans (Richardson), the record being based upon 4 specimens (Nos. 7972 to 7975) collected by Dr. W. H. Jones, and now in the Philadelphia Academy. During the Agassiz South Pacific expedition of the Albatross in 1899-1900, 2 examples of this species were taken in the surface towing net at 8 p. m., September 8, 1899, at latitude 10° 57′ N., longitude 137° 35′ W., southeast of the Hawaiian Islands. These

2 specimens are doubtless identical with those recorded by Mr. Fowler, and are apparently distinct from *R. coruscans*, the type of which came from between St. Helena and Ascension Islands and others from between Australia and New Zealand. They are near *R. andrew* Lütken, from which they seem to differ in the blunter snout, the more slender tail, and in having the posterolateral photophore somewhat before the adipose fin.

Type, No. 50622, U. S. N. M. (field No. 05805), 1.3 inches long, collected by the *Albatross* at 8 p. m., September 8, 1899, at the surface at 137° 35′ W., 10° 57′ N.; cotype, No. 2736, U. S. F. C., same size, collected at same time and place.

Rhinoscopelus coruscans, Fowler, Proc. Ac. Nat. Sci. Phila. 1900, 498 (near the Sandwich Islands); not of Richardson.

Family SYNGNATHIDÆ. The Pipe-fishes and Sea-horses.

10. Hippocampus fisheri Jordan & Evermann, new species.

Eye 2.8 in snout; snout 2 in head; D. 18, on 4 rings; A. 4; P. 15; rings 12 + 34.

Tail longer than head and trunk; trunk rather deep, compressed, its width 1.7 in depth; eye small, equal to interorbital width; interorbital space concave; gill-opening small, high; spines on head and body rather high, sharp; 2 rings on trunk between each pair of larger spines; tail with 3 rings between each pair of larger spines; coronet well developed, with 5 spines; spines over eye blunt; base of dorsal about equal to snout; anal small, long; pectoral broad, rays rather long.

Color in life, trunk below middle row of rings yellowish golden, above middle row blackish brown on orange ground; knobs orange; lower portion of knobs on 8 to 11 rings spotted with dark brown; side and top of tail same as back of trunk; ventral side pale dirty orange; head, crown and snout dirty dark brown; an orange band across snout and one before eyes; pale brownish golden over gills; chin orange; iris yellowish golden with 8 reddish streaks radiating from pupil; fins pale; a red spot before each eye at each side of preorbital spine.

Color in alcohol, pale brown, upper surface with dark brown marblings; side with small roundish dark spots.

The above description is from the type, No. 50625, U. S. N. M. (field No. 03835), a specimen 2.6 inches long, obtained at Kailua, Hawaii, where the species was new to the natives. We have 5 other examples, each about 3 inches long, taken from the stomach of a dolphin (*Coryphæna* sp.) which was captured at Hilo, July 18, 1901.

When fresh, No. 03507, a male, was pink or pale cardinal along and near the keels; plates on back and above middle row of knobs on side mottled blackish on pale red ground; plates below middle row of knobs and on belly porcelain white; egg-pouch uniform pale cardinal-red, paler than rest of body; tail same pink or pale cardinal, mottled with blackish blotches; top of head and crown blackish on pale red; cheek, jaw, and snout pink. Some examples had ventral side of tail and portion behind fourth prominent spine of tail uniform pale cardinal-red.

This species is named for Mr. Walter V. Fisher, of Stanford University.

We have the following specimens:

Field No.	Length.	Locality.	Final disposition of specimen.
	Inches.		
03835	2.6	Kailua....	Type, No. 50625, U. S. N. M.
03507	....	Hilo.......	Cotype, No. 7450, L. S. Jr. Univ. Mus.
....	....	...do....	Cotype, No. 2760, U. S. F. C.
....	....	...do....	Cotype, No. 3306, Field Col. Mus.
....	....	...do....	Cotype, No. 1687, Bishop Mus.
....	....	...do....	Cotype, No. ——, M. C. Z.

11. Hippocampus hilonis Jordan & Evermann, new species.

Eye about 4 in snout; snout 2 in head; D. 16, on 3 rings; rings 12 + 35. Tail a little longer than head and trunk; trunk rather deep, compressed, its width 2 in depth; eye small, equal to interorbital width, which is concave, broader posteriorly; gill-opening high, rather large; spines on head and body very blunt, rounded, or obsolete, though forming knobs of more or less equal size along tail; coronet with rounded knobs, before which is a short keel or trenchant ridge; base of dorsal about 1.35 in snout.

Color in alcohol, dark or blackish brown, more or less uniform.

This species is known to us only from the example described above. It is closely related to the Japanese *Hippocampus aterrimus* Jordan & Snyder, but on comparison with the type of that species was found to differ in the presence of the keel on the top of the head and in other minor characters. It is also close to *H. ringens*.

Type, No. 50626, U. S. N. M. (field No. 03832), a specimen 6 inches long, presented to us by Mr. A. M. Wilson, of Hilo, Hawaii, where he obtained the specimen.

Family ATHERINIDÆ. The Silversides.

12. Atherina insularum Jordan & Evermann, new species.

Head 4 in length; depth 4.75; eye 3 in head; snout 4; interorbital 2.8; maxillary 2.5; mandible 2.2; D. vi-i, 11; A. 17; scales 46, 6 rows from anterior base of anal upward and forward to spinous dorsal.

Body oblong, compressed; head triangular, the sides compressed, top flat; mouth large, oblique, maxillary reaching front of pupil, lower jaw included; teeth in rather broad villiform bands on jaws, vomer, and palatines; interorbital space very broad and flat; snout broad, truncate; origin of spinous dorsal slightly posterior to vertical at vent, slightly nearer tip of snout than base of caudal; longest dorsal spine about 2.4 in head, reaching nearly to vertical at front of anal; distance between spinous and soft dorsals equal to distance from tip of snout to middle of pupil; edge of soft dorsal concave, anterior rays somewhat produced, their length 1.9 in head; last dorsal ray about one-half longer than one preceding; base of soft dorsal 1.8 in head; origin of anal considerably in advance of that of soft dorsal, the fins similar, anterior rays about 1.7 in head, base of anal 1.3 in head; caudal widely forked, the lobes equal; ventral short, barely reaching vent; pectoral short, broad, and slightly falcate, its length about 1.4 in head. Scales large, thin, and deep, 19 in front of spinous dorsal, 6 rows between the dorsals and 9 on median line of caudal peduncle.

Color when fresh, clear olive green with darker edges to scales; lateral stripe steel blue above, fading into the silvery belly; fins uncolored.

Color in alcohol, olivaceous above, silvery on sides and below; scales of back and upper part of side with numerous small round coffee-brown specks, disposed chiefly on the edges; median line of back with a darkish stripe; middle of side with a broad silvery band, plumbeous above, especially anteriorly, more silvery below; top of head and snout with numerous dark brownish or black specks; side of head silvery, opercle somewhat dusky, sides and tip of lower jaw dusky; dorsals and caudal somewhat dusky, other fins pale; pectoral without dark tip.

This small fish is common inside the reef in shallow bays everywhere in the Hawaiian Islands. Many individuals were seen off the wharf at Lahaina on Maui. Our collections of 1901 contain 20 specimens from Kailua, from 1.5 to 3.5 inches long; 43 from Hilo, 1.5 to 2.25 inches long; and 1 from Honolulu, 2.25 inches in length. Numerous specimens were obtained by the *Albatross* at Honolulu in 1902, 1 of which is taken as our type and 3 others as cotypes.

Type, No. 50819, U. S. N. M., 4.25 inches long, obtained by the *Albatross* at Honolulu. Cotypes No. 2741, U. S. F. C., 3.9 inches long; No. 2302, Am. Mus. Nat. Hist., 3.9 inches long; and No. 4063, Field Col. Mus., 3.5 inches long, all collected at Honolulu by the *Albatross*.

Family HOLOCENTRIDÆ. The Squirrel-fishes.

13. Myripristis berndti Jordan & Evermann, new species.

Head 2.8 in length; depth 2.4; eye 2.7 in head; snout 4.7; maxillary 1.7; interorbital 4.9; D. x-i, 16; A. iv, 14; P. i, 14; V. i, 7; scales 4-32-7.

Body elongate, deep, compressed, its greatest depth at base of ventral; head large, compressed, its depth less than its length; snout short, blunt, convex, its width about twice its length; upper profile of head straight from above nostril to occiput; eye large, high, its diameter a little less than posterior part of head, and its upper rim hardly impinging upon upper profile of head; mouth very large, oblique; mandible slightly projecting, the maxillary not reaching posterior margin of eye; distal expanded extremity of maxillary 1.7 in eye; several enlarged, blunt teeth on outer front edges of jaw and sides of mandible; teeth in jaws fine, in broad bands, also on vomer and palatines; tongue thick, pointed, and free in front; suborbital rim narrow, finely serrate; lower posterior margin of maxillary with blunt denticulations; lips rather thick and fleshy; nostrils close together, posterior very large, close to front rim of orbit; bones of head all finely serrate; opercle with well-developed spine; gill-opening large, filaments rather large; gillrakers long, fine, longest longer than longest gill-filaments; pseudobranchiæ very large, outer portions free for half their length; dorsal spines slender, first 3.4 in head, second 2.6, third 2.2, fourth 2.2, tenth 6.4, and last 3.5; anterior dorsal rays elevated, produced into a point, first 1.8, second 1.7, and last 8; first and second anal spines short, third 2.6, and fourth 2.8; soft anal similar to soft dorsal, anterior ray 1.75, third 1.8, and last 6; caudal forked, lobes pointed, 1.2; pectoral rather small, pointed, 1.4; ventral 1.6, reaching .65 distance to anus; caudal peduncle elongate, compressed, 2.2 in head, its depth 3.25; scales large, ctenoid, deep; lateral line slightly arched, running obliquely down on side along upper part of caudal peduncle; 4 rudimentary caudal rays above and below, slender, sharp-pointed, and graduated.

Color in life (No. 03370), deep red, with silvery luster; no stripes on side, a blood-red band across gill-opening and base of pectoral; fins deep red, without white edgings, distal half of spinous dorsal shading into orange.

Color in alcohol, pale straw-color, fins plain and paler; upper margin of opercle blackish, and axil of pectoral black; anterior margins of soft dorsal and anal whitish.

We take pleasure in naming this species for Mr. Louis E. Berndt, superintendent of the Honolulu market. Described from an example (No. 03346) taken at Honolulu, where the species is rather common. Our collections contain the following specimens:

Field No.	Length	Locality.	Final disposition of specimen.	Field No.	Length.	Locality.	Final disposition of specimen.
	Inches.				Inches.		
03346	9	Honolulu.	Type, No. 50627, U. S. N. M.	04870	9	Honolulu.	Cotype, No. 3947, Field Col. Mus.
03370	8	....do....	Cotype, No. 16818, Bishop Mus.				
04834	8	....do....	Cotype, No. ...., M. C. Z.	04872	8	....do....	Cotype, No. 24212, Ac. Nat. Sci. Phila
04836	7	....do....	Cotype, No. 2701, U. S. F. C.				
04849	7	....do....	Cotype, No. 2282, Am. Mus. Nat. Hist.	04873	8.5	....do....	Cotype, No. 7451, L. S. Jr. Univ. Mus.
04850	7	....do....	Cotype, No. 1489, Cal. Ac. Sci.	04874	7	....do....	
04856	7	....do....	Cotype, No. 50628, U. S. N. M.	04878	7	....do....	
04857	8.5	....do....	Cotype, No. 9801, Mus. Ind. Univ.	04882	7	....do....	
				04885	8	....do....	
				04891	9	....do....	

14. Myripristis chryseres Jordan & Evermann, new species.

Head 2.75 in length; depth 2.5; eye 2.4 in head; snout 5.5; maxillary 1.9; mandible 1.8; interorbital 5; D. x-i, 14; A. iv, 12; scales 4-34-6.

Body short, stout, and compressed; dorsal profile evenly convex from tip of snout to origin of soft dorsal; ventral outline nearly straight to origin of anal whose base is equally oblique with that of soft dorsal; caudal peduncle short but slender, and not greatly compressed, its length from base of last dorsal ray to first short spinous caudal ray 1.3 in eye, its least width about 3 in its least depth which is 1.8 in eye; head heavy, short; mouth moderately large, the gape in closed mouth reaching vertical of middle of eye; maxillary very broad, triangular, reaching nearly to vertical of posterior line of eye, with a broad, curved supplemental border; surface of maxillary roughly striated, anterior edge near the angle strongly dentate; lower jaw strong, somewhat projecting, the tip with 2 rounded rough prominences fitting into a distinct notch in upper jaw; teeth short, in narrow villiform bands in jaws and on palatines, a small patch on vomer, none on tongue; eye very large, orbit exceeding postocular part of head; lower edge of eye on level with axis of body; snout short, 2 in orbit; interorbital space nearly flat, strongly rugose; 2 long ridges from preorbitals to nape; outside of these a short ridge beginning above front of pupil, extending backward and branching upon nape; supraocular ridge spinescent posteriorly; suborbital narrow, strongly dentate below, upper edge in front somewhat roughened; opercular bones all strongly toothed; opercular spine short and obscure (stronger in most of the cotypes); dorsal spines slender, fifth longest and strongest, its length 2.5 in head; first dorsal spine somewhat posterior to base of pectoral, its length 2 in eye, spines gradually shorter from fifth; space between dorsals very short, about equal to length of tenth spine; dorsal rays long, length of longest a little greater than orbit, last equal to pupil; first anal spine very short, second short and triangular, its length about 1.5 in pupil; third anal spine long, strong and straight, longer than fourth, its length equal to diameter of orbit; fourth anal spine slender, its length 1.3 in orbit; anal rays longer than those of dorsal; caudal widely forked, lobes equal, their length 1.5 in head; pectoral long and narrow, its length 1.4 in head, the tip reaching past tips of ventrals; ventrals slender, pointed, nearly reaching vent and nearly as long as pectoral.

Scales smaller than in *M. murdjan*, number in lateral line 34 in type, 35 to 38 in some of the cotypes; scales strongly dentate, and striate near the edges; a strongly dentate humeral scale.

Color in life, bright scarlet, centers of the scales paler; a blackish-red bar behind, and on edge of, opercle, continued as red (not black) into the axil; first dorsal golden, with red basal blotches on membranes; second dorsal golden, with crimson at base, spine and first ray white; caudal golden, first ray white above and below; anal golden, the spines and first ray white; all the vertical fins narrowly edged with red; ventrals mostly pink, with golden wash on first rays; pectoral plain crimson; axil light red.

Color in alcohol, yellowish or orange white, the edges of the scales paler; some of the scales with small brownish dustings on the edges; edge of opercle black; opercle and cheek somewhat silvery; fins all pale yellowish, without dark edges. In some individuals the general color is more silvery, and in one example (No. 04860) the axil of the pectoral is somewhat dusky. In life the color is more scarlet than in *M. murdjan* and the fins yellow, not red as in *M. murdjan* and all other Hawaiian species.

Myripristis chryseres is related to *M. murdjan*, from which it differs in the smaller scales, larger eye, less black in the axil, and the absence of black edges to the dorsal and anal fins as in the life colors already noted, the yellow fins being the most conspicuous character in life. It reaches a length of 9 or 10 inches and appears to be moderately abundant at Honolulu and Hilo.

Type, No. 50629, U. S. N. M. (field No. 03463), a specimen 8 inches long, obtained at Hilo, Hawaii. The numerous cotypes and the museums in which they have been deposited are indicated in the following tabular list of specimens:

Field. No.	Length.	Locality.	Final disposition of specimen.	Field No.	Length.	Locality.	Final disposition of specimen.
	Inches.				*Inches.*		
2558	4	Honolulu.	Type, No. 50629, U. S. N. M.	04862	9	Hilo	Cotype, No. ——, M. C. Z.
03463	8	Hilo......		04863	9	.....do....	Cotype, No. 9802, Ind. Univ. Mus.
04823	8.5	.....do....	Cotype, No. 7452, L. S. Jr. Univ. Mus.	04867	8	Honolulu.	Cotype, No. 2702, U. S. F. C.
04827	8	.....do....	Do.	04868	7	.....do....	Cotype, No. 1689, Bishop Mus.
04833	6	Honolulu.	Cotype, No. 2283, Am. Mus. Nat. Hist.	04869	8	.....do....	Cotype, No. 1490, Cal. Ac. Sci.
04839	7	.....do....	Cotype, No. 24273, Ac. Nat. Sci. Phila.	04887	8	.....do....	
				04889	7	.....do....	
04860	9.5	Hilo	Cotype, No. 50630, U. S. N. M.	04890	8	.....do....	
04861	9	.....do....	Cotype, No. 3948, Field Col. Mus.	04892	8	.....do....	

15. Myripristis argyromus Jordan & Evermann, new species.

Head 3.5 in length; depth 2.75; eye 2.4 in head; snout 5; maxillary 1.8; mandible 1.6; interorbital 3.75; D. x–i, 15; A. iv, 13; scales 4–33–5.

Body rather long and compressed, dorsal and ventral outlines about equally and evenly convex from snout to origins of anal and soft dorsal fins; head rather large but short; mouth moderate, maxillary reaching vertical at posterior edge of pupil, the exposed portion broad, triangular, the upper edge concave, the end rounded and the anterior edge with short blunt teeth, strongest at angle; tip of upper jaw with a shallow notch roughened at its outer edges; jaws equal, lower fitting into the notch of upper and with 2 patches of strong blunt tooth-like tubercles at its tip; eye large, its middle above level of tip of upper jaw; interorbital space wide and slightly convex; 2 low, nearly parallel median ridges from tip of snout to nape, diverging slightly at their middle, another low ridge from above orbit backward to nape, and another backward around orbit; ridges on nape divergent; suborbital narrow, dentate on both edges; opercular bones all striate and dentate at the edges; opercle with a short, flat, triangular spine; scales large, rough, striate near the edges which are finely toothed; a series of 4 or 5 large modified scales across nape, and a series of triangular scales along bases of dorsal and anal; about 10 scales in front of dorsal; origin of dorsal about over lower base of pectoral; dorsal spines slender, the first 3.2 in head, third and fourth longest, about equal to orbit; interval between dorsals very short; anterior dorsal rays somewhat produced, their length equal to snout and eye; edge of fin concave, last rays nearly 3, or equal to pupil; anal spines graduated, the first very small, second short but stout, third much longer and stoutest, its length 1.3 in eye, fourth still longer and more slender; anterior anal rays produced, their length about equal to that of longest dorsal rays, free edge of fin concave; caudal evenly forked, the lobes equal to length of head; pectoral long and pointed, reaching beyond tips of ventrals, about 1.3 in head; ventrals shorter, 1.6 in head, their tips equally distant between their bases and that of first anal ray.

Color in alcohol, pale yellowish-white, brightest above, more silvery on side and belly; opercular bones with fine round brownish specks; edge of opercle not black, scarcely dusky; axil dusky inside but not showing above fin; fins pale yellowish-white without any dark on edges.

Type, No. 50631, U. S. N. M. (field No. 04829), a fine specimen 9.5 inches long, obtained by us at Hilo, Hawaii.

M. argyromus is related to *M. berndti*, but is distinguished by the more slender body, the absence of black on the opercle, and the paler axil. It does not appear to be abundant and is represented in our collections by only 8 specimens. All the other specimens are taken as cotypes. The data regarding each will be found in the following list:

Field No.	Length.	Locality.	Final disposition of specimen.
	Inches.		
04829	9.5	Hilo	Type, No. 50631, U. S. N. M.
04830	9.0	Honolulu	Cotype, No. 7453, L. S. Jr. Univ. Mus.
04835	6.5	....do	Cotype, No. 1690, Bishop Mus.
04840	6.5	....do	Cotype, No. 2284, Am. Mus. Nat. Hist.
04877	9.0	Honolulu	Cotype, No. 2703, U. S. F. C.
04879	8.0	....do	Cotype, No. ——, M. C. Z.
04880	7.5	....do	Cotype, No. 9863, Ind. Univ. Mus.
04881	9.0	....do	Cotype, No. 3949, Field Col. Mus.

16. Myripristis symmetricus Jordan & Evermann, new species.

Head 3.2 in length; depth 2.4; eye 2.2 in head; snout 5; interorbital 3.8; D. x-i, 15; A. iv, 14; P. i, 14; V. i, 7; scales 4–36–6.

Body elongate, deep, compressed, greatest depth about midway between origin of ventrals and anal; upper and lower profiles evenly convex; head compressed, as long as deep, its width 1.7 in its length; snout short, broad, blunt, and steep; upper profile of head straight from above nostril to occiput; eye very large, high, hardly impinging upon the upper profile of head, its diameter greater than postocular region; mouth very large, oblique; mandible slightly projecting, and reaching posteriorly to below posterior rim of pupil; distal expanded extremity of maxillary 2.35 in eye; several enlarged blunt teeth on outer front edges of mandible; teeth in jaws, on vomer, and palatines very fine, in bands; tongue thick, pointed, free; suborbital rim narrow, finely serrate; lower posterior margin of maxillary smooth; lips rather thick and fleshy; nostrils close together, posterior very large, close to front rim of orbit; bones of head all finely serrate; opercle with well-developed spine; gill-opening large, filaments large; gillrakers long, fine, longest longer than longest gill-filament; pseudobranchiæ very large; dorsal spines slender, sharp, first 2.75 in head, second 2.1, third 2, fourth 1.9, tenth 6, and last 3.6; soft dorsal with anterior rays elevated, produced into a point which projects beyond tip of posterior rays when fin is depressed, first ray 1.4 in head, third 1.35, and last 3.75; anal spines graduated to last, third enlarged, 2.5 in head, fourth 2.0; soft anal similar to soft dorsal, anterior rays produced, first 1.4, third 1.3, and last 4.6; caudal elongate, deeply forked, the lobes pointed, 1.2 in head, and reaching slightly behind tips of ventrals; ventrals sharp-pointed, 1.4 in head, spine 2.2; caudal peduncle elongate, compressed, its length 1.8 and its depth 3.2; scales large, finely ctenoid, deep on middle of side; lateral line running obliquely back, slightly curved at first, and posteriorly along upper side of caudal peduncle; 4 rudimentary, slender, sharp-pointed, graduated rays along upper and lower edges of caudal; scales narrowly imbricated along middle of side.

Color in alcohol, pale straw-color; fins paler, except the anterior dorsal and anal rays, which are grayish; margin of opercle above blackish; axil of pectoral black.

This species was found both at Honolulu and Hilo, but does not appear to be abundant. Only 4 specimens are in our collections:

Field No.	Length.	Locality.	Final disposition of specimen.
	Inches.		
04866	5.5	Hilo	Type, No. 50632, U. S. N. M.
04864	5.5	....do	Cotype, No. 7454, L. S. Jr. Univ. Mus.
04865	5.5	....do	Cotype, No. 2704, U. S. F. C.
04924	5.0	Honolulu	Cotype, No. 3950, Field Col. Mus.

17. Flammeo scythrops Jordan & Evermann, new species.

Head (measured to end of flap) 2.75 in length; depth 3; eye 3 in head; snout 4; maxillary 2.1; mandible 1.8; interorbital 5; D. xi, 13; A. iv, 9; scales 5–48–7, 5 rows on cheek; Br. 7.

Body oblong, rather slender; dorsal outline gently and rather evenly curved from tip of snout to origin of soft dorsal, more nearly straight from tip of snout to nape; ventral outline less convex; head long; snout long and pointed; maxillary broad, with a strong supplemental bone whose lower edge forms a broad angle; end of maxillary slightly concave; lower jaw long, much projecting, tip prominent; mouth large, not greatly oblique; maxillary nearly reaching vertical at posterior line of pupil; lips broad, rounded, and soft; eye large, lower edge of pupil on axis of body; interorbital space with a broad, shallow groove between low ridges, one on each side; space between ridge and eye with short, curved ridges; nape on each side with a group of 8 or 10 short, sharp ridges, diverging backward and ending in short, sharp spines; posterior part of supraocular with a patch of short spines; suborbital dentate on its lower edge; preorbital with 2 blunt prominences in front, a strong, recurved spine below, ridges and spines on its upper surface; opercular bones all strongly striate, the striæ ending in short spines; entire surface of interopercle striate; opercle with 2 strong spines, the lower the stronger, its length 1.6 in orbit; preopercle with a very strong spine at angle, its length nearly equaling diameter of orbit, its surface striate, and its base with a series of small spines; under surface of dentary somewhat roughened; surface of articular bone much rougher; jaws each with a broad band of villiform teeth, the outer series on upper jaw stronger; a narrow series on each palatine and a patch on vomer; scales moderate, the surfaces usually nearly smooth, the edges finely toothed; a series of strongly striate scales across nape, and a strong, striated plate at shoulder; lateral line well developed, little arched, with about 45 pores; bases of soft dorsal and anal each with series of modified triangular scales; caudal with small scales on base and fine scales on membranes, extending well toward tips of outer rays; origin of spinous dorsal in advance of base of pectoral or over middle of upper opercular spine; dorsal spines in a broad, deep groove, moderately strong, middle one longest, 2.3 in head, first a little shorter than snout, tenth more than half eye; dorsal rays longer than spines, longest 2.2 in head; first anal spine very short, second about 3 times as long; third anal spine very long and strong, but little curved, reaching past base of anal, its length 1.5 in head; fourth anal spine shorter and more slender, its length 2.25 in head and equaling longest anal rays; last anal ray much shorter, 1.6 in eye; pectoral long and slender, 1.2 in head, the tip nearly reaching vent; ventrals shorter, equal to snout and eye; caudal forked, the lobes equal, not strongly divergent, their length about equaling that of third anal spine; rudimentary caudal spines 5 above, 4 below, strong and sharp.

Color in life, head red above, paler on sides, nearly white below; tips of jaws rich red; side of body with about 10 or 12 narrow yellow stripes separated by red or rosy stripes of about same width, those below paler and somewhat purplish; under parts purplish or pinkish white; the stripes beginning at edge of opercle and ceasing at base of caudal peduncle, which is rich red above, becoming paler on side and below; membranes between the first and third dorsal spines rich blood-red, those between other spines white at base, each with distal portion lemon-yellow in front and red behind, last 2 or 3 membranes with little or no yellow; dorsal spines pale rosy, nearly white; soft dorsal, anal, pectoral, and ventral with rays rosy, membranes pale; ventral with a little yellow at base; anal spines somewhat dusky; caudal rich blood-red, paler distally; eye red, a narrow yellow ring around pupil.

Another example (No. 03041), much faded, was bright red; stripes on side equally bright golden; fins red; edges of dorsal membranes pale; no markings evident on fins.

Color in life of another example (No. 03451), side with 10 or 11 longitudinal golden or yellow bands; spinous dorsal more or less white; membranes between first and third dorsal spines more or less deep vermilion, except the upper marginal portion behind second spine, which is white; a red blotch along margin of membranes just before each of the other dorsal spines.

Color when fresh of another specimen (No. 03490), violet-rose with 10 stripes of bright golden on side; dorsal red, mottled with golden, the first two spines deep red; soft dorsal and other fins rather light red without edgings, and scarcely darker behind third anal spine; pectoral and ventrals pink; a red dash across cheek, space above and below whitish; temporal region deep red; iris red.

All these colors fade in alcohol and the fish becomes a pale yellowish white, the longitudinal lines on side showing faintly as duller and brighter stripes of yellowish white; fins all whitish or yellowish white, membranes of spinous dorsal whiter.

The above description from the type, No. 50633, U.S.N.M. (field No. 03488), a specimen 9 inches long obtained by us at Honolulu. An examination of our large series of cotypes shows but slight variations, the characters appearing quite stable. In some examples the upper opercular spine is the larger, in others the 2 are equal; in 2 examples we find 3 opercular spines.

This species has been several times called *Holocentrum argenteum*. The species described under that name by Quoy & Gaimard from New Guinea resembles this in the slender body and general coloration, but differs in having the lower jaw included, eye much smaller, mouth smaller, and the preopercular spine weaker. It was probably intended for some species with the lower jaw included.

Holocentrus ticre Lesson, from Tahiti, is more likely to be the present fish. It is figured as elongate, with the spinous dorsal low and the opercular spines equal. The plate is, however, too rough to permit certain identification and approaches almost as closely to *Holocentrus diploxiphus* as to *Flammeo*.

This is one of the most abundant species in the markets at Honolulu and Hilo. It reaches a length of 8 to 10 inches. Our collections contain the following specimens, all of which, except the first, are taken as cotypes:

Field No.	Length.	Locality.	Final disposition of specimen.	Field No.	Length.	Locality.	Final disposition of specimen.
	Inches.				Inches.		
03488	9.25	Honolulu.	Type. No. 50633, U. S. N. M.	04965	8.50	Honolulu.	Cotype, No. 7456, L. S. Jr. Univ. Mus.
03041	8.00	.....do....	Cotype.	04966	8.50	.....do....	Cotype, No.2705,U.S.F.C.
03451	8.50	.....do....	Cotype. No. —. M. C. Z.	04967	7.50	.....do....	Cotype, No.2705,U.S.F.C.
03490	8.50	Hilo....	Cotype. No. 7455, L. S. Jr. Univ. Mus.	04968	8.25	.....do....	Cotype, No. 24214, Ac. Nat. Sci., Phila.
04925	8.25	.....do....	Do.	04969	7.50	.....do....	Cotype. No. 2285, Am. Mus. Nat. Hist.
04926	10.00	.....do....	Do.				
04927	9.00	.....do....	Cotype, No. 1491. Cal. Ac. Sci.	04970	8.50	.....do....	Cotype, No. 7456, L. S. Jr. Univ. Mus.
04928	9.25	.....do....	Cotype.				
04949	6.25	Honolulu.	Cotype.No.3967, Field Col.Mus.	04971	7.75	.....do....	Do.
04954	8.75	.....do....	Cotype, No. 7456, L. S. Jr. Univ. Mus.	04972	7.75	.....do....	Do.
				04973	8.75	.....do....	Do.
04956	8.75	.....do....	Cotype, No. 9804. Ind. Univ. Mus.	04974	8.75	.....do....	Do.
				04975	8.00	.....do....	Do.
04957	7.25	.....do....	Cotype.	04976	5.00	.....do....	Do.
04958	8.50	.....do....	Cotype. No.1691, Bishop Mus.	04977	7.00	.....do....	Do.
04959	7.50	.....do....	Cotype.	04988	9.00	.....do....	Do.
04964	9.00	.....do....	Cotype. No. 50634, U. S. N. M.				

Holocentrum argenteum, Steindachner, Denks. Ak. Wiss. Wien, LXX, 1900, 492 (Honolulu and Laysan); not of Cuvier & Valenciennes.

18. Holocentrus xantherythrus Jordan & Evermann, new species.

Head 2.8 in length; depth 3; eye 3 in head; snout 4; maxillary 2.7; interorbital 5; D XI–14; A. IV, 10; scales 4–47–8.

Body elongate, compressed, greatest depth about base of ventral; upper profile steep; lower profile nearly horizontal; head compressed, its depth about 1.2 in length, width 2.25; eye large, high, impinging upon upper profile in front, anterior, and a little less than postocular region; snout short, pointed, its upper profile obliquely straight; jaws rather large, subequal; maxillary reaching beyond front margin of pupil or to first third of eye, its distal expanded extremity 2.7 in eye; supplemental maxillary large; lips rather thick, fleshy; teeth small, short, in rather broad bands in jaws and on vomer and palatines; tongue elongate, pointed, free in front; nostrils close together, posterior, a deep cavity in front of middle of eye; interorbital space broad, very slightly concave; preorbital with a large spine in front, its margins serrate; suborbital narrow, with finely serrate margin; preopercle with a large dagger-like spine at lower angle; opercle with 2 similar spines on upper margin, upper one much the larger; bones of head with serrate margins; gill-opening rather large, filaments and pseudobranchiæ well developed; gillrakers short, compressed, few, and much shorter than longest filaments; fleshy axillary flap small; dorsal spines sharp-pointed, first 3.2 in head, second 2.8, third 1.9, last 7; anterior dorsal rays high, second 2.4 in head, third 2.2, last 6.5; third anal spine very large, not reaching beyond soft rays, 1.7 in head, fourth 2.25; anterior anal rays longest, first 1.75 in head, second 1.9, last 6; caudal rather small, deeply forked; pectoral small, 1.6 in head; ventral sharp-pointed, 1.4, spine 2; caudal peduncle elongate, compressed, its length 2.1 in head, depth 4; scales rather large, ctenoid; lateral line nearly straight, running obliquely down along upper side of caudal peduncle.

Color in life (No. 02989), bright red, belly more or less silvery; about 10 narrow longitudinal silvery stripes; uppermost pinkish; side of head silvery with pinkish shades; a white stripe from preorbital to base of preopercular spine; spinous dorsal deep red without streaks or black marking, a white spot behind first and second spines at base, tips of third to seventh spines whitish; soft dorsal, anal, caudal, and pectoral plain pink; anal with membrane of third spine and first soft ray deep red; ventral pink, spine and first soft ray white, second soft ray deep red anteriorly, posteriorly whitish.

Another example (No. 03161), was rose red when fresh, with about 10 very faint light rosy streaks along rows of scales, these much less distinct than in other species; cheek rosy with one broad oblique white band; dorsal plain red, the membrane fading to white, no light stripes on dark areas; other fins plain light red; membrane of fourth anal spine not darker; iris pink.

Another example (field No. 03467), deep crimson when fresh, with 10 narrow, sharply defined, white stripes along rows of scales; an oblique white stripe below eye from snout to base of preopercular spine; dorsal clear deep red, clouded with darker; soft dorsal, caudal, and anal light bright red; membrane between third and fourth anal spines blood-red; pectoral deep red; ventrals red, spines white, their membranes blood red.

Color in alcohol, pale brown or brownish white, washed more or less with silvery or brassy white; side with 9 or 10 longitudinal white stripes; fins pale.

This species is related to *Holocentrus ensifer*, differing mainly in the presence of two well-developed spines on the upper margin of the opercle. It is one of the most abundant of the family in Hawaiian waters. It is represented in our collections by 40 examples, as follows:

Field No.	Length.	Locality.	Final disposition of specimen.	Field No.	Length.	Locality.	Final disposition of specimen.
	Inches.				*Inches.*		
05999	6	Honolulu.	Type, No. 50635, U. S. N. M.	04227	6.5	Honolulu.	Cotype, No. 2845, Ind. Univ. Mus.
02934	6.5	...do....	Cotype, No. 50836, U. S. N. M.				
02983	5.5	...do....	Cotype, No. 2706, U. S. F. C.	01936	5.75	...do....	Cotype, No. 1492, Cal. Ac. Sci.
02989	4.5	...do....	Cotype, No. 3951, Field Col. Mus.	01958	6.25	...do....	
				01959	6.25	...do....	
03160	5.25	...do....	Cotype, No. 7457, L. S. Jr. Univ. Mus.	01960	5.5	...do....	
				01961	5.5	...do....	
03161	6.25	...do....	Do.	01940	5.5	...do....	
03294	5.75	...do....	Cotype, No. 1692, Bishop Mus.	01990	5.75	Kailua	
03295	5.75	...do....	Cotype, No. ——, M. C Z.	01234	5.25	Honolulu.	
03293	5	...do....	Cotype, No. 2286, Am. Mus. Nat. Hist.	01230	5.5	...do....	
				01241	3.25	Kailua	
03467	5.75	...do....	Cotype, No. 21215, Ac. Nat. Sci. Phila.	01978	3.75	...do....	
				01226	5.75	Honolulu.	
03468	6.5	...do....	Cotype, No. 2753, U. S. F. C.	05905	5.6	...do....	

And 15 other examples from Honolulu, ranging in length from 4 to 6.5 inches.

19. Holocentrus ensifer Jordan & Evermann, new species.

Head 3 in length; depth 2.7; eye 3 in head; snout 3.5; maxillary 2.25; interorbital 5; D. xi, 15; A. iv, 11; P. i, 14; V. i, 8; scales 4–17–8.

Body elongate, compressed, greatest depth at ventral fin; upper profile decidedly more convex than lower; head compressed, much longer than deep, pointed, its width a little more than half its length; eye moderate, about 1.2 in postocular part of head, and slightly impinging upon upper profile; snout pointed; mouth moderate, oblique; maxillary broad, with large supplemental bone distally, equal to half diameter of eye; lips thick, fleshy; teeth minute, in broad bands in jaws, and on vomer and palatines; tongue pointed, free in front; nostrils close together, posterior a large cavity with several small spines projecting over; preorbital with 2 large strong spines and about 6 strong serrations on its margin; suborbital rim narrow; bones of head all more or less finely serrate, the opercle above and preopercle below each with a long, strong, dagger-like spine; interorbital space broad, very slightly concave; a fleshy axillary flap; gill-opening large, filaments moderately long, much longer than gill-rakers which are compressed and not very numerous; pseudobranchiæ large; spinous dorsal long, membrane between spines not much incised, first 2.2, second 2.1, third 2, last 4.2; anterior dorsal rays longest, fourth 1.8, last 7.5; third anal spine largest, 1.75, fourth 2.3; soft anal similar to soft dorsal, third spine not reaching beyond rays; caudal rather small, forked; pectoral 1.3; ventral 1.4, spine 2; caudal peduncle compressed, its length 2.2, depth 4; scales rather large, ctenoid; lateral line arched a little at first and running down obliquely on upper side of caudal peduncle.

Color in life, bright red; side with about 8 yellow longitudinal bands; spinous dorsal vermilion tinged with yellow; soft dorsal rosy with front margin white and behind this above, red; anal whitish with red between third spine and first ray; caudal red, margined above, and along the emargination with whitish; pectoral whitish with red lines; ventral rosy with front margin white.

Another example (field No. 03454), in life had yellow and red longitudinal bands above and yellow and white below; spinous dorsal vermilion, other fins red with white borders. Another (field No. 03472), was brilliant scarlet red with 11 golden streaks along rows of scales, upper 4 broadest, and third and fourth most distinct and oblique; a white or golden streak across cheek; fins plain scarlet without dark patches. Color, when fresh, of another specimen (field No. 03494), bright red verging to scarlet; side red, with 4 golden stripes along back and 6 silver stripes below these, golden and silver, very bright; head crimson; a white band on cheek; spinous dorsal deep scarlet with crimson edge; soft dorsal light crimson with a white, then a dark crimson edge; caudal blood red, edged above and below with white, posterior part of fin abruptly pale; anal with pale spines, then blood red, then pinkish; ventral with white spine, then dark red, then pink; pectoral light red, axil deep red.

Color in alcohol, pale brown or brownish white, the longitudinal bands on sides, together with scales on cheeks and opercle, silvery; fins pale.

This species was obtained by us at Honolulu and Kailua, and appears to be common at the former place. The collections contain the following specimens:

Field No.	Length	Locality	Final disposition of specimen.	Field No.	Length.	Locality.	Final disposition of specimen.
	Inches.				*Inches.*		
03448	6	Honolulu.	Type. No. 50837, U. S. N. M.	04951	9.5	Honolulu.	Cotype. No. 24216. Ac. Nat. Sci. Phila.
03472	6	Kailua ...	Cotype. No. 7458, L. S. Jr. Univ. Mus.	03454	9.75	......do....	Cotype. No. 9806. Ind. Univ.Mus.
04929	8.75	Honolulu.	Cotype. No. 2708. U. S. F. C.	03494	8.5	......do....	Cotype. No. 3952, Field Col.Mus.
04930	8.5	......do....	Cotype. No. 1483, Bishop Mus.				
04931	8	......do....	Cotype. No. ——, M. C. Z.				
04950	8.75	......do....	Cotype. No. 2287, Am. Mus. Nat. Hist.				

Family CARANGIDÆ. The Pampanos.

20. Carangus elacate Jordan & Evermann, new species.

Head 3.6 in length; depth 3.4; eye 4.5 in head; snout 3.8; interorbital 3.8 in snout; maxillary 2.1; preorbital 8.5; mandible 1.9; D. vii-i, 19; A. ii-i, 16; scutes 28.

Body slender, compressed, not greatly elevated; snout rather short, profile ascending to nape in a gentle curve, slightly trenchant; mouth large, slightly oblique; lower jaw somewhat projecting; maxillary reaching posterior edge of orbit, its width at tip 1.5 in orbit; supplemental maxillary well developed, its width 3.25 in entire width; gape reaching vertical of posterior edge of pupil; villiform teeth on vomer, palatines and tongue, those on jaws in a single row, small and somewhat canine-like; eye large, anterior; adipose eyelid strongly developed behind; supraocular region with two ridges, extending to humeral region, the lower the stronger; posterior half of body, beginning at origin of soft dorsal, long and gently tapering to caudal peduncle; caudal peduncle much depressed, its least depth scarcely half its least width; distance from base of last dorsal ray to origin of caudal fin equal to snout and pupil; fins small; origin of spinous dorsal posterior to base of pectoral by a distance equal to eye; longest dorsal spine slightly greater than snout; anterior rays of soft dorsal somewhat produced, about 1.8 in head; anal similar to soft dorsal, its origin under eighth soft dorsal ray, anterior ray produced, but scarcely equaling longest soft dorsal rays; caudal widely forked, lobes apparently equal; pectoral long and falcate, reaching past origin of anal, exceeding head in length by 0.65 diameter of eye; ventrals short, 2.4 in head; scales rather large, a low sheath at base of soft dorsal and anal anteriorly; breast entirely scaled; lateral line strongly arched above pectoral, joining straight portion under sixth dorsal ray, chord of arched portion 1.6 in straight part.

Color in alcohol, rusty olivaceous above, paler on side below lateral line, belly white; top of head dark olive, side and lower jaw lighter, with strong brassy tinge on postocular and on lower portions of opercle; lower jaw profusely covered with fine brown points; a black spot at upper end of opercular opening; axil black; vertical fins all more or less dark; produced part of soft dorsal almost black, low part of soft dorsal black at base, then lighter, narrowly tipped with dark; anal dark brown, with a subterminal stripe of yellowish white along edge of fin; pectoral and ventrals pale.

The above description based upon the type, No. 50638, U. S. N. M. (field No. 04452), a large example 27 inches long, from Honolulu.

This species somewhat resembles *Caranguss marginatus*, from which it differs in the much more slender body, larger eye, and dark anal fin. The type is the only example obtained.

Family SERRANIDÆ.   The Sea-basses.

21. Pikea aurora Jordan & Evermann, new species.

Head 2.5 in length; depth 3; eye 5 in head; snout 4; interorbital 6.2; maxillary 2.25; D. VIII, 13; A. III, 8; scales 5–55–22; Br. 7; gillrakers short and rather weak, about 9 + 5.

Body moderately stout, the back slightly elevated, head rather long and pointed; snout depressed, the anterior profile nearly straight from tip of snout to occiput; mouth large, maxillary reaching posterior margin of pupil, supplemental bone not developed, the tip broad, 1.5 in orbit; mouth somewhat oblique, the lower jaw strongly projecting; teeth in broad villiform bands on jaws, vomer, and palatines; tongue naked; eye moderate, high up, chiefly above axis of body; anterior nostril in a short tube at edge of prenasal; posterior nostril small, round, near upper anterior edge of orbit; edge of preopercle slightly dentate, especially on lower arm; opercle ending in a broad flap with a weak, flat spine; pseudobranchiæ rather small; interorbital low, very little convex; caudal peduncle stout, compressed, and very deep, the depth equaling snout and eye; fins rather small; origin of dorsal posterior to that of pectoral, slightly nearer base of last ray than tip of snout; dorsal spines low and weak, the third longest, 3.6 in head; soft portion of dorsal somewhat elevated and pointed, with longest ray 1.9 in head; anal similar to soft dorsal but smaller and somewhat posterior, fifth ray 2 in head; caudal truncate or slightly lunate; ventrals short, not nearly reaching vent, their length 1.75 in head; pectoral rather long and slender, reaching origin of anal, its length about 1.4 in head; scales rather small, finely ciliate, somewhat loose; entire head, except interorbital, snout, and under parts, scaled; lateral line well developed, with a strong arch above the pectoral and distinctly decurved under last dorsal ray.

Color in life (field No. 03342), top of head, upper half of anterior part of body, and whole posterior half of body pale rosy; lower part of head, and lower parts of anterior half of body white with faint rosy wash; top of head and back in front of dorsal vermiculated with greenish yellow lines; middle portion of upper jaw yellow with a broad sulphur-yellow stripe from it to eye, then back of eye to opercular opening; a narrow sulphur stripe on posterior edge of maxillary and continued interruptedly downward and backward across cheek to opercle; a few small yellow spots across cheek between the two stripes; tip of lower jaw yellow; yellow of back in about 6 indefinite lines; dorsal pale rosy, spinous part greenish yellow at base, this extending toward tip posteriorly and forming a submarginal yellow stripe on soft part, narrowly bordered above by rosy; rest of fin rosy; caudal dark rosy, paler toward tip, then with blackish red edge, a greenish yellow stripe along upper and lower margins narrowly edged with rosy; anal yellow anteriorly, rest of fin pale rosy; pectoral and ventrals pale rosy; yellow of lower jaw bounded by rosy, rest of jaw and chin whitish; some examples with posterior half of side with scattered small greenish yellow spots, these extending on caudal; eye with a broad brown bar through the middle, white above and below.

Color in alcohol, pale yellowish white, lighter below; body, especially posteriorly, caudal, and soft dorsal fins with numerous small distinct brown spots; head pale, a white line extending along upper edge of maxillary and across cheek to opercular opening, a similar but less distinct white line from eye to upper edge of gill-opening; between these 2 a few white specks; all the fins except caudal and soft dorsal plain yellowish white.

Four specimens of this interesting and handsome species were obtained by us, 2 at Honolulu and 2 at Hilo. Four others are in the collection made at Honolulu in 1898 by Dr. Wood.

Field No.	Length.	Locality.	Final disposition of specimen.	Field No.	Length.	Locality.
	Inches.			O. P. J.	Inches.	
05232	6.2	Hilo	Type, No. 50675, U. S. N. M.	403	4.5	Honolulu.
05233	4.8	....do	Cotype, No. 3971, Field Col. Mus.	682	5.25	Do.
05231	5.75	Honolulu.	Cotype, No. 7184, L. S. Jr. Univ. Mus.	687	6	Do.
03342	5.2	....do	Cotype, No. 2734, U. S. F. C.	6074	6	Do.

22. Anthias kelloggi Jordan & Evermann, new species.

Head 2.5 in length; depth 2.5; eye 4.5 in head; snout 3.6; maxillary 2; interorbital 5.4; D. xi, 15; A. iii, 7; P. 15; scales 4–36–10; gillrakers 16 + 4.

Body short, deep, and compressed; dorsal outline greatly arched, profile from origin of spinous dorsal to tip of snout nearly straight, being gently concave over interorbital space; ventral outline nearly straight; caudal peduncle compressed, its greatest depth 3 in head; head longer than deep; snout bluntly pointed, lower jaw prominent, slightly the longer; mouth large, nearly horizontal; a narrow band of small, sharp, conic teeth on palatines, a small patch on vomer, a band of cardiform teeth on upper jaw, a narrower band in lower jaw; several large canine teeth in each jaw anteriorly, 3 of these close together on middle of each side of lower jaw, these hooked backward; 6 or 8 large pores on lower side of mandible and several on upper part of snout; maxillary reaching to posterior edge of orbit, its greatest width 1.5 in eye; edge of preopercle above angle and edge of opercle below the upper middle of base of pectoral denticulate; 2 broad opercular spines, the upper the larger; eye anterior, its lower edge on line with upper base of pectoral; fins large, the second soft dorsal ray and upper rays of upper caudal lobe being produced each as a filament, the dorsal filament being produced half its length beyond rest of fin; dorsal spines stout and strong, the first spine 2.3 in third, the fifth being the highest, 2.5 in head; base of spinous dorsal 1.15 in head; base of soft dorsal 2.3 in head, its fourth ray 3.5 in head, the last ray 1.4 in fourth; caudal truncate, the lower rays produced slightly as a filament, but not nearly so long as the upper lobe; second anal spine longest, 2.5 in head; second soft ray longest, 2 in head; pectoral very long and large, reaching to origin of soft anal, the eighth and ninth rays from the top the longest, 1.4 in head; scales large, finely ctenoid, in regular series; entire body and head scaled; basal portion of all fins except spinous dorsal with small scales; lateral line strongly convex, not concurrent with the dorsal profile, becoming straight on middle of caudal peduncle; one row of scales behind tip of last dorsal ray.

Color in alcohol, pale brown, the fins lighter; in life, red.

Only three specimens of this species were obtained, all having been taken with the hook in deep water off Kailua, in southwestern Hawaii. It is allied to *Anthias japonicus* Steindachner & Doderlein.

Named for Dr. Vernon Lyman Kellogg, professor of entomology in Stanford University.

Field No.	Length.	Locality.	Final disposition of specimen.	
	Inches.			
05278	7.75	Off Kailua	Type. No. 50642, U. S. N. M.	
03703	8	......do....	Cotype, No. 7400, L. S. Jr. Univ. Mus.	
05277	8.5	......do....	Cotype, No. 2711, U. S. F. C.	

Family APOGONIDÆ. The King of the Mullets.

23. Apogonichthys waikiki Jordan & Evermann, new species.

Head 2.4 in length; depth 3; eye 3.2 in head; snout 4.6; interorbital 6; maxillary 2; D. vii–i, 8; A. ii, 7; scales 2–24–5.

Body short, stout, and compressed; dorsal outline strongly arched from tip of snout to posterior base of soft dorsal; ventral outline comparatively straight from tip of mandible to origin of anal; vent immediately in front of origin of anal; caudal peduncle deep and compressed; head rather large; mouth large, slightly oblique, jaws equal, maxillary reaching posterior edge of pupil; eye rather small, slightly above axis of body; interorbital space narrow, little convex; opercular and preorbital bones entire; a band of small villiform teeth in each jaw, and on vomer and palatines; fins moderate, origin of spinous dorsal nearer base of last soft ray than tip of snout; first dorsal spine very short, second about half length of third, which is equal to eye and snout; base of soft dorsal equal to depth of caudal peduncle; longest dorsal rays 2.25 in head; caudal rounded, its length 1.75 in head; origin of anal slightly posterior to that of soft dorsal, its longest rays 2.4 in head; pectoral slender, reaching past origin of anal, its length 1.5 in head; ventrals short, barely reaching origin of anal, their length nearly 2 in head; scales large, weakly ctenoid, firm and somewhat deeper than long; lateral line strongly developed, following outline of back until under last dorsal ray, where it curves downward, following middle line of caudal peduncle to base of caudal fin.

Color in alcohol, head and body rather dark brownish, a lighter crossband around body at nape and across opercles; another light band surrounding body between the 2 dorsal fins; 3 dark brown lines radiating from the eye, the first downward across cheek to tip of maxillary, the second backward across cheek toward base of pectoral, the third upward and backward to origin of lateral line; spinous dorsal blackish, especially on last spine; soft dorsal, anal, and caudal dusky, narrowly edged with white; pectoral pale, crossed by about 6 obscure brownish crossbars; ventrals black or very dark brown, the outer rays somewhat paler.

The above description is based upon the type, No. 50639, U. S. N. M. (field No. 20), a specimen 1.5 inches long, obtained from the coral rocks in front of Waikiki, near Honolulu, August 22, 1901.

This species is distinctly related to *A. alutus* of the coast of Florida, from which it differs markedly in color and in the more slender body. Only one specimen was obtained.

Genus FOWLERIA Jordan & Evermann, new genus.

Fowleria Jordan & Evermann, new genus of *Apogonidæ* (*aurita*).

This genus differs from *Apogonichthys* only in the character of the lateral line, which is developed only on the anterior part of the body.

Several species occur in crevices of coral rock in the South Seas. All of them are of very small size and some are brightly colored.

This genus is named for Mr. Henry Weed Fowler, of the Academy of Natural Sciences of Philadelphia.

24. Apogon snyderi Jordan & Evermann, new species.

Head 2.7 in length; depth 3.1; eye 3.7 in head; snout 3.7; interorbital 4.5; maxillary 2.2; mandible 2; gape 3; D. vii–i, 9; A. ii, 8; C. 17; P. 10; scales 2–25–5; Br. 6.

Body short and stout, moderately compressed, the dorsal and ventral outlines about equally curved; head rather large, conic; snout conic, the anterior profile very slightly curved from tip of snout to origin of spinous dorsal; mouth oblique, jaws subequal, the lower slightly included; maxillary long, reaching not quite to posterior edge of pupil, its width at tip 2 in eye, supplemental bone well developed; interorbital space rather broad, slightly convex, preorbital narrow, least width 3 in eye; teeth on vomer and jaws, the latter in villiform bands; none on palatines; gillrakers slender, 10 on lower limb of first arch; caudal peduncle compressed and deep, the least width about 4 in its depth; scales large, deep, closely imbricated, strongly ctenoid and loose; lateral line beginning at upper end of gill-opening, nearly straight to base of caudal fin, 4 scales in front of spinous dorsal; nape with a striated shield; edge of opercle thin and smooth; both margins of preopercle and edge of interopercle serrate, teeth strongest at angles; a series of moderately strong teeth along lower edge of orbit; origin of spinous dorsal nearer snout than base of last dorsal ray; first dorsal spine very short, fourth longest, about 2 in head, second 2 in the fourth, seventh 2 in second; first soft rays longest, 1.8 in head; caudal deeply emarginate, longest rays about 1.6 in head; anal similar to soft dorsal, somewhat smaller, its origin under last rays of soft dorsal; ventrals pointed, scarcely reaching vent, 1.9 in head; pectoral reaching vertical of vent, 1.7 in head.

Color in alcohol, pale yellowish brown, darkest above; a darker brownish band extending from upper edge of opercle along side, just above lateral line, to posterior edge of soft dorsal; another broader, more distinct brown band from tip of snout through eye and along middle of side to base of caudal fin, covering lateral line on caudal peduncle; caudal peduncle at base of caudal fin with a broad dusky crossbar, usually darkest on upper half, sometimes obscure, sometimes with a darker blotch or spot in the upper portion; upper parts of head covered with fine dark brown punctulations; lower jaw similar, but somewhat paler; membranes of anterior 2 or 3 dorsal spines black, others finely punctulate; soft dorsal pale at base, above which is a broad indistinct dark crossband, the color confined chiefly to the interorbital membranes, this color extending to near tip of last rays; outer part of soft dorsal pale; anal similar to soft dorsal, the black bar narrower and nearer base of fin, rest of fin white; caudal dusky on membranes of outer 1 or 2 rays, the fin otherwise white, with a few fine punctulations on the interradial membranes; ventrals pale; distal parts of the first and second rays and their connecting membrane black; axil and base of pectoral somewhat dusky.

Color in life (field No. 198, O. P. J.), pale red; 2 longitudinal pearly lines on body; first dorsal with a dusky olivaceous anterior border; white lines along fourth, fifth, sixth, and seventh spines, the membrane olivaceous; second dorsal with many white and some olivaceous spots; anal with a dusky line along base, the distal part red; base of caudal dusky, rest of fin pale red; ventral with a white spot near tip; pectoral pink; iris yellow.

Another example (field No. 03499) was coppery brown when fresh, with trace of dusky band along side; a faint black bar at base of caudal, forming a black spot above end of lateral line; some dusky on opercle; first dorsal dusky; second dorsal brownish red with some dark; anal same with a basal flesh-colored bar below it; caudal reddish brown; ventrals same, with first ray pinkish and dusky behind it; some dusky on opercle.

This species reaches a length of about 6 inches. It was obtained by Garrett in the Hawaiian, Society, and Paumotu islands. Our collections contain numerous specimens from Honolulu and Hilo. We have examined also 12 specimens in the collection made by Dr. O. P. Jenkins.

This species closely resembles *Apogon menesemus*, from which it differs chiefly in coloration; the black caudal crescent, which is such an excellent distinguishing mark in *A. menesemus*, is wholly absent in this species; moreover, the 2 silvery lateral bands, which become dark brown in spirits, are not found in *A. menesemus*; and the black on the anal and soft dorsal is less conspicuous in *A. snyderi*. It belongs to the subgenus *Pristiapogon* of Klunzinger, having both limbs of the preopercle serrate.

This species is figured by Bleeker, Day, and Günther, the figures of Bleeker and Günther being colored. The best figure is that of Günther in Fische der Südsee, who calls it *Apogon frenatus*, but the species originally thus named seems to be quite different, as Bleeker has already noticed.

Named for Mr. John O. Snyder, assistant professor of zoology in Stanford University.

Our collection contains the following specimens of this species:

Field No.	Length.	Locality.	Final disposition of specimen.	Field No.	Length.	Locality.	Final disposition of specimen.
	Inches.				*Inches.*		
03072	5.25	Honolulu.	Type, No. 50640. U. S. N. M.	13	3.4	Honolulu.	
03065	3.8	.....do....	Cotype. No. 50641, U. S. N. M.	14	3	.....do....	
03067	4.8	.....do....	Cotype, No. 7459. L. S. Jr Univ. Mus.	21	2	Hilo......	
02941	4.5	.....do....	Do.	16	3.25	Honolulu.	Cotype. No. 9807. Mus. Ind. Univ.
03079	4.4	.....do....	Cotype, No. 2709, U. S. F. C.	05145	4.8	Hilo	Cotype. No. 9808, Mus. Ind. Univ.
03215	4.6	.....do....	Cotype, No. 2710, U. S. F. C.	05146	4.8	.....do....	Cotype. No. 3956, Field Col. Mus.
03216	4.5	.....do....	Cotype, No. ——, M. C. Z.	05147	4.6	.....do....	Cotype. No 3957, Field Col. Mus.
03217	4.25	.....do....	Cotype, No. ——, M. C. Z.	05148	4.75	.....do....	Cotype. No. 1493, Cal. Ac. Sci.
03218	4.8	.....do....	Cotype, No. 2288, Am. Mus. Nat. Hist.	05149	4.4	.....do....	Cotype. No. 1494, Cal. Ac. Sci.
03219	4.75	.....do....	Cotype, No. 2289, Am. Mus. Nat. Hist.	05150	4.75	.....do....	Cotype. No. 1694, Bishop Mus.
02499	5	.....do....	Cotype, No. 24217, Ac. Nat. Sci. Phila.	05151	4.5	.....do....	Cotype. No. 1695, Bishop Mus.
1	4.8	.....do....	Cotype, No. 24218, Ac. Nat. Sci. Phila.	O. P. J. 198	4.75	Honolulu.	
2	4.75	.....do....		688	5	.....do....	
3	4.75	.....do....		411	5	.....do....	
4	4.5	.....do....		409	5.4	.....do....	
5	4.5	.....do....		406	4.75	.....do....	
6	4.25	.....do....		408	4.5	.....do....	
7	4.5	.....do....		417	4.4	.....do....	
8	4	.....do....		a 428	4.25	.....do....	
9	4	.....do....		a 680	4.25	.....do....	
10	3.5	.....do....		b 430	4	.....do....	
11	3.8	.....do....		b 431	3.75	.....do....	
12	3.5	.....do....		b 673	5.5	.....do....	

a Dr. Wood.   b Jordan & Snyder.

Apogon frenatus, Günther, Fische der Südsee, I, 19, taf. 19, fig. A, 1873 (Hawaiian, Society, and Paumotu islands); Steindachner, Denks. Ak. Wiss. Wien, LXX, 1900, 484 (Honolulu), not *Apogon frenatus* Valenciennes, Nouv Ann. Mus. Hist. Nat. 1832, 57, pl. 4, fig. 4, nor of Klunzinger

Family PRIACANTHIDÆ. The Catalufas.

25. Priacanthus alalaua Jordan & Evermann, new species. "*Alalaua.*"

Head 3.2 in length; depth 2.65; eye 2.4 in head; snout 3.6; maxillary 2; interorbital 3.8; D. x, 14; A. III, 15; scales 13–85 to 90–45, 70 pores; Br. 6; gillrakers, about 22 on lower arm.

Body short, deep, compressed, ovate; upper profile of head nearly straight; snout very blunt; mandible prominent, produced; mouth very oblique; teeth small, sharp, in bands on jaws, vomer and palatines; tongue rounded, free in front; maxillary reaching almost to front margin of pupil, its greatest width 2 in eye; edge of preopercle finely serrate, with a sharp, flat, serrated spine directed backward at angle; margins of interopercle, subopercle, and opercle entire; opercle with an obscure flat spine; interorbital space slightly convex; eye very large, its lower edge a little above base of pectoral and in line with axis of body; nostrils small, close together, the anterior with elevated rim; posterior nostril oblong, with broad flap; gillrakers rather slender, about 22 on longer arm of first arch, longest about 3 in eye; origin of spinous dorsal over upper base of pectoral; dorsal spines rather uniform, the longest about equal to orbit; soft portion of dorsal somewhat elevated, rounded, fourth ray 1.7 in head; anal spines rather stronger than those of dorsal, third the longest, 1.1 in orbit; soft portion of anal similar to that of soft dorsal, rays of about equal length; caudal truncate, the middle rays slightly greater than orbit; pectoral short, bluntly pointed, not reaching tip of ventral, length 1.4 in head; ventrals longer, just reaching base of second anal spine, their length 1.2 in head; ventral spine about 1.25 in longest ray, or 1.7 in head; scales small, firm and rugose, those of lateral line somewhat enlarged; entire head, as well as body, densely scaled; lateral line rising abruptly for 6 or 7 pores from gill-opening, thence concurrent with back to caudal peduncle.

Color in life, silvery, light olive above, somewhat flushed with red in irregular blotches; chin red; spinous dorsal olive-yellowish, especially on edge; ventrals black, rays whitish; fins unspotted.

Young of 4 inches in length are dirty gray, browner above, with no trace of red in life; some brown spots along lateral line; fins dusky, anal and ventral darkest; iris a little brownish red.

Color in alcohol, plain yellowish-white: spinous dorsal and anal somewhat dusky; ventral membranes black, the rays white, other fins pale yellowish-white. In some examples the color is much more flushed with red, especially above; the red paler and more evanescent than in the other species; fins red, unspotted; the spinous dorsal edged with golden; upper lip golden; ventral membrane black, pectoral pale.

There seems to be but little variation in this species; the younger individuals appear to be more brightly colored or with more evident wash of red than was shown in the type.

We have examined the following specimens:

Field No.	Length.	Locality.	Final disposition of specimen.
	Inches.		
01170	8.25	Honolulu	Type, No. 50643, U. S. N. M.
03395	7	......do......	Cotype, No. 7461, L. S. Jr. Univ. Mus.
03420	6	......do......	Cotype, No. 2712, U. S. F. C.

Family LUTIANIDÆ.   The Snappers.

Genus BOWERSIA Jordan & Evermann, new genus.

Bowersia Jordan & Evermann, new genus of *Lutianidæ* (*violescens*).

Body long, rather slender and moderately compressed; top of head evenly rounded, the supraoccipital crest extending forward on cranium; jaws equal, lower not projecting; bands of villiform teeth on both jaws, the outer series somewhat enlarged and canine-like; villiform teeth on vomer, palatines, and tongue; maxillary slipping for its entire length under the rather broad preorbital; eye large; opercle entire, ending in 2 flat, obscure spines, the space between them deeply emarginate, but filled by soft membrane; preopercle scarcely dentate; dorsal fin continuous, the last ray produced nearly twice length of preceding one.

This genus is related to *Apsilus*, with which it agrees in the presence of villiform teeth on the vomer and palatines, but from which it differs in having well-developed teeth on the tongue, and in the produced last dorsal and anal ray. Two species are known.

We take much pleasure in naming this new genus for the Hon. George M. Bowers, United States Commissioner of Fish and Fisheries, in recognition of his active and intelligent interest in promoting scientific work, especially the investigation of the aquatic resources of the Hawaiian Islands.

  a. Scales rather large, about 60 in lateral line; preorbital broad, 7.75 in head .... *violescens.*
  aa. Scales smaller, about 68 in lateral line; preorbital narrow, 10 in head .. .. *ulaula.*

26. Bowersia violescens Jordan & Evermann, new species. "*Opakapaka.*"

Head 3.25 in length; depth 3.5; eye 4.4 in head; snout 3; maxillary 2.6; mandible 2; interorbital 3; preorbital 7.75; scales 8-60-15; D. x, 10; A. iii, 8; Br. 7; gillrakers 5+14.

Body long, rather slender, moderately compressed, tapering gradually into the rather long caudal peduncle; head large, longer than deep; snout moderate, rather bluntly conic; mouth large, maxillary reaching anterior third of pupil, slipping for its entire length under the thin edge of the rather broad preorbital, the width of its tip 2 in eye; mandible strong, but not projecting; broad bands of villiform teeth on jaws, vomer, palatines, and tongue, the outer series in the jaws slightly enlarged and canine-like; eye large, its lower edge in line with axis of body; interorbital broad, gently convex; anterior profile but slightly curved from tip of snout to nape, thence more strongly arched to origin of dorsal, descending in a long, low curve to caudal peduncle; ventral outline but slightly convex; caudal peduncle rather long, 2 in head, its least width about 1.6 of its least depth, which is 1.8 in its length, measured from base of last dorsal ray to base of supporting caudal rays; gillrakers few, rather strong and short, the longest about 2.6 in eye; opercle smooth, ending in 2 flat, obscure spines (more strongly developed in each of the cotypes); preopercle obscurely serrate at the angle (more distinctly so in the cotypes); fins moderately developed, the dorsal fin continuous, without notch, its origin over base of pectoral and equally distant from tip of snout and base of fourth ray, length of entire base of fin and to tip of last ray twice length of head; first dorsal spine moderately short, closely bound to the second, whose length exceeds it by about one-half; seventh dorsal spine longest, its length equal to that of snout; last dorsal ray produced, its length about 1.7 times that of the preceding; anal similar to soft dorsal, its origin under base of third or fourth dorsal ray; first anal spine very short, third longest and strongest, its length equaling diameter of eye; last anal ray produced, its length equaling that of produced dorsal ray; caudal rather widely forked, lobes about equal, their length, measured from base of first supporting ray, equaling head; ventrals pointed, their tips not reaching vent, length 1.4 in head; pectoral long, slightly falcate, the tip about reaching tips of ventrals, its length about 1.2 in head; scales large, deeper than long and rather loose; cheek and opercles scaled, 5 rows on cheek; a large bony humeral scale, from which extends to nape a series of somewhat modified scales, in front of which is a patch of ordinary scales; lateral line complete and well developed, beginning at lower edge of humeral scale and following curvature of back to base of middle caudal rays; the pores little or not at all branched.

Color in life (field No. 03404), light rosy olive, with violet shades, pale below; center of each scale of back shining violet; dorsal reddish flesh-color, its base anteriorly yellowish olive; caudal flesh-color, rosy along the edges; anal similar, its edge light lavender gray; ventrals pale, shaded with light orange; pectoral flesh-color, violaceous at base; snout violet, iris light yellow. A flesh-colored violaceous fish without color markings anywhere. Another specimen (field No. 03417) freshly dead, had the body, head, and caudal light rosy; ventrals white; outer margin of spinous dorsal golden, the membranes with irregular golden areas; pectoral and anal not distinctly colored; iris yellow.

Color in alcohol of type (field No. 03018), above dusky silvery, bases of scales brown; sides and under parts silvery, with pale greenish-yellow tinge; top of head somewhat olivaceous, sides rusty silvery; axil of pectoral dusky; fins all pale or yellowish-white.

This species reaches a length of about 2 feet and is an important food-fish.

Only 4 specimens were secured:

Field No.	Length.	Locality.	Final disposition of specimen.
	Inches.		
03018	24	Honolulu..........	Type, No. 50690, U. S. N. M.
05040	....	....do	Cotype, No. 7473, L. S. Jr. Univ. Mus.
03404	....	....do	Cotype, No. 2721, U. S. F. C.
03417	....	....do	Cotype, No. 8813, Ind. Univ. Mus.

27. Bowersia ulaula Jordan & Evermann, new species. "*Ulaula.*"

Head 3.6 in length; depth 3.8; eye 3.8 in head; snout 3.8; maxillary 2.9; mandible 2.4; interorbital 10; preorbital 10; scales 8-68-14; D. x, 11; A. iii, 8; Br. 7; gillrakers 21+5.

Body long and slender, the dorsal outline in a low, gentle curve from tip of snout to base of caudal, the ventral outline but gently convex; head moderate, bluntly conic; snout rather short; mouth moderate, somewhat oblique, the jaws equal; maxillary moderate, slipping for its entire length under the narrow, thin preorbital, its width at tip 2.8 in eye; bands of villiform teeth on vomer, palatines, tongue and jaws, those of outer series in the latter scarcely enlarged; opercle ending in 2 obscure, flat spines, the space between them deeply emarginate but filled by membrane; preopercle rather distinctly serrate, the teeth very short; eye rather large, its lower border in line with axis of body; preorbital very narrow, much narrower than in *B. violescens;* interorbital space narrower than in the preceding species, slightly convex; caudal peduncle long, its length from base of last dorsal ray to first supporting rays of caudal 1.7 in head, its least width about 2.1 in its least depth, which is 2.1 in its length; gillrakers rather numerous, close-set, the longest about 2.2 in eye; fins moderately developed, the dorsal continuous, without notch, its origin slightly behind base of pectoral and equally distant between tip of snout to base of fifth or sixth dorsal ray; head 2 in distance from origin of anal to middle of last dorsal ray; first dorsal spine rather short, about 1.9 in length of second; fifth dorsal spine longest, its length equal to distance from tip of snout to pupil; last dorsal ray produced, its length about 1.8 times that of the preceding; anal similar to soft dorsal, its origin under base of third dorsal ray; first anal spine very short, the third longest, its length 1.2 in diameter of eye; soft anal similar to soft dorsal, the last ray produced and of equal length with that of dorsal; caudal densely scaled and widely forked, lobes equal, their length, measured from base of first supporting rays equaling that of head; ventrals not pointed, their tips not reaching vent, their length 1.6 in head; pectoral long, slightly falcate, its tip reaching vent and much beyond that of ventral, its length equaling that of head; scales rather small, closely imbricated, deeper than long, their edges finely ciliated; cheek and opercles scaled, 6 rows on cheek; a large bony humeral scale from which extends a series of modified scales to nape, and in front of which is a patch of ordinary scales; lateral line complete and well developed, beginning at lower edge of humeral scale and following contour of back to base of middle caudal rays, the tubes little branched.

Color in alcohol, brownish or purplish olivaceous above, paler on side; under parts nearly plain white; each scale of back and upper part of side with a darker brown spot, these forming indistinct rows, about 6 above lateral line; side below lateral line with less distinct horizontal lines; upper parts of head olivaceous brown, lower parts paler, spines of dorsal fin purplish, the membranes white, purplish at tips; soft dorsal with rays whitish, membranes purplish; caudal slightly dusky, other fins plain whitish.

This species is related to *B. violescens,* from which it differs chiefly in the shorter snout, larger eye, shorter maxillary, shorter mandible, narrower interorbital space, decidedly smaller scales, more numerous gillrakers, and more posterior insertion of dorsal fin. Only one specimen known, type No. 50661, U. S. N. M. (field No. 04104). 14.25 inches long, from Hilo, Hawaii Island.

28. Etelis evurus Jordan & Evermann, new species.

Head 3.2 in length; depth 3.6; eye 3 in head; snout 3.9; maxillary 2.2; interorbital 3.6; D. x, 11; A. iii, 8; scales 5–50–11; Br. 6; gillrakers 15 + 6, longest about 2 in eye.

Body rather long, tapering, moderately compressed; dorsal outline slightly convex, ventral outline nearly straight; head considerably longer than deep, compressed, subconic; snout bluntly pointed, less than eye, equal to portion of eye anterior to posterior edge of pupil; mouth large, oblique; small bands of villiform teeth on vomer, palatines, and anterior part of each jaw; a single row of small, wide-set, slender canine teeth on the outer edge of each jaw, those in upper jaw slightly larger and more wide-set; a single larger canine tooth on the side of each jaw in front, those in the upper jaw the larger; maxillary extending to middle of pupil; eye very large, its lower edge slightly below axis of body; preopercle finely serrate; opercle with 2 broad, flat spines, not produced, the upper rather obscure; fins moderately developed; origin of spinous dorsal slightly posterior to base of pectoral, its distance from tip of snout equaling that to base of sixth dorsal ray; dorsal fin deeply notched, almost divided; first dorsal spine short, its length but slightly greater than diameter of pupil; third dorsal spine longest, 2.1 in head; ninth spine short, its length 2.75 in third; soft dorsal not elevated, the rays about equal, the last 1.75 in third spine; anal similar to soft dorsal, the first spine very short, the third about 1.8 in third dorsal spine, last anal ray about equal to last dorsal ray; caudal deeply notched, the lobes much produced, the upper the longer, its rays greatly exceeding length of head, or about 2.4 in body; ventrals long, but not reaching vent by a distance equaling half diameter of pupil, their length 1.5 in head; pectoral long, reaching vent, the upper rays somewhat produced, their length 1.2 in head; scales moderate, firm, covering body, nape, opercles, and breast; a large humeral scale; lateral line beginning at lower edge of humeral scale and following contour of back to base of caudal fin.

Color in life, of a specimen (field No. 03481) 14 inches long, brilliant rose-red, the side from level of eye abruptly silver, with rosy shades; snout, jaws, eye, and inside of mouth red; fins all rose-color, the dorsal and caudal bright; ventrals and anal pale, the former washed with red on center; axil pale pink; pectoral pale rosy.

Color in alcohol, uniform yellowish white, paler below; fins all pale yellowish white, the caudal lobes somewhat dark.

This species is related to *Etelis oculatus* of the West Indies, from which it differs in the somewhat larger scales, much longer caudal lobes (9.5 times length of middle rays instead of 4 times, as in *E. oculatus*), and larger eye. From *E. carbunculus* Cuvier & Valenciennes, from Isle de France, it seems to differ in having only 16 instead of 20 scales in a transverse series and in the coloration.

This species, one of the handsomest of all Hawaiian fishes, is thus far known only from Hilo, Hawaii, in the market of which we obtained 13 fine examples, measurements of which are given in the following table:

Field No.	Length.	Locality.	Final disposition of specimen.	Field No.	Length.	Locality.	Final disposition of specimen.
	Inches.				Inches.		
03482	12.5	Hilo	Type, No. 50662, U. S. N. M.	04086	13	Hilo	Cotype, No. 24224, Ac. Nat. Sci., Phila.
04082	13.25	...do	Cotype, No. 2722, U. S. F. C.	04087	12.75	...do	Cotype, No. 3968, Field Col. Mus.
03481	14	...do	Cotype, No. 50663, U. S. N. M.				
04080	15	...do	Cotype, No. 7174, L. S. Jr. Univ. Mus.	05287	13	...do	Cotype, No. 9814, Ind. Univ. Mus.
04081	15.25	...do	Do.				
04083	15.5	...do	Cotype, No.1701, Bishop Mus.	05288	11.6	...do	Cotype, No.1498, Cal. Ac. Sci.
04084	12.75	...do	Cotype, No. ——, M. C. Z.	05289	12 6	...do	
04085	16.5	...do	Cotype, No. 2296, Am. Mus. Nat. Hist.				

Family KYPHOSIDÆ. The Rudder-fishes.

29. Sectator azureus Jordan & Evermann, new species.

Head 4 in length; depth 3; eye 5 in head; snout 3.65; maxillary 4; interorbital 2.4; D. xi, 15; A. iii, 13; scales 14–81–20.

Body elongate, ovoid, greatest depth about at tip of pectoral; head slightly longer than deep, compressed; snout very bluntly convex; jaws about equal, maxillary not reaching front of eye; mouth small, horizontal; teeth very small, compressed, in a single series in each jaw; minute villiform teeth on vomer, palatines and tongue; tongue broad, rounded and free in front; preopercle entire, posterior edge very oblique; lower edge of eye on a line with upper base of pectoral, posterior margin well in front of middle of head; interorbital broad, strongly convex, a deep groove in front of eye to nostril; caudal peduncle rather long, 1.9 in head; origin of spinous dorsal slightly in front of base of ventrals, well behind pectoral, its distance from tip of snout slightly greater than depth of body; longest dorsal spine 3 in head, last dorsal ray elongate, being one-fourth longer than other rays, its length 3.4 in head; third anal spine longest, 4.9 in head; first anal ray longest, 3.4 in head; base of anal 1.8 in base of dorsal; caudal deeply forked, lower lobe the longer, 3.5 in body; pectoral short, slightly longer than ventrals, 1.8 in head, the spine more than half length of longest ray; scales cycloid, present on head except on jaws and in front of eye, very minute on all the fins except ventrals; lateral line concurrent with dorsal outline; peritoneum dark gray.

Color in life, dark steel-blue, becoming paler below; a definite deep blue stripe from snout below eye widening on opercle, and thence straight to center of base of caudal; below it a narrow bright golden stripe from angle of mouth to lower part of caudal, and then a fainter blue stripe below this; a blue stripe from eye to upper part of gill-opening, interspace shaded with green; a deep blue stripe from upper part of eye along each side of back to base of upper caudal lobe; upper fins dusky golden or olivaceous; ventrals yellow; anal and lower lobe of caudal dirty golden; pectoral translucent.

Color in alcohol, deep steel gray, brown above, each scale with a very pale spot, the edge pale; lower surface whitish silvery; a pale streak of gray behind eye to edge of opercle; dorsal fin gray brown like the back; caudal and pectoral whitish; inside of ventrals dusky orange; ventrals and anal dusky; inside of pectoral blackish brown.

Type, No. 50864, U. S. N. M. (field No. 03363), a specimen 15.25 inches long, taken off the shore near Heeia, Oahu Island.

This species must be very rare, being unknown to the fishermen and only the single specimen having been obtained by us.

Family MULLIDÆ. The Surmullets.

30. Mulloides flammeus Jordan & Evermann, new species.

Head 3.6 in length; depth 4; eye 4.3 in head; snout 2.25; interorbital 3.5; maxillary 2.6; mandible 2.1; shortest distance from eye to upper edge of maxillary 1 in eye; D. vii-9, longest dorsal spine 1.75 in head, longest dorsal ray 2.6; A. 7, longest ray 2.7; scales 3–41–6; pectoral 1.5; ventral 1.4.

Body oblong, not much compressed; head heavy, broad, the interorbital space broad and slightly convex; snout rather long and pointed, not abruptly decurved; mouth rather large, somewhat oblique, the lower jaw but slightly included; maxillary broad, slipping for most of its length under the thin preorbital, its tip not reaching orbit by diameter of pupil; eye rather large, high, slightly posterior; gillrakers 18 + 7, the longest about 2 in eye, serrate; opercular spine obscure in adult, more plainly developed in the young; origin of dorsal a little nearer posterior base of soft dorsal than tip of snout; distance between dorsals considerably less than snout, about 2.6 in head; anal similar to soft dorsal, its origin somewhat more posterior; ventrals rather long, reaching slightly beyond tip of pectoral; caudal deeply forked, the lobes equal, about 1.2 in head.

Color in life (field No. 03459), bright rose-red, with 5 broad crossbands of darker clear rose, which vanishes very soon after death; a very faint yellow lateral streak, with yellow shades on scales below; lower side of head rose, snout and lips very red; 2 wavy golden streaks from below eye to angle of mouth, lower conspicuous; first dorsal clear red; second dorsal deep red on the lower half, fading above; caudal deep red at base, fading outward; anal pink, pectoral light yellow; ventral creamy red; barbels red, paler toward tip; iris silvery.

A color note on specimens, field Nos. 03054 and 03055, says that they were rosy in life.

Color in alcohol, pale dirty olivaceous above, yellowish white on sides and belly; head-yellowish olive above, pale on cheek and below; a yellowish band from snout under eye; fins all colorless, the spinous dorsal slightly dusky, all with slight yellowish tinge; ventrals with the middle membranes blackish. Smaller examples show considerable rosy on the sides, indicating that the fish in life was probably red or rosy in color.

This species somewhat resembles *Mulloides auriflamma*, from which it differs in the smaller eye, larger, more oblique mouth, longer maxillary, the longer, less decurved, more pointed snout, and fewer gillrakers. It bears some resemblance to *M. pflugeri*, but has the eye larger and the snout longer and more pointed. Compared with *M. samoensis*, it has a much larger and more oblique mouth, and a considerably longer maxillary, as well as a different coloration. It does not agree with any of the plates of Day, Günther, or Bleeker, nor with any current descriptions. In life its banded coloration gives it a very handsome appearance. It is found in deeper water than most of the other species.

M. flammeus seems to be fairly abundant and is represented in our collections by the following 9 specimens, the first of which is taken as the type and all the others as cotypes:

Field No.	Length.	Locality.	Final disposition of specimen.	Field No.	Length.	Locality.	Final disposition of specimen.
	Inches.				Inches.		
03740	9.5	Kailua ...	Type, No. 50665, U. S. N. M.	03995	11.25	Hilo	Cotype, No. 24225, Ac. Nat. Sci. Phila.
03029	6	Honolulu.	Cotype, No.1702, Bishop Mus.				
03054	6.5	.....do	Cotype, No. 7475, L. S. Jr. Univ. Mus.	03934	9	.....do	Cotype, No. 9815, Ind. Univ. Mus.
03055	6	.....do	Cotype, No. ——, M. C. Z.	03937	12	.....do	Cotype, No. 3853, Field Col. Mus.
03778	9.75	.....do	Cotype, No.2723, U. S. F. C.				
03459	10	Hilo	Cotype, No. 2297, Am. Mus. Nat. Hist.				

31. Pseudupeneus chrysonemus Jordan & Evermann, new species.

Head 2.8 in length; depth 3.4; eye 5.3 in head; snout 1.7; interorbital 3.5; maxillary 2.3; D. viii-9; A. i, 7; scales 3–30–7.

Body slender, not greatly compressed, the back gently and rather uniformly elevated from tip of snout to dorsal; ventral outline slightly convex; head moderate; snout long; bluntly pointed; mouth moderate, slightly oblique, the lower jaw included; maxillary broad at tip, falling short of vertical of orbit by diameter of pupil; interorbital space convex; eye small, in posterior half of head; teeth rather large, in a single band in each jaw; barbels long, 1.2 in head, reaching nearly to base of ventrals; opercular spine small; fins rather large; third dorsal spine longest, 1.5 in head, or equal to distance from tip of snout to middle of pupil, third ray longest, 3.2 in head; base of spinous dorsal 1.4 in third spine; base of soft dorsal 1.4 in longest spine; origin of spinous dorsal nearer last dorsal ray than tip of snout by longitudinal diameter of pupil; distance between dorsals 1.5 in eye; length of caudal peduncle 1.5 in head; pectoral long, pointed, slightly falcate, 1.4; ventrals slightly longer, 1.3; last anal ray 2.9, equal to base of fin; caudal shallowly forked, lobes 1.3 in head, middle rays 2.75 in upper lobe; scales finely ctenoid and obscurely dendritic; lateral line concurrent with the back, the pores with few branches, the number usually not exceeding 5 or 6; 2 scales between the dorsals, 8 on dorsal side of caudal peduncle; peritoneum somewhat silvery.

Color when fresh, deep scarlet red, especially a shade from snout through eye toward tail; first dorsal plain scarlet, second paler golden with oblique stripes of scarlet and yellow edge; caudal orange; reddish at base, yellowish at tip; anal like second dorsal; pectoral pale orange; ventrals deep red; barbels bright yellow; iris red. In life, a pale streak backward from eye to middle of side parallel with back; side with 2 blotches of deep red; a row of dark spots along bases of both dorsals; young of 3 inches, from the rock pools, in life, dark olive green above with a dark olive streak along lateral line and 3 dark shades under first dorsal, second dorsal, and back of caudal peduncle; tip of first dorsal cherry-red, edged with white; second dorsal and caudal translucent, scarcely reddish; ventrals and anal bright cherry-red, former mesially dusky; barbels golden.

Color in alcohol, pale yellowish; each scale below dorsal with brownish edgings, generally most distinct in young and often entirely disappearing with age; a series of smaller obscure spots along median line from opercle to tip of pectoral; sides and under parts with faint traces of rosy.

This species may be known by the series of dusky blotches along each side of the dorsal fin and by the simple structure of the lateral line. In life it is at once known by its golden barbels.

The above description based upon a specimen (field No. 03929), 8 inches long, obtained at Honolulu, in 1898, by Dr. Wood. We have examined 4 other specimens of approximately the same size obtained at the same time, and numerous examples collected by us at Honolulu and Hilo.

The following is our list of specimens of this species:

Field No.	Length.	Locality.	Final disposition of specimen.	Field No.	Length.	Locality.	Final disposition of specimen.
	Inches.				*Inches.*		
03476	6.75	Hilo	Type. No.50666, U. S. N. M.	03989	8.25	Hilo	Cotype, No. 2298, Am. Mus. Nat. Hist.
03920	4.5	Honolulu		03990	7.5	do	Cotype, No. 24226, Ac. Nat. Sci. Phila.
03928	7	Hilo		03991	8	do	Cotype, No. 9846, Ind. Univ. Mus.
03930	8.5	do		03992	8	do	Cotype, No. 3955, Field Col. Mus.
03931	6.5	do		03993	8.5	do	Cotype, No. 1499. Cal. Ac. Sci.
03932	7	do		03994	7.75	do	Cotype, No. 50676, U. S. N. M.
03933	8.5	do		04005	4	Honolulu	
03935	8.5	do		04006	5	do	O. P. J.
03936	8.5	do		03929	*a*8	do	
03938	5.25	do			*a*5.5	do	
03939	4.75	do			*a*5.5	do	
03983	5	do	Cotype, No. 7476, L. S. Jr. Univ. Mus.		*a*5.5	do	
03984	5	do	Cotype, No. 7476, L. S. Jr. Univ. Mus.		*b*5.75	do	
03985	5.25	do	Cotype, No.2724, U. S. F. C.				
03986	6	do	Cotype, No.2725, U. S. F. C.				
03987	6.75	do	Cotype, No. 1703, Bishop Mus.				
03988	6.5	do	Cotype. No. ——, M. C. Z.				

a Collected by Dr. Wood, 1898.   *b* Collected by Jordan & Snyder in 1900.

32. Upeneus arge Jordan & Evermann, new species. "*Weke*" or "*Weke Puco.*"

Head 3.75 in length; depth 4.1; eye 5 in head; snout 2.25; interorbital 3; maxillary 2.3; shortest distance between maxillary and eye 1.25 in longitudinal diameter of eye; D. viii-9, second spine 1.5 in head; A. ii, 6, longest anal ray 1.9 in head; pectoral 1.5; ventrals 1.45; scales 3–40-7.

Body oblong, compressed, deepest through the spinous dorsal; head moderate, compressed, profile arched from origin of the spinous dorsal to tip of snout, steepest on snout; snout bluntly rounded; lower jaw included; mouth moderate, slightly oblique; tongue short, rounded anteriorly, not broad nor thick, and not free; teeth in villiform bands on each jaw and on vomer and palatines; maxillary moderate, reaching anterior edge of eye, moderately broad and sheathed for more than half of its length; eye rather small, high, median, adipose eyelid well developed; barbels not reaching edge of gill-opening; pseudobranchiæ well developed; gillrakers 16 +6, finely serrate, last 5 or 6 on longer limb very blunt and short, pupil of eye contained 1.5 in longest; spinous dorsal 1.5 in depth, first 2 spines even, longer than the others and longer than base; distance from snout to origin of spinous dorsal one-third distance from snout to last scale on caudal; distance between dorsals slightly less than base of soft dorsal; soft dorsal slightly concave; caudal deeply forked, upper lobe longer; anal similar to soft dorsal, inserted slightly behind the latter; ventrals reaching slightly beyond pectoral, rays of pectoral slightly the longer; lateral line concurrent with dorsal outline; scales large, finely ctenoid; entire body and head scaly.

Color in life, pale green, changing to white below; edges of scales on back and down to lateral line purplish brown, giving the appearance of 3 rather distinct stripes of purplish brown, with greenish centers on the scales; side with 2 broad yellow stripes, the upper beginning on opercle at level of eye and running to caudal just above lateral line, the latter being crossed under soft dorsal; second beginning on base of pectoral and running to base of caudal just below lateral line, this stripe less distinct and narrowing posteriorly; opercle bright rosy; top of head dusky; cheek white with some rosy; lower jaw white; barbels yellow; dorsal fins pale, each crossed by 2 or 3 brownish rosy bars; caudal white, upper lobe with 4 broad brownish red bars running downward and backward, 1 at base narrow; lower lobe with similar but much broader black bars running upward and backward, 2 of them more distinct than the others; 2 longish dark spots on inner rays; anal, ventrals, and pectoral pale, ventrals rather pale yellowish; iris yellowish, pink above.

Color in alcohol, above, bluish olivaceous, the side becoming lighter, almost white on belly; borders of scales dusky; first dorsal spine with 3 or 4 dark spots, and the upper posterior edge of membranes with dark spots; soft dorsal with 3 dark spots on anterior edge and similar spots on upper part of fin; caudal fin with dark bands, upper lobe with about 6, those on lower lobe 4, much broader; other fins pale.

This species resembles *Upeneus vittatus* (Forskål), described from Djidda, Arabia, but the latter has the belly abruptly deep yellow in life.

This is an abundant and important food fish at Honolulu, where we obtained 10 specimens and where 4 others were collected by Dr. Jenkins in 1889. It is equally common at Hilo and in Pearl Harbor. It lives in shallow water along quiet shores, and is known as "Weke" or "Weke Puco."

The following is our list of specimens:

Field No.	Length.	Locality.	Final disposition of specimen.	Field No.	Length.	Locality.	Final disposition of specimen.
	Inches.				*Inches.*		
02999	8.5	Honolulu.	Type, No. 50667, U.S.N.M.	03795	10.25	Honolulu.	Cotype, No.3954, Field Col. Mus.
03019	10	....do....	Cotype, No. 7477, L. S. Jr. Univ. Mus.	03796	10	....do....	Cotype, No.9817, Ind. Univ. Mus.
03148	10.5	....do....	Cotype, No.2726, U.S.F.C.	03797	8.5	....do....	Cotype, No. 1500, Cal. Ac. Sci.
03288	8	....do....	Cotype, No. 1704, Bishop Mus.				
03791	10	....do....	Cotype, No. M.C.Z.	O.P.J.			
03793	9.25	....do....	Cotype, No.2299, Am. Mus. Nat. Hist.	135	9	....do....	
03794	12.5	....do....	Cotype, No.24227, Ac. Nat. Sci. Phila.		8	....do....	
					8	....do....	
					10.5	....do....	

Upeneoides vittatus, Streets, Bull. U. S. Nat. Mus., No. 7, 71, 1877 (Honolulu); not of Forskål.

Family POMACENTRIDÆ. The Demoiselles.

33. Glyphisodon sindonis Jordan & Evermann, new species.

Head 3.5 in length; depth 1.75; eye 3.4 in head; snout 3.5; maxillary 3.4; interorbital 2.8; D. XII, 19; A. II, 15; scales 4-28-9, 22 pores.

Body short and deep, dorsal outline evenly arched from tip of snout to soft dorsal; head deeper than long, compressed; snout short and conic; mouth small, horizontal, lower jaw slightly shorter; maxillary reaching to anterior edge of orbit; a single row of small, rather blunt, slightly compressed teeth on each jaw; preopercle entire, opercle ending in 2 small flat spines, upper very small and obscure; eye anterior, high, its lower edge above upper base of pectoral; interorbital broad, steep and convex; fins large, origin of dorsal over base of ventrals, its distance from tip of snout equal to distance from base of last ray to tip of upper caudal lobe; spines strong and long, first 0.7 of fourth, which is 1.9 in head and of same length as following spines; middle dorsal rays produced, longest ray 1.25 in head; anal similar to soft dorsal, longest ray 1.25 in head, second spine longest, 2 in head; caudal forked, upper lobe the longer; ventrals reaching past vent, outer rays longest, about equal to head; pectoral broad, upper rays longest, equal to head; scales large, ctenoid, covering entire body and head except lower jaw and snout anterior to eye; lower limb of preopercle scaled; large scales covering nearly all of dorsal spines, smaller scales covering as much of soft dorsal and anal and nearly all of caudal; very minute scales on base of pectoral, none on rays of ventrals; lateral line concurrent with dorsal outline, on 22 scales, ending 3 rows of scales short of posterior base of dorsal, then dropping 3 rows of scales and continuing obscurely on middle of caudal peduncle to base of caudal fin.

Color in alcohol, uniform very dark brown, nearly black; 2 narrow wavy bands of white on side, first beginning about under fourth dorsal spine and extending under about middle of pectoral, thence curving slightly backward toward vent, rather indistinct below pectoral; second band beginning under last dorsal spine and first ray, extending toward middle of anal, rather obscure, indistinct for 2 or 3 scales before reaching anal; fins all black, pectoral slightly lighter than others; a large black ocellated spot with a narrow white border on back and lower part of soft dorsal, larger than eye, just back of last white bar.

This species agrees with typical *Glyphisodon* in all respects except that none of the teeth appears to be emarginate. It agrees with *Chrysiptera* in the entire preopercle and preorbital and naked snout, but differs from the type of that genus in having the teeth in a single series.

The above description based on the type, No. 50669, U. S. N. M. (field No. 04524), a specimen 3.75 inches long, from Honolulu. One other specimen obtained. It is taken as a cotype and is No. 2727, U. S. F. C. reserve series (field No. 03732), a specimen 2.75 inches long, from Kailua, where it was first discovered by Michitaro Sindo, for whom the species is named.

34. Pomacentrus jenkinsi Jordan & Evermann, new species.

Head 3.4 in length; depth 1.8; eye 3.3 in head; snout 4; maxillary 3.2; interorbital 2.75; D. XIII, 16; A. II, 13; scales 4–29–11; Br. 4.

Body ovate, deep, compressed, dorsal outline rather steep, evenly curved from tip of snout to soft dorsal, following edge of scales on spinous dorsal; head deeper than long, compressed subconic; snout bluntly conic, jaws equal; maxillary reaching anterior edge of eye; mouth small, horizontal; a single row of close-set, incisor teeth in each jaw; posterior edge of preopercle roughly serrate; opercle ending in 2 short flat spines, the upper very obscure; interorbital wide, strongly convex; fins rather large; origin of dorsal over ventral, origin of each equally distant from tip of snout; first 2 or 3 dorsal spines shorter than others; others about of equal length, shorter than the longest dorsal rays, the median rays being longest, 1.5 in head; caudal forked, lobes rounded, upper the longer; anal rounded, longest ray 1.5 in head, second spine rather stout and strong, 2.2 in head; ventrals long, reaching vent, 1.1 in head; pectoral broad, upper rays the longer, 1.2 in head; scales large, finely ctenoid; body and head, except lower jaw and snout, scaled, scales on top of head small; bases of all the fins except ventrals well covered with fine scales, those on spinous dorsal larger; lateral line concurrent with dorsal outline to a line under base of third or fourth dorsal ray, where it drops 3 rows of scales to middle of caudal peduncle, whence it continues to base of caudal fin, the detached portion little developed.

Color in life, ground color dark drab; central portion of scales olivaceous, each one with black on lower part of posterior edge forming vertical bands on body; axil black; outer border of dorsal fin, above scaled part, black; pectoral dusky olivaceous, black at base; ventral and anal black; caudal dusky with posterior border lighter; iris bright yellow.

Color in alcohol, dark brown, edges of scales darker; a dark stripe on upper edge of membranes of spinous dorsal, broadest and most distinct anteriorly; rest of dorsal, and caudal and pectoral dark brownish; ventrals and anal dark, almost black; a black blotch at upper base of pectoral, continuous with the black axil.

This is a very abundant species among the Hawaiian Islands. Numerous specimens were obtained at Honolulu in 1889 by Dr. Jenkins, and others by Dr. Wood in 1898 and Dr. Jordan in 1900. Our own collections, made in 1901, contain numerous specimens, the localities represented being Honolulu, Hilo, and Kailua.

The above description is based chiefly upon a specimen (field No. 04526) 4.8 inches long, obtained by us at Honolulu.

The field numbers and lengths of a few of our specimens are given in the following table:

Field No.	Length.	Locality.	Final disposition of specimen.
	Inches.		
03331	3.75	Honolulu	No. 7479, L. S. Jr. Univ. Mus.
04516	4.75	Hilo	No. 7480, L. S. Jr. Univ. Mus.
04517	4.5	......do	No. 3959, Field Col. Mus.
04518	4.2	......do	No. 50671, U. S. N. M.
04519	4.2	......do	No. 2728, U. S. F. C.
04520	4	......do	No. 2729, U. S. F. C.
04521	3.8	......do	No. 1705, Bishop Mus.
04522	5.25	Kailua	No. ——, M. C. Z.
04523	4.8	......do	No. 24228, Ac. Nat. Sci. Phila.
04526	4.8	Honolulu	No. 50670, U. S. N. M.
04527	4.5	......do	No. 2300, Am. Mus. Nat. Hist.

Pomacentrus nigricans, Quoy & Gaimard, Voyage Uranie, Zool., 399, 1824 (Sandwich Islands); Cuvier & Valenciennes, Hist. Nat. Poiss., v, 425, 1830; Günther, Cat , iv, 34, 1862 (Sandwich Islands); not *Holocentrus nigricans* Lacépède, Hist. Nat. Poiss., iv, 332 and 367, 1803, locality unknown, collected by Commerson.

Eupomacentrus marginatus Jenkins, Bull. U. S. Fish Comm. for 1899 (June 8, 1901), 391, fig. 5, Honolulu (Type, No. 49700, U. S. N. M., Coll. O. P. Jenkins); not *Pomacentrus marginatus* Rüppell.

Family LABRIDÆ.   The Wrasse-fishes.

35. Lepidoplois strophodes Jordan & Evermann, new species.

Head 2.75 in length; depth 2.75; eye 4.65 in head; snout 3.25; mouth 3.1; interorbital 4; D. xii, 10; A. iii, 12; scales 7-34-13.

Body oblong, compressed; head longer than deep; upper and lower profiles evenly and slightly convex; snout long, pointed, rounded above; jaws produced, pointed, about equal; mouth large, maxillary reaching beyond front of eye; teeth strong, forming a sharp cutting edge on sides of jaws, front of each jaw with 4 large canines; eye rather large, anterior, high in head; posterior margin of preopercle very finely emarginate; interorbital space rather broad, convex; nostrils small, anterior in short tube; dorsal spines pungent, longest 3 in head, last 3.5; third anal spine longest, 2.8; third anal ray 1.9 in head; pectoral rounded, 1.7; ventrals pointed, 1.4; caudal broad at base, truncate; caudal peduncle broad, compressed, its depth 2; scales large, thin, those on front of dorsal, along its base and that of anal, small; lateral line concurrent with back, sloping down at caudal, then running straight to its base.

Color in life, pale rosy white; upper parts of the snout, nape, and side to base of about ninth dorsal spine, lemon-yellow, extending down on side to level of upper edge of pupil; side of head very pale rosy, 2 irregular broken lines of wine-colored spots across snout and through eye to posterior edge of opercle, a similar row of 4 oblong spots from angle of mouth downward and backward to edge of opercle; cheek and side of lower jaw with numerous small irregularly placed orange spots; side with about 16 brighter rosy longitudinal lines, those above less distinct on account of the deeper rosy ground color, those below more distinct, the ground color being more white; side between anal and soft dorsal fins with a broad sooty black spot extending irregularly upon both fins and fading out upon body anteriorly, the posterior edge being nearly vertical and well defined; caudal peduncle and base of the caudal fin whitish, with a slight tinge of rosy, a pale rosy band separating this from the black lateral area; region in front and below the pectoral with about 4 series of small reddish brown spots; pectoral region and the under parts somewhat bluish; dorsal fin rich lemon-yellow, the tips of the soft rays whitish, and a small, round, black spot on middle of membrane of second spine; base of soft rays and last dorsal spines rosy from intrusion of the rosy wash on side of body; last dorsal rays sooty black at the base from extension of the black spot on the side; caudal pale lemon-yellow; anal pale rosy in center, lemon on spines and along tip of fin, base of fin sooty black from intrusion of black spot on side of the body, the black extending farthest down on the interradial membranes; pectoral very pale rosy; ventrals pale rosy, the membranes bluish, the tip of second ray blackish.

Color in alcohol (field No. 04291), gray-brown, gradually darker posteriorly; space between soft dorsal and anal abruptly black, the color extending forward in darker streaks along the rows of scales and forming a large black blotch on soft dorsal and anal; top of head and space before dorsal abruptly pale; posterior part of caudal peduncle also abruptly pale: a black blotch on dorsal between second and third spines, not involving third and fourth, as in *L. bilunulatus;* dorsal and caudal otherwise pale; a pale blotch at base of posterior dorsal rays; side with narrow dark brown longitudinal lines, coalescing posteriorly with the black blotch; 2 narrow brown streaks from lip to front of eye, then back across side of head above; edged with narrow, darker, wavy lines; a wavy streak from corner of mouth toward base of pectoral, lower side of head with small brown spots or blotches; ventral fin mostly dusky.

This species is very close to *Lepidoplois bilunulatus,* differing chiefly in the dark zone on posterior part of body and in the smaller size of the dorsal spot. Our specimens are all young, but we have the young of *L. bilunulatus* scarcely larger and showing the markings of the adult.

Our collection contains the following specimens of this species, all from Honolulu, where it is not uncommon:

Field No.	Length.	Locality.	Final disposition of specimens.
	Inches.		
04291	4.7	Honolulu..........	Type. No. 50872. U. S. N. M.
03520	3.8	.....do............	Cotype. No. 2730. U. S. F. C.
03532	4.5	.....do............	Cotype. No. 7481. L. S. Jr. Univ. Mus.
04292	3.75	.....do............	Cotype. No. 176. Bishop Mus.
04293	3.75	.....do............	Cotype. No. 3963. Field Col. Mus.

VERRICULUS Jordan & Evermann, new genus.

Verriculus Jordan & Evermann, new genus of *Labridæ (sanguineus).*

Body elongate, subfusiform, compressed, with rather long, pointed snout: snout rather large, with anterior canines strong, 1 to 1: posterior canines present: lateral teeth short, confluent in a serrated cutting edge; cheek and opercle scaly: preopercle entire, both limbs more or less scaly; scales moderate, about 40 in lateral line; lateral line continuous; D. xii, 10; A. iii, 12; dorsal spines low, pungent; soft dorsal and anal not elevated, their bases without scales; caudal subtruncate; pectoral short.

This genus is allied to *Verro* and *Nesiotes.* From its nearest relative, *Nesiotes,* it differs in the presence of a posterior canine tooth. The single known species is brilliantly colored.

36. Verriculus sanguineus Jordan & Evermann, new species.

Head 2.9 in length; depth 3.5; eye 6.2 in head; snout 3.1; mouth 2.8; interorbital 4.75; D. xii, 10; A. iii, 12; scales 5–40–13.

Body elongate, compressed, oblong; head long, pointed, conic, its depth 1.7 in its length; eye small, its posterior margin in middle of length of head; snout long, pointed, rounded; jaws produced, equal; mouth large, nearly horizontal, corner reaching below front rim of eye; lips thick, fleshy; teeth strong, those on sides short, close-set, forming a sharp cutting edge on side of jaw; 5 canines in front of upper jaw, 4 in front of lower, a posterior canine on each side of upper jaw; tongue long, pointed, free in front; preopercle not serrate; interorbital space broad, convex; nostrils small, anterior in short tube; dorsal spines strong, sharp-pointed, longest in middle and posteriorly; last dorsal spine 4 in head; anal spines strong, last spine longest, 3.75; seventh anal ray 3; caudal rounded; dorsal and anal fins scaled at base; pectoral rounded, 1.9 in head; ventrals short, spine strong, pointed, two-thirds longest ray, which is 2 in head; caudal peduncle broad, deep, 2.2 in head; scales small, thin, cycloid; head with very small thin cycloid scales on occiput, cheek, greater part of opercle, behind eye, and on opercles, otherwise naked; lateral line slightly curved in front, then obliquely down to base of caudal.

Color in life, deep red, edge of upper jaw and lower tip golden; a long stripe from eye along back to base of caudal golden, with a red shade, a vertical black bar edged with golden above, on opercular region; a long blackish area covering it from eye to above pectoral, with some blackish before, behind and above; a black spot at base of caudal; dorsal and caudal golden, first dorsal edged with violet and with the lower half violet; anal entirely deep blood-red; ventrals golden; pectoral reddish, golden at base.

Color in alcohol, very pale brown; a dusky band from snout across back of head and on side, fading out indistinctly posteriorly; a blackish spot at middle of base of caudal; opercle posteriorly with black vertical blotch; fins all pale or light brown.

Described from the type, No. 50677, U. S. N. M. (field No. 03489), an example 7.5 inches long, taken at Hilo, with hook and line, in deep water with *Etelis evurus*, *Eteliscus marshi*, *Erythrichthys schlegeli*, *Antigonia steindachneri*, and *Anthias fuscipinnis*.

We have examined only one example, the one described above.

37. Pseudocheilinus evanidus Jordan & Evermann, new species.

Head 3 in length; depth 3.8; eye 4.5 in head; snout 3; preorbital 6.2, interorbital 5.5; D. ix, 11; A. iii, 9; scales 2–25–6.

Body short, deep and compressed; head long, conic; snout long, sharply conic; anterior profile rising in a relatively straight line from tip of snout to nape, thence gently convex to base of caudal peduncle; ventral outline less convex; mouth large, horizontal, below axis of body, gape reaching anterior line of orbit; upper jaw with 3 pairs of anterior canines, outer strongest, curved outward and backward; lower jaw with a single pair at tip, similar to inner above; jaws laterally with a single series of smaller conic teeth; preorbital narrow, oblique; eye high up, its lower border on axis of body; interorbital space rather broad and flat; depth of caudal peduncle about 2 in head; scales large, surfaces finely striate; head, nape, and breast with large scales; lateral line following contour of back until under base of sixth dorsal ray, where it is interrupted, reappearing 2 rows farther down and continuing on 6 or 7 scales to base of caudal fin; fins rather large; dorsal spines somewhat greater than eye in length, spines with a sheath of large scales reaching nearly to their tips; soft dorsal and anal with a lower sheath; soft dorsal elevated, rays equal to snout and eye; anal similar to soft dorsal, second spine strongest, nearly as long as snout; anal rays somewhat longer, equaling those of soft dorsal; caudal rounded, its length 1.3 in head, its base covered with very large, thin scales.

Color in life, according to Mr. Sindo, body dull brick-red; belly and base of anal pale purplish; about 17 thin, thread-like longitudinal yellowish streaks along side anteriorly; dark greenish blotches above eye and on snout; a bluish horizontal bar on cheek, below which is a yellow bar; median line of throat and tip of snout brick-red; edges of opercle and preopercle bright purple; a purple stripe with reddish edges through middle of dorsal fin, below which the color is dull brick-red, like that of body, and above which the spinous dorsal is orange-yellow, the margin of the membranes bright cardinal-red; above the purple streak in the soft dorsal is a bright yellow streak, above which the fin is cardinal-red, fading gradually upward; dorsal rays purplish; tip of soft dorsal somewhat red; caudal rays purple, the membranes immediately next to the rays yellow, middle part dull brick-red; anal same as caudal; ventrals pale purplish; pectoral pale; iris scarlet-red.

The same specimen after having been in spirits more than a year has the body light brownish-blue; a pale streak along each row of scales, but no trace of the narrow yellowish streaks above noted; top of head and upper part of cheek dusky blue; opercle and edge of preopercle rich blue; dorsal, anal, and caudal fins bright blue, the soft dorsal pale on outer two-thirds, dorsal rays bright blue; ventrals and pectoral light blue, latter darker blue at base. The color of this specimen in spirits is wholly different from that which it possessed in life, and it would be difficult to believe that such changes had taken place except that the specimen was carefully tagged in the field when the color note in life was taken.

Since writing the above, we have noticed similar changes in the Samoan species, *P. hexataenia*. The blue shades are permanent in spirits, while the pink or crimson wash soon vanishes in spirits.

The 17 thread-like streaks, mentioned in Mr. Sindo's field notes above, have vanished entirely in the original type. A number of specimens taken by Mr. Snyder at Laysan, while on the *Albatross*, retain these traits, the streaks being almost white, like white threads, covering most of the side anteriorly. This is a very peculiar color mark, which should well distinguish the species in life.

A single specimen, type, No. 50678, U. S. N. M. (field No. 05757), was taken by Mr. Sindo in Henshaw's pool near Hilo, a deep tide-pool in the lava rocks.

38. Hemipteronotus baldwini Jordan & Evermann, new species.

Head 3.25 in length; depth 3; eye 5.75 in head; snout 1.75; maxillary 3; preorbital 2.2; interorbital 4.8; D. ii–viii, 13; A. iii, 13; scales 3–27–9.

Body moderately short and deep, greatly compressed; head slightly deeper than long; anterior profile nearly vertical from mouth to front of eye, sharply cultrate; dorsal outline gently convex, sloping to the deep caudal peduncle; ventral outline less convex; caudal peduncle very narrow, the depth 2.25 in head; mouth small, horizontal, the maxillary nearly reaching vertical of orbit; the jaws equal, each provided anteriorly with a pair of strong curved canines and laterally with a single row of short close-set conic teeth; lower jaw strong, its outline very convex; preorbital nearly vertical and very deep; preopercle and opercle smooth, with membranous edges, the latter produced somewhat in a broad rounded flap; origin of dorsal but little posterior to orbit, far in advance of base of ventrals; first 2 dorsal spines somewhat removed from third but connected to it by a low membrane, their length scarcely greater than the gape of mouth; remaining dorsal spines short and weak, scarcely equaling gape; soft dorsal low, the rays slightly longer than the spines; anal similar to soft dorsal, rays somewhat longer; caudal slightly convex, rays 2 in head; outer ray of each ventral somewhat produced, not reaching vent, the length about 1.9 in head; pectoral broad, the longest rays 1.7 in head; scales large, thin, smooth, firmly attached, those on breast somewhat reduced; head naked, except about 4 series of small scales extending from eye downward to level of mouth; lateral line curving abruptly upward from upper end of gill-opening, following contour of back to the scale under third dorsal ray from last where it drops 3 rows and continues to base of caudal, the pores simple, unbranched.

Color in life (field No. 03123), pale, yellowish white over head and body; a diffuse lemon-yellow blotch under and above pectoral; a large brownish-black blotch on lateral line under seventh to tenth dorsal spines; dorsal fin yellowish-white, tip of detached part with a jet-black crescent (this marking variable in position, it sometimes being farther posterior), rest of fin faintly mottled with yellowish and olive, the latter in narrow oblique lines; caudal yellowish white; anal yellowish white, with narrow, wavy, pale-blue lines, and a large jet-black spot bordered with blue on membrane of last 5 rays; iris whitish.

Color in alcohol, creamy yellowish white; head somewhat orange on cheek and opercles; faint rosy lines downward from eye to mouth and on preopercle; median line of anterior profile bluish; middle of back with a large black or brownish black blotch lying on lateral line, beneath which is a large white blotch under and above pectoral fin; anterior part of spinous dorsal blackish at edge, the color ocellated, rest of dorsal yellowish white with narrow purplish cross-lines; anal similar, with a large jet-black spot on last 4 rays; caudal color of soft dorsal; pectoral and ventrals yellowish white.

Color in alcohol, of one of the female cotypes (No. 03372), pale olivaceous, the general color that of the male; dorsal with black spots on membranes of second, third to fourth, and eighth spines, the latter ocellated; a series of about a dozen small black spots back of the dorsal blotch on side above lateral line; no black spot on anal.

The above description based upon the type, No. 50644, U. S. N. M. (field No. 03414), a male example, 8.5 inches long, obtained at Honolulu.

Another specimen, also a male (field No. 03371), was in life, livid gray; each scale posteriorly with a vertical spot of violet; anterior line of profile bright violet; a violet line downward from eye with a whitish area behind it on cheek; an oblique violet line downward and backward from opercular flap to behind axil; behind this a vague yellow area, behind which is an ovate white spot, each scale around which has a vertical bar of bright violet; above this a large black blotch washed with brick-red; dorsal bluish-gray, the rays posteriorly with an increasing amount of orange, where the blue is reduced to oblique crossbands, an intermarginal line of violet, a small black spot on last ray; membranes of second to fourth dorsal spines with a terminal black ocellus; anal pale golden, with oblique bluish stripes, a large jet-black ocellus bordered with blue on last rays; caudal pale orange, crossed by bluish lines; ventrals and pectoral pale.

Still another male example (field No. 03004) was described as follows: General color very pale smoky white, edges of scales pale bluish, beneath seventh to ninth dorsal spines a large blotch, brick-red above, pale rosy below, all irregularly overlaid with black or brown; median line from tip of snout to base of first dorsal spine bright blue; a narrow bright blue line downward from anterior part of eye to angle of mouth; region above pectoral pale lemon yellow, a short oblique pale blue line above base of pectoral; dorsal pale flesh-color, with short vertical bluish lines, with 3 jet-black spots at tips of first, second, and fourth spines; anal pale yellowish, a black spot on distal half of last 3 rays; caudal pale, with obscure bluish cross-lines; pectoral and ventrals white; iris yellowish, red at lower posterior angle.

Another example, a female (field No. 03372), 7.5 inches long, from Honolulu, which is taken as a cotype, differed in life coloration from the male in lacking the black ocellus on the anal and in having more violet on the white lateral spot, also more golden before it; violet lines and spots obscure, but present; 3 to 8 small blackish points above lateral line behind black dorsal blotch; a small black ocellus on second to third dorsal spines and one on seventh dorsal spine, these wanting in some females; fins otherwise colored as in the males, but the blue fainter and the orange of dorsal brighter.

Another female example (field No. 03005) differed in color from field No. 03004 only in the absence of black on the dorsal and anal fins, the paler blue lines on head, the paler caudal fin, and in having black spots on the back. Another female example (field No. 03271), 7.5 inches long, in life had the head and body smoky white; a large bluish white spot under tip of pectoral; snout bluish around border and surrounded by a broad pale yellow space involving nearly all of anterior half of side below level of eye; a large black spot under fifth to sixth dorsal spines, crossing lateral line and penetrating yellow of side, nearly reaching white spot; back of this a series of about a dozen small black specks, scattered along side above lateral line to near end of dorsal fin; median line of snout and head blue; dorsal pale, with wavy yellow cross-lines, pinkish toward margin; caudal pale; anal pale, with about a dozen pale yellow crossbars; pectoral and ventrals pale; iris yellow and red.

This beautiful and abundant species is represented in our collection by 41 specimens, 19 of which are males and 22 females. The differences in coloration of the two sexes are very marked. The male, in all specimens examined, has the jet-black spot upon the last rays of the anal, a marking which is not present in any of the females examined. The female always has a series of small black specks on the side above lateral line posterior to the large lateral blotch. These markings, the small black spots on the side of the female and the large jet-black spot on the anal of the male, would apparently always serve to distinguish the two sexes.

The extent of variation in color among individuals of the same sex is indicated in the color descriptions given above. We should have added that occasionally there is a small jet-black spot upon the last rays of the dorsal.

This is one of the most abundant and beautiful species found among the Hawaiian Islands. It appears to be related to *H. melanopus* of Bleeker, but differs from it markedly in the presence of the large black lateral blotch and in the absence of the large red lateral blotch shown in Bleeker's figure.

This species is named for Mr. Albertus H. Baldwin in recognition of his paintings of American and Hawaiian fishes.

Field No.	Length.	Sex.	Locality.	Final disposition of specimens.	Field No.	Length.	Sex.	Locality.
	Inches.					*Inches.*		
03414	8.5	Male	Honolulu	Type, No. 50644, U. S. N. M.	05603	8.5	Male	Honolulu.
03123	8.8	Male	...do	Cotype, No. 2713, U. S. F. C.	05604	8.5	...do	Do.
03372	7.5	Fem	...do	Cotype, No. 2714, U. S. F. C.	05605	8.5	...do	Do.
03371	7.5	Male	...do	Cotype, No. 7462, L. S. Jr. Univ. Mus.	05606	7.75	...do	Do.
03004	7	Male	...do	Cotype, No. 3951, Field Col. Mus.	05609	6.75	Fem	Do.
03005	6.25	Fem	...do	Cotype, No. 3962, Field Col. Mus.	05627	9.2	Male	Do.
03271	7.25	Fem	...do	Cotype, No. 7462, L. S. Jr. Univ. Mus.	05628	7.6	Fem	Do.
03084	6.5	Fem	...do	Cotype, No. 50645, U. S. N. M.	05629	8.12	Male	Do.
05438	8.25	Male	Hilo	Cotype, No. ——, M. C. Z.	05630	8.5	...do	Do.
05589	7.25	Fem	Honolulu	Cotype, No. ——, M. C. Z.	05631	8.5	...do	Do.
05439	7.5	Male	Hilo	Cotype, No. 2290, Am. Mus. Nat. Hist.	05632	8.6	Fem	Do.
05591	6.2	Fem	Honolulu	Cotype, No. 2291, Am. Mus. Nat. Hist.	05633	7	...do	Do.
05465	9.25	Male	...do	Cotype, No. 24219, Ac. Nat. Sci. Phila.	05634	7	...do	Do.
05599	6.8	Fem	...do	Cotype, No. 24220, Ac. Nat. Sci. Phila.	05635	7	...do	Do.
05586	8.25	Male	...do	Cotype, No. 9809, Mus. Ind. Univ.	05636	6.25	...do	Do.
05600	6.5	Fem	...do	Cotype, No. 9810, Mus. Ind. Univ.	05637	6.5	...do	Do.
05587	8.5	Male	...do	Cotype, No. 1495, Cal. Ac. Sci.	05639	6.25	...do	Do.
05607	7.4	Fem	...do	Cotype, No. 1496, Cal. Ac. Sci.	05640	6.25	...do	Do.
05568	8	Male	...do	Cotype, No. 1696, Bishop Mus.	05641	6.2	...do	Do.
05608	7	Fem	...do	Cotype. No. 1697, Bishop Mus.	05642	5.75	...do	Do.
03124	8.75	Male	...do		O. P. J.			
					n 624	6.5	Male	Honolulu.

a Collected by Jordan & Snyder.

39. Xyrichthys niveilatus Jordan & Evermann, new species.

Head 3.3 in length; depth 2.4; eye 6.2 in head; snout 1.8; preorbital 2; maxillary 3; interorbital 4.7; D. II-VII, 12; A. III, 12; scales 3-28-8.

Body short, deep, and very greatly compressed; anterior profile nearly vertical from tip of upper jaw to front of eye, thence in a parabolic curve to dorsal fin; anterior dorsal outline very trenchant; body tapering rather evenly from head to caudal peduncle, which is greatly compressed and very deep, depth at middle equaling preorbital; head short; snout very short and blunt; mouth small, horizontal, the maxillary nearly reaching anterior edge of orbit; jaws equal, each with a pair of strong curved canines in front, and a single series of smaller, conic teeth laterally, the canines of lower jaw most prominent and extending in front of upper jaw; eye small, high up; the interorbital space narrow and trenchant; opercles smooth, without spines or serrations, ending in thin flexible edges; preorbital vertical and very deep; origin of dorsal fin above posterior line of orbit, far in advance of base of ventrals; first 2 dorsal spines somewhat removed but not detached from third, the membrane between second and third spines moderately notched, length of second spine about 2.7 in head, remaining dorsal spines subequal, weak, about equal to gape; dorsal rays low, the last few somewhat produced, their length 3 in head; anal similar to soft dorsal; caudal short, slightly convex, rays about equal to preorbital; outer ray of ventral somewhat produced, not reaching vent, its length equaling depth of preorbital; pectoral broad, its tip reaching vent, its length equaling distance from snout to edge of preopercle.

Scales large, thin, and with membranous edges, those on breast somewhat smaller; head entirely naked, except for a few small scales below the eye; lateral line beginning at upper end of gill-opening following closely the curvature of back to the scale under the last dorsal ray but 2, where it drops 3 scales and continues to base of caudal, the pores simple, rarely branched.

Color in life, grayish; each scale of posterior half of body with a large violet spot, more narrow and brighter near middle of body, the edge of each scale broadly golden-olive; a large golden area, anteriorly deep orange, above pectoral and on edge of opercle; behind this a large quadrate pure white area extending to tip of pectoral; a few scales in golden area with bright violet markings; head shaded with violet, a bright violet stripe downward from eye to behind angle of mouth; a lunate black area shaded with red just below front of soft dorsal; spinous dorsal violet-gray, edged with reddish; soft dorsal golden, with violet vermiculations at base, its edge orange; anal golden, with bluish vermiculations; caudal similar, with the bluish markings; pectoral faintly reddish; ventrals dirty white.

One of the cotypes, a male (field No. 03373), agreed in life coloration with the type except that behind the opercle is a golden area with the bright violet stripes across anterior basal part; behind this a large milk-white patch beyond tip of pectoral; a violet border was around the white and blackish above the yellow.

Color in alcohol, dirty yellowish white, dusky above; head with some purplish reflections; a thin purplish line downward from anterior edge of orbit to tip of maxillary, a similar but less distinct line from humeral region downward to subopercle; a yellowish white blotch on side above base of pectoral, in the base of which are 2 or 3 small purplish spots; a large white area on middle of side under and above tip of pectoral, separated from the yellowish blotch by purplish brown on 2 or 3 scales; a black spot covering the larger part of 3 scales on side above lateral line under base of first 3 dorsal rays, back at base of last dorsal rays somewhat dusky; anterior portion of dorsal fin dusky olivaceous, soft dorsal, anal and caudal pale yellowish crossed by narrow, wavy, pale purplish lines; ventrals and pectoral plain yellowish white.

This handsome fish is rather common about Honolulu.

Field No.	Length	Locality.	Final disposition of specimen.
	Inches.		
03345	9.5	Honolulu......	Type, No. 50646, U.S. N. M.
05164	9.75	......do......	Cotype, No. 3960, Field Col. Mus.
03373	9.25	......do......	Cotype, No. 2715, U. S F.C.
05597	8.2	......do......	Cotype, No. 7463, L. S. Jr. Univ. Mus.
05598	7.4	......do......	Cotype, No. ——, M. C. Z.
05590	5.5	......do......	Cotype, No. 2292, Am. Mus. Nat. Hist.

Family SCARIDÆ. The Parrot-fishes.

40. Scarus jenkinsi Jordan & Evermann, new species.

Head 3 in length; depth 2.5; eye 6.5 in head; snout 2.6; preorbital 4.7; interorbital 3; D. IX, 10; A. III, 9; P. 13; scales 2-24-7.

Body short, very deep and greatly compressed; head short, nearly as deep as long, snout short and blunt; mouth small; each jaw with 1 or 2 blunt canines; dorsal and ventral outlines about equally convex; anterior profile rising rather irregularly from tip of snout to origin of dorsal; caudal peduncle deep, its least depth 2 in head. Scales large, deeper than long; 2 rows of large scales on cheek and 1 row on subopercle; a row of thin modified scales at base of dorsal and anal; a few very large, thin scales on base of caudal; lateral line ceasing under last dorsal ray, reappearing 2 rows lower down and continuing to base of caudal, the pores with 2 or 3 irregular branches; dorsal rays soft and flexible, not pungent; dorsal spines somewhat elevated posteriorly, longest a little more than 2 in head; first ventral spine obscure, the others soft and flexible; anal rays somewhat shorter than those of dorsal;

caudal shallowly lunate, the outer rays not greatly produced; ventrals moderate, 1.6 in head, not reaching to origin of anal by a distance equal to two-fifths their length; pectoral broad, 1.2 in head.

Color of a nearly fresh specimen, bright blue-green, brightest on posterior half of body, each scale broadly edged with reddish brown; lower anterior part of body reddish brown, with traces of blue-green; top of head brownish red or coppery, a broad deep blue-green band on the upper lip, extending on side of head to below eye; lower lip with a narrow brighter blue-green band connecting at angle of mouth with the one from upper lip; chin with a broad coppery-red bar, followed by a broader bright blue-green one; caudal green, median part pale, banded with green spots; dorsal bright green at base and tip, the middle pale greenish, translucent; anal similar, the distal band broader; pectorals and ventrals deep vitriol-green with whitish markings.

Color in alcohol, dirty greenish, side with about 8 longitudinal series of greenish blotches; head olivaceous above, paler on cheeks; upper lip broadly pea-green at edge, this color continued backward to under eye; edge of lower lip pale green, continued around angle of mouth uniting with the same color from upper lip; chin with a broad, pale crossbar behind which is a broader, pale green one which extends up on cheek nearly to orbit; back of this is a still broader, white crossbar interrupted in the middle by greenish; subopercle and lower edge of preopercle with a large, irregular, green patch; a median green line on breast to base of ventrals; dorsal green at base and along edge, the middle portion paler; anal similar to dorsal, the green border broader; caudal bright pea-green on the outer rays, the inner ones pale with 4 or 5 cross-series of green spots, tips of rays darker; ventrals pale green, the edges dark pea-green; pectoral pale green, darker green on the upper rays.

This species is related to *Scarus gilberti* from which it differs in the greater depth and the somewhat different coloration. It is also related to *Scarus lauia*, but differs in the much greater depth, the less produced caudal lobes, the greater width of the green head markings, and the color of the fins.

Only one specimen was obtained, type, No. 50647, U. S. N. M. (field No. 02944), 14 inches long, obtained at Honolulu, June 6. Named for Dr. Oliver P. Jenkins.

41. Scarus lauia Jordan & Evermann, new species. *"Lauia."*

Head 2.8 in length; depth 2.7; eye 6.75 in head; snout 2.6; preorbital 4.8; interorbital 2.8; D. ix, 10; A. iii, 9; P. 13 on one side, 14 on other; scales 2-25-6.

Body short, stout and compressed; head heavy; snout rather short, bluntly rounded; dorsal and ventral outlines about equally arched, anterior profile slightly concave before the eyes; nape strongly convex; mouth small, nearly horizontal, in axis of body; upper jaw with 1 or 2 moderately strong, backwardly directed canines; a similar but smaller canine sometimes present on lower jaw; cutting edge of upper jaw fitting outside that of lower; teeth small, entirely above axis of body; opercle with a broad short flap. Scales large, their surface with fine lines and granulations; nape and breast with large scales; cheek with 2 rows of large scales, about 7 scales in each; subopercle and lower limb of preopercle each with a row of scales; opercle with large scales; lateral line broken under last dorsal ray, reappearing one row lower down and continuing to caudal fin, the pores with 2 to 4 branches; a series of these oblong scales along base of dorsal and anal; base of caudal with 3 or 4 very long, thin scales. Dorsal spines soft and flexible, not pungent, the longest about 2.7 in head; soft portion of dorsal somewhat higher, especially posteriorly where the rays are about 2.4 in head; anal spines soft and flexible, the first obscure, the third about 4.3 in head; anal rays higher, the last but one longest, 3 in head; caudal deeply lunate, the 3 or 4 outer rays above and below produced, length of middle rays 2.3 in head, or 2 in outer rays; ventrals moderate, not reaching vent, 1.9 in head; pectoral broad, the free margin oblique, length of longest rays 1.3 in head.

Color in life, head brownish yellow before eyes, the jaws lighter yellow; cheek washed with brownish and blue, throat greenish; nuchal and opercular regions brownish orange; body salmon-color above, the belly lighter yellow, most of the scales with an edging of greenish blue; a deep blue line from nostril before and behind upper part of eye; upper lip deep blue, the streak forming an interrupted line before eye; lower jaw with 2 blue cross-lines, 1 marginal; a dark blue spot behind angle of mouth; deep blue blotches on interopercle; dorsal deep blue with a peculiar jagged stripe of light brownish yellow; anal with blue spots at base, then light yellow, then deep blue, then green with blue edge; caudal brownish yellow, with bright blue edgings and a median area of bright golden green; ventrals golden, trimmed with bright blue; pectoral golden with deep blue above and greenish blue on lower rays, a salmon streak across base with greenish blue behind it.

Color in life of another example (No. 03040, 10 inches long), pale coppery rosy, darker on first 3 rows of scales; the center of each scale in the first 5 rows greenish blue; under parts pale rosy, with orange wash; head pale rosy, a small postocular blue spot, a short blue line forward from eye, and a second of same color on upper lip and across cheek to eye, where it has a slight break, then continues under eye as a greenish-blue bar; under lip with narrow blue edge; chin faded salmon with a double blue crescent; space from chin to isthmus bright blue; an oblong bright blue spot on subopercle, behind which is a smaller irregular one bordered above by a broad greenish-blue space; dorsal greenish blue, with a broad submedian orange band, the lower greenish-blue band made up of large, scarcely connected, bluish spots, the upper half continuous with a narrow bright blue border; a small orange blotch on base of last dorsal ray; caudal pale rosy at base, then with a greenish bar, followed by a broad rosy bar, then by a broad terminal greenish-blue bar, dark blue in front, greenish in middle and pale blue on outer third; upper and lower edges of caudal blue, below which is a broad rosy orange stripe; anal greenish blue at base, then a broad orange stripe, the outer half greenish blue with narrow bright blue edge; pectoral orange anteriorly, pale bluish behind, the anterior border blue; ventrals orange, anterior edge and tip blue; iris pale orange.

Color in spirits, light dirty grayish white, lighter below; a narrow pea-green stripe on edge of upper lip, breaking up into irregular spots from angle of mouth to lower edge of orbit, a similar stripe from nostril to eye and slightly beyond upper posterior border of eye; these lines sometimes continuous and unbroken; lower jaw edged with green, a broader pea-green cross-stripe at anterior edge of branchiostegal opening; subopercles each with a broad green stripe; line of union of gill-membranes broadly green; dorsal with a series of large olive-green spots at base and a broad band of similar color on distal half, these separated by a paler band and cut by intrusions from it both above and below; dorsal fin with a very narrow paler border; and with a series of greenish spots at base, then a broad pale yellowish white line, bounded distally by an indefinite, wavy, black line shading off into the greenish of the distal half; last ray of anal dusky on its outer third; caudal greenish-olive at base and on produced outer rays, edges of fin above and below green; middle rays with a broad lunate area of pale green, scalloped proximally by dark green, separated from the lighter green base by a broad whitish interspace, the upper and lower edge also darker green; ventrals creamy white, the outer edge pale greenish; pectoral whitish, the upper edge dusky.

There is some variation in the width of the green markings on the head, sometimes the stripes on the lower part of the head being very broad.

This species is related to *Scarus gilberti*, from which it differs in the more strongly produced caudal lobes, in the narrower lines on the snout, the broader green lines on the throat, the absence of a green median line on the breast, and in the very different coloration of the fins.

Field No.	Length.	Locality.	Final disposition of specimen.
	Inches.		
03485	14	Hilo	Type, No. 50648, U. S. N. M.
04152	14	Honolulu	Cotype, No. 7464, L. S. Jr. Univ. Mus.
03040		....do	Cotype, No. 2716, U. S. F. C.
04353		....do	Cotype, No. 9811, Ind. Univ. Mus.
04352		....do	Cotype, No. 3964, Field Col. Mus.

Also one specimen (No. 12046), 9 inches long, obtained by Dr. Wood at Honolulu.

42. Scarus barborus Jordan & Evermann, new species.

Head 3.2 in length; depth 3.2; eye 6.6 in head; snout 2.9; interorbital 2.9; preorbital 4; D. ix, 10; A. iii, 9; P. 14; scales 2-25-6.

Body oblong, not very deep nor greatly compressed; head about as long as deep, conic, compressed; snout short, blunt and rounded; upper jaw produced, its lip double, covering entire dental plate; lower lip covering half of dental plate; no canine teeth; eye anterior, high, its lower border considerably above upper base of pectoral; caudal peduncle short and deep, its depth 2 in head.

Origin of dorsal over upper base of pectoral, spines flexible, short, not quite as long as rays; longest ray 2.1 in head; longest anal ray 2.2 in head; caudal truncate; ventrals 1.9 in head, not reaching vent by half their length; pectoral 1.5 in head. Scales large and thin, very slightly roughened by radiating lines of granulations extending to margins of scales; lateral line interrupted, the pores being on 18 scales, then dropping 2 rows to row of scales under posterior base of dorsal, and continuing to base of caudal on middle of caudal peduncle, 7 pores in the shorter part, which begins on the row following the row on which the upper part ends, there not being 2 pores in the same row; the scales extend well out on the caudal, the last scale of lateral line, very large and thin, being the largest scale on the fish; 4 scales in median line before dorsal; 2 rows of scales on cheek, 5 scales in upper row and 2 to 4 in lower, sometimes only 2 on posterior part; 2 rows on opercle, and 1 on lower margin.

Color in alcohol, grayish leaden brown, lighter below; no markings on fins different from corresponding parts of body evident.

The above description is based on the type, No. 50649, U. S. N. M. (field No. 04316), a specimen 7.75 inches long, from Honolulu; cotype, No. 2735, U. S. F. C. (field No. 04354), 7.5 inches long, from Honolulu; cotype, No. 7465, L. S. Jr. Univ. Mus. (field No. 650), 5.5 inches long, from Honolulu.

Family TEUTHIDIDÆ. The Tangs.

43. Teuthis atrimentatus Jordan & Evermann, new species.

Head 3.8 in length; depth 1.9; eye 4.2 in head; snout 1.2; interorbital 3; D. ix, 27; A. iii, 25.

Body deep, compressed, ovoid, the upper profile steeper than lower, evenly convex; jaws low, not produced, lower inferior; mouth small, inferior; teeth broad, compressed, edges crenulate; nostrils close together, anterior larger, with small fleshy flap; anterior dorsal spines graduated to posterior, the longest 1.5 in head; fourth dorsal ray 1.4; third anal spine longest, 1.9; first anal ray 1.5; caudal large, emarginate, upper and lower rays produced in sharp angular points, upper much longer than lower; pectoral about 3.5 in body; ventrals sharp-pointed, 3.6 in body, spine half the length of fin; caudal peduncle compressed, 2 in head; caudal spine large, depressible in a groove, 3.1 in head; scales very small, ctenoid, few, and very minute on vertical fins; lateral line high, arched, at first descending under fifth dorsal spine, then straight to below middle of soft dorsal, finally falling down and running along side of caudal peduncle to tail.

Color in life (No. 02996), coppery brown, crossed by numerous, very narrow, pale blue lines, those above axis of body running somewhat upward and backward, and with short broken lines of same interspersed, those below more regular but less distinct; cheek brassy, with about 5 narrow pale blue lines from eye to snout; a conspicuous jet-black spot on caudal peduncle at base of last dorsal ray, each of these extending slightly upon pale rusty, each with 5 or 6 narrow brassy lines parallel with margin; edge of each blackish; last rays of dorsal and anal more brassy; caudal dark, blackest on outer part of middle rays; pectoral pale lemon; ventrals dusky, blacker toward tips; iris brownish, white on posterior part. Another example (No. 03474) was dull olive-gray, unmarked, save a faint whitish band across nape and back part of head; fins plain dusky gray.

Color in alcohol, very dark chocolate brown; side with about 40 narrow irregular or incomplete series of indistinct dark slaty longitudinal lines; cheek with similarly colored lines running obliquely downward; fins, except pectoral, all more or less blackish or dusky; dorsal with about 5 blackish longitudinal bands; anal with several similar indistinct blackish bands; base and axils of last dorsal and anal rays blackish; pectoral brown.

This common species is well distinguished from *Teuthis dussumieri* and other streaked species by the black ink-like spot in the axil of the dorsal and anal fin. It has several times been recorded under the erroneous name of *Acanthurus lineolatus*, but the species originally called by that name must be something else. Numerous specimens were obtained by us at Honolulu, where it was also secured by Dr. Jenkins and Dr. Wood. We have examined the following examples:

Field No.	Length.	Locality.	Final disposition of specimen.	Field No.	Length.	Locality.	Final disposition of specimen.
	Inches.				*Inches.*		
05481	9.5	Honolulu.	Type, No. 50673, U.S.N.M.	05378	6.2	Honolulu.	Cotype, No. 1501, Cal. Ac. Sci.
02996	6.8	....do....	Cotype, No. 7482, L. S. Jr. Univ. Mus.	....	6.1	....do.....	Cotype.
				05491	5.2	....do.....	Cotype.
03146	5.7	....do....	Cotype, No. 2731, U. S. F. C.	05486	5.75	....do.....	Cotype.
03205	5.75	....do....	Cotype, No. ——, M. C. Z.	05484	6.7	....do.....	Cotype.
03474	4.5	....do....	Cotype, No. 2301, Am. Mus. Nat. Hist.	05498	5.5	....do.....	Cotype.
03729	6.75	....do....	Cotype, No. 24229, Ac. Nat. Sci. Phila.	Dr. Wood. 3	....	....do.....	Cotype.
05018	7	....do....	Cotype, No. 3965, Field Col. Mus.	....	5.3	....do.....	Cotype.
				O. P. J.	6.6	....do.....	Cotype.
05020	5.2	....do....	Cotype, No. 9818, Ind. Univ. Mus.	....		....do.....	Cotype.
05365	5.6	....do....	Cotype, No. 1707, Bishop Mus.	141	4.4	....do.....	Cotype.
				....	4.6	....do.....	Cotype.

Acanthurus lineolatus, Günther, Fische der Südsee, i, 112, taf. lxxiii, fig. A, 1873 (Society Islands); Steindachner, Denks Ak. Wiss. Wien, lxx, 1900, 493 (Honolulu); not of Cuvier & Valenciennes.

Family BALISTIDÆ. The Trigger-fishes.

44. Pachynathus nycteris Jordan & Evermann, new species.

Head 3.5 in length; depth 1.9; eye 5 in head; snout 1.25; interorbital 2.6; preorbital 1.5. D. III – 33; A. 29; scales about 80. Body short, stout, deep, and greatly compressed; head short; dorsal and ventral profiles about equally curved; caudal peduncle short, compressed, its least depth about twice diameter of eye, its least width about equal to diameter of eye; a short horizontal groove in front of eye below nostrils; nostrils small, close together in front of upper part of eye; teeth broad, close-set, forming a continuous plate, the teeth, however, not united; lips thin; mouth small, horizontal, in axis of body; lower jaw very slightly the longer; gill-opening short, nearly vertical; a group of bony scutes under pectoral back of gill-opening, one of these considerably enlarged; scales regularly arranged in rows, their surfaces granular; lateral line beginning at posterior edge of eye, ascending to within 7 scales of spinous dorsal and continuing to near origin of soft dorsal, where it disappears; scales on posterior portion of body and on caudal peduncle each with a slightly raised crest at its center, these forming series of ridges along the side. First dorsal spine strong, blunt, and rough, its length about 2 in head; second dorsal spine shorter and much weaker, its length scarcely more than one-third that of first; third dorsal spine remote from the second and very short, not extending above dorsal groove; soft dorsal gently rounded, its rays of approximately equal length, the longest equaling the distance from tip of snout to posterior edge of eye; base of soft dorsal slightly greater than distance from tip of snout to posterior base of first dorsal spine or equaling distance from tip of snout to lower base of pectoral axil; anal similar to soft dorsal, the rays somewhat longer than those of soft dorsal, the base somewhat shorter; caudal short and rounded, the rays about 1.75 in head; pectoral short, the upper rays longest, about 3 in head.

Color in alcohol, rich brownish or velvety black; spinous dorsal black; soft dorsal pale yellowish or whitish, margined with black, the lower half crossed by 4 narrow parallel black lines; anal similar to soft dorsal, but with only 2 narrow black lines on its basal half; caudal dusky, yellowish at tip; pectoral yellowish.

Only one specimen. Type, No. 50821, U. S. N. M. (field No. 05089), 6.25 inches long, Honolulu.

Family TETRODONTIDÆ. The Puffers.

45. Lagocephalus oceanicus Jordan & Evermann, new species.

Head 2.8 in length; depth 3.6; eye 4.5 in head; snout 2.4; interorbital 3.2; depth of caudal peduncle 6; D. 12; A. 12; C. 10; P. 14.

Body rather elongate, moderately compressed, greatest depth at vertical of pectoral; head long; snout long, blunt at tip, the sides flattened; anterior profile from tip of snout to vertical of pectoral in a long, low, even curve; ventral outline little convex when not inflated; mouth small; teeth pointed at median line, the cutting edge sharp; nostrils separate, not in tubes, the anterior somewhat the larger, their distance from eye about half their distance from snout or about half the interorbital space; gill-opening vertical, 1.2 in eye, extending a little above base of pectoral, inner flap entirely hidden by outer; eye rather large, wholly above axis of body; interorbital space very little convex; cheek long; caudal peduncle nearly round, tapering, its length from anal fin equaling snout; back, upper parts of sides and entire head entirely smooth, no spines or prickles evident; belly covered with small 4-rooted spines, most prominent when belly is inflated, spiniferous area not extending on throat anterior to eye, nor on side above base of pectoral, but in front of anal extending upward to level of lateral fold; a line of very small mucous pores curving above eye on interorbital space; a strong cutaneous fold on lower part of side of caudal peduncle from above anterior base of anal to lower base of caudal fin; no dermal fold on head or anterior part of body; mucous pores inconspicuous; dorsal fin somewhat anterior to anal, pointed, anterior rays produced, their length equal to that of snout; anal similar to dorsal, its rays somewhat longer; caudal lunate, outer rays about 2 in head; pectoral broad, its length a little greater than snout, 2.3 in head.

Color in life, back blackish, fading into deep steel-blue on side; side and below from level of upper edge of eye abruptly silvery-blue; sides of belly white, with round black spots about as large as pupil, these most distinct about pectoral, before, below, and behind the fin; upper fins dusky; caudal mottled black, tipped with white; pectoral black above and behind, pale below; anal pale, broadly tipped with blackish.

Color in alcohol, bluish black above; side from upper level of eye abruptly bluish silvery; back crossed by 7 or 8 narrow darker cross-streaks; belly white, with a series of about 9 to 12 small roundish black spots, chiefly below the pectoral; cheek dusky; pectoral, dorsal, and caudal dusky, tips of the latter paler; anal whitish, a little dusky at tip. A somewhat smaller example (4.5 inches long) has larger dark spots along middle of side above level of pectoral.

This species is known to us from 2 small examples obtained in the market of Honolulu. It is related to *Lagocephalus stellatus* (Donovan) of Europe (*Tetrodon lagocephalus* of Günther, not of Linnæus), but differs in the much shorter pectoral, more conspicuous spots, and rather greater extension of the prickly region of the breast. The types of *Tetrodon lagocephalus* Linnæus are reputed to have come from India. According to Linnæus, this species had 10 dorsal and 8 anal rays. It may have been based on *Lagocephalus sceleratus* or some other East Indian species, but there seems to be no evidence that it was identical with the European *Lagocephalus stellatus*. In any event, the Hawaiian form seems different from any other yet known.

Type, No. 50820, U. S. N. M. (field No. 03379), 5 inches long, obtained at Honolulu; cotype, No. 7784, L. S. Jr. Univ. Mus. (field No. 534, paper tag), 4.5 inches long, also from Honolulu.

Family OSTRACIIDÆ.   The Trunk-fishes.

46. Ostracion oahuensis Jordan & Evermann, new species. "*Momo Awaa.*"

Head 3.9 in length; depth 2.9; eye 2.9 in head; snout 1.2; preorbital 1.6; interorbital 1; D. 9; A. 9; P. 10; C. 10.

Body 4-sided; dorsal side of carapace evenly convex, its greatest width one-fourth greater than head; lateral dorsal angles not trenchant, slightly convex anteriorly, then evenly convex; snout blunt, the anterior profile ascending abruptly then strongly convex in front of eyes; interorbital space nearly flat; cheek flat; side of body concave, its width about equal to head; ventral keel prominent, evenly convex; ventral surface nearly flat posteriorly, but little convex anteriorly, its greatest width 1.4 times length of head, its length just twice its width; gill-opening short, not exceeding two-thirds diameter of eye; least width of anterior opening of carapace 1.75 in interorbital, or 1.5 times diameter of orbit, the depth nearly twice orbit; mouth small; teeth rich brown; least depth of posterior opening of carapace much less than width of anterior opening, equaling distance from lower edge of preorbital to pupil; length of caudal peduncle less than that of head, its depth 2.2 in its length; no spines anywhere. Dorsal fin high, its edge obliquely rounded, its length 1.3 in head; anal similar to dorsal, the edge rounded, its length 1.2 in dorsal; caudal slightly rounded, its rays nearly equal to head; pectoral with its free edge oblique, the rays successively shorter, length of fin equal to height of dorsal.

Color in life, dark brown with blue tinges; interorbital space showing more or less golden; small whitish spots profusely covering entire dorsal surface; no spots on side of body or on face; no spots on ventral surface except a faint one of a slightly darker color than general gray color of surface; one longitudinal row of golden spots on each side of upper part of caudal peduncle from carapace to base of caudal fin; pectoral, anal, and dorsal fins with transverse rows of faint spots; caudal bluish black at base, white on posterior half; a broad light or yellowish area below eye; iris golden.

Color in alcohol, rich brown above, the sides darker, and the ventral surface paler, brownish about margins, dusky yellowish within; entire back with numerous small, roundish, bluish-white spots; upper half of caudal peduncle with similar but larger spots; forehead and snout dark brown; lips brownish black; cheek dirty yellowish; sides and ventral surface wholly unspotted; base of caudal blackish, paler distally, the dark extending farthest on outer rays; other fins dusky, with some obscure brownish spots.

This species is related to *O. camurum* Jenkins, from which it differs in the smaller, more numerous spots on back, the entire absence of spots on side, the smaller size of the spots on the caudal peduncle, and the brighter yellow of the suborbital region. Only 2 specimens known, both from Honolulu.

Type, No. 50668, U. S. N. M. (field No. 03443), a specimen 5.6 inches long, obtained by us at Honolulu, July 25, 1901. Cotype, No. 7478, L. S. Jr. Univ. Mus. (field No. 2156), an example 5.25 inches long, collected at Honolulu, in 1898, by Dr. Wood.

Family SCORPÆNIDÆ.   The Rockfishes.

47. Pterois sphex Jordan & Evermann, new species.

Head 2.4 in length; depth 2.65; eye 3.8 in head; snout 3.2; interorbital 5.2; maxillary 2 35; mandible 2; D. xiii, 11; A. iii, 7; P. 16; V. i, 5; scales 10–56–13.

Body elongate, compressed, greatest depth at first dorsal spines; back only slightly elevated; snout rather short, rounded; mouth large, oblique; maxillary reaching below anterior rim of orbit, its distal expanded extremity 1.75 in eye; teeth fine, in bands in jaws and on vomer; lips rather thin, fleshy; tongue pointed, compressed and free in front; jaws nearly equal; eighth dorsal spine longest, equal to head; penultimate spine 4; fifth dorsal ray 1.75; third anal spine longest, 2.2; third anal ray longest, 1.5; caudal rounded, elongate, 1.4; pectoral long, the rays more or less free for at least half their length; ventral 1.3 in head, reaching beyond origin of anal; ventral spine 2.1; caudal peduncle compressed, its depth 3.75; nasal spine very small; preocular, supraocular and post-ocular spines present, the upper bony ridge over eye being serrate; tympanic, coronal, parietal, and nuchal spines present, coronal very small and close together and parietal with 4 serrations; a finely serrate ridge from behind eye over opercle to suprascapula; a finely serrated ridge over preorbital and cheek to margin of preopercle, ending in a strong spine, below this 2 other spines; preorbital with a strong spine over maxillary posteriorly, and with fine serrations above; scales ctenoid, present on top of head, cheeks, and opercles, head otherwise naked; tubes of lateral line single, in straight line to base of caudal; several fleshy flaps on head, 1 above eye, 1 from lower preorbital spine, and 2 from along margin of preopercle.

Color in alcohol, very pale brown, whitish beneath; side with 9 broad, deep brown bands alternating with narrow brown bands on trunk and posterior portion of head; narrow brown bars from below penultimate dorsal spine with a narrower brown line on each side above lateral line; lower surface of head whitish, without crossbands; spinous and soft dorsal and caudal each with 4 dusky brown crossbands; base of anal with 2 broad similar bands, and soft portion of anal with 3 series of irregular crossbands; axil of pectoral above with white blotch; pectoral whitish with 10 blackish crossbands; a brown band in front of base of pectoral extending on lower pectoral rays; ventral with dusky blotch at base, outer portion with about 5 dusky crossbands.

The only example we have seen of this species is the type, No. 50650, U. S. N. M. (field No. 05030), 6 inches long, obtained by us at Honolulu.

48. Scorpænopsis catocala Jordan & Evermann, new species.

Head 2.1 in length; depth 2.75; eye 7.25 in head; snout 3.1; interorbital 4.3; maxillary 1.8; D. xii, 10; A. iii, 5; P. 18; V. i, 5; scales 9–42–22.

Body elongate, greatest depth at first dorsal spines; back elevated, swollen, or convex, below first dorsal spines; snout rather long, with an elevated prominence; mouth large, oblique; maxillary large, expanded extremity broad, 6.5 in head; teeth in broad villiform bands in jaws, those on vomer small; no teeth on palatines; tongue small, pointed, free in front; lips rather thick, fleshy; eye small, a little in front of middle of length of head; a deep pit below eye; top of head with deep square pit just behind interorbital space; anterior nostril with broad fleshy flap; posterior large, without flap; four spines on side of snout above anterior nostril; preocular, supraocular, postocular, tympanic, parietal, and nuchal spines present; a series of spines running across cheek below eye; several large spines on lower part of preopercle; several spines on opercle; side of head above with many small spines; suprascapular with several small spines; dorsal spines rather strong, third longest, 3.75 in head; last dorsal spine 3.8; second dorsal ray 2.7; second anal spine enlarged, a little longer than the third, 3.4 in head; first anal ray longest, 2.4 in head; caudal rounded, 2 in head; pectoral large, lower rays thick, fleshy, curved inward; sixth pectoral ray 1.7 in head, lowest 3.7; base of pectoral broad, 2.25; ventral spine strong, 3.1 in head, second ray longest, 1.9; the innermost ray joined by a broad membrane to belly; caudal peduncle compressed, its depth 4 in head; head and body with many fringed fleshy flaps; scales moderately large, ctenoid.

Color in life (field No. 03382), excessively mottled, streaked, and spotted; body dark purplish brown or claret shaded, the spaces gray tinged with sulphury yellow; head all dull brown, flaps colored like the space about; belly to axillary region whitish with reticulations and irregular marks of yellowish olive; axillary region wine-brown, finely mottled with yellowish white in streaks and spots; a few round black spots behind and in axil; inside of pectoral with a large jet-black blotch at upper part of base, bordered with orange; around this a large yellow area, then 6 oblong black spots on the membranes of upper rays above middle, then a broad rose-red band, fading into violet below, the rim gray; ventrals bright brown and gray, red shaded on inner face; inside of branchiostegals salmon-color, striped with white, the membranes yellow; membranes of upper jaw salmon-color mottled with light yellow; tip of upper jaw orange with a golden ridge dividing a triangular spot of indigo-blue between vomer and premaxillary; a golden line on each side in front of palatines; tip of tongue light yellow; a triangular indigo-colored spot behind teeth of tip of lower jaw; a golden streak behind it on membrane before tongue; lower lip salmon-color especially behind where hidden.

Color in alcohol, dark purplish, beautifully mottled with dusky and darker; head mottled above with dusky; fins with many fine dusky and brown wavy lines; base of pectoral both outside and inside brownish, the latter variegated with white and blackish brown; outer portion of inside of pectoral covering first 5 rays with a series of broad blackish spots; ventrals more or less brownish variegated with gray and whitish; body whitish, mottled with pale brown; edges of buccal folds, inside of mouth, deep yellow; a deep blue blotch directly behind teeth in front of each jaw.

This species is related to *Scorpæna gibbosa* (well figured by Günther in Fische der Südsee), from which it differs in the much rougher and less depressed head, much larger flaps on opercles and mandible, and the presence of a very large fringed flap on the anterior nostril, this being obsolete in *S. gibbosa*. We have compared our specimens with examples of *S. gibbosa* from Apia.

This species was obtained both at Honolulu and Hilo, and appears to be not uncommon. Our collections contain 8 excellent examples, as follows:

Field No.	Length.	Locality.	Final disposition of specimen.
	Inches.		
05298	9.5	Honolulu.........	Type, No. 50051, U. S. N. M.
03382	9.3	.....do...........	Cotype, No. 7466, L. S. Jr. Univ. Mus.
03521	6.8	.....do...........	Cotype, No. 2717, U. S. F. C.
05294	7.75	.....do...........	Cotype, No. 1698, Bishop Mus.
05295	6	.....do...........	Cotype, No. ——, M. C. Z.
05296	8.2	Hilo.............	Cotype, No. 3966, Field Col. Mus.
05299	7.5	Honolulu.........	Cotype, No. 2293, Am. Mus. Nat. Hist.
05299	6.3	Hilo.............	Cotype, No. 24221, Ac. Nat. Sci., Phila.

49. Dendrochirus hudsoni Jordan & Evermann, new species.

Head 2.5 in length; depth 2.5; eye 3.4 in head; snout 3.3; interorbital 5; maxillary 2.1; mandible 1.8; D. XIII, 10; A. III, 6; P. 18; V. I, 5; scales 8-52-13.

Body elongate, compressed, rather deep, the greatest depth at fifth dorsal spine; profiles of trunk above and below more or less even; head compressed; snout short, rounded; mouth large, maxillary nearly reaching below middle of eye, its distal expanded extremity equal to half eye; minute teeth in bands in jaws and on vomer; lips thin; tongue pointed, compressed, free in front; jaws nearly equal; anterior nostrils each with a small fleshy flap; interorbital space deeply concave; fifth dorsal spine longest, 1.25 in head; penultimate spine 5.2; second anal spine longest, 2.1; third anal ray longest, 1.3; caudal rounded, 1.25; pectoral 2.4 in trunk, reaching below middle of base of soft dorsal, rounded, and only membranes between lower rays slightly incised; ventral rounded, reaching base of first anal ray; caudal peduncle compressed, its least depth 3.5 in head; nasal spines very small; preocular, postocular, tympanic and coronal spines present; parietal and nuchal spines forming a single ridge; a ridge of spines behind eye above opercle; a ridge of spines below eye, ending in a spine on margin of preopercle; 2 spines below this also on margin of preopercle; no opercular spines; margin of preopercle with spine projecting down and back; skinny flap above eye equal to its diameter, and another from preorbital spine; scales small, ctenoid; head naked except some scales on opercle, cheek, and side above; lateral line running obliquely down to base of caudal.

Color in alcohol, pale brown or whitish; side with 3 pairs of deep brown vertical bands, first on posterior part of head preceded by a deep brown streak from below eye, second on middle and posterior part of spinous dorsal, and third extending out on soft anal and basal portion of soft dorsal; soft dorsal, caudal, and anal pale or whitish; membranes of dorsal spines deeply incised in front, each spine with 3 brown crossbands; pectoral grayish with a blackish brown basal blotch and 5 blackish crossbands; ventral blackish with 2 whitish or grayish blotches.

This species is especially characterized by the unspotted soft dorsal, anal, and caudal. From *Dendrochirus barberi* Steindachner, it is distinguished by the longer pectoral which reaches to below the posterior dorsal rays.

Named for Capt. C. B. Hudson, in recognition of the excellence of his paintings of Hawaiian fishes.

We have examined 5 specimens of this species, as follows:

Field No.	Length.	Locality.	Final disposition of specimen.
	Inches.		
03547	1.8	Waikiki	Type, No. 50632, U. S. N. M.
651	1.9	Reef near Honolulu	Cotype, No. 7467, L. S. Jr. Univ. Mus.
652	1.9	Honolulu.........,..	Cotype, No. 2718, U. S. F. C.
O.P.J. 301	4	.....do	
	3.5	.....do	

Family GOBIIDÆ.   The Gobies.

QUISQUILIUS Jordan & Evermann, new genus.

Quisquilius Jordan & Evermann, new genus of *Gobiidæ* (*eugenius*).

Allied to *Asterropteryx*. Body robust, covered with large, ctenoid scales; snout blunt; mouth large, very oblique, with 2 series of sharp teeth in jaws, the inner depressible; side of head with several series of short papillary fringes; ventrals separate, their rays i, 5, joined at base by a narrow frenum; dorsals short, the first with 6 spines, the second with 12 short rays.

The genus is distinguished from other small Eleotrids by the papillary fringes on preorbital, jaws, and opercles.

50. Quisquilius eugenius Jordan & Evermann, new species.

Head 2.8 in length; depth 3.8; eye 3.25 in head; snout 4.25; width of mouth 2.4; interorbital 2 in eye; D. vi-12; A. 10; V. i, 5; scales 25,-12.

Body robust, compressed, greatest depth about middle of belly; head large, elongate, broad, depth 1.4 in its length, width 1.25; snout short, blunt, rounded above; jaws large, lower projecting; mouth large, very oblique, its posterior margin reaching below front of eye; upper jaw with 2 series of teeth, sharp-pointed, outer larger, the inner depressible; mandible with teeth similar to those in upper jaw; no teeth on vomer and palatines; tongue truncate, front margin not notched; eye large, high, anterior; nostrils separated, anterior in small tube, posterior close to upper front margin of eye; interorbital space narrow, very deeply furrowed; a series of fringe-like papillæ running from preorbital along upper margin of maxillary down behind corner of mouth where it joins another series running along under surface of mandible, and continued back and upward on margin of preopercle; anterior margin of opercle with a small vertical series of papillæ, each papilla a little shorter than diameter of eye; gill-opening large, continued forward till nearly below posterior margin of eye; spinous dorsal rather small, spines flexible, with tips produced in short filaments; soft dorsal high, median rays rather longer than others; anal more or less similar to soft dorsal, posterior rays very long; caudal rather large, round; pectoral broad, round, equal to head; ventrals small, 1.25 in head, sharp-pointed, and joined at base of inner rays by a narrow frenum; caudal peduncle compressed, its length 1.6 in head, depth 2.4; scales large, ctenoid, those on upper part of head very small; snout, interorbital space, jaws, and lower surface of head naked; no lateral line.

Color in life (field No. 03554), body with transverse bands of dark brown with olivaceous tinge alternating with dirty white; edges of scales in dark brown portions lighter; dorsal, anal, and caudal dark brown, edged in part with white; pectoral light reddish brown.

Color in alcohol, brown; 12 dark brown crossbands on side, the last 6 very broad, much broader than the pale interspaces; vertical fins dark slaty; pectoral pale slaty; ventral pale on outer posterior portion, blackish slaty on inner.

We have examined the following examples:

Field No.	Length.	Locality.	Disposition of specimens.
	Inches.		
	1.4	Waikiki	Type. No. 50674, U. S. N. M
03554	.8	Honolulu............	Cotype, No. 7483, L. S. Jr Univ. Mus.
	1.2	......do	Cotype, No. 2732 U. S. F. C.
	1.2	......do	Cotype, No. 1708, Bishop Mus.
	.9	......do	Cotype, No. ......, M. C. Z.
	1.1	......do	Cotype, No. 3970, Field Col. Mus.
	.9	......do	Cotype, No. 24230, Ac. Nat. Sci. Phila.

51. [a] Gnatholepis knighti Jordan & Evermann, new species.

Head 3.5 in length; depth 4.25; eye 3.8 in head; snout 3.6; width of mouth 2.5; interorbital 2.25 in eye; D. vi–12; A. 12; P. 16; V. 5.5; scales 32,–9.

Body elongate, compressed, not depressed in front, greatest depth at the middle of belly; head elongate, its depth 1.25 in its length, its width 1.5; snout oblique, blunt, broad; upper profile of the head obtuse, with a prominence over eye in front; mouth rather broad, the maxillary not reaching posteriorly to below front rim of orbit; lips rather thin; teeth small, sharp, in narrow bands in jaws with an outer enlarged series; no teeth on vomer or palatines; interorbital space very narrow, level; nostrils small, close together in front of eye, anterior with flap of very short, fleshy cirri; eye high, small, a little anterior; gill-opening restricted to side, nearly vertical, its length 2.25 in head; scales large, finely ctenoid, and becoming much larger on posterior side of trunk; scales small on belly in front of ventrals, cycloid; scales moderately large, cycloid on the upper part and side of head, head otherwise naked; dorsal fins well separated, spines flexible and with extremities of most free and filamentous; first 1.6 in head, fifth 1.7, last 2.7; soft dorsal long, last rays longest, first 1.7, last 1.25; anal similar to the dorsal, but lower, first ray 2.8, last 1.25; caudal rounded, the median rays very long, a little longer than head; pectoral with upper median rays longest, all rather fine, about equal to length of caudal; ventrals rather large, frenum uniting in front, rather broad, length equal to pectoral; caudal peduncle compressed, length 1.2 in head, depth 2.25.

Color in life, pale flesh-color, upper parts with dark brownish spots and blotches; a series of about 8 brownish blotches along middle of side; a small dark spot on base of pectoral; opercle dusky; fins all pale, spinous dorsal with brown edge; iris bluish white.

Color in alcohol, pale brown, side with numerous small dark brown spots and 7 large dark brown blotches; a dark brown streak below eye, and another across opercle; spinous dorsal very pale brown with about 3 blackish brown cross-lines, very distinct on first spines, running somewhat obliquely, and becoming indistinct posteriorly; soft dorsal with the spines pale or whitish brown and membranes between blackish brown; anal more or less dark gray brown; caudal very pale brown or whitish, spotted in cross-series with brown; pectoral pale brown; ventrals dark brown, paler along edges.

Color when fresh, of example from Hilo, olive-green, rather pale, and with 7 blackish crossbands; caudal spot small and inconspicuous; black bar below eye, narrow and very distinct; back crossbarred with many spots of dusky olive; side with longitudinal streaks of dark brown spots along rows of scales, these irregular and variable, mixed, especially behind, with spots of pale sky-blue; dorsal, anal, and caudal dotted finely with dark olive; pectoral pale olive; ventrals blackish; anal plain blackish, paler at base. In most examples examined the head was finely dotted with bright pale blue on cheeks and opercles.

This small but interesting species is generally common in brackish water about Hilo and Honolulu. Our collections contain a total of 123 specimens; 15 of these have been tagged and their measurements are given in the table; 101 other specimens from Hilo range in length from 1.1 to 2.5 inches, the average length being 1.81 inches. From Waianae we have 5 specimens, 1.3 to 1.8 inches in length, the average being 1.62 inches. From the pond at the Moana Hotel at Waikiki, we have 2 examples, each 0.8 of an inch long. The average length of our 123 specimens is 1.81 inches.

The species is named for Master Knight Starr Jordan, who first noticed it in the pond at the Moana Hotel at Waikiki Beach near Honolulu.

---

[a] The genus *Gnatholepis* Bleeker seems to be equivalent to *Hazeus* of Jordan & Snyder.

The following are some of the specimens examined:

Field No.	Length.	Locality.	Final disposition of specimen.
	Inches.		
2150	2.25	Hilo	Type, No. 50653, U. S. N. M.
778	1.75	.....do	Cotype, No. 7468, L. S. Jr. Univ. Mus.
783	2.5	.....do	Do.
784	2.2	.....do	Cotype, No. 2719, U. S. F. C.
785	2.2	.....do	Cotype, No. ——, M. C. Z.
788	2.25	.....do	Cotype, No. 2294, Am. Mus. Nat. Hist.
789	2.2	.....do	Cotype, No. 24222, Ac. Nat. Sci. Phila.
790	2.2	.....do	Cotype, No. 9812, Ind. Univ. Mus.
792	2.1	.....do	Cotype, No. 3969, Field Col. Mus.
793	1.9	.....do	Cotype, No. ——.
794	2	.....do	Cotype, No. 1497, Cal. Ac. Sci.
795	1.8	.....do	Cotype, No. 1699, Bishop Museum.
797	1.75	.....do	
799	1.7	.....do	
800	1.5	.....do	

Acentrogobius ophthalmotænia, Streets, Bull. U. S. Nat. Mus., No. 7, 60, 1877 (coral reefs at Oahu); not of Bleeker.

52. Gobiopterus farcimen Jordan & Evermann, new species.

Head 3.25 in length; depth 3.5; eye 3.2 in head; snout 3.5; D. vi–11; A. 9; scales 28 (27 to 29)–10.

Body rather robust, compressed, greatest depth at gill-opening; head rather large, depth 1.25 in length, width 1.4; upper profile of head evenly convex from tip of snout to origin of dorsal; jaws large, mandible very large, slightly produced; mouth large, very oblique, maxillary extending beyond front margin of eye; teeth in jaws uniserial, rather large, somewhat canine-like; two small depressible canines on posterior part of bone behind anterior series; lips large, thick, fleshy; tongue not emarginate, large, thick, rounded; nostrils close together, posterior very large, in front of upper margin of orbit with elevated rim; interorbital space very narrow, concave; scales large, ctenoid; a large pore behind and above base of pectoral; gill-opening large, continued forward below; spinous dorsal small, flexible, spines ending in filaments, beginning behind base of pectoral; soft dorsal high, rays of nearly uniform length; anal with posterior rays elongate, much longer than anterior; caudal elongate, rounded; pectoral broad, round, equal to head; ventrals long, equal to head, broad, without any frenum in front; caudal peduncle compressed, its length 1.5 in head, depth 2.25.

Color in alcohol, pale brown, trunk covered all over with very pale minute brown dots; fins very pale brown, dorsals dusky, especially the spinous; 3 vertical pairs of pale brown cross-lines over side of head.

Described from an example 1.1 inches long, taken at Hilo. Type, No. 50654, U. S. N. M.

VITRARIA Jordan & Evermann, new genus.

Vitraria Jordan & Evermann, new genus of *Gobiidæ*, subfamily *Luciogobinæ* (*clarescens*).

Body elongate, translucent, covered with very small thin scales; mouth small, oblique; teeth minute; gill-opening rather narrow; dorsals small, the rays v–11; pectoral rather long; ventrals small, united in a circular disk. Small gobies of the coral reefs, allied to the Japanese genus *Clariger*, but with the first dorsal of 7 small spines instead of 3.

53. Vitraria clarescens Jordan & Evermann, new species.

Head 4.6 in length; depth 6.7; eye 3.5 in head; snout 4.5; D. viii–11; A. i, 10.

Body elongate, slender, compressed, greatest depth between dorsal fins; head elongate; pointed, conic, depth 1.75 in its length, width 2; snout rather long, rounded; jaws prominent, upper slightly produced; mouth oblique, maxillary reaching a little beyond anterior margin of eye; teeth not evident; tongue broad, truncate; snout above, interorbital space, and top of head more or less flattened; nostrils well separated, anterior nearly midway in length of snout, posterior close to front of eye; eye rather large, anterior; gill-opening restricted to side, rather small; scales very small; dorsal spines flexible, first dorsal small, the last three spines very small (minute stubs, broken in the type) the fin beginning behind tip of ventrals; soft dorsal beginning a little nearer base of caudal than tip of snout, about over insertion of anal, and anterior rays of both fins longest, those of anal gradually smaller behind, the last 2 minute and close together; caudal emarginate, lobes rounded; pectoral rather long, lower rays longest; ventrals small, united to form a small round disk whose diameter is 2.25 in head; caudal peduncle compressed, elongate, its length equal to head, its depth 2.25 in head.

Color in alcohol, very pale translucent brown, 7 V-shaped pale brown markings on upper side of body united over back; fins whitish.

We have examined 7 examples obtained at Hilo, each about 1.2 inches in length:

Locality.	Final disposition of specimen.	Locality.	Final disposition of specimen.
Hilo.....	Type, No. 50655, U. S. N. M.	Hilo......	Cotype, No. 2295, Am. Mus. Nat. Hist.
Do....	Cotype, No. 7469, L. S. Jr. Univ. Mus.	Do......	Cotype. No 24223, Ac. Nat. Sci. Phila.
Do....	Cotype, No. 2720, U. S. F. C.	Do......	Cotype, No. 1700, Bishop Museum.
Do....	Cotype, No. ——, M. C. Z.		

Family PTEROPSARIDÆ.

OSURUS Jordan & Evermann, new genus.

Osurus Jordan & Evermann, new genus of *Pteropsaridæ* (*Parapercis schauinslandi* Steindachner). This genus is allied to *Parapercis*, from which it differs in having the caudal fin deeply forked instead of truncate.

Family FIERASFERIDÆ.

54. Fierasfer umbratilis Jordan & Evermann, new species.

Head 10.2 in length; depth 15.2; eye 5 in head; snout 4.8; mouth 2.6; interorbital 4.5.

Body very elongate, compressed; tail very long and tapering gradually in a long point; head elongate, conic, its depth 2 in length, width 2.25; snout rather broad, conic, and produced beyond mandible; mandible broad, flattened below; mouth nearly horizontal, broad, the gape reaching below posterior margin of eye; eye rather small, anterior, without eyelid, and placed about first quarter of head; nostrils well separated, anterior with elevated rim, posterior a short, crescent-like slit; interorbital space rather broad, convex; gill-opening low, inferior, rather long; gill-membrane free from isthmus, its angle nearly an eye diameter distant from posterior margin of eye; dorsal fin almost rudimentary, very low and thin; anal rather broad, in middle its height is about 0.75 in eye, from which point it gradually decreases to tip of tail, where it is rudimentary, like dorsal; tail ending in a fleshy point, caudal fin apparently absent; pectoral small but relatively large, 3.1 in head, rays very minute; lateral line distinct, running down along middle of side on posterior half of tail; no scales.

Color when fresh (field No. 03506), pale olivaceous, with pale greenish spots; a pale bluish streak in each spot over lateral line; pale purplish oblong spots on lower half of body; head greenish-olive, with pale green spots closely set on cheek and jaw; pale purplish dots on upper part of cheek and behind eye; first dorsal same as body, but the spots yellowish; a black spot behind first and second rays, tips pale; rays of second dorsal checked alternately with yellowish-green and white; caudal same as second dorsal, but margin yellowish; anal, yellowish-olive; tip blackish; pectoral and ventrals pale; iris greenish-yellow; dull red streaks radiating from pupil.

Color in alcohol, brown; head and end of tail dark sooty or blackish brown, the color formed of dark points; greater part of anal fin, lower surface of body anteriorly and pectoral and branchiostegal membranes, pale straw color; lower surface of trunk more or less blotched with pale brown.

Our collection contains but 2 specimens of this species, both obtained at Hilo. Type, No. 50656, U. S. N. M. (field No. 03506), a specimen 7.6 inches long; cotype, No. 7470, L. S. Jr. Univ. Mus. (field No. 528), an example, 6.4 inches long.

Fierasfer umbratilis occurs also in the South Seas, and is readily distinguished from most related species by its dark, non-translucent coloration.

F. boraborensis from Borabora, briefly described by Kemp, has the pectoral 6 to 7 times in head.

Family PLEURONECTIDÆ. The Flounders.

55. Engyprosopon hawaiiensis Jordan & Evermann, new species.

Head 3.8 in length; depth 1.75; eye 3.25 in head; snout 4.25; interorbital 6.3; maxillary 2.8; D. 79, A. 56; P. i, 10; V. i, 5; scales 14–46–15.

Body elongate, deep, rather ovoid, greatest depth about end of pectoral; head very deep, its length 0.7 in depth; upper profile very convex in front, steep; snout short, obtuse; jaws small, produced a little, the mandible slightly projecting; lips rather thin; mouth curved a little, very oblique, the small maxillary reaching a little beyond front margin of eye; teeth in jaws very small, sharp-pointed; eyes well separated, lower anterior, placed in first third of head, the upper about two-fifths an eye diameter posterior; nostrils close together, with elevated rims; interorbital space a little more than half an eye diameter in width, deeply concave; gill-opening small; gillrakers rather short; scales large, finely ctenoid, very small on rays of vertical fins; lateral line strongly arched at first for first two-ninths its length, then straight to base of caudal; dorsal beginning on snout, the anterior rays free for only a short portion of their extremities, first 5 in head, fiftieth 2.1, this the highest region of the fin; anal more or less similar, first 3.25, thirtieth 2; caudal rounded, middle rays longest, 1.1; pectoral short, pointed, 1.4; ventrals rather broad, base of left 3, first and last rays about equal; right ventral smaller; caudal peduncle compressed, its depth 1.9.

Color in alcohol, dark olivaceous brown, fins dark gray-brown, each ray finely specked with olivaceous brown; left pectoral specked with dark brown, right pectoral dull creamy or brownish white like the right side of body.

Type, No. 50657, U. S. N. M., taken at Hilo, the only example we have seen, 3 inches long.

56. Engyprosopon arenicola Jordan & Evermann, new species.

Head 3.6 in length; depth 1.9; eye 4.3 in head; maxillary 3; D. 78; A. 57; P. i, 11; V. 5; scales 14–36–17. Body elongate, very deep, rather ovoid, the greatest depth at tip of pectoral; head much deeper than long, the upper profile steep, strongly convex; snout obtuse; jaws very oblique, mandible slightly projecting; maxillary very oblique, reaching below anterior margin of eye; lips rather thin, fleshy, fringed along margins; teeth in jaws minute, sharp-pointed; eyes close together, lower anterior placed about first third of length of head; upper eye about one-third an eye diameter posterior; nostrils well separated, with raised fleshy rims forming a flap; interorbital space very narrow, concave; gill-opening rather small, restricted to side; gillrakers small, short, few; scales large, finely ctenoid; lateral line strongly arched for anterior fourth of its length, then straight to base of caudal; anterior dorsal rays free distally for one-half their length, first ray 3 in head, forty-fifth 1.8, which is the highest region of the fin; anal similar to dorsal, but anterior rays not free for half their length; first ray 3.5, thirtieth 1.8; caudal elongate, median rays longest, equal to head; pectoral short, pointed, 1.5; ventrals rather large, the left with its base 5 in head, first ray 3.6, last 2.6, almost entirely in front of the right, which is much smaller; caudal peduncle broad, compressed, its depth 2.2 in head.

Color in alcohol, very pale brown; side marked with many large incomplete rings of blackish or dusky and with a number of dusky spots in between; fins whitish, the vertical or unpaired with large blackish spots on membranes between rays and similar small ones scattered about, those of caudal forming about 4 crossbands; several dusky spots at base of pectoral; right side whitish.

We have seen but 2 examples, both taken at Hilo: Type, No. 50658, U. S. N. M., 2.5 inches long. Cotype, No. 7471, L. S. Jr. Univ. Mus., 1.9 inches long.

Family ANTENNARIIDÆ.

57. Antennarius drombus Jordan & Evermann, new species.

Head (to end of opercle) 2.5 in length; depth 1.75; eye 5 in head; snout 4; width of mouth 2; D. 1-1-12; A. 7; P. 12; V. 5.

Body very deep, compressed, back elevated; head deep, with blunt conic profile in front, somewhat oblique above; snout broad, obtuse, surface uneven; mouth broad, large, nearly vertical; maxillary concealed under skin, reaching below anterior part of eye; lips fleshy; teeth in jaws minute, in narrow bands; teeth on palatines rather large, sharp-pointed, none on vomer; tongue broad, thick;

mandible large, with fleshy knob at symphysis, projecting; nostrils circular, well separated, with rounded fleshy rims; interorbital space convex, roughened; top of head with rather large concave pit; eye high, anterior; bait rather short, only reaching a little beyond first spine, with fleshy caruncle at extremity; dorsal spines short, first free, rough, depressible in pit on top of head; second dorsal spine twice length of first, equal to width of mouth, depressible, and united with skin of back to its tip; posterior dorsal rays longest, and the last, like that of anal, united to caudal peduncle by a membrane; anal similar, rounded, elongate, 1.5 in head; pectoral broad; ventral small, rounded; caudal peduncle small, compressed, its depth equal to interorbital space; body rather rough, mucous pores on head and in lateral line with excrescences; side of body with many pointed cutaneous flaps; second dorsal spine and first dorsal ray very rough, also with cutaneous flaps; lateral line very convex, running down toward middle of base of anal.

Color in alcohol, pale plumbeous gray, more or less spotted or mottled with darker; belly and lower surface rather pale, the spots distinct; fins all more or less pale with dark spots, some at basal portions of dorsal and anal darker; iris blackish with radiating lines of golden.

The above description is from the type, No. 50659, U. S. N. M. (field No. 541), taken at Waikiki, near Honolulu.

Another example (field No. 539) shows some differences: Head (to end of opercle) 2.5 in length; depth 1.7; eye 3 in head; maxillary 1.8; width of mouth 1.7; interorbital 3.7; D. 1-1-12; A. 7; P. 12; V. 5.

Body very deep, compressed, back elevated; head deep, gibbous, with blunt conic profile in front, somewhat oblique above; snout broad, obtuse, short, surface uneven; mouth large, obliquely vertical; maxillary large, reaching a little beyond front portion of eye; lips fleshy; teeth in jaws minute, sharp, in bands; teeth on roof of mouth large, sharp-pointed; tongue large, broad, thick; mandible large, with knob at symphysis, projecting; nostrils well separated, close to end of snout, each with elevated fleshy rims, the anterior higher; interorbital space broad, elevated, uneven; top of head with rather large pit; eye high, anterior; bait short, reaching tip of first dorsal spine, with caruncle at extremity; dorsal spines short, depressible; first dorsal spine half length of second, free, depressible in pit on top of head; second dorsal spine large, joined by skin to its tip; dorsal rays of about equal height, seventh 1.3 in head, and the last, like lower portion of last anal ray, adnate to caudal peduncle by a membrane; anal rounded; caudal elongate, rounded; pectoral broad; ventral small; body rather rough, mucous pores on head and lateral line with excrescences; along the lateral line and anterior region of dorsal are many cutaneous flaps; lateral line convex, running down to above middle of anal.

Color in alcohol, dark gray-brown; edges of vertical fins whitish, the pale border rather broad and very distinct along posterior, dorsal, anal, and caudal rays; side with about 6 large round blackish spots; caudal with some pale or indistinct mottlings; pectoral and ventral with rather broad margins, median portion dusky; iris more or less silvery.

A. drombus seems nearest related to *A. nummifer* Cuvier & Valenciennes, originally described from Malabar. Probably the specimens from the South Seas referred to the latter belong rather to *A. drombus*. *A. nummifer* is said to be red in color with dark spots, and, as figured by Dr. Day, differs in several respects from *A. drombus*. Both these species differ from *A. commersoni* and its numerous allies or variants (*A. niger A. leprosus*, *A. rubrofuscus*, and *A. sandvicensis* from Hawaii) in the shortness of the first dorsal spine or fishing rod. This is scarcely longer than the second spine in *A. drombus*, but in *A. commersoni* it is twice as long.

Our collections contain but 2 examples of this species, the type, No. 50659, U. S. N. M. (field No. 541), and cotype, No. 7472, L. S. Jr. Univ. Mus. (field No. 539), both taken on the reef at Waikiki, near Honolulu.

DESCRIPTIONS OF A NEW GENUS AND TWO NEW SPECIES OF FISHES FROM THE HAWAIIAN ISLANDS.

BY DAVID STARR JORDAN AND BARTON WARREN EVERMANN.

Since the publication of our recent paper[a] on new species of fishes from the Hawaiian Islands, further studies of our large collections have resulted in the discovery of an interesting new species of *Tropidichthys* and a remarkable new genus of *Scorpænidæ*. These are described in the present paper. Illustrations of both species will be given in our final report.

Tropidichthys psegma Jordan & Evermann, new species.

Head 3 in length; depth 2; eye 4.5 in head; snout 1.5; interorbital 2.3; D. 11; A. 11; C. 8; P. 16.

Body short, stout, moderately compressed; snout long, conic; anterior dorsal profile rising evenly to region above gill-opening, at which point the body is deepest; interorbital flat; gill-opening nearly vertical, short, its length less than diameter of eye; mouth low, below axis of body; teeth strong, convex, cutting edge sharp; eye small, supraorbital rim not prominent; caudal peduncle deep, its least depth about 2 in head, its least width 4 in its least depth; length of caudal peduncle from dorsal fin to base of caudal fin 1.3 in head; from base of anal fin 2 in head; dorsal prominence equally distant between tip of snout and posterior base of caudal; base of dorsal 1.5 in height of fin, which latter is 2 in head; anal similar to dorsal, its edge rounded; caudal truncate, or very slightly convex, 1.2 in head; pectoral broad, its base 2.6 in head, free edge oblique, posterior rays 1.5 in anterior ones; body mostly smooth; interorbital space and snout above and on side with small prickles; belly with a few prickles; a scattered patch also on side above pectoral.

Color in alcohol, dark brown above, paler below, 3 or 4 short black lines running forward from orbit, and same number backward; lower part of side, especially posteriorly, and lower part of caudal peduncle, with small roundish black spots; snout and interorbital space crossed by about 12 narrow black lines, these extending down on side of snout; side of snout with 3 or 4 narrow black lines from chin toward eye, separated by paler lines; posterior to these small irregular black spots covering entire cheek, dotted over with fine white specks; ends of spines, pectoral, dorsal, and anal pale whitish, their bases largely brownish black; caudal dark brownish or black.

This species is known to us only from the type, No. 50885, U. S. N. M. (field No. 2561) 3.75 inches long, obtained by us at Honolulu in 1901.

We have compared this specimen with examples of *T. coronatus* Vaillant & Sauvage, obtained by the *Albatross* in 1902, and find them quite distinct.

IRACUNDUS Jordan & Evermann, new genus.

Iracundus Jordan & Evermann, new genus of *Scorpænidæ* (*signifer*).

Allied to *Hebedceus* and *Pontinus*. Body rather elongate, compressed, covered with small, weakly ctenoid scales; fins not scaly; head not depressed; formed as in *Sebastodes*, the spines moderately developed; head and body with dermal flaps; teeth on jaws and vomer, none on palatines; dorsal fin deeply divided, the spines 11 in number, the fourth much elongate; pectoral rays undivided; anal rays III, 5; ventral rays I, 5; caudal rounded; vent at base of first anal spine; air bladder obsolete.

---

[a] Descriptions of new genera and species of fishes from the Hawaiian Islands. <Bull. U. S. Fish Comm. 1902 (April 11, 1903), pp. 161-208.

Iracundus signifer, Jordan & Evermann, new species.

Head 2.4 in length; depth 3.2; eye 4 in head; maxillary 2; D. x, 1, 9; A. iii, 5; P. 17; V. i, 5; scales about 9-55;30, about 45 pores.

Body rather elongate, moderately compressed, the head conic, not depressed; mouth large, oblique, the lower jaw slightly projecting, the maxillary reaching to opposite posterior margin of pupil; teeth in moderate bands in the jaws, the inner teeth in the upper jaw slightly largest; vomer with small teeth; palatines toothless; interorbital area deeply concave, little wider than pupil; spines on top of head low and rather sharp, much as in *Sebastodes*; preocular, supraocular, postocular, tympanic, occipital, and nuchal spines present; a ridge with 2 spines outside the tympanic spine; preorbital moderate, about as wide as eye, with a sharp spine turned forward and a blunt spine turned backward; suborbital stay a narrow, simple ridge, reaching base of preopercular spine, which is straight and very short; 3 lower preopercular spines reduced to blunt points; opercle with 2 slender diverging spines, the upper the larger, their points not reaching edge of membrane; head with numerous, broad, fleshy flaps; a fringed flap at the nostril, 2 on edge of preorbital, 2 on lower limb of preopercle, and a high fringed flap above eye, about as long as pupil; small simple flaps on the cheek, the end of the maxillary, and elsewhere on head; large pores on lower jaw, under suborbital stay, and elsewhere; opercle and upper part of cheek with rudimentary, embedded scales; jaws naked; top of head scaleless, occiput covered with thin skin and scarcely depressed; gillrakers very short, thickish, and feeble, all but about six reduced to mere rudiments; no slit behind last gill; body covered with small, close-set scales, which are slightly ctenoid; scales on nape small, on breast minute; lateral line conspicuous, provided with dermal flaps; numerous scattered flaps on sides of body.

Dorsal fin very deeply notched, the spines rather slender, pungent, the first a little longer than eye, the second and third subequal, about half longer, the fourth greatly elevated, 1.5 in head, almost twice height of third and fifth, which are subequal; sixth, seventh, and eighth slightly longer than fifth, tenth very short, eleventh half length of fourth; soft dorsal high, the longest rays nearly half head; rays of all the fins scaleless; caudal long, rounded, 1.4 in head; anal high, the spines graduated, the third a little longer than second, which is 2.6 in head; longest soft rays 1.8 in head; pectoral with the rays all simple, the longest 1.2 in head, lowest rays shortened and thickened; ventral fins inserted below axis of pectoral, rather long, 1.6 in head, not quite reaching anal, inner rays well free.

Color, pale in alcohol, doubtless vermilion red in life, the flaps on body pinkish; a single jet-black spot about half diameter of pupil near tip of membrane between second and third spines of dorsal.

The only example known is the type, No. 50886, U. S. N. M. (field No. 655), a specimen 4.2 inches long, taken by us on the coral reef at Honolulu.

THE FRESH-WATER FISHES OF WESTERN CUBA.

By C. H. EIGENMANN,

Professor of Zoology, University of Indiana.

SAN JUAN RIVER, LOOKING UPSTREAM FROM THE FIRST BEND ABOVE THE UNITED
HABANA RAILROAD BRIDGE

SAN JUAN RIVER FROM ABOVE THE UNITED HABANA RAILROAD BRIDGE.

THE FRESH-WATER FISHES OF WESTERN CUBA.

By C. H. EIGENMANN,

Professor of Zoology, University of Indiana.

During March, 1902, the writer, accompanied by one of his students, Mr. Oscar Riddle, as assistant and interpreter, made a series of collections in the fresh waters of western Cuba, in the streams accessible by the Western Railway and the United Habana Railways. Attempts to reach waters remote from the railways were abandoned on account of the expense, both in time and money. Sumidero was reached by horse from Pinar del Rio, and the caves about Cañas on foot and by volante. The original and chief object of the visit to Cuba was to secure material for a study of the eyes of the blind fishes, *Stygicola* and *Lucifuga*. In this I was successful. The fresh-water fishes proved also of considerable interest. As might have been expected, many of the more abundant and larger species had been previously described by Poey. Nothing, however, was known about the distribution of fresh-water fishes, and there were found a surprising number of new species.

I wish here to express my thanks to Mr. A. P. Livesey and Mr. J. E. Wolfe, the managers of the Western and United Habana Railways, who did all in their power to make the available time profitable from a scientific standpoint. I am also under obligations to Mr. Philip Hammond, the chief engineer of the United Habana Railroads, for suggestions and various favors. Mr. Pascual Ferreiro, of the Cuban railway postal department, kindly acted as guide to the Pedregales caves, and Mr. Francisco Martinez and his brother to the caves about Cañas. The success of the expedition was largely due to my companion, Mr. Oscar Riddle, whose previous stay in Porto Rico and trip to Trinidad and the Orinoco had familiarized him with the language of the country and enabled him to deal with the natives.

The drawings illustrating this paper were made by Mr. Clarence Kennedy.

In his "Memorias sobre la Historia Natural de la Isla de Cuba," tomo 2, pp. 95–114, 1856, Poey describes two species of blind fishes, *Lucifuga subterraneus* and *Lucifuga dentatus*, from caves on the southern slopes of the jurisdiction of San Antonio, Guanajay, and San Cristobal." They were first brought to notice by the surveyor, D. Tranquilino Sandalio de Noda. Specimens were secured for Poey by Dubroca, Fabre, and Layunta.

---

[a] Jordan & Evermann, in their Fishes of North and Middle America, III, p. 2501, give a number of localities for which I can find no authority in Poey. Thus "San Antonio, Cuba (Coll. D. Tranquilino); Sandalio de Noda (Coll. D. Juan Antonio Fabre)" and cave at the "Castle of Concord." I can not find the authority for the locality San Antonio, Cuba; the collector given as D. Tranquilino is probably D. Tranquilino Sandalio de Noda: the second locality Sandalio de Noda is probably the latter section of the name of the man who first called attention to them. Castle of Concord should probably read "Coffee plantation La Concordia."

The localities from which Poey secured *Lucifuga subterraneus* are (1) Cave of Cajio, 5.2 miles south of La Guira de Melena (Noda, 1831); (2) cave of the coffee plantation La Industria, halfway between Alquizar and Guanimar (Dubroca); (3) cave of Ashton, near San Andres (Fabre); (4) cave of the Dragon, on the cattle farm San Isidro, near Las Mangas (Fabre); (5) cave on the coffee plantation La Concordia, 5.2 miles from Alquizar (Layunta); (6) cave near the beehouse of the coffee plantation La Paz (Dubroca); a well near the tavern Frias (Fabre).

Lucifuga dentatus was secured in Nos. 1, 2, and 3. Those from Nos. 1 and 2 were said to be without the least vestiges of eyes; those from No. 3 with vestiges.

I visited a number of the caves from which Poey secured his specimens, intending to obtain specimens from as many of Poey's localities as possible, but especially from those from which he secured both species of the blind fishes. The towns Guira de Melena, Alquizar, San Andres (now Cañas), mentioned by Loey, are successive stations along the Western Railroad, and Las Mangas is a town a short distance off the railroad beyond Cañas. We made our first stop at Alquizar, hoping to be in the middle of the cave region. The driver we engaged at Alquizar to take us to La Industria showed his independence and originality by taking us over a rough road during a half day's drive parallel to the railroad to Frias and Ashton near the station Cañas, but we were not inclined to quarrel with him, as we at once secured fishes in Ashton, and caves were reported to us as very abundant in the whole region south of Cañas. Many of these caves were visited by us, as well as that at Cajio (by Mr. Riddle) and others in a widely distant part of the island. Those from which we secured specimens I propose briefly to describe here. The cave on La Industria we did not visit, and in fact, except while we were engaging our volante, no one about Alquizar seemed to know the plantation La Industria. It is possible that the name has been changed in recent years. Bearing in mind our experiences with the volantemen at Alquizar we made no attempt to find La Concordia, which is also reached from Alquizar.

The "caves" about Cañas can best be described after a few words concerning the country in general in which they occur. The territory about Cañas is entirely drained by underground streams. The streams rising in the hills and mountains forming the watershed between north and south drainage run above ground for a distance and then disappear underground. The Ariguanabo River thus runs into a bank at San Antonio de los Baños and disappears among fallen rocks. A few yards away from its "sumidero" the water can be seen running in its underground channel through an opening in the thin roof of the channel. A few yards farther on a dry channel leads down to the water which at the end of the channel disappears among fallen rocks. Other rivers disappear in a similar manner. Their waters reappear, in part at least, in a number of "ojos," some near the coast south of San Antonio. The region drained by underground streams is flat, with frequently no indications of surface streams and their erosion, and extends westward to near San Cristobal, where the first permanent surface stream is observed. At Artemisa and Candelaria stream beds contained pools of water at the time of our visit.

From San Cristobal to Pinar del Rio there were many small but perennial streams. Eastward from Cañas the cave region has an unknown extent. Poey limited it to the jurisdiction of Guanajay, but it certainly extends as far east as the meridian of Matanzas, and from reports probably beyond Cienfuegos. East of Rincon there are, however, frequent river beds, dry during the time of our visit. This main region, belonging to the southern slope, sends a point northward from Rincon to the Almendares River in the northern watershed. Aside from the "Ojos de agua" along the edge of the cienegas skirting the southern coast there are two notable places where undergound rivers find an exit. One at Vento supplies the entire city of Habana with its water, the other serves to make the region about Guines a garden, its waters being used for irrigation. Others in all probability have a subaqueous exit to the south.

The large spring at Vento, sufficient to supply the city of Habana with its water, is the only one which issues on the northern slope, so far as I know. The origin of the supply from the Vento spring has not been traced. It issues but a few feet from the Almendares River, and it is very remotely possible, though not at all probable, that it derives its water from the upper courses of the Almendares. At the time of our visit the water of the spring was 1 degree warmer than that of the Almendares River. The region north of that river, across the river from Vento, being shut out from a possible contributing source, it undoubtedly derives its water from the system of underground streams mentioned above. An examination of the best available map and the levels of the Western and United Habana Railroads makes it seem quite certain that the Vento springs derive their water from the region immediately south of Vento and north of Rincon and Bejucal—that traversed by the two railroads mentioned. This region contains various sinks without surface outlets, as well as dry sink-holes, and is the northward-projecting point of the cave region mentioned above. A notable sink-hole in this region is that at Aguada, on the United Habana Railroad. This is very broad, shallow, and dry during the dry season, but the water rises to stand over 10 feet deep on the railroad track during some of the wet seasons.

THE CAVES VISITED.

The soil over the region under consideration is thin, the surfaces of the very irregularly corroded rocks jutting out in numerous places. This, together with the fact that the water of the underground streams is but a few feet underground, gives the region an entirely different aspect from the sink-hole and underground-stream region in Indiana, Kentucky, etc. It is, in the first place, impossible to enter the underground streams, and there are no funnels on the surface to indicate the location of an underground stream or its tributary. In places the thin limestone roof of an underground chamber has given way and enables one to get to the water, which in all the numerous places we visited was stationary, not flowing. With one or two exceptions the water was covered with a continuous crust of carbonate of lime, due to the evaporation and discharge of carbon dioxide from the surface of the perfectly quiet lime water. When the water is disturbed flakes of variable size break loose and gradually sink to the bottom. All of the so-called caves about Cañas, with one possible exception, were sink-holes formed by the breaking of the thin roof of a larger or smaller underground chamber. In all of the caves where stalagmites and stalactites were noticed these extended for 3 feet or more into the water. Inasmuch as they could not have formed under water, the latter must have risen since their formation. Usually the walls of the sink-hole retreat downward and sidewise, suggesting that a channel filled with water extends down and out from the sink-hole. The impression is irresistibly made that there exists here an extensive series of drowned caves. As our visit to the caves occurred near the end of the dry season, this excess of water could not have been due to recent excessive rains. The water in the caves was, however, still falling and fell several inches during March.

The condition could have been brought about (1) by the rising of the southern coast of Cuba, resulting in a changed incline in the underground streams and a consequent drowning of the caves; (2) by a blocking of the outlets of the streams; (3) by an increase of the rainfall above that provided for in the past. I was at first inclined to favor the first view, since caves as widely different as those of Cañas and Alacranes showed the same characters; but a cave at Matanzas, on the northern slope, showed exactly the same conditions. A local blocking of the outlets is scarcely conceivable on account of the wide separation of caves showing the same conditions. Concerning the rainfall I am not able to speak, but any other cause, since the caves are above the level of the ocean, does not occur to me.

The detailed description of various caves may begin with Modesta No. 1. The caves, for the most part, do not have distinctive names, but are called after the Finca on which they occur. I have added the numbers 1, 2, etc., for convenience. Modesta No. 1 is an ideal cave of which all the rest are modifications. There is nothing on the surface to distinguish it when one is a few feet away. The cave is bell-shaped, with an opening 10 to 15 feet at the top. A tree growing at its margin sends vertical roots down to the bottom. On these roots notches have been cut, and the descent is made by means of them. At the water level at the time of our visit the cave was oval in section, 30 by 45 feet in extent. In the middle of the bell, and immediately under the opening, there was a large pile of rocks, cemented together in places by stalagmitic material and rising but a few inches above the water. The water, beautifully clear, became rapidly deeper in all directions and could be seen to extend out in at least two directions in deepening channels filled to their top with water. The roots descending from the opening at the top to the island, a distance of about 15 feet, here divided suddenly into a tuft of innumerable rootlets, most of them in the water. Such roots were found in almost all the caves, and the young blind fishes were always found in among the rootlets; the big ones among the rocks.

Modesta No. 2, also called Hawey, is a cave of the same type, except that the central mass of fallen rock forms an arch over the water, and that it can be reached from one side by an inclined plane, also formed of fallen material. Rootlets were very abundant here and small blind fishes equally so. The water was probably not more than 10 feet below the surface and at one edge was very deep—how deep we had no means of determining. Part of the opening had at one time been walled in and the cave was used as a well. It is possible that this is the well mentioned by Poey as being near Frias and containing blind fishes. But a number of other wells in the neighborhood reach caves.

Modesta No. 3 contained no blind fishes.

San Isidro No. 2 is a duplication of Modesta No. 1, but with different proportions. The central mass is higher and holds a number of stalagmites. It is only partially surrounded by water. The entrance is by roots.

PLATE 20.

RIO SAN DIEGO AT PASO REAL LOOKING UPSTREAM FROM BELOW WESTERN RAILROAD
BRIDGE

RIO DEL PINAR, LOOKING UPSTREAM FROM BRIDGE.

San Isidro No. 1 is an underground chamber containing 2 to 3 feet of water in places and muddy in the bottom. It is reached by an inclined plane from the side and is more like a cave as ordinarily understood than the caves of the Modesta type.

La Frias No. 1 is a cave of the Modesta type, but larger and with the roof of one side fallen, so that the descent to the water is made by an inclined plane, and the water forms a crescent about 15 feet wide under the retreating walls of the side opposite to the entrance.

La Frias No. 2 is another cave of the Modesta type, 100 feet across at the bottom and with the water 40 feet from the surface. The roof has fallen in at one side and the central mass of rocks rises nearly to the level of the surrounding region, so that the descent can be made by a winding inclined plane. Part of the roof, very thin and worn through in places, is still standing, supported by stalagmite-stalactite columns. The roots of trees wind about stalagmites or descend as straight and unsupported stems 15 and more feet through the cave to the water, where there is the usual breaking up into rootlets.

Ashton No. 1 is a cave of the Modesta type, with a large amount of fallen material at one side, where one can descend to the water by means of steps. The diameter of the cave is possibly 100 feet. A large tree grows in the center of the fallen material. At the deeper part of the cave, opposite the entrance, the walls retreat downward. Stalactites and stalagmites are present, and there is a crescent of pools of water of variable width and depth. Channels filled with water are seen to lead off from the bottom of the pools. This is one of the rare places where green plants are growing in the water. They are confined to the dextral pools as one enters.

Ashton No. 2 differs from all the other caves. It is more open; the roof has fallen in, so that there are no retreating walls, as in the other caves. The fallen material slopes gradually to the water, which is shallow and densely covered with duckweed. A channel leads off from the left end of the water as one enters, and we went into this with our canoe, but the walls soon came down to the water. The place was sultrily hot and smelled disagreeably of decaying plants, so that we were very glad to get out. This was one of the few caves we visited in which there were no fishes. This cave practically joins Ashton No. 1, the dividing wall being but a few yards wide.

Ashton No. 3, called Los Baños, differs from the others in that considerably more of the roof of the original chamber remains standing and the amount of territory covered by the water is also considerably larger. The depth of the water differs from a few inches to "overhead." This cave is occasionally used as a bath. At the time of our visit the water was covered with the usual crust of lime.

Cajio, Jaiguan, and La Tranquilidad were visited by Mr. Riddle.

Cajio, 6 or 8 miles southeast of La Guira de Melena, differs considerably in one particular from all the other caves. The water lies at a much greater distance from the entrance than in the others. The entrance is an ordinary sink-hole 10 to 12 feet deep. Instead of finding the water at the bottom of this sink, as in the Modesta type, it is perfectly dry. Leading from this, however, there is a dark, narrow passageway, 100 feet long, which leads to a very large chamber with a crescent-shaped body of water. This long channel is not an inclined plane, but runs parallel to the surface. The thin roof of the main chamber has a hole blasted into it, through which light reaches the water. The floor is of fallen rock, as in other caves

The cave of Jaiguan, 2 or 3 miles east of Cajio, is a chamber 100 feet long by 40 or 50 feet wide. It probably contains more water than any other cave visited, with the possible exception of Pedregales. There is an opening in the center of the roof affording considerable light. The entrance is a small sink-hole at the edge of the cave, which is provided with steps cut in the rock. The roots of a tree reach from the central opening down to the water, a distance of 18 feet. The rootlets, very abundant here, were found to shelter many of the young blind fishes. In this cave is a central, rocky islet, formerly completely surrounded by water, as in Modesta No. 1. A narrow bridge of rocks has been built, which unites it with the entrance to the cave.

La Tranquilidad No. 1, in the Cañas region, 3 miles west of Ashton, is of the Modesta type. It is a large cave, entered only by means of a rope from an opening in the roof 4 feet in diameter. The water here presents an extensive surface, but nowhere is it deep. Large specimens were taken here. Side channels allow one to follow the water farther in this cave than in any other visited. The greater part of the cave is very dark.

La Tranquilidad No. 2 is a small cave with much light. It is in the center of a large sink probably 600 feet in diameter. The water is not deep and is easily accessible to cattle and swine. The bottom is formed of very soft, deep mud.

The number and species of fishes taken in these caves are listed under the head of the various species. In many of the cases 2 blind crustaceans, both of them new species, were found to be abundant. One of them is a very graceful *Palæmonetes*, the other, a *Cirolana*, is much more abundant and forms a large part of the food of the blind fishes. [a]

My attention was called by Messrs. Wolfe and Hammond to the Pedregales caves near Alacranes in Matanzas Province, 60 miles east of the easternmost of the caves from which Poey recorded blind fishes. They were visited more for the sake of visiting all the caves that might possibly contain blind fishes than with the expectation of finding any. When near these caves we inquired whether any of them contained fishes and were told "yes, but they don't amount to anything; they don't have any eyes." After this remark we felt thoroughly comfortable in a place where certainly nothing else contributed to comfort.

Pedregales cave, about 3 miles from Alacranes, differs in some respects from the caves of the Cañas region. The cave slopes down as steeply as can conveniently be descended from a narrow opening, but once inside it widens out, descending continually. The floor was formed of a section of a cone that recalls the central masses in the bell-shaped caves of Cañas. The floor and the roof were elaborately decorated with stalagmites and stalactites, some of them united and ranging from 3 feet in diameter to a fraction of an inch. These were pure white when broken, but tinted a red on the outside by the coral earth. When struck they gave out a clear bell-like tone, and the striking of various-sized columns by different members of the party produced a pleasing chime-like effect. Among the caves that I have visited this is approached in elaborateness of decoration only by the cave of the fairies in Colorado.

[a] See W. B. Hay, in U. S. Nat. Mus., XXVI, pp. 429-435.

At a vertical depth, judged by the depth of a near-by well to be about 75 feet, water was encountered in the form of the usual crescent. There was no indication that we had reached the bottom of the cave and it is not known how deep the water is, for roof and floor continue to slope down with the same incline and stalagmites rise from a depth of at least 3 feet beneath the level of the water at the time of our visit, when it was about 3 feet below its maximum height. The water was covered with a crust of lime and no fishes were seen.

An account of the Pedregales caves will soon be issued by Mr. Pascual Ferreiro, of the Cuban railway postal service, a member of the international copyright commission.

About half a mile beyond Pedregales is the M cave, so called from the M-shaped path that leads from the surface to the water. The descent in this cave is much steeper than in Pedregales and the stalagmitic decorations much less elaborate. A dim light penetrates to the water. Here, as in Pedregales, the cave floor continues to descend for an unknown distance below the level of the water. Fishes were more abundant here than in any other cave visited. They were all of one species.

We visited another cave in Matanzas Province, about 20 miles north of the Pedregales, at the edge of the city of Matanzas. It was essentially like the M cave, but contained no fishes.

THE STREAMS VISITED.

Between the western end of the island and Union, south of Matanzas, a number of streams run by independent courses from the watershed to the sea. Those west of San Cristobal are perennial. Those immediately east of San Cristobal consisted at the time of our visit of a series of independent pools. East of Artemisa the streams run above ground only part of their course, then enter caves and continue their course to the sea underground, or reappear as "ojos de agua" a short distance from the ocean. In the western part of this region, from Cañas to at least Guira de Melena, there are no dry beds or other surface indications of drainage. Farther to the west dry beds of streams, narrow and crooked, were seen, but with one exception there is no perennial stream between San Cristobal and Union except along the coast. The one exception is the stream formed by the large springs near Guines. East of Union we did not go.

The Rio Sabanalamar at San Cristobal is about 20 feet wide and varied from 1 foot to 2 inches in depth in cross-section. The water is in places very swift and shallow; in others "over head" and in pools. The banks of the river are clay; the bottom is grass-grown except in the deep pools and over riffles. We seined up and down from the railroad bridge and also in an old channel of the river containing a muddy pool entirely cut off from the river. The water of the river was clear and at 10 a. m. had a temperature of 23 C.

The Rio Palacios at Los Palacios varied from 5 to 40 feet in width. The water was clear, the bottom alternately gravel, mud, and weed-grown. Temperature of the water 23 C. We seined up and down the ford at the end of the main street. Between Los Palacios and Paso Real the country is in part swampy, with lily ponds.

The Rio San Diego at Paso Real de San Diego is 15 to 40 feet wide with steep banks about 20 feet high. The water was clear, in pools and riffles, and 23° C. We seined near the railroad bridge. There are several dry beds of tributaries in the neighborhood. At Herradura a small creek, 5 feet across and with banks 20 feet deep, was crossed. No fishing was attempted.

Between Las Ovas and Golpe were ponds with white water-lilies. No fishing was attempted here.

The Rio del Pinar is, at the town of the same name, a broad shallow stream with alternating riffles, pools, and weedy patches. Collections were made above and below the wagon road near the ice factory.

The Rio Cuyaguateje is the most western river of any size that flows to the south. Collections were made near Sumidero. The river near this town tunnels twice through rocky walls several hundred feet high, and in one instance probably not much more than 200 feet thick, and in the other probably several times as thick.

The Almendares River is a deep and swift stream about 40 feet wide emptying into the ocean near Habana. It was scarcely possible to seine in the river itself. Collections were made above and below a dam at Calabazar and in a small tributary just above the dam at Calabazar. The temperature was not taken at the time of seining. A few days later it was 25 at Vento.

The Ariguanabo is of special interest, inasmuch as it is one of the rivers that disappears in a cave. Collections were made just above its entrance to the cave. It is a clear, swift stream running through the town of San Antonio de los Baños. Above the town a dam at the ice factory has deepened the water so that a small steamer can run up to Laguna Ariguanabo. At the time of our visit the river was for a long distance above the town blocked with water hyacinths and other water plants. A much smaller though similar brook which runs through Guanajay also disappears in the ground several miles below the town. No attempt was made at seining.

Collections were also made in the outlet of the Yumuri at Matanzas, but we did not succeed in ascending to fresh water, and no fresh-water fishes were secured.

We ascended the San Juan River from Matanzas to the head of tide water, where a shallow ford occurs. Collections were made in the ford, above and below the ford, and in pools of spring water. Immediately above the ford the surface of the stream was covered with water hyacinths, and the stream was 4 and more feet deep. At the ford the water had a maximum depth of about 18 inches, and in places formed shallow riffles. Below the ford the banks become steep and the water is too deep for a collecting seine.

PECULIARITIES AND ORIGIN OF THE CUBAN FISH FAUNA.

There are recorded in the present paper 36 species and subspecies. These belong to 25 genera and 13 families. A number of other species have been taken in the same region, notably *Lepisosteus tristœchus*. Of the 37 species and subspecies (including the last-named species) but 4, aside from members of the *Gobiidæ*, are found in fresh water elsewhere. They are the species of *Lepisosteus, Symbranchus, Agonostomus,* and *Anguilla. Lepisosteus tristœchus* is found in the fresh waters of Mexico and the southern United States. *Symbranchus marmoratus* is generally distributed through the fresh waters of the tropics of America. *Anguilla chrysypa* is also found in the streams of eastern North America, and *Agonostomus monticola* is found in the fresh waters of the West Indies and Mexico. Of the remaining species only those of the genus *Heros* belong to a strictly fresh-water family. The genus *Heros* is generally distributed in South and Central American waters, one of its members entering the United States. The members of the marine family of *Gobiidæ* are found in the streams and brackish water of tropical America generally, and their presence in Cuba is not so significant as their absence would be. A number of the species enumerated are marine, and their presence in the rivers may be looked upon as purely fortuitous; these are *Tarpon atlanticus, Doryrhamphus lineatus, Centropomus undecimalis, Lutianus jocu, L. griseus, Eucinostomus meeki, Gobius soporator, G. boleosoma*, and *Lophogobius cyprinoides*.

Two species, in many ways the most interesting fishes found in the region examined, are members of the deep-sea family *Brotulidæ;* they are the blind-fishes *Stygicola dentatus* and *Lucifuga subterraneus*. These have evidently worked their way up the underground streams and are now becoming readapted to the light in the upper courses of the streams. No other members of the family are found in fresh water anywhere. *Atherina* is a marine genus with the peculiar Cuban species as its sole fresh-water representative. The remaining species are all members of the *Pœciliidæ*, a family inhabiting brackish water and coastwise streams. Of the *Pœciliidæ* 2 genera, *Girardinus* and *Toxus*, are peculiar to Cuba.

The origin of the Cuban fauna is then not far to seek. We have, as mentioned above, a number of marine species, more or less regular visitors of the fresh water. We have species widely distributed in the brackish water and coastwise streams whose presence is predicable (*Gobiidæ*), and we have local modifications of families with a wide distribution in the brackish and fresh waters of the tropics of America (*Pœciliidæ*). We have, furthermore, local adaptations of marine species to fresh water (*Brotulidæ* and *Atherina*). The origin of all the above is simple of explanation. The species whose presence is of greatest interest are the strictly fresh-water species of *Lepisosteus*, evidently belonging to the North American fauna, and *Symbranchus* and *Heros* as evidently members of the South American fauna. The presence of the eel in the fresh waters of Cuba is to be expected, inasmuch as it very probably breeds in the ocean near Cuba. The presence of *Symbranchus, Heros*, and *Lepisosteus tristœchus* and *Agonostomus monticola* shows that the fresh-water fauna of Cuba has a greater affinity for that of Mexico than for that of Florida, and that these forms probably reached Cuba by way of Yucatan.

SYSTEMATIC LIST OF FISHES COLLECTED, WITH DETAILS OF DISTRIBUTION.[a]

SYMBRANCHIDÆ.

Symbranchus marmoratus Bloch. T. Pinar del Rio.

ANGUILLIDÆ.

Anguilla chrysypa Rafinesque. San Juan, near its first ford; Paso Real.

ELOPIDÆ.

Tarpon atlanticus (Cuvier & Valenciennes). M.

Pinar del Rio, 4 specimens, 20, 119, 182, and 192 mm., from a deep pool beneath the wagon bridge, many miles from the sea. They are locally known as "sadina," and we had been told that we should find them in this spot.

PŒCILIIDÆ.

The members of this family are everywhere abundant, especially in streams bordering the cave region.

Key to the genera of Cuban Pœciliida.

a. Anal fin of male similar to that of female, oviparous: intestinal canal short, little convoluted; teeth little movable; dentary bones firmly connected; lower jaw strong and usually projecting beyond upper.
  b. Teeth all pointed, in villiform bands.
    c. Air-bladder well developed, no caudal ocellus; gill-openings not restricted above, opercles free from shoulder girdle; dorsal and anal nearly equal; origin of dorsal in advance of anal...............*Fundulus*
    c. Air-bladder wanting; a black ocellus at root of caudal in male; dorsal smaller than anal, its origin behind that of anal...*Rivulus.*
  b. Teeth tricuspid in one row, no villiform band of teeth; body short and deep, compressed; dorsal short, of 10 to 12 rays, first ray slender and rudimentary...*Cyprinodon*
aa. Anal fin in male placed well forward and modified into a sword-shaped intromittent organ.
    d. Intestinal canal short, little convoluted; teeth all pointed, in bands; eye normal; jaws not produced; dorsal short, of 6 to 10 rays, behind origin of anal; mouth wide, chin low..................*Gambusia.*
    dd. Intestinal canal elongate, much convoluted.
      e. Teeth compressed, entire, without lateral cusps.
        f. Anal process in male very long, serrate behind near tip and with finger-like claspers (the prepuce being modified into a pair of clasps); dorsal in both sexes behind origin of anal.
          g. Dentaries and intermaxillaries firmly united; teeth of outer row fixed, a band of minute teeth behind them.
            h. Teeth of outer row much expanded at tip, broadly spade-shaped in upper jaw, close-set, their margins overlapping; teeth near middle of lower jaw asymmetrically expanded, lateral lobes prolonged and ending in a point...*Glaridichthys.*
            hh. Teeth of outer row wide-set, scarcely expanded, spear-shaped, those near middle of lower jaw in two irregular series..*Toxus. h not*
          gg. Dentaries and intermaxillaries loosely joined, teeth of outer row movable, inserted on lips, a few teeth behind them or none, those of outer row wide-set, scarcely expanded, spear-shaped; those of middle of lower jaw in two irregular series..*Girardinus.*
        ff. Anal process comparatively short, a leaf-shaped prepuce attached to the anterior surface covers the tip, tip without claspers; dorsal in female in advance of origin of anal.
          i. Tip of anal process in male ending in a simple antrorse hook, no serra on its posterior surface...*Pœcilia*
          ii. Teeth all pointed, origin of dorsal behind that of anal....................................*Heterandria.*

Fundulus cubensis Eigenmann, new species. *

I am somewhat in doubt as to the generic position of this species. Its short intestine, double row of teeth, unrestricted gill-openings, and position of its dorsal in relation to its anal, and similarity of sexes (at least one of the three specimens is a male) seem to indicate that it is a species of *Fundulus.*

Type: No. 9667, Ind. Univ. Mus., 29 mm. long; Pinar del Rio.

Cotypes: Two specimens, 26½ mm. long; Pinar del Rio, at the ford just above wagon bridge.

Head 3.6, about equal to depth; D. 11 or 12; A. 10 or 1, 10; scales 24. Origin of dorsal very slightly nearer tip of snout than base of middle caudal rays and over the eighth scale of lateral line; origin

---

a The following characters are used to indicate the general distribution of the genera and species enumerated.
t, genus peculiar to Cuba; *, species peculiar to Cuba; T, generally distributed in tropical fresh waters; M, marine species.
b refer = a quiver full of arrows.

of anal under eleventh scale. Dorsal and anal high, angular behind, last ray somewhat produced, 4.3 in length to base of caudal; tip of last anal ray reaching base of caudal, tip of last dorsal ray nearly so; ventrals reaching to origin of anal; pectorals about to middle of ventrals; mouth very small and very oblique; eye longer than snout, 3 in head, very little less than interorbital; profile straight in front, slightly convex toward dorsal. Teeth all conical, in at least two series, those of outer series enlarged.

Coloration brilliant. Ventrals and anal without pigment, caudal with two or more cross-series of spots; a dark humeral spot just above origin of pectoral; dorsal dusky; a dark band through lower part of eye forward below chin; a dark band from eye to tail, its median half much darker than edges, bordered above and below by pigmentless bands bright orange in life, the upper one beginning below origin of dorsal; back above upper light band thickly covered with pigment spots; below lower light band a dusky band of equal width consisting of a series of contiguous V-shaped markings from axil of pectoral to tail; below this another pigmentless band extends from axil of pectoral to just behind anal; below this is a narrower band of dusky from axil to last anal ray; below this another short pigmentless band; edge of branchiostegal membranes and a line forward from their union to chin black; ventral surface otherwise without pigment except a faint line between ventrals.

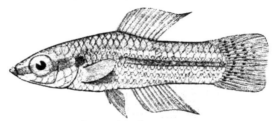

FIG. 1 *Fundulus cubensis* Eigenmann, new species.

Gambusia punctata Poey *

Everywhere abundant. The following specimens were collected: San Antonio, 3 females, the largest 92 mm. long; Modesta No. 2, 3 females, the largest 84 mm.; Modesta No. 3, 4 females 37 to 72 mm.; San Cristobal, 50 females 31 to 84 mm.; 11 males 40 to 53 mm.; Palacios, 18 females 54 to 88 mm.; 5 males 50 to 61 mm.; Paso Real, 2 females 50 to 55 mm.; Sumidero, 3 females and 1 male; Pinar del Rio, 9 females, the largest 84 mm.; San Juan near Matanzas, 30 females, the largest 70 mm.; 5 males, the largest 42 mm.; Calabazar, 7 females, the largest 75 mm.

Gambusia puncticulata Poey. *

This species is shorter and deeper than *Gambusia punctata*, which it greatly resembles. The dorsal in the specimens taken counts 8 to 10, not 11, as Garman found in his specimens.

San Antonio, 3 females, the largest 58 mm., D. 9; San Cristobal, 4 females, the largest 48 mm., D. 8, 9, and 10; Palacios, 1 female 39 mm., D. 9; Pinar del Rio, 4 females, the largest 47 mm., and 3 males, the largest 35 mm.

GLARIDICHTHYS Garman

Intestinal canal elongate; males with anal fin modified into a very long intromittent organ; jaws much more firmly united than in *Girardinus*, each with a series of close-set, broad-tipped, entire or but slightly crenulate teeth, these teeth not movable, a narrow band of smaller, broad-tipped, conical or tricuspid teeth behind them; fins small, anal in advance of dorsal in both sexes.

Most nearly allied to *Goodea* and *Girardinus*, and differing from them in character of teeth, *Goodea* having tricuspid teeth, *Girardinus* having movable loosely-set, and *Toxus* having hastate teeth.

Glaridichthys uninotatus Poey. *

Abundant. Fifty specimens preserved from San Cristobal, 47 to 84 mm. long; a number of specimens have, in addition to the lateral spot, a spot on either side of anus and sometimes a black streak connecting the two; in one instance the lateral spot on one side is entirely replaced by the anal spot; males (13) 38 to 47 mm.

At Palacios 24 specimens were preserved, 50 to 81 mm. Fluctuations in the lateral spots of female as in San Cristobal specimens; in one specimen only a small spot on one side, none on other.

Paso Real, 2 females; Pinar del Rio, 4 females.

Inner series of teeth in upper jaw small, with sharply triangular or tricuspid teeth in 4 or 5 rows; inner series of lower jaw little expanded at tip. Teeth of outer row near middle of jaws irregularly expanded, lateral lobes prolonged into a point; lateral teeth of lower and of upper jaw equally expanded.

Glaridichthys falcatus Eigenmann, new species. *

Type, No. 9664, Ind. Univ. Mus., a female, 82 mm. long, from San Cristobal.

Cotypes: Eight females from an old river channel at San Cristobal, the smallest 60 mm. long, the largest 85 mm.; 4 females 50 to 53 mm., from Palacios, taken in a muddy pool in the river bed at the ford; 8 females and 2 males from Rio del Pinar, the females 38 to 47 mm., the males 29 and 37 mm. This species reaches its maximum size and is most abundant in warm, muddy pools.

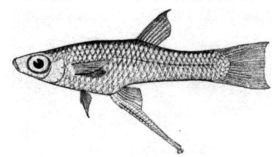

Fig. 2. *Glaridichthys falcatus* Eigenmann, new species. Male.

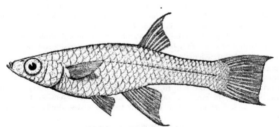

Fig. 3. *Glaridichthys falcatus* Eigenmann, new species. Female.

Body long, slender, little compressed; head 4; depth 4 (in pregnant females 3.5); D. 9; A. 11; scales 29; head broad, wedge-shaped in profile, with lower jaw very oblique, projecting; eye very large, longer than snout, 2.6 in head, 1.4 in interorbital; mouth very oblique, small; interorbital divided into 3 distinct regions by longitudinal grooves, central portion convex; origin of dorsal equidistant from base of middle caudal ray and origin of pectoral; dorsal and anal falcate; second rays sickle-shaped, each extending for one-third its total height beyond tip of last ray when folded back, little less than length of head; caudal emarginate, some of outer rays prolonged; origin of anal in female about equidistant from base of middle caudal ray and anterior margin of eye, its seventh ray under origin of dorsal; ventrals usually reaching to anal; pectorals about to middle of ventrals in female, to base of anal in male.

San Cristobal specimens very pale; a dusky streak from nape along middle of back to caudal, scales above lateral line faintly edged with black; a black line along middle of sides composed of a single series of chromatophores, a black streak along ventral surface from anal to caudal; otherwise colorless.

Palacios specimens colored like those from San Cristobal, except one in which each scale of the side below lateral line is edged with a series of chromatophores and there is a faint hint of 8 dark spots along median black line; region above lateral line dusky.

Pinar del Rio specimens colored like the darker Palacios specimen, sometimes a black streak on either side of anus and forward to ventrals. Male with modified portion of the anal very long, with retrorse spines behind and a little clasper on tip of longest rays.

Glaridichthys torralbasi Eigenmann, new species. *

Type, No. 9662, Ind. Univ. Mus., male, 45 mm. long, from Pinar del Rio, represents apparently a new species. In general appearance it very greatly resembles the males of *Girardinus metallicus* and *Glaridichthys uninotatus*. From the former it differs in possessing bands of teeth in jaws behind spatulate row, first row of teeth fixed, and dentary and premaxillaries much more firmly united; from the second it differs in coloration, having no lateral spot and a conspicuous dorsal band. There are no other species which it resembles.

D. 9; A. 10; scales 28; head 3.6; depth 5.3; body compressed, elongate; mouth small, subterminal, lower jaw not prominent; eye greater than snout, less than 3 in head, equal to interorbital; teeth in outer series overlapping, those of lower jaw more pointed; a band of minute teeth behind at least the front row in upper jaw; dorsal small, its first ray equidistant from base of middle caudal rays

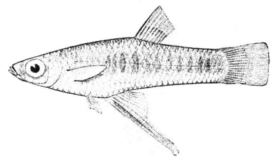

Fig. 4. *Glaridichthys torralbasi* Eigenmann, new species.

and from eye, its highest ray nearly equal to distance of pectoral from anterior margin of eye; caudal truncate, equal to distance of pectoral from tip of snout; anal process long, with serrae behind, and a clasper at end of anterior rays; ventrals small, not much longer than eye; pectoral reaching to anal; dorsal with an arched band reaching from full length of first ray to base of last; caudal, ventral, and pectoral colorless; first ray of anal process largely black; sides with about 10 dark cross-streaks; all scales margined with black, dorsal ones not heavily so; a dark dorsal streak, a black ventral line behind anal, a dusky band around head just in front of eyes.

I take pleasure in naming this species for Prof. José I. Torralbas, of the chair of zoology in the University of Habana.

Girardinus metallicus Poey. †

San Cristobal, very abundant, largest female, 77 mm. long; largest male, 45 mm. A few females with black on anal.

Palacios; largest female, 69 mm.; largest male, 41 mm.; ventral band from chin to tail black in one male; one male blotched with black.

San Antonio; largest female, 79 mm.

Ashton, females usually with black on anal; largest female, 51 mm.; largest male, 38 mm.

Pinar del Rio; largest female, 68 mm.; largest male, 41 mm.; a number of males with a black streak of varying intensity and width along the ventral surface.

Girardinus garmani Eigenmann, new species. † *

Type, No. 9661, Ind. Univ. Mus., one male, 35 mm., Pinar del Rio. Cotypes, one male, 35 mm., Pinar del Rio; one male, 38 mm., Palacios.

D. 9; A. 9; scales 29; depth 3.4 to 3.6; head 3.6 to 3.8. Body compressed; head truncated, lower jaw nearly vertical; mouth very small, lips thick, teeth in a single series in each jaw, sloped as in *G. metallicus*, very movable, intermaxillaries and dentaries not united; eye as long as snout, 2.5 in head, equal to interorbital; origin of dorsal a little nearer pectoral than base of middle caudal rays; dorsal rounded, small; dorsal, caudal, and pectoral of about equal height, equal to distance of pectoral from eye; anal process 2.5 in length, serrate near its tip behind and with a clasper at its end; ventrals very small, reaching to the anal; ventral surface colorless except a black line from anal to caudal; scales of side with a dark margin of increasing width toward back; a dusky dorsal streak; head in front of eye dark above and below; region below eye colorless; a well-defined black spot on base of last 5 dorsal rays; first dorsal ray black; anal process blackish on basal half, a small indistinct black spot on distal half of last anal membrane and extending at times on neighboring regions; sides without streaks or bars.

This species differs from *G. metallicus* and *G. denticulatus* in being entirely without streaks or spots or bars on the sides, in the number of anal rays, and in other characters.

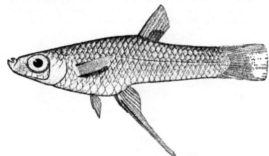

FIG. 5. *Girardinus garmani* Eigenmann, new species

TOXUS Eigenmann, new genus. †

Toxus Eigenmann, new genus of *Poeciliida* (*riddlei*).

This genus differs from *Glaridichthys* in its narrow teeth, from *Girardinus* in having its jaws firmly joined.

Toxus riddlei Eigenmann, new species. †*

Type, No. 9656, Ind. Univ. Mus., a female, 59 mm. long, from San Cristobal.

Cotypes: 1 female 66 mm. long, and 2 males 33 and 34 mm. long, from San Cristobal. Head 4; depth 3.4; D. 9; A. 10; scales 28; origin of dorsal midway between base of middle caudal rays and origin of upper pectoral ray and over thirteenth scale of lateral line; origin of anal below eleventh scale; fins moderate; longest dorsal ray equaling length of head without snout; eye equaling snout, little more than 3 in head; interorbital convex, equaling snout and eye; profile slightly curved; outer row of teeth movable, spear-shaped, not very closely set, brown-tipped, a band of minute teeth behind them; a dark lateral band crossed by about 6 indistinct dark crossbands; scales of sides with a light center and a narrower or broader margin of dark, forming reticulations; a dark streak extending down between eye and angle of mouth; rest of lower side of head and belly white; pectoral colorless; caudal faintly dusky; anal with a faint dark band through the middle, the tips and base colorless; dorsal dusky.

Male much smaller, the color contrasts sharper; about 7 well-marked dark crossbands in the larger specimen; dorsal tipped with dusky, a black band from base of last dorsal rays forward toward basal third of fourth dorsal ray; in the larger specimen a series of dark spots on dorsal rays on a level with tip of first one; a black band through middle of anal, most intense upon and entirely covering last rays; first fully developed ray black in the large specimen, colorless in the smaller; origin of anal under eighth scale.

I take pleasure in naming this species for Mr. Oscar Riddle, to whom much of the success of the expedition to Cuba is due.

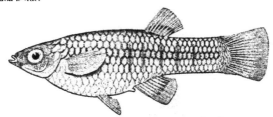

Fig. 6. *Tozus riddlei* Eigenmann, new species. Female.

Fig. 7. *Tozus riddlei* Eigenmann, new species. Male.

Pœcilia vittata Guichenot. *

San Cristobal, over 50 females, the largest 94 mm. long; a number of them with peculiar black blotches; the small ones with 3 yellow stripes below the lateral line and frequently with one or more series of black dots along lower part of side; 32 males, the largest 70 mm.

San Antonio, 5 females, the largest 110 mm.; 2 males, the largest 73 mm.

Los Palacios, 70 females, the largest 100 mm., showing great variability in the intensity of coloration; 35 males, the largest 65 mm.

Calabazar, 9 females, the largest 102 mm.; 4 males, the largest 70 mm.

Paso Real, 3 females, the largest 80 mm.; 1 male, 53 mm.

Pinar del Rio, 17 females, the largest 83 mm.; 9 males, the largest 63 mm.

Sumidero, 2 females.

Heterandria cubensis Eigenmann, new species. *

Type, No. 7663, Ind. Univ. Mus., a female, 59 mm., from Los Palacios.

Cotypes, one female with young, 53 mm., from Los Palacios, and one female, 38 mm., from Pinar del Rio. These specimens agree well with the characters of the genus *Heterandria*, as restricted by Garman, except that the outer series of teeth are movable. Head 5 to 5.2; depth 3.3 to 3.7; D. 9; A. 10; scales 29. Body elongate, little compressed, general shape that of *Fundulus;* profile regularly curved from dorsal fin to eyes, flattened over eyes and forward; mouth small, opening upward, the lower jaw projecting; bones of jaw loosely united; eye longer than mouth, 2.5 to 2.7 in head, very little less than interorbital; origin of dorsal a little nearer head than base of middle caudal rays, over first third of anal; dorsal and anal both falcate, the anterior rays extending considerably beyond tip of last when laid back; highest dorsal ray slightly shorter than highest anal ray, about equal to length of head; caudal a little longer than head; ventral reaching to anus; pectoral reaching to ventral.

Scales of the mid-dorsal line with their dorsal halves dusky, those of entire side margined with black, most distinct above lateral line; a series of 12 narrow dark vertical bands about as wide as pupil and as high as eye; an irregular black streak along middle of side; dorsal tipped with dusky, the first membrane black; a black line and a few chromatophores along each anal ray; a black line from anal to caudal; chin dusky; pectoral, ventral, and belly colorless.

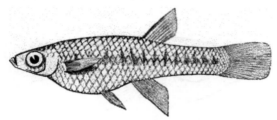

Fig. 8. *Heterandria cubensis* Eigenmann, new species.

SYNGNATHIDÆ.

Doryrhamphus lineatus (Valenciennes). M.

Two specimens from the Rio San Juan below the ford.

ATHERINIDÆ.

Atherina evermanni Eigenmann, new species. *

Type No. 9657, Ind. Univ. Mus., 45 mm., from San Cristobal. Cotypes, 35 specimens, 36 to 50 mm., from San Cristobal, and 5 specimens, 28 to 38 mm. long, from Pinar del Rio.

D. v-i, 9-11; A. i, 12 to 15; head 3.5 to 4; depth 4.5 to 5; scales 32; head rather pointed; mouth oblique, the lower jaw projecting; maxillary reaching a little beyond front of eye; eye 2.75 in head, equal to distance from tip of snout to anterior margin of pupil; little wider than interorbital; teeth minute; spinous dorsal inserted behind tips of ventrals, its origin equidistant from tip of snout and middle caudal rays, or a little nearer caudal; caudal peduncle very slender, its least depth less than

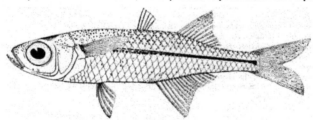

Fig 9. *Atherina evermanni* Eigenmann, new species.

length of eye; caudal little less than length of head; anal inserted in advance of origin of dorsal; ventrals small, not reaching anal; pectoral reaching tips of ventrals; a conspicuous lateral band most intense on caudal peduncle, gradually fading out under pectoral; region above this in all cases thickly peppered with black cells, most thickly so along median line; region below this in many cases similarly but less intensely spotted; lower side of head and breast white; ventrals nearly free from pigment; all the other fins with pigment, cells of greater or less intensity along the rays. This species is readily distinguishable from the other species of *Atherina* by the smaller number of scales.

I take pleasure in naming this species for Dr. Barton Warren Evermann, in recognition of his valuable work on the fishes of the West Indies, especially his work on the fishes of Porto Rico.

MUGILIDÆ.

Agonostomus monticola (Bancroft). T.

Rio San Juan, near its first ford, abundant, the largest 140 mm. long; Sumidero, abundant, the largest 170 mm.; Pinar del Rio, abundant, the largest 160 mm.

CENTROPOMIDÆ.

Centropomus undecimalis (Bloch). M.

Rio San Juan, at its first ford, abundant, varying in length from 55 to 250 mm.

LUTIANIDÆ.

Lutianus jocu (Bloch & Schneider). M.

Two specimens from the Rio San Juan, just below the ford.

Lutianus griseus (Walbaum). M.

One small specimen from the Rio San Juan, just below the ford.

GERRIDÆ.

Eucinostomus meeki Eigenmann, new species. M.

Type, No. 9090, Ind. Univ. Mus., a specimen 135 mm., from San Juan River, just below its first ford. General appearance of *Ucirma lefrogi*, differing from all other species of the genus *Eucinostomus* in having but 2 anal spines.

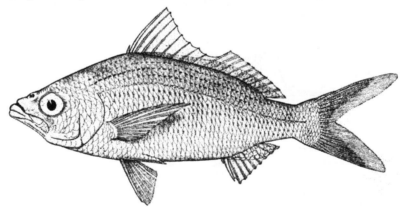

Fig. 10 *Eucinostomus meeki* Eigenmann, new species

Head 3.25; depth 3; D. IX, 10; A. II, 8; scales 4–46–9; eye 1 in snout, 3 in head, 1 in interorbital. Body elongate, little compressed or elevated, the dorsal profile but little more elevated than the ventral; snout pointed, the profile from snout to dorsal gently arched; mouth narrow, terminal, but little above the lower margin of the eye; maxillary reaching to vertical from front of eye, 3.4 in head, its exposed part boat-shaped, a trifle more than twice as long as wide, 5 in head; intermaxillary groove entirely naked, its width 5 in the interorbital; preopercle and preorbital entire; dorsal spines slender, the second longest, 6 in the length; ventrals short, reaching half way to anal; pectoral long, 3.5 in the length, reaching beyond tips of the ventrals, but not to vent. First anal spine minute, the second equal to the length of the eye.

Color, ashy gray, with some metallic reflections; dusky lines along the rows of scales; sides and back everywhere punctate with minute dots; vertical fins dusky; ventrals and pectorals lighter.

Named for Dr. Seth Eugene Meek, assistant curator of zoology, Field Columbian Museum, in recognition of his excellent work on Mexican fishes.

CICHLIDÆ.

Individuals of the genus *Heros* are as numerous in the streams of Cuba as individuals of the *Centrarchidæ* are in the streams of equal size in the Ohio valley. They were found by us down to tide water, but not in it. Only a single species has been recorded from Cuba, and nothing has been said either concerning its distribution or its variation. No one, except possibly Poey, has before this compared numbers of specimens from different places or even from the same place. Such a comparison is therefore very desirable, and the material collected far surpasses all other collections made before. We have altogether 236 specimens from various localities. An examination of all of these proves either the presence of several instead of a single species on the island or a remarkable variation with localities. A definition of the variations has proved very elusive. The numbers of fin rays and scales are uniform, so that the differences exist in the proportions and the color. But the coloration also has a certain underlying uniformity. There is a spot near the middle of the side, another at the base of the caudal, and an obscure third above the gill-opening. There are numerous small spots on the fins and on scales of the sides, especially below and on the opercles, and sometimes on the cheeks. There is also a longitudinal streak from the eye through the lateral spot to the caudal spot, and a

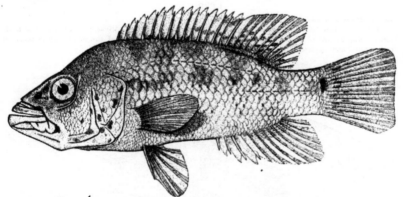

Fig. 11. *Heros tetracanthus torralbasi* Eigenmann, the type of new subspecies.   (Type A.)

definite number of crossbars, both streak and bars most conspicuous in the young and in light-colored adult individuals. This uniformity of underlying structure makes defining of species or varieties a difficult proceeding. The polymorphism is further complicated by instances like the following: The specimens from San Antonio are readily referable to a certain form found at Calabazar, although they differ from Calabazar specimens in quite readily distinguishable features; but one of them differs notably from all other specimens collected at San Antonio, and would unhesitatingly be considered a species distinct from the other specimens from the same locality. But at Palacios the same form branching from the Calabazar form approaches the characters of the single specimen from San Antonio.

I venture to describe here certain of the most aberrant forms as new, without, however, feeling sure that they are really distinct varieties or species or that some of the other forms referred to *H. tetracanthus* are not also new.

Heros tetracanthus Cuvier & Valenciennes. *

Heros tetracanthus torralbasi Eigenmann, new subspecies.

(25 specimens, 60 to 181 mm. long, from Calabazar.)

These specimens come from the Almendares River, and as this flows near to Habana it is very probable that the type of *Heros tetracanthus* came from the same river. Cuvier & Valenciennes say that Poey's drawing, on which their *tetracanthus* was based, resembled *Ambloplites* in outline, and possessed spots in the angles of the scales. This very well describes some specimens I have (figs. 12 and 13).

D. xv, 11; A. iv, 9; depth greatest below first dorsal spine, 2.5 to 2.7; depth of caudal peduncle 2.5 to 2.25 in head; scales 27 to 29; pores 17 to 21 - 10 to 13; body heavy forward, tapering from the shoulders to the caudal peduncle; jaws heavy, lips thick; snout 2.75 to 2.5 in head; eye 4.5 to 5 in head (3.5 in young), 1.5 in interorbital; no pore in upper angle of gill-opening (except on one side of one individual.); highest dorsal and anal rays reaching base of caudal; highest dorsal ray 4.5 to 5.3 in

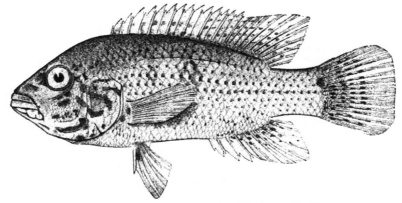

FIG. 12. *Heros tetracanthus tetracanthus* Cuvier & Valenciennes   (Type B.)

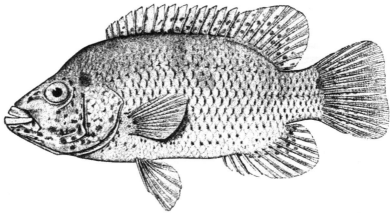

FIG. 13. *Heros tetracanthus tetracanthus* Cuvier & Valenciennes   (Type C.)

the length; last dorsal ray 3 in the longest; lateral and caudal spots conspicuous in young, which have a series of light crossbars; two light bars usually confluent over the lateral spot; fins dusky, the vertical ones lighter-edged and with some spots on their bases; no spots on head or body in the youngest; in larger ones spots appear about the base of the pectoral, opercle, angle of preopercle and mandible. There is great variation in the distinctness of the lateral bands.

In the larger specimens there are 2 types of coloration; type A has more or less distinct vertical bars, alternating light and dark; the lateral and caudal spots are distinct, the dark crossbars are darkest in a line between the two; a dark streak extends from the eye to second dark bar, this with the darker areas on the crossbands forming an interrupted lateral band; cheeks unspotted, opercles and mandible with dark spots; ventral surface plain; a few scattered spots along the sides; vertical fins more or less spotted at the base.   Type B shows no crossbands, each scale of the side with a dark spot forming longitudinal series; cheeks as well as opercles and mandible spotted or the spots confluent into lengthwise streaks; vertical fins more conspicuously spotted; cheeks in the young of both types unspotted; sides of the young of type B less regularly spotted than in the adult.   Figs. 11 and 12 are drawn from males, of the same size and with reproductive organs in the same stage of development. Fig. 12, type B, evidently represents the variety figured by Poey, and is the original *tetracanthus*. Fig. 11, representing type A, may be termed *H. tetracanthus torralbai* Eigenmann, var. nov. (No. 9672, Ind. Univ. Mus.).

(78 specimens, 44 to 160 mm. long, from San Cristobal.)

Most of these were taken out of a muddy lagoon near the river, and all were very pale in color, the crossbars showing well.   In these paler specimens there are no indications of a longitudinal stripe. The largest, which also came out of the lagoon, is nearly uniform light ashy, there being but faint indication of crossbars and spots; there are faint spots on fins and opercle.   Other smaller specimens are everywhere profusely spotted.   In the darker specimens from the river there is a dark lateral band.

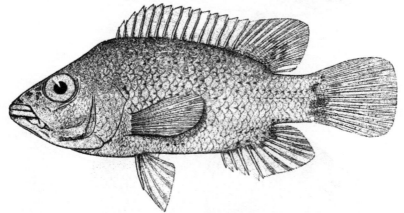

Fig. 14. *Heros tetracanthus griseus* Eigenmann, new subspecies.

(70 specimens, 47 to 190 mm. long, from Los Palacios.)

These specimens are of types A and B from Calabazar, with some distinct features   The dark spots (of B) along the rows of scales are, in some of the lighter individuals, nearly faded out; in the darkest ones they spread nearly over entire margin of scales (fig. 13).   The cheeks are spotted or streaked in some of the largest specimens and not in others.   The lateral band in some specimens of Type A is as well developed as in some San Antonio specimens; vertical bars vary also very much in intensity.   One dark specimen resembles in almost all respects *Heros griseus* from San Antonio.   Depth, 2.25 to 2.5.

(17 specimens, 62 to 155 mm. long, from Pinar del Rio, are of A and B from Calabazar.)

(4 specimens, 45 to 240 mm. long, from Sumidero.)

All these specimens are dark, the smaller one nearly uniform, with but faint crossbars; lateral and caudal spots distinct.   The largest one is a male, very dark above, without distinct markings, and with black streaks and spots on cheeks, opercles, and lower sides.   Depth 2; head 2.6; eye 5 in head   These may referred to *tetracanthus*.

Among 32 specimens from San Antonio there are 3 distinct types, one of which may simply be the adult of one of the others. They are all elongate, the depth being 2.3 to 2.5 in length. There are, in the first place, 4 adults measuring 160 to 190 mm. in length, of type B from Calabazar. They differ from the Calabazar specimens in having the spots along the scales larger and less regular. In two of the specimens the cheeks are spotted, in two others the spots are confluent into vertical or longitudinal streaks. In the largest some of the dorsal rays are prolonged, reaching to near middle of caudal. These are probably the adult of 17 specimens from 92 to 157 mm. long. In all of these, even in the smallest, there are spots on the cheeks, more numerous in the larger, and confluent into streaks in the largest. Sides irregularly spotted; lateral band more or less conspicuous as the fish is lighter or darker; there are traces of the usual light and dark bands in some individuals. All of these are evidently typical *H. tetracanthus*.

There are 11 specimens, 90 to 123 mm. long, evidently modifications of *H. tetracanthus torralbasi* from Calabazar, in which, except one very dark specimen, the lateral band is very conspicuous. Vertical bands quite well marked except in darkest specimens. Cheeks are unspotted except in one individual which has faint spots; sides without small dark spots.

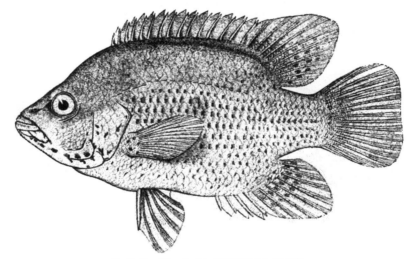

FIG. 15. *Heros tetracanthus latus* Eigenmann, new subspecies

Heros tetracanthus griseus Eigenmann, new subspecies *

Type: No. 9670, Ind. Univ. Mus.; a specimen 117 mm. long, from San Antonio.

D. xv, 11; A. iv, 8; scales 27; pores 17 - 10; depth a little more than 2.5; depth of caudal peduncle 2.5 in head; head 2.7 in the length. Shape and general characters of *Heros tetracanthus* from the same place, differing in the color and the notably larger eye, as compared with specimens of *H. tetracanthus* of the same size. Eye 3.7 in head (4.6 in *H. tetracanthus* of the same size); 1 in interorbital (1.5); 1.25 in snout (1.7); preorbital five-sevenths of eye; snout 3 in head; highest dorsal and anal reaching base of caudal; highest dorsal 4.7 in the length; highest anal 5; ventrals reaching to vent. No lateral spot; a faint caudal spot; sides ashy with irregular dark spot; a few whitish streaks through some of the scales above lateral line; cheeks plain, a few spots on opercles; soft portions of vertical fins spotted at base; no traces of dark crossbars.

Heros tetracanthus latus Eigenmann, new subspecies. *

Type: No. 9669, Ind. Univ. Mus., a specimen 160 mm. long, from San Juan.

This is a narrow, deep fish with projecting lower jaw, a pointed snout, and a depression in the profile over eyes. D. xv, 11; A. iv, 9; depth 2; head 2.7; depth of caudal peduncle 2 in head; scales 28; pores, 18 + 12; snout pointed 2.75 in head; eye 4.5 in head, 1.5 in interorbital; no foramen in upper angle of gill-opening; maxillary reaching vertical from front of orbit; highest dorsal and anal rays reaching to end of basal two-fifths of caudal; highest dorsal ray twice as high as last ray, equaling longest caudal ray, equaling length of head without opercle; highest anal ray slightly shorter; ventrals to vent.

Ashy gray, darker above to light below; each scale of lower parts of side with black spot on tip, extending over to next scale and forming distinct series; lower part of opercle, preopercle, lower part of cheek, and lower jaw with dark spots; vertical fins dusky; soft dorsal and anal spotted; ventrals dark, inner ray light; pectorals pale.

Heros tetracanthus cinctus Eigenmann, new subspecies. *

Type: No. 9671, Ind. Univ. Mus., a specimen 129 mm. long, from Paso Real.

Four specimens were taken at Paso Real measuring a 68, b 72, c 129, and d 136 mm. long, respectively, and differing in coloration from those taken at any other point. The two larger are very dark, one having very distinct markings.

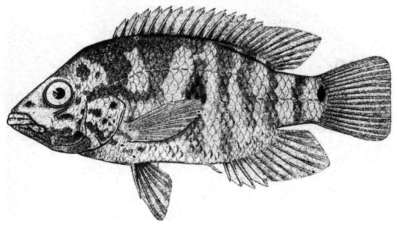

Fig. 16. *Heros tetracanthus cinctus* Eigenmann, new subspecies.

The specimens approach *H. nigricans* but have a normal lateral line: c (the type) is most aberrant in its coloration, d approaches the coloration of *tetracanthus torralbasi* from the Almendares, a and b are indistinguishable from other young except in the band through the lateral spot.

D. xv, 11 (c); xvi, 10 (d); xv, 10 (b); xvi, 11; A. iv, 8; scales 28, pores 15 + 9 (a); 16 + 8 (b); 15 + 9 (c); 19 + 11 (d); head 2.6 to 2.7; depth 2.25 to 2.2; depth at end of opercle 2.4 to 2.43; snout pointed, 2.75 to 3 in head; mouth horizontal, maxillary nearly concealed when the mouth is closed; maxillary about reaching vertical from front of orbit; highest dorsal and anal rays reaching little beyond base of caudal 5 in length, last dorsal ray 2.3 in the longest; ventrals reaching vent; gill-opening with a supplementary pouch above, reached by a larger or smaller foramen.

Dark; a black lateral and a black caudal spot; side with 7 light crossbars; a light streak across nape from upper angle of gill-opening, another across from behind eye, another between eyes; snout light; spaces between light bars form dark bars of about equal width except first two on body, which are much wider at the top; a black streak from eye across upper angle of gill-opening to the second dark bar; cheeks and opercles with black spots and streaks, ventral surface spotted; vertical fins and ventrals dark; soft dorsal and anal with obscure spots; pectoral pale, unspotted. The dark band in which the lateral spot is located continued to the back.

Young much lighter colored, a few dark spots along opercle and below cheek; ventral surface unspotted, lateral and caudal spots conspicuous.

A fifth specimen, 152 mm. long, from Paso Real, is a typical *tetracanthus*.

Heros nigricans Eigenmann, new species.

Type, No. 9668, Ind. Univ. Mus., a specimen 192 mm. long, from Pinar del Rio

One specimen was secured. It is the most prominent of the aberrant forms, and I should unhesitatingly describe it as a distinct species if I had obtained more than one specimen.

D. xiv, 11; A. iv, 10; scales 28; pores about 15 - 0; head 2.6; depth a trifle less than 2 in the length; depth at end of opercle 2 in length ⅓ diameter of eye; depth over middle of eye equals length of head, less one ocular diameter; snout pointed, upper and lower profiles nearly equally inclined to behind eye; upper profile gibbous behind eye; maxillary very little exposed when mouth

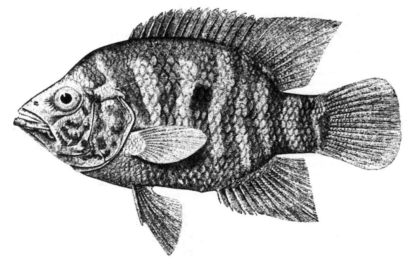

Fig. 17. *Heros nigricans* Eigenmann, new species.

is closed; eye 1.5 in interorbital. Gill-cavity, with a small supplementary pocket at its upper angle, entered by a large foramen; lateral line irregularly developed on left side; no pores on either side of tail; soft dorsal falcate, fifth ray as long as caudal, 3 times as long as last ray, its tip reaching nearly to middle of caudal, 3.6 in length, longer than head, less opercle; caudal broadly rounded; anal falcate, fourth ray longest, equaling head, less opercle; ventrals reaching to vent.

Color everywhere nearly black, with 7 lighter crossbars, 2 of which are on caudal peduncle; an additional light streak from angle of gill-opening across nape; cheeks and opercles with lighter marking; fins nearly uniform black.

GOBIIDÆ.

Philypnus dormitator (Lacépède). T.

Rio San Juan, and seen in Vento springs near Havana; very abundant in the San Juan, where specimens 46 to 256 mm. long were obtained; young with black stripe from tip of lower jaw to caudal.

Dormitator maculatus (Bloch). T.

Rio San Juan, at its mouth and at the first ford; very dark; lagoon at San Cristobal, very light.

Eleotris pisonis (Gmelin). T.   Rio San Juan, at its mouth and at the first ford.

Lophogobius cyprinoides (Pallas). M.   Rio San Juan.

Gobius soporator Cuv. & Val.   M.   Mouth of Rio San Juan at Matanzas and at its first ford.

Gobius boleosoma Jordan & Gilbert.   M.   Mouth of Rio San Juan and at its first ford.

Awaous taiasica (Lichtenstein).   T.   Sumidero.

BROTULIDÆ.

Stygicola dentatus (Poey). †*

This blind fish was taken in the M cave near Alacranes, 20 specimens; Jaiguan, 5 specimens; Frias, 2 specimens; Modesta, 4 specimens; and Tranquilidad, 7 specimens; caves near Cañas without specific locality, 5 specimens. In all, 43 specimens were secured, ranging from 60 to 152 mm. long. Poey records this species from Cajio and Ashton, in which it was not found by us. He also records it from La Industria, which is said to be between Cajio and the Cañas caves.

The males of *Stygicola dentatus* are distinctly larger than the females. Average length of the 20 females caught is 97 + mm., the largest one 120 mm.   The average size of the 23 males is 113 + mm., the largest one being 152 mm. long. The males were in excess of the females in the ratio of 100 females to 115 males. There is but an appreciable difference in the averages of the fins, as far as these could be counted, the average formula for the females being D. 91.4, A. 74; and for the males D. 91.1, A. 73.6; or the average for the two are D. 91.2, A. 73.6.

Lucifuga subterraneus Poey. †*

This species was taken in all but one of the caves in which *Stygicola dentatus* was taken, and in several others besides. The localities are Ashton, 13 specimens; Los Baños, 5 specimens; Cajio, 3 specimens; Hawey, 16 specimens; San Isidro, 2 specimens; Jaiguan, 18 specimens; Las Frias, 5 specimens; Modesta, 2 specimens; Tranquilidad, 3 specimens; Cañas, without specific locality, 9 specimens; total, 76 specimens, ranging from 24 to 94 mm. long. The females of *Lucifuga subterraneus* are distinctly larger than the males. In making the average for the size of the sexes individuals less than a year old were not considered, because the differences in the sexes would, if present, be but very slight, and because in such young the sex could not always be determined with certainty. An examination of all specimens makes it probable that at the end of a year after birth the young are about 50 mm. long. In obtaining the average size of the sexes only specimens over 50 mm. were considered. The males above this size measure 59.7 mm. on an average, with a maximum of 94 mm. The females measured 71.1 mm. on an average, with a maximum of 93 mm. Of the specimens over 50 mm. long 23 were males and 22 females, or 100 females for every 104.5 males. The fin formula to the nearest decimal for those of the individuals over 50 mm. which would be counted is, males, D. 82.1, A. 67.4; females, D. 81.9, A. 68. The average formula for those less than 50 mm. long is D. 83, A. 67.2, or for all together, D. 82.6 +, A. 67.5.

While the average number of rays differs considerably in the two species, the number in each varies so much that the numbers in individual cases overlap, individuals of *Lucifuga* reaching as high as 88 dorsal rays and individuals of *Stygicola* as low as 87. The same is true regarding the anal.

A female of this species (fig. 3, pl. 21), 65 mm. long, contained four young about 20 mm. long.

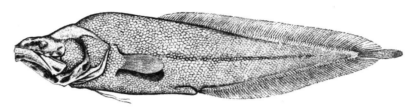

STYGICOLA DENTATUS.

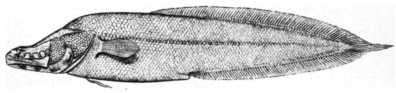

LUCIFUGA SUBTERRANEUS.

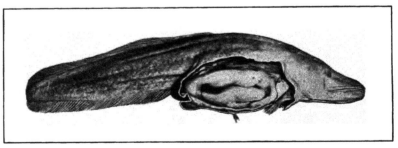

LUCIFUGA. A BLIND FISH CONTAINING UNBORN YOUNG WITH WELL-DEVELOPED EYES.

Contributions from the Biological Laboratory of the U. S. Fish Commission
at Woods Hole, Massachusetts.

THE ORGAN AND SENSE OF TASTE IN FISHES.

By C. JUDSON HERRICK,

Professor of Zoology in Denison University

237

CONTENTS.

238

Contributions from the Biological Laboratory of the U. S. Fish Commission,
Woods Hole, Massachusetts.

THE ORGAN AND SENSE OF TASTE IN FISHES.

By C. JUDSON HERRICK.

Professor of Zoology in Denison University.

INTRODUCTION.

The practical problems connected with the fisheries have been attacked (and in large measure successfully solved) by a rough-and-ready application of the method of trial and error, and the scientific investigator has merely to follow after and explain why a given form of trap or method of lure is successful with one species of fish and not with another. But there remain many unsolved problems of great economic importance, and it is the function of scientific research to contribute to the solution of these problems in a more orderly and economical manner, even though it often happens that the investigator best qualified to solve the scientific problem has not the practical knowledge of fishery matters necessary to apply his own results to economic problems, and so his facts have to be worked over from the other point of view before they become practically useful.

We are, in fact, profoundly ignorant of the senses and instincts of the fishes, even those connected with their feeding habits, which are of so direct importance to all commercial fisheries. Nearly all which one finds in the scientific literature bearing on the senses of fishes is merely inference of function based on a study of the structure of the organs—a most precarious pathway for scientific research. My own studies on the nerve components of fishes have led me to certain inferences regarding the functions and the distribution of the organs of taste in fishes, and the present study is an attempt to follow out these inferences by the determination of more exact facts regarding the pathways of gustatory stimuli as anatomically demonstrable, together with sufficient direct physiological experiment to furnish definite information of the function served by this system of sense organs and of their nervous paths in the fishes.

Neurologists have always paid a great deal of attention to the conduction paths within the central nervous system, and in recent years special efforts have been made to isolate the various functional systems of neurones, tracing the exact path of the sensory impulses from the peripheral organ to the primary sensory center, thence to the various secondary centers and return reflex paths. This motive underlies the recent studies on the nerve components and, indeed, much of the best morphological work on the nervous system in all times.

Some years ago I formulated the following definition of such a functional system of neurones, with special reference to the peripheral members of the system:

The sum of all the nerve fibers in the body which possess certain physiological and morphological characters in common so that they may react in a common mode. Morphologically each system is defined by the terminal relations of its fibers, by the organs to which they are related peripherally, and by the centers in which the fibers arise or terminate. The fibers of a single system may appear in a large number of nerves repeated more or less uniformly in a metameric way (as in the general cutaneous system of the spinal nerves), or they may all be concentrated into a single nerve (as in the optic nerve).

Now, if we add to this the secondary paths related to the primary central end stations referred to above, and the chief reflex arcs directly associated therewith, we shall have a picture of the system in its entirety.

The functional system with which we are especially concerned in the present research is that known to comparative anatomy as the communis system, including (1) unspecialized visceral sensory fibers ending free in the mucous surfaces of various viscera without special sense organs—probably phylogenetically the more primitive elements—and (2) specialized sensory fibers always ending in connection with highly differentiated sense organs in the mouth, pharynx, lips, or outer skin, known as taste buds, terminal buds, or end buds, and in general serving the function of taste. These specialized elements are probably of more recent phylogenetic origin than the first group, and the term "gustatory system" will be used to designate these organs, wherever placed on the body surface, together with their nervous pathways toward and within the brain. In other words, the gustatory system is that portion of the communis system of neurones which serves the sense of taste, as distinguished from those communis neurones which serve less highly specialized visceral sensations.

These two groups of fibers can easily be distinguished peripherally of the brain, but centrally they have not as yet been successfully analyzed. Hence in treating of the central gustatory path we can not be sure that we do not include the unspecialized visceral system also. But since in some fishes the gustatory fibers preponderate many fold over the unspecialized fibers of the communis system, there is no ambiguity arising from this central confusion of the two elements so far as the gustatory system is concerned, since the secondary paths as clearly traceable in these fishes must be made up chiefly of gustatory fibers.

The central gustatory path is not definitely known either in man or in any other vertebrate, so far as shown by the available literature. I have therefore studied with some care the brains of some fishes in which this system is enormously developed, in the hope that they would throw light on this unsolved problem of vertebrate anatomy. And in this I have not been disappointed, though my study of the central paths is not yet sufficiently advanced for publication.

As intimated above, sense organs belonging to the communis system and presumably serving the function of taste are found in the mouths of all fishes (" taste buds"). They are frequently found upon the lips, and in some cases they are found likewise plentifully distributed over extensive areas of the outer skin of the head and trunk. In this latter case they are commonly termed terminal buds or end buds (Endknospen, Becherorgane, of the Germans). They must in all cases be sharply distinguished from the neuromasts, or organs of the lateral-line system (German,

Nervenhügel), though these latter occur in the skin of fishes in a great variety of forms, often resembling the terminal buds very closely. The innervation and functions of the two systems of organs are, however, wholly different, and they really have nothing to do with each other. I shall illustrate more fully in a later section of this paper the structure of the terminal buds and the details of their innervation. I here call attention merely to the important fact that both in structure and in nerve supply they resemble most closely the taste buds of the mouth. From this one naturally infers for them a gustatory function. Since, however, inferences are not in order when facts are available, I have undertaken to determine experimentally the function of these cutaneous sense organs of the communis system.

The experiments which I have made are of an exceedingly simple nature, the attempt being to put the fish while under observation in as nearly normal conditions as possible and to utilize the ordinary feeding and other instinctive reactions so far as possible in the accumulation of the data. These are the methods of the old-time observational natural history, it is true, as contrasted with the methods of precision of the modern physiological laboratory. They have, however, proved sufficient for their purpose, which was merely to determine the class of stimuli to which the terminal buds are sensitive, or the sensational modality which they serve, rather than to contribute to the chemical physiology of taste in general.

The chief obstacle to experiments of this sort, and one which many observers seem to have made no serious efforts to overcome, is the natural timidity or shyness of wild creatures when kept in the confined and unnatural quarters necessary for close observation. The rôle played by fear in animal behavior has been vividly brought to our notice by Whitman ('99), and, like this observer, I find that young animals which have been reared in captivity are much more approachable and tractable under experimental conditions than adults which have been reared in their natural freedom. In fact, with several species I quite failed to get the adults to take food at all in captivity, though they were under observation for long periods.

REVIEW OF LITERATURE.

Surprisingly little attention has been paid to the physiology of taste in fishes, and this literature is very scanty. On the other hand, the anatomical investigation of these sense organs has been extensively followed for nearly a century, though often in a blind and profitless way. The history of opinion upon the significance of these sense organs has been quite fully given by Merkel ('80) in his great monograph published in 1880, and the earlier phases of this history need not be again reviewed further than to mention a few salient features.

In 1827 Weber observed the taste buds on the peculiar palatal organ of the carp and correctly interpreted their function. He also figured the brain of the carp, illustrating the enormous vagal lobes from which these taste buds receive their innervation. Leydig discovered in 1851 the terminal buds of the outer skin of fishes and gave a detailed account of their structure, which subsequent research has shown to be in some respects inaccurate. In 1863 F. E. Schulze gave a more accurate description of the "*becherförmigen Organe*" of fishes, in which he distinguished the specific sensory cells from the supporting cells. He also correctly inferred their function to be similar to that of taste buds within the mouth, viz, the perception of chemical stimuli.

In 1870 the same author (F. E. Schulze, '70) made a further important contribution to the problem of the terminal buds by the demonstration that they differ structurally from all neuromasts, or organs of the lateral-line system. The neuromasts are commonly sunken below the skin in canals, tubes, or pits, but in some cases they are strictly superficial and resemble in external form the terminal buds very closely—a feature which led Leydig ('51, '79, '94) and others to assume that the two classes of organs are mere varieties of a common type. Schulze showed that the neuromasts can in all cases be differentiated from the terminal buds by the fact that their specific sensory cells (pear cells) extend only part way through the sensory epithelium and fail to reach the internal limiting membrane, while in the terminal buds both specific sensory cells and supporting cells pass through from external to internal limiting membrane.

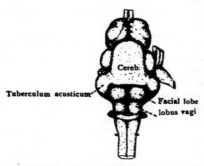

FIG. 1.—Dorsal view of the brain of the yellow cat-fish (*Leptops olivaris* Raf.). The olfactory bulbs with most of their cerum have been removed, also the membranous roof of the fourth ventricle, exposing the facial and vagal lobes. This ventricle is bounded behind by a transverse ridge containing the commissura infima Halleri and the commissural nucleus of Cajal. × 2.

This distinction was confirmed by Merkel ('80), who, with curious inconsistency, while recognizing the structural dissimilarity of the two classes of organs, nevertheless, as we shall see below, ascribes to both essentially the same function, touch. This matter was put to the decisive test in my contribution on *Ameiurus* ('01), a type which possesses both terminal buds and neuromasts in great abundance and diversity of forms. Schulze's contention is supported both by the structure of the organs and by their innervation, for I have shown that all neuromasts of whatever form are innervated by acustico-lateralis nerves from the tuberculum acusticum of the brain, while all terminal buds, whether within the mouth or in the outer skin, are innervated by communis nerves related centrally to a single center within the brain. This center is bilobed, the lobus vagi receiving most of the communis fibers from the mouth cavity by way of the vagus and glossopharyngeus and the lobus facialis the communis fibers from the terminal buds of the outer skin by way of the facial nerve (cf. fig. 1).

Similar terminal buds have been found in the outer skin of many species of Teleostomes and in Cyclostomes, but, so far as certainly known, nowhere else among vertebrates (save on the lips of some other classes). Their distribution among the fishes is very irregular, being most abundant among the siluroids, cyprinoids, ganoids, and cyclostomes, in general bottom fishes of sluggish habit, often living in mud and rarely belonging to the predaceous types which find their food chiefly by the sense of sight. The following list of fishes which have been shown to possess terminal buds on the outer skin is by no means complete, but will serve to illustrate the wide range of species which have acquired this peculiarity:

Fishes possessing terminal buds on the outer skin.

Accrina. On fins and body (Merkel, '80).
Acipenser sturio, sturgeon. On barbel (Merkel, '80). Also other sturgeons.
Agonus cataphractus, pogge. On the villiform tentacles beneath the head (Bateson, '90).
Ameiurus melas, cat-fish, and other North American Siluridæ. On barblets and nearly the whole body surface (Herrick, '01).
Amia calva, bowfin. On skin of head and other parts (Allis, '97).
Anguilla vulgaris, eel. On the fins, lips, and anterior nostril (Merkel, '80; Bateson, '90).
Aspius alburnus (Merkel, '80).
Barbus fluviatilis. On barblet (F. E. Schulze, '63).
Branchiostoma lanceolatum = Amphioxus lanceolatus, lancelet. On the oral cirri (Merkel, '80).
Carassius auratus, gold-fish. On the whole body (numerous authors; Herrick).
Cephalacanthus = Cobitis fossilis, flying gurnard (Merkel, '80).
Cottus scorpius, sculpin. On fins (Merkel, '80).
Cynoscion = Corvina (Merkel, '80).
Cyprinus carpio, carp, and other cyprinoids. On whole body (Merkel, '80 and others).
Dactylopterus (Merkel, '80).
Discognathus lamta, Indian carp. Over the whole body surface (Leydig, '94).
Enchelyopus = Motella, four-bearded rockling. On barblets and pelvic fins (Bateson, '90).
Gadus callarias, cod. On lips, barbel, fins, and body (Merkel, '80; Herrick, '00).

Gadus luscus, pouting. On the lips, barblet, and pelvic fins (Bateson, '90).
Gadus merlangus, whiting. On lips (Bateson, '90).
Gadus pollachius, pollack. On lips (Bateson, '90).
Gaidropsarus = Motella, three-bearded rockling. On all the barblets and pelvic fins (Zincone, '78; Bateson, '90).
Gobius, goby. On fins (Merkel, '80).
Hippocampus, sea horse (Merkel, '80).
Leptocephalus conger, conger eel. On the outer and inner lips (Bateson, '90).
Leuciscus delineatus. On the body generally (Leydig, '94).
Leuciscus dobula (Leydig, '57).
Lota vulgaris, ling. On barblet (Merkel, '80).
Mullus barbatus, mullet. On barblet (Zincone, '78; Merkel, '80).
Petromyzon fluviatilis, lamprey. On skin of whole body (Merkel, '80, and others).
Pygosteus = Gasterosteus pungitius, stickleback (Merkel, '80).
Rhodeus amarus. On the body generally (Leydig, '94).
Scorpæna (Merkel, '80).
Silurus glanis, cat-fish (Merkel, '80).
Solea vulgaris, sole. "Contrary to the natural presumption, the villi on the lower (left) side of the head do *not* bear sense organs, though, as Mr. Cunningham informs me, such organs are found between the villi" (Bateson, '90).
Tinca vulgaris, tench. On barblet (Merkel, '80).

As already suggested, our knowledge of the functions of all of the sense organs of fishes is very imperfect, since speculation based upon structure has seemed more attractive to most authors than accurate physiological research. The monograph of Merkel ('80), with its great wealth of accurate anatomical data on the structure and distribution of terminal buds in all classes of vertebrates, gives an excellent illustration of the dangers in the path of even so skillful an observer when he goes beyond the bounds of observed fact and enters the field of speculation. This author recognizes the close structural resemblance between these organs and the undoubted organs of taste in the human body. He controverts, however, the clear argument of F. E. Schulze for their gustatory function on merely theoretical grounds. His first objection is based on their innervation. Instead of being supplied by a single gustatory nerve, the glossopharyngeus, they may be supplied, he says, by any other body nerve. This objection has been totally removed by the discovery (compare especially my own *Ameiurus* paper, already referred to, published in October, 1901) that all terminal buds, no matter where located on the body and no matter from what nerve branches their innervation seems to come, are in reality supplied by nerves of a single physiological system, terminating in the brain in a single center—the communis nerves.

Again, he objects to Schulze's theory that the terminal buds serve to localize gustatory stimuli on the various parts of the body, on the ground that an organ of chemical sense stimulated by substances in solution in the environing fluid could not receive a sufficiently circumscribed stimulation. It is unnecessary to follow the argument in detail, for the experiments which I shall describe shortly show conclusively that when the sapid substance is brought into contact with these organs or very near to them the stimulus is accurately and very promptly localized, and in fact some of the fishes studied habitually find their food by this very power, the gustatory stimulus calling forth an immediate reflex movement toward the point stimulated. It is probable that the local sign is not given by the gustatory (communis) nerves, but by the accompanying tactile (general cutaneous) nerves of the corresponding cutaneous area (which general cutaneous nerves Merkel, curiously enough, denies to the fishes altogether, whereas, in fact, they are plentifully supplied to all parts of the skin), though my experiments do not decisively answer this question.[a] Weak stimuli, especially when uniformly diffused through the water, are, it is true, not at all localized; but strong stimuli are unquestionably localized by one method or another.

In fact, Merkel agrees with Jobert that the terminal buds of the outer skin are tactile in function. This is based largely on the erroneous belief, referred to above, that there are no free tactile nerve endings in the skin of fishes, and also on the observed tactile sensibility of the barblets and other parts of the body known to be most plentifully supplied with terminal buds. But I have shown that all of these parts of the body receive, in addition to communis nerves for the specialized sense organs, a most liberal general cutaneous innervation for tactile sensibility; and the experiments which follow go to show practically that these two functions commonly cooperate in setting off the reflex of seizing food, though they may be experimentally isolated.

Merkel now proceeds to carry his argument to its logical conclusion (and likewise to a *reductio ad absurdum*) by denying the gustatory function to all terminal buds, even those within the mouth supplied by the glossopharyngeal nerve, of all vertebrates below the *Mammalia*.

He finally concludes that both the neuromasts of the lateral-line system and the terminal buds are tactile organs, the buds being the more delicate; but if these are deficient, then the neuromasts may be elevated to a more delicate functional value; both of which conclusions, in the light of our present knowledge, illustrate the dangers attending an attempt to determine function on the basis solely of observed structure, without adequate physiological control.

The general works contain numerous references to the subject, but usually chance observations or speculative conclusions. Günther says, under the caption "Organ of taste":

Some fishes, especially vegetable feeders, or those provided with broad molar-like teeth, masticate their food; and it may be observed in carps and other cyprinoid fish that this process of mastication frequently takes some time. But the majority of fish swallow their food rapidly and without mastication, and therefore we may conclude that the sense of taste can not be acute. The tongue is often entirely absent, and even when it exists in its most distinct state it consists merely of ligamentous or cellular substance, and is never furnished with muscles capable of producing the movements of extension or retraction, as in most higher vertebrates. A peculiar organ on the roof of the palate of cyprinoids is perhaps an organ adapted for perception of this sense; in these fishes the palate between and below the upper pharyngeal bones is cushioned with a thick, soft, contractile substance, richly supplied with nerves from the Nervi vagus and glossopharyngeus.

---

[a] On this point, see the further experiments recorded in the Addendum, pp. 270-271.

Regarding the peculiar palatal organ of the cyprinoids, it has been known since Weber's account in 1827 that this is plentifully supplied with taste buds, and Weber himself brought forward strong indirect evidence that its function is gustatory. The following observations (and many similar ones might be cited from the literature of sport) are taken from the section on "The Trouts of America," by William C. Harris, in the American Sportsman's Library.

The angler can not resist the belief that the senses of smell and taste are well developed in trout. They eject the artificial fly, if the hook is not fast in the flesh, at the instant they note its nonedible nature, or when they feel the gritty impact of the hook. They will not eat impure food, and they have the faculty of perceiving odors, and various scents attract or repel them. This has been verified from the earliest days of our art, when ancient rodsmen used diverse and curious pastes and oils, which were seductive to fish; in Walton's day, and long after, this practice was followed and the records tell us of its success. When I was a boy and the Schulkill River was swarming with the small white-bellied cat-fish, than which no more delightful breakfast food ever came out of the water, the only bait used to catch them was made of Limburger cheese, mixed with a patch of cotton batting to hold it firm on the hook. No other lure had the same attraction for them because, no doubt, of the decided odor of the cheese.

The problems connected with the relative significance of the several sense organs of the fishes have been treated both anatomically and experimentally in the excellent paper of Bateson ('90). After anatomical remarks, based largely on his own careful studies, on the eyes, olfactory organs, and gustatory organs, he recounts a series of admirable and well-considered experiments made to test the parts played by these organs in the normal feeding of various kinds of fishes.

These observations are grouped under two chief heads, viz., "Senses of fishes which seek their food by scent" and "The senses of fishes which seek their food by sight." Though the taste buds in the mouth and outer skin are described and correctly interpreted in the anatomical part of the paper, these organs are scarcely considered at all in the physiological part, and this is really the greatest weakness of the paper. Since my own observations in part follow so closely in the footsteps of Bateson (though completed in the main before his paper was accessible to me), and since they are in general confirmatory of his, it will be of interest to review portions of his paper at this time.

He gives the following list of fishes which he has observed "to show consciousness of food which was unseen by them, as, as will hereafter be shown, there is evidence that they habitually seek it without the help of their eyes":

Protopterus annectens, mud-fish.	*Motella mustela*, five-bearded rockling.
Scyllium canicula, rough dog-fish.	*Nemacheilus barbatula*, loach.
Scyllium catulus, nurse-hound.	*? Lepadogaster gouanii*, sucker.
Raja batis, skate.	*Solea vulgaris*, sole.
Conger vulgaris, conger eel.	*Solea minuta*, little sole.
Anguilla vulgaris, eel.	*Acipenser ruthenus*, sterlet.
Motella tricirrata, three-bearded rockling.	

He says: "To this list may almost certainly be added the remainder of the *Raiidæ*, together with the angel-fish (*Rhina squatina*) and *Torpedo*." Unfortunately, however, Bateson in his list does not distinguish between those fishes in which smell obviously plays the leading part and those in which taste or touch or both are used to compensate for the reduction of vision, and it is this defect which it is hoped that the present contribution may in part correct.

Most of the forms in the list above are more or less nocturnal animals, but they differ much in this regard. The part attributed to the sense of sight and smell in Bateson's studies is so similar to my own conclusions in many respects that it seems fitting to quote the greater part of his description, especially since the species observed by us are in all cases different. He says:

None of these fishes ever start in quest of food when it is first put into the tank, but wait for an interval, doubtless until the scent has been diffused through the water. Having perceived the scent of food, they swim vaguely about and appear to seek it by examining the whole area pervaded by the scent, having seemingly no sense of the direction whence it proceeds. Though some of these animals have undoubtedly some visual perception of objects moving in the water, yet at no time was there the slightest indication of any recognition of any food substance by sight. The process of search is equally indirect and tentative by day and by night, whether the food is exposed or hidden in an opaque vessel, whether a piece of actual food is in the water or the juice only, squeezed through a cloth, and, lastly, whether (as tested in the case of the conger and the rockling) the fish be blind or not. * * * The perceptions, then, by which these animals recognize the presence of food are clearly obtained by means of the olfactory organs and apparently exclusively through them. I was particularly surprised to find no indication of the possession of such a function by the sense organs of the barbels and lips or by those of the lateral line. As has been already described, the pelvic fins and barbels of the rocklings (*Motella*) and the lips, etc., of most fishes bear great numbers of sense organs closely comparable in structure with the taste buds of other vertebrates. No one who has seen the mode of feeding of the rockling or pouting (*Gadus luscus*) can doubt that these organs are employed for the discrimination of food substances; but the fact already mentioned, that the rockling in which the olfactory organs had been extirpated did not take any notice of food that was not put close to it, points to the conclusion that they are of service only in actual contact with the food itself.

Bateson gives also a considerable list of fishes which he has observed to get their food chiefly by the sense of sight, and he is doubtless correct in asserting that the majority of fishes belong to this class. None of these sight-hunting fishes while living in his tanks appeared able to see their food by night, or even in twilight. None of the fishes which he enumerates as belonging to this class showed symptoms of interest when the juice of food substances was put into the water, and other evidence is brought forward to show that the sense of smell plays little or no part in helping them to discover their food.

I have not studied any of the species mentioned by Bateson, but for the forms studied by me, which have an extensive supply of terminal buds on the outer skin, I fully confirm most of the statements quoted above, save that in determining the part played by sight I did not blind any of my fishes and save that the statement that in fishes of his first group "at no time was there the slightest indication of any recognition of any food substance by sight" is strictly true of none of my fishes except *Ameiurus*, though in some of the other cases it is approximately true.

The only important respect in which my observations are not in harmony with those of Bateson is in connection with the part played by the sense of taste in some of these types of fishes. I have studied the gustatory reactions of fishes closely allied to the rockling and having the same arrangement of terminal buds on the barblets and pelvic fins, and am convinced that Bateson's failure to get clear gustatory reactions from these organs was due to the insufficiency of his methods of experiment rather than to the absence of the function. In general, it may be stated that the part played by the gustatory reflex in the case of fishes having an extensive supply of terminal buds on the outer skin is of vastly greater importance than Bateson appears to have recognized.

The only other paper of importance dealing with the sense of taste in the fishes experimentally which has come to my notice is the great monograph on the senses of taste and smell by Nagel ('94). He investigated the sense of taste in the following fishes:

(1) FRESH-WATER TYPES: *Anguilla anguilla* (old and quite young); *Cyprinus carpio; Barbus fluviatilis; Leuciscus cephalus; Gasterosteus aculeatus; Gobius fluviatilis; Silurus glanis* (young specimen); *Cobitis fossilis*.
(2) MARINE TYPES: *Pristiurus; Scyllium catulus* and *S. canicula; Syngnathus acus; Uranoscopus scaber; Lophius piscatorius*.

Nagel tested all the fresh-water fishes mentioned in this list by bringing bitter, sour, sweet, and salty solutions in contact with the skin, without getting any response to the stimulus. Thus, the carp, wels (*Silurus*), and stickleback did not respond to a stimulation of the skin of the body with quinine, though the last-named fish gave an immediate response when the solution touched the lips. He concludes:

In the fresh-water fishes, according to my observations, the power of taste is completely lacking in the outer skin; or, more precisely, in no part except the head is there gustatory sensibility.

For such of these forms as possess no terminal buds on the skin of the body this is doubtless true; but for the other fishes, including, doubtless, *Silurus* and *Cyprinus*, it is certainly a mistake. In gadoid fishes I got a clear reaction against quinine solution when it was applied to the free fin rays, which are known to be supplied with terminal buds, but not from other parts of the skin.

Among the elasmobranch fishes Nagel found *Scyllium catulus* and *S. canicula* to be sensitive to very dilute solutions of vanilla all over the body and fins. Bitters were not perceived thus, nor oil of rosemary, but they are very sensitive to creosote. He controverts Schwalbe's argument that the terminal buds of the outer skin of fishes probably have a gustatory function by reason of the similarity of their structure with that of taste buds in the mouth, and concludes:

A real sense of taste, such as man and many other animals have in the mouth, appears to be absent in the outer skin of all fishes and Amphibia.

It will appear from the following pages that this conclusion is erroneous. I will merely add here that if Nagel had worked with sapid solutions, with which his fishes were presumably already familiar, instead of with substances like sugar and vanilla, toward which no clearly established reflexes had been established in the natural environment of the fishes, his conclusions might have been different.

TERMINAL BUDS AND THEIR INNERVATION.

The terminal buds of the fishes tabulated above, and doubtless many others which might be mentioned, are of the same type and presumably provided with similar innervation by communis nerves, for cutaneous branches of the communis root of the facial nerve are known to reach the areas provided with the buds in all cases which have been adequately studied. These organs may therefore all be defined morphologically as belonging to the communis system of sense organs, along with the taste buds of the mouth cavity and as distinct from the lateral-line organs and all other types of sense organs. In order to support this position there remains merely the proof that the terminal buds and taste buds have a similar function. This evidence is presented subsequently in this paper.

The terminal buds of fishes have been often described and figured, and I have little to add to the classical descriptions save in the matter of distribution and innervation. Those in the mouth are supplied by branches of the x, ix, and vii pairs of cranial nerves, the first two nerves supplying those in the gill regions and the pretrematic branch of the glossopharyngeus also running forward to supply those on the hyoid arch (tongue). The communis root of the facialis (= portio intermedia of human anatomy) and its geniculate ganglion supply the taste buds on the palate by the r. palatinus facialis (= great superficial petrosal nerve of man), and other buds on the lining of the cheek, on the jaws, and on the lips by other branches, some of which are secondarily associated with branches of the trigeminus and most of which have no homologues in mammalian anatomy, though some one or more of them probably represent the chorda tympani.

In *Ameiurus* I have shown ('01) that terminal buds occur in the skin of practically

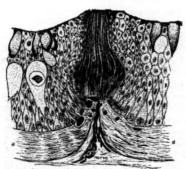

FIG. 2.—Section through the skin of the top of the head of *Ameiurus nebus*, showing a terminal bud - 375. (From the *Journal of Comparative Neurology*, vol. XI, No. 3, Oct., 1901, plate XVII, fig. 11.) At *d* is the dermis, which is raised into a low papilla under the sense organ and whose center is pierced by the nerve for the organ.

the whole body surface, most abundantly on the barblets and diminishing in frequency toward the tail. These buds (see fig. 2) rest on a low papilla of the dermis, quite different from that figured by Merkel ('80, plate v, fig. 1) for the terminal buds of *Silurus*. His figure shows a much smaller organ, resting upon a greatly elongated papilla in an epidermis which is apparently thicker than in *Ameiurus*. Merkel states ('80, p. 72) that terminal buds always occur on such a dermal papilla. While this is certainly the general rule, we find occasionally instances where the papilla is absent, as on the filiform fins of the hake, where I find the buds imbedded in the epidermis and extending only part way through it, with a layer of unmodified epidermal cells between the bud and the dermis.

All parts of the body of *Ameiurus* which are supplied with terminal buds are reached by branches of the facial nerve from the geniculate ganglion. In other words, the rami from the communis root of the facialis are distributed to nearly the whole outer body surface. On the distal side of the ganglion these rami usually join themselves to other cutaneous branches which are phylogenetically older, belonging to the general cutaneous and lateral-line systems. Even the great recurrent branch into the trunk, the ramus lateralis accessorius, which passes out of the cranium as a practically pure communis nerve, anastomoses with the spinal nerves at their ganglia and its fibers are ultimately distributed along with the general cutaneous fibers from these spinal ganglia. Fig. 3 illustrates the courses of the chief cutaneous branches of the communis system in *Ameiurus melas*, the nerves of all other systems being omitted from the sketch.

Proximally of the geniculate ganglion the communis root of the facialis pursues an uncomplicated course to the primary gustatory center within the medulla oblongata. In most fishes this root passes back close to the floor of the fourth ventricle as the fasciculus communis (= fasc. solitarius of mammals) to terminate in the vagal lobe of the same side, and receives in its course the communis root of the glossopharyngeus nerve. But in siluroids and cyprinoids, where the very abundant terminal buds of the outer skin are all innervated from the communis root of the facial nerve, the consequent increase in the size of this root has resulted in a great enlargement of the cephalic end of the gustatory center (vagal lobe) which appears on the dorsal surface of the oblongata as the facial lobe. This structure is paired in siluroids and was formerly called the lobus trigemini, an inadmissible term, since it has nothing whatever to do with the trigeminus nerve. In cyprinoids it is unpaired and is referred to in the older literature as the tuberculum impar.

The cyprinoid fishes also have long been known to have terminal buds (*Becherorgane*) widely distributed over the outer body surface; but neither the innervation of these organs nor the exact composition of the cranial nerves has ever been worked out in any cyprinoid fish. A cursory examination of a series of sections prepared

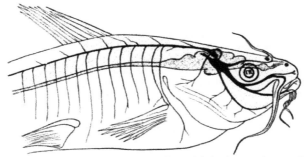

FIG. 3.—A projection of the cutaneous branches of the communis root of the facial nerve in *Ameiurus melas*, as seen from the right side. The outline of the brain is indicated by the stippled area and the positions of the eye and anterior and posterior nostrils are indicated. The projection is reconstructed from serial sections, but is not drawn accurately to scale. More detailed reconstructions of the cranial nerves and lateral line sense organs of this fish are given in the *Journal of Comparative Neurology*, vol. XI, No. 3, plates XIV and XV (Herrick '01).

by the Weigert method through the entire head and body of a small gold-fish (*Carassius auratus*) has convinced me that the same conditions in general prevail in the cyprinoids as in the siluroids. That is, the enormous size of the vagal lobes of cyprinoids is explained by the fact that these are the terminal centers for the vast numbers of nerve fibers entering the brain by way of the IX and X nerves from the palatal organ, this remarkable structure being crowded over its entire extent with taste buds and probably serving to filter food particles out of the mud taken into the mouth.

On the other hand, the tuberculum impar, or facial lobe, receives the entire communis root of the facial nerve. This root receives fibers from practically all parts of the outer surface of the body, and we may infer by analogy with other fishes that these fibers connect with the terminal buds in these cutaneous areas, though we have as yet no actual demonstration of this fact. The terminal buds of the skin of the head are supplied mainly, as in *Ameiurus*, by way of the infraorbital trunk. The terminal buds in the skin of the body of the gold-fish are not, however, supplied by a ramus lateralis accessorius, or recurrent facial nerve, as in *Ameiurus* and the gadoid fishes, for this nerve, as has long been known, is absent in the cyprinoids.

There is, however, in these fishes an intracranial anastomosis between the v+vii ganglionic complex and the ix+x complex, the composition of which has thus far remained unknown. This proves to be the recurrent branch of the facialis, carrying communis fibers from the geniculate ganglion into the trunk. The details of the peripheral distribution of these fibers have not been fully worked out, but the main path in the gold-fish is as follows:

The geniculate ganglion of the facialis is clearly separable from all other ganglionic masses of the trigemino-facial complex and is composed of two portions, each of large size. The more dorsal portion corresponds to the greater part of the ganglion in other teleosts and distributes its fibers chiefly by way of the infraorbital trunk. The more ventral portion sends cephalad a very large palatine nerve, and caudad a still larger nerve which represents morphologically, though not topographically, the r. recurrens facialis of the siluroids, etc., or the facial root of the r. lateralis accessorius as found in the cod.

This nerve passes back along the lateral side of the great auditory root and at the level of the superficial origin of the ix nerve it divides into several strands, one of which passes dorsally of the ix root, the others ventrally. These latter, however, pass upward so as to lie, farther back, dorsally of all of the vagus roots except that of the lateralis branch of the vagus. All of these communis fibers now join themselves to the r. lateralis vagi and, passing through the ganglion of the latter nerve, both components enter the body of the fish bound up in a single nerve trunk in which the fine communis fibers are for a time completely surrounded by the coarse lateralis fibers. The communis fibers go off in successive branches along with lateralis fibers. The details of the distribution have not been worked out, though I think it would not be difficult to do so with the material at hand. It is highly probable that the communis fibers are for the terminal buds sparsely distributed over the skin of the body and that the terminal buds of the trunk are all innervated from these communis fibers in the r. lateralis vagi, just as the buds in the skin of the head are innervated by other communis fibers from the geniculate ganglion of the facialis, an arrangement substantially identical in morphological plan with that of the siluroid fishes.

The conditions here, so far as studied, confirm essentially the conjectures to which I was led from a study of the literature (Herrick, '99, p. 400), and accord so completely with the morphological interpretation there proposed that we merely refer the reader to that passage in the Menidia paper.

FUNCTIONS OF TERMINAL BUDS.

EXPERIMENTS ON SILUROID FISHES.

The cat-fish (*Ameiurus nebulosus*) upon which this series of experiments was conducted (except a few experiments specifically designated) were hatched in the open at Cranville in the spring of 1901. In October of that same year they were taken to the laboratory and kept through the following winter in tanks. Microscopic examination of the skin and barblets shows that their skin and cutaneous sense organs at this age are practically in the adult condition. During the winter they were fed on various kinds of meat chopped fine, sometimes cooked, but usually raw.

In one small aquarium were kept half a dozen cat-fish, several ordinary "shiners" (*Notropis* sp."), and some small "spotted suckers" (*Minytrema melanops* Rafinesque). Casual observations made during the winter while feeding showed that the shiners use the eyes chiefly in capturing their food. A bit of meat dropped into the water will usually be seized instantly and devoured before it has time to sink to the bottom of the tank. After it has fallen to the bottom it is apt to be long overlooked unless the fish happens upon it in its aimless wanderings, or unless its attention is called to it by the movements of other fishes which may be eating it. These fishes, when observed, are usually swimming about in the mid-depths of the tank, not resting near the bottom. I have observed the same behavior in *Menidia* and other large-eyed species.

The behavior of the suckers was totally different. These fishes lie on the bottom most of the time unless disturbed, though if frightened they are very active, swimming powerfully and leaping out of the water. When food is thrown in they never pay the slightest attention, nor are they attracted by the sight of other fishes struggling for the meat. They are exceedingly shy and rarely eat when under observation. They lie quietly much of the time or swim slowly about, dragging the fleshy lips of the highly protrusible mouth over the bottom of the tank. If they thus happen upon a bit of meat this is sucked into the mouth, worked over with the pharyngeal teeth apparently, and then often ejected forcibly from the mouth, to be again taken, perhaps, and the process repeated—a behavior very characteristic of the way they take the bait, I am told by fishermen.

The cat-fish, like the suckers, keep strictly to the bottom of the tank. They are often quiet in the darkest corners or lying under débris, but much of the time are slowly dragging the mental and post-mental barblets along the bottom. The nasal barblets are held projecting well upward, and the maxillary barblets are directed outward and backward, their tips trailing the bottom or waving gently back and forth. They appear never to use their eyes directly for catching food to the slightest degree under the conditions of these experiments. No attention is paid to particles of food thrown into the water, even though they settle down within a few millimeters of the nose or barblet of the fish. The only case observed by me in which the eyes seem to serve in finding food is when a large piece of meat is thrown in and one fish begins to "worry" it. His movements may attract others until as many fish as can reach it are all tugging at it at once. If, however, a shadow is caused to fall upon the water, as by hovering the hand over the aquarium, the fishes are greatly disturbed and dart wildly about. They always seek the darkest corners of the tank and lie under dead leaves resting on the bottom of the tank for the most part, showing that the eyes are not by any means functionless and the fishes are strongly negatively phototactic.

If the cat-fishes in the course of their aimless movements along the floor of the aquarium touch a bit of meat with the lips or barblets, it is instantly seized and swallowed. Food in the immediate neighborhood of the fish is not discovered at once, but after a time appears to affect the fish in some way, probably through the sense of

---

a Notropis has very small tuberculum impar and vagal lobes, the latter scarcely larger than in the cod, *Menidia*, and physoclistous fishes generally. From this one may safely infer that cutaneous terminal buds are not as highly developed in this form as in the larger cyprinoids.

smell, as the maxillary barblets begin to wave about more actively and finally the fish becomes restless. He does not find the food, however, unless in the course of his movements it actually touches some part of the body.

During May and June, 1902, more systematic experiments were undertaken with these fish, and since these experiments are typical of those subsequently performed on other species of fishes I shall recount them in some detail. At first a few specimens were taken out in a shallow tray and the attempt made to feed them in various ways under close observation. They were, however, so much frightened by the exposure to bright daylight and by the proximity of the observer, in spite of all precautions, that no reactions could be obtained which were at all satisfactory. A bit of fresh meat on a long-handled needle could be thrust slowly toward the fish as he lay quietly on the bottom, rubbed over his body or on the barblets, and even over the lips, without evoking a movement of any kind in response. The same observation was made with the spotted suckers. The fishes in both cases had been without food for several days and were very hungry, but were obviously too much frightened to respond to the food stimulus.

On another occasion the same conditions were prepared, except that a few dead leaves were littered over the bottom of the tray. The fish when placed in the tray immediately sought the shelter of the leaves, and, after a suitable interval to enable them to become accustomed to the place, the feeding experiments were repeated. Selecting a fish which was entirely concealed under a large leaf, save for a projecting barblet, a bit of meat on a slender wire was gently passed down into the water in such a way as to touch the projecting barblet. It was instantly seized and swallowed. This was repeated many times with several of the fishes.

In subsequent experiments the fish were not removed from their own tank, but the water was drawn off so that it was only about six inches deep. Here they would lie under the leaves and the experiment could be continued with a minimum of disturbance to the fishes. The experiment of touching the barblet with meat was repeated hundreds of times with an almost invariable result that the fish instantly turned and snapped up the morsel. If the meat was merely held very close to the barblet it usually produced no response. The reaction was obtained equally well, no matter which barblet was touched.

In a later series of experiments I found that the fish would almost always turn and seize the meat if he were touched at any point on the head or body. If the tail of the fish projected out from under a leaf and the skin near the root of a tail fin were touched with meat the fish would turn and seize the meat. This reaction was not so uniformly made at first as that from the barblets, but after a dozen or so of trials it followed with equal promptness and uniformity, the fish apparently requiring a little practice to learn the movement perfectly.

The experiments last described were repeated the next day and by this time it was found that the fishes had become so tame that they would take the meat if offered to them in the open, without the shelter of the dead leaves, though not so certainly as when under the cover of the leaves, often taking fright from the shadow of the observer's hand or from some other cause.

In none of these cases did the fishes appear to see the bait or to perceive it in any way other than by actual contact with the skin at some point. If the bait were held a moment in front of them and then moved slowly away they would not follow it. If, however, it touched a barblet and then moved rapidly away before the fish had time to seize it, then the fish would sometimes follow it a short distance.

At this point the relations of vision and smell to these reactions should receive some further consideration. These young fishes, like their adults, spend much of their time buried under the débris of the bottom, with perhaps a barblet or a portion of the tail only projecting. Under these circumstances it is easy to apply the stimulus to various parts of the skin with the assurance that the contact is wholly invisible to the fish. Many such experiments show decisively that the reaction takes place in the same way whether the fish is able to see the stimulus applied or not. The visual factor being so conclusively ruled out, I have not thought it necessary to blind the fish for further control.

This conclusion of course must be limited strictly to fish of the species and age under investigation. It by no means follows that they may not subsequently learn to use their eyes in finding food, as well as in escaping from their enemies. Indeed, during the later experiments of this series, after the fishes had been fed for several weeks almost daily with meat on the end of a wire, I saw some slight evidence that they took note of the bait by the sense of sight, but the observations were in no case conclusive. Whether the adult *Amiurus nebulosus* ever uses the eyes in the capture of food I have no definite information, though from the habit of spending much of the time during the day completely buried in the mud and of feeding chiefly at night it is very improbable that they do so. With the channel cat-fish, *Ictalurus*, the case is certainly different.

Mr. I. A. Field tells me that while fishing for bass in the Black River, Ohio, he has sometimes caught large specimens of *Ictalurus* with live minnows as bait. The current was swift and the minnows were kept off the bottom of the river and in motion all the time. At the meeting of the American Association for the Advancement of Science, at Pittsburg, July 1, 1902, in the course of a brief report upon these experiments, I asked the question whether anyone ever caught a cat-fish on a spoon hook. Dr. L. L. Dyche stated that he has occasionally caught the channel cat (*Ictalurus*) on a spoon in a small lake, but only in bright sunlight. Dr. Eigenmann stated that *Ictalurus* has much better eyes than *Amiurus*. They are not only larger, but the retinal pattern is more nearly like that of other fishes, while that of *Amiurus* is decidedly degenerate.

The part played by the sense of smell is much more difficult to determine. As intimated above, I have evidence that the gustatory organs of the skin can function only in contact with the sapid substance. The most highly flavored food can be held within a millimeter or two of the barblet or lips without calling forth the characteristic instantaneous reflex. I will narrate one experience which was many times repeated in a variety of modifications. Three fishes were lying quietly under a small water-soaked leaf. A bit of rather stale beefsteak, with a strong odor, was held on the tip of a fine wire over the edge of the leaf under which they were lying and separated by a centimeter or two from the nostrils of the fishes. The leaf was considerably corroded by decay, and doubtless the odor could freely permeate it, though it was nearly or quite opaque. After some ten seconds the fishes began to move restlessly about in circles under the leaf, which was soon swept away by their movements.

As a rule the fishes swam in narrow circles close to the bottom and for a long time failed to find the meat, though they seemed to be aware of its general position for they never circled far away. If the meat were very slowly moved across the aquarium the fish could be drawn in this way after it for a considerable distance, though the meat was never found unless in the course of their apparently aimless movements one of the fishes came in contact with it, when it was instantly snapped up.

This aimless circling movement may be termed provisionally the *seeking reaction*, since it is so different from the characteristic movement made when the stimulus is in contact with the body—a sharp turn of the body and instantaneous seizing of the bait—which I shall term the *gustatory reaction*. Unfortunately, I have not had opportunity as yet to carry out extirpation experiments on *Ameiurus* to determine decisively the part played by the olfactory organ in this reaction. (Compare the experiments on the tomcod narrated below.)

The fishes upon which these experiments were performed have unfortunately been lost. At the present time I have a fresh lot of *Ameiurus* fry under observation, and have already verified many of the conclusions reached with the first lot. But this second collection of fishes has not, at the time when this report is submitted, been in captivity long enough to become sufficiently accustomed to their new surroundings to feed freely and fearlessly. After some months of further preliminary observation, I hope to carry on experiments which may shed some light on the sense of smell in these fishes. But this must be reserved for a later report. A few subsequent observations are noted on pages 270–271.

We must content ourselves at the present time, then, with the inference that the sense of smell plays at least a small part in these reactions, for the animals became slightly restless in the proximity of the stimulus, though they were not in contact with it; this, however, appears never to provoke a definite reaction of seizing the food, but merely a vague reaction in search of food. On the other hand, physical contact with the irritating substance causes a definite and precise reaction which is practically constant. This points either to touch or to taste.

To test the relative part played by stimulation of these two sets of sense organs, the following series of experiments was performed. A half dozen fish in an aquarium were tested a score of times with fresh meat on the tip of a wire, as in the previous cases. The reaction was obtained uniformly, no matter what part of the body or head was touched. Half an hour after the close of these experiments a bit of cotton wool was wound around the tip of a wire and the fishes were tested with this exactly as they had been with the meat. For the first six trials the barblets only were touched. The fish in each case turned and seized the cotton as promptly as the meat had been taken. The cotton would be immediately dropped. After a few more trials the fishes would generally turn when touched, but would check their movement before the cotton was actually taken into the mouth. Several specimens were now tested on the trunk with the cotton. One or two turned completely around and took the cotton, but generally there was a slight movement only toward the cotton, which was checked before the cotton was reached. After a few further tests, the fishes would usually pay no attention to a contact with the cotton on the skin of the body and the reaction by the barblets became uncertain, until finally the cotton could be freely rubbed over the barblets or lips of some of the individuals without producing any response.

These experiments were many times repeated, sometimes using white cotton, sometimes red cotton, and sometimes fresh meat. The reaction was uniformly obtained with the meat. If at the close of a few experiments with the meat a minute pledget of cotton was substituted for the meat, there was feeble or no response from rubbing the body with the cotton, though upon touching the barblets the fish would usually turn and often would seize the cotton and drop it again at once. After several repetitions, the fish became wholly indifferent to the cotton, no matter how it was applied, or they would if touched upon a barblet turn toward it without biting it. They were now again tested with bits of meat. This they took as eagerly and as precisely as before, showing that they were still hungry.

After the interval of a day or two the fishes would still appear to remember the cotton, and I rarely, after the first trials, got a prompt "gustatory" reflex with the cotton. If they noticed it at all, they would turn slowly and touch it with the lips or a barblet in a tentative or inquiring manner, only to turn away again without taking it into the mouth. This deliberate movement may be designated, for reasons to appear immediately, as the *tactile reflex*, as distinguished from the instant seizing of food, the "gustatory reflex."

These experiments seem to show that in the reactions to the meat, both from the barblet and from the skin of the body, the senses of taste and touch both participate. This is in accord with the known innervation of the skin and barblets, for all parts of the body surface receive general cutaneous (tactile) nerves, and all parts are plentifully provided with terminal buds (taste buds) which are innervated by communis (gustatory) nerves. The experiments further suggest that these two sensory factors can be experimentally isolated by training.

The fishes having become accustomed by brief training to make the simple reflex of seizing the food under the stimulus applied to any part of the barblets or skin, and doubtless utilizing both gustatory and tactile sensations, the gustatory factor is eliminated by the substitution of cotton wool for the meat. The tactile sensation alone proves to be sufficient to set off the reflex after the training previously given. The stimulus is, however, never followed by satisfaction and is soon given up, the fishes after further practice not reacting to the tactile stimulus alone. If, however, the gustatory sensation is added, by the substitution of meat for the cotton, the original reflex is given as promptly as before. This would seem to indicate that, while the tactile sensation alone is not sufficient to maintain the reflex, the addition of the gustatory element is sufficient, and therefore that the gustatory element is the essential element in setting off the reflex. This hypothesis was tested by an extensive series of experiments similar in plan to those last described.

In general there was no noticeable difference between the reaction to the white cotton and that to the red, though in some cases, especially toward the end of the series of experiments, after the fishes had learned to pay no attention to white cotton when touched at any point by it, they would sometimes turn and touch the red cotton with the lips or a barblet, immediately to turn away again without biting the cotton as they did at first. The reaction is not the quick turn and instant seizing of the bait, which I have termed the "gustatory reaction," but a more deliberate movement similar to what I termed above the "tactile reaction." This occurred only when the cotton was in plain view at the time of the contact and is probably in this case partly a visual response, called forth by the similar appearance of the red cotton and bits of beefsteak on which they were habitually fed. It was not by any means constant, for, in general, after the first few days, contact with neither color of cotton called forth any response whatever.

After this result was reached, I dipped the pledgets of white cotton in the filtered juice of fresh beef and touched the body surfaces and barblets with them in the same way as before. In all cases I got a typical "gustatory" reaction exactly the same as with the meat, and this reaction persisted after many trials with no diminution. The cotton was taken instantly into the mouth and tugged vigorously. No amount of training served to eradicate or to weaken this reflex.

I next prepared a small bulb syringe, with the delivery tube drawn out to a very fine point. This was filled with the water in which the fishes were and a fine jet directed against their bodies. They either paid no attention or were disturbed and swam away. I now substituted for the water in the syringe the juice of raw beef pressed out and strained. When a jet of this fluid was directed against the side of the body, the fish always instantly turned and tried to take the end of the syringe. The reaction was identical with that produced when a corresponding part of the body is touched with raw meat. I invariably got the reaction, both from the sides of the body as far back as the root of the tail fin and from the skin of the head and barblets.

I also tested the fishes with bits of red brick held in forceps. The forceps seemed to frighten the fishes. They either paid no attention to the contact with the brick (when touched in such a way that they could not see the point of contact), or else the harsh contact seemed to frighten them. I then touched them on various parts of the body and the barblets with bits of brick which had been soaked in raw meat juice. In most cases they would turn and touch the brick with the lips or take it into the mouth, but often they seemed frightened and would swim away. I then gave them a few bits of meat with the forceps and found that they took it eagerly, being very hungry, but it had to be given more cautiously than with the wire, as they were afraid of the forceps if they saw them clearly.

Next I dropped bits of brick which had been soaked in meat juice in front of the fishes as they lay under leaves with the barblets projecting beyond the edges of the leaves. In all such cases, upon touching the brick with a barblet, they seized the brick and bit at it viciously. Often they would return to it a second or third time and try to bite it. I dropped similar bits of brick which had not been soaked in meat juice in front of them in the same way, but they paid no attention to them, or in a few cases they would touch them with the barblets and then swim away again ("tactile" reaction). They never attempted to bite them. Clearly they taste the meat juice in the bricks when they are touched by a barblet, and the experiment when the body was touched by a similar brick held in forceps shows that they taste the juice by the body also.

On one occasion I tested the fishes with pieces of cooked meat that had been long boiled so that nearly all of the extractives were drawn out. The experiments were conducted just like those with the raw meat, but the fishes gave by no means so clear reactions to it. Upon touching the sides of the body, the fishes usually paid no attention to the stimulus, treating it just as they did cotton. I then touched the barblets a few times, and to this they would generally react by turning and taking the meat, but not always nor so promptly as with fresh meat. Upon testing the sides of the body again after this experience I got a reaction. The fishes would turn and touch the meat with the barblet or lips before taking it, rarely giving the quick reaction characteristic of fresh meat. Evidently the cooked meat has less taste to the fishes than fresh meat and this interferes with the reaction. They eat the cooked meat when they are sure that it is edible.

These experiments, all of which were many times repeated and controlled, I think show conclusively that practically the whole cutaneous surface of *Ameiurus* is sensitive to both tactile and gustatory stimuli, and that the latter call forth characteristic reflexes which are of the greatest value to the fish in procuring food. The fish normally reacts to contacts on the body by both types of stimuli - to the mere tactile stimulus (if at all) by a tentative movement calculated to bring the doubtful substance into contact with the more highly sensitive barblets or lips, but to the tactile stimulus accompanied by the gustatory by an immediate, rapid, and precise movement calculated to seize the food. This latter reflex is unvarying and is very persistent under a great variety of forms of stimulation. The former ("tactile") reflex is less stable, and may be readily eliminated by a simple course of training. Clearly the gustatory element of the sensation complex resulting from a contact with a sapid substance is more important than the tactile element.

It is clear that in order to call forth the characteristic "gustatory" reflex the stimulus must be quite strong and rather sharply localized. For when there is only a small amount of meat juice diffused through the water, as by the presence of a piece of fresh meat near the fish, he is not able to localize it accurately, but exhibits only the "seeking reaction." I have not as yet been able to convince myself whether the fish could accurately localize a strong and sharply localized gustatory stimulus with no tactile element. In all the experiments in which meat juice was directed against the body with a pipette or syringe there was doubtless some tactile effect produced by the impact of the jet. We know from the experiments that pure tactile stimuli can be accurately localized on the skin, and there can be no doubt that under normal conditions these assist in the localization of the food object. Compare the further discussion in the Addendum, pages 270-271.

EXPERIMENTS ON GADOID FISHES.

The preceding experiments were all carried on in the zoological laboratory of Denison University; the experiments on marine fishes which follow were made during the summer of 1902 at the U. S. Fish Commission laboratory at Woods Hole. The feeding reactions of three types of gadoids were studied, viz. young pollock (*Pollachius virens*), about 10 cm. long; hake (*Urophycis tenuis*), about 20 cm. long, and young adult tomcod (*Microgadus tomcod*).

As is well known, the hake and tomcod have a mental barblet which is known to be abundantly set with terminal buds and which receives both communis and general cutaneous innervation. In all three types the lips are freely supplied with terminal buds and there is a recurrent branch of the facial nerve, the ramus lateralis accessorius, which carries communis fibers into the trunk to supply terminal buds found on the fins, especially the free rays of the ventral or pelvic fins. These fins are far forward under the throat. In the pollock they are but little modified; in the tomcod two rays are about twice as long as the others and for about half their length they project freely below the rest of the fin. In the hake all of the rays of this fin are suppressed save these modified free rays, so that the fin is filliform, branched at the end. Microscopic examination shows that the terminal buds are more abundant on the more highly modified fins. The hake also has a free filament on the dorsal fin produced by the extension of the third and fourth rays beyond the others. I have not examined this free filament microscopically, but know that it receives communis fibers from the r. lateralis accessorius, and have no doubt that it also has numerous terminal buds, as the experiments show it to be very sensitive to gustatory stimuli. The pollock have very large eyes and are excellent visualizers. When food is thrown into the water, they dart for it and in general they take their food by the visual reflex. So keen is the vision that it would be difficult to carry on any experiments, such as I have done with the other two species, without first blinding the fish. Nor do they habitually drag the bottom with the free ventral fin rays as the others do. I have, therefore, not devoted much attention to this species, preferring to study more carefully those species in which the gustatory reflex plays the greater part in the life of the fish.

The hake (*Urophycis tenuis*). These fishes, like the tomcods, readily adapt themselves to life in captivity, and are easily experimented upon in small tanks. They are excellent visualizers, though not so much so as the pollock. When bits of meat are thrown into the water they usually catch them before they fall to the bottom, and their keen vision makes difficult such experiments as I carried on with the cat-fishes. They do not seem to recognize by sight food lying on the bottom, but only when it is in motion. But bits of meat, fish, or clam lying on the bottom are usually found by the aid of the free ventral fins. These fishes spend much of their time in slowly swimming in an apparently aimless manner close to the bottom of their tank. During these movements the filamentous pelvic fins are so held that their tips drag the bottom. These fin rays are quite long, and they are usually directed obliquely forward, outward, and downward, with the two branches of each fin widely divaricated, so that the four tips touch the ground in a line transverse to the body axis at about the level of the mental barblet. In this way the bottom under the fish and for a short distance on either side is thoroughly explored as the fish swims over it, and all food particles with which the barblet or free fin rays come in contact are taken by a quick and precise movement similar to that set off in the siluroids by contact with their barblets. Bits of meat or clam on the end of a slender wire could be laid on the bottom of the tank and then slowly moved up under or behind the fish and the reflex from the ventral fins tested in this way. Such experiments, however, had to be made with great caution and many times repeated to rule out possible visual sensations which likewise call forth an immediate reflex.

Bateson ('90, *a*) records similar reactions with the rockling (*Motella*), a gadoid fish with the same general structure and distribution of terminal buds as the hake, but with better developed barblets. (On the structure of the pelvic fins of *Motella* compare Bateson's account on p. 214 with that on p. 234 of the same volume.) Bateson, moreover, got the same reflex with fishes which had been blinded, and I have not thought it necessary to repeat this experiment, for my fishes give sufficiently clear evidence that this reflex from the fins is wholly independent of vision. We have, however, to investigate the parts played by tactile, gustatory, and olfactory sensations.

Bateson's remarks ('90, *a*, p. 214) in this connection on the rockling may be quoted here. The three-bearded and the five-bearded rockling are nocturnal and lie still all day.

Generally, both the animals take no notice of food until it has lain in the water some minutes, when they start off in search of it. The rockling searches by setting its filamentous pelvic fins at right angles to the body, and then swimming about feeling with them. If the fins touch a piece of fish or other soft body, the rockling turns its head round and snaps it up with great quickness. It will even turn round and examine uneatable substances, as glass, etc., which come in contact with its fins, and which presumably seem to it to require an explanation. The rocklings have great powers of scent, and will set off in search of meat hidden in a bottle sunk in the water. Moreover, a blind rockling will hunt for its food and find it as easily as an uninjured one.

The above, taken in connection with other passages, shows that this author considers that the food is found largely by scent, and that the fin reaction is essentially tactile, though he has seen the sense organs on the pelvic fins and recognized their resemblance to taste buds.

Examination of stomach contents shows that the normal food of these hake is largely crustaceans, particularly shrimps. I fitted up a tank with some seaweed and put into it a large number of prawns (*Palæmonetes*), mostly living, but some dead. Upon putting the hake into this tank, they immediately ate some of the dead prawns from the bottom and afterwards caught the live ones, but very slowly and with many failures. The response seems to be wholly visual. The fishes would repeatedly pass directly over living prawns, touching them with the fins or being brushed by their antennæ, but so long as the crustaceans were quiet they seemed not to notice them. If, however, a prawn was killed and crushed and thrown back into the water, it was immediately found. Upon another occasion I put a live clam into the tank with the hake, where it remained for several days, with siphons greatly extended. The fishes repeatedly brushed over this siphon with their free fins, but never paid any attention to it, though if a similar siphon were cut off from a live clam, so as to allow some of the juices to escape, it would be immediately taken and eaten. Evidently live food is not clearly located by the gustatory organs of the fins.

Besides observing as fully as possible the normal feeding habits of the hake, I experimented upon the reactions to stimuli applied to both the pelvic and the filamentous dorsal fins. As mentioned by Bateson, the pelvic fins are freely used to explore all manner of substances which may attract the notice of the fish, whether edible or not. After these fishes have become accustomed to being fed small bits of meat or clam or mussel (*Modiola*) in their tank, they immediately swim toward any small unfamiliar body with the pelvic fins thrust forward to touch it before the mouth reaches it. Sometimes the tips of these fins close over it with a movement strongly suggestive of grasping, though of course this they can not do.

Upon testing by contact with meat or other bait, the free dorsal filament is found to be quite as sensitive to gustatory stimuli as the filamentous ventrals. The reflex in this case is very characteristic and constant—the fish upon touching a savory morsel checks its forward movement and immediately "backs water" so as to reverse the movement of the body until the object is directly above the mouth, when it is taken at once. This reflex usually (though not so invariably) follows a contact of meat upon any part of the dorsal fin, as well as the free filament. The reflex rarely fails when any one of the filamentous fins is touched by freshly cut meat. After meat has been in the water for fifteen minutes or more it seems to lose its savor and the fins may be repeatedly dragged over it without calling forth a response, and the same is true of the barblet and lips.

I tested the filamentous fins with a wisp of cotton wool on a fine wire, as I did the cat-fishes. It was rarely noticed at all by the pelvic fins, but at the first contact with the filamentous dorsal the fish reacted just as he did to meat with which he had been tested immediately before. Upon repetition, the response was soon discontinued. For a few tests the fish would pause, and perhaps back up slowly so as to smell the suspicious object or touch it with the barblet, but it was not taken into the mouth. After from two to ten tests no further attention was paid to the cotton, or the fish would pause a moment without backing up. This experiment was many times repeated in the course of the first day of its trial and daily thereafter for some time. If three or four hours intervened between two series of about twenty tests, the first one or two tests of the second series might be followed by an incomplete reaction, but after that usually no notice was taken of the cotton. The fishes apparently remembered the preceding tests. But if more than twenty-four hours intervened between tests, the process of training usually had to be gone over again.

The fact that the hake does not appear to remember the difference between the pure tactile stimulus and the tactile plus the gustatory for so long a time as the cat-fish does is probably to be explained by the fact that the number of taste buds on the filamentous fins of the hake is much less than that on the barblets of the cat-fish, and therefore the gustatory element in the sensation complex is doubtless much less in the hake. The whole course of the experiments indicates that the response is in fact much more strongly tactile in the hake.

During the course of these experiments I often alternated bits of meat with the cotton wool, and at other times substituted cotton that had been soaked in clam juice. In these cases I always got the characteristic gustatory reaction by all of the filamentous fins, no difference being observable between the reaction to meat of clams or fish and that to cotton soaked in filtered clam juice.

I also tested the hake with gelatin which had been soaked up in cold water. Shreds of the well-softened gelatin were fastened to the end of a wire and brought into contact with the body surface. The reactions were identical with those obtained with white cotton. The gelatin shreds are very nearly colorless and absolutely tasteless to my tongue. But to the sense of touch they are almost exactly the same as the bits of fresh clam meat with which most of these experiments have been conducted. The hake at first would take the bait when the filamentous dorsal was touched, but if the gelatin was taken into the mouth it would be immediately rejected, and after a few trials the fish would no longer respond to the stimulus. He acted in the same way when the pelvic fins were stimulated. Shreds of the softened gelatin falling through the water were sometimes noticed, but rarely taken into the mouth, and if so, were immediately rejected. Similar shreds lying on the bottom were neglected, even though the barblet and filamentous fins dragged over them repeatedly.

I next took small clam shells that had been lying long in the tanks containing the fish and were thoroughly cleaned of fleshy matter and which the fishes had not paid any attention to for days. These I dried and warmed and then filled with melted gelatin which had been previously softened up in cold water. Upon cooling there results a mass, colorless, tasteless, and odorless, which feels almost exactly like the flesh of the clam, which has often been fed to the fishes in this way. Upon dropping these shells into the water, the fishes eagerly snatch them up, feel of them with the lips or barblet, and then bite into the gelatin. They immediately reject the gelatin and they never repeat the process. Even if they draw the fins or barblets repeatedly over the shells and the contained gelatin, they never again pay any attention to them.

I also repeated with the hake the experiments which I had previously carried out upon the cat-fish, using a fine-pointed pipette and sapid solutions. The fishes were in all cases first tested with sea water taken from the tank in which they were swimming. On one occasion (the first test made) a jet of water directed against the filamentous dorsal was followed by the characteristic backward movement of the fish, so that he finally received the jet in the face. He turned and tried to take the point of the pipette in his mouth – a purely tactile reflex apparently. This response I never got again with this or any other fish, though occasionally the fish would stop, hesitate a moment, and then swim on, paying no further attention to the stimulus. If the jet of water is directed against the pelvic fin while it is extended and searching the bottom for food, the fin is usually quickly withdrawn and pressed against the side of the body.

The pipette was then filled with the freshly prepared and strained juice of the mussel (*Modiola*), and this was directed against the fish in the same way. The fishes responded instantly, just as when stimulated by meat, whether the jet was directed against the filamentous dorsal, or the dorsal fin at any part, or the side of the body, or the free pelvic fin. The reflex is immediate and unmistakable, more sharply defined than I usually get by contact with the meat of the same mussel. The experiment was many times repeated, always with the result that the jet of water was ignored or avoided, while the jet of mussel or clam or crab juice was eagerly sought, the fish usually snapping at the end of the pipette.

I have carried out no systematic chemical experiments to determine the gustatory preferences of the fishes, having shaped my experiments so far as possible along the lines of the normal feeding habits of the species studied. Nagel and some other previous students of these problems have relied chiefly on reactions to unpleasant stimuli, and the reader is referred to their works, though I consider this a less satisfactory line of inquiry than the study of normal reactions to food substances. The few fragmentary observations which I have made with chemical stimulants I shall, however, record in their appropriate places.

Specimens of hake were tested with a 0.2 per cent solution of hydrochloric acid made up in distilled water, the acid being directed against the body by means of a fine pipette. The dorsal and ventral fins, the sides of the body, and the lips were tested. When first tested on the fins one hake turned and tried to take the pipette, much as he did with the clam juice. Afterwards this fish, as well as all the others from the first, seemed rather to dislike the acid and would swim slowly away. There is no constant reaction, however, and in fact the fishes act very much as they do when a jet of simple sea water is directed against them. They do not appear to dislike the acid intensely. Later I tested these fishes with a 1 per cent solution of hydrochloric acid in sea water. This is decidedly unpleasant and is uniformly avoided.

The experiments recorded seem to show clearly that the hake receives both tactile and gustatory stimuli by means of the free fin rays and to some extent doubtless by other parts of the outer body surface. What rôle may be played by the sense of smell remains obscure. To test the powers of locating concealed food the following experiments were tried:

In a tank containing two hake which were very hungry I placed a piece of fresh clam meat concealed between two small, old, and thoroughly clean clam shells which had been lying for some time in the bottom of the tank. The fishes did not seem to smell the meat at a distance and so be attracted to the spot where the shells were, but if in the course of their aimless movements along the bottom of the tank they passed over the shells, they generally stopped a moment, smelled around, and then passed on, first feeling over the whole area of the shell with their free fins. As time passed, this reaction became less clear until after some fifteen minutes they generally passed over the shells without paying any attention. They never found the meat. This experiment was many times repeated with the same result. The sense of smell can play no strong part in the locating of their food. It may play some small part, though I incline to believe that the interest which the fishes show in the concealed bait is excited by a vague stimulus to the terminal buds on the fins. Compare the experiments made after extirpation of the olfactory organs in the tomcod described below.

The tomcod (Microgadus tomcod). These fishes are much less active than the hake, spending most of the time lying quietly on the bottom of their tank. They have not so keen sight as the hake and pollock, but still obtain much of their food by this sense, catching food thrown in before it reaches the bottom. They do not catch live prawns in captivity so well as the hake do, yet prawns and other active crustaceans are found in the stomachs of specimens taken with the seine. The dorsal fin lacks the free filamentous rays and is not especially sensitive to gustatory stimuli. The ventral fins are, however, very efficient in locating sapid substances lying on the bottom. They are shorter than those of the hake and are not thrust forward, but incline slightly backward. Like the hake, the tomcods spend much time in slowly exploring the bottom, though they assume a very different position, with the head directed downward at an angle of some 30 to 45 with the bottom, so that the tips of the barblet and ventral fins just drag the bottom. When food particles are located they are snapped up by a quick lateral movement similar to that of the cat-fishes. Sometimes, however, stimulus of the ventral fins is followed by a reversed swimming movement, the fish backing up to take the bait. At other times the fish when exploring the bottom swims slowly backward, so that no change of direction is necessary when food is located.

I made a series of tests with cotton wool and cotton dipped in clam juice similar to those described for the hake, and with the same results. I also repeated the tests made with sea water and with strained clam juice by the aid of a pipette, with identically the same results as with the hake. After a few tests the fishes ignore sea water and plain cotton, but invariably respond to cotton soaked in clam juice and to the juice itself as they do to meat. The tomcod reacts to bits of clear gelatin soaked up in water essentially as the hake does.

I also tested the tomcod with hydrochloric acid, 0.2 per cent in distilled water and 1 per cent in sea water. Both are obviously avoided. I filled a fine pipette with a solution of quinine sulphate in sea water, about 0.1 per cent—a very bitter solution. The tomcod swims away immediately if applied either to the lips or to the pelvic fins, but appears not to notice it if applied to other parts of the body.

Within two old clam shells, which had been lying in the tank with the tomcods for several days and had remained unnoticed, was placed a piece of fresh clam. They were then closed together and laid on the bottom of the aquarium containing a tomcod. Shortly the fish passed near it, appeared to perceive it, turned from his course, and passed and repassed the spot until the shell was located, apparently by smell, by a method of "trial and error." Then he rooted at the shell vigorously until the two halves were separated and he could get the meat. I repeated this with a piece of squid within the shells with the same result. I tried two empty shells in the same way. He saw me put them into the water, came up to investigate, smelled (?) of the shells and went away without so much as touching them, and never came back to them again.

These experiments were repeated in many forms many times. In most of these cases the efficient organ in discovering the presence of the food was almost certainly the pelvic fin. At least, this alone located it, for the fish swam about (possibly feebly smelling something good), but did not make a definite movement toward the bait until the fins were dragged over the crack between the two shells containing it, from which the juices were doubtless being diffused out into the surrounding water. Then he backed up in the typical way. If the bait was not found within a very few minutes it was left unnoticed, even though subsequently uncovered.

These fishes almost invariably find a concealed bait, though the hake rarely does so. The hake seems to perceive the odor or savor of the food, for he lingers about the spot where it is concealed, but never makes a movement to uncover it. The tomcod, on the other hand, actively pushes things about with his snout until the bait is discovered. But, unlike the gadoid fishes which Bateson describes, these fishes do not get the scent of the food at any considerable distance and then search for it. They do not notice the bait until within a few centimeters of it, and there is no evidence that the sense of smell assists at all in the localization.

To test this point the olfactory organ was extirpated in several tomcods which had given the reaction last described clearly. Several ways of performing this operation were tried. The most successful method was to etherize the fish sufficiently to keep him quiet and then operate in a shallow tray with the mouth kept under water, cutting off the olfactory nerves or crura with a sharp scalpel. The wound suppurated badly, but appeared to give the fish no serious trouble, as they feed normally from the second day onward. Without going into the details of the observations, I may say that after the third or fourth day the fishes took their food in all respects like uninjured fishes, so far as could be observed. They gave all of the characteristic reflexes that have been mentioned above, including the discrimination between cotton wool and cotton dipped in clam juice and between sea water and clam juice applied with a pipette, etc. The operated fish would locate a concealed bait by means of the pelvic fins exactly as the normal fish does, and he would similarly root it out and eat it. In short, the gustatory reflexes, so far as I have observed them, were absolutely unmodified by the operation. That the olfactory apparatus was totally destroyed was verified by autopsy dissections made after the close of the observations.

OTHER FISHES.

The sea-robin (*Priotonus carolinus*). --The three finger-like rays of the pectoral fins of the gurnards have long attracted the attention of zoologists, and the American species of *Prionotus* have been made the subject of a careful research by Morrill ('95). He finds that, as in the closely related European *Trigla*. the free rays are totally devoid of terminal buds or other specialized sense organs and that the sensory nerves with which these free rays are so abundantly supplied end free, like tactile nerves in general.

He also made some interesting physiological experiments. The normal food of these species, so far as known, is small fish, young clams, shrimps, amphipods and other small crustacea, squid, lamellibranch mollusks, annelids, and seaweeds. (Linton, 1901, p. 470.) They are constantly feeling about the sand, turning over stones and feeling under them, etc., with these free rays, and undoubtedly find their food largely in this way, especially the annelids, mollusks, and crustacea; but in captivity the eyes are used chiefly in securing the food. Morrill writes further:

In order to test the use of the free rays independently of sight the crystalline lens and cornea were removed from some fish, and in other cases the cornea was covered with varnish, balsam, or tar. The repeated experiments were negative in their result, as the fish paid no attention to the food, even when it was placed in contact with the free rays.

Morrill concludes "that the free rays have been modified for tactile purposes, and that they are mainly, if not altogether, used in searching for food."

Morrill's dissections leave it uncertain whether the free rays of the pectoral fins receive communis nerves, as they should do, of course, if these organs had given evidence of gustatory powers. The only source of communis fibers for this fin would be through the ramus lateralis accessorius (r. recurrens facialis). Stannius (1849, p. 49) did not find this nerve in *Trigla gurnardus* and *T. hirundo*. I dissected a specimen of *Prionotus carolinus* and found the same to be true here, so that it can be taken as assured that no communis nerves reach the pectoral fin in this species.

After an examination of the feeding habits of the adult sea-robin and of young specimens about 10 cm. long I quite agree with Morrill that the reaction to food particles by the free fin rays is tactile only, with no gustatory element. When adults are fed with fresh clams or mussels, the shells split open to expose the meat, they turn and bite out the meat as soon as a free ray touches the soft flesh. Young fishes did not give this reaction so invariably, and evidently relied much more on sight. Clean clam shells filled with melted gelatin were reacted to like the fresh clams once or twice by each fish, but usually were thereafter ignored.

The free rays constantly stir up the sand and gravel of the bottom. If soft edible particles are touched the head may be turned to snap them up, especially with old fishes. With younger ones this usually does not happen unless the particle is seen while in motion. In fact, with these younger fishes the purpose of the activity of the free rays seems to be in the main the agitation of particles on the bottom to bring them into the range of vision. Almost any unfamiliar object, such as a bit of coal or a brightly colored pebble or any soft particle, if seen while in motion, will be apt to be taken into the mouth. The analysis is done here—not by the peripheral cutaneous organs. All small objects thrown into the water are taken into the mouth as they fall; bits of filter paper, gelatin, etc., will be taken and immediately rejected. The same bit of paper or excrement may be taken and rejected a half dozen times in rapid succession, the reflex following in a perfectly automatic way as soon as the moving object is seen. Small worms when thrown into the water would be captured before they had time to reach the bottom, but if placed on the bottom they would seek shelter under pebbles and remain unnoticed until they were stirred up and sent floating off, when they would be seen and taken at once. The free fin ray was observed to touch the worm when concealed without evoking a response. A moment later the worm was set in motion and taken at once.

I got no evidence that the fishes smell or otherwise detect the presence of food at a distance or concealed from sight and touch. Meat inclosed between clam shells, which a tomcod would have secured within a minute or two, remained unnoticed, though the outsides of the shells were repeatedly fingered over by the free rays and similar bits of meat were taken at once if in motion near the fish.

The young sea-robins eat crab meat well. I made a strong extract of crab meat and filtered it. Now with a fine pipette a jet of clean sea water was directed against the free pectoral-fin rays. There was no response, or if the jet was strong the fin was folded against the body. The extract of crab applied in the same way with the pipette gave the same result. Even when the jet is directed against the lips the fish usually pays no attention or is disturbed and swims away. This would seem to indicate that the sense of taste is absent or very feeble on all of the exposed parts of the body. Thus the absence of special gustatory sense organs, of communis nerves, and of gustatory reactions from the free rays of the pectoral fins serve as mutual controls.

The king-fish (Menticirrhus saxatilis).—These fishes have a short, thick mental barblet, and they were studied to compare their reactions with those of the siluroid and gadoid fishes. Most of the types of experiment made previously on the latter fishes were repeated on the king-fish. Without going into details, the experiments seemed to show in general that the king-fish is not a pure visualizer, though vision is somewhat used in finding food. This seems to be in the main a tactile reaction, as most of the food taken was by contact and nonnutritious substances were generally taken if they felt like food. For instance, colorless gelatin is taken at the first contact and repeatedly thereafter for an indefinite number of times, though in each case it is at once rejected as soon as it enters the mouth. The sense of taste seems to be limited to the mouth, and I found no evidence of a gustatory reaction by the barblet, though the experiments were not sufficiently numerous or varied to be conclusive. They do not find a concealed bait.

The toad-fish (Opsanus tau).—These fishes were experimented upon at the same time as the hake and tomcod, and by the same methods. The toad-fish never found a concealed bait and never seemed to get food by any other reflex path than the visual or tactile. The fleshy, cutaneous appendages of the skin were especially tested to bring out possible gustatory reactions, but with negative results save for those bordering on the lips, where it was impossible to exclude the participation of taste buds on the lips. This agrees with the anatomical findings of Miss Clapp (1899), whose careful study of the skin of this fish failed to reveal any terminal buds on these appendages or elsewhere away from the buccal cavity. A jet of sea water directed against these appendages or the body surface in general usually disturbs or frightens the animal merely, if it is noticed at all. A jet of clam juice similarly applied calls for the same reaction unless it is so directed as to reach the lips, in which case the fish reacts to it just as the hake and tomcod do, attempting to take the tip of the pipette in the mouth. The following solutions were applied in the same way by a fine pipette to various parts of the body surface: 0.2 per cent hydrochloric and 1 per cent hydrochloric acid in sea water, and 0.1 per cent quinine sulphate in sea water. In all cases the fishes paid no attention to the stimulus unless the substance was so applied as to come into contact with the lips. The experiments lead me to conclude that the toadfish can taste only within the mouth and on the lips, and that if the cutaneous appendages have any sensory function it is tactile only.

CONCLUSION.

The morphological and physiological significance of the terminal buds of fishes is a problem which has exercised some of the ablest morphologists for over half a century. The methods of the older anatomy have signally failed to yield concordant results. Not until the innervation of the cutaneous sense organs was worked out from the standpoint of nerve components was this confusion relieved. The older morphologists (Schulze, Merkel, and others) discovered a morphological criterion, the "hair cells," by which the terminal buds could be distinguished from cutaneous sense organs belonging to the lateral-line system. But this fact attained its significance only when it was discovered that the organs of the lateral-line system, or neuromasts, which possess the "hair cells," are always innervated by lateralis nerves related centrally to the tuberculum acusticum, while terminal buds, which lack the "hair cells," are always innervated by communis nerves which are related centrally to the primary gustatory centers of the vagal and facial lobes.

Presumably, then, lateral-line organs and terminal buds have different functions; and, further, the function is probably not tactile in either case, since all parts of the skin receive general cutaneous nerves in addition to the special sensory components, and these general cutaneous nerves are related proximally to different centers from either of the others. The lateral-line organs are known to be used in the maintenance of bodily equilibrium and the perception of mass motion of the water. (Compare the recent works of Lee and Parker.) On the other hand, the terminal buds are related in structure and innervation to undoubted taste buds of the mouth, and hence the inference that their function is taste. This inference is abundantly confirmed by the experiments here recorded, and the function and morphological rank of the terminal buds are at last definitely fixed.

It may be regarded as established that fishes which possess terminal buds in the outer skin taste by means of these organs and habitually find their food by their means, while fishes which lack these organs in the skin have the sense of taste confined to the mouth. The delicacy of the sense of taste in the skin is directly proportional to the number of terminal buds in the areas in question.

Numerous unrelated types of bony fishes from the siluroids to the gadoids which possess terminal buds have developed specially modified organs to carry the buds and increase their efficiency. These organs may take the form of barblets or of free filiform fin rays. The free rays of the pelvic and dorsal fins of gadoid fishes are thus explained, and indeed this is possibly the motive for the migration into the jugular position of the pelvic fins of the gadoids.

In all cases where terminal buds are found on barblets or filiform fin rays gustatory nerves belonging to the communis system are distributed to them. These barblets and free fin rays likewise receive a very rich innervation of tactile or general cutaneous nerves, so that they merit their popular designation—"feelers." Both sets of end organs undoubtedly cooperate in the discrimination of food, and the animal has the power of very accurate localization of the stimulus. Whether the gustatory stimulus alone can be localized apart from its tactile accompaniment can not at present be stated. A purely tactile stimulus with no gustatory element can be localized precisely, and I have as yet no conclusive evidence that a pure gustatory stimulus, even when strong, can be located by the fish. It is certain that feeble and widely diffused gustatory stimuli can not be accurately located by the fishes which I have experimented with, either by the terminal buds or by any other organs.

The fishes in which the cutaneous terminal buds are most highly developed are in general bottom feeders of rather sluggish habit, and in some cases they are nocturnal feeders. The high development of this sense is compensated for in some fishes by the reduction of others. The visual power of the fishes is especially apt to suffer degradation. This degradation may be organic, a positive degeneration of the visual apparatus, as in *Amiurus*, or it may be merely functional. In the latter case, though the organs of vision are not necessarily modified, these organs are not actually used in procuring food, the fish being unable to effect visual reflexes toward food substances or to correlate visual stimuli with the movements necessary to react toward food substances. The fish may be perfectly able to effect other visual reflexes, but is apparently unable to understand the significance of food when perceived by the sense of sight only. This particular central reflex path has never been developed, or has atrophied from disuse. Nature has here effected for the species something similar to what is accomplished in individual men occasionally by disease, in the production of certain aphasias.

The number of reflex activities habitual to an animal with a nervous system as simply organized as the bony fish is probably far smaller than is commonly supposed, and these activities are in general characterized by but little complexity of organization. It is probably quite within the range of possibility to determine by observation and experiment for any given species of fish, to a high degree of accuracy, what these habitual activities are and to work out by histological methods the reflex arc within the nervous system for each of them; and since the human nervous system is built up on the same general plan as the piscine nervous system it follows that such a thorough and systematic correlation of function with structure would be profitable from many points of view.

ADDENDUM.

During the winter and spring of 1903 some further observations have been made with the purpose of answering (among others) the question raised above, whether fishes can localize a sensation received by the terminal buds alone with no tactile accompaniment; or, in other words, whether gustatory sensations may be provided with a local sign as tactile sensations are. (This question, of course, does not necessarily involve the more general one as to the essential nature of the local sign, whether it is due to a "specific energy" of the peripheral nerve or sense organ or to central differentiation in the terminal nucleus.)

Some recent clinical observations suggest that in human beings such a localization of gustatory sensations is possible. Cushing (Johns Hopkins Hospital Bulletin, vol. xiv, No. 144, 1903, p. 77) reports after destruction of the Gasserian ganglion and total paralysis of general sensation on the anterior part of the tongue, that the gustatory sensibility remains unimpaired, and that in this case the gustatory sensations can be localized. It is not, however, absolutely certain that it is the gustatory fibers which effect the localization, for the chorda tympani, which was uninjured, may carry also a certain number of fibers for general sensation from the facialis root in addition to gustatory fibers, as Cushing assumes is the case with the chorda from some of his results and from those of Köster.

My own observations were made on the young of *Ameiurus* from 5 to 8 cm. long, received from the State fish hatchery, at London, Ohio, in October, 1902, and kept under observation in tanks during the following winter. These fishes prove to be more shy and less teachable than the smaller *Ameiurus* fry (about 3 cm. long) hatched by wild parents, upon which the experiments reported in the preceding pages were made.

I have verified on these fishes most of the observations made on the smaller fishes last year. The most noticeable difference in their behavior is the evidently greater visual power in these fishes. As soon as they began to feed freely in the presence of the observer (which required several months of training) they began to show evidence of visual recognition of a moving bait, if very near them, and provided they had just previously been fed with the same food in the same way. They never under any circumstances notice visually a still bait, and their recognition of a moving bait is at best very imperfect and only an occasional occurrence.

Upon putting a concealed bait in a tank with the fishes I found no evidence that they are able to locate it by the sense of smell or otherwise from a distance, provided the water is still. If, however, they swim near enough to the capsule containing the bait (beef liver, cheese, etc.) to pass the barblets into the strong diffusion currents emanating directly from the bait, it is located instantly. The reactions here are essentially like those by which the tomcod localizes a concealed bait, though I have not completed the experiment by extirpation of the nose to determine what part, if any, is played by the sense of smell. So far as my experiments have gone these fishes will not locate a concealed bait in still water unless they pass within 5 cm. of it.

In running water, however, the case is quite different. I constructed a long, narrow tank, so arranged that a slow stream of water can pass through it from end to end. By covering the lower end of the tank and illuminating moderately the upper end, it can be so arranged that the negative phototaxis will counteract any positive rheotaxis and the fishes will remain in the lower end of the tank. If now liver or other strong bait is placed above them, the fishes will promptly swim up the current and locate the meat.

The experiments seem to indicate that concealed food can not be located by these fishes from a distance in quiet water (cf. Nagel, 1894), but that if the fish passes within a few centimeters of it the diffused juices are recognized and the food located promptly. In running water, however, the fishes will follow the diffused juices up the stream for considerable distances and so find the food—a fact well known to every fisherman. Tactile sensations are clearly not involved; it lies between the senses of smell and taste, and I have not as yet gone far enough with this series of experiments to decide finally the part played by the sense of smell.

I have, however, tested the sensitiveness of the barblets to diffused savors more fully. Raw meat or beef liver was minced, extracted in a little water, and strained. A wisp of cotton was wound on the end of a slender wire, dipped in the meat juice, and gently lowered so as to lie a few millimeters from the tip of a barblet of a cat-fish which was otherwise entirely concealed under a large leaf. The fish was unable to see the cotton and actual contact with the barblet was carefully avoided. Within a few seconds the fish became conscious of the savor and turned *toward the cotton*. Again, I filled a glass tube, of about 3-mm. bore, with the meat juice, closed the upper end with the finger, and carefully lowered the open end down over a projecting barblet, as in the previous case. The specific gravity of the meat juice is slightly greater than that of the water, and from the lower end of the tube (the upper end being kept closed) the juice slowly diffused downward enveloping the tip of the barblet, without, however, any noticeable current being produced in the water. The fish locates the stimulus and turns toward the source of it. In other cases I colored the juice with a little blood, so that the course of the diffusion currents could be observed, and it is evident that the reaction follows the stimulus of the *barblet* only, and not the organ of smell, for the movement is made before the diffusion currents have had time to reach the nostril.

These reactions are not as prompt or precise as those given after a *contact* with a sapid substance where a tactile sensation accompanies the gustatory, and in a large percentage of the cases there is no definite reaction toward the point stimulated, but merely the more vague "seeking reaction" to which reference has been made above. Nevertheless they indicate on the whole that pure gustatory stimuli, if very strong and applied to a small area of the percipient organ, *can be localized in space, or have a* "*local sign.*"

May 30, 1905.

LITERATURE CITED.

ALLIS, E. P., JR.
'97. The cranial muscles and cranial and first spinal nerves in *Amia calva*. *Journ. Morph.*, XII, 3.

BATESON, W.
'90. The sense-organs and perceptions of fishes, with remarks on the supply of bait. *Journ. Marine Biol. Assoc., London*, I, pp. 225–256.
'90a. Sense of touch in the rockling. *Ibid.*, p. 214.

CLAPP, CORNELIA M.
'99. The lateral line system of *Batrachus tau*. *Journ. Morph.*, XV, 2.

GRABER, V.
'85. Vergleichende Grundversuche über die Wirkung und die Aufnahmstellen chemischer Reize bei den Tieren. *Biol. Cnt.*, Bd. V, Nos. 13, 15, 16.
'89. Ueber die Empfindlichkeit einiger Meertiere gegen Riechstoffe. *Ibid.*, Bd. VIII, pp. 743–754.

GÜNTHER, A. C. L. G. '80. An introduction to the study of fishes. *Edinburgh*.

HARRIS, WM. C. '02. Salmon and trout American Sportsman's Library. *N. Y., The Macmillan Co.*

HERRICK, C. JUDSON.
'99. The cranial and first spinal nerves of *Menidia*; a contribution upon the nerve components of the bony fishes. *Journ. of Compar. Neurology*, vol. 9, pp. 153–455.
'00. A contribution upon the cranial nerves of the cod-fish. *Ibid.*, X, 3.
'01. Cranial nerves and cutaneous sense organs of North American Siluroid fishes. *Ibid.*, XI.

JOBERT.
'72. Études d'anatomie comparée sur les organes du toucher chez divers mammifères, oiseaux, poissons et insectes. *Ann. sc. nat.*, 5 Ser., T. XVI.

KAHLENBERG, L.
'98. The action of solutions on the sense of taste. *Bul. Univ. of Wisconsin, Science Series*, vol. II, 1, pp. 1–31.

LEE, FREDERIC S.
'92. Ueber den Gleichgewichtssinn. *Centralbl. f. Physiol.*, Bd. 6, pp. 508–512.
'93. A study of the sense of equilibrium in fishes. *Jour. of Physiol.*, vol. 15, pp. 311–343.
'94. A study of the sense of equilibrium in fishes, Part II. *Jour. of Physiol.*, vol. 17, pp. 192–210.
'98. Functions of the ear and lateral line in fishes. *Amer. Jour. of Physiol.*, I, pp. 128 144.

LEYDIG, FR.
'51. Ueber die äussere Haut einiger Süsswasserfische. *Zeits. wiss. Zool.*, III.
'79. Neue Beiträge zur anatomischen Kenntniss der Hautdecke und Hautsinnesorgane der Fische. *Festschr. z. 100 Jähr. Naturf. Ges. zu Halle*.
'94. Integument und Hautsinnesorgane der Knochenfische. Weitere Beiträge. *Zool. Jbr. Abt. f. Anat. u. Ontogen.*, VIII, 1, pp. 1–152.

LINTON, E. '01. Parasites of fishes of the Woods Hole region. *Bull. U. S. Fish Com. for 1899*.

MERKEL, FR. '80. Ueber die Endigungen die sensiblen Nerven in der Haut der Wirbelthiere. *Rostock*. Contains extensive bibliographies.

MORRILL, A. D. '95. Pectoral appendages of *Prionotus* and their innervation. *Journ. Morphology*, XI.

NAGEL, W. A.
'94. Vergleichend physiologische und anatomische Untersuchungen über den Geruchs- und Geschmackssinn und ihre Organe, mit einleitenden Betrachtungen aus der allgemeinen vergleichenden Sinnesphysiologie. *Bibliotheca Zoologica, Stuttgart*, Heft 18. Contains a bibliography of 335 titles.

PARKER, G. H.
'03. Sense of hearing in fishes. Abstract of a paper read before the Am. Assoc. for the Advancement of Science. *Science, N. S.*, vol. XVII, No. 424.
'03a. The sense of hearing in fishes. *Am. Naturalist*, XXXVII.
'03b. Hearing and allied senses in fishes. *Bull. U. S. Fish Commission for 1902*.

RICHARDS, T. W.
'98. The relation of the taste of acids to their degree of dissociation. *Am. Chemical Journal*.

SCHULZE, F. E.
'63. Ueber die becherförmigen Organe der Fische. *Zeits. f. wiss. Zool.*, XII, 2.
'67. Epithel- und Drüsenzellen. *Arch. f. mikr. Anat.*, III.
'70. Ueber die Sinnesorgane der Seitenlinie bei Fischen und Amphibien. *Arch. f. mik. Anat.*, VI.

STANNIUS, H. '49. Das peripherische Nervensystem der Fische, anatomisch und physiologisch untersucht. *Rostock*.

TODARO, F. '73. Les organes du goût et la muqueuse bucco-branchiale des Sélaciens. *Arch. Zool. Expérim.*, II, pp. 534–558.

WEBER. '27. Ueber das Geschmacksorgan der Karpfen. *Meckel's Archiv f. Anat.*

WHITMAN, C. O. '99. Animal behavior. Biological lectures, Woods Hole, session of 1898. *Boston*.

ZINCONE, A. '78. Osservazioni anatomiche su di alcune appendici tattili dei pesci. *Rend. Accad. Napoli*, XV.

CONTRIBUTIONS TO THE BIOLOGY OF THE GREAT LAKES.

ROTATORIA OF THE UNITED STATES.

II. A MONOGRAPH OF THE RATTULIDÆ.

·

By H. S. JENNINGS,

Assistant Professor of Zoology in the University of Michigan.

CONTRIBUTIONS TO THE BIOLOGY OF THE GREAT LAKES.

ROTATORIA OF THE UNITED STATES.

II. A MONOGRAPH OF THE RATTULIDÆ.

By H. S. JENNINGS.

Assistant Professor of Zoology in the University of Michigan.

INTRODUCTION.

There is perhaps no need so great in American zoology as to have the different groups of invertebrates thoroughly described and set in order, so that the worker in ecology, physiology, variation, or morphology can determine them without becoming a professional systematist. As matters stand at the present time, most of our aquatic invertebrates can not be determined without the study of much scattered literature, ancient and modern, and much wearisome and unprofitable sifting of synonymy.

Happily there is at present a strong movement toward remedying this state of affairs. What is needed is a set of studies comprising monographic treatments of the various groups—each account of a group complete in itself, so far as the American species are concerned, so that any species of the group can be determined without reference to other literature. This can be accomplished if different investigators select circumscribed groups of not too great extent, perhaps a single genus, and set this thoroughly in order, describing and figuring all species likely to occur in America, and bringing the names into consonance with recognized rules of nomenclature. It is such a study of one of the families of the Rotatoria that is herewith presented.

The *Rattulidæ* are a family of free-swimming Rotatoria, containing altogether about 40 to 45 species. Their chief general interest lies in their peculiar unsymmetrical structure, most of them having the organs so disposed as to give the impression that the body has been twisted, while the primitive bilateral symmetry is still further disturbed by a number of the organs becoming rudimentary on one side. They are found as a rule amid aquatic plants in the quiet parts of lakes, ponds, and streams. Only one of them (*Rattulus capucinus* Wierz. & Zach.) can be said to be limnetic—that is, commonly found free-swimming at a distance from the vegetation of the shores and bottoms. A few occur in swamps; but clear water, amid actively growing vegetation, is the place where the *Rattulidæ* abound. In such regions they are often among the most abundant of the Rotifera.

The classification of the *Rattulida* has fallen into great confusion. This statement could be made of almost any of the larger groups of Rotatoria, but it is perhaps more strikingly true of this family than of any other. Many species have been described under several different specific and generic names, while in other cases several different species have been described under a single name. The twisted, unsymmetrical structure has always been more or less of a puzzle to systematists, making it difficult to determine even what were properly to be considered dorsal and ventral surfaces, and the great difference in appearance between contracted and extended animals has further tended to favor confusion. It has seemed to the writer that there is no group of the Rotatoria so much in need of a thorough revision as this one. For this reason it has been taken up first.

In the following paper I attempt to give an account of the structure and movements of these animals, paying especial attention to the asymmetry and its biological significance, and to furnish as far as possible full descriptions and figures of all known species. A large majority of the known species I have myself been able to study, and in these cases the descriptions and figures are based on my own observations. I have attempted to make these so detailed that further mistakes in the identification of these species will hardly be possible. In the case of species which I have not been able to examine myself I give the figures and descriptions which have been published by other authors. Many of these descriptions are very unsatisfactory, as comparison with a large number of species is necessary for bringing out the important characteristics, and such comparison has, in the absence of preserved material, been almost wholly lacking until very recent times.

In the preparation of this paper I have been especially indebted for assistance of the most essential character to Mr. Charles F. Rousselet, of London, England, and to Mr. F. R. Dixon-Nuttall, of Eccleston Park, North Prescot, England. Mr. Rousselet placed at my disposal his valuable mounted collection of the *Rattulida*, including a number of species which I did not have in my collection, and has assisted me throughout the work with valuable notes and suggestions. Mr. Dixon-Nuttall sent me his notes and drawings of a considerable number of species of *Diurella*, which he had long been studying, and gave me permission to make use of some of his excellent figures, a number of which are given on plates IV and XIII. The continued cooperation of these two careful investigators has added much to the completeness and accuracy of this paper, and has made it possible, by comparison of specimens, to be certain that my determinations of doubtful species agree with those of the best European authorities.

I am indebted also for specimens of *Rattulida* from Lake Bologoe in Russia to the kindness of Dr. Romuald Minkiewicz, of the University of Kasan, Russia; to Herr Max Voigt, of Plön, I am under obligations for specimens of his new species, *Diurella rousseleti*. For notes and other assistance I am indebted to Herr Oberförster L. Billinger, of Stuttgart, Germany; to Prof. Dr. Otto Zacharias, director of the Freshwater Biological Station at Plön, Germany, and to Prof. Dr. Karl Eckstein, Eberswalde, Germany. It is a pleasure to express here my thanks to these gentlemen.

METHODS.

The use of preserved mounted specimens has been the basis for the present work. It is only through the methods devised within the last decade by Mr. Charles F. Rousselet, of London, England, that the use of such preserved material has become possible in the study of the Rotatoria. Hence the complete lack hitherto of type or reference specimens among these animals. This has been one of the prime causes of the great confusion in the classification of the Rotatoria. A few of the genera have been worked over in the last few years with the use of preserved specimens by Mr. Rousselet and his collaborators in England. It is not too much to say that it will be necessary to go over the entire group of Rotatoria in the same manner before order can be brought out of the present confusion.

Killing and preservation.—The collections of preserved material on which the following paper is based were made as follows: The Rotifera were taken in various ways—by towing with the tow net in water free from vegetation, by washing aquatic plants in jars of water, by bringing into the laboratory quantities of aquatic plants together with some of the water about them, etc. Most of the Rotatoria come after a time to the lighted side of the vessel in which the material collected is placed. These are transferred in large numbers to a watch glass and placed beneath a simple microscope or low power of the compound microscope, where the movements of the organisms can be observed.

Now a considerable quantity of Rousselet's narcotizing fluid is mixed with the water in the watch glass. One-fourth as much narcotizing fluid as there is water, or a larger or smaller proportion, may be used, as seems desirable from observation of the movements of the animals. Rousselet's narcotizing fluid consists of 2 per cent solution of hydrochlorate of cocaine, 3 parts; methyl alcohol, 1 part; water, 6 parts. This causes the animals to swim slowly and gradually to settle to the bottom; they will soon die, and if allowed to die unfixed will be quite worthless for study, destructive changes taking place in the tissues at the moment of death or perhaps even before. As soon, therefore, as most of the rotifers have sunk to the bottom, as much of the water as possible is drawn off from above them with a pipette. Then a small amount of 0.25 per cent osmic acid is introduced, which kills and fixes the rotifers at once. Now remove the osmic acid as quickly as this can be done without taking up too many of the rotifers (within a minute or two if possible), and wash several times in distilled water. In thus fixing the rotifers in large numbers at once, it is usually impossible to draw off the osmic acid as soon as would be best, so that the animals become much blackened. But the blackening may be removed later with hydrogen peroxide. If the osmic acid has been used at the right time usually a majority, or at least many, of the rotifers will be found to be fixed well extended. But as the time required for narcotization varies with different species as well as with different individuals of the same species, many of the animals will be found contracted or with the structure partly obscured by degenerative changes. With practice, however, it will become possible to secure a sufficiently large percentage killed in good condition to make the collection very valuable.

For study of the loricate Rotifera it is advisable to kill some part of every collection directly by means of osmic acid, without previous narcotization, for in the loricate rotifers some of the most important distinctive characters can best be seen in contracted specimens.

After several washings the collections are preserved in 3 to 6 per cent formalin (3 to 6 parts commercial formalin to 100 parts water). They can not be preserved in alcohol without causing extensive shrinkage, rendering them useless for further study.

These collections may later be examined under a lens in order to study the rotifers belonging to any family, genus, or species, and the specimens desired picked out by means of a pipette drawn to a capillary point. The different species are sorted into different watch glasses, and the blackening due to the osmic acid is removed by drawing off most of the formalin and adding a few drops of hydrogen peroxide for a few minutes. As soon as the desired degree of bleaching is reached the hydrogen peroxide is replaced by formalin. The formalin should be changed several times and allowed to stand several hours before mounting the specimens, otherwise bubbles of oxygen may appear under the cover glass after it is sealed.

Specimens which have not been in osmic acid long enough to require bleaching are better in some respects than those that have been bleached by the hydrogen peroxide, as the latter removes the pigment from the eye, as well as the blackening due to the osmic acid.

The specimens are then mounted in hollow-ground slides. The slides should be thin and the concavities shallow, so that it will be possible to use high powers of the microscope. The specimens are transferred to the concavities along with some of the formalin, and covered with a circular cover glass. It is best not to leave any bubbles of air beneath the cover. The superfluous fluid is withdrawn from the edge of the cover with a bit of filter paper, and the cover is then sealed.

It is, of course, necessary to use some sealing material that will not allow water to evaporate through it. Mr. Rousselet, the originator of this method of mounting rotifers, recommends the following for sealing the mounts: After fixing the cover with a ring composed of a mixture of two-thirds gum damar with one-third gold size, there are added two coats of pure shellac, followed by three or four coats of gold size, allowing twenty-four hours for each coat to dry.

The following account of the *Rotifida* is based on the study of 101 collections, made as above, and representing about half as many different stations. These collections were mostly made about the shores of Lake Erie, during the summers of 1898, 1899 and 1901, while the writer was connected with the biological work on the Great Lakes carried on by the United States Fish Commission. The following regions were examined with special thoroughness:

1. The region about the islands in the western part of Lake Erie.
2. The south or Ohio shore of Lake Erie, in the region known as East Harbor, some distance from Sandusky, Ohio.
3. The lake shore and river at Huron, Ohio.
4. The region about Erie Harbor, Pennsylvania, including the swamps and ponds on Presque Isle.
5. Long Point, on the Canadian shore of Lake Erie.
6. Many collections have also been made about Ann Arbor, Mich., in the Huron River, and in a number of small streams and ponds in the neighborhood.

These collections have been supplemented by specimens and notes furnished by a number of investigators in Europe, as mentioned in the introduction.

STRUCTURE OF THE RATTULIDÆ.

The *Rattulidæ* are Rotatoria, usually of small size, in which the cuticle of the body has become stiffened to form a sort of shell, called a lorica. At the anterior end is a ciliated area or corona, by means of which the animal swims; this may be retracted within the lorica. At the posterior end is a small separate joint, known as the foot (*f*, figs. 1, 27, 46, etc.). To the foot are attached one or two bristle-like structures, which are called the toes. The internal organs comprise an alimentary canal, nervous, muscular, excretory, and reproductive systems, and certain mucus glands. In the following account these sets of organs will be taken up in order.

I. EXTERNAL FEATURES.

(1) *General form.*—The more usual form of the body in the *Rattulidæ* is that of a cylinder, or long oval, frequently curved. In some cases the body is much elongated, as in *Rattulus elongatus* Gosse (pl. XII, fig. 102), or *Diurella insignis* Herrick (pl. II, fig. 15); in other cases it is short and plump, as in *Diurella porcellus* Gosse (pl. II, figs. 19–21). In a few cases (*Rattulus latus* Jennings, pl. VII, figs. 65, 66; *R. multicrinis* Kellicott, pl. VI, figs. 55–57) the body is broad and ovoid in form.

A striking feature of the animals is their tendency to asymmetry in shape. This shows itself in many ways. The body with the toes usually forms a curve, concave to the right, convex to the left (figs. 1, 8, 16, 28, 46, 95, 99, 102, etc.). The curve is often not simple, but is of such a nature that the body forms a segment of a spiral. This is perhaps best seen in fig. 1, of *Diurella tigris* Müller; it is a characteristic which is difficult to represent in a drawing, although often very noticeable in the animal itself. As will be seen later, the asymmetry shows itself in the form and arrangement of many organs.

(2) *Lorica*—The body is covered with a hardened cuticula, known as the lorica. The lorica covers the body completely, being without openings at the sides, but it is open anteriorly for the projection of the corona, and posteriorly for the protrusion of the foot. The lorica is not so stiff and unyielding in the *Rattulidæ* as in many of the Rotifera, usually permitting considerable change of form. Compare, for example, the extended form of *Rattulus longiseta* Schrank (pl. VIII, fig. 67) with the contracted form in the same species (pl. VIII, fig. 70). In some species the lorica is stiffer, not permitting such marked changes in shape.

Head-sheath.—The anterior part of the lorica is usually set off from the remainder of the body by a slight constriction. This anterior portion, covering the head, may be known as the head-sheath (*h. s.*, figs. 1, 3, 8, etc.). It presents a number of interesting characteristics, and some that are very important in classification. Only in *Rattulus latus* Jennings is it impossible to distinguish a head-sheath from the remainder of the lorica.

The head-sheath frequently has longitudinal plaits, if they may be so designated, which serve for permitting the folding of the head-sheath when the head is retracted within the lorica. These are well seen in figs. 3, 4, 58, 59, and 62. These plaits seem to be due to alternate longitudinal strips of hard, stiff material, and of soft, yielding cuticula. On the inner surface of the head-sheath are many fine transverse muscle fibers (shown especially in figs. 58 and 59, pl. VI). When the head is drawn within the lorica, these longitudinal folds are brought together by the yielding of the soft strips between them, and partly slip over one another, so that the size of the head-sheath is greatly reduced and the anterior opening nearly or quite closed. It is possible to withdraw the head, at least partly, in most species without causing the complete folding of the head-sheath; evidently a supplementary contraction of the fine transverse muscle fibers is necessary to bring this about.

In some species (notably *Diurella tigris* Müller, pl. I, figs. 3, 4; *Diurella rousseleti* Voigt, pl. IV, fig. 37; *Rattulus multicrinis* Kellicott, pl. VI, fig. 58; *Rattulus capucinus* Wierz. & Zach., fig. 59, and *Rattulus cylindricus* Imhof, pl. VII, fig. 62) the head-sheath falls when contracted into very regular folds. In *D. tigris* Müller, *D. rousseleti* Voigt, and *D. intermedia* Stenroos, and perhaps in other species, the number of these folds is nine. In some other species, as, for example, in *Rattulus gracilis* Tessin, pl. V, fig. 48, the folds are very irregular. In still other species no such folds are present, and the lorica may remain widely open when the head is retracted. This is the case, for example, in *Rattulus scipio* Gosse, pl. V, fig. 52.

On the anterior dorsal margin of the head-sheath there are in certain species of the *Rattulidæ* a number of teeth. In *Diurella rousseleti* Voigt there are nine well-marked teeth; in other species there are but one or two. Leaving out of consideration for the present the case of *Diurella rousseleti*, we may classify the teeth in other species into two categories:

(a) In *Rattulus multicrinis* Kellicott (pl. VI, figs. 55 and 58), *Rattulus capucinus* Wierz. & Zach. (pl. VI, figs. 59–61), and *Rattulus cylindricus* Imhof (pl. VII, fig. 62), there is a single nearly median projection of the dorsal lorica edge, extending over the head. In *Rattulus cylindricus* Imhof this is prolonged into a long hook, curved downward over the anterior opening of the lorica. In these cases the tooth seems to be nearly or quite in the middle line.

(b) In a number of other species there is either one tooth (*Rattulus gracilis* Tessin, figs. 45–48; *Rattulus scipio* Gosse, figs. 50–52; *Diurella tigris* Müller, figs. 1, 3, 4; *Diurella tenuior* Gosse, figs. 7, 8; *Diurella weberi*, figs. 12–14 and 116–117; *Diurella intermedia* Stenroos, figs. 108, 109)—or two teeth (*Rattulus longiseta* Schrank, figs. 67–70; *Diurella insignis* Herrick, figs. 15, 16; *Diurella porcellus* Gosse, figs. 19, 20; *Diurella stylata* Eyferth, figs. 27–30), which seem of a different character. These lie distinctly to the right of the dorsal middle line (so far as that can be defined), and form prolongations of one or both edges of the "striated area" of the lorica, hereafter described. When there are two of these teeth they are usually unequal in size, the right one being longer. (Only in *Diurella stylata* Eyferth are they nearly or quite equal in length.) In most species they are merely short teeth, but in *Rattulus longiseta* Schrank and *Diurella stylata* Eyferth they are long spines. The position of these teeth on the right side is one of the markedly unsymmetrical characters of the *Rattulidæ*. A further account of these teeth may best be deferred until the "striated area" has been described.

Many of the species have no teeth at the anterior edge of the lorica. The anterior opening of the lorica is usually oval, with a slight notch near the ventral middle line. In some few cases the edge of the head-sheath projects farther on the left side of the opening than on the right. This is notably the case in *Diurella weberi* n. sp. (pl. XIII, figs. 116–117); it is slightly so in *Diurella tenuior* Gosse and *Diurella brachyura* Gosse, and perhaps in other species.

In some cases three or four or more teeth have been described by different authors at the anterior edge of the lorica. In many cases this is due to the optical effect of the longitudinal folds in the head-sheath above described or to the slight rounded projections of certain parts of the head-sheath mentioned in the last paragraph. Sometimes the folds of the head-sheath project as sharp teeth. An example of this condition is found in *Diurella rousseleti* Voigt, where there are nine of these teeth. These, however, are of different character from the one or two teeth which I have described above. These latter are structures to a certain extent *sui generis*, and I shall, as a rule, restrict the use of the term *teeth* in this connection to them.

Stenroos (1898) has described a new species. *Mastigocerca (Rattulus) rosea*, which is said to have two long teeth or spines, like those of *Rattulus longiseta* Schrank, at the *ventral* margin of the lorica. In other respects the animal resembles *Rattulus longiseta* Schrank. As this peculiar position of the teeth is unknown in any other of the *Rattulidæ*, and is entirely out of harmony with the structure and behavior of the *Rattulidæ* in other respects (as will appear later), it seems possible that there was an error of observation in this case.

Striated area, Ridge.—One of the most peculiar characteristics of the *Rattulida* is the presence on the lorica of a dorsal longitudinal area, striated transversely, which extends from the anterior edge some distance backward on the body. This area shows the most varied differentiations in different species—in some appearing as a single high ridge, in others as two ridges, in others as a depression, while in still other cases there is no change in the surface of the lorica at this region except the transverse striations. This peculiar area is so characteristic for the *Rattulida*, and plays such a part in determining their forms, that it must be treated in full.

The area is unsymmetrical in position, usually beginning at the anterior margin of the lorica, to the right of the mid-dorsal line, and passing obliquely backward and toward the left side. Its sides are, as a rule, rather sharply defined, frequently appearing as thickenings or ridges. This area shows in *Rattulus elongatus* Gosse a condition which will serve as a useful point of departure for an understanding of the various differentiations which it undergoes in other species. In *R. elongatus* Gosse (pl. XII, fig. 102) the area begins at the anterior edge as a broad, shallow furrow, with well-marked sides. This furrow lies a little to the right of the position of the eye, as seen from above. From the sides of the furrow transverse striations pass toward its middle (and a little forward). The striations are not continuous from one side to the other, but meet in the middle of the furrow in a sort of rhaphe.

The furrow proper extends backward for a distance only somewhat greater than the diameter of the lorica. Near its posterior end, in its middle line, is situated the dorsal antenna. Though the furrow or depression below the general surface ceases at the point above indicated (shown at *x*, figs. 102 and 105), the striated area continues, with well-defined edges, for about one-third the length of the lorica.

In *Rattulus longiseta* Schrank (pl. VIII, fig. 67) the striated area is of very nearly the same character as in *Rattulus elongatus* Gosse, save that it exists as a depression throughout its entire length, reaching to the middle of the lorica. In this species we have another characteristic feature added—the relation of the striated area to the two anterior teeth or spines. *The two teeth are continuations of the thickened edges of the striated furrow.* This appears to be true in all species where the teeth exist. The tooth or spine which forms the continuation of the right edge is much longer than the left one.

What is the function of this striated area and what are the transverse striations which mark it? The striated furrow, as we find it in *Rattulus longiseta* Schrank, bears much resemblance to one of the longitudinal folds in the head-sheath of such species as *Rattulus capucinus* Wierz. & Zach. and *Rattulus multicrinis* Kellicott (pl. VI, figs. 58 and 59); and these folds are cross-striated, just as in the case of the furrow. The striations in the folds of the head-sheath are evidently fine muscular bands, which have the office of bringing the folds together when the head is withdrawn.

In the case of the dorsal striated area, it seems beyond question that the striations are of the same nature—that they are muscular bands. They are clearly not surface markings, but are internal bands. This is seen with especial ease in such forms as *Rattulus carinatus* Lamarck and *Rattulus bicristatus* Gosse, in which the striated area rises in the form of one or two ridges. Moreover, the two edges of the furrow may be closely approximated, when the animal is strongly retracted, as in pl. VIII, fig. 70. When the head is extended the bases of the two teeth (on the opposite sides of the furrow) are a considerable distance apart; but when the animal is contracted to a maximum degree the two are almost in contact.

The striated area therefore represents a longitudinal flexible portion of the lorica, permitting an increase or decrease in the circumference of the body. The striations are muscle fibers, by means of which the approximation of the two sides is brought about. These fibers are attached at the middle and at the two thickened edges of the area.

In *Rattulus mucosus* Stokes (pl. X, fig. 86) the two edges of the striated area are raised as pronounced ridges, leaving a broad and deep furrow between them. The striations (muscle fibers) pass from the summit of the ridges to the bottom of the furrow. Stokes (1896) states that he has seen the two ridges drawn toward each other, and I believe that I have observed the same thing.

In *Rattulus bicristatus* Gosse the two edges of the area reach their highest development, rising as two very high prominent ridges with a broad, deep furrow between them (pl. IX, figs. 77 and 78). The muscles are grouped in pronounced bundles, which pass from near the summit of the ridges to the middle of the broad groove between them. In a squarely side view of the ridges the ends of the muscle bundles are seen as irregular areas.

In another series of species, of which *Rattulus carinatus* Lamarck (pl. XI, figs. 95, 97), *Rattulus lophoessus* Gosse (figs. 98, 99), and *Diurella tigris* Müller (fig. 1), may be taken as types, only the right edge of the striated area is elevated into a ridge, the left not rising above the general surface of the body. Thus a single ridge is produced, having its edge toward the right, and sloping gradually to the left. The left edge of the striated area may usually be recognized as a sharp, well-defined line, but not at all elevated. The muscle fibers run from the summit of the ridge (on the right) to the base of the ridge, at the left boundary of the area. The interruption of the fibers in the middle of the area can usually still be made out (though it is not indicated in all the figures).

Thus we have produced the peculiar condition found in many of the *Rattulidæ* and well shown in fig. 1 and fig. 95 (pl. XI)—a high, sharp ridge passing on the right side of the body obliquely backward. Why the right ridge should thus have developed rather than the left one we shall try to bring out in our general discussion of the asymmetry of the *Rattulidæ*.

In addition to the types already described the striated area is present, in a considerable number of species, neither in the form of a well-defined ridge nor as a well-defined groove, but merely as a flexible area with marked transverse striations. This is the case, for example, in *Rattulus rattus* Müller (pl. XI, figs. 100, 101). In this organism the striated area is in some cases apparently swollen out to form a slight rounded ridge; in other cases it seems to lie at the general level of the lorica surface, while in still other specimens it seems to form a slight depression. It is probable that these are functional differences, due to the state of contraction or extension of the specimen. Almost every intergradation is found, from the furrow of *R. elongatus* Gosse to the high ridge of *R. carinatus* Lamarck. In perhaps the majority of species (especially in *Diurella*) the striated area is merely slightly elevated at its right edge, forming a low ridge, not conspicuous in most views.

The area in which the transverse striations can be seen usually passes from the anterior edge to the middle of the length of the body, or to a point some distance behind the middle. The ridge formed by the elevation of the area sometimes continues back farther than the striations, and may extend to the beginning of the foot (as in *Rattulus lophoessus* Gosse, pl. XI, figs. 98, 99).

Among the species which I have studied with care only *Rattulus latus* Jennings and perhaps *Rattulus multicrinis* Kellicott and *Rattulus capucinus* Wierz. & Zach. show no sign of the striated area.

A word further should be said about the relation of the striated area to the teeth or spines at the anterior edge of the lorica. Those of the second category mentioned on page 280 are formed as outgrowths of the thickened edges of the striated area. Where two teeth are present both the edges project, that formed by the right edge being usually the longer. When only one tooth is present it is formed by a projection of the right edge of the area.

The anterior projections of the first category mentioned on page 280, found only in *Rattulus cylindricus* Imhof, *Rattulus capucinus* Wierz. & Zach., and *Rattulus multicrinis* Kellicott, are formed in a somewhat different way. The initial stage in the production of such a projection is found in *Rattulus elongatus* Gosse (pl. XII, fig. 102); the entire width of the striated area projects at the anterior edge as a rounded lobe. In *Rattulus capucinus* Wierz. & Zach. and *R. multicrinis* Kellicott the projection has developed into a large triangular tooth. In *Rattulus cylindricus* Imhof (pl. VII, fig. 62) the tip of this tooth has further developed into a long hook, curved down over the corona. The three species showing this peculiar differentiation occupy a different position from most of the other species in many other respects also.

(3) *Foot.*—The foot is a short, conical structure attached to the body at the posterior end. The foot shows little variation in structure, except in size and form, being in some cases short and thick, in others slender. In a few cases (*Diurella porcellus* Gosse, *D. sulcata* Jennings, etc.), the foot is very small, so as to be hardly recognizable as a separate structure. In some of these cases it is usually held completely retracted within the body. Sometimes the foot shows one or two faint annulations which have at times been described as joints.

The most peculiar thing about the foot in the *Rattulidæ* is its usually unsymmetrical attachment to the body. The joint between the foot and the body is commonly oblique, extending farther back on the left (or left dorsal) side than on the right. This is well shown in fig. 86 (pl. X), fig. 99 (pl. XI), and fig. 103 (pl. XII). In some cases the posterior edge of the lorica projects backward some distance over the foot on the left side, but not on the right. The foot is thus attached to the lorica in such a way that it can bend to the right, but not to the left.

(4) *Toes.*—The toes form perhaps the most peculiar characteristic of the *Rattulidæ*. Most of the Rotifera have two short posterior appendages attached to the foot, placed side by side, and, like most paired organs, similar in form and size. But in the *Rattulidæ* we find the two toes in the majority of cases unequal, sometimes excessively so, and no longer side by side. In some species one of the toes has almost disappeared, while the other has become immensely developed, forming a straight rod as long as the body (in *Rattulus cylindricus* Imhof, for example, pl. VII, fig. 62).

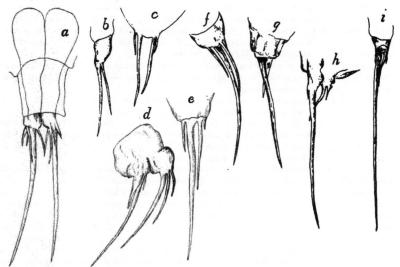

Fig. 1. - Dorsal views of the toes in a number of species of *Rattulidæ*, showing gradual reduction of the right toe.
(a) *Diurella tigris* Müller; (b) *D. stylata* Eyferth; (c) *D. brachyura* Gosse; (d) *D. porcellus* Gosse; (e) *D. insignia* Herrick; (f) *D. tenuior* Gosse; (g) *Rattulus gracilis* Tessin; (h) *R. lophoessus* Gosse; (i) *R. elongatus* Gosse.

The steps in the series of changes by which this is brought about may be clearly followed by comparing the toes of different species. In a few species (*Diurella tigris* Müller, *D. sulcata* Jennings, *D. intermedia* Stenroos, etc.) the two toes are still equal, as in other rotifers. One of these will serve best as a starting-point. We will select *Diurella tigris* Müller, whose toes are shown in text-figure 1, at *a*.

The toes form two long, curved, pointed, spine-like rods of equal size. At the base of each are four small flattened spines (so-called substyles), which usually lie closely applied to the base of the toes. The use of these substyles was pointed out by Plate (1886), and will be readily appreciated when one of the habits of the animals is understood. The posterior part of the body contains two large glands (pl. I, figs. 3, 4, *m. g.*), which secrete a quantity of mucus, which is stored up in two large sacs (figs. 3, 4, *m. r.*). These sacs open one at the base of each toe, and discharge the mucus out upon the surface of the toe. Thence it trails behind the animal as a long thread, by means of which the rotifer attaches itself to various external objects and hangs in the water, as a spider by its thread. The mucus passes out of the sac between the substyles and the main toes, and the four substyles serve to direct its course out along the surface of the toe.

But the two toes in *Diurella tigris* Müller are not placed exactly side by side, as in most rotifers, but they partake of the prevailing asymmetry of the animal. The attachment of the toes to the foot is oblique, like that of the foot to the body, so that the right toe lies at a higher level than the left. The arrangement will be

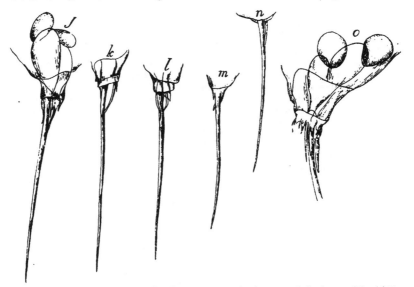

Fig. 1.—Dorsal views of the toes in a number of species of *Rattulus*, showing gradual reduction of the right toe. (*j*) *R. longiseta* Schrank; (*k*) *R. scipio* Gosse; (*l*) *R. carinatus* Lamarck; (*m*) *R. multicrinis* Kellicott; (*n*) *R. pusillus* Lauterborn; (*o*) *R. bicristatus* Gosse (base of toe only).

best understood if one conceives it to have been brought about as follows: The toes, originally concave downward, have been twisted at their attachment to the foot, so that their concavity now faces to the right (pl. I, fig. 1), and the right toe lies above the left, as the animal creeps along the bottom. The toes and foot can therefore now bend only to the right, not toward the ventral side, as in most rotifers.

Now, as a result of the condition above described, the two toes no longer have the same relation to the environment as they have in a bilaterally symmetrical animal. This similar relation to the environment is usually assigned as a reason for the similarity of paired organs, and the lack of such similar relation to the environment may become an equally good ground for the loss in similarity of two organs no longer having this relation. The lower, originally left, toe is now next to the bottom when the animal is creeping, and will more often come in contact with it than will the right toe. Moreover, when a thread of mucus hangs from the toes and catches on some object on the bottom, it will more often be that from the lower (left) toe.

So, perhaps as a consequence of this change of position and of relation to the environment, the right or upper toe begins to degenerate. The steps in degeneration are easily traceable and are shown in text-figure 1. In *Diurella stylata* Eyferth (*b*) and *D. brachyura* Gosse (*c*) the toes are almost equal, but the left is a little longer. In *Diurella porcellus* Gosse (*d*) the difference is greater. In *Diurella insignis* Herrick (*e*) and *D. tenuior* Gosse (*f*) the right toe is about half as long as the left. In *Rattulus gracilis* Tessin (*g*) it is about one-third the length of the left. The right toe now forms a small spine, which has its tip bent toward the main or left toe, and lying against the latter. *Rattulus lophoessus* Gosse (*h*) shows a still farther step; *R. elongatus* Gosse (*i*), *R. longiseta* Schrank (*j*), *R. scipio* Gosse (*k*), *R. carinatus* Lamarck (*l*), *R. mullicrinis* Kellicott (*m*), and *R. pusillus* Lauterborn (*n*) still farther ones in the reduction of the right toe and corresponding increase in the left one. In the species last named (*i* to *n*) the rudimentary right toe has usually been classed with the substyles; it can generally be recognized, however, by its form and position, as well as, at times, by the fact, shown in *j*, that one of the mucus reservoirs opens at its base. Finally, there are certain species, as *Rattulus bicristatus* Gosse (*o*), *R. mucosus* Stokes, and others, in which it is very difficult, or impossible, to distinguish between the rudimentary right toe and the substyles.

It is probable that this degeneration of one of the toes is related primarily to the habit, so common in the *Rattulidæ*, of becoming suspended from foreign objects by a thread of mucus attached to the tip of the toe, and then revolving on the long axis. It is evident that a single, long rod is much better fitted to serve as a pivot than two toes side by side. These would impede the revolution by furnishing resistance to the water.

The substyles are present all through the series. Their number varies; in most cases each of the two toes seems to have two, three, or four. In *Rattulus bicristatus* Gosse (text-figure 1, *o*) at least eight can be seen about the base of the main toe; among these the rudimentary right toe can hardly be distinguished from the others.

Hand in hand with the reduction of the right toe goes a reduction of the mucus reservoir which is connected with it. The reduction of the mucus reservoir is not so extensive as that of the toe, and it never completely disappears. Indeed, in some cases where the toes are very unequal, the two reservoirs remain of the same size. This is true in *Rattulus stylatus* Gosse (fig. 92, pl. X.) Unequal reservoirs are shown in text-figure 1, *j* and *o*.

Apparently, in some species at least, the toes are of the full length when the animal is hatched from the egg, while the body is much smaller than it later becomes. Thus in young specimens the toe is much longer in proportion to the body than in adults. This is well shown by comparing figs. 18 (pl. II), 51 (pl. V), 90 (pl. X), representing young specimens respectively of *Diurella insignis* Herrick, *Rattulus scipio* Gosse, and *R. mucosus* Stokes, with the other figures representing adult specimens of these species. This is a point worthy of special note, as it may easily lead to error in specific determinations.

(5) *Corona.*—The truncate anterior end is unprotected by the lorica and bears in the *Rattulida*, as in other rotifers, cilia by means of which the organism moves and by which it obtains its food. Partly surrounded by the cilia are usually also a number of antenna-like organs. This whole complex of structures at the anterior end is known as the corona. In the *Rattulida* the structure of the corona does not vary a great deal in the different species. The main features of the corona will be seen by an examination of that of *Diurella stylata* Eyferth (pl. III, fig. 31). Partly surrounding the corona, especially in the dorso-lateral region, are two curves of cilia (*a*), forming together nearly a semicircle. These two curves are not continuous with one another, but there is a gap between them in the middle dorsal region. These cilia are the organs of locomotion of the animal.

At the sides of the mouth (*m*) are two other curves of shorter cilia (*b*). These are connected with food-taking, and may be said to belong strictly to the mastax or pharynx. When the mastax is pushed far beyond the surface of the head, as sometimes happens, these cilia are seen to be borne upon its end. This is well shown in fig. 56 (pl. VI), in *Rattulus multicrinis* Kellicott.

In the dorsal part of the corona, in the median line, is a thick dorsal projection (pl. III, fig. 31, *c*). In many *Rattulida* this is more slender than in *Diurella stylata* Eyferth. At the sides of this process, but lying a little ventral to it, are two smaller prominences (*d*) bearing cilia. Just above the dorsal projection, shown in fig. 31, is another thick dorsal process, shown in side view in fig. 27, *e*.

The four curves of cilia described above (*a* and *b*) are present in all the *Rattulida*. There is also almost invariably a single, thick dorsal process (*c*). In the other antenna-like structures there is more variation. In *Rattulus multicrinis* Kellicott (pl. VI, fig. 57) the upper median process (*e*) is very long, while the lower one (*c*) is short. There are two long lateral processes (*d*) on each side. In *Rattulus latus* Jennings (pl. VII, fig. 65) the corona is similar to that of *R. multicrinis* Kellicott. The lower dorsal process (*e*) bears on its end two small processes. The parts of the corona which can be seen easily in most species of the *Rattulida* are the cilia and the large dorsal process. The latter lies, as a rule, a little to the left of the end of the striated area of the lorica.

The functions of the curious club-shaped or antenna-like organs of the corona are not known beyond the general probability that they are sense organs.

In addition to the (probable) sense organs on the corona, there are three other structures which doubtless have sensory functions. These are the so-called dorsal and lateral antennæ.

Dorsal antenna.—The dorsal antenna is found, as in most rotifers, on the dorsal surface, some distance back of the anterior end of the head. It usually lies a very little behind the constriction which separates the head-sheath from the remainder of the lorica. It consists, in well-developed cases, of a small club-like structure, projecting through an opening in the lorica and bearing one or more fine setæ. It is best developed in *Rattulus cylindricus* Imhof, where the seta which it bears is very long and conspicuous (pl. VII, figs. 62, 63, 64).

From the antenna there may often be traced a fine cord running to the brain. This has, just within the lorica, a spindle-shaped thickening.

In many species no setæ can be observed on the dorsal antenna, and often the only trace of it is the opening in the lorica, through which it should pass. This opening is probably to be found in all the *Rattulidæ*.

The position of the dorsal antenna perhaps indicates the position of the dorsal median line. This is true at least in other rotifers, and in the *Rattulidæ* the surface which bears it is above when the animal is creeping on the bottom. Its place with reference to the striated area is therefore of interest. In *Rattulus elongatus* Gosse, *R. cylindricus* Imhof, and *R. bicristatus* Gosse, and, indeed, as a rule in the species in which the striated area forms a furrow with its two sides equally developed, the dorsal antenna lies in the middle of the furrow. But in *Rattulus mucosus* Stokes this is not true. The dorsal antenna in this species lies to the left of the striated area, in a notch in the outer side of the left ridge.

A similar position is found in almost all species in which the striated area is developed as a single ridge. The antenna lies to the left of the ridge, usually at about the left edge of the striated area. (See pl. XI, figs. 95 and 100.)

Lateral antennæ.—In most free-swimming rotifers (as in the *Notommatidæ*, from which the *Rattulida* are without doubt derived) the two lateral antennæ are situated one on each side, in the posterior third of the body, symmetrically with relation to one another. Many species of the *Rattulida* have preserved nearly this primitive position, though usually with slight variations. There may be a tendency for the right antenna to be a little farther forward (as in *Rattulus scipio* Gosse, pl. V, fig. 52, and *Rattulus carinatus* Lamarck, pl. XI, fig. 95), or to be a little nearer the dorsal side, as in *Rattulus elongatus* Gosse (pl. XII, figs. 102, 103, 105), or the opposite tendencies may be shown. But in some cases there is a very remarkable asymmetry in the positions of the two antennæ. In *Rattulus cylindricus* Imhof, for example, the left antenna is at about the middle of the length of the body (pl. VII, fig. 63), while the right antenna is very far back, at the place where the lorica is joined by the foot. In *Diurella stylata* Eyferth the left antenna is still farther forward (figs. 28, 29, pl. III), while the right one is on the posterior part of the body.

There are probably no species of *Rattulida* in which the lateral antennæ can not be found by careful search.

INTERNAL ORGANS.

The internal organs partake, to a considerable degree, of the asymmetry so characteristic of the external anatomy of the *Rattulida*. Otherwise the internal structure in this group does not present a great deal that is different from what is found in most of the related Rotifera, so that I shall treat of it only briefly. The *Rattulidæ* are not a favorable group for a study of the characteristic internal structure of the Rotifera.

(1) *Alimentary canal.*—The alimentary canal shows the following parts: The mouth opens into a muscular pharynx known as the mastax, containing chitinous jaws or *trophi*. From the mastax a short, narrow tube, the œsophagus, passes backward to widen into the large, thick-walled stomach. The stomach narrows to form the intestine, which passes straight back to the anus. The entire course of the alimentary canal is well shown in fig. 63, pl. VII (*Rattulus cylindricus* Imhof), and fig. 77, pl. IX (*Rattulus bicristatus* Gosse).

Mouth.—The mouth opens on the truncate anterior end, or corona, near its ventral side (pl. III, fig. 31, *m*). At its sides are two curves of cilia (*b*) which serve the purpose of carrying small particles of food to the mouth.

Mastax.—The mastax is a muscular, pharynx-like structure, forming the first part of the alimentary canal (*mx.*, pl. VII, fig. 63). Its anterior end forms a nearly circular area on the corona, within which lies the mouth (pl. III, fig. 31, *m*). The two curves of cilia above spoken of, at the sides of the mouth, are really borne on the end of the mastax, as appears when the latter is pushed far out (pl. VI, fig. 56).

The mastax is large, filling up a considerable portion of the anterior part of the lorica. It is composed chiefly of a mass of muscles, which act upon the chitinous jaws. Only the anterior part of the mastax is hollow and receives the food, the posterior three-fourths or more being a solid mass of muscle. The transverse muscles are often very evident as striations (pl. XIII, figs. 108, 111, 115, 119).

In consequence of the asymmetry of the trophi the mastax frequently shows an unsymmetrical form, as, for example, in fig. 61, pl. VI. The œsophagus opens into the mastax on its dorsal side near its anterior part, as shown in fig. 63, pl. VII, and fig. 105, pl. XII. In many species the œsophageal opening is clearly somewhat on the right side of the mastax.

The mastax frequently has connected with it one or more prominent glands. One on the left side is especially marked in *Rattulus multicrinis* Kellicott (pl. VI, fig. 57) and *Rattulus latus* Jennings (pl. VII, fig. 65). These glands are apparently not present in all species.

Trophi.—The chitinous jaws or trophi vary a great deal among the different species, and usually show a considerable degree of asymmetry. The trophi of *Rattulus carinatus* Lamarck were well described by Gosse (1856) in his classical paper on the "Structure, Functions, and Homologies of the Manducatory Organs in the class Rotifera." The trophi of *Rattulus longiseta* Schrank (pl. VIII, figs. 71, 72) furnish a good example of the typical structure. Following Gosse, we may distinguish three main portions—the two lateral parts, known as *mallei*, and a central structure, the *incus*. Each of these is composed of several portions.

The malleus consists of two chief parts, a long distal rod, the *manubrium* (*mn.*), and a shorter proximal portion, the *uncus* (*u.*). The two mallei are unequal in size, the left one being the larger. The left uncus bears teeth, while the right one is merely a straight rod without teeth.

The incus or central portion consists of three main parts. There is a long curved median rod, the fulcrum (*fu.*), which, as the side view (fig. 71) shows, lies at a level nearer the ventral surface than do the manubria. In side view the fulcrum is seen to consist of two rods, the ventral one being very thin and united to the other by membrane. The fulcrum bears at its proximal end two large structures known as *rami* (*ra.*). These articulate with the fulcrum and inclose a space between them. At their proximal ends they, like the unci, bear a number of teeth. The rami have their lower or distal ends produced into a long process for the attachment of muscles. These, with Gosse, we may designate as the alulæ (*al.*). The left alula is considerably longer than the right.

In addition to these chief portions there are a number of chitinous rods forming a framework which lies on the dorsal side of the proximal end of the trophi (pl. VIII, fig. 71, *su.*). These arise from the manubria and are connected with the rami. The function of this framework is not very clear. In some cases it seems to support a sort of chitinous fringe about the mouth (pl. XIII, fig. 118).

The typical parts of the apparatus are the manubria, unci, fulcrum, and rami, and in our account of the variations of the trophi among different species we shall take only these into consideration.

In the small group of related species comprising *Rattulus multicrinis* Kellicott, *R. capucinus* Wierz. & Zach., *R. cylindricus* Imhof, and *R. latus* Jennings, the trophi are nearly or quite symmetrical. The manubria are approximately of the same length and the alulæ seem not strongly developed.

Most of the remaining species of the genus *Rattulus* have the trophi moderately unsymmetrical, the left manubrium being considerably larger than the right. This is the case, for example, in *R. elongatus* Gosse (pl. XII, fig. 107), *R. longiseta* Schrank (pl. VIII, fig. 72), *R. bicuspes* Pell (pl. VIII, fig. 76), and *R. bicristatus* Gosse (pl. IX, fig. 80). In *R. mucosus* Stokes (pl. X, fig. 91) there is a much greater asymmetry, and the trophi have a very peculiar character. The left manubrium and uncus, the fulcrum and rami, are heavy and massive, while the right manubrium and uncus are reduced to mere slender rods. There appear to be no teeth, the trophi seeming to be designed rather for crushing than biting.

In *Diurella* the asymmetry of the trophi is on the whole much more pronounced than in *Rattulus*, most of the species of *Diurella* having jaws fully as unsymmetrical as those last described, or even more so. In *Diurella porcellus* Gosse (pl. II, fig. 22) the right manubrium is as long as the left, but is excessively slender—a mere bristle. In *D. sulcata* Jennings (pl. II, fig. 26) and *D. tennior* Gosse (pl. I, fig. 10) the reduction of the right manubrium has gone still farther; it has become much shorter than the left one. In *D. tigris* Müller (pl. I, fig. 2), finally, the culmination of asymmetry is reached; the right malleus is a minute rudiment, while the left one is massive.

Gosse (1856) described *Diurella porcellus* as having the right manubrium quite lacking. This is not the case with the specimens of that species which I have examined, though it is much reduced. I have found none of the *Rattulidæ* in which the right manubrium could not be discovered.

It is striking that the trophi are most unsymmetrical as a rule in the species of the genus *Diurella*, though the toes in this genus are less unsymmetrical than in *Rattulus*. This is probably due to the fact that in *Diurella* the body is as a rule more slender and more curved than in *Rattulus*. As the curve is of such a nature that the right side is concave, there is much less space on this side than on the convex left side, so the internal structures on the right side are reduced. This is especially noticeable in the trophi. In *Rattulus*, where the body is usually more swollen and less curved, there is not so much occasion for the reduction of the right side.

Œsophagus.—The œsophagus is merely a short, slender passageway with thin walls, which begins on the dorsal side of the mastax, on its anterior one-fourth. It is well shown in fig. 63 (pl. VII), fig. 77 (pl. IX), and fig. 105 (pl. XII).

Stomach.—The stomach is an enlarged sac, with thick, apparently glandular walls, forming a direct continuation of the œsophagus. In the broad-bodied species, such as *Rattulus latus* Jennings (pl. VII, fig. 65), it lies on the right side.

At the anterior end of the stomach are the two gastric glands, one on each side. These are small solid structures, often lobulated and showing a number of prominent nuclei. They are well shown in fig. 77 (pl. IX), fig. 87 (pl. X), and fig. 102 (pl. XII).

Intestine.—The stomach narrows at its posterior end to form the intestine (*in.*, pl. VII, fig. 63; pl. XII, fig. 102). The walls of the intestine are usually thinner and less colored than those of the stomach, but there is no precise line of demarcation between the two. The intestine narrows rapidly to end at the *anus*, which lies beneath the edge of the lorica, just above the beginning of the foot, a little to the left of the middle line.

Food of the Rattulidæ.—The food of the *Rattulidæ* seems to consist chiefly of small particles suspended in the water, which are brought to it by its cilia, or of the floccose material covering the surface of water plants. The animals may often be seen creeping over the stems or leaves of water plants with the corona against the surface, as if they were feeding, but it is very rarely that one sees any definite namable thing devoured. In one case, and one only, I have seen a *Rattulus* display predatory tendencies. A *Rattulus gracilis* Tessin seized a young *Diurella tenuior* Gosse, which happened to be near it, pierced the lorica with its jaws, tore out a piece from the side of its prey, and devoured it. The jaws of many other species seem better fitted for carnivorous habits than do the comparatively weak ones of *Rattulus gracilis* Tessin, but I have seen no other instances of the character just described.

(2) *Brain.*—The brain in the *Rattulidæ* is usually a large, oblong body, rather prominent, which lies on the dorsal side of the mastax, in the anterior part of the body. In front the brain has no defined boundary, merging into the mass of substance which supports the corona. Its main mass frequently lies to the left of the striated area or ridge on the lorica. The brain is usually somewhat shorter than the mastax, but in a few cases—notably in *Diurella stylata* Eyferth—it forms a very large sac, extending backward more than half the length of the lorica (*br.*, pl. III, fig. 27). In such cases the brain is seen to be made up of large cells, whose outlines are clearly distinguishable.

In the *Rattulidæ* the brain has no opaque, chalky mass at its posterior end, such as is found in many of the *Notommatadæ*. Gosse (1889) described as *Rattulus cimolius* an animal in which the brain has such a chalk mass, but from Gosse's description and figure (see p. 342 and fig. 138, pl. XV) it seems clear that this animal was not one of the *Rattulidæ*; it should rather be classed with the *Notommatadæ*.

Connected with the brain is the single *eye*. This is a hemisphere of red pigment, usually attached to the posterior end or under side of the brain. In a number of species the brain is divided at its posterior end into two unequal lateral lobes, the left one being smaller and bearing the eye at its tip. This condition is shown in fig. 99, pl. XI (*Rattulus lophoessus* Gosse), and in figs. 102 and 103, pl. XII (*R. elongatus* Gosse); it is present in a number of other species also. The dorsal antenna is connected with the brain by a slender cord, which is very evident in *Rattulus cylindricus* Imhof (pl. VII, fig. 63). It is probable that the lateral antennæ are thus connected with the brain also. From each of these there passes forward a slender cord, but I have not succeeded in tracing this to the brain.

(3) *Excretory organs.*—The excretory organs do not differ essentially from those found in other rotifers and are not strikingly developed in the *Rattulidæ*, so that this group is not a favorable one for their study. For this reason I have not paid especial attention to the excretory system. It consists essentially of the well-known lateral canals, one on each side, which open at their posterior ends into a small bladder-like structure, the contractile vacuole. These parts are shown in fig. 24 (pl. II) and fig 32 (pl. III).

The lateral canals are two slender tubes, in some species considerably convoluted, in others much less so, which begin in the anterior part of the body and run backward, one along each side, to the contractile vacuole. The lateral canal bears on each side four or five small evaginations, each containing a long cilium; these are the so-called flame cells or vibratile tags (pl. II, fig. 24).

The contractile vacuole (fig. 24, *c. r.*) is unusually small and inconspicuous in the *Rattulida*. It is a spherical vesicle lying just beneath or at the side of the intestine near its posterior end. It is situated above the large mucus reservoirs (*m. r.*), which are sometimes mistaken for the contractile vacuole.

The contractile vacuole usually pulsates rather rapidly, perhaps in consequence of its small size. Twenty times per minute seems a not uncommon rate; in *Diurella brachyura* Gosse, according to Stokes (1896), there are 40 pulsations per minute. The contractile vacuole in most rotifers opens into the intestine near its posterior end; this matter has not been especially investigated in the *Rattulida*.

(4) *Reproductive organs.*—The male seems to be quite unknown in the *Rattulida*. No member of this family is given in Mr. Charles Rousselet's list of male rotifers hitherto described (1897), and I have myself seen nothing of a male in any of the species studied.

The ovary (*ov.*) is an irregular, frequently somewhat lobular, organ, differing in no important manner from the same organ in most of the related free-swimming rotifers. As Plate (1886) has shown, the ovary in most of the Rotatoria consists of two parts, a vitelline portion ("Dotterstock") and a germinal portion ("Keimstock"). The latter is smaller than the former; from it the eggs are directly produced. The germinal portion seems to lie at the right side or right anterior corner of the vitelline portion, in the *Rattulida*. The vitelline portion contains a small number (usually if not always eight, in this family) of large, conspicuous nuclei.

The ovary lies on the ventral side of the alimentary canal, usually mostly to the left of the median line. In *Rattulus latus* Jennings (pl. VII, fig. 65) it lies entirely to the left of the alimentary canal, not on the ventral side of the latter at all.

The eggs are formed in the germinal portion, to the right of the main body of the ovary. When the egg has reached a considerable size, it usually occupies a large space on the right side of the body, as in fig. 32 (pl. III).

In the *Rattulida*, so far as known, the egg, after extrusion, is carried attached to the lorica in only one species, *Rattulus cylindricus* Imhof. In this case the animal is frequently found carrying the egg attached to the posterior end of the lorica, above the foot (pl. VII, fig. 62).

(5) *Mucus glands and reservoirs.*—The glands and reservoirs for supplying the tenacious mucus-like substance, by which the animals attach themselves to various objects, are unusually well developed in the *Rattulida*. The reservoirs especially form a large, clear, oval sac, or a pair of sacs, filling a considerable part of the hinder portion of the body. There are typically two of the glands and two of the reservoirs in the *Rattulida*. They are well shown in fig. 92 (pl. X); fig. 79 (pl. IX), and fig. 69 (pl. VIII). The two glands are rounded or irregular granular bodies, lying near the ventral surface, just behind the ovary.

The two reservoirs are usually pressed close together or even united, so that it is perhaps just as correct to speak of a single reservoir divided into two chambers by a longitudinal partition, as of two reservoirs. Into these chambers passes the secretion from the glands; it may often be found in preserved specimens as a solid mass. In living specimens the reservoirs are entirely clear, and have often been taken for the contractile vacuole, occupying as they do the position usually taken by the vacuole in other rotifers.

One of the two reservoirs opens at the base of the right toe, the other at the base of the left toe (see fig. 6, pl. I, and fig. 69, pl. VIII). The tenacious secretion passes out between the base of the toe and the substyles, being directed by the latter down along the surface of the toe. From the tip of the toe it trails off into the water, like a spider's web, and attaches itself to any object with which it comes in contact. The animal then remains suspended in the water, like a spider from its thread (though of course the rotifer, owing to the movement of the cilia, may hang upward or horizontally, as well as downward). The animal spins about on its long axis, remaining nearly in the same position, or it may of course move in the circumference of a circle about the object to which it is attached.

While thus attached, the action of the cilia brings food to the mouth, just as in the Rotifera that are permanently fixed by their posterior ends. The free-swimming Rotifera which have this secretion of mucus have thus the advantage of being able to temporarily change their roving method of life into a fixed one.

But the thread produced by the mucus is not so strong, apparently, but that by an extra effort it may be broken at any moment. Often a specimen will be seen to swing about from its point of attachment for a considerable time, then suddenly to start rapidly forward, swimming with the most complete freedom. Often too the mucus seems to act merely as a yielding thread—moderating the course of the animal a little—but being drawn out as the animal progresses.

Sometimes the mucus becomes a trap which results in the death of the animal. A specimen will sometimes bring the *base* of its toes against some solid object, as the glass slide on which it is undergoing examination, at the moment when a large quantity of the mucus has been given out. It thereupon sticks fast, perhaps by the entire length of the toe, to the glass and can not escape. It then remains attached at this point till it dies. It is probable that such an accident rarely occurs except when the animal is under such unusual and cramped conditions as it finds between the slide and the cover-glass.

In *Diurella tigris* Müller, in which the two toes are equal, the two reservoirs are also equal (pl. I, fig. 6). But in most species in which the right toe has become rudimentary, the right reservoir has likewise much decreased in size. This is the case, for example, in *Rattulus bicristatus* Gosse (pl. IX, fig. 79) and *R. longiseta* Schrank (pl. VIII, fig. 69). In *R. stylatus* Gosse, however, the two reservoirs are still nearly or quite equal in size (pl. X, fig. 92), although the right toe has nearly disappeared.

THE ASYMMETRY OF THE RATTULIDÆ AND ITS BIOLOGICAL SIGNIFICANCE.

The writer has already given in a separate paper[a] a general discussion of the significance of asymmetry in a number of lower organisms, so that only the salient points, with their application to the *Rattulidæ*, will be set forth here.

All the *Rattulidæ* are more or less unsymmetrical in their structure. If we seek for a general statement which shall express the nature of this asymmetry we shall find it most fully set forth as follows: Conceiving the middle to be a fixed point, the

---

[a] Asymmetry in Some Lower Organisms and its Biological Significance. Mark Anniversary Volume, N. Y., 1903.

anterior part of the body seems to be twisted over to the right, the posterior part over to the left. This will perhaps best be appreciated by examining fig. 1 (*Diurella tigris* Müller). The anterior part of the ridge is far to the right of the middle line. The single tooth is also to the right of the middle; in those species where there are two teeth (as *Rattulus longiseta* Schrank, *Diurella porcellus* Gosse, etc.) the left tooth is nearly in the middle line, the right tooth much to the right of that line.

At the posterior end, on the other hand, the indications are that what was primitively dorsal has passed to the left, while the right-hand one of the paired structures has taken a dorsal position. The dorsal projection of the lorica over the foot has become shifted to the left (shown particularly well, for example, in the figure of *Rattulus lophoessus* Gosse, pl. XI, fig. 99). The right toe has come to lie nearly on the dorsal side of the left one, so that the concavity of the toes (originally ventral) has become directed to the right (fig. 1). Thus the foot and toes can bend only to the right, not to the left.

The body has become not merely twisted on its primitive straight axis, however, but is often bent at the same time so as to form a segment of a spiral (seen especially well in fig. 1, of *Diurella tigris* Müller). As a result of this the left side has become convex, the right side concave. (Compare the following dorsal views in which this is evident: Figs. 1, 16, 29, 46, 52, 75, 78, 95, 99, 103.) These features are, of course, much more marked in some species than in others.

These general changes have induced certain secondary ones. The originally right toe, which has become dorsal, gradually degenerates until it has become in many species a mere rudiment. The right mucus reservoir is likewise involved in this change, becoming smaller than the left. Owing, perhaps, to the enlargement of the left side as a result of its convexity, and the diminution of the right side owing to the concavity falling here, there is a tendency for the internal organs to be better developed on the left side than on the right. This is most strikingly brought out in the structure of the trophi. The right half of the trophi, as shown in the account of these organs, is almost invariably smaller than the left, and in many cases is quite rudimentary.

Altogether, we may say that the body in the *Rattulidæ* tends to take the form of a segment of a spiral, and that this change from the primitive bilateral symmetry has induced also a considerable number of subsidiary changes.

What is the significance of this peculiar condition in the *Rattulidæ?*

The key to the asymmetry of this group is to be found in a study of the movements and behavior of the animals. The unsymmetrical structure is, of course, not a primitive condition, but these animals were originally bilaterally symmetrical. The fundamental plan of structure is still that of bilateral symmetry; certain parts have been reduced or changed in position so that asymmetry has resulted, but the bilateral ground plan is easily traceable. The nearest relatives of the *Rattulidæ* are still bilaterally symmetrical. Probably no one familiar with the Rotatoria will be inclined to question the view that the *Rattulidæ* are derived from the *Notommatadæ*. The *Notommatadæ* are typically creeping forms. They live among the weeds, on the surfaces of which they creep about by means of their cilia, keeping the mouth, as a rule, against the surface.

The differentiations shown by bilaterally symmetrical organisms are usually brought into relation theoretically with their methods of movement, and doubtless very justly. Anterior and posterior ends differ because they come into different relations with the environment, owing to the forward movement. In the same way dorsal and ventral surfaces differ because they come into different relations with the environment—the ventral side being more commonly in contact with a surface, the dorsal side not thus in contact, but subjected to the light and other influences coming from above. On the other hand, the right and left sides are in a similar relation to the environment, there being no influence which acts on one differently from the way it affects the other; hence they remain alike.

Analogous considerations apply to the radially symmetrical form. But there is another type of structure, having an equally definite relation to the method of life and movement—a type which has not been hitherto recognized, at least not as having a definite relation to a widespread method of locomotion and life. This is what may, in general, be characterized as a spiral type of structure, or at least as a one-sided type. This type of structure is found in many organisms which swim freely through the water in a spiral course. Its typical representatives are the Infusoria—the Ciliata and Flagellata.

The spiral course may be characterized as the simplest device to enable an organism to make progress in a given direction through the free water without fulfilling the difficult condition of making all sides identically alike, or of making the differences exactly balance each other.[a] In the spiral course the organism continually keeps one side toward the outside of the spiral. In other words, it is in reality always turning toward one side. The tendency to deviate thus caused is compensated by a revolution on the long axis, which continually brings the side in question into a new position. The path thus becomes a spiral, while if revolution on the long axis did not occur it would be a circle.

Now, the organisms which habitually make use of this method of progression have a form which is adapted to it. In the ciliate and flagellate Infusoria, which move in this manner, the form is usually unsymmetrical, often clearly spiral; and here the spiral form seems to be primitive; at least it was not developed from an originally bilateral form. But in the *Rotulida* we have a group of animals, fundamentally bilateral, which are taking on this spiral, unsymmetrical form as an adaptation to their method of movement.

Movements of the Rotulida.—If we examine in detail the movements of one of the *Rotulida*, taking, for example, *Diurella tigris* Müller (fig. 1), we find that it swims through the water in a spiral, of such a course that its twisted body forms a segment of the spiral path (text figure 2). The animal revolves to the right and swerves toward its dorso-dextral side, while it at the same time progresses. The result is a path almost exactly that which would be produced if the animal were moving on the inside of a hollow cylinder and the dorso-lateral spiral ridge ran in a groove on the inner surface of the cylinder, which fitted it precisely and had the same curvature. The effect is the same as that produced by the spiral grooves on the inner surface of a rifle barrel, giving the ball a rotary motion about the axis of flight. The result is here, as in the rifle ball, to make the axis of progression a straight line.

---

[a] For the grounds on which this statement is based, as well as a general discussion of spiral movement and unsymmetrical structure, see the paper on *Asymmetry*, etc., already cited (p. 263); also a paper by the present author on *The Significance of the Spiral Swimming of Organisms*, in the American Naturalist, vol. 35, 1901, pp. 369–378.

It is evident, therefore, that the general form of the body is adapted to the path which the body follows through the water. And this general form is produced by a twisting of the body from its original bilateral symmetry into the condition already minutely described. The reason why only the right half of the striated area is, as a rule, elevated into a ridge, which slopes to the right, and why the ridge has an oblique course, are entirely evident in the light of the method of movement. The course is always a right spiral, and the single oblique ridge, sloping to the right, greatly favors the spiral movement, while if the left ridge were developed, it would act in opposition to the spiral course. The reason why the right side is concave, the left convex, with the consequent asymmetry of some of the internal organs (notably the trophi) is equally evident. All these things are adaptations to the spiral movement, and, specifically, to movement in a right spiral.

But there are some points which still need elucidation. Why has the foot become twisted into such a position that the toes can be bent only to the right? Why does the right toe degenerate? And why are the teeth at the anterior dorsal margin of the lorica confined to the right side?

These points will be better understood if we examine the behavior a little further. As we have seen, the animals continually swerve, while swimming, toward the dorso-dextral part of the body—that which bears the ridge. This result is due to two components, (1) a tendency to swerve toward the dorsal side, as when lifting the body from the bottom (a tendency which is present in almost all free-swimming rotifers), and (2) the revolution toward the right. The resultant of these two components is a turning toward the dorso-dextral region.

Now, as in the Infusoria,[a] the usual reaction to a stimulus in the *Rattulidæ* is closely related to the method of locomotion. When a *Diurella* or *Rattulus* while swimming freely through the water meets an obstacle it alters its course simply by turning still farther than usual toward the side to which it is already swerving—that is, toward the dorso-dextral side. If the obstacle is small it is thus at once avoided. If the obstacle on the other hand is large, such as a flat surface, which prevents further movement in the

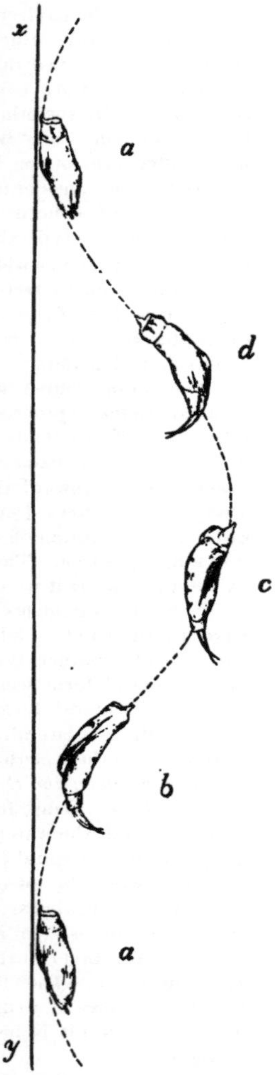

─────────────────────
[a] See Jennings, *On the Movements and Motor Reflexes of the Flagellata and Ciliata.* Am. Jour. Physiol., vol. 3, pp. 229-260.

Fig. 2.—Spiral path followed by *Diurella tigris* Müller, showing that the animal continually swerves toward the dorso-dextral side.

axis of the spiral, the animal continues to swerve toward the dorso-dextral side till its general direction is completely changed. Text figure 3 represents such a reaction in *Diurella tigris* Müller.

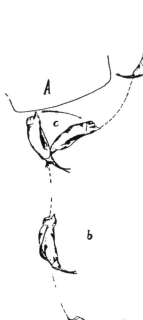

The animal may be stimulated in other ways; the usual result is to induce it to swerve farther toward the dorso-dextral side. If there is really no obstacle the path becomes merely a wider spiral than usual for some distance.

Now, it is evident that if the animal, when thus turning, strikes against any object, it will be the dorso-dextral angle of the head which receives the shock. The corona is of course not covered by the lorica, as is the rest of the body, so that it might easily be injured in such cases. But at the point where the corona would strike—at the dorso-dextral angle—is the tooth (or teeth). This takes the blow which would otherwise fall upon the delicate corona.

Sometimes the animal swims forward into a small angle, where it can not directly turn, as between the surface film of water and the bottom of a watch glass. In this case the animal begins, as usual, to turn toward the dorso-dextral side, but as a result it may merely "bump" its head against the bottom. It nevertheless perseveres trying to turn in the same direction, while at the same time it revolves on its long axis. Thus the head will be dragged and "bumped" along the surface until in time the dorso-dextral angle (through the revolution) becomes directed toward the free water. No one who has seen this peculiar performance (which is not at all uncommon) can remain in doubt as to what is the significance of the tooth or teeth at the dorso-dextral side of the anterior end of the lorica. These teeth take all the "bumping" while the animal is turning, in place of its falling upon the delicate corona. The teeth are placed just where they will serve to protect the delicate head when the anterior end comes in contact with anything. Owing to the invariable swerving toward the dorso-dextral side, the head, if it ever strikes against obstacles at all, will strike on this dorso-dextral angle, where the teeth are ready to protect it.

Fig 3.- Diagram of a reaction to a stimulus in *Diurella tigris* Müller. A represents an obstacle. The animal turns toward the dorso-dextral side which bears the tooth and ridge

The striking against objects is by no means rare even in the ordinary swimming

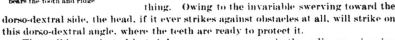

of the animal. It often swims along with its spiral path tangent to a surface, almost every turn bringing the animal against the surface. Such a tangent surface may be represented by the line *x–y* in text figure 2. But, as the figure shows, of course it is always the dorso-dextral angle which comes in contact with the surface, and the tooth or teeth protect the soft head from injury. No teeth are present on the left side, because they would serve no purpose in that position.

Finally, the twisting of the foot and toes, so that they can turn only to the right, finds its explanation along the same line. The entire animal is constructed on the plan of turning to the right, and the arrangement of the toes is merely another adaptation to this. If the toes were so arranged as to bend downward, a sudden stroke with them would turn the organism toward the ventral side, quite in opposition to the other tendencies of the animal. But with the toes turning to the right, their action is brought into harmony with the rest of the behavior of the animal. On getting to a place where it can go forward no further, or as a result of other strong stimulation, the animal turns its toe or toes suddenly and strongly to the right and forward. By this the usual swerving of the animal to the right is strongly accentuated; the path of the animal is thus suddenly changed.

I have attempted to give an explanation of the decrease in size of the right toe in the general account of the toes (p. 284), which may be referred to here to complete the account of the factors which result in the production of asymmetry in this group.

A few other points should be mentioned in regard to the movements of these animals. There are a few of the smaller species of *Rattulidæ*, with short thick bodies, such as *Diurella porcellus* Gosse and *D. brachyura* Gosse, which do not invariably swim in a spiral, though they do usually. In some cases one of these animals will be seen to swim for a short distance in the following manner: With ventral side down (or up), the body swings on the long axis from side to side, giving it a peculiar rocking motion, but without revolving completely. After swimming for a short distance in this manner the animal may suddenly begin to revolve and continue its course in a spiral path like the other *Rattulidæ*.

The *Rattulidæ* not infrequently creep along the substratum with the coronal face against the surface. Under these circumstances the animal of course does not revolve. But the unsymmetrical structure produces its effect even in this case. The animal very rarely moves in a straight line, but usually follows the curve indicated by the form of the body, thus circling continually to its right. That this might perhaps be expected will be seen by examining the figures of *Rattulus lophoessus* Gosse (pl. XI, fig. 99) and *Diurella tigris* Müller (fig. 1), as seen from above.

The habit which these animals have of affixing themselves to foreign objects by a string of mucus has already been described (p. 293).

The above account of the movements of the *Rattulidæ* has been drawn from a study of a considerable number of species. Indeed, all through the work on the group special attention was paid to this matter. I have studied especially in this connection *Diurella tigris* Müller, *Diurella porcellus* Gosse, *Diurella brachyura* Gosse, *Diurella stylata* Eyferth, *Rattulus rattus* Müller, *Rattulus carinatus* Lamarck, *Rattulus bicristatus* Gosse, *Rattulus mucosus* Stokes, *Rattulus bicuspes* Pell, and *Rattulus elongatus* Gosse. In all these the behavior is essentially as set forth above.

CLASSIFICATION.

The classification of the *Rattulidæ* has been in a very confused condition. There is little agreement as to the division of the family into genera or as to the names which are to be used for the genera. The specific names are in an equally unsatisfactory condition.

In the present paper the writer attempts to use the names, both generic and specific, which are in accordance with the rules of nomenclature adopted by the International Congresses of Zoology. In view of the approaching publication, by the German Society of Zoologists, of a systematic review of the entire animal kingdom, " Das Thierreich," in which these rules are to be applied, it seems impossible that any names not in accordance with these rules can long persist. While it is of course inconvenient to be compelled to change some names that have come into rather general usage, the confusion so caused will not last long, and it will be a great advantage to get the nomenclature once established on a generally recognized basis. In the case of the *Rattulidæ* the confusion is already so great that the adoption of the names required by the recognized rules of nomenclature can scarcely be called even an inconvenience.

I shall give in the following a brief historical review of our knowledge of the *Rattulidæ*, with the purpose of showing the generic names which must be used.

Historical review.—The first of the *Rattulidæ* to be described was *Rattulus rattus*, by Eichhorn, in 1775. Eichhorn called it the " Water Rat " (" Die Wasser-Ratte "). In 1776 Müller gave it the name *Trichoda rattus* The genus *Trichoda* included a heterogeneous group of microscopic organisms, of which the animal at present under consideration by no means formed the type, so that the genus *Trichoda* does not belong to the *Rattulidæ*. Müller described also *Rattulus carinatus* under the name " *Trichoda rattus vesiculam gerens*," and a third species of the *Rattulidæ*, *Trichoda lunaris*, which it seems impossible now to recognize. In the same year (1776) Schrank gave the name *Brachionus cylindricus* to *Rattulus rattus* Müller. The name *Brachionus* had already been used for the rotifers which still bear that name, so that it is not available for the *Rattulidæ*, and the specific name *cylindricus* is a synonym of *rattus*.

In 1786 Müller described *Diurella tigris* as *Trichoda tigris*—the specific name *tigris* thus of course having priority for this species.

Schrank next described *Rattulus longiseta*, at first (1793) under the name *Brachionus rattus*; then (1802) under the name *Vaginaria longiseta*. The type of Schrank's genus *Vaginaria* was not one of the *Rattulidæ*, so that we may leave this name out of consideration. The specific name *longiseta* evidently has priority for the animal under consideration, however, in place of Ehrenberg's name *bicornis*.

In 1816 Lamarck founded for Müller's *Trichoda rattus* and "*Trichoda rattus vesiculam gerens*" the genus *Rattulus*. This generic name therefore evidently has priority over any other for the *Rattulidæ*, and must take the place of the commonly used name *Mastigocerca* for the genus to which Müller's species (*R. rattus* and *R. carinatus*) belong.

In 1820 the same forms were placed by Goldfuss in the genus *Trichocerca*. This name is of course merely a synonym, so far as the *Rattulidæ* are concerned.

In 1824 Bory de St. Vincent founded for these same animals the genus *Monocerca*, giving them both together the name *Monocerca longicauda*. Both the generic and specific names are thus of course synonyms, and must be dropped.

Bory de St. Vincent at about the same time (1824) described under the name *Diurella tigris* the animal which I describe below as *Diurella porcellus* Gosse. As the specific name *tigris* had already been used by Müller for a member of the same genus, it can not persist. But the generic name *Diurella* is the first one given to one of the *Rattulidæ* having nearly equal toes; hence this name has the priority for the genus so distinguished.

We find, therefore, that the generic name *Rattulus* is to be used for the species having one very long toe; *Diurella* for those having two short, nearly equal, toes.

In 1830 Ehrenberg founded the genus *Mastigocerca* for *Rattulus carinatus*, while placing the species *rattus* in Bory's genus *Monocerca*. The names *Mastigocerca* and *Monocerca* have since been much used, owing to Ehrenberg's great authority; they are both, however, merely synonyms of *Rattulus*. The name *Rattulus* Ehrenberg restricted to a small organism which he identified with Müller's *Trichoda lunaris*, and to which he attributed, rather emphatically, two eyes, a character not known at present to be possessed by any of the *Rattulidæ*.

Eichwald (1847) founded the genus *Bothriocerca* for one of the *Rattulidæ*, apparently belonging to the genus *Diurella*, though his account is so vague that the animal can not be identified. In any case *Bothriocerca* is merely a synonym.

Dujardin (1841) included *Diurella tigris* Müller in his new genus *Plagiognatha*, a genus containing a heterogeneous group of organisms, supposed to resemble each other in their jaws. This genus was not a natural one and must be given up.

Schmarda (1859) founded the genus *Heterognathus* for certain species, part of them at least belonging to the *Rattulidæ*—apparently species of the genus *Diurella*. . The type of this new genus had two *equal* toes, and was probably *Diurella tigris* Müller. If we are to classify in a genus by themselves the species having equal toes (thus following Gosse), this genus would have to receive, according to the laws of priority, the name *Heterognathus* Schmarda.

In 1886 Tessin gave it as his opinion that the *Rattulidæ* could not be distinguished into well-defined natural genera, *Rattulus gracilis* Tessin forming a connecting link between *Monocerca* (*Rattulus*) and *Diurella*. He therefore united all the species in the new genus *Acanthodactylus*. The giving of a new name was of course an unjustifiable proceeding, even granting the truth of Tessin's contention. If all the *Rattulidæ* are to be united into a single genus, the name *Rattulus* undoubtedly has the priority. Moreover, the name *Acanthodactylus* was already preoccupied, in the Reptilia (See Hoffman, in Bronn's Klassen und Ordnungen des Thierreichs, Bd. 6, Abth. 3, p. 1089).

Finally, in 1889, Gosse, in Hudson & Gosse's Monograph of the Rotifera, distinguished genera as follows: To the species with one long toe was given Ehrenberg's name *Mastigocerca*. The genus *Rattulus* was given an entirely new sense, different from that in which either Lamarck, Bory, or Ehrenberg had used it. In it were placed the species having two equal toes (including some species which clearly do not belong to the *Rattulidæ* at all). Finally, the genus *Cœlopus* was founded on the basis of a peculiar structural characteristic, which Gosse thought he had discovered in some of the species of *Diurella*. Gosse thought that the toes in *Diurella porcellus* Gosse, *D. tenuior* Gosse, *D. caria* Gosse, and *D. brachyura* Gosse consisted of "one broad plate with another laid upon it, in a different plane," and on this feature he founded the genus *Cœlopus*. As has been repeatedly pointed out of late, Gosse was quite mistaken upon this point; these species have two equal or unequal bristle-like toes. This genus *Cœlopus* must then of course be dropped, as a synonym of *Diurella*. *Mastigocerca* as used by Gosse is equivalent to *Rattulus* in its primitive application, so that it must be replaced by *Rattulus*. Finally, the group of species which Gosse distinguished as *Rattulus* can not well be separated from *Diurella*, and should be included in that genus.

Lord (1891) founded a new genus, *Elosa*, said to belong to the *Rattulidæ*, but lacking a foot. From the description which he gives of this animal, it seems clear that it is not one of the *Rattulidæ*, but belongs rather with *Asconorpha*.

Division of the Rattulidæ into genera.—We may now inquire a little more fully into the basis for classification in this family. On what grounds can the *Rattulidæ* be divided into genera?

The characters which have been used by previous authors are mainly two: (1) the presence or absence of a lorica; (2) the number and relative length of the toes.

As to the first point, Ehrenberg distinguished two genera, *Monocerca*, without a lorica, and *Mastigocerca*, with a lorica. In the former he placed *Rattulus rattus* Müller; in the latter, *Rattulus carinatus* Lamarck. Now, these two are so closely related that it is doubtful whether they should not be considered one and the same species, and both have the cuticula stiffened to form a lorica. The same is true of the other species, *Rattulus longiseta* Schrank, included by Ehrenberg in his non-loricate genus *Monocerca*. In fact, the distinction between *Monocerca* and *Mastigocerca* had no basis in reality, and it is quite impossible to divide the family in this manner, for all have a lorica.

As to the second point, Bory de St. Vincent (1824) included in *Monocerca* the species having a single long toe, while *Diurella* had two evident toes. This distinction, in one form or another, has been kept up and is in use at present.

Tessin (1886) held that this was not a good basis for division into genera, for he saw in *Rattulus gracilis* Tessin (pl. v, figs. 45–49), a species which, with its shorter toe about one-third the length of the main toe, formed a transition from the single-toed to the two-toed forms. He therefore united all the *Rattulidæ* in a single genus.

There can be no question but that Tessin was right in believing that intermediate stages could be found between the two-toed and one-toed forms. In fact, as I have shown in the account of the toes, almost every gradation can be found between the condition with two equal toes, and that where only a single toe can at first be detected, and all the species can be shown to have two toes, though the right one is in many cases a mere rudiment.

If, therefore, we are to consider the genus a natural group, including only species that are more closely related to each other than to any species of another genus, I believe there is no escape from the necessity of classifying the *Rattulidæ* all in one genus. I have made many attempts to group them into what seemed natural genera on other bases than the toes, but found that all had the same defects; some of the species within the genus were apparently not so closely related to each other as they were to some species outside the genus.

Perhaps the nearest to a natural group within the family would be made by separating off *Rattulus capucinus* Wierz. & Zach., *R. cylindricus* Imhof, and *R. multicrinis* Kellicott as a separate genus. But *R. elongatus* Gosse is very closely related to the two former, and seems itself closely related to *R. rattus* Müller. If the last-named species should be included in the new group it would have to carry with it *R. carinatus* Lamarck, *R. lophoessus* Gosse, etc. Moreover, *Rattulus latus* Jennings and *R. bicuspes* Pell seem related to *R. multicrinis* Kellicott. On the whole, such a group could not be separated off without being open to all the objections which may be made to the classification on the basis of the toes.

But the idea that a genus must represent a well-defined natural group, all the species of which are more closely related to each other than to any outside species, has as a matter of fact been largely given up in practice. Generic divisions are more commonly made on artificial grounds, to break up an otherwise unwieldy group into convenient divisions.

On this basis it seems to me that we may properly retain the old genera based upon the toes. In one group may be classed, as heretofore, those species which make the general impression of having a single long toe. This group must receive the name *Rattulus*. In the other group will be placed those that make clearly the impression of having two toes, and to this group the name *Diurella* belongs.

Then arises the question as to how we shall define our two genera so as to decide in doubtful cases in which genus the species shall go. None of the definitions heretofore given will suffice, for they have been made upon false grounds and without a knowledge of the real structure and amount of variation in the toes.

We shall probably do best to frame our definition so that it shall retain in the genus *Diurella* those that have heretofore been looked upon generally as having two toes, while *Rattulus* shall include those that have generally been considered one-toed forms. This will be best accomplished if we define the two groups as follows: *Diurella* includes those species in which the smaller of the two toes is more than one-third the length of the larger; *Rattulus*, the species in which the smaller toe is but one-third or less of the length of the larger, or seems to be lacking.

The history of our knowledge of the group shows that where, as in *Diurella tenuior* Gosse and *D. insignis* Herrick, the smaller toe is nearly (though not quite) one-half the length of the larger, the animal is naturally classified with the two-toed species. For this reason it is better to make the dividing point come at the proportion of one-third rather than at one-half. The division thus obtained is perhaps the most natural of any that could be made. The chief place where it fails is of course in the species that are near the dividing line, in separating such closely related species as *Diurella tenuior* Gosse and *Rattulus gracilis* Lessin.

Another genus, to include species having the toes exactly equal (answering to Gosse's genus *Rattulus*), might be recognized. But this seems to me hardly advisable. The equality of the toes is only one point on a long scale of variation and seems, in the present case at least, not worthy of being so strongly marked. In our American *Rattulidæ* this would separate from all others *Diurella tigris* Gosse, *D. intermedia* Stenroos, *D. sulcata* Jennings, and *D. caria* Gosse, which certainly do not form a group well marked off from the other species of *Diurella*. If such a genus should be recognized it would have to receive the name *Heterognathus* Schmarda (1859), as this was the first genus founded for equal-toed forms, its type species, *Heterognathus macrodactylus* Schmarda, being without much doubt none other than *Diurella tigris* Müller.

Specific distinctions. —As to the distinction of species, this seems not intrinsically so difficult in the *Rattulidæ* as in some other groups of the Rotatoria, notably the *Anuræadæ* and *Brachionidæ*. In the latter families individual variation is so great that it is often almost impossible to determine whether two considerably differing specimens should or should not be considered different species. In the *Rattulidæ* variation does not extend to such lengths as this, and with good specimens the species may usually be recognized with much certainty. Of the 29 species which are described in the following from my own observation, there are, I am convinced, only one or two cases where later investigation may possibly unite two into one. One of these is that of *Rattulus rattus* Müller and *Rattulus carinatus* Lamarck, which has always been considered doubtful. Two very distinct forms are certainly found—one with a ridge, the other without—and I have not succeeded in showing that the two are really identical. Further, the species grouped about *Diurella tenuior* Gosse are rather critical as to specific distinctions.

Points to be noted in descriptions of the Rattulidæ.—From many of the descriptions of the *Rattulidæ* given in the literature, it is exceedingly difficult to determine the animal in question, but this is due to the fact that the characteristic distinguishing features of the animal have not been noted. It will be well to point out, therefore, the features that are of especial importance in distinguishing species, and that should be included, if possible, in every description. I give them in the order of their importance.

1. The teeth at the anterior margin of the lorica, their absence or presence, their number and relative size, if present. Those at the dorsal or dorso-dextral margin should be clearly distinguished from others due to the folding of the head-sheath when retracted.

2. The toes, their length relatively to the body and to each other; their position.

3. The general form of the body.

4. The absence or presence of the longitudinal folds in the head-sheath; their form, especially when the lorica is retracted, and any other characteristics of the anterior margin of the lorica.

5. The "striated area," whether developed as a single or double ridge, a furrow, a smooth area, or not developed at all.

As many other features should, of course, be added as possible, but the above are the most important ones and should not be omitted from descriptions of any of the *Rattulidæ*. Of course, an accurate figure or figures (showing the above points, as well as others) is perhaps even more important than a good description.

Several of the most important points above mentioned, notably the presence, number, and relative size of the teeth at the anterior margin of the lorica, and the relative length of the toes, have very usually been omitted from specific descriptions of the *Rattulidæ*: this makes it very difficult to recognize the animals.

The following systematic account of the *Rattulidæ* is arranged thus: I first give the characteristics of the family. This is followed by a key to the genera and species, which may be of assistance in locating quickly a given species; though for a determination, of course, the entire description and the figures should be studied. Some of the poorly described or doubtful species, which I have not myself seen, could not be taken into the key owing to the uncertainty as to important technical characters; these, however, are referred to at appropriate places in the key.

Then follows a description of all the well-founded species of *Rattulidæ* under the two genera. Under each genus I divide the descriptions into two parts, the first including the species which I am able to describe from my own observations, thus making the accounts full on all points important for classification; the second, species which I have not myself seen. Of these latter I have compiled descriptions from other authors and have copied the best figures I could find.

The species which are described from my own observations number twenty-nine, including all that have been found in America as well as a number of others. The remaining well-established species number seven.

Finally, I have added a list and notes on doubtful species, species that are insufficiently described for recognition, and animals that have been wrongly classified with the *Rattulidæ*. These are in many cases accompanied by copies of the original figures.

Family RATTULIDÆ.

Loricate rotifers, with the structure somewhat unsymmetrical in certain features. Lorica usually cylindrical and curved, or ovate or ovoid; closed all around, with an opening at each end for the protrusion of the head and the foot. The anterior portion of the lorica usually set off from the remainder as a head-sheath, by a slight constriction. On the dorsal surface of the lorica, usually if not always somewhat to the right of the median line, a longitudinal area which is transversely striated; this striated area is generally striated, and may be developed as a single or double ridge, a furrow, or remain smooth. (It is absent in only two or three species.) Eye single, occipital, attached to the brain. Foot short, frequently attached unsymmetrically to the lorica. Toes bristle-like, their place of attachment usually twisted so as to bring the right toe somewhat to the dorsal side of the left; the toes sometimes equal, but the right or dorsal toe usually shorter; sometimes quite rudimentary. Minute, bristle-like "substyles" at the base of the toes. Trophi usually unsymmetrical, the right manubrium smaller than the left; sometimes rudimentary. Mucus glands and reservoirs much developed.

Key to the genera and species.

A. The two toes equal or the shorter toe more than one-third length of longer   I. DIURELLA Bory de St Vincent.
B. A single long toe with usually or always an inconspicuous shorter one, the latter not more than one-
      third the length of the longer one .... II. RATTULUS Lamarck.

I. DIURELLA Bory de St. Vincent.

a1. Toes equal.
   b1. With a single tooth at the dorsal or dorso-dextral anterior margin of the lorica
      c1 The toes about one-third as long as the body; body elongated, cylindrical, curved   ....1. *D. tigris* Müller.
      c2. Toes shorter, less than one-third the body length; body shorter and very small; no visible ridge.
                                                                                        4 *D. intermedia* Stenroos.
         (See also 9 *D. sulcata* Jennings.)
   b2. No tooth (or tooth very inconspicuous), hardly noticeable, at the anterior dorsal margin of the lorica.
      c1. Foot minute, usually retracted within the lorica; toes very short; lorica ending behind in a sharp
         angle; two deep grooves surrounding the body nearits middle   ..   ...   9. *D. sulcata* Jennings.
      c2. Very small; lorica much swollen and rounded behind and above, bringing the foot entirely on the
         ventral surface. Otherwise much as in the last   ..   ....10. *D. curta* Gosse.
      c3. Body projecting much above and behind the foot; the toes wide apart at base, about one-third the
         length of the body........ ...   .   13 *D. sejunctipes* Gosse.
      c4. Body projecting behind and above the small foot; a ring-like fold or collar surrounding the lorica in
         front of its middle; toes slender, about half as long as the lorica........... ... .14. *D. collaris* Rousselet.
         (See also 17. *D. brevidactyla* Daday.)
a2. Toes unequal (the right one shorter).
   b1. A single tooth at the anterior dorsal margin.
      c1 Body elongated, cylindrical, curved; the right toe about half as long as the left, or a little less than
         half .... 2. *D. tenuior* Gosse.
      c2. The body short, thick, curved, with a very prominent ridge; toes short, the right one a very little
         shorter than the left .... 3. *D. weberi*, n. sp.
         (See also *D. uncinata* Voigt, page 319, note.)

b². With two teeth or spines at the dorsal or dorso-dextral anterior margin of the lorica.
  c¹. The two teeth short, the right one larger than left; body long, slender, curved; right toe about half
    as long as the left, or less...5. *D. insignis* Herrick.
  c². The two teeth short, the right one larger than the left; body short, thick; the foot turned beneath
    the ventral surface; the right toe a little shorter than the left...............6. *D. porcellus* Gosse.
  c³. The two teeth developed into long, sharp, slender spines, nearly or quite equal in length; body long,
    conical in shape, scarcely curved; the two toes very close together and nearly equal (the left a
    little longer)...7. *D. stylata* Eyferth.
b⁴. With many (nine) teeth at the anterior margin of the lorica.........8. *D. consuetti* Voigt.
b⁵. Without teeth at the anterior margin of the lorica; foot at the posterior end of the body.
  c¹. Toes very nearly equal...11. *D. brachyura* Gosse.
  c². Right toe about two-thirds as long as the left...............................12. *D. dixon-nuttalli*, n. sp.
    (See also *D. brevidactyla* Daday, *D. marina* Daday, and *D. helminthodes* Gosse, which could not
    be taken into the key because certain important technical characters, particularly in regard to
    the anterior margin of the lorica, are unknown.)

II RATTULUS Lamarck.

a¹. With a single tooth or projection at the anterior dorsal (or dorso-dextral) margin of the lorica
  b¹. The single, not very conspicuous, tooth situated at dorso-dextral part of anterior margin of lorica,
    forming a prolongation of the ridge (the ridge may be inconspicuous).
    c¹. Body elongated, nearly cylindrical, not strongly curved; head-sheath sharply set off from the
      remainder of the lorica and very contractile; main toe about half the length of the lorica, accom-
      panied by a shorter toe one-third the length of the main one...............18. *R. gracilis* Tessin.
    c². Body elongated, nearly cylindrical, not strongly curved; the head-sheath not sharply set off nor
      very contractile; toe two-thirds the length of the lorica, or still longer............19. *R. scipio* Gosse.
    c³. Body elongated, nearly cylindrical, strongly curved.............................35. *R. curvatus* Levander
    c⁴. Body elongated, fusiform, not strongly curved; anterior tooth very inconspicuous; toe one-half to
      two-thirds the length of the lorica; a small spur projecting backward from the base of the toe
      when the latter is turned forward.................................20. *R. macerus* Gosse.
      (See also *R. uniden̄s* Stenroos (?) and *R. cuspidatus* Stenroos (?).)
  b². The single large tooth forming a triangular projection from the median dorsal part of the anterior
    margin of the lorica, overhanging the corona.
    c¹. The lorica oval or ovoid in form...21 *R. multicrinis* Kellicott.
    c². The lorica elongated, cylindrical.
      d¹. The anterior tooth prolonged as a long, slender hook bending over the corona, but not always
        visible; toe almost or quite as long as the body.........................22. *R. cylindricus* Imhof.
      d². The large anterior tooth forming a hood-like projection over the corona; body somewhat curved;
        toe about half the length of the body.................................23. *R. capucinus* Wierz. & Zach
        (See also 36. *R. dubius* Lauterborn.)
a². With two long teeth or spines at the anterior margin of the lorica
  b¹. The two teeth at dorso-dextral margin of lorica; the right longer than the left............24 *R. longiseta* Schrank.
  b². The two teeth at the ventral margin of the lorica (?).................................*R. roseus* Stenroos (?)
a³. Without teeth at the anterior margin of the lorica.
  b¹. Lorica with two prominent ridges on its dorsal (or dorso-dextral) surface.
    c¹. Very large; the two ridges very high and extending two-thirds length of body...............25. *R. bicristatus* Gosse.
    c². Smaller; the two ridges lower, reaching back only about half the length of body...............26. *R. mucosus* Stokes.
  b². A single very prominent thin ridge on the dorso-dextral surface of the lorica
    c¹. The ridge high and thin, extending about one-half the length of the lorica...............27. *R. carinatus* Lamarck.
    c². The ridge high and thin, extending nearly or quite the entire length of the lorica...............29. *R. lophoessus* Gosse.
  b³. The ridge either not prominent or lacking (a low ridge can be detected in most of these species, on
    careful examination).
    c¹. Body broad, ovate, very unsymmetrical at the posterior end; no trace of ridge...............30. *R. latus* Jennings
    c². Body short, thick, arched dorsally; toe longer than the lorica; lateral antennæ protected by pro-
      jecting spines..31. *R. bicuspes* Pell.
    c³. Large; body long, slender, straight, tapering posteriorly; toe two-thirds the length of the lorica.
        ....32. *R. elongatus* Gosse.
    c⁴. Body short, irregular, somewhat conical; toe less than half the length of the lorica...............33. *R. stylatus* Gosse
      (See also *R. brachydactylus* Glasscott.)
    c⁵. Very small; body truncate in front, gently arched dorsally; toe about the length of the lorica; sub-
      styles very inconspicuous.................................34 *R. pusillus* Lauterborn.
    c⁶. Body oval, much larger than in the last; toe about the length of the lorica; substyles easily seen.
        ....28 *R. rattus* Müller.

DESCRIPTIONS OF GENERA AND SPECIES.

I. DIURELLA Bory de St. Vincent.

Generic characters.—Two toes, either equal, or the shorter more than one-third the length of the longer. The longer toe less than one-half the length of the body. Body nearly cylindrical; curved or twisted.

A. DESCRIPTIONS OF THE SPECIES STUDIED BY THE AUTHOR.

1. Diurella tigris Müller (pl. 1. figs. 1-6).

Synonyms: *Trichoda tigris* Müller (1786); *Notommata tigris*, Ehrenberg (1833, 1838); *Heterognathus macrodactylus* Schmarda (1859); *Monommata tigris*, Bartsch (1870); *Rattulus tigris*, Hudson & Gosse (1889).

Distinguishing characters.—This species may be known by the two equal toes (fig. 6), the single tooth at the dorsal anterior edge of the lorica, and the nearly cylindrical curved body. It has a striking resemblance to *D. tenuior* Gosse, from which it is distinguished by the equality of the two toes. It differs also in the usually greater size, the somewhat greater prominence of the ridge, and the slightly greater slenderness and distinctness of the foot.

External features.—The body is elongated and curved, appearing to be cylindrical, in a cursory view. Really the body rises to a ridge on the right side, so that in section it has the form shown in fig. 5. Preserved specimens usually lie, owing to the form of the body, in such a position that the ridge does not appear in profile; hence it is very easily overlooked; in living specimens it is more conspicuous.

The head-sheath is rather distinctly set off from the remainder of the lorica by a constriction. It is marked by nine longitudinal plaits (fig. 3), at which the head-sheath folds when the head is withdrawn, thus closing the anterior opening completely (fig. 4). On the right side the anterior edge bears a single prominent tooth.

The ridge (fig. 1) begins as a backward prolongation from the base of the tooth. It extends backward and to the left, seeming to have a slightly spiral course, and reaching almost to the foot. Along its left side are transverse striations, similar to those so prominent in many species of *Rattulus*, but less conspicuous.

The degree of development of the ridge varies greatly in preserved specimens. In some it can scarcely be seen at all. In others it is visible in the anterior part of the body, but seems to extend only half the length of the lorica or less. These differences are perhaps due only to optical difficulties resulting from the position of the specimen, but I am inclined to believe that there are really such differences in the development of the ridge in different specimens. These differences are perhaps functional, depending upon the degree of contraction of the animal (see the general account of the striated area, p. 281). In view of these facts the size and length of the ridge can not be considered a distinguishing character in this species.

As a whole, the body may be seen to form a segment of a spiral, a spiral that is further accentuated by the position of the toes (*q. v.*).

Corona.—The corona bears a single club-shaped frontal process; its other features have not been studied especially.

Antennæ.—The dorsal antenna (fig. 1, *d. a.*) lies just to the left of the ridge, a very little behind the constriction separating off the head-sheath. The lateral antennæ are in the usual position, on the posterior third of the body (fig. 1).

Foot.—The foot is rather slender and sharply set off from the body. The joint between the foot and body appears to lie in a transverse plane, without the asymmetry which is so marked in many of the species of *Rattulus*.

Toes (fig. 6).—The two equal toes are stout, curved rods, about one-third the length of the body. They are attached to the foot in such a way that the base of the right toe lies above that of the left, and the concavity of the toes faces to the right (fig. 1). When the toes bend (at their attachment), they bend to the right. Each toe has at its base a number (at least four) of short, sharp spines or substyles (fig. 6). At the base of each toe opens one of the two mucus reservoirs (fig. 6, *m. r.*).

Internal organs.—The eye lies at the posterior end of the brain and appears in dorsal view to the left of the ridge (fig. 1, *e*). The two mucus reservoirs are of equal size, and each opens separately at the base of one of the toes. The trophi are well developed, and in this species their asymmetry reaches perhaps its highest development. The left manubrium is long and heavy; the right one a mere rudiment—a short, slender spicule (fig. 2). The other internal organs call for no special remark.

Measurements.—Length of body, 0.175 to 0.225 mm.; of toes, 0.050 to 0.075 mm.; total, 0.225 to 0.300 mm.

Movements.—For an account of the movements of this species, see the general discussion of the movements of the *Rattulidæ*, p. 295.

History.—In the systematic and faunistic literature this species has long been confused in a very curious way with another to which it bears very little resemblance, namely, with *Diurella porcellus* Gosse (*D. tigris* Bory). This is due to the fact that both received the name *tigris*, one from Müller, the other from Bory de St. Vincent, and they have often since been supposed to be identical. Ehrenberg (1838) confused the two, citing Bory's *D. tigris* as a synonym of his *Notommata tigris*. Gosse, in Hudson & Gosse's Monograph (1889), describes the present animal as *Rattulus tigris*, but notes in a rather perplexed way that Eckstein's account (1883) of the animal does not agree with his own. Eckstein had described under this name Bory's species (*Diurella porcellus* Gosse). Bilfinger (1894, p. 51) seems to have been the first to set forth clearly the fact that Ehrenberg's *Notommata tigris* and Bory's *Diurella tigris* are two distinct animals. Attention has been called to the same fact by Weber (1898, p. 513) and probably by others. It will be well to give here a list of the animals mentioned by different authors under the specific name *tigris* (assigned to various genera), specifying in each case which of the two animals, *Diurella tigris* Müller or *D. porcellus* Gosse, was really meant, so far as that can be determined.

Trichoda tigris Müller, 1786 = *Diurella tigris* Müller.
Diurella tigris Bory de St. Vincent, 1824 = *D. porcellus* Gosse.
Notommata tigris, Ehrenberg, 1838, 1888 = *D. tigris* Müller.
Notommata tigris, Perty, 1852 = *D. tigris* Müller.
Notommata tigris, Pritchard, 1861 = *D. tigris* Müller.
Monommata tigris, Bartsch, 1870, 1877 = *D. tigris* Müller.
Diurella tigris, Eckstein, 1883 = *D. porcellus* Gosse.
D. tigris, Herrick, 1885 = *D. porcellus* Gosse.
D. tigris, Eyferth, 1885—The figure seems to represent *D. tigris* Müller, but the description applies best to *D. porcellus* Gosse.
D. tigris, Plate, 1886 = *D. porcellus* Gosse.
D. tigris, Blochmann, 1886 = *D. porcellus* Gosse.
Acanthodactylus tigris, Tessin, 1886 = *D. porcellus* Gosse.
Rattulus tigris, Hudson & Gosse, 1889 = *D. tigris* Müller.[a]
Rattulus tigris, Wierzejski, 1893 = *D. porcellus* Gosse (?).
Rattulus tigris, Levander, 1894 = *D. porcellus* Gosse.
Rattulus tigris, Hood, 1895 = *D. tigris* Müller.
Rattulus tigris, Scorikow, 1896—The description does not agree with either of the species under consideration.
Rattulus tigris, Stenroos, 1898 = *D. tigris* Müller.
Rattulus tigris, Jennings, 1900, 1901 = *D. tigris* Müller.

Distribution.—In America *Diurella tigris* Müller is very common in aquatic vegetation in the quiet parts of streams and lakes. I have recorded its presence in the following localities: Put-in Bay Harbor and East Harbor, Lake Erie; Huron River at Ann Arbor, Mich.; Portage River, Ohio. It has also been recorded from Bangor, Me., by "J. C. S." (1883), and from the neighborhood of Cincinnati, Ohio, by Turner (1892), but it is impossible to say in these cases which of the two species that have gone under this name (*D. tigris* Müller or *D. porcellus* Gosse) was meant.

In Europe: England (Gosse, 1889); Ireland (Glascott, 1893; Hood, 1895); Germany, near Tübingen (Bartsch, 1870), and in Württemberg (Bilfinger, 1894); Tyrol (Dalla Torre, 1889); Hungary (Bartsch, 1877; Kertesz, 1894); Lake Nurmijärvi, in Finland (Stenroos, 1898).

Also in India, near Calcutta (Anderson, 1889); New Guinea (Daday, 1901); Ceylon (Daday, 1898); Natal, South Africa (Kirkman, 1901).

[a] In a previous paper (Jennings, 1900) I was inclined to believe that the animal described and figured by Gosse was not the real *Notommata tigris* of Ehrenberg, owing to the disproportionately large size of the anterior end in Gosse's figure, as well as to the unusual form of the body. But after studying many specimens of this and other *Rattulidæ* I am convinced that Gosse's figure is a poorly drawn representation of a much contracted specimen of this species.

2. Diurella tenuior Gosse (pl. 1, figs. 7-10).

Synonyms: *Coelopus tenuior* Gosse (1889); *Mastigocerca flectocaudatus* Hilgendorf (1898).

Distinguishing characters.—This species is to be known by its elongated curved body, with a single tooth at the anterior margin of the lorica, and the unequal toes (pl. 1, fig. 9), the right toe being only about one-half as long as the left or a little less than one-half as long.

It has much resemblance to *D. tigris* Müller, from which it differs in the unequal toes and in certain other characters mentioned in the account of that species. It also greatly resembles *Rattulus gracilis* Tessin, from which it differs in the following particulars: The body in *D. tenuior* is regularly curved, so that no straight outlines appear, as in *Rattulus gracilis*; the second toe is longer in proportion to the main one than in *Rattulus gracilis*, where the lesser toe is only about one-third the length of the main toe, while here it is about one-half the length of the latter; the head is much less sharply set off from the remainder of the lorica than in *Rattulus gracilis*; the tooth is more pronounced and the ridge less prominent than in the last-named species; the foot is shorter and less prominent than in *Rattulus gracilis*.

Diurella tenuior Gosse also has a striking resemblance to *Diurella insignis* Herrick in the form and in the toes, but the latter species has two teeth at the anterior edge of the lorica in place of one, is much larger, and there are other differences in details.

This species is related, finally, to *D. intermedia* Stenroos, but *D. tenuior* is usually much larger than *D. intermedia* and is longer in proportion to its diameter. The two differ especially, however, in the toes, those of *D. intermedia* being equal.

External features.—The body is long and cylindrical, much as in *D. tigris* Müller. There is a low oblique ridge on the right-hand side, passing backward from the point of origin at the anterior tooth to about the middle of the length of the lorica. This is striated transversely, as in other species. The head-sheath is marked off from the remainder of the lorica by a slight constriction. It has longitudinal folds somewhat similar to those of *D. tigris* Müller, though perhaps hardly so prominent; by these the anterior opening of the lorica can be nearly closed when the head is withdrawn. On the dorso-dextral part of the anterior margin there is a tooth, perhaps hardly so prominent as that of *D. tigris* Müller, but rather more pronounced than in *Rattulus gracilis* Tessin.

The corona has not been especially studied.

Antennæ.—The dorsal antenna (figs. 7, 8) lies in the usual position, a little to the left of the ridge. The right lateral antenna has the usual position on the posterior third of the body (fig. 8). The left lateral antenna I have not seen.

Foot.—Rather broadly conical, not so sharply set off from the lorica as in *D. tigris* Müller.

Toes (fig. 9).—The two toes are unequal in size, the right toe being about half, or a little less than half, the length of the left toe. The main (left) toe is a curved, pointed rod, about half the length of the lorica. The right toe is much more slender and is so curved that its tip usually lies against the main toe at about the middle of the length of the latter. The right toe seems to be, as a rule, a trifle less than half the length of the main toe. This species, in its technical characters, is on the boundary line between *Rattulus* and *Diurella*, and is as closely related to *Rattulus gracilis* Tessin as to any of the species of *Diurella*.

Just outside the base of the main toe is a substyle which is nearly as long as the right toe. There is also a minute substyle just outside the base of the right toe.

Internal organs.—These call for no special remark, except in the case of the trophi (fig. 10). The trophi are very similar to those of *Diurella tigris* Müller and *Rattulus gracilis* Tessin—the right side being very rudimentary as compared with the left.

Measurements.—Different specimens of this species vary excessively in size. Two specimens drawn to the same scale are shown in figs. 7 and 8 (pl. 1). The length of body varies from 0.135 to 0.21 mm.; length of toes, from 0.055 to 0.08 mm.; total, from 0.19 to 0.29 mm.

History.—This species was first described by Gosse in Hudson & Gosse's Monograph of the Rotifera (1889). Like many of Gosse's descriptions, that of this species is inaccurate in some details. For example, he states that the head is defended by two or three projecting points. Weber (1898) has likewise given a description and figure of this species, repeating Gosse's statement that there are three or four points at the anterior edge of the lorica, though his figure shows but one. It is probably the longitudinal folds in the head-sheath that have given rise to the impression that there were several teeth. These, when the head is contracted, often give an appearance as of projecting teeth, though a close examination reveals the incorrectness of this.

Hilgendorf (1898) described as *Mastigocerca flectocaudatus* n. sp. a rotifer which, from his description and figure, bears much resemblance to *Diurella tenuior* Gosse. Apparently the author himself concluded that the supposed new species was *D. tenuior* Gosse, for in my copy of Hilgendorf's paper the name *Mastigocerca flectocaudatus* is crossed out and "*Coelopus tenuior*" substituted, apparently by the hand of the author. Hilgendorf gives no measurements, his figures are not very detailed and are apparently not made with the camera, so that it is difficult to form an independent judgment as to the identity of the animal. It will be best, therefore, to accept the view that this was *D. tenuior* Gosse.

Distribution.—In America: This species is not rare in the vegetation of lakes and streams. I have found it in the following localities: Old Channel, between Round Lake and Pine Lake, Charlevoix, Mich.; Put-in Bay Harbor and East Harbor, Lake Erie; Long Point, Canada, near "The Cottages"; swamps on North, Middle, and South Bass islands in Lake Erie; Portage River, Ohio; Huron River at Ann Arbor, Mich.; East Sister Lake, Ann Arbor, Mich.; ditch in tamarack swamp region, near Ann Arbor, Mich. Doubtfully reported by Kellicott (1888) from the Shiawassee River at Corunna, Mich.; Sandusky Bay, Lake Erie (Kellicott, 1896); waters connected with the Illinois River at Havana, Ill. (Hempel, 1898).

In Europe: England (Gosse, 1889); Ireland (Glasscott, 1893); Gr. Plöner See, Germany (Zacharias, 1893); Switzerland (Weber, 1898); Bohemia (Petr, 1890); Hungary (Kertesz, 1894).

Also in New Guinea (Daday, 1901); Ceylon (Daday, 1898); New Zealand (?) (Hilgendorf, 1898, as *Mastigocerca flectocaudatus*).

3. Diurella weberi, n. sp. (pl. I, figs. 11-14; pl. XIII, figs. 116 and 117).

Synonym: *Coelopus porcellus* Weber (1898), in part.

Distinctive characters.—*Diurella weberi* is to be distinguished from its nearest relative, *Diurella porcellus* Gosse, by the single tooth at the anterior edge of the head-sheath, and by the broad, rounded projecting plate (see fig. 14) at the left side of the anterior opening—as well as by the high, thin ridge. It differs from *Diurella brachyura* Gosse in the presence of the anterior tooth and of the ridge; from *Diurella sulcata* Jennings in the prominent tooth, the inequality of the toes, and the presence of the ridge; from *Diurella intermedia* Stenroos in the unequal toes and the presence of the ridge. *Diurella tenuior* Gosse, which in technical characters resembles this, is easily distinguished from it in practice by the high, thin keel, the shorter body, and the shorter, only slightly unequal toes of *Diurella weberi*.

External features.—The body is short, and curved in the arc of a circle, much as in *Diurella porcellus* Gosse, though it is not so thick. The head-sheath is indistinctly set off from the rest of the lorica by a slight constriction. At the anterior margin of the lorica, to the right of the dorsal median line, is a single sharp, prominent tooth. From this tooth there runs backward a high, thin ridge, which is transversely striated and extends about two-thirds the length of the lorica (pl. I, fig. 12). This ridge is much more prominent than the ridge of *Diurella porcellus* Gosse. One of the most peculiar characteristics of this species is the large, rounded projection from the left side of the anterior margin of the lorica. This is especially noticeable in a retracted specimen (see figs. 13 and 14); but gives form to the head even in extended animals (see figs. 12 and 117, and compare Weber (1898), fig. 2, pl. 20). In retracted specimens a number of folds may at times be seen in that part of the head-sheath not formed by the plate just mentioned.

Corona.—The corona has not been thoroughly studied. It bears a thick dorsal process.

Antennæ.—The dorsal antenna lies in the usual position, to the left of the ridge. The right lateral antenna is in the usual place on posterior third of the body; the left lateral antenna is much farther forward, only a little behind middle of body and near dorsal side (figs. 14 and 116).

Foot.—The foot is not quite so nearly inclosed within the lorica as in *Diurella porcellus* Gosse and is not situated so far forward on the ventral side.

Toes (fig. 11).—The two toes are nearly equal, but the left toe is a little longer than the right. Possibly the difference in length is a little less in this species than in *Diurella porcellus* Gosse. The length of the main toe is about equal to the diameter of the body. Three or four inconspicuous substyles are found at the base of the toes; these are much less conspicuous than in *D. porcellus* Gosse.

Internal organs.—The internal organs offer nothing of especial interest. The trophi differ from those of *D. porcellus* Gosse in being straighter and more slender and in not showing in side view the long transverse piece which makes the left manubrium so conspicuously "crutch-shaped" in the latter species. (Compare the trophi as shown in figs. 12 and 21.)

Measurements.—Length of body, 0.09 to 0.12 mm.; of toes, 0.03 to 0.04 mm.; total, 0.12 to 0.16 mm.

History.—This species was figured by Weber (1898, pl. 20, figs. 2-4) as *Cœlopus porcellus* Gosse (*Diurella porcellus* Gosse). Weber's description confuses the two species; thus, the two teeth at the anterior edge, mentioned by Weber, but not figured, belong to the real *D. porcellus* Gosse, not to the species which he figures. That the figure represents the present species is shown by the general form, the high, sharp ridge, the single tooth, the form of the trophi, and the extension of the anterior edge of the lorica on the left side, all points which are characteristic of the present species and distinguish it clearly from *D. porcellus* Gosse.

Distribution.—*Diurella weberi* is not very common. I have recorded it from the following localities: East Harbor, Lake Erie; swamps on North and South Bass islands in Lake Erie, and on Presque Isle near Erie, Pa.; Huron River at Ann Arbor, Mich.; a ditch in the tamarack swamp region, near Ann Arbor, Mich.

Weber (1898) found this species in the Botanical Gardens at Geneva, Switzerland, and at St. Georges, Switzerland. Mr. F. R. Dixon-Nuttall informs me that he finds it in ponds in England. Mr. Charles F. Rousselet has sent me a sketch of what is evidently this species, made from specimens found in New Zealand.

4. Diurella intermedia Stenroos (pl. XIII, figs. 108-110).

Synonym: *Cœlopus intermedius* Stenroos (1898).

Distinguishing characters.—This species is to be distinguished from *Diurella brachyura* Gosse, which it much resembles, by the tooth at the dorsal anterior edge of the lorica, and by the equal toes. From *D. weberi*, with which it agrees in the single tooth, it differs in the absence of the conspicuous ridge and in the equality of the toes. From *D. porcellus* Gosse it differs in having but a single tooth at the anterior edge. From *D. sulcata* Jennings, finally, it differs markedly in the absence of the furrows surrounding the body about the middle.

External features.—The body is nearly cylindrical, not so short as in *D. porcellus* Gosse, and curved. The head-sheath is set off by a slight constriction from the remainder of the lorica. It has nine longitudinal plaits for folding when the head is withdrawn. At the dorsal edge, a little to the right of the median line, is a single well-marked tooth.

The lorica bears no distinct ridge, though a faintly striated area, in some cases apparently a little depressed, extends backward from the base of the tooth about half the length of the body.

Corona.—Corona of usual character. It bears a single sharp dorsal process (pl. XIII, fig. 108).

Antennæ.—The dorsal antenna lies a little in front of the constriction which separates the head-sheath from the rest of the lorica. The right lateral antenna lies in the usual position on the posterior one-fourth of the body. The left lateral antenna I have not found.

Foot.—Very short, not pushed so far forward on ventral side as in *D. porcellus* Gosse.

Toes (pl. XIII, fig. 110).—The two toes are equal, or so nearly so that one can not be certain of a difference in length. There are two substyles, one a little longer than the other, each more than half the length of the main toes.

Internal organs.—The trophi have not been minutely studied. Their general appearance is shown in pl. XIII, fig. 108. The gastric glands are very small and fastened to the stomach only by slender, thread-like ducts. The other internal organs call for no special remark.

Measurements.—Total length, about 0.13 to 0.16 mm.; toes, about 0.03 to 0.04 mm.

History.—This species was recently described by Stenroos (1898). I have found but a few specimens, and most of our detailed knowledge of the animal is derived from the notes and figures of Mr. F. R. Dixon-Nuttall, which he has with great kindness placed at my disposal. His figures are reproduced in figs. 108 and 110.

Distribution.—I have found only a few specimens, from the Huron River at Ann Arbor, Mich. Stenroos (1898) found the animal in Lake Nurmijärvi, Finland. Mr. Dixon-Nuttall informs me that examples have often been sent him from Dundee, Scotland, by Mr. John Hood.

5. Diurella insignis Herrick (1885) (pl. II. figs. 15–18).

Distinctive characters.—This species may be known by the long, slender, curved body; the two slightly unequal teeth at the anterior margin of the lorica, and the two unequal toes, the longer one (in adults) a little less than one-half the length of the lorica. It shows much resemblance to *Diurella tenuior* Gosse, but is distinguished from it by the possession of two teeth at the anterior margin and by the more elongated form. The toes are almost identical in the two species.

External features.—The body is more elongated and slender than in any other species with which I am acquainted, and is gently curved. It tapers slightly from a point not far from the anterior end to the foot. The anterior portion of the lorica, the head-sheath, is set off from the remainder by a constriction, and is provided with a number of longitudinal plaits at which folding takes place when the head is retracted. At its anterior edge, a little to the right of the dorsal line, it bears two teeth. These are of unequal size, the right one being considerably the longer. The length of this right tooth varies considerably in different specimens; it seems to be especially prominent in young specimens (pl. II. fig. 18). The two teeth are separated by a considerable interval. The left tooth is small and is very easily overlooked, especially when the corona is extended, so that specimens of this species are likely to be thought to have only a single tooth if a careful examination is not made.

From the teeth a low ridge extends backward, having its edge to the right, even with the right tooth, and sloping gradually to the left. The ridge is very inconspicuous and easily overlooked. It extends backward for three-fourths of the length of the lorica and is marked as usual by transverse striations.

Corona.—The corona is of the usual character, having two marginal curves of cilia and two about the mouth. There is a short, thick, dorsal process, and two lateral projections bearing cilia (as in *Diurella stylata* Eyferth, pl. III. fig. 31); other processes I have not seen.

Antennæ.—The dorsal and left lateral antennæ I have not been able to find in this species; the right lateral antenna is in the usual position on the posterior fourth of the body (pl. II. fig. 16).

Foot.—The foot is of the usual short conical form, obliquely attached to the lorica, so that it may turn to the right, but not to the left.

Toes.—The toes (fig. 17) are almost identical with those of *Diurella tenuior* Gosse, save that they are longer. The left toe in an adult animal (pl. II. figs. 15 and 16) is a little less than one-half the length of the body, while in a young specimen (fig. 18) it is considerably more than one-half the length of the body. The right toe is one-half the length of the main toe, or a little less. Just outside the base of the main (left) toe there is a substyle, which is one-half the length of the right toe. At the base of the right toe there is a minute, rudimentary substyle.

Internal organs.—The trophi (see fig. 18) are very unsymmetrical, as in *Diurella tigris* Müller and *D. tenuior* Gosse. The right malleus is very small and slender, though perhaps not quite so much reduced as in the two species last mentioned. The single eye is attached to the brain near its posterior end; in a dorsal view it lies considerably to the left of the ridge or striated area on the lorica (fig. 16). The other internal organs call for no special remark.

Measurements.—Total length, 0.32 to 0.37 mm.; main toe, 0.10 to 0.12 mm.; shorter toe, about 0.05 mm.

History.—This species was described by Herrick in 1885. Herrick's description was brief and his figure extraordinarily poor, and as the species has not hitherto been found again, it has usually been relegated to the limbo of "doubtful species." But Herrick's description fits very well the specimens which I have, while his figure looks as if it had been drawn from memory. The animal has not again been mentioned since Herrick's paper.

Distribution.—Herrick found *Diurella insignis* in Minnesota. I have found it to be rather rare, but somewhat widely distributed amid the vegetation of lakes, ponds, and streams. My records show it to have been observed in the following localities: Put-in Bay Harbor and East Harbor, Lake Erie; inlet on Starve Island, close to South Bass Island, in Lake Erie; swamp on Presque Isle near Erie, Pa.; East Sister Lake near Ann Arbor, Mich.

This species has not been found in Europe.

6. Diurella porcellus Gosse (pl. ii, figs. 19–23).

Synonyms: *Diurella tigris* Bory de St. Vincent (1824); *Monocerca porcellus* Gosse (1851); *Acanthodactylus tigris* Tessin (1886); *Coelopus porcellus* Hudson & Gosse (1889).

Distinctive characters.—This species is to be known by the short, plump, curved body; by the two toes, one a little longer than the other, usually kept folded beneath the body, and especially by the two teeth at the anterior margin of the lorica on the dorsal side. From all the other closely related species it differs in the presence of these two teeth, the others having one or none.

External features.—The body is short and thick, and strongly curved, so that the back forms an arc of a rather small circle. The posterior end is broad and rounded, the opening for the foot being on the ventral surface. The head-sheath is marked off from the remainder of the lorica by a slight constriction; it bears at its anterior margin, a little to the right of the middle line, two teeth, which are very similar to those of *Diurella insignis* Herrick. The right one of these is the longer, and is separated from the left by a slight interval. Ventrally the anterior margin has a broad, shallow notch. When the lorica is strongly contracted the two sides of this notch project as two decided points, one of which is seen in fig. 20. These two points might be called teeth, and this animal is therefore sometimes said to have four anterior teeth, two dorsal and two ventral. These two ventral teeth, due to the folding of the head-sheath, are of a different character from the dorsal ones, however, and are not to be noticed when the head is fully extended.

Extending backward from the larger one of the two dorsal teeth is a ridge, having its edge directed to the right. It is striated transversely from near its summit to a line some distance to the left of it. The ridge is not prominent, and in some specimens there is a decided depression just to the left of the ridge, so that the ridge appears merely as the edge of the depression. In other cases the back seems nearly smooth, only the striated area being visible, with perhaps a marked line at its right edge. These differences are probably functional changes due to the varying states of contraction of the specimens, though I have not been able to demonstrate this.

Corona.—The corona has a short, median, club-shaped process. It has not been fully studied in other respects.

Antenna.—The dorsal antenna is just to the left of the ridge, in the depressed area, when the depression is present. It is situated a little behind the constriction which sets off the head-sheath. The lateral antennae are in the usual position on the posterior one-fourth of the body, the left one somewhat in advance of the right.

Foot.—The foot is very small and partly inclosed within the lorica.

Toes.—There are two unequal toes, the left one being about equal in length to the diameter of the body, while the right one is a little shorter (fig. 23). Each of the toes is accompanied at its base by two substyles, one of them in each case being more than half the length of the shorter toe. The right toe usually lies with its tip against or across the longer left toe. This gives an appearance which Gosse (1889) interpreted as being due to two flat, spoon-shaped toes, the one lying within the other. The inner sides of the two toes were supposed to be the outlines of the smaller toe; the outer sides those of the larger toe. On the basis of this supposed structure the genus *Coelopus* was founded.

Internal organs.—The trophi are unsymmetrical, though the right manubrium is not lacking, as represented by Gosse (1855). It is a very slender rod, a mere bristle, but of the same length as the left manubrium. The latter is markedly "crutch-shaped" in side view (fig. 21), though this is not noticeable in a dorsal or ventral view. The remainder of the internal organs call for no special mention.

Measurements.—Length of body without toes, 0.14 to 0.15 mm.; toes about 0.05 to 0.06 mm.

History.—This species was first described by Bory de St. Vincent in 1824, as *Diurella tigris*. Since the name *tigris* had been given by Müller to another species, Bory's name can not be retained for this species. It has been used, however, by many investigators since Bory's time. For a list of accounts of this animal under the specific name *tigris*, see the list given in the account of *Diurella tigris* Müller, above. Gosse (1851) described this animal as a new species, under the name *Monocerca porcellus*; this specific name *porcellus* is therefore the correct one to use, under the accepted rules of nomenclature. In Hudson & Gosse's Monograph (1889) Gosse founded a new genus, *Coelopus*, for this and a number of related species. As set forth in the general account of the taxonomy (p. 300), this genus was founded on a mistaken idea and was without justification, so that the species must be reunited with *Diurella*. Weber (1898) has given an extended description, supposedly of this species, under Gosse's name *Coelopus porcellus*. But rather curiously, his figures (pl. 20, figs. 2 and 3) do not represent this species, but afford an excellent picture of another species, which I have called *Diurella weberi* (q.v.). This has but one tooth in place of two at the anterior margin of the lorica. Weber's description introduces characters from both the species concerned. The two anterior dorsal teeth, of which he speaks (but which he does not figure) belong to *Diurella porcellus*, but the high ridge belongs to *D. weberi*. The fact that Weber had before him *D. weberi*, not *D. porcellus*, is perhaps the reason why he could not find the prominent substyles at the base of the toes, as represented by Plate and others, for these are much less prominent in the former species.

Distribution.—This species is one of the most common amid the vegetation of swamps, lakes, and streams. I have recorded its presence in the following localities: Lake St. Clair; Lake Erie (Put-in Bay Harbor, East Harbor, Long Point, Erie Harbor, and various other parts of Lake Erie); Crooked Lake, Newaygo County, Mich.; Old Channel, Charlevoix, Mich.; swamp on South Bass Island in Lake Erie; pools at Hanover, N. H.; Graveyard Pond, Presque Isle, near Erie, Pa.; Huron River at Ann Arbor, Mich.; ditch in the tamarack swamp region near Ann Arbor, Mich. Other observers have recorded it in America as follows: Ohio and Minnesota (Herrick, 1885, as *Diurella tigris*); Shiawassee River at Corunna, Mich. (Kellicott, 1888); Sandusky Bay, Lake Erie (Kellicott, 1896); waters connected with the Illinois River at Havana, Ill. (Hempel, 1898.)

Also taken in many parts of Europe, some of the more characteristic localities in Europe and elsewhere being: Germany (Plate, 1886); common in England (Gosse, 1889); Ireland (Glascott, 1893; Hood, 1895); near Basel, in Switzerland (Ternetz, 1892); Finland (Levander, 1894, as *Rattulus tigris*; Stenroos, 1898); near Kharkow, Russia (Scorikow, 1896).

In Natal, South Africa (Kirkman, 1901).

7. Diurella stylata Eyferth (1878) (pl. III, figs. 27–31).

Synonyms: *Rattulus bicornis* Western, 1893; *Coelopus similis* Wierzejski (1893) (?); *Rattulus bicornis* n. sp. Scorikow (1896); *Mastigocerca bicornis* Minkiewicz, 1900.

Distinctive characters.—This species is to be known by the two very slender, nearly or quite equal spines at the dorsal edge of the anterior margin of the lorica, by the conical form, and by the two short, unequal toes, the longest being little more than one-third the length of the lorica.

External features. The body is elongated conical, the thickest portion being near the anterior end or somewhat back of the anterior end (pl. III, figs. 27–30). Thence the body tapers regularly backward to the base of the toes. The head-sheath is set off from the remainder of the lorica by one or two marked constrictions. The head-sheath falls into many folds (fig. 30), when the head is retracted. At the anterior margin of the head-sheath, apparently a little to the right of the middle line, are the two long, slender spines which form the most characteristic features of this animal. These spines are nearly equal in length, though in most if not all specimens the right one is a trifle shorter than the left—a condition not found in any other species of the *Rattulidæ*. The length of the spines is usually about equal to the diameter of the lorica at its thickest point, though there is considerable variation. The spines are not absolutely fixed in position, but can be bent down over the corona for some distance when the latter is retracted. At times one of the spines may cross the other at its tip (fig. 28).

Extending backward from each spine is a ridge-like thickening, the two ridges including between them a narrow, transversely striated area (fig. 29). The entire area seems a little elevated above the general surface of the lorica, the side ridges being a little more elevated than the part between them.

Corona (fig. 31).—The corona, in its main features of the usual character, consists of the following parts: (1) Two semicircles of large cilia raised on elevations at the sides of the head (fig. 31, *a*). In many specimens when alive there is a prominent red spot at the dorsal or inner ends of these elevations (fig. 31, *r. s.*), almost as brightly red as the eye. (2) Two semicircles of cilia on slight elevations at the sides of the mouth (fig. 31, *b*). (3) A large, central dorsal, fleshy projection (*c*). (4) Above this (fig. 27, *c*) a smaller dorsal projection. (5) Two small short lobes, apparently crowned with cilia, at the sides (and ventrad) of the central dorsal projection (fig. 31, *d*).

Antennæ.—The dorsal antenna is situated in the striated area, a little back of the constriction separating off the head-sheath. The lateral antennæ are remarkably unsymmetrical in their position, recalling the condition found in *Rattulus cylindricus* Imhof. The left lateral antenna (fig. 28, *l. a.*) is far forward, considerably in advance of the middle of the body, while the right lateral antenna is far back, near the base of the foot (fig. 28, *r. a.*).

Foot.—The foot is very slender, conical in form, and so attached to the lorica that it may bend to the right, but not to the left. In other words, the posterior part of the body is constructed as if it had been twisted over to the left. The position of the toes is likewise such as would be explained by such a twist.

Toes.—The two toes lie very close together, as a rule, so that in a cursory examination they might be taken for one. They are unequal, the right toe being a little shorter than the left. The longest toe is about one-third the length of the lorica. In correspondence with the position of the foot, as set forth above, the right toe lies with its base above the left, further adding to the difficulty of distinguishing the two toes in a dorsal view. This position is such as would be attained by a twist of the posterior part of the body to the left, as mentioned in the account of the foot. At the base of the toes there are, according to Western (1894), three small substyles; these I have not been able to see.

Internal organs.—The brain is immensely enlarged, in some specimens extending farther backward than the middle of the body and·taking up a large share of the space within the lorica (see fig. 27, *br.*). In such cases the brain can be seen to be made up·of large cells, the outlines of which can be clearly traced (see the figure just referred to). If it is possible that this immense size is a transitory condition, not always present. The eye is attached to the dorsal surface of the brain, in front of the middle of the latter. In a dorsal view the eye underlies the thickening or ridge which runs backward from the base of the left anterior spine (fig. 29). The trophi are rather slender, and are somewhat unsymmetrical, the left manubrium being much better developed than the right (fig. 31). The mucus reservoir is divided longitudinally into two equal halves, one of which opens at the base of each toe. The other internal organs call for no special remark.

Measurements.—Total length, about 0.275 mm.; length of toes, 0.05 to 0.06 mm.; of anterior spines, 0.035 to 0.045 mm.

Movements.—*Diurella stylata* Eyferth swims in a rather wide spiral, in an awkward manner. The animal continually rotates over to the right as it swims, and at the same time it swerves continually toward the side which bears the spines; thus the spiral is produced.

When suddenly stimulated, as by swimming against an obstacle, or by the striking of some other organism against it as it swims, the animal usually reacts as follows: The cilia are partly withdrawn and the dorsal spines are bent down a little over the entrance to the lorica. If the stimulus is very strong the cilia are completely retracted and the animal remains quiet. If the stimulus is not so strong the cilia are only partly retracted and immediately begin operations again. But now they act in such a way as to turn the organism toward the side which bears the spines. The organism therefore swerves in the direction so indicated. This is, of course, the same direction in which the swerving occurs in the usual movement, only after a stimulus the swerving is more pronounced, so that the entire course of the animal is changed. For some time after the stimulation has occurred the swerving toward the side bearing the spines is much more marked than usual, so that the path followed becomes a much wider spiral.

History.—*Diurella stylata* was described by Eyferth in 1878. This description (1885, p. 111) was not clear in its account of the anterior spines, one of which he says arises from the "neck," while he seems to imply that the other rises from the brain. The folds in the anterior edge of the lorica he described as "one or two short spines" on the ventral side. Otherwise his description is good and his figure is at once recognizable as identical with the organism I have described above. Western (1893) redescribed this species as *Rattulus bicornis*. In his first description he described and figured the two toes as equal in length, a mistake which he afterward corrected (Western, 1894, p. 7). Scorikow (1896) described this species as a new one; by a rather curious coincidence he selected the same name (*Rattulus bicornis*) as Western had done. He also made the same mistake as Western in describing and figuring the two toes as equal. In the same year as Western, Wierzejski (1893) described what seems to be the same species under the name *Cœlopus similis*. In Wierzejski's figure the two equal anterior spines are shorter than usual, and the body is thicker and shorter, while the character of the toes is not clearly shown (nor described in the text). It is possible, therefore, that Wierzejski's species is not the same as *Diurella stylata* Eyferth.

Hood (1895) gives good figures and a description of this animal under the name *Rattulus bicornis*. Minkiewicz (1900) again describes this species as new, under the name *Mastigocerca birostris*.

Distribution.—This species is not common, though it sometimes occurs in large numbers in swampy ponds. I have recorded its presence in the following places: East Harbor, Lake Erie (near Sandusky, Ohio); pond near United States fish-hatchery, Put-in Bay, Ohio; Portage River, Ohio.

In Europe: England (Western, 1893); Ireland (Hood, 1895); Württemberg, Germany (Bilfinger, 1892); bayous of the Rhine (Lauterborn, 1898); Austrian Poland (Wierzejski, 1893, as *Cœlopus similis*); near Kharkow, Russia (Scorikow, 1896); Lake Bologoe, Russia (Minkiewicz, 1900).

8. Diurella rousseleti Voigt (pl. IV, figs. 37–39).

Synonym: *Cœlopus rousseleti* Voigt (1901)

Distinctive characters.—This small species is at once known by the nine projecting points at the anterior margin of the lorica, the upper right-hand one being a little larger than the others. The animal does not closely resemble other species of *Diurella*; its closest relatives seem to be *Rattulus stylatus* Gosse and *Rattulus pusillus* Lauterborn. But the shorter toe is about half the length of the longer one, so that it is necessary to place the animal with the *Diurellas*.

External features.—The small body is usually rather short and thick (fig. 37), slightly bent, and tapering backward in conical fashion to the toes. The proportions of the body vary considerably, as will be seen by comparing fig. 37 and fig. 39. The large head-sheath is marked off from the remainder of the lorica by a slight constriction. The entire circumference of the head is set with large projecting points or teeth, there being nine of these in all. They are somewhat larger on the right side than on the left, and the dorso-dextral one is a little larger than any of the others. These teeth are formed as projections of the plaits of the head-sheath, and are represented in much less pronounced form, as mere rounded projections, in some other species. The larger dorso-dextral tooth evidently corresponds to the single tooth of *Diurella tigris* Müller and other single-toothed species. Between the teeth the lorica in *D. rousseleti* forms furrows which are flexible. These fold when the head is strongly retracted, so that the teeth are brought into close contact. Between the two dorsal teeth is a somewhat larger furrow, which passes backward to the constriction which separates the head-sheath from the body. This furrow perhaps represents the "striated area" of other species.

Corona.—This bears, according to Voigt (1901), a very long central dorsal process, bent upward and showing wavy lines on its lower side.

Antenna.—The dorsal antenna is in the usual position on the dorsal side, near the constriction which separates off the head-sheath. The lateral antennæ have not been observed.

Foot.—The foot is a short, conical structure, of the usual form.

Toes.—There are two toes (fig. 38), the right one being about one-half the length of the left. The two toes are very close together, and the right one is very slender, so that it is easily overlooked; the impression is then received that the animal has but a single toe. The longer toe is about one-third the length of the body, or a little less than one-third. It is very slightly curved, the concave side of the curve being dorsal.

Internal organs.—According to Voigt (1901) there is a large red eye on the posterior end of the large brain. Trophi large, unsymmetrical.

Measurements.—Length of body, 0.095 mm.; of toe, 0.03 mm. Length of the long anterior dorsal process of the corona, when extended, 0.0195 mm.

History.—This species was described by Voigt, without a figure, in 1901. Through the kindness of Herr Voigt I have received a quantity of material containing specimens of the animal, and have thus been able to study it at first hand. The figures herewith given are the first published. Fig. 39 is due to Mr. Dixon-Nuttall; the others I have myself made.

Distribution.—Not yet been found in America. It should be looked for in small ponds.

In Europe: Plankton of the Schöh-See, Heiden-See, and Schluen-See, near Plön, Germany (Voigt, 1901). Also found in England.

9. Diurella sulcata Jennings (pl. II. figs. 24–26, and pl. XIII, figs. 113, 118, 119).

Synonyms: *Rattulus sulcatus* Jennings (1894); *Cœlopus brachiurus (?)* or *Rattulus cryptopus* Bilfinger (1894).

Distinguishing characters.—This species is to be known by the short, curved body, with two prominent constrictions about it: by the very short, equal toes, attached to a foot which is usually withdrawn into the body, and by the unarmed anterior edge of the lorica. It bears some resemblance to *Diurella brachyura* Gosse, but is distinguished from the latter by the equal toes, the constrictions, and the general form. It also resembles *Diurella caria* Gosse, but is much larger, and the posterior part of the lorica ends in an entirely different manner in the two species.

External features.—The body is nearly cylindrical and strongly curved, the dorsal line forming nearly an arc of a circle. The foot is usually retracted within the lorica (fig. 25), so that the ventral line meets the dorsal in a sharp angle, giving a very characteristic feature of this animal; when the foot is extended, however, as in fig. 24, this angle does not appear. Surrounding the body a little in front of the middle are two pronounced grooves, which separate off the anterior part of the lorica from the remainder. There are no pronounced teeth nor spines at the anterior margin of the lorica, though the dorsal edge projects a little farther than the ventral, so that this might perhaps be described as a very slightly marked tooth. On the dorsal surface of the lorica (apparently in the middle line, though this is very difficult to determine in an animal of this form, in which a dorsal view is rarely obtained) there is a shallow furrow, between two slight ridges, extending back about to the first transverse groove. This furrow is striated transversely. In some specimens this furrow seems scarcely to exist at all. The ridge to the right of the furrow is a little higher than the one to the left.

Corona.—The corona has the usual two sets of cilia—those about the outer edge and a small curve on each side of the mouth. In addition to these, the following structures may be distinguished: (*a*) A large fleshy dorsal process, pointed in side view (fig. 24), but nearly rectangular from above: (*b*) two small antenna-like processes, one on either side of and below the dorsal process; (*c*) a large, rounded, central projection of the coronal surface below the dorsal process (fig. 24).

Antennæ.—The dorsal antenna is in the median furrow, about halfway back to the first circular groove. The lateral antennæ, very minute, are in the usual position, on the posterior third of the lorica.

Foot.—The foot is scarcely distinguishable as a separate structure, since it is small and is habitually retracted within the lorica (fig. 25). It can be extended, however (fig. 24), and is a very short joint of the usual form.

Toes.—The two toes are equal in length, very short, and are usually concealed for half their length within the lorica, the tips projecting downward (fig. 25). When the foot is extended, the toes point forward (fig. 24.) Each toe is accompanied on its outer side by a substyle about one-third its own length.

Internal organs.—The mastax is very large, and contains large, well-developed trophi. These are unsymmetrical, the right manubrium being a mere slender bristle, much smaller than the left (fig. 26). The mucus reservoir is large, and divided by a longitudinal partition into two equal halves. The contractile vacuole is very small, lying above the mucus reservoir, on the right side of the intestine. The remainder of the internal organs call for no special mention.

Measurements.—Length of body without toes, 0.17 to 0.18 mm.; length of toes, 0.03 to 0.035 mm.

History.—This species was described by the present author in 1894 as *Rattulus sulcatus*. In the same year Bilfinger (1894) described and figured it, considering it to be possibly Gosse's *Cœlopus* (*Diurella*) *brachyurus*. Since that time it has been mentioned, with notes, by Stenroos (1898) and figured by Jennings (1901).

Distribution.—*Diurella sulcata* Jennings is very common in summer in the vegetation of our lakes. I have found it in the following localities: Lake St. Clair: Old Channel, Charlevoix, Mich.; West Twin Lake, 6 miles from Charlevoix, Mich.; Put-in Bay Harbor and East Harbor, Lake Erie. Kellicott (1896) found this species in Sandusky Bay, Lake Erie.

In Europe: Württemberg, Germany (Bilfinger, 1894): Lake Nurmijärvi in Finland (Stenroos, 1898).

10. Diurella cavia Gosse (pl. III. figs. 35 and 36).

Synonym: *Coelopus cavia* Gosse 1889.

Distinctive characters.—This species is to be recognized by the very small, plump body, without teeth at the anterior margin of the lorica, and the projection of the foot, so that the foot arises from the ventral surface of the body, and by the short, equal toes. It differs from *D. porcellus* Gosse and *D. intermedia* Stenroos in the absence of teeth at the anterior margin of the lorica and in the equal toes, together with differences in general form. From *D. brachyura* Gosse it differs in the short, thick body and the large, rounded backward projection over the foot and in the equal toes. From *D. sulcata* Jennings, its nearest relative, it differs in its much smaller size and in the great posterior enlargement projecting as a large, rounded protuberance over the foot, giving the animal an entirely different appearance from the last-named species.

External features.—The lorica is short and thick, arched dorsally and only slightly curved ventrally (fig. 36). The lorica projects backward as a large, hollow protuberance, extending considerably back of foot. There are one or two slight constrictions about the middle of the body, much as in *D. sulcata* Jennings, but less marked. The anterior margin of the lorica is without teeth. Extending back from the anterior margin to nearly the middle of the lorica, a little to the right of the middle line, is a depressed, striated area, its two edges being a little elevated.

Corona.—The corona bears the usual thick dorsal process; in other respects it has not been specially studied. The antennæ I have not seen.

Foot.—The foot is very small, scarcely noticeable as a separate joint. It is situated considerably in front of the posterior end, on the ventral surface.

Toes.—The two toes are equal in length, the length being somewhat less than the diameter of the lorica. In the specimen studied by the author the two toes extended backward and were crossed (fig. 35). This is doubtless by no means the rule. In Mr. Gosse's specimens the toes were turned forward, as in *D. porcellus* Gosse. There is a substyle at the base of each of the toes (not shown in the figures). The internal organs seem to offer nothing exceptional. In the single specimen at my disposal I was not able to make out the trophi.

Measurements.—Length of body without toes, 0.115 mm.; length of toes, 0.032 mm.

History.—This species was described as *Coelopus cavia* by Gosse in Hudson & Gosse's Monograph of the Rotifera (1889). It has not since been described or figured.

Distribution.—I have found but a single specimen of this species, from the northern swamp on Middle Bass Island, in Lake Erie.

In Europe: Epping Forest, England (Gosse, 1889); Ireland (Glascott, 1893); Austrian Poland (Wierzejski, 1893).

11. Diurella brachyura Gosse (pl. III. figs. 32–34, and pl. XIII. figs. 114 and 115).

Synonyms: *Monocerca brachyura* Gosse 1851; *Diurella rattulus* Eyferth 1878 and 1885; also Eckstein (1883); *Acanthodactylus rattulus* Tessin 1886; *Coelopus brachyurus* Hudson & Gosse 1889; *Rattulus palpitatus* Stokes 1896.

Distinctive characters.—This species is to be known by the small, curved body, less plump than in *D. porcellus* Gosse and *D. cavia* Gosse; the lack of teeth at anterior margin of lorica; the fact that the foot is not on the ventral surface; and the nearly equal toes, of length about equal to diameter of lorica. It is nearest to *D. cavia*, from which it is distinguished by the more slender body, tapering to the posterior end, and the fact that the foot is not on the ventral surface.

External features.—The body is cylindrical in form, much more slender than in *D. porcellus* Gosse, and tapers toward the posterior end. In extended specimens the thickest part of the body is the middle, the head region being a little narrower (fig. 32). The body is curved, so that the dorsal line forms nearly an arc of a circle. Together with the toes, which continue the curve of the body (when not bent up against the lorica), a full semicircle is thus formed. The head-sheath is not sharply set off from the remainder of the lorica, though a slight constriction between the two is evident. There are no teeth at the anterior margin of the lorica. The head-sheath may be folded longitudinally when the head is retracted, as in many other species. At such times one of the folds on the left extends a little beyond the others, forming thus a slight rounded, very inconspicuous, projection (fig. 33). This projection disappears when the head is fully extended. Usually no ridge is apparent, though on some specimens there is evidently a slight elevation of the lorica in the usual position of the ridge, to the right of the dorsal line. Careful examination of favorable specimens shows that a striated area, such as marks the ridge when it exists, is always present (fig. 33). This is broad, and extends back to about the middle of the body; it has two sets of the striations, meeting each other along a central rhaphe.

Corona.—The corona is of the usual character. It bears a single thick dorsal process (fig. 32), and apparently two very slender lateral processes, though of these I could not be quite certain.

Antennæ.—The dorsal antenna is in the striated area, a little behind the constriction setting off the head-sheath. The lateral antennæ are in the usual position, on posterior third of body.

Foot.—The foot forms a continuation of the tapering body at its posterior end, not being pushed forward on the ventral side, as in *D. caria* Gosse and *D. porcellus* Gosse. It is of the usual short conical form.

Toes (figs. 34 and 114).—The two toes are very nearly equal, the left being a very little longer than the right. Frequently the tip of the right toe lies against the left, but this is by no means always true. The longest toe is about equal in length to the diameter of the body. At the base of each toe on its outer side is a single substyle, about one-third the length of the toe.

Internal organs.—The trophi are of the usual character, the right manubrium being much reduced. The small contractile vacuole (fig. 32, *cv.*) lies above the mucus reservoir and contracts very rapidly (according to Stokes (1896) 40 times per minute). The rest of the internal organs call for no special remark.

Measurements.—Length without toes, 0.10 to 0.13 mm.; length of toes, about 0.03 mm.

History.—This species was described by Gosse in 1851 as *Monocerca brachyura*. Eyferth (1878) proposed, for a form which he said was much smaller than *D. stylata*, the name *Diurella rattulus*, but he gave no further account of the animal. Eckstein (1883) described and figured the animal under the name proposed by Eyferth. Tessin (1886) gave a few notes on the animal under the name *Acanthodactylus rattulus*. In Hudson & Gosse's Monograph (1889) this species was transferred to Mr. Gosse's new genus *Cœlopus*, receiving the name *Cœlopus brachyurus*. As this genus was based on an error, the species must of course go back to *Diurella*. Finally, Stokes (1896) described this as a new species, under the name *Rattulus palpitatus*, the specific name relating to the rapidity of the pulsations of the contractile vacuole. Stokes's description and figure apply in every detail to *D. brachyura*, so that there was no reason for giving the animal a new name.

Figures of this species have also been given by Jennings (1900 and 1901).

Distribution.—This species is not very common and seems as a rule to inhabit swampy ponds. I have recorded it from East Harbor, Lake Erie, near Sandusky, Ohio; from the Huron River at Ann Arbor, Mich.; from pools near Hanover, N. H.; and from marshy ponds on North, Middle, and South Bass islands, and on Presque Isle, all islands in Lake Erie. Kellicott (1888) reported its presence in the Shiawassee River at Corunna, Mich.; Stokes (1896, as *Rattulus palpitatus*) found it near Trenton, N. J.

In Europe: England (Gosse, 1889); Ireland (Glasscott, 1893); near Rostock, Germany (Tessin, 1886); Württemberg, Germany (Bilfinger, 1892, as *D. rattulus*); Finland (Levander, 1894, as *D. rattulus*).

12. Diurella dixon-nuttalli n. sp. (pl. IV, figs. 40 to 44).

Distinctive characters.—This species is to be known by the absence of teeth at the anterior margin of the lorica and by the two toes, one about two-thirds the length of the other. It is closely related to *D. brachyura* Gosse, from which it differs in the greater inequality of the two toes, as well as in general form. (Compare the figures of the two species.) From *D. sulcata* Jennings and *D. caria* Gosse this species differs in having unequal toes.

External features.—The body is nearly cylindrical, somewhat curved, and tapers toward the posterior end. The dorsal line is convex, the ventral line nearly straight, or concave. The head-sheath is set off from the remainder of the lorica by a constriction, and has a number of longitudinal folds, where it yields when the head is retracted. It is without teeth at its anterior edge. On the dorsal surface of the lorica a short furrow extends backward from the anterior margin to a point some distance behind the constriction which separates the head-sheath from the remainder of the body (fig. 40). This evidently corresponds to the striated area of other species.

Corona.—The corona bears the usual median dorsal club-shaped process, as well as a number of other prominences (fig. 40). Otherwise it seems to be of the usual character.

Antennæ.—The dorsal antenna is situated in the dorsal furrow mentioned above, a little behind the constriction which separates off the head-sheath. The lateral antennæ are in nearly the usual position on the posterior third of the body, but the left anterna is considerably farther forward than the right. (Compare figs. 41 and 44.)

Foot.—A short, thick joint, which can apparently be retracted within the lorica (see fig. 40).

Toes.—There are two toes, very close together. The longer left toe is about half as long as the body of the animal, while the right toe is about two-thirds the length of the left (fig. 43). There are two minute substyles at the base of the main toes (fig. 43).

Internal organs.—The prominent brain bears a large red eye at its posterior end. The trophi (fig. 42) are unsymmetrical, the right malleus being much reduced.

Measurements.—Total length, 0.15 to 0.18 mm., of which the toe forms about one-third.

History.—This species has not been described before. It was drawn by Mr. Dixon-Nuttall some years ago, and copies of his drawing have been distributed to many workers on Rotifera under the name *Cœlopus brachyurus*, but he agrees with me that this is not really the *Cœlopus brachyurus* of Gosse (see the account of *Diurella brachyura*), so that it is necessary to give it a new specific name. I name it therefore after the investigator who first figured it. The figures herewith presented (figs. 40–44) are all by Mr. Dixon-Nuttall, and my description is based upon them.

Distribution.—This species has not been found in America. According to Mr. Dixon-Nuttall it is common in ponds in England.

DESCRIPTIONS COMPILED FROM OTHER AUTHORS.[a]

13. Diurella sejunctipes Gosse (pl. xiv, figs. 120, 121).

Synonym: *Rattulus sejunctipes* Gosse (1886).

Distinguishing characters.—" Body projecting much above and behind the foot; toes, two, coequal, slender, decurved, set side by side, wide apart " (Gosse, 1889, p. 66). The body is said to be stout, plump, and curved; the foot is short and thick. Gosse described what is evidently the mucus reservoir as "a great basal bulb, wholly internal," forming part of the foot. The toes are two equal acute slender styles, so curved as to continue the outline of the body, and are wide apart at the base. The trophi were figured by Gosse "conjecturally."

Described by Gosse from notes by Dr. F. Collins. Found by the latter in a pool near Wellington Military College, Birks, England.

Stenroos (1898) found this species in Lake Nurmijärvi, in Finland, and gave a figure (fig. 121) and measurements. Length of body, 0.109 mm.; thickness, 0.03 mm.; length of toes, 0.03 mm.

Distribution.—As above and in Bohemia (Petr, 1890).

14. Diurella collaris Rousselet (pl. xiv, fig. 127).

Synonym: *Rattulus collaris* Rousselet (1896).

I give herewith Mr. Rousselet's description of this species in his own words:

"In shape the body is roughly cylindric, slightly curved behind; the lorica is finely pitted or stippled, giving it a roughened appearance; it has no dorsal ridge and is fairly stiff, except in the neck region, where the integument is more flexible and frequently forms a thickened collar when the animal is bending or retracting, and from this characteristic peculiarity the animal derives its specific name. The foot opening is oblique, nearly ventral, and the lorica overhangs the foot dorsally in a marked degree. The head is elongated, truncate in front, and somewhat tapering anteriorly, and it is furnished with a simple wreath of cilia; it contains a conical brain mass, with a red eye at the tip, and long jaws of the *Rattulus* type. The long, thin œsophagus is attached to

---

a *Diurella uncinata* Voigt. While this paper was passing through the press, Voigt published a brief diagnosis of a new species of *Diurella*, under the name *Cœlopus uncinatus* (Zoologischer Anzeiger, Bd. 25, 1902, p. 679). For the sake of completeness I append a translation of his description: "Body short, curved. Anterior edge of the lorica slightly denticulate. Somewhat to the right of the middle line, when the animal is viewed from the dorsal side, arises a long, rapidly narrowed, somewhat curved process. The short foot shows two unequal curved toes. Jaws unsymmetrical. A large red eye-spot. Length of the body without the process, 0.095 mm. Length of the largest toe 0.02 mm. Length of the frontal process, 0.027 mm. Occurrence. November, 1900 and 1901 in the Schluen-See and Schöh-See, amid *Potamogeton* and *Phragmites*. Specimens few."

The diagnosis is not accompanied by a figure. A full description, with figures, is promised for the forthcoming (ninth) Heft of the Forschungsberichte aus der Biol. Station zu Plön.

the antero-dorsal part of the mastax and widens into the large saccate stomach and intestine. Rounded gastric glands are attached to the anterior part of the stomach in the usual way. The ovary is an oval plate with large nuclei embedded in its granular substance, and it has generally a large maturing egg attached to it. Lateral canals, with flame cells attached, and a contractile vesicle are present. The dorsal antenna emerges from a small depression in the head just behind the tip of the brain, and the lateral antennae are situated in the lumbar region, on each side of the body. The foot emerges nearly ventrally; it consists of two short joints and is furnished with two very long, thin, narrow, glassy toes, about half the size of the body in length. The toes are nearly straight for about half their length, then they are decurved; one or two very small substyles are present at the base of each toe. In swimming the animal moves slowly, as if the small ciliary wreath were not powerful enough to move the comparatively large body, and I always found it at the bottom of my tanks among the sediment.

"Length: Total, with toes, $\frac{1}{3}$ inch (0.317 mm.); of body alone, $\frac{1}{12}$ inch (0.212 mm.); of toes alone, $\frac{1}{12}$ inch (0.105 mm.). Habitat, Sandhurst, Berks."

Stenroos (1898) found this species in Lake Nurmijärvi, in Finland, and gives a description and figure. Stenroos's specimens were larger than those of Rousselet, the body shorter and thicker, the projection of the lorica back of the foot larger, and the foot consisted of but a single joint, instead of two, as described by Rousselet. (This last-named difference probably arises merely from a variation in interpretation as to what should be called a "joint.")

15. Diurella helminthodes (Gosse (pl. xiv, fig. 122).

Synonym: *Rattulus helminthodes* (Gosse (1889).

Distinguishing characters.—"Body very slender, especially in front; the width less than one-fifth the length; toes without accessory styles at base; brain clear." (Gosse, 1889, p. 65.) This species was described by Gosse from a single dead specimen. He says that it approaches *Diurella tigris* Müller in form, in the slenderness and in the comparative length of the toes, but it is much more elongated and the anterior part especially is more slender than in *D. tigris* Müller. He thinks there is a low dorsal ridge, beginning insensibly near the middle of the length and ending in an oblique angle near the foot. Gosse thought that no substyles were present, but was not absolutely certain of this. Whether or not a tooth is present at the anterior edge, as in *D. tigris* Müller, Gosse does not say. Length to tips of toes, 0.25 mm.; of toes, 0.066 mm.; width and depth of body, 0.05 mm.

Glasscott (1893) lists this species from Ireland and states that the anterior part was of the same diameter as the posterior.

Scorikow (1896) has given a description of a rotifer which he identified doubtfully as this species, without a figure, but his account adds nothing of importance to that of Gosse.

Distribution.—Gosse (1889) found *D. helminthodes* in a pool near Birmingham, England; Glasscott (1893) in Ireland; Scorikow (1896) near Charkow, Russia; Wierzejski (1893) in Austrian Poland.

16. Diurella marina Daday (1889) (pl. xiv, figs. 123-126).

This species was described by its author in the Magyar language, so that I am unfortunately unable to make use of his description. His figures are reproduced in pl. xiv, figs. 123-126.

In a brief note in German, Daday (1890) says that *Diurella marina* most resembles *Diurella tigris* of Ehrenberg, but is distinguishable from it by the structure of the mastax and the peculiar border of the head-sheath of the lorica. What these peculiarities are must be judged from the figures. The figures do not show whether the toes are equal or unequal.

This species is marine and was found by Daday in the Bay of Naples.

17. Diurella brevidactyla Daday (1889) (pl. xiv, fig. 128).

This species, like the last, was described in the Hungarian language, so that I can not use the description. In a brief résumé Daday (1890) says that this species is distinguished from *D. marina* Daday by the simple anterior edge of the lorica, that its toes are very short, and that its mastax is different from that of *D. marina*. It is likewise a marine species and was found in the Bay of Naples.

II. RATTULUS Lamarck.

Generic characters.—One long toe, which is usually accompanied by another (the right toe), which is rudimentary, being not more than one-third length of main toe. The main toe usually more than half the length of body. Body cylindrical, oval, or ovoid; usually less curved than in *Diurella*.

DESCRIPTIONS OF THE SPECIES STUDIED BY THE AUTHOR

18. Rattulus gracilis Tessin (pl. v, figs. 45–49).

Synonyms: *Acanthodactylus gracilis* Tessin 1886; *Mastigocerca ternix* Gosse (1889).

Distinguishing characters.—This species is to be distinguished by the elongated, only *slightly* curved body, with the head-sheath sharply set off, and with many longitudinal folds, the main (left) toe about one-half to two-thirds the length of the body and the rudimentary (right) toe about one-third the length of the main one. Its nearest relative is *Diurella tenuior* Gosse, from which it differs in having the shorter toe only one-third the length of the longer, in the less curved form, and the head-sheath sharply set off from the body. From *Rattulus scipio* Gosse it differs in having a shorter main toe, with the right toe longer in proportion, and in the marked folds of the head-sheath when the head is retracted.

External features.—The body is elongated and shaped much as in *R. scipio*, save that it is a little more curved, the dorsal line being markedly convex, while the ventral line is nearly straight. In a dorsal view the sides of the body frequently appear nearly straight, as shown in fig. 46. There is a ridge on the dorso-dextral side, extending from the foot to the head. On the head-sheath the ridge is less prominent; it ends anteriorly in a minute tooth (figs. 45, 47). The ridge is transversely striated, the striations extending some distance to the left on the lorica. These striations are very inconspicuous, owing to the opaqueness of the internal organs, so that they can be seen only in especially favorable specimens.

The head-sheath is sharply set off from the rest of the body by a deep constriction, and is as a rule much narrower than the rest of the lorica. It has many longitudinal folds, by which it can be folded into very small compass and the anterior opening almost completely closed when the head is strongly retracted (fig. 48). These folds almost disappear when the head is unusually extended (fig. 45). The dorsal portion of the head-sheath projects considerably beyond the ventral portion when the head is strongly retracted.

On its right side, in the continuation of the ridge of the lorica, the head bears a single tooth (figs. 46–48). This is very minute, so that it is easily overlooked; it is not mentioned by Tessin (1896) nor Gosse (1889), though it was observed by Bilfinger (1894).

As to the general form, it is perhaps possible to distinguish two varieties of this species. Those which were sent me by Mr. Rousselet from Prescot, England, differed from the specimens found in America in the more slender body, perhaps a little curved, and with the head-sheath not so sharply set off from the rest of the lorica. This English form is shown in figs. 45 and 47, while American specimens are shown in figs. 46, 48, and 49. The differences do not seem to me sufficient to justify considering these different species. In other characteristics than those mentioned the specimens are alike.

Corona.—The corona bears a prominent dorsal process; otherwise it has not been minutely studied.

Antennæ.—The dorsal antenna lies to the left of the ridge, at the junction of the head-sheath with the rest of the lorica (fig. 46, *d. a*). The two lateral antennæ are in the usual position on the sides, on the posterior fourth of the body (fig. 46).

Foot.—The foot is short and thick as compared with that of *Rattulus scipio*, and the lorica does not project over it in a free edge on the left dorsal side, as in the last-named species.

Toes (fig. 46, 47).—The main or left toe is from one-half to two-thirds the length of the lorica. The smaller or right toe (*r. t.*) is about one-third the length of the main one, and its distal end lies across the latter. At the left side of the main toe is a large substyle, about one-half or more of the length of the right toe. On the outer side of the right toe is a similar but very minute substyle. The larger of the two mucus reservoirs is connected with the main or left toe; the smaller with the rudimentary right toe.

Internal organs.—These offer nothing of unusual interest. The stomach is usually large and very opaque, making it difficult to study the internal structure. The eye lies considerably to the left of the ridge. The jaws are rather weak and are unsymmetrical, the right manubrium being a short, very slender rod, while the left one is stout and very much larger.

Measurements.—Length of lorica, without toes, 0.17 mm.; of longest toe, 0.08 mm.; of shorter toe, 0.03 mm.; total length, 0.25 mm.

Movements.—*Rattulus gracilis* Tessin is a slow swimmer. As it moves through the water it revolves upon its long axis to the right, so that the path becomes a spiral. The dorso-dextral ridge is always directed toward the outside of the spiral. In other words, the animal swerves continually toward the ridge, the latter serving thus to cut the water. When stimulated suddenly, as by coming in contact with an obstacle, the animal swerves strongly toward the dorso-dextral side—that is, toward the ridge.

A specimen of this species was seen to feed upon a young specimen of *Diurella tenuior* Gosse. The jaws of *Rattulus gracilis* Tessin were extended far out (as in fig. 49) and seized the side of the prey; a piece of the *Diurella* was then torn out and devoured.

History.—This species was first described by Tessin (1886) under the name of *Acanthodactylus gracilis*. In 1889 Gosse described the same animal in the supplement to Hudson & Gosse's Monograph under the name *Mastigocerca iernis*. Bilfinger (1894) has given a better description of this animal than either Tessin or Gosse, but did not give a figure.

Distribution.—In America: East Harbor, Lake Erie; Graveyard Pond, on Presque Isle, near Erie, Pa.; Huron River at Ann Arbor, Mich. (abundant in *Ceratophyllum*).

In Europe: Near Rostock, Germany (Tessin, 1886); lakes in England (Gosse, 1889); Ireland (Hood, 1893); Württemberg, Germany (Bilfinger, 1894).

19. Rattulus scipio Gosse (pl. v, figs. 50–52; pl. XIII, figs. 111–112).

Synonyms: *Mastigocerca scipio* Gosse (1889); *Mastigocerca unidens* Stenroos (?) (1898); *Mastigocerca cuspidata* Stenroos (?) (1898).

Distinguishing characters.—This species is distinguished by the usually somewhat prismatic lorica, sometimes curved, widely open in front, with the head-sheath not sharply set off from the rest of the lorica; the single tooth near the anterior margin, and the single long toe three-fourths or more of the length of the lorica, accompanied by a short "substyle" (the right toe).

External features.—The lorica is elongated, often with nearly straight sides (fig. 50), though sometimes curved (fig. 111). In adult specimens (figs. 50 and 52) the diameter of the body is nearly uniform for three-fourths of the length, being very little narrower in the head region, and at the posterior end tapering in conical fashion to the foot. In young specimens (fig. 51) the largest part of the body is nearer the anterior end, and the lorica tapers thence regularly backward to the foot. The form of the lorica is not greatly changed in fully retracted specimens.

Considerably to the right of the middle line the lorica rises to a pronounced dorsal ridge, which aids much in giving the body a prismatic appearance. The ridge inclines sharply to the right and extends from the anterior edge fully three-fourths of the length of the body. It is marked with the usual transverse striations; these extend for a considerable distance to the left of the ridge (fig. 52). At the anterior end the ridge bears a tooth, which is fairly prominent though not large. The tooth is not at the very anterior margin of the lorica, but arises from a little behind this. In a retracted specimen (fig. 52) it projects slightly beyond the edge of the lorica, while in extended specimens (fig. 51) its tip may not reach the edge.

The head-sheath is not very sharply marked off from the rest of the lorica, though a slight constriction between the two may be detected, especially marked on the ventral side. The head-sheath does not show longitudinal folds or flutings, such as are prominent in *Rattulus longiseta* Schrank and *R. gracilis* Tessin, and does not constrict or change its form greatly when the head is fully retracted. This gives one of the most striking characteristics of this species. At the anterior edge the head-sheath flares a little (fig. 51) and the anterior aperture remains widely open, even when the head is retracted (fig. 52).

Corona.—The corona bears a thick, in dorsal view somewhat triangular, dorsal process.

Antennæ.—The dorsal antenna lies to the left of the ridge, at left edge of striated area. The two lateral antennæ are in the usual position, the right one being a little in advance of the left.

Foot.—The foot is rather sharply set off from the rest of the body. The posterior dorsal edge of the lorica projects on the left side some distance over the point of attachment of the foot, so that the latter can not bend to the left, but bends almost directly to the right (fig 52).

Toes.—The single main toe (representing the left toe of *Diurella*) is nearly or quite as long as the lorica: the right toe (figs. 50 and 111, *r. t.*) is rudimentary and small. The main toe apparently does not grow during the life of the animal, while the remainder of the body does, so that in a young specimen the toe is as long as the entire body (fig. 51), while in adult specimens (figs. 50, 52, and 111) it is only about three-fourths or less the length of the body.

Internal organs.—These offer nothing especially noteworthy. The trophi (fig. 51) are very unsymmetrical, the left manubrium being long, stout, and curved; the right one, a slender, straight rod about three-fourths the length of the left.

Measurements.—Length of adult body, 0.2 mm.; of toe, 0.15 mm.; total, 0.35 mm.

History.—This species was first described by Gosse in 1889, on page 61 of Hudson & Gosse's Monograph, vol. 2. Like many of Gosse's descriptions, the account of this species is somewhat inexact, the figure and description not agreeing in all points. It is on account of this inexactness in Gosse's descriptions that I have considered it justifiable to identify the species here described with that described by Gosse. The resemblance in general appearance and form of the body is great, as will be seen by comparing Gosse's figure with my fig. 51. But Gosse describes the animal as having three spines at the anterior margin of the lorica, each running back some distance as a sharp ridge. His figure shows but one of these spines, and no rotifer is known which would answer to this description. Gosse probably took the profile of the flaring edges of the lorica for two of the spines.

A more important difference is in the length of the toe. Gosse describes and figures the main toe as a little less than half the length of the lorica, while in the species here described it is much longer. Gosse's notes and figures are often inaccurate, however; for example, he states that in this species the mastax occupies more than half the body length, while in his figure it does not occupy one-third the body length. I have thought it best, therefore, to give this species Gosse's name, at east until one corresponding more exactly to Gosse's description is found.

No description or figure of this species, except that of Gosse, has been published.

Distribution.—In America: Put-in Bay Harbor and East Harbor, Lake Erie; Graveyard Pond, Presque Isle, near Erie, Pa.; near "The Cottages," Long Point, Canada, on north shore of Lake Erie.

In Europe: England (Hudson & Gosse, 1889); Ireland (Glascott, 1893); Württemberg, Germany (Bilfinger, 1892); Gr. Plöner See, Germany (Zacharias, 1893); near Basel, Switzerland (Ternetz, 1892); Lake Nurmijärvi in Finland (Stenroos, 1898); Bohemia (Petr, 1890). Also in Ceylon (Daday, 1898).

20. Rattulus macerus Gosse (pl. v, figs. 53, 54).

Synonyms: *Mastigocerca macera* Gosse (1889); *Mastigocerca fusiformis* Levander (1894).

Distinguishing characters.—This species is to be known by the elongated fusiform body (sometimes a little curved); the toe one-half to two-thirds the length of the body, *the short spur* (figs. 53, 54, *sp.*) *projecting backward from the base of the toe* when the latter is bent forward, and the single, very small and inconspicuous tooth at the right anterior edge of the lorica.

External features.—The body is elongated and fusiform, the dorsal surface much more convex than the ventral. In some specimens (fig. 53) the body is slightly curved. The head-sheath is marked off, as usual, by a slight constriction. It bears at its anterior margin, to the right of the dorsal middle line, a small, very inconspicuous tooth. This tooth is very easily overlooked, being hidden commonly by the fleshy head; it was not observed by Gosse or Levander. What corresponds to the ridge or striated area is not strongly marked; in contracted specimens (fig. 53) it may be noticed as a broad, elevated area extending backward from the region of the tooth. In fully extended specimens it can hardly be seen at all.

The corona has not been studied.

Antennæ.—The dorsal and lateral antennæ are in the usual positions, the former a little behind the constriction separating off the head-sheath; the latter on the posterior fourth of the lorica, at the sides.

Foot.—The foot is slender and cylindrical. It bears at its tip a spur, which is described in the account of the toes.

Toes.—The main toe is a nearly straight rod; in the adult (fig. 54) about half the length of the lorica; in a young specimen (fig. 53) it is about two-thirds as long as the lorica. There are two "substyles" (one doubtless representing the right toe); the longer of these is about one-fourth the length of the main toe.

The most peculiar feature of the foot and toes in this species is a spur-like point which extends backward from the distal end of the foot, at the base of the toes. It is shown in figs. 53 and 54 (*sp.*) as well as in the figure of this species given by Levander (1894). This spur is not found, so far as I am aware, in any other species. When the toe is extended straight back from the body the spur is not visible.

Internal organs.—The few specimens at my command did not permit a study of the trophi. Otherwise the internal organs seem to offer nothing worthy of special mention.

Measurements.—Length of adult without toe, 0.3 mm.; length of toe, 0.14 mm.; total, 0.44 mm.

History.—*Rattulus macerus* was described by Gosse (1889) from a single, partly disorganized specimen. He did not notice the anterior tooth on the lorica (which is very inconspicuous), and he describes and figures the lorica as thicker in its posterior part. But in his specimen the head and part of the internal organs had flowed out in a disorganized mass, leaving the anterior part of the lorica collapsed, so that it is natural that the posterior half should have been a little thicker than the anterior. Otherwise his description agrees well with the specimens I have found.

Levander (1894) described this animal as *Mastigocerca fusiformis* n. sp., and gave a very characteristic outline figure. His figure shows the spur at the base of the toe, but he did not notice the inconspicuous tooth at the right anterior margin of the lorica (his figure shows a view from the left side, where this would not be seen). It seems to me that there is not sufficient difference between the accounts of Gosse and Levander to justify considering them as describing different species. Gosse's description is a little less full than Levander's, though both give only brief general descriptions—neither of them mentioning (for example) the tooth or the very characteristic spur (though Levander shows the latter in his figure).

Scorikow (1896) describes this species without a figure. He considered *R. macerus* Gosse and *R. fusiformis* Levander to be the same, but incorrectly included *R. gracilis* Tessin as a synonym.

Distribution.—I found four specimens of this species in material taken from the marshy part of Lake Erie about "The Cottages," on Long Point, Canada. Gosse (1889) met with it in water from Woolston Pond, Hants, England; Levander (1894) in ponds and pools in Finland; Stenroos (1898) in Lake Nurmijärvi in Finland; Scorikow (1896) in a swamp near Kharkow in Russia.

21. Rattulus multicrinis Kellicott (pl. vi, figs. 55-58).

Synonym: *Mastigocerca multicrinis* Kellicott (1897).

Distinguishing characters.—This species is distinguished at once from all others by its broad, regularly ovate form. From *Rattulus latus* Jennings, the only species which resembles it at all in general appearance, it is markedly distinguished by the symmetrical form of the posterior part of the lorica. (Compare fig. 57 of *R. multicrinis* with fig. 65 of *R. latus*.)

External features.—The lorica is broadly ovate in dorsal or ventral view, widest in the middle region, narrowing a little in front to the capacious head-sheath and tapering rapidly and regularly behind to the foot. The form is remarkable for the almost complete lack of the asymmetry which is so striking in most of the *Rattulidae*.

In side view (fig. 56) the lorica swells out strongly on both the dorsal and ventral sides. There is in this species nothing comparable to the usual ridge or striated area.

The anterior part of the lorica or head-sheath is not sharply separated from the rest of the lorica, though there is a wide, shallow constriction in its base. The head-sheath is marked by numerous longitudinal folds and the anterior edge is crenate, each fold projecting a little as a rounded point. In the dorsal middle line there is a large, prominent triangular point projecting considerably beyond the rest of the lorica. When the head is retracted (fig. 58) the head-sheath becomes folded and the anterior opening nearly closed. The anterior part of the lorica then has a striking resemblance to the same portion in *Rattulus capucinus* Wierz. & Zach. (pl. vi, fig. 59) and *R. cylindricus* Imhof (pl. vii, fig. 62), indicating that *R. multicrinis* is closely related to these.

Corona.—This species shows the more complicated type of rattulid corona with especial clearness (figs. 56, 57). The following parts may be distinguished: (1) Two large half circles of cilia about the dorsal and lateral margins of the corona (a). (2) Two smaller arcs of cilia, one on each side of the mastax (b). When the mastax is protruded far out, it appears that these two arcs are actually borne on the tip of the mastax itself (fig. 48, b). (3) A long, blunt central process (e) borne on the dorsal margin of the corona. (4) A short, pointed central process (c) just below the last. (5) On each side of the last named two straight, slender processes (d), the inner one in each pair being pointed, the outer one blunt. All these processes were well described by Kellicott (1897). The corona resembles considerably that of *Rattulus latus* Jennings, and is still more like that of *R. capucinus* Wierz. & Zach., as described by Wierzejski & Zacharias (1893).

Antennæ.—The dorsal antenna is nearly or quite in the middle line, just in front of the line separating the head-sheath from the remainder of the body (fig. 55). The lateral antennæ are situated one on each side, about half way back from the middle of the body (fig. 55, *l. a.*).

Foot.—The foot forms a short cone, tapering rapidly to the toes.

Toes.—The main (left) toe is not quite so long as the body, and is nearly straight. The right toe is rudimentary, forming a short spine which lies obliquely across the base of the left or main toe. Between this and the main toe is a minute substyle, and there is a similar one at the left of the base of the main toe.

Internal organs.—The eye is attached to the large brain some distance in front of its posterior end. The trophi (fig. 57) are stout and almost symmetrical, a condition found in only a few of the *Rattulidæ*, but occurring in *R. capucinus* Wierz. & Zach., evidently the nearest relative of *R. multicrinis* Kellicott. The mastax can be protruded far out from the lorica, as shown in fig. 56. The remainder of the internal structure calls for no special remark beyond the statement that this species furnishes an excellent opportunity for a study of the characteristic internal organs of the *Rattulidæ*, these being particularly well displayed in the broad body of this animal.

Measurements.—Length of body, 0.18 to 0.20 mm.; of toe, 0.09 to 0.10 mm.; total, 0.27 to 0.30 mm.

History.—This species was described by Kellicott in 1897, from Sandusky Bay, Lake Erie. It has not been reported by anyone else until the present time.

Distribution.—*Rattulus multicrinis* Kellicott has thus far not been found elsewhere than in Lake Erie. I found it in East Harbor, Lake Erie; Kellicott (1897) in Sandusky Bay, Lake Erie.

22. Rattulus cylindricus Imhof (pl. VII. fig. 62–64).

Synonyms: *Mastigocerca cylindrica* Imhof (1891); *Mastigocerca setifera* Lauterborn (1893); *Mastigocerca hamata* Zacharias (1897); *Mastigocerca hamata*, var. *bologoensis* Minkiewicz (1900).

Distinguishing characters.—This species is to be distinguished by the median anterior curved hook which hangs down over the anterior opening of the lorica (not always visible); by the longitudinal folds of the head-sheath when the head is retracted; by the very long, prominent dorsal antenna (not always visible); by the nearly cylindrical body, usually highest a little in front of the foot; by the long toe, nearly or quite equaling, or sometimes exceeding, the length of the body, and by the habit of carrying the egg attached to the posterior end of the lorica.

External features.—The body is nearly cylindrical in form, but in many specimens it rises gradually toward the posterior end, its highest point lying just in front of the foot (fig. 62). Here the body falls off steeply to the foot (figs. 62 and 63). In some specimens, however, the body tapers gently backward to the foot (fig. 64). It was from such specimens, only still more slender than fig. 64, that Zacharias's species *Mastigocerca hamata* was described. The dorsal line shows in side view a characteristic slight depression just behind the dorsal antenna, rising again back of this region. The ventral line is very nearly straight.

The head-sheath is not sharply set off from the rest of the lorica, though there is usually a gentle, shallow constriction where the lorica passes onto the head. The head-sheath has longitudinal folds similar to those found in *Rattulus multicrinis* Kellicott and *R. capucinus* Wierz. & Zach. By means of these folds the anterior opening of the lorica can be quite closed (fig. 62). The median dorsal part of the anterior edge projects as a triangular point (as in the two species just mentioned), but in *R. cylindricus* Imhof the tip of this point is prolonged to form a hook, which bends downward over the anterior opening of the lorica (fig. 62, 64). This hook is thickened just distad of the place where it joins the lorica. *R. cylindricus* Imhof is distinguished by this hook from all other species of *Rattulidæ*. It is important to note, however, that when the head is fully extended this hook is frequently not visible; it seems to be either turned back or hidden by the cilia. Such a case is shown in fig. 63. This fact, if not carefully noted, is likely to lead to incorrect determinations of specimens not showing the hook. There is a slight furrow passing back from the base of the hook in the dorsal median line for about one-third the length of the lorica, and this furrow is marked with faint cross-striations. We have in this species, therefore, a slightly marked " striated area," which seems to be quite lacking in its nearest relatives, *R. multicrinis* Kellicott and *R. capucinus* Wierz. & Zach.

Corona.—The corona has not been studied thoroughly. The preserved specimens at my disposal did not permit of such study. There are two slender, lateral, antenna-like appendages, however, as seen in fig. 63.

Antenna.—The dorsal antenna is long and usually very prominent, as shown in figs. 62 and 63. In other cases it is merely a bundle of short, hair-like processes (fig. 64), while in still other preserved specimens I have not been able to see it at all. In these cases the antenna may have been injured.

Lauterborn (1893) described this species as new under the name *Mastigocerca setifera*, merely because Imhof (1891) did not mention the dorsal antenna in his original description. Lauterborn held that owing to the prominence of the dorsal antenna it could not have been overlooked if Imhof had really had this species before him. This is negatived by the fact, just stated, that specimens are often met with in which the antenna is inconspicuous or invisible. Minkiewicz (1900) described this same species anew—again without mention of the dorsal antenna—though specimens of his species, received through his courtesy, show the prominent antenna clearly. It is thus evident that because the antenna is not mentioned in a description one can not conclude that it is nonexistent nor even that it is inconspicuous. The name *Mastigocerca setifera* has therefore no foundation and must be considered a synonym of *R. cylindricus* Imhof.[a]

The lateral antennae, as Bilfinger (1894) has shown, are strikingly unsymmetrical in position. The left is on the flank, at about the middle of the length of the lorica (fig. 63), while the right is far back, almost exactly at the junction between the lorica and foot.

Foot.—The foot is very small, and not clearly marked off from the rest of the body.

Toes.—In this species the disproportion between the right and left toes has reached its maximum. The right is a mere, minute, scale-like bristle, hardly noticeable, while the left (forming the " toe proper") is a long, straight rod, almost or quite as long as the entire body of the animal. There is a small substyle on the outer side of the main toe, nearly as long as the rudimentary right toe (fig. 64). The latter lies, as usual, across the base of the main toe.

Internal organs.—The eye is situated at about the middle, or a little behind the middle, of the long brain (fig. 64). The trophi have not been thoroughly studied. The specimens which I have had at hand have not shown these clearly. According to Bilfinger (1894), they are nearly symmetrical. The ovary (fig. 64, *or.*) may be seen to be connected behind with the cloaca. The egg is carried in this species attached to the posterior part of the lorica, above the foot (fig. 62). No other species of the *Rattulidæ* is known which thus carries the egg with it. The other internal organs call for no special remark.

Measurements.—Length of body, 0.26 mm. to 0.31 mm.; of toe, 0.23 mm. to 0.32 mm.; total, 0.49 mm. to 0.63 mm.

History.—This species was described briefly, without a figure, by Imhof (1891). In 1893 Lauterborn redescribed it, at first identifying it with Imhof's species, but in a postscript to his paper giving it a new name (*Mastigocerca setifera*), because Imhof had failed to mention the prominent dorsal antenna. (See the account of the dorsal antenna above.) The first figure of this species was given by Bilfinger (1894), together with a good description. Zacharias (1897) redescribed the animal under the name *Mastigocerca hamata*. The specimen figured by Zacharias shows a more slender form than any of the other figures given, and the body slopes even more gradually to the foot than in my fig. 64. But as these points are clearly very variable (compare fig. 62 and fig. 64), and Zacharias's specimens agreed in other points with this strikingly characterized form, especially in the hook and the very long toe, it seems beyond doubt that his species is the same as *R.*

---

[a] If this were not done, we should be forced to the absurdity of identifying as *R. cylindricus* those specimens in which for any reason we could not see the antenna, while others would receive the name *R. setifera*.

cylindricus Imhof. Finally, Minkiewicz (1900) redescribed and figured this species under the name *Mastigocerca hamata* var. *bologoensis.*

Distribution.—I have found this species in East Harbor, Lake Erie, near Sandusky, Ohio. Specimens were also sent me by Prof. E. A. Birge from inland lakes in Wisconsin.

In Europe this species has been recorded as follows: Bayous of the Rhine (Lanterborn, 1893); Württemberg, Germany (Bilfinger, 1894); Lake Bologoe in Russia (Minkiewicz, 1900); a pond in Germany (Zacharias, 1897).

23. Rattulus capucinus Wierzejski & Zacharias (pl. vi, figs. 59–61).

Synonyms: *Mastigocerca capucina* Wierzejski & Zacharias (1893); *Mastigocerca hudsoni* Lanterborn (1893).

Distinguishing characters.—This species is at once distinguished from its nearest relative, *R. multicrinis* Kellicott, by the elongated, cylindrical form of the body. From all other known species it is distinguished by the large triangular projection of the lorica above the head, a character which it shares with *R. multicrinis* Kellicott alone.

External features.—The body is an elongated cylinder, somewhat curved toward the ventral side, as shown in fig. 60. There appears to be considerable variation in the proportions of the body. Those studied by the author (from Germany, obtained through the kindness of Mr. C. F. Rousselet) were of the proportions shown in figs. 60 and 61; but the figures of Wierzejski & Zacharias (1893) show a much shorter animal, while the figure of Lanterborn (1893) is still shorter, and the ventral surface forms almost a straight line. The specimens which I have examined agree more nearly in their proportions with those found by Levander (1894) and Stenroos (1898) in Finland. The lorica seems to have nothing which can be compared with the ridge or the striated area in most *Rattulidæ.*

The head-sheath is set off by a marked constriction from the remainder of the lorica. There are many longitudinal seams at which the sheath folds when the head is retracted (fig. 59). Between these seams the parts of the head-sheath project at the anterior margin, so that the edge is crenate. The dorsal part of the head-sheath runs out to a strong triangular point, projecting far over the ventral edge of the lorica. This gives the retracted head the appearance of a capucin cap, whence the specific name. The whole structure is almost identical with that of *R. multicrinis* Kellicott.

Corona.—The corona, according to Wierzejski & Zacharias (1893), is very similar to that described above for *R. multicrinis* Kellicott. There are two central antenna-like processes, the more dorsal one being longer, and two lateral processes on each side. There are likewise two dorsolateral semicircular wreaths of cilia. These are shown in fig. 61.

Antennæ.—The dorsal antenna has not been observed in this animal. The lateral antennæ are in the usual position on the flanks, about half way back from the middle of the body (fig. 60).

Foot.—The foot is a short, conical structure. It is overhung on its dorsal surface by a roof-like backward projection of the lorica (fig. 60).

Toes.—The main (left) toe is a nearly straight rod about half the length of the body. The rudimentary right toe is one-fourth to one-third the length of the main toe, and lies across the proximal part of the latter. A small, scale-like substyle lies against the side of the main toe.

Internal organs.—These call for no special remark, save in the case of the trophi. These, according to Wierzejski & Zacharias (1893), are not unsymmetrical.

Measurements.—Length of body, 0.30 mm.; of toes, 0.125 mm.; total, 0.425 mm.

History.—This species was described in 1893 by Wierzejski & Zacharias, under the name *Mastigocerca capucina.* In the same year Wierzejski (1893) gave a description (in Polish) and repeated the figures given by Wierzejski & Zacharias (1893), while Zacharias (1893) also gave a brief description and a new figure. In the same year Lanterborn (1893) described this animal as *Mastigocerca hudsoni.* A figure and brief description were given by Levander in 1894, and notes by Stenroos (1898).

Distribution.—In America: Lake St. Clair and West Twin Lake, near Charlevoix, Mich.

In Europe: Gr. Plöner See, in Germany (Zacharias, 1893); bayous of the Rhine (Lanterborn, 1893); Württemberg, Germany (Bilfinger, 1894); Galicia, Austro-Hungary (Wierzejski, 1893); Lake Nurmijärvi in Finland (Stenroos, 1898); Lohijärvi-See in Finland (Levander, 1894); River Oudy near Kharkow, Russia (Scorikow, 1896); Lake Bologoe in Russia (Minkiewicz, 1900).

24. Rattulus longiseta Schrank (pl. VIII, figs. 67-72).

Synonyms: *Brachionus rattus* Schrank (1793): *Vaginaria longiseta* Schrank (1802); *Monocerca bicornis* Ehrenberg (1830, 1838); *Monocerca cornuta* Eyferth (1878); *Acanthodactylus bicornis* Tessin (1886); *Mastigocerca bicornis* Hudson & Gosse (1889).

Distinguishing characters.—The characteristic features of this animal are the two long spines at the anterior dorsal edge of the lorica. Of these the right is much longer than the left. The only species which at all resembles this is *Rattulus roseus* Stenroos, which is said to have the spines at the ventral anterior margin instead of the dorsal (but see the account of that species, p. 341).

External features.—The body is usually fusiform in shape; when well extended it is elongated, widest at about the middle or a little in front of the middle, and tapering thence regularly backward to the foot. But the form of the lorica varies greatly with the degree of extension of the animal, as well as with the age of the individual. Young specimens are often broadest near the anterior end (especially when the head is retracted, fig. 68), and the body is slender and tapers rapidly to the foot. The lorica is flexible and permits great changes of form in one and the same individual. When strongly retracted the animal is shorter and thicker, and the body becomes almost oval in form (fig. 70). The anterior portion of the lorica or head-sheath is not distinctly marked off from the rest in this species, though a notch at the point of separation can usually be detected on the ventral side when the animal is retracted (fig. 70).

The lorica is marked on the dorsal surface a little to the right of the middle line by a shallow longitudinal furrow, passing backward from the anterior end to about the middle of the length of the body (fig. 67). The direction of the furrow is slightly oblique, its anterior end lying a trifle farther to the right than its posterior end. The furrow is marked by transverse striations, really muscle fibers, attached within the two ridges which form the boundaries of the furrow. These two bounding ridges project at the anterior margin of the lorica as two long spines, forming the most characteristic feature of this animal. The right one of the two spines is the longer, usually twice as long or more than twice as long as the left. (Weber, 1898, figures the left spine as the longer, and Gosse, 1855, states that the left spine is the longer in this species. It is, of course, possible that there is variation in this matter, but I examined a large number of preserved specimens with this matter in mind, and found that in all cases the right spine was longer.)

The head-sheath has longitudinal plaits or flutings, where folding takes place when the head is retracted. The anterior margin of the lorica differs exceedingly in the contracted and expanded conditions. In a fully extended living individual (fig. 67), the anterior part of the lorica is wide open, and the margin shows, in addition to the two long dorsal teeth, four or more minute points, lateral and ventral. There is usually no trace of the longitudinal folds so prominent in the retracted individual. In retracted specimens (figs. 68, 70), on the other hand, the anterior opening is much smaller, and many ridges and grooves are visible, owing to the folding of the lorica. Each of the longitudinal ridges runs out to form a small point or tooth, so that the anterior margin seems to bear many teeth.

Corona.—The corona bears a large dorsal frontal process: otherwise it has not been thoroughly studied.

Antennæ.—The dorsal antenna lies within the striated furrow, a short distance from the anterior end (fig. 67). The two lateral antennæ are in the usual position, one on each side, about one-fourth the body length in front of the foot. In specimens where an exact comparison between the two was possible, the left antenna was situated a little anterior to the right.

Foot.—The foot is a short, conical structure, attached to the body only slightly obliquely, so that its movement is not so nearly limited to a turning to the right as we find it to be in many of the *Rattulidæ.*

Toes.—The right toe (fig. 69, *r. t.*) has nearly disappeared, so that it is customary to speak of the left one as *the* toe, while the right is classed merely with the substyles. The main toe is usually about two-thirds the length of the body. The substyles are small scales, one of which lies on each side of the main toe and the rudimentary right toe. The latter lies a little above the main toe, with its tip against it.

Internal organs.—The eye is attached to the brain, and in a dorsal view lies usually considerably to the left of the dorsal furrow on the lorica. The trophi (figs. 71 and 72) are unsymmetrical, the right malleus being much more slender than the left. (For a full description of the trophi in this species see the general account of the trophi, p. 289.)

Measurements.—Length of body without toe or anterior spines, 0.30 mm.; length of toe, 0.20 mm.; length of longest anterior spine, 0.06 mm.; total, 0.56 mm.

History.—This is one of the best known of the *Rattulidæ* and has, ever since the time of Ehrenberg, gone under the specific name *bicornis*, though Ehrenberg admitted that Schrank's name *longiseta* was the first one given.

The animal was first described by Schrank in 1793. He confounded it at that time with *Rattulus rattus*, and gave it, therefore, the name *Brachionus rattus*. In 1802 Schrank recognized the distinction between this and *Rattulus rattus*, and gave the present animal the specific name *longiseta*, placing it, along with a heterogeneous group of organisms, in the genus *Vaginaria*. The specific name *longiseta* must, according to the rules of priority, be used for this animal in place of Ehrenberg's name *bicornis*. Ehrenberg (1830) recognized this animal as Schrank's species, but changed the name to *bicornis*, because he thought this name more appropriate than *longiseta*. This proceeding is, of course, not a justifiable one according to the rules of nomenclature.

The only synonym which has been added for the specific name since the time of Ehrenberg is the *Monocerca cornuta* of Eyferth (1878). There can be no question, it seems to me, that this is the same species as *Rattulus longiseta* (*Monocerca bicornis*). Eyferth himself seemed of that opinion, saying that even if this is the same species as Ehrenberg's *bicornis*, the name must be changed to *cornuta*, and giving the new name *cornuta* with a mark of interrogation. The new name was based upon the number of teeth at the anterior edge of the lorica, but Eyferth's account (1885) of these agrees very closely with what may be observed in *Rattulus longiseta*, save that he considered one of the two long spines to be an antenna, an error similar to that which he made in the case of *Diurella stylata* (q. v.).

Eyferth says that in *Monocerca cornuta* there is a dorsal ridge, ending in a spine, and that at the sides and on the "chin" at the anterior edge there are two pairs of smaller points, statements which are true for *R. longiseta*, as shown in fig. 67. He did not recognize the doubleness of the ridge nor of the spine, errors of a character which were frequently made at that time and which occur repeatedly in Eyferth's account of the *Rattulidæ*. Hudson & Gosse (1889, Supplement, p. 35) have made quite unnecessary difficulties for the recognition of *Monocerca cornuta* Eyferth as *Rattulus longiseta* (*Monocerca bicornis* Ehrenberg) by altering and adding to Eyferth's description, though their account is supposedly taken from that of Eyferth. They say that the two lateral teeth are half the length of the dorsal spine, though Eyferth makes no such statement. In Eyferth's figure the exact position of the real anterior edge of the lorica is not discernible, so that the relative length of the spines can not be judged from this. Hudson & Gosse add "no substyles," though Eyferth states exactly the contrary. A comparison of fig. 68 of the present paper with Eyferth's fig. 24, Taf. VII (1885), will show at once how such a figure as that of Eyferth could be made from the present species.

The following figures or descriptions of this animal have been given (doubtless the list could be increased): Schrank (1793, 1802, 1803); Ehrenberg (1830, 1838); Dujardin (1841); Perty (1852); Leydig (1854); Pritchard (1861); Bartsch (1870, 1877); Eyferth (1878, 1885); Blochmann (1886); Tessin (1886); Hudson & Gosse (1889); Bilfinger (1892); Bergendal (1892); Glasscott (1893); Wierzejski (1893); Eckstein (1895); Scorikow (1896); Stenroos (1898); Weber (1898); Jennings (1900, 1901).

Distribution.—In America: *Rattulus longiseta* Schrank is common amid plants in lakes and streams, though it rarely occurs in large numbers. I have recorded its presence in the following places: Put-in Bay Harbor and East Harbor, Lake Erie; marsh on the shores of Lake Erie at "The Cottages," Long Point, Canada; Huron River at Ann Arbor, Mich.; Lake St. Clair; Chippewa Lake, Mecosta County, Mich.; Round Lake and Pine Lake at Charlevoix, Mich; pools at Hanover, N. H. Other observers have recorded it as follows: Pond near Bangor, Me. (J. C. S., 1883); waters connected with the Illinois River at Havana, Ill. (Hempel, 1898).

In Europe (only typical localities given): Germany (Ehrenberg, 1838, and many other authors); England (Gosse, 1889); Ireland (Glasscott, 1893); Hungary (Bartsch, 1877); Greenland (Bergendal, 1892); Bohemia (Petr, 1890); Russia (Scorikow, 1896); Finland (Levander, 1894); Tyrol (Dalla Torre, 1889); Austrian Poland (Wierzejski, 1893); Switzerland (Weber, 1898); Roumania (Cosmovici, 1892).

25. Rattulus bicristatus Gosse (pl. IX, figs. 77-80).

Synonym: *Mastigocerca bicristata* Gosse (1889).

Distinguishing characters.—This very large species is distinguished from all others by the two high dorsal ridges passing from the anterior end backward for about three-fourths the length of the lorica. From *Rattulus mucosus* Stokes, the only other species which has two prominent ridges, it is distinguished by the greater height and length of the ridges; these in the last-named species extend only about one-half the length of the lorica, or less, and they are much lower than in *R. bicristatus*. The two species differ in many other respects also.

External features.—The body in side view (fig. 77) is oblong, two to three times as long as wide, the dorsal line forming a nearly regular arch from head to foot, the ventral line less convex and notched at the junction of the head-sheath with the rest of the lorica. The body is not so thick from side to side as it is dorso-ventrally, a dorsal view (fig. 78) showing an oblong form, the length about three times the width.

The whole appearance of the animal is dominated by the two great longitudinal ridges. These begin at the anterior end, some distance apart, and extend backward and a little to the left, ending about one-third the length of the body from the beginning of the foot. The ridges are high, thin at the edges, and grow thicker toward their bases. They inclose between them a wide V-shaped trough (fig. 78).

Within the ridges are broad, well-defined bands of muscle fibers, passing from the upper part of each ridge to the floor of the furrow between them. These bands evidently correspond to the transverse striations occurring in other species; the fibers are not usually grouped into distinct bands, as they are in *R. bicristatus*.

The head-sheath is marked off from remainder of lorica merely by a slight constriction, largely confined to ventral side. The ridges continue on head-sheath to anterior margin. There is a slight notch between the ends of the two ridges (fig. 78), and a very slight one on the ventral side.

Corona.—The corona is of the usual character. It bears two slender lateral processes (fig. 77); a medial dorsal process has not been observed.

Antennæ.—The dorsal antenna is situated at the bottom of the groove midway between the two ridges (fig. 78). In the retracted specimen the eye appears just in front of it. The lateral antennæ (fig. 78) are in the usual position, a short distance in front of the base of the foot.

Foot.—The foot is rather large, conical, and so attached to the lorica that it can bend to the right and ventrally, but not to the left nor dorsally.

Toes.—There is a single, very long, curved toe, accompanied by numerous substyles (fig. 79). One of these substyles is longer than the others and may represent the rudimentary (right?) toe. But the primitive arrangement seems entirely lost, so that it is not possible to demonstrate this or to show certainly which of the two original toes is represented by the main one. The length of the main toe is in young individuals equal to or greater than that of the body; in larger specimens the toe is somewhat shorter than the body. As many as eight substyles can be counted, in favorable specimens, at the base of the main toe; possibly the number is still greater. They are much more prominent than in most species, some of them standing at a considerable distance from the base of the toe (fig. 79).

Internal organs.—The eye is situated sometimes at the middle of the brain, sometimes at its posterior end. It lies beneath the groove between the two ridges. The trophi are very large and strong, and bear many teeth. The left side is less developed than the right (fig. 80).

Rattulus bicristatus Gosse, owing to its great size, is unusually favorable for a study of the viscera, but there are no special features to add to the description given in the general account of the anatomy of the *Rattulidæ* (p. 288).

Measurements.—Length of body, 0.25 to 0.30 mm.; of toes, 0.24 to 0.25 mm.; total, 0.49 to 0.55 mm. Greatest height of body, 0.10 to 0.14 mm. This species was described by Gosse (1889) in the supplement to Hudson & Gosse's Monograph of the Rotifera. Figures or descriptions of it are also given by Glasscott (1893, poor), Stenroos (1898), and Jennings (1900 and 1901).

Distribution.—In America: This species is not uncommon amid the vegetation of rivers, lakes, and ponds. I have found it in the following localities: Put-in Bay Harbor and East Harbor, Lake

Erie: West Twin Lake near Charlevoix, Mich.; Huron River at Ann Arbor, Mich. Hempel (1898) records it from waters connected with the Illinois River at Havana, Ill.

In Europe: Dundee, Scotland (Gosse, 1889); Ireland (Glasscott, 1893; Hood, 1895); Württemberg, Germany (Bilfinger, 1894); Basel, Switzerland (Ternetz, 1892); Lake Nurmijärvi in Finland (Stenroos, 1898); Galicia, Austro-Hungary (Wierzejski, 1893).

26. Rattulus mucosus Stokes (pl. x, figs. 86-91).

Synonyms: *Mastigocerca mucosa* Stokes (1896); *Mastigocerca rectocaudatus* Hilgendorf (1898)?

Distinguishing characters.—This species shares with *Rattulus bicristatus* Gosse the peculiarity of having two well-marked dorsal ridges with a furrow between them. But in *R. mucosus* Stokes the ridges are lower and extend back only to about the middle of the length of the lorica; the entire animal is considerably smaller and of a different form and there are many other points of difference. *Rattulus mucosus* bears much resemblance in general form to *R. rattus* Müller, *R. carinatus* Lamarck, and *R. lophoessus* Gosse, but the presence of the two ridges distinguishes it at once from these.

External features.—In side view (fig. 87) the body of an adult specimen is broadly oblong, the length being little more than twice the depth. The head-sheath is marked off from the remainder of the lorica by a slight constriction, most marked on the ventral side. The anterior margin is without teeth or spines. On the ventral side there is, when the head is retracted, a deep, narrow, longitudinal fold, which looks like a gap (fig. 89), and on the dorsal side there is a slight notch between the two ridges.

The two ridges extend from the anterior edge backward and a little to the left to a point a little behind the middle of the lorica (fig. 86). They are not so high as in *R. bicristatus* Gosse, and the furrow between them is not so wide. For this reason the two ridges may appear in side view as but a single one (figs. 87, 88), one completely hiding the other, or they may be entirely overlooked. They are marked with transverse striations, similar to those of *R. bicristatus* Gosse.

In young specimens (fig. 90) the body is narrower behind than in adults and the toe is longer in proportion to the length of the body.

Corona.—The corona bears the usual wreaths of cilia about its outer margin, two small arcs at the sides of the mouth, and three antenna-like processes. One of these is dorsal and club-shaped; the other two are very slender lateral rods.

Antennae.—The dorsal antenna is situated, not within the groove between the ridges, as in *R. bicristatus* Gosse, but to the left of the left ridge. It projects from a rounded depression on the left side of this ridge. (This depression, though not the antenna, is indicated in fig. 86.) The lateral antennae occupy the usual position on the sides, considerably behind the middle of the lorica.

Foot.—The foot, of the usual short, conical form, is joined to the body in a more unsymmetrical manner than usual. The lorica projects far over it on the left side (fig. 86), but not on the right. Thus the foot and toe can be bent to the right side (fig. 88), but not to the left.

Toes.—There is a single, long, main toe, accompanied by three (possibly more) substyles. The rudimentary right toe is here hardly distinguishable from the substyles. The main toe is nearly straight, and is frequently carried for long periods bent up against the right side (as in fig. 88); the animal then swims about as if it had no toe. In a young specimen (fig. 90) the toe is about as long as the body; in older specimens it is shorter than the body.

Internal organs.—These offer nothing of especial interest, except in case of the trophi (fig. 91). These are very massive, but the right manubrium has almost disappeared, persisting merely as a short, slender rod. The trophi are thus more unsymmetrical than is usual in the genus *Rattulus.*

Measurements.—Length of body, 0.18 to 0.20 mm.; of toe, 0.12 to 0.15 mm.; total, 0.30 to 0.35 mm.

History.—This species was first described by Stokes, in 1896, as *Mastigocerca mucosa.* It had been observed by various investigators and referred, doubtfully, to *Rattulus bicristatus* Gosse. Thus, Jennings (1894, p. 19) and Kellicott (1897, p. 50) mention finding a species which has two ridges, but does not agree with accounts of *R. bicristatus* Gosse. In addition to the description and figure of Stokes, this species has been figured by Jennings (1900 and 1901).

Hilgendorf (1898) describes as *Mastigocerca rectocaudatus* a species which resembles the present one in many respects. Hilgendorf does not mention the two ridges, but says there is a "median dorsal cleft" which is especially noticeable when the head is withdrawn; this may well have been the furrow between the two ridges. It would be very easy to interpret the structure as merely a furrow or cleft, the ridges being considered merely the sides of the cleft. Hilgendorf's figures are not detailed, so that it is difficult to be certain of the identity of his species; it certainly resembles the present one, and his account hardly justifies the founding of a new species.

Distribution.—This species is one of the most abundant of the Rotatoria amid the vegetation of the shallower parts of the lakes. I have found it in the following places: Lake St. Clair; Chippewa Lake, Mecosta County, Mich.; Crooked Lake, Newaygo County, Mich.; Round Lake at Charlevoix, Mich.; pond at Hanover, N. H.; Huron River at Ann Arbor, Mich., and at the following stations on Lake Erie: Put-in Bay Harbor, East Harbor, Long Point (Canada, near "The Cottages").

Stokes (1896) found this animal in a pond near Trenton, N. J.; Kellicott (1897, under the name *Mastigocerca bicristata*) in Sandusky Bay, Lake Erie.

New Zealand (?), Hilgendorf (1898) as *Mastigocerca rectocaudatus*.

This species has not yet been recognized in Europe.

27. Rattulus carinatus Lamarck (pl. XI, figs. 95–97).

Synonyms: *Trichoda rattus vesiculam gerens* Müller (1786); *Rattulus carinatus* Lamarck (1816); *Trichocerca rattus* Goldfuss (1820); *Monocerca longicauda* Bory de St. Vincent (1824); *Mastigocerca carinata* Ehrenberg (1830, 1838), and most subsequent authors; *Monocerca carinata* Eyferth (1885); *Acanthodactylus carinatus* Tessin (1886).

Distinguishing characters.—This species is at once known by the high, thin keel or ridge extending somewhat more than half the length of the lorica. It differs from *R. lophoessus* Gosse in the fact that the ridge does not reach the entire length of the lorica. There are also differences in the toes and in other features.

External features.—The body is a long oval, widest near the middle and tapering toward both ends. The dorso-ventral and lateral diameters are nearly the same (compare figs. 95 and 97). The part of the lorica enveloping the head is marked off from the remainder by two slight constrictions. The anterior edge of the lorica is quite unarmed, and forms a gentle curve, bounding the anterior opening, without teeth, angles, or notch. In the ventral region the curve forms a slight shallow concavity (fig. 97); at this point the lorica can be folded inward when the head is strongly retracted, forming what appears to be a deep, narrow gap.

The most striking feature of this organism is the very high, narrow ridge. This begins at the anterior end, considerably to the right of the median line, and extends obliquely backward and to the left (fig. 95), stopping a little behind the middle of the body. The ridge is inclined strongly to the right and is marked with transverse striations. These striations appear to be muscle fibers passing from the top of the ridge (at its right edge) to the left and downward. When the ridge is seen from the side the ends of the fiber bundles show the arrangement given in fig. 97. The striations extend on the surface of the lorica some distance to the left of the ridge (fig. 95).

Corona.—The corona has the usual two arcs of cilia about its dorso-lateral margin and two at sides of mouth. There is a short, thick median dorsal process and two slender lateral ones (fig. 97).

Antennæ.—The dorsal antenna lies on the left side of the ridge, just behind the head-sheath (fig. 95). The lateral antennæ are in the usual position on the posterior part of the body, the right one considerably in advance of the left (fig. 95).

Foot.—The foot is of the usual short, conical form. The lorica projects over it dorsally much farther on the left side than on the right (fig. 95), so that the foot may be bent to the right or down, but not to the left nor up.

Toes (fig. 96).—The single main toe (l. t.), representing the left toe of the genus *Diurella*, is an almost straight rod of nearly or quite the length of the lorica. It is accompanied at its base by a number of short scales and spines, one of which, curved so that the tip lies against the main toe, seems (by comparison with a number of other species) to represent the right toe (fig. 96, r. t.). This is in the present species shorter than one of the substyles proper.

Internal organs.—The internal organs offer nothing of especial interest except the trophi. These are decidedly unsymmetrical, the right malleus being considerably smaller than the left. The trophi are essentially like those of *Rattulus longiseta* Schrank (pl. VIII, fig. 72), but with the right malleus perhaps a little smaller.

Measurements.—Length of body, 0.16 to 0.17 mm.; of toe, 0.14 to 0.15 mm.; total, 0.30 to 0.32 mm.

History.—*Rattulus carinatus* Lamarck seems to have been the second species of the *Rattulidæ* that was described. Müller (1786) considered it a variety of *Rattulus rattus* Müller, and described it under the name of *Trichoda rattus vesiculam gerens*. He thought that the ridge was an egg sac: hence, the above name. Lamarck (1816) founded the genus *Rattulus* for this species and *R. rattus*, considering them as one species—a view which is, perhaps, not fully disproved yet. To this species he gave the name *Rattulus carinatus*, thus evidently basing his description on the characteristics of the species having the ridge. The name *carinatus* has therefore been properly used ever since for this species. The animal has been repeatedly transferred from one genus to another (see synonymy), the name *Rattulus* becoming completely supplanted by Ehrenberg's name *Mastigocerca*. According to the recognized rules of nomenclature this species must be restored to the first genus (*Rattulus*) which was founded to contain it.

Distribution.—In America *Rattulus carinatus* Lamarck occurs widely distributed in ponds, lakes, swamps, and rivers, but is usually taken in small numbers. I have found it as follows: Put-in Bay Harbor and East Harbor, Lake Erie; swamps on North, Middle, and South Bass islands, in Lake Erie; Huron River at Ann Arbor, Mich.; Portage River, Ohio; Lake St. Clair, Lake Michigan, Round Lake, and Pine Lake, near Charlevoix, Mich.; pools at Hanover, N. H.; ditch 5 miles south of Ann Arbor, Mich.; and in the following inland lakes of Michigan: West Twin Lake, Muskegon County; Crooked Lake, Newaygo County, and Chippewa Lake, Mecosta County.

By other observers it has been recorded as follows: Pond near Bangor, Me. (J. C. S., 1883); Shiawassee River at Corunna, Mich. (Kellicott, 1888); Sandusky Bay, Lake Erie (Kellicott, 1896); waters connected with the Illinois River at Havana, Ill. (Hempel 1898).

In Europe: Common in England (Gosse, 1889); Ireland (Hood, 1895); Württemberg, Germany (Bilfinger, 1892); Gr. Plöner See (Zacharias, 1893); near Rostock, Germany (Tessin, 1886); near Basel, Switzerland (Ternetz, 1892); near Geneva, Switzerland (Weber, 1898); Tyrol (Dalla Torre, 1889); Finland (Levander, 1894, and Stenroos, 1898); Galicia, Austro-Hungary (Wierzejski, 1893); Hungary (Kertesz, 1894; Daday, 1897); Bohemia (Petr, 1890); Livland, Russia (Eichwald, 1847); Kharkow, Russia (Scorikow, 1896). (Many other authors have listed this animal.)

Also at Sandringham, Australia (Anderson and Shepherd, 1892); in Ceylon (Daday, 1898); in New Guinea (Daday, 1901).

28. Rattulus rattus Müller (pl. XI, fig. 100, 101).

Synonyms: *Trichoda rattus* Müller (1776 and 1786); *Brachionus cylindricus* Schrank (1776); *Trichoda cricetus* Schrank (1803); *Rattulus carinatus* Lamarck (1816, in part); *Trichocerca rattus* Goldfuss (1820); *Monocerca longicauda* Bory de St. Vincent (1824, in part); *Monocerca rattus* Ehrenberg (1830, 1838); *Mastigocerca rattus* Hudson & Gosse (1889).

This animal is practically identical with *Rattulus carinatus* Lamarck, save that it lacks the ridge which forms so conspicuous a feature of that species. Some observers (Müller, Lamarck, Bory, Dujardin, etc.) have held that *R. rattus* and *R. carinatus* are merely varieties or variations of the same species. It is possible that this is true; in the lack of positive evidence that one may be transformed into the other it will be more convenient to retain the separate names, however, so that they may be recorded separately when desirable.

After study of a large number of specimens of these two species from many different localities I am convinced that there is no sharp distinguishing character except the presence or absence of the ridge. Yet in general the specimens of *R. rattus* Müller which I have seen have been larger than those of *R. carinatus* Lamarck and the body not so strikingly fusiform in shape, but more equal in diameter throughout.

In place of the high ridge of *R. carinatus* the present species has a broad longitudinal area, not elevated, which is marked with transverse striations and occupies the same position as the ridge in *R. carinatus* (fig. 100). The dorsal antenna lies in a notch on the left side of this striated area; a little behind it lies the eye (fig. 100).

The foot and toes are identical with those of *R. carinatus* Lamarck.

Measurements.—Length of body, about 0.17 to 0.18 mm.; of toe, 0.13 to 0.16 mm.; total, 0.30 to 0.32 mm.

History.—This seems to have been the first of the *Rattulidæ* observed. It was discovered by Eichhorn in 1775 and was called by him the Water-Rat ("die Wasser-Ratte"). Müller in 1776 gave it the name *Trichoda rattus*. Müller, Lamarck, Bory de St. Vincent, Dujardin, and various others have considered this species to be identical with *R. carinatus* Lamarck. This view has some evidence in its favor, though it can not be considered established. If it should ever be shown conclusively that the two are only forms of the same species, the name *Rattulus rattus* would prevail over *R. carinatus* as being the older name.

Descriptions or figures of this species are to be found in Ehrenberg (1838), Perty (1852), Eyferth (1885), Tessin (1886), Plate (1886), Hudson & Gosse (1889), Levander (1894), Scorikow (1896), Stenroos (1898).

Distribution.—In America this species is very common amid vegetation in quiet waters. I have found it in the following localities: Put-in Bay Harbor, Lake Erie; Long Point, Canada, near "the Cottages," on the north shore of Lake Erie; pools and swamps on South Bass Island in Lake Erie. It has been recorded by other observers in the United States as follows: New York (Ehrenberg, 1843); near Minneapolis, Minn. ("J. W.," 1883); near Cincinnati, Ohio (Turner, 1892); Sandusky Bay, Lake Erie (Kellicott, 1897).

In Europe: England (Gosse, 1889); Ireland (Glascott, 1893; Hood, 1894); Germany (Bartsch, 1870; Plate, 1886; Tessin, 1886); Tyrol (Dalla Torre, 1889); Bohemia (Petr, 1890); Finland (Levander, 1894; Stenroos, 1898); Switzerland (Ternetz, 1892); Hungary (Toth, 1861; Bartsch, 1877; Kertesz, 1894; Daday, 1897); Livland, Russia (Eichwald, 1847); near Kharkow, Russia (Scorikow, 1896); also in Ceylon (Daday, 1898); abundant in Greenland (Bergendal, 1892).

29. Rattulus lophoessus (Gosse (pl. XI. figs. 98, 99).

Synonym: *Mastigocerca lophoessa* (Gosse (1889).

Distinguishing characters.—The distinctive features of this animal are the long, high ridge, reaching from the anterior end to the very foot and inclined far over to the right; the fusiform body, the unarmed anterior margin of the lorica, and the short rudimentary right toe, one-fourth the length of the main toe. It closely resembles *R. carinatus* Lamarck in many respects, but differs from it in the length of the ridge and in the toes.

External features.—The body is fusiform in shape, and somewhat more elongated than in *R. carinatus* Lamarck. The head-sheath is marked off from the remainder of the lorica by an evident constriction. The anterior edge of the lorica is without teeth or spines; it has a shallow depression in the ventral middle region (fig. 98).

The ridge is nearly as high as in *R. carinatus* and extends from the anterior edge of the lorica to the very foot. It is situated considerably to the right of the middle line and is inclined far over to the right, resembling thus the right one of the two ridges in *R. bicristatus* Gosse. It is striated, as in *R. carinatus* Lamarck, and the striations extend considerably to the left of the ridge, ending at a well-defined line (fig. 99). Gosse (1889) and Weber (1898) describe the ridge as being interrupted, so as to form two or more arches. This was not the case in the specimens which I studied nor in those described by Bilfinger (1894).

Corona.—The corona has the usual dorsal club-shaped process, according to Bilfinger (1894); it has not been studied otherwise. In the preserved specimens at my disposal the corona was partly withdrawn.

Antennæ.—The dorsal antenna is situated considerably to the left of the ridge at the edge of the striated area. The lateral antennæ I have not been able to see in the preserved specimens. The right one is figured by Bilfinger (1894) in the usual position on the posterior third of the body.

Foot.—The foot is a short cone, of the usual character. The lorica projects far back over its base on the left side (fig. 99), so that the foot is free to bend to the right, but not to the left.

Toes (fig. 98).—The left or main toe (*l. t.*, fig. 98) forms a long, nearly straight rod, about two-thirds as long as the body. Above its base, separated from it by a well-marked gap, is the right toe (*r. t.*), about one-fourth to one-third as long as the main toe. The right toe is bent toward the main toe, its tip overlying the latter. At the base of both the main toe and the smaller one are one or two scale-like substyles. The larger lobe of the mucus reservoir (*m. r.*) opens at the base of the main toe; the smaller lobe at the base of the rudimentary right toe.

Internal organs.—The eye lies on the left side of the large brain (fig. 99). The trophi could not be studied in detail in the specimens at my command. As noted above, the mucus reservoir is divided into two unequal lobes, opening at the bases of the right and left toes, respectively. The other internal organs offer nothing of especial importance.

Measurements.—Length of body, 0.23 mm.; of toe, 0.15 mm.; total, 0.38 mm.

History.—This species was described by Gosse, in Hudson & Gosse's Monograph of the Rotifera, as *Mastigocerca lophoessa*. Figures and descriptions have since been given by Bilfinger (1894) and Weber (1898).

Distribution.—*Rattulus lophoessus* Gosse has not yet been found in America. For the opportunity of studying it I am indebted to Mr. Charles Rousselet, of London, England, who sent me excellent mounted specimens of this rare species.

In Europe: England and Scotland in pools (Gosse, 1889); Ireland (Hood, 1895); Württemberg, Germany (Bilfinger, 1894); Tyrol (Dalla Torre, 1889); Switzerland (Weber, 1898); Lake Nurmijärvi in Finland (Stenroos, 1898).

30. Rattulus latus Jennings (pl. VII, figs. 65, 66).

Synonym: *Mastigocerca lata* Jennings (1894).

Distinguishing characters.—This species is at once distinguished from all others by the broad ovate lorica, coupled with the striking lack of symmetry at the posterior end. This asymmetry distinguishes it at once from *R. multicrinis* Kellicott, the only species that resembles it at all in general appearance.

External features.—The lorica is broadly ovate in dorsal or ventral view, the width being about five-eighths of the length. The dorso-ventral measurement is about two-thirds that of the width, so that the animal is dorso-ventrally somewhat depressed. When seen in side view the dorsal line is a uniform curve from the front of the head to the base of the large foot. The ventral line is a similar but less convex curve from the junction of the head-sheath with the lorica to the base of the toe, so that the two curves are not symmetrically placed. The lorica is peculiarly unsymmetrical in a dorsal or ventral view, for the posterior part of the body, bearing the foot, is a thick, truncate cone lying not in the middle line, but on the left side (fig. 66). On the right side there is a blunt projection corresponding in position to that bearing the foot, but a little smaller. Between it and the left projection is a well-defined notch.

There is no sign of a ridge or striated area in this species.

The head-sheath is scarcely marked off from the remainder of the lorica at all; only a slight angle on the ventral side marks where it begins. In front the ventral edge of the lorica ends in a broad notch, at the bottom of which is a projecting tooth (fig. 65). Dorsally the anterior edge of the lorica is a slightly uneven curve, with neither a distinct notch nor a projecting tooth (fig. 66). The form of the lorica is not changed appreciably when the head is retracted.

Corona.—The corona consists of the following parts: (1) A dorsal and lateral fringe of cilia, forming about two-thirds the circumference of the head and interrupted in the dorsal middle region; (2) at the middle of the dorsal edge a flattened non-setigerous column, truncate at the end; (3) below the last, a similar flattened process, bearing at its free end a pair of minute styles. Below and at the side of this are (4) a pair of somewhat club-shaped processes curving ventrad. At either side of the middle of the coronal disk are (5) four small papillae, the two inner of which, at least, bear long setae. These are partially surrounded by (6) an incomplete circle of cilia.

Antennæ.—The dorsal antenna appears as a small tube, reaching the dorsal surface of the lorica near the dorsal middle line, some distance from the anterior margin of the lorica (fig. 66, *d. a.*). The lateral antennæ are visible in a dorsal view, lying one above each of the two posterior projections of the lorica (fig. 66, *l. a.*).

Foot.—The foot is very short, scarcely distinguishable as a separate joint. It is borne by the left one of the two projections in which the lorica ends.

Toes.—The main toe (representing the left toe) is a slender, pointed rod, continuing the curve of the left side of the lorica. It is about four-fifths as long as the lorica. It is accompanied by three short, unequal substyles, the longest (representing the right toe) about one-fifth the length of the main toe, the others much shorter. The toe is united to the body in such a way that it can be turned to the right, but not to the left.

Internal organs.—The mastax is oblong, truncate at either end; its circular end appears in a ventral view in front of the broad pectoral notch of the lorica. To its sides are attached two projecting glandular bodies (fig. 65, *gl.*). The trophi are nearly or quite symmetrical. The internal organs partake in their arrangement of the peculiar asymmetry that appears in the lorica. The stomach lies to the right of the middle; its walls contain many large, spherical, light-yellowish, refractive granules. The ovary lies to the left of the stomach, not ventral to it. The lateral canals of the left side lie ventral to the ovary and present three flame cells, one at the side of the posterior end of the mastax, one at the side of the anterior end of the stomach, and one just in front of the contractile vesicle. The two halves of the mucus reservoir (fig. 65, *m. r.*) are pushed widely apart, the left one being much the larger. The brain is of the usual form; on its dorsal surface it bears the eye, formed of a large clear sphere, embedded in a deep red cup.

Measurements.—Length of body, 0.17 to 0.18 mm.; of toe, 0.12 mm.; total, 0.29 to 0.30 mm.

History.—This species was described by the present author in 1894. A description and figure are given by Stenroos (1898) and a figure by Jennings (1901).

Distribution.—In America: I have found this species in the following localities: East Harbor, Lake Erie, near Sandusky, Ohio; Lake St. Clair; West Twin Lake near Charlevoix, Mich.; Graveyard Pond on Presque Isle near Erie, Pa.; Huron River at Ann Arbor, Mich. Kellicott (1896) records this species from Sandusky Bay, Lake Erie; Hempel (1898) from waters connected with the Illinois River at Havana, Ill.

In Europe: Lake Nurmijärvi, Finland (Stenroos, 1898).

31. Rattulus bicuspes Pell (pl. VIII, figs. 73–76).

Synonyms: *Mastigocerca bicuspes* Pell (1890); *Mastigocerca spinigera* Stokes (1897).

Distinguishing characters.—This very peculiar species may be known by its short, plump body, high arched dorsally; by the unarmed anterior edge of the lorica; the very prominent lateral antennæ, protected by stout spines; and by the long toe, longer than the body.

External features.—The body is very short and thick, the ventral line nearly straight, while the dorsal line is a high arch. The highest part of the body is a little behind the middle; thence it falls off suddenly to the foot, there being in some cases even an inward curve just above the foot (fig. 73). The anterior part of the lorica, or head-sheath, is not strongly marked off from the remainder, though a slight constriction can sometimes be seen behind it (fig. 73). There are no teeth or spines at the anterior margin. Considerably to the right of the dorsal middle line is a ridge, which is fairly prominent in living individuals, but seems less noticeable in preserved specimens. It is rather broad and reaches from near the anterior edge to a point some distance behind the middle of the lorica (fig. 75). It is transversely striated (i. e., it contains transverse muscle bands).

Corona.—The corona is of the usual character, the only point deserving especial mention being the antenna-like processes. There are five of these, as in *R. latus* Jennings, *R. multicrinis* Kellicott, and *R. capucinus* Wierz. & Zach. The median dorsal process is stout and club-shaped; on each side of this, and nearer the ventral side, are two slender processes close together (fig. 75).

Antennæ.—The most peculiar feature of this species are the lateral antennæ. These are in the usual position, on the posterior part of the lorica, near the dorsal side. Each forms (or is accompanied by) a large, sharp spine, with its basal part enlarged. According to Stokes (1897), fine setæ may be seen just in front of these spines, close against them; these I have not seen.

The dorsal antenna is in the usual position, a short distance behind the anterior edge of the lorica. It lies just to the left of the striated area (fig. 75).

Foot.—The foot is a small, short structure, arising as a continuation of ventral part of body.

Toes.—There is one main toe (representing the left toe of *Diurella*); this is longer than the body and is nearly straight. The rudimentary right toe is very small and curves toward the main toe, its tip lying against the latter.

Internal organs.—This species is not a favorable one for the study of the internal organs, these being crowded together in the short, thick lorica in such a way as to make it very difficult to disentangle them. There seems in any case to be nothing calling for special remark. The trophi (fig. 76) are rather stout, and of the usual unsymmetrical form, the left side being much better developed.

Measurements.—Length of body, 0.12 mm.; of toe, 0.13 to 0.14 mm.; total, 0.25 to 0.26 mm.

History.—This species was described by Pell in 1890 as *Mastigocerca bicuspes.* Stokes redescribed it in 1897 as *Mastigocerca spinigera.* It has been figured by Jennings (1900 and 1901).

Distribution.—I have found this species as follows: East Harbor, Lake Erie, near Sandusky, Ohio; Graveyard Pond, Presque Isle, near Erie, Pa.; Huron River at Ann Arbor, Mich. Pell (1890) does not note where he found this species. Stokes (1897) described it from a pool near Trenton, N. J. It has not yet been recorded from Europe.

32. Rattulus elongatus Gosse (pl. XII. figs. 102–107).

Synonyms: *Mastigocerca elongata* Gosse (1886); *Mastigocerca grandis* Stenroos (1898).

Distinguishing characters.—This species may be known by its large size, its elongated form tapering toward the posterior end, the very long main toe accompanied by a " substyle" one-sixth to one-fourth its length, and the unarmed anterior edge of the lorica.

External features.—The body is long and slender, somewhat larger at or near the anterior end and tapering back to a slender foot. The dorsal line is somewhat arched, the ventral line nearly straight (fig. 105). The anterior portion of the lorica, or head-sheath, is not marked off from the remainder of the lorica by a constriction. The anterior edge is without spines or teeth.

The dorsal surface of the lorica bears a broad, transversely striated area extending backward about one-third the length of the lorica. The anterior part of this striated area (a little less than one-half of its length) is depressed, so that there is a rather broad furrow extending from the anterior edge backward for a distance somewhat greater than the diameter of the body (to the point *x*, figs. 102 and 105). Near the posterior end of this furrow is the opening for the dorsal antenna (fig. 102, *d. a.*). That part of the lorica forming the bottom of the furrow projects a little at the anterior edge of the lorica (fig. 102).

When the head is extended it is covered with a somewhat stiffened membrane which lies in transverse folds, giving this region a wrinkled appearance (figs. 103 and 104).

Corona.—The corona seems of the usual character. There is a large club-shaped dorsal process (not shown in the figures), but lateral processes have not been observed with certainty.

Antennæ.—The dorsal antenna is in the dorsal furrow near its posterior end. The lateral antennæ are in the usual position (figs. 102 and 105) save that the right one lies much farther toward the dorsal side than does the left, if the position of the dorsal antenna is taken as indicating the dorsal middle line. In a dorsal view only the right one of the lateral antennæ shows (figs. 102 and 103).

Foot.—The foot is more slender than usual, but is otherwise of the ordinary form. The lorica projects farther over it on the left side than on the right, so that the foot may bend to the right, but not to the left.

It will be noticed that there are a number of features in this region which seem to indicate that the posterior part of the animal is to be considered as twisted, so that the primitively dorsal side is now turned to the left, the primitively right side being dorsal. The attachment of the foot to the body is one of these features; usually in rotifers the body projects over the foot on the dorsal side, not on the left side. The position of the lateral antennæ indicates the same thing; the left one is now far over toward the ventral side, the right one nearly dorsal (fig. 103). Still more striking, from this point of view, is the attachment of the toes to the foot. The two toes are no longer side by side, but the primitively right (rudimentary) toe lies almost directly above the left (main) toe (fig. 105). If the hinder part of the body could be twisted about 90 to the right all these structures would regain their usual positions.

Toes (fig. 106).—The main toe (*l. t.*), representing the left toe of the genus *Diurella*, is a long, nearly straight, tapering rod, two-thirds to four-fifths the length of the lorica. The right toe (*r. t.*) is rudimentary, but is nevertheless much better developed than in many species of the genus *Rattulus;* it is a crooked spine from one-sixth to one-fourth the length of the main toe. From its base it curves ventrally toward the main toe, which it crosses (fig. 105). As has been set forth in the account of the foot, the base of this, primitively the right toe, lies in this animal almost directly dorsal to the main (left) toe. Both the main toe and the rudimentary toe are accompanied at the base by scale-like substyles, each toe having at least two of these (fig. 106).

Internal organs.—The internal organs are beautifully and clearly displayed in *R. elongatus* Gosse, rendering this perhaps the most favorable species that exists for a study of the anatomy. Beyond what was given in the general account of the anatomy of the *Rattulidæ* (p. 288), only two or three points need special mention. The brain is unsymmetrical, there being a broad right lobe or division, and a narrow left lobe, shorter than the right one. This left lobe bears at its end the eye (fig. 103). The trophi are well developed and decidedly unsymmetrical, the left half being much larger and stronger than the right (fig. 107).

Measurements.—Length of body, 0.38 to 0.46 mm.; of toe, 0.28 to 0.35 mm.; total, 0.66 to 0.81 mm.

History.—This species was described by Gosse (1880) in Hudson & Gosse's Monograph of the Rotifera, as *Mastigocerca elongata.* A description (in Russian), without figure, is given by Scorikow (1896), and figures are given by Jennings (1900 and 1901). Stenroos (1898) redescribed this animal (with a figure of the anterior end) as *Mastigocerca grandis* n. sp. His description of *Mastigocerca grandis* fits *Rattulus elongatus* Gosse perfectly, the figure of the anterior end showing the form when the head is partly extended.

Distribution.—In America: *Rattulus elongatus* Gosse is not rare in the quiet parts of rivers, ponds, and lakes, though large numbers are not usually found together. I have taken this species in the following localities: Put-in Bay Harbor and East Harbor, Lake Erie; pools at Hanover, N. H.; Portage River, Ohio; Huron River at Ann Arbor, Mich. It has also been found in Sandusky Bay, Lake Erie (Kellicott, 1897), and in waters connected with the Illinois River at Havana, Ill. (Hempel, 1898).

In Europe: England and Scotland (Gosse, 1880); Ireland (Hood, 1895); Württemberg, Germany (Bilfinger, 1892); bayous of the Rhine (Lanterborn, 1893); near Basel, Switzerland (Ternetz, 1892); Galicia, Austria-Hungary (Wierzejski, 1893); Lake Nurmijärvi in Finland (Stenroos, 1898, as *Mastigocerca grandis*); Bohemia (Petr, 1890); near Kharkow, Russia (Scorikow, 1896).

Also in Ceylon (Daday, 1898).

33. Rattulus stylatus Gosse (pl. x, figs. 92–94).

Synonyms: *Monocerca stylata* Gosse (1851); *Mastigocerca stylata* Gosse (1889).

Distinguishing characters.—This species is to be distinguished by the very short toe, one-half or less the length of the body; by the irregular form of the body, with the head-sheath strongly set off from the remainder of the lorica, and by the unarmed anterior edge. There is no other species which bears a very close resemblance to it.

External features.—In this species the body is more irregular in form than in perhaps any other. In dorsal or ventral view the body is ovate in general shape, truncate in front, but tapering rapidly behind to the toe. The posterior half of the body is thus conical, the apex of the cone being surmounted by the toe. In side view the ventral line is nearly even, while the dorsal line is arched (fig. 93). The head-sheath or anterior portion of the body is smaller than the remainder, and set off from it by a broad and deep furrow, so that in side view a prominent "hump" is observed just behind the furrow (fig. 93). The furrow does not run uninterruptedly around the body, but is rather an irregular fold; it is farther forward on the ventral side than on the dorsal. The anterior margin of the lorica is without teeth or projections of any kind.

Of this species I was able to study only mounted specimens (and these through the kindness of Mr. Charles F. Rousselet), and none of those available presented a directly dorsal view. It was therefore difficult to tell whether there was anything corresponding to the ridge or striated area or not. There seemed no sign of a ridge, but apparently there is a dorsal depression running backward from the median dorsal anterior margin of the lorica; this could not be determined with certainty, however. The animal is clearly not so markedly unsymmetrical as are many of the *Rattulidæ.*

Corona.—The corona bears the usual club-shaped dorsal process; it has not been minutely studied in other respects.

Antennæ.—Dorsal and lateral antennæ in the usual positions, as shown in figs. 92 and 94.

Toes.—The main toe is very short, as compared with that of most of the species of *Rattulus,* being usually somewhat less than one-third the length of the body. It is slightly curved, the concavity of the curve being on the dorsal side, as shown in figure 93. Closely appressed against the main toe is a substyle (perhaps the rudimentary right toe), which tapers very rapidly. Owing to its slenderness and close approximation to the main toe it is very difficult to determine the length of the substyle. In one case it appeared to be from one-third to one-half the length of the main toe. It is very easily overlooked.

Internal organs.—The trophi are a little unsymmetrical. The mucus reservoir is divided into two equal halves, in spite of the fact that one of the toes has become rudimentary.

Measurements.—Length of body, 0.18 mm.; of toe, 0.05 mm.; total, 0.23 mm.

History.—This species was described by Gosse (without a figure) in 1851. It is figured and described by Gosse (1889) in Hudson & Gosse's Monograph. No description or figure of this species has been given since, though Bilfinger (1894) and Eckstein (1895) have a few notes on it. Eyferth (1885) thought it might be identical with *Diurella stylata* Eyferth, though the two seem not to have the remotest resemblance.

Distribution.—This species has not been found in the United States, and my figures and description are from specimens kindly sent me by Mr. C. F. Rousselet, of London, England.

In Europe: England (Gosse, 1851, 1889); Ireland (Hood, 1895); Württemberg, Germany (Bilfinger, 1894); Müggel-See, Germany (Eckstein, 1895); bayous of the Rhine (Lauterborn, 1898).

34. Rattulus pusillus Lauterborn (pl. IX, figs. 81–85).

Synonym: *Mastigocerca pusilla* Lauterborn (1898).

Distinguishing characters.—This species may be known by its minute size, the short, plump body without teeth or spines, and the rod-like toe about four-fifths the length of the body.

External features.—The body is short and thick, without striking external features of any sort. The head-sheath is marked off by a light constriction from the remainder of the body. Considerably to the right of the dorsal median line is a small, shallow furrow running obliquely backward and corresponding to the striated area of other *Rattulidæ*. This is very inconspicuous, so that it may very easily be overlooked. In some specimens it appears to extend less than half the length of the animal; in others it passes to a point considerably back of the middle, while in still others it is scarcely observable at all. Possibly the furrow disappears when the lorica is distended, as by strong contraction. The form of the lorica shows considerable variation, as will be seen by a comparison of the figures given on plate IX. There are no teeth, spines, or notches at the anterior edge of the lorica.

Corona.—This is of the usual structure. There are two curves of cilia at the sides of the coronal disk and two about the mouth. A median dorsal club-shaped process exists, but no lateral processes are to be observed.

Antennæ.—The left lateral antenna is in the usual position (fig. 83), but the others have not been observed.

Foot.—The foot is very small; it shows no peculiarities in other respects.

Toes.—There is a single bristle-like toe, usually about four-fifths as long as the lorica, but varying. In some cases it is little more than half the length of the lorica. It is nearly straight, though there is a slight bend a short distance from its base, like that to be observed in *Rattulus stylatus* Gosse. Closely appressed to the base of the main toe, so as to be very inconspicuous, is a short substyle, about one-sixth the length of the main toe.

Internal organs.—These offer nothing peculiar. The trophi are of the usual type, the left side being considerably more developed than the right.

Measurements.—Length of body, 0.085 to 0.11 mm.; of toe, 0.06 mm.; total, 0.14 to 0.17 mm.

History.—Lauterborn (1898) lists this species under the name *Mastigocerca pusilla*, but does not give a description. Through the kindness of Dr. Lauterborn I received a sketch of his animal, which shows that it is identical with the rotifer which I have found in the Great Lakes. I have therefore used Lauterborn's specific name.

Distribution.—*Rattulus pusillus* is rare. I have found it in East Harbor, Lake Erie, near Sandusky, Ohio, and in ponds on Middle and South Bass islands, in Lake Erie. Mr. Rousselet has sent me specimens collected at Hanwell, in England. Lauterborn (1898) found it in the bayous of the Rhine.

DESCRIPTIONS COMPILED FROM OTHER AUTHORS.

35. Rattulus curvatus Levander (pl. XIV, fig. 129).

Synonym: *Mastigocerca curvata* Levander (1894).

This species is distinguished by the long, cylindrical, slender body, very strongly **curved, and** by the prominent movable spine at the anterior margin of the lorica. The rather long **head-sheath** is set off from the body, as in most species, by a slight constriction. The toe is about one-**third the** length of the body. Other details are not given by Levander.

Measurements.—Length of body, 0.176 mm.; thickness of body, 0.033 mm.; length of **toe**, 0.055 mm.

Distribution.—This species was found by Levander in Finland along the shores of **the island** Löfö.

36. Rattulus dubius Lauterborn (pl. XIV, fig. 133).

Synonym: *Mastigocerca dubia* Lauterborn (1894).

This marine species was described briefly by Lauterborn from a single specimen found **near** Helgoland. The body is rather short, with a slight constriction between the head portion **and the** remainder. At the anterior end, both dorsally and ventrally, a triangular projection of the **lorica**. The toe about half the length of the body; without substyles. Length of the body, 0.11 **mm.**; thickness in front, 0.032 mm.; length of toe, 0.058 mm. Other details not given.

DOUBTFUL SPECIES, SPECIES INSUFFICIENTLY DESCRIBED, AND SPECIES WRONGLY ASSOCIATED WITH THE RATTULIDÆ.

Rattulus unidens Stenroos (?) (pl. XV, fig. 135).

Synonym: *Mastigocerca unidens* Stenroos (1898).

I give in the following a translation of Stenroos's description of this animal. His **account of** the internal organs is omitted, since these show nothing peculiar.

"The body is elongated, cylindrical, scarcely narrowed posteriorly. The head portion **is set** off from the body by a line, and its length is somewhat less than its width. On the ventral **side it** appears to be cleft backward to the boundary. On the dorsal side there is a chitinous **ridge or** dorsal keel, which is extended forward as a tooth. At its broad basis, at the boundary of the **head** portion, a small opening is visible. Dorsal setæ, on the other hand, I have not found. **The foot** joint is about as long as broad, narrowed posteriorly, and furnished with three spines or toes, **the** medial one of which is more than half as long as the body, while the two lateral ones are **bristle-like**, equal in length, and somewhat curved.

"Length of the body with the foot-joint, 0.26 mm.; thickness of the body, 0.06 mm.; **length of** the toe, 0.125 mm.; of the substyles, 0.034 mm.

"The rotatory organ is furnished with about four short and thick finger-like papillæ. **The** large, crescent-shaped red eye is furnished with a refractive lens, and lies upon the large, oval **brain.**" (Stenroos, 1898, p. 145).

It will be noticed that this species agrees with *Rattulus gracilis* Tessin in the single **tooth** at the anterior end, and approximately in the body form and in the toes. Possibly it should **be** considered a synonym of the species just named. Or it may have been described from **specimens** of *Rattulus scipio* (Gosse). It was found by Stenroos in Lake Nurmijärvi in Finland.

Rattulus cuspidatus Stenroos (?) (pl. XV, fig. 136).

Synonym: *Mastigocerca cuspidata* Stenroos (1898).

This species bears much resemblance to *Rattulus scipio* Gosse, and should probably **be** identified with that species. It was described by Stenroos from a camera sketch made some **time** before the description was written.

Stenroos notes its resemblance to *R. scipio* Gosse, but considers the two distinct. The **body** is broadest in front, and gradually narrowed toward the rear. The head-sheath is set off by **a** slight constriction, and is furnished with a tooth which is said to rise from the right ventral **side;**

the figure shows it to be in the same position as the tooth of *Rattulus scipio* Gosse (q. v.). The main toe is not quite half the length of the body, and is furnished with two substyles, of which one is long and S-shaped.

The peculiarly formed head-sheath, the sharp lateral spine at the anterior end, and the long substyle are considered the characteristic features of this species.

Length of body, 0.30 mm.; thickness, 0.067 mm.; length of toe, 0.138 mm.; of the longest substyle, 0.05 mm.

The animal was found in Lake Nurmijärvi in Finland.

Rattulus roseus Stenroos (?) (pl. XV, fig. 137).

Synonym: *Mastigocerca rosea* Stenroos (1898)

This species has a close resemblance to *Rattulus longiseta* Schrank (*Mastigocerca bicornis* Ehrb.), from which it is said by Stenroos to differ chiefly in two points. One is in the form of the body—in this species broadest in the anterior half of the body, while *Rattulus longiseta* is said by Stenroos to be broadest in the middle. This, however, is a character which will by no means always hold for *R. longiseta*; it is not at all rare to find specimens that are broadest near the head. The chief point of difference is, however, that the two teeth at the anterior edge of the lorica are said to be on the ventral side, in place of on the dorsal side, as in *R. longiseta*.

This difference is unquestionably sufficient to justify the formation of a new species. But when we consider the relation of the anterior teeth in other *Rattulidæ* to the dorso-dextral striated area or ridge, and to the method of movement of the animal, as described in the first part of this paper, it is difficult not to question the presence of two teeth like those of *Rattulus longiseta* Schrank on the *ventral* anterior margin of the lorica. In Stenroos's figure (reproduced in fig. 137, pl. XV), which seems clearly a dextro-ventral view, if we suppose that the shorter tooth is seen through the transparent head of the animal (which may sometimes be done), the figure would agree throughout with *R. longiseta* Schrank. Stenroos states, however, that he has compared the two species, and that they are different.

Length of body, 0.336 mm.; thickness, 0.086 mm.; length of toe, 0.218 mm.; of the substyle, 0.05 mm.

This species was found by Stenroos in Lake Nurmijärvi, in Finland.

Rattulus brachydactylus Glascott (pl. XIV, fig. 130).

Synonym: *Mastigocerca brachydactyla* Glascott (1893)

I copy herewith the entire account (as well as the figure) of this species.

"*Sp. ch.*—Body irregularly cone-shaped; head lumpish; toe style-like, short, straight, no substyles, no ridge.

"Allied to *M. stylata* [*Rattulus stylatus* Gosse], but broadest at the head; body an irregular cone, puckered into constrictions, but not gibbous in the middle; the toe straight and finely pointed, only one-fourth the length of the body; no substyles; gait wobbling.

"*Habitat.*—A pond, County Wexford."

It is, of course, not entirely evident from the above description that this animal is really one of the *Rattulidæ* at all, and it is doubtful if the description and figure are sufficient to permit of its recognition if found again.

Rattulus antilopæus Petr (1890).

This may have been *Diurella tigris* Müller, though from the figure and description it is impossible to be certain. The description given by Petr is as follows:

"Body cylindrical, somewhat narrowing toward both ends; foot broad, one-jointed, ending in two toes bent toward each other sickle-wise; these about half the length of the body. At the base they are provided with two pairs of little spines." (Petr, 1890, p. 222).[a]

The figure resembles a poorly drawn contracted specimen of *Diurella tigris* Müller, with the toes strongly bent in opposite directions and crossing one another. This condition of the toes, on which the new species seems to be based, is almost certainly due to distortion. A species so inadequately described and figured can only be dropped.

---

[a] For the translation of this diagnosis from the original Czech language, in which it was written, I am indebted to Prof. George Rebec, of the University of Michigan.

"Rattulus lunaris" (pl. XIV, figs. 131, 132).

(*Trichoda lunaris* Müller, 1776; *Rattulus lunaris* Ehrenberg, 1838; *Mastigocerca lunaris* Weisse, 1847.)

"*Rattulus lunaris*" was evidently described from one of the smaller species of *Diurella*, but the descriptions have been so inexact, and perhaps erroneous, that it seems impossible to recognize the animal with certainty. In fact, it is probable that several different species of *Diurella*, inaccurately observed, have been given this name, so that it was really a collective designation, which can not be restricted to any particular species.

Judging from the description and figure of Ehrenberg (1838), *Rattulus lunaris* has a striking resemblance to *Diurella brachyura* Gosse. The form of the body, the position of the foot, the length of toe relatively to that of the body, and the unarmed anterior edge of lorica, are striking points of similarity. But Ehrenberg assigns to this animal two eyes, which, if credited, prevents the identification of the two. Ehrenberg decidedly emphasizes the presence of the two eyes; names the animal "Brillenratte" on account of them, and discusses their position, in a way that makes it difficult to believe he could have been entirely mistaken as to their existence. A number of other investigators, notably Eichwald (1847), Perty (1852), Bartsch (1877), and Wierzejski (1893), have reported finding Ehrenberg's *Rattulus lunaris*, and Eichwald especially mentions that it can be distinguished from closely related species by the two eyes. Ehrenberg states also that the toe is simple and styliform. Two of his figures are reproduced in figs. 131, 132.

Weisse (1847) describes as *Mastigocerca lunaris* what he considers to be Müller's original species. Weisse's animal had *but one eye*, and he seems to incline to the belief that the assignment of two eyes to this animal by other observers was a mistake. He notes that he himself had reported finding Ehrenberg's *Rattulus lunaris*, but that after once noticing the single eye he was never again able to find specimens with two. But Weisse's description does not help greatly in deciding what animal should be called *lunaris*, owing to the fact that his description was evidently based on observation of at least two different animals. He says that some specimens had the toe about one-third the length of the body, while in others the toe was full half the length of the body. The former is represented in his figs. 4 and 5, the latter in his fig. 6. Judging from his description and figures, the former may have been *Diurella brachyura* Gosse, the latter *Diurella tenuior* Gosse. But there is no statement as to the presence or absence of teeth at the anterior edge of the lorica, and the figures and descriptions are in other respects also so general in character that it is quite impossible to be certain in the matter.

Taking all together, it seems necessary to let the name *lunaris* drop, it being impossible to recognize any definite species as corresponding to the description given.

Distemma setigerum Ehrenberg (pl. XIV, fig. 134).

This animal, from the structure of its toes and its general appearance, seems to belong to the *Rattulidæ*, where it would be assigned to the genus *Diurella*. But the assignment to it of two eyes by Ehrenberg prevents its identification with any known species. Ehrenberg's specific characters are: Body oblong-ovate; the two eyes red; the toes seta-like and decurved. He mentions the fact that it might easily be confounded with *Rattulus (lunaris)*.

Bartsch (1877) reports this species from Hungary.

Monocerca valga Ehrenberg (1838).

As noted by Hudson & Gosse (1889), this was apparently a male rotifer of some sort.

"Rattulus cimolius" Gosse (1889) (pl. XV, fig. 138).

There is nothing in Gosse's description to indicate that this animal belongs to the *Rattulidæ*, except, possibly, the unsymmetrical trophi. But this is a character which is not at all rare, as Lund (1899, p. 70) has observed, in various *Notommatadæ*. For the rest, all the characters mentioned by Gosse are quite foreign to the *Rattulidæ*, but are characteristic for some of the *Notommatadæ*. The skin is flexible (there being, so far as can be judged from description or figure, no sign of a lorica); brain opaque; toes blade-like; there are no substyles; apparently no eye; auricles present on corona. None of these characters are found in the *Rattulidæ*, so that it seems that there is absolutely no ground for including this species in the present family.

Gosse found one specimen at Sandhurst, Berks, England; another specimen in a pool near Birmingham, England. Glasscott (1893) reports finding it in Ireland.

"Rattulus calyptus" Gosse (1889) (pl. xv. fig. 139).

The case stands with this species as with *"Rattulus cimolius."* There is nothing in Gosse's figure or description that gives the least indication that this organism is one of the *Rattulidæ*. It is without a lorica; toes blade-shaped; brain clear; no eyes; face furnished with "pendent veil-like lobes of flesh." The animal was marine, being found in the tide pools on the Scottish coast.

Found also in Ireland (Hood, 1895).

Cœlopus (?) minutus Gosse (1889) (pl. xv. fig. 144).

This species was described from a single specimen, which was nearly dead. Only its general appearance gave it any claim to be placed among the *Rattulidæ*, for in its other characteristics it gave little indication of belonging to this family. It had two eyes, wide apart; apparently no mastax or rotatory organ. The toe appeared to be a single short, slender spine. In place of mastax and trophi there was a tube leading from the anterior end to the stomach. The body was thick, short, and rounded; the foot short and thick. The animal was excessively minute, being but 0.05 mm. long. It was found in Black Loch, near Dundee, Scotland.

Glasscott (1893) reports finding a dead specimen of this animal in Ireland, and says that the toes were two broad, decurved blades, exactly alike, and stretched out in a line with the body. If her specimen was really the same as Gosse's, this account of the toes, of course, removes all reason that may have existed for considering this one of the *Rattulidæ*.

Elosa worrallii Lord (1891) (pl. xv. fig. 140-143).

This rotifer was assigned by Lord to the *Rattulidæ*, so that an account of it should perhaps be given here. The animal is without a foot and toes, but in some respects, notably in its asymmetry, it perhaps does show some resemblance to the *Rattulidæ*. I should consider that it belongs rather with the genus *Ascomorpha*; others of this genus have some points of resemblance with the *Rattulidæ*. Lord's description (Lord, 1891, p. 324) is as follows (somewhat abridged):

"Lorica ovate, widest behind, trilobate in optical section; eyes two, one frontal, one cervical; trophi unsymmetrical; foot and toes absent. The lorica, which, as stated, is three-lobed, is on the dorsal aspect oval, widest behind, with a posterior rounded projection, the continuation of the dorsal lobe. There is a peculiar crescentic opening posteriorly on the left under side, visible both on dorsal and ventral aspects. * * * On the sides of the head are two triangular, movable pieces, the points of which can be made to meet and protect the retracted corona, much as in *Cœlopus porcellus*, an evident approach to the more perfect defensive armature of *Dinocharis*. The cervical eye is dark and rather large; the frontal one, which is to the right of the median line, is small and pale, and in many of the specimens can be easily overlooked. The mastax is long, pear-shaped, and three-lobed; the trophi are protrusile and asymmetrical; there is a long fulcrum, with a terminal knob. The left manubrium is nearly as long, while the right one is short and rudimentary. The stomach is generally filled with brown alimentary matter, and there is a distinct intestine, which in newly collected specimens is invariably of a pale-green color; neither salivary nor gastric glands were discoverable, and I think they would hardly have escaped notice had they been present."

Bothriocerca affinis Eichwald (1847) (pl. xv. fig. 145).

This animal was evidently a species of *Diurella*, but what species it is impossible to decide, owing to the indefiniteness of Eichwald's figure and description. In fact it was probably described from observation of more than one of the smaller species of *Diurella*, for he says that specimens found in pools near Kaugern differed from those found in the Drixe, in the presence of a small tooth at the dorsal and ventral anterior margin of the lorica. Eichwald had thus evidently at least two different species before him, though they were described as one.

Eichwald says that the "foot" (meaning what is now called the toe) had a longitudinal furrow; this appearance was due of course to the space between the two toes. Altogether it is evident that both the generic and specific names must be dropped; the former as synonymous with *Diurella*, the latter because the species is unrecognizable.

Bothriocerca longicauda Daday (1889) (pl. xv. fig. 146).

This marine organism Daday apparently classes with the *Rattulidæ*. As the description is in Hungarian, I am unable to make use of it. In a brief German résumé Daday (1890) says that this species differs from *Bothriocerca affinis* Eichwald in the fact that the anterior edge of the lorica has several excavations, and that the toe is very long. Other characteristics must be judged from the figure (pl. xv. fig. 146.)

Heterognathus brachydactyla Schmarda (1859).

This species is so inadequately described as to be quite unrecognizable, so that it will have to be dropped completely.

Heterognathus notommata Schmarda (1859).

Insufficiently described for recognition. The figure bears a slight resemblance to *Diurella tenuior* Gosse.

Heterognathus diglenus Schmarda (1859).

This was not one of the *Rattulidæ*, but a *Diglena*, apparently *Diglena catellina* Ehr.

DISTRIBUTION OF THE RATTULIDÆ.

As an examination of the foregoing list will show, of the 36 well-established species 25 have been found in America and 32 in Europe. Four species described from America have not yet been found in Europe; these are *Diurella insignis* Herrick, *Rattulus multicrinis* Kellicott, *R. mucosus* Stokes, and *R. bicuspes* Pell. Eleven European species have not been found in America, namely: *Diurella rousseleti* Voigt, *D. dixon-nuttalli* Jennings, *D. sejunctipes* Gosse, *D. collaris* Rousselet, *D. helminthodes* Gosse, *D. marina* Daday, *D. brevidactyla* Daday, *Rattulus lophoëssus* Gosse, *R. stylatus* Gosse, *R. curvatus* Levander, and *R. dubius* Lauterborn.

It is not improbable that all the species found in Europe will in time be found in this country, and that future workers in Europe will detect the four American species there. It was only within a few years that *Rattulus latus* Jennings was described from the United States; it was soon after found by Stenroos in Finland. Several of the species described in this paper for the first time are shown to be distributed in both Europe and America. Ten of the species here listed from America were not known hitherto to exist in this country.

Some of the better-known species are shown to have a very wide distribution. For example, *Diurella tigris* Müller has been found widely distributed in Europe and America and in India, New Guinea, Natal, and Ceylon. *Diurella tenuior* Gosse has been found in New Guinea, Natal, and New Zealand, as well as in Europe and America. *Rattulus carinatus* Lamarck is recorded from all parts of Europe and the United States, and from New Guinea, Ceylon, and Australia. It is probable that the group as a whole will be found to have a cosmopolitan distribution.

Of the 25 American species 23 have been found in Lake Erie. A characteristic feature in the distribution of these animals is the fact that many species may be found in a single restricted area. Thus, from a small pool not more than 30 feet across, in the Huron River at Ann Arbor, Mich., 14 species were taken, 6 species of *Diurella* and 8 of *Rattulus*. From East Harbor, Lake Erie, 20 species have been taken; this, however, is a rather extensive region.

LITERATURE CITED.

[The papers named have all been consulted in the original, except those marked with an asterisk. The references in the text to those thus marked are based upon abstracts or citations by other authors.]

ANDERSON. H. H.
    1889. Notes on Indian Rotifers.   Journ. Asiatic Soc. Bengal. vol. 68, pt. 2. No. 4. pp. 345–358.
ANDERSON. H. H., AND SHEPHERD, J.
    1892. Notes on Victorian Rotifers.   Proc. Roy. Soc. Victoria. n. s., vol. 4, pp. 69–80.
BARTSCH, SAMUEL.
    1870. Die Rädertiere und ihre bei Tübingen beobachteten Arten.   Stuttgart.
    1877. Rotatoria Hungariæ.   Budapest.
BERGENDAL. D.
    1892. Beiträge zur Fauna Grönlands.   Ergebnisse einer im Jahre 1890 in Grönland vorgenommenen Reise.   I. Zur Rotatorien-Fauna Grönlands.   Sep.-Abd. aus: Kongl. Fysiograf. Sällskapets Handlingar, ny följd, bd. 3, pp. 1–180.
BILFINGER, L.
    1892. Ein Beitrag zur Rotatorienfauna Württembergs.   Jahreshefte des Vereins f. vaterl. Naturkunde in Württemberg. 1892. pp. 107–118.
    1894. Zur Rotatorienfauna Württembergs.   Ibid., 1894. pp. 35–65
BLOCHMANN, F.
    1886. Die mikroskopische Thierwelt des Süsswassers.   122 pp.
BORY DE ST. VINCENT.
    1824 a. *Encyclop. méthod.   Vers.
    1824 b. *Dict. class. d'hist. nat.
COSMOVICI, LÉON C.
    1892. Rotifères.   Organisation et faune de la Roumanie.   Le Naturaliste, vol. 14.
DADAY, EUG. VON.
    1889. A Napolyi öböl Rotatoriái.   Értekezések a Természettudományok Köréből. Kiadja a Magyar tud. Akademia. 1889. pp. 1–52.
    1890. Die Räderthiere des Golfes von Neapel.   Math. und Naturw. Berichte aus Ungarn, Bd. 8, pp. 349–353.
    1897. Rotatorien.   From: Resultate der Wiss. Erforschung des Balatonsees, Bd. 2, Th. 1, Sect. 5.
    1898. Mikroskopische Süsswasserthiere aus Ceylon.   Anhangsheft zum XXI. Bande der Természetrajzi Füzetek. pp. 1–119.
    1901. Mikroskopische Süsswasserthiere aus Deutsch-Neu-Guinea.   Természetrajzi Füzetek, 24. pp. 1–56.
DALLA TORRE. K. W. VON.
    1889. Studien über die mikroskopische Thierwelt Tirols.   I. Theil: Rotatoria. Zeitschr. des Ferdinandeums für Tirol und Vorarlberg.   Dritte Folge. Hft. 33. pp. 239–252.
DUJARDIN. F.
    1841. Histoire naturelle des zoophytes: Infusoires.   Suites à Buffon.   Paris. 1841.   684 pp.
ECKSTEIN. KARL.
    1883. Rotatorien der Umgegend von Giessen.   Zeitschr. f. wiss. Zool., Bd. 39. pp. 343–443.
    1895. Die Rotatorienfauna des Müggelsees.   Zeitschr. f. Fischerei u. deren Hilfswissenschaften, 1895. Hft. 6. pp. 1–5.
EHRENBERG. C. G.
    1830. Beiträge zur Kenntnis der Organisation der Infusorien und ihre geographischen Verbreitung, besonders in Sibirien.   Physik. Abhandl. der Königl. Akademie der Wissenschaften zu Berlin.
    1833. Dritter Beitrag zur Kenntniss grosser Organisation in der Richtung des kleinsten Raumes.   Physik. Abhandl. der Königl. Akademie der Wiss. zu Berlin.
    1838. Die Infusionsthierchen als vollkommene Organismen.   Leipzig.
    1843. Verbreitung und Einfluss des mikroskopischen Lebens in Süd- und Nord-Amerika.   Berlin.
EICHHORN. J. C.
    1775. Wasserthiere, die mit keinem blossen Auge nicht können gesehen werden und die sich in den Gewässern um Danzig befinden.   Danzig (2nd edit., 1781. Berlin.)
EICHWALD.
    1847. Erster Nachtrag zur Infusorienkunde Russlands.   Bull. de la Soc. Imp. des Naturalistes de Moscou. t. 20. Seconde partie. pp. 285–366.
EYFERTH. B.
    1878. Die einfachsten Lebensformen des Thier und Pflanzenreiches.   Braunschweig.
    1885. Ibid., second edition.
GLASSCOTT. Miss L. S.
    1893. A list of some of the Rotifera of Ireland.   Scientific Proc. Roy. Dublin Soc., vol. 8 (n. s.), part 1. no. 6. pp. 29–86.
GOLDFUSS.   1820. *Handbuch der Zoologie.

GOSSE, P. H.
    1851. A catalogue of Rotifera found in Britain, with descriptions of 5 new genera and 32
        new species. Ann. and Mag. Nat. Hist., sec. ser., vol. 8, 1851, pp. 197–203.
    1856. On the structure, functions, and homologies of the manducatory organs in the class
        Rotifera. Phil. Trans., 1856, pp. 419–452.
    1889. See Hudson and Gosse, 1889.
HEMPEL, A.
    1898. A list of the Protozoa and Rotifera found in the Illinois River and adjacent lakes at
        Havana. Ill. Bull. Ill. State Lab. of Nat. Hist., vol. 5, art. 6, pp. 301–388.
HERRICK, C. L.
    1885. Notes on American Rotifers. Bull. Sci. Lab. Denison University, vol. 1, pp. 43–62.
HILGENDORF, F. W.
    1898. A contribution to the study of the Rotifera of New Zealand. Trans. New Zealand
        Institute, vol. 31, pp. 107–134.
HOOD, J.
    1895. On the Rotifera of the County Mayo. Proc. Roy. Irish Academy, third ser., vol. 3,
        pp. 664–706.
HUDSON, C. T., AND GOSSE, P. H.
    1889. The Rotifera, or Wheel Animalcules. Two volumes, with supplement. London.
IMHOF, O. E.
    1891. Ueber die pelagischen Fauna einiger Seen des Schwarzwaldes. Zool. Anz., 14. Jahrg.,
        pp. 33–38.
JENNINGS, H. S.
    1894. The Rotatoria of the Great Lakes and of some of the inland lakes of Michigan. Bull.
        Mich. Fish Commission, no. 3, pp. 1–34.
    1896. Report on the Rotatoria.   A biological examination of Lake Michigan in the Tra-
        verse Bay region. Bull. Mich. Fish Commission, no. 6, pp. 85–93.
    1900. Rotatoria of the United States, with especial reference to those of the Great Lakes.
        Bull. U. S. Fish Commission for 1899, pp. 67–104.
    1901. The Rotatoria. Synopses of North American invertebrates, no. 17. American Natural-
        ist, vol. 35, pp. 725–777.
KELLICOTT, D. S.
    1888. Partial list of the Rotifera of the Shiawassee River at Corunna, Mich. Proc. Am. Soc.
        Microscopists, vol. 10, pp. 1–13.
    1896. The Rotifera of Sandusky Bay. Proc. Am. Soc. Microscopists, vol. 18, pp. 155–164.
    1897. The Rotifera of Sandusky Bay. (Second paper.) Ibid., vol. 19, pp. 43–54.
KERTESZ, KALMAN.
    1894. Budapest és Környéknék Rotatoria-faunája. Készult a Kir. tud. Egyet. Allattani és
        összehasonlito boneztani intezetben. Budapest, 1894.
KIRKMAN, T.
    1901. List of some of the Rotifera of Natal. Journ. Roy. Micr. Soc. London, 1901, pp. 229–241.
LAMARCK, J. B. P. A. DE.
    1816. Histoire naturelle des animaux sans vertèbres. T. deuxième. Histoire des Polypes.
        (Also second edition, 1836.) Paris.
LAUTERBORN, ROBERT.
    1893. Beiträge zur Rotatorienfauna des Rheins und seiner Altwasser. Zool. Jahrbücher,
        Abth. f. Syst. Geograph. und Biologie. Bd. 7, Hft. 2, pp. 254–271.
    1894. Die pelagischen Protozoen und Rotatorien Helgolands. Sonderabdr. aus: Wiss. Meeres-
        untersuch., herausgegeben v. d. Kommission z. Unters. d. Deutschen Meere in Kiel
        u. d. Biol. Anstalt auf Helgoland, neue Folge. Bd. 1, pp. 208–213.
    1898. Ueber die zyklischen Fortpflanzung limnetischer Rotatorien. Biol. Centralblatt, Bd. 18,
        pp. 173–183.
LEVANDER, K. M.
    1894. Materialien zur Kenntniss der Wasserfauna in der Umgebung von Helsingfors, mit
        besonderer Berücksichtigung der Meeresfauna. Acta Soc. pro Fauna et Flora
        Fennica, XII. No. 3, pp. 1–72.
LEYDIG, F.
    1854. Ueber den Bau und die systematische Stellung der Räderthiere. Zeitschr. f. wiss.
        Zool., Bd. 6, pp. 1–120.
LORD, J. E.
    1891. A new Rotifer (Elosa worrallii Lord). International Journal of Microscopy and
        Natural Science, ser. 3, vol. 7, no. 11, pp. 323–325.
LUND, C. WESENBERG.
    1899. Danmarks Rotifera. I. Grundtrækkene i Rotiferernes Økologi, Morfologi og Syste-
        matik. København, 1899.
MINKIEWICZ, R.   1900. Petites études morphologiques sur le "limnoplancton." Zool. Anz.,
        Jahrg. 23, pp. 618–623.
MÜLLER, O. F.   1776. *Prodromus zoolog. danicæ. Addenda.
        1786. *Animalcula Infusoria.

PELL, A. 1890. Three new Rotifers. The Microscope, vol. 10, pp. 143–145.
PERTY, M. 1852. Zur Kenntniss kleinster Lebensformen. 228 pp.
PETR, F. 1890. Vírníci (Rotatoria) vysočiny českomoravské. Zvláštní otisk z Věstníka královske
    české společnosti náuk, pp. 215–225.
PLATE, L. 1886. Beiträge zur Naturgeschichte der Rotatorien. Jen. Zeitschr., Bd. 19, pp. 1–120.
PRITCHARD, A. 1861. A history of infusoria. London.
ROUSSELET, C. F.
    1896. Rattulus collaris n. sp. and some other Rotifers. Journ. Quekett Micr. Club, ser. 2,
        vol. 6, no. 39, pp. 265–269.
    1897. On the male of Rhinops vitrea. Journ. Roy. Micr. Soc., 1897, pp. 4–9.
"J. C. S." 1883. Pond life in winter. Am. Monthly Micr. Journ., vol. 4.
SCHMARDA, L. K.
    1859. Neue wirbellose Thiere beobachtet und gesammelt auf einer Reise um die Erde, 1853 bis
        1857. Erster Band.
SCHRANK.
    1776. * Beiträge zur Naturgeschichte.
    1793. * Naturforscher. XXVII, p. 26.
    1802. * Briefe naturhist. Inhalts an Nau, p. 383.
    1803. * Fauna boica. III, 2.
SCORIKOW, A. S.
    1896. Rotateurs des environs de Kharkow (Russian). Contributions of the Society for the
        Investigation of Natural History of the University of Kharkow. pp. 1–168.
    1898. Note sur quelques Rotateurs des environs de Kharkow. Ibid., pp. 1–6.
STENROOS, K. E.
    1898. Das Thierleben im Nurmijärvi-See. Acta Soc. pro Fauna et Flora Fennica, Bd. 17, No. 1
        pp. 1–259.
STOKES, A. C.
    1896. Some new forms of American Rotifera. Ann. and Mag. of Nat. Hist., ser. 6, vol. 18,
        pp. 17–27.
    1897. Some new forms of American Rotifera. II. Ibid., ser. 6, vol. 19, pp. 628–633.
TERNETZ, CARL.
    1892. Rotatorien der Umgebung Basels. 54 pp. Basel.
TESSIN, G.
    1886. Rotatorien der Umgegend von Rostock. Arch. d. Freunde d. Naturgesch. zu Mecklen-
        burg, Bd. 43, pp. 133–170.
TOTH, ALEX.
    1861. Die Rotatorien und Daphnien der Umgebung von Pest-Ofen. Verhdlg. der Kaisl.-
        Königl. zool.-bot. Gesellsch. in Wien, Bd. 11, pp. 183–184.
TURNER, C. H.
    1892. Notes upon the Cladocera, Copepoda, Ostracoda, and Rotifera of Cincinnati, with descrip-
        tions of new species. Bull. Sci. Lab. Denison Univ., vol. 6, pt. 2, pp. 57–74.
VOIGT, MAX.
    1901. Diagnosen bisher unbeschriebener Organismen aus Plöner Gewässern. Zool. Anz., Bd.
        25, pp. 35–39.
"J. W." 1883. To the Editor. Am. Monthly Micr. Journ., vol. 4, p. 18.
WEBER, E. F.
    1898. Faune rotatorienne du bassin du Leman. Revue suisse de zool., t. 5, pp. 263–785.
WEISSE, J. F.
    1847. Dorococcus globulosus Ehr., nebst Beschreibung dreier neuer Infusorien, welche bei St.
        Petersburg in stehenden Wässern vorkommen. Bull. Acad. Imp. des Sciences de
        St. Pétersbourg. Classe physico-math., sér. 2, t. 5, no. 15, pp. 226–230.
WESTERN, G.
    1893. Notes on Rotifers, with description of four new species and of the male of Stephanoceros
        eichhornii. Journ. Quek. Micr. Club, ser. 2, vol. 5, pp. 155–160.
    1893b. Notes on Rotifers. Ibid., ser. 2, vol. 5, p. 308.
    1894. Some foreign Rotifers to be included in the British Catalogue. Ibid., ser. 2, vol. 5, pp.
        420–426.
WIERZEJSKI, A.
    1893. Rotatoria (wrotki) Galicyi. 106 pp. Krakow.
WIERZEJSKI, A., UND ZACHARIAS, O.
    1893. Neue Rotatorien des Süsswassers. Zeitschr. f. wiss. Zool., Bd. 56, pp. 236–244.
ZACHARIAS, O.
    1893. Faunistische und biologische Mittheilungen am Gr. Plöner See. Forschungsberichte aus
        der biol. Station zu Plön. Theil 1.
    1897. Neue Beiträge zur Kenntniss des Süsswasserplanktons. Ibid., Theil 5. pp. 1–9.

INDEX TO GENERIC AND SPECIFIC NAMES.

DESCRIPTION OF PLATES.

PLATE I.

Diurella tigris Müller.

1. Dorsal view, showing the ridge (· 350).
2. Ventral view of trophi, a little from the left (· 665).
3. Ventral or ventro-sinistral view (· 350).
4. Ventral or ventro-sinistral view of strongly retracted specimen (· 350).
5. Optical section of body from rear (· 350).
6. Ventral view of foot and toes, with mucus reservoirs (m. r.) (· 600).

Diurella tenuior Gosse.

7. Small specimen, dorsal or dorso-dextral view (× 350).
8. Larger specimen, dorsal or dorso-dextral view (· 350).
9. Ventral view of toes (· 600).
10. Trophi, dorsal view (· 600).

Diurella weberi n. sp.

11. Ventral view of toes under pressure (· 665).
12. Right side view (· 600).
13. View of anterior edge of lorica, from the right (· 350).
14. Lorica, from left side (· 350).

PLATE II.

Diurella insignis Herrick.

15. Ventro-dextral view, head retracted (· 350).
16. Dorsal view, retracted (· 350).
17. Toes, pressed out, ventral view (· 350).
18. Young specimen, ventro-dextral view (· 350).

Diurella porcellus Gosse.

19. Dorsal view (× 350).
20. Lorica, from right side (· 350).
21. Left side of extended specimen (· 350).
22. Trophi, ventral view (cf. trophi in fig. 21) (· 665).
23. Foot and toes, dorsal view (the toes were bent beneath body, as in fig. 21; animal viewed from ventral side) (· 665).

Diurella sulcata Jennings.

24. Left side, foot extended (· 350).
25. Right side, foot retracted (× 350).
26. Trophi, dorsal view (· 665).

PLATE III.

Diurella stylata Eyferth.

27. Right side (· 385).
28. Dorsal view of retracted specimen (× 385).
29. Dorsal view of extended specimen (· 385).
30. Left side of retracted specimen (· 385).
31. Corona (· 600).

Diurella brachyura Gosse.

32. Right side (· 600). c. v. for c. v. = contractile vacuole.
33. Left side of lorica, showing projecting point on left side of anterior margin (· 600).
34. Toes, dorsal view (· 665).

Diurella caria Gosse.

35. Dorsal view (· 350).
36. Right side, showing sac-like protrusion of lorica above and behind toes (· 350).

PLATE IV.

The figures on this plate, except figs. 37 and 38, were drawn by Mr. F. R. Dixon-Nuttall, of Eccleston Park, near Prescot, England, who kindly placed them at the disposal of the author for the present paper.

Diurella rousseleti Voigt.

37. Side view of retracted specimen (· 600).
38. Toes, ventral view (· 665).
39. Side view of retracted specimen.

Diurella dixon-nuttalli n. sp.

40. Right side.
41. Right side of another specimen.
42. Trophi, dorsal view.
43. Toes, dorsal view.
44. Left side.

PLATE V.

Rattulus gracilis Tessin.

45. Right side of much-extended specimen (from Prescot, England) (· 350).
46. Dorsal view (from Huron River, Michigan) (· 350).
47. Dorsal view of lorica (from Prescot, England) (· 350).
48. Dorsal view of anterior part of retracted specimen (· 350).
49. Ventral view, trophi extended (· 350).

Rattulus scipio Gosse.

50. Ventro-dextral view of extended specimen (· 350).
51. Dorso-dextral view of extended specimen, young (· 350).
52. Dorsal view of retracted specimen (· 350).

Rattulus macerus Gosse.

53. Right side of young specimen, retracted (· 285). sp. spur.
54. Right side of adult specimen, extended (· 285). sp. = spur.

PLATE VI.

Rattulus multicrinis Kellicott.

55. Dorsal view of lorica (· 350).
56. Side view, with head and mastax extended (· 350).
57. Ventral view of extended specimen (× 350).
58. Side view of anterior part of lorica when the head is strongly retracted (× 350.)

Rattulus capucinus Wierz. & Zach.

59. Side view of anterior part of lorica when the head is strongly retracted (· 285).
60. Right side (· 285).
61. Ventral view. The toe extends obliquely toward the observer, so that its full length is not seen (· 285).

PLATE VII.

Rattulus cylindricus Imhof.

62. From Lake Bologoe, Russia, left side of retracted specimen, bearing egg (· 245).
63. Left side of extended specimen (from Germany) (· 285).
64. Left side of specimen from East Harbor, Lake Erie (· 245).

Rattulus latus Jennings.

65. Ventral view of extended specimen (· 350).
66. Dorsal view of lorica (· 285).

PLATE VIII.

Rattulus longiseta Schrank.

67. Right side view (· 285).
68. Ventral view of young specimen, retracted (· 285).
69. Ventral view of posterior part of body, with toes, mucus glands and reservoirs (· 800).
70. Specimen strongly retracted (· 285).
71. Trophi, from left side (· 965).
72. Trophi, dorsal view (· 965).

Rattulus bicuspes Pell.

73. Left side (· 350).
74. Young specimen, right side (· 350).
75. Dorsal view (· 350).
76. Trophi, dorsal view (· 965).

PLATE IX.

Rattulus bicristatus Gosse.

77. Right side (· 350).
78. Dorsal view, to show furrow and ridges (· 285).
79. Posterior part of body, foot and base of toes showing the mucus reservoirs and substyles (· 965). (The guide line from *m. r.* should extend farther to the left, to reach the reservoirs.)
80. Trophi, flattened by pressure.

Rattulus pusillus Lauterborn.

81. With head partly retracted, from Wehrle's Pond, Middle Bass Island, Lake Erie (· 350).
82. With head extended, from same locality as the last (· 350).
83. Dorsal view (· 965).
84. Side view (· 965).
85. From Hanwell, England (· 350).

PLATE X.

Rattulus mucosus Stokes.

86. Dorso-dextral view of lorica, to show ridges (· 350).
87. Side view (· 350).
88. Side view of contracted specimen (· 350).
89. Ventral view of lorica (· 350).
90. Young specimen (· 350).
91. Trophi, dorsal view (· 965).

Rattulus stylatus Gosse.

92. Ventral view (· 285).
93. Left side (· 285).
94. Dorso-sinistral view (· 350).

PLATE XI.

Rattulus carinatus Lamarck.

95. Dorsal view (· 350).
96. Foot and toes, dorso-sinistral view (· 600).
97. Right side (· 350).

Rattulus lophoessus Gosse.

98. Right side (· 285).
99. Dorsal view (· 285).

Rattulus rattus Müller.

100. Dorsal view (· 350).
101. Left side (· 350).

PLATE XII.

Rattulus elongatus Gosse.

102. Retracted, from East Harbor, Lake Erie, dorsal view (· 285).
103. Extended, from England, dorsal view (· 190).
104. Ventral view of anterior end with head extended (· 180).
105. Left side (· 285).
106. Dorso-sinistral view of foot and toes (· 350).
107. Trophi, ventral view (· 600).

PLATE XIII.

The figures on this plate, except fig. 108, were drawn by Mr. F. R. Dixon-Nuttall, of Eccleston Park, near Prescot, England, who kindly placed them at the disposal of the author for the present paper.

Diurella intermedia Stenroos.

108. Right side (from England).
109. From Huron River, Mich. (· 350).
110. Toes pressed out.

Rattulus scipio Gosse.

111. From a living specimen.
112. Toes pressed out.

Diurella sulcata Jennings.

113. Toes pressed out.

Diurella brachyura Gosse.

114. Toes pressed out.
115. Right side.

Diurella weberi n. sp.

116. Left side.
117. Right side.

Diurella sulcata Jennings.

118. Trophi.
119. Left side.

PLATE XIV.

Diurella sejunctipes Gosse.

120. Dorsal view, after Gosse (1889).
121. Left side (· 350), after Stenroos (1898).

Diurella helminthodes Gosse.

122. After Gosse (1889).

Diurella marina Daday.

123. After Daday (1890).
124. Ventral view of anterior edge of lorica, after Daday (1890).
125. Mastax, after Daday (1890).
126. Anterior end, strongly contracted, after Daday (1890).

Diurella collaris Rousselet.

127. After Rousselet (1896).

Diurella brevidactyla Daday.

128. After Daday (1899).

Rattulus curvatus Levander.

129. (× 390), after Levander (1894).

Rattulus brachydactylus Glascott.

130. After Glascott (1893).

"*Rattulus lunaris*" Ehrbg.

131. Dorsal view, after Ehrenberg (1838).
132. Side view, after Ehrenberg (1838).

Rattulus dubius Lauterborn.

133. After Lauterborn (1904).

Diatemma setigerum Ehr.

134. After Ehrenberg (1838).

PLATE XV.

135. *Rattulus unidens* Stenroos (× 296), after Stenroos (1898).
136. *Rattulus cuspidatus* Stenroos (× 450), after Stenroos (1898).
137. *Rattulus roseus* Stenroos (× 230), after Stenroos (1898).
138. "*Rattulus cimolius*" Gosse, after Gosse (1889).
139. "*Rattulus calyptus*" Gosse, after Gosse (1889).
140. *Elosa worrallii* Lord, side view, drawn by F. R. Dixon-Nuttall.
141. *Elosa worrallii* Lord, trophi, drawn by F. R. Dixon-Nuttall.
142. *Elosa worrallii* Lord, dorsal view, after Lord (1891).
143. *Elosa worrallii* Lord, section of body, after Lord (1891).
144. "*Cyclops* (?) *minutus*" Gosse, after Gosse (1886).
145. *Bothriocerca affinis* Eichwald, after Eichwald (1847).
146. *Bothriocerca longicauda* Daday, after Daday (1899).

Abbreviations used in the plates.

al. alula.	*in.* intestine.	*œ.* œsophagus.
br. brain.	*l.* left.	*ov.* ovary.
c. v. contractile vacuole.	*l. a.* left lateral antenna.	*r.* right.
d. a. dorsal antenna.	*l. mu.* left manubrium.	*ra.* ramus.
e. eye.	*l. t.* left toe.	*r. mu.* right manubrium.
f. foot.	*m.* mouth.	*r. t.* right toe.
fu. fulcrum.	*m. g.* mucus gland.	*sp.* spur.
g. g. gastric gland.	*m. r.* mucus reservoir.	*st.* stomach.
gl. gland.	*mu.* manubrium.	*u.* uncus.
h. s head-sheath.	*mx.* mastax.	

PLATE I

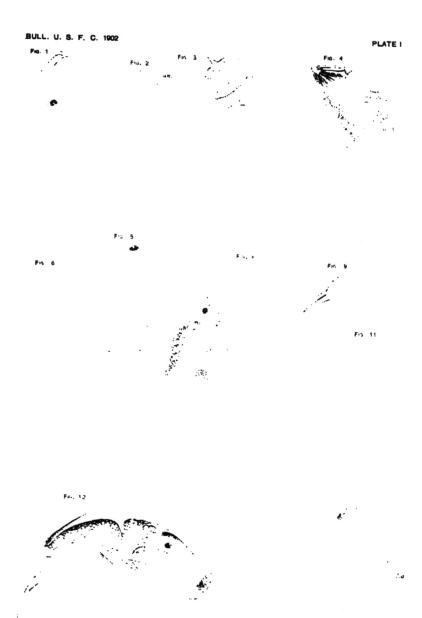

FIG. 1

FIG. 2   FIG. 3   FIG. 4

FIG. 5

FIG. 6   FIG.   FIG. 9

FIG. 11

FIG. 12

M. S. J.

FIG. 15   F.i. 16   F.i. 1   FIG. 18

FIG. 2.

FIG. 23

H. S J.

PLATE III

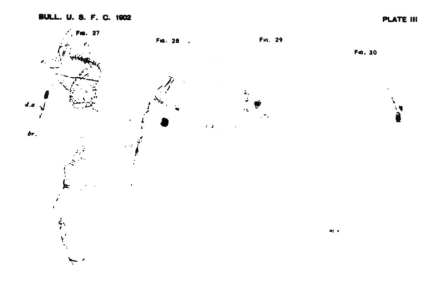

Fig. 27

Fig. 28

Fig. 29

Fig. 30

Fig. 31

Fig. 35

Fig. 32

Fig. 33

Fig. 36

Fig. 34

H. S J

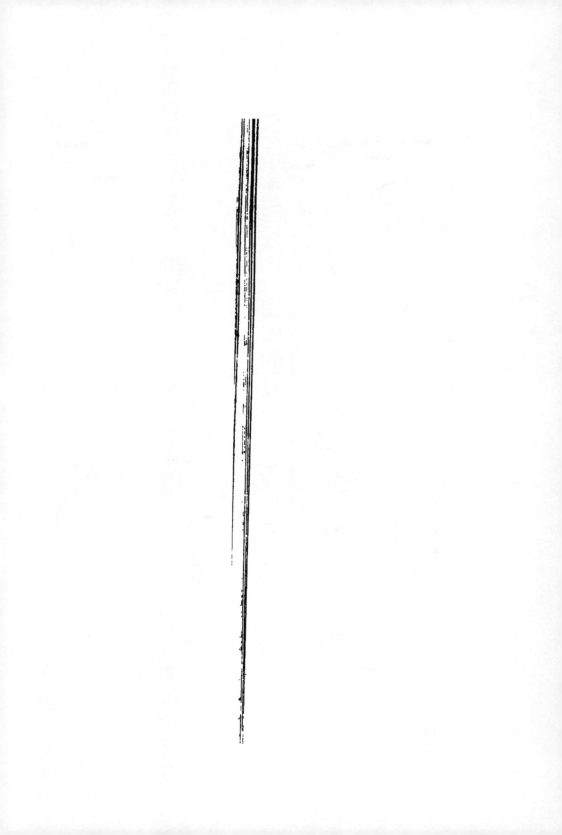

PLATE IV

Fig. 37

Fig. 39

PLATE V

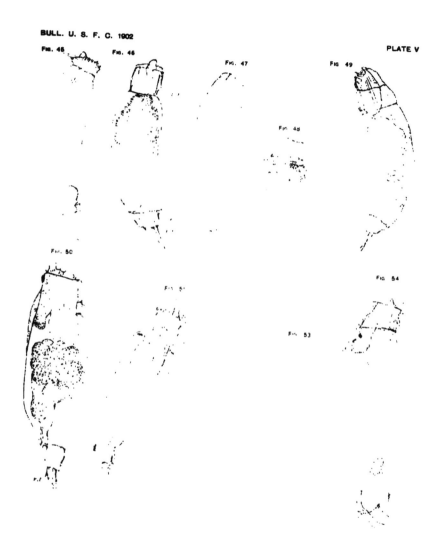

Fig. 45

Fig. 46

Fig. 47

Fig. 49

Fig. 48

Fig. 50

Fig. 51

Fig. 54

Fig. 53

FIG 15

PLATE VII

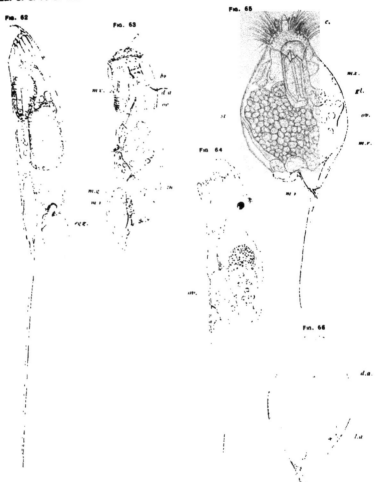

Fig. 62

Fig. 63

Fig. 65

Fig. 64

Fig. 66

PLATE VIII

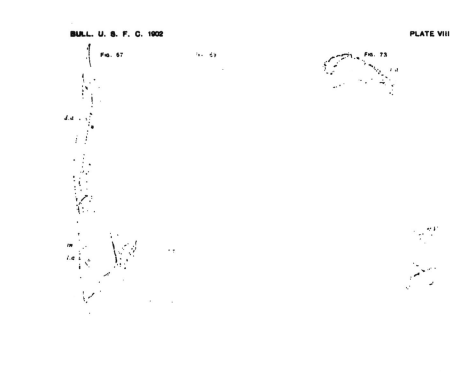

Fig. 67

Fig. 69

Fig. 73

PLATE IX

FIG. 77

FIG. 81

FIG. 82

78

FIG. 83

FIG. 84

PLATE X

FIG. 86

FIG. 87

FIG. 89

FIG. 90

FIG. 91

FIG. 88

H. S. J.

BULL. U. S. F. C. 1902

PLATE XI

FIG. 95   FIG 97

FIG. 99

FIG. 101

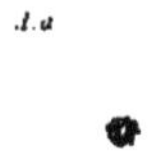

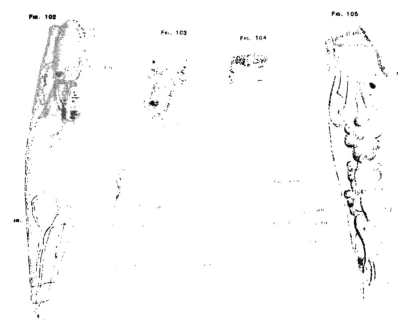

Fig. 102

Fig. 103

Fig. 104

Fig. 105

H. S J del

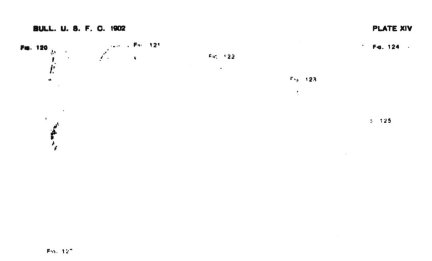

Fig. 120　　　Fig. 121　　　Fig. 122　　　Fig. 124

Fig. 123

Fig. 125

Fig. 127　　　Fig. 126

Fig. 128

Fig. 134

PLATE XV

Fig. 135

Fig. 136

Fig. 137

Fig. 138

Fig. 139

Fig. 140

NOTES ON SOME FRESH-WATER FISHES FROM MAINE,

WITH

DESCRIPTIONS OF THREE NEW SPECIES.

By WILLIAM CONVERSE KENDALL,

Assistant, United States Fish Commission.

NOTES ON SOME FRESH-WATER FISHES FROM MAINE, WITH DESCRIPTIONS OF THREE NEW SPECIES.

BY WILLIAM CONVERSE KENDALL,

Assistant, United States Fish Commission.

About forty years ago Ezekiel Holmes published a list of the fishes of Maine,[a] chiefly compiled and containing but few fresh-water species. Over thirty years later the present writer published a report[b] upon an investigation of the fresh waters of Washington County, which contained about the first record of observations upon Maine fresh-water fishes since Holmes's publication. Prior to this time there had been no systematic collecting in the inland waters of the State. In the four years immediately following some small collections were made, upon which there has been no detailed report.

In 1898 the United States Fish Commission, realizing that knowledge directly valuable to fish-culture and indirectly to the public could be derived from a study of the landlocked salmon and its native habitat, detailed the author to make such an investigation of Sebago Lake basin. Since then up to the present time the fresh waters of Maine have received considerable attention and a large amount of important information has been obtained.

In ten years 22 salt and fresh water species not previously recorded from the State, 12 of which are fresh-water[c] forms and 3 of which are new to science, have been found. This raises the list of native fresh-water (including anadromous) fishes from 35 to 47 species. Others have had their recorded range considerably extended in the State, and some which have not been recorded since their description, or known, perhaps, from only a single locality, have been found widely distributed. These statements are not astonishing when the great extent of the fresh waters in the State and the small amount of work done there are taken into consideration. There still remains a large unexplored area, and doubtless other forms new to the State and perhaps new to science may be discovered.

It is not the aim of this paper to enter into the details of the results of this work, this being reserved for a future more comprehensive paper, but to call attention to a few interesting fresh-water species of Maine fishes and put on record some observations regarding them.

---

a Second Annual Report upon the Natural History and Geology of Maine, 1862.

b Notes on the Fresh-water Fishes of Washington County, Me. Bulletin U. S. Fish Commission 1899, 13.)

c The additions to the fresh-water faunal list other than those mentioned in this paper are (1) *Chrosomus erythrogaster*, (2) *Semotilus atromaculatus*, (3) *Notropis muskoka*, (4) *Couesius plumbeus*, (5) *Fundulus diaphanus*, (6) *Eucalia inconstans*. Of marine species there are a number recorded in general range but mentioned from no definite locality in Maine, of which no account has been taken. The salt-water additions not recorded as extending so far north as Maine are (1) *Narcine occidentalis*, (2) *Gasterosteus gladiunculi*, (3) *Mugil curema*, (4) *Stenotomus chrysops*, (5) *Centropristes striatus*, (6) *Menticirrhus saxatilis*, (7) *Prionotus carolinus*, (8) *Spheroides maculatus*, (9) *Lophopsetta maculata*, and (10) *Macrourus bairdi*, of which *M. saxatilis* and *M. bairdi* have not previously been noted.

For courtesies, advice, and assistance the writer is indebted to Dr. David Starr Jordan, president of Leland Stanford Junior University; Dr. Barton W. Evermann and Dr. Hugh M. Smith, of the U. S. Fish Commission; Dr. Tarleton H. Bean and Mr. Barton A. Bean; the commissioners of inland fish and game, of Maine; Mr. Elmer D. Merrill, of the U. S. Department of Agriculture; Mr. Daniel Cummings, proprietor of a sportsman's camp and fish and game warden at Square Lake, and Mr. John A. Story, of the State fish-hatchery at Caribou, Me.

The drawings of the new species are by Mr. A. H. Baldwin.

Catostomus catostomus (Forster). *Northern Sucker.*

This species is known elsewhere by various other names, such as "small-scaled sucker," "long-nosed sucker," "red-sided sucker," etc. The only published intimations of its occurrence in Maine are by Prof. C. H. Hitchcock,[a] who said that a red-sided sucker was peculiar to Rangeley lakes, and Dr. A. Leith Adams,[b] who incidentally refers to the above locality and says the fish occurs in the St. Croix Lake waters. These are not undoubted references to this species, however, for the common sucker (*C. commersonii*) often has red sides during the breeding season. Dr. Adams's reference is definite regarding the Skiff Lake Stream specimens identified by Dr. Günther, but Skiff Lake Stream, though tributary to the St. Croix lakes, is in New Brunswick, and Dr. Adams's identification of it in Grand Lake Stream, where he procured his "silvery trout," may have been a mistake, owing to the common red-sided character of breeding *C. commersonii*. The inference is, however, that this species may be found in the St. Croix waters in Maine.

The first positive evidence of its occurrence in Maine was furnished by Mr. Elmer Merrill, who collected it at Craig Brook and sent it with some other species to the U. S. National Museum. Mr. Merrill gave the writer the following note regarding it:

"Mr. Atkins says it is the common sucker of Alamoosook and Toddy ponds, where it is abundant. In Craig Brook, in June, males were seen devouring the eggs as fast as extruded by the females."

This species has since been collected by the writer and Mr. Thomas B. Gould in Glasier Lake, which is an expansion of the St. Francis River, tributary to the St. John River, and in the "thoroughfare" connecting Long Lake with Cross Lake, of the Eagle Lake system, Aroostook County, in October, 1901. In this thoroughfare white-fish were spawning, and this sucker and *C. commersonii* were feeding upon the eggs.

This species is doubtless more widely distributed than it seems, but being an inhabitant of the cooler depths of the lakes, it is seldom seen unless in the breeding season when it ascends the streams to spawn, or in the fall when it follows trout, salmon, and white-fish to their spawning-grounds to feed upon their eggs. It is probably never recognized by the inexperienced observer, who sees in it only "a sucker." Like its congener, the common sucker, it varies in adult size; in some localities a length of 18 inches or more is attained, and in others only 6 to 10 inches. It may be distinguished from the common sucker by its longer nose, thicker lips, and smaller scales.

Leuciscus neogæus (Cope).

Prior to its discovery in New Brunswick in 1888, and again in 1895 by Philip Cox,[c] this species had not been recorded east of Wisconsin and Michigan. It was found by Kendall & Gould in Bill

---

[a] "The *Salmo oquassa* Girard, or blueback trout, an uncommon variety of dace, and a red-sided sucker are peculiar to these waters." (Geology of Maine, Second Annual Report, Natural History and Geology of the State of Maine, 1862, 328.)

[b] "During June, when the silvery salmon trout is plentiful in the streams connecting the various lakes, there is found associated with it a red banded sucker, 5 or 6 inches in length, with a brilliant red bar extending lengthwise down its sides. I examined several specimens of this fish, kindly procured for me from Skiff Lake Stream of the eastern Schoodic chain of lakes, by Major Monk and Captain Wolseley. It seems to be also found in the upper waters of the Androscoggin in the state of Maine, but further there are no accounts of its presence north of the state of Vermont, where it was discovered by Le Sueur and named by him *Catostomus longirostris*. The specimens above referred to were examined by Dr. Günther, who informs me that they differ only from this species in the length of the anal fin, which varies according to sex and season." (Field and Forest Rambles, with Notes and Observations on the Natural History of Eastern Canada, 1873, 252.)

[c] History and present state of the Ichthyology of New Brunswick, with a catalogue of its fresh-water and marine fishes. (Bulletin XIII, Natural History Society of New Brunswick, 1895, 41.)

Fish Brook, a tributary of the east branch of the Penobscot River about a mile below Matagamon Lake, in Second or Matagamonsis Lake in October, 1900; in these same places and in a small pond near Hale Pond, West Branch of the Penobscot, in August; Smith Brook, outlet of Haymock Lake, a tributary of Eagle Lake of Allagash River, in September, and Cross Lake Thoroughfare, Aroostook County, in October, 1901. In September, 1902, it was obtained by the same collectors in Lunkasooe Lake, which empties into the east branch of the Penobscot.

This is a beautiful little minnow. The small examples are not easily distinguished by color from *Chrosomus erythrogaster*, which usually occurs with it. An examination of the pharyngeal teeth, however, will quickly decide the matter. *Chrosomus* having but a single row on each bone, while the other has a double row. Some differences occur in the Maine specimens from those from farther west, to which, though not to separate them as distinct species, it is desirable to call attention. These will be shown in the accompanying table, from which it will be seen that there is considerable variation amongst them. Cox mentions still more in New Brunswick specimens.

The writer was privileged to examine some of the New Brunswick specimens, which were considerably deeper and more "stubby" fish than any of ours, with very steep foreheads and exceedingly projecting lower jaws. These were doubtless breeding males. Some of our specimens indicate the same form but to a lesser degree. This seems to be the first record of this species from Maine.

Table of proportional measurements of Leuciscus neogæus from Maine.

[Bill Fish Brook, September 7, 1901.]

Total length in millimeters.	Head in length without tail.	Depth in length.	Eye in head.	Snout in head.	Maxillary in head.	Mandible in head.	Pharyngeal teeth.	Scales.	Dorsal rays.	Anal rays.	Height of dorsal in head.	Height of anal in head.	Pectoral in head.	Ventral in head.	Sex.
76	4	5.24	4.25	3.77	4.25	3.40	1.4–5.2	82	8	8	1.54	1.70	1.70	2.26	Fem.
66	4.23	5	3.71	3.25	3.25	2.60	1.5–5.1	83	8	8	1.31	1.44	1.44	1.85	Fem.
63	4	4.72	4.33	4.33	4.33	3.25	2.5–5.2	83	8	8	1.44	1.62	1.44	2	Fem.
65	4.15	4.69	3.71	3.71	3.25	2.60	1.5–4.1	83	8	8	1.44	1.62	1.62	1.85	Fem.
65	4.50	4.99	4	4	4	3	1.5–5.0	83	8	8	1.20	1.33	1.33	1.71	Male.
63	4.33	4.72	4	4	4	3	1.5–4.1	85	8	8	1.20	1.50	1.50	1.71	Male.

[Southards Pond, August 19, 1901. A small pond near Hale Pond, West Branch Penobscot waters, having connection with Hale Pond only during stages of high water.]

Total length in millimeters.	Head in length without tail.	Depth in length.	Eye in head.	Snout in head.	Maxillary in head.	Mandible in head.	Pharyngeal teeth.	Scales.	Dorsal rays.	Anal rays.	Height of dorsal in head.	Height of anal in head.	Pectoral in head.	Ventral in head.	Sex.
89	4	5.06	4.75	3.80	3.16	2.71	2.4–5.2	77	8	8	1.58	1.72	1.72	2.11	Fem.?
89	4.11	4.95	4.50	3.60	3.60	2.57	2.4–4.2	82	8	8	1.50	2	2	2.37	Fem.?
91	4.22	5.01	4.50	3.60	3.60	2.57	2.4–4.12	77	8	8	1.63	1.80	1.80	2.25	Fem.?

[Matagamon Lake, October 3, 1901.]

Total length in millimeters.	Head in length without tail.	Depth in length.	Eye in head.	Snout in head.	Maxillary in head.	Mandible in head.	Pharyngeal teeth.	Scales.	Dorsal rays.	Anal rays.	Height of dorsal in head.	Height of anal in head.	Pectoral in head.	Ventral in head.	Sex.
72	4	4.61	4.25	4.25	5	3	1.5–4.2	83	9	1.70		1.50	2.14	Fem.
69	4.13	4.64	4	3.50	3.51	2.80	1.5–4.1	80+	8	8	1.40	1.75	1.44	2	Fem.
69	3.73	4.30	3.75	3.70	3.75	3	1.5–4.1	80	8	8	1.50	1.50	1.50	2.14	Fem.
69	3.86	4.45	4.28	3.75	3.75	3	1.5–5.1	80+	8	8	1.50	1.66	1.50	2.14	Fem.
69	4.14	5.27	4.66	4.66	4.66	2.80	1.5–4.2	80+	8	8	1.33	1.55	1.55	1.75	Fem.
69	4.14	5.21	4.66	3.50	3.50	2.80	1.5–4.1	80+	8	8	1.40	1.40	1.55	1.75	Fem.

[Cross Lake Thoroughfare, October 23, 1901.]

Total length in millimeters.	Head in length without tail.	Depth in length.	Eye in head.	Snout in head.	Maxillary in head.	Mandible in head.	Pharyngeal teeth.	Scales.	Dorsal rays.	Anal rays.	Height of dorsal in head.	Height of anal in head.	Pectoral in head.	Ventral in head.	Sex.
73	4	5	5	3.50	3.50	2.50	1.5–4.1	About 80	8	8	1.50	1.66	1.66	2.14	Fem.
73	4.06	5.08	5	4.28	4.28	3	1.5–5.1	About 80	8	8	1.57	1.87	1.66	2.14	Fem.
73	4.06	5.08	5	3.75	3.75	3	1.5–4.1	About 80	8	8	1.87	1.87	1.87	2.14	Fem.
65	3.79	4.41	4.66	3.50	3.50	2.80	2.5–4.2	About 80	8	8	1.40	1.40	1.40	1.80	Male.
63	4	4.33	3.25	3.25	3.25	2.60	2.5–4.2	Over 80	8	8	1.30	1.30	1.30	1.62	Male.
62	3.84	4.16	3.71	3.25	3.25	2.60	1.5–4.2	Over 80	8	8	1.30	1.52	1.30	1.75	Male.

Leuciscus carletoni Kendall, new species. *Chub Minnow.*

Head 4.35 in length; depth 4.83; eye 4.44 in head; snout 3.33; D. i, 8; A. i, 8; scales 12–73–18; teeth 2, 5–4, 2. Body elongate, rounded, back little elevated; head blunt, the profile moderately steep; mouth terminal, oblique; maxillary 3.33 in head, with small barbel just above its extremity; jaws subequal; mandible 2.85 in head; lateral line slightly decurved, nearly continuous, absent only on last scale; scales rather small. Dorsal fin inserted behind front of ventral, its height 1.25 in head; anal 1.53 in head; pectoral and ventral of moderate length, the former 1.53 and the latter 1.81 in head. Peritoneum pale.

Coloration, above, dusky olive, somewhat speckled with brown; an irregular dusky stripe along the lateral line to base of caudal, ending in a small black spot; below lateral line creamy white with brownish spots on side; dorsal and pectoral dusky; other fins pale. Colors after preservation in formalin and later in alcohol very much intensified.

Type, No. 50832, U. S. N. M., an individual 102 millimeters long, one of numerous specimens collected by W. C. Kendall and Thomas B. Gould in Bill Fish Brook, a tributary of the East Branch of the Penobscot River about a mile below Matagamon Lake, September 7, 1901. Cotypes, No. 2744, U. S. F. C.

Named for Hon. Leroy T. Carleton, chairman of commissioners of inland fish and game of Maine.

It will be seen from the accompanying table that there is considerable individual variation among the specimens from this one locality. There is also found a locality variation which, though possibly accidental, seems from its constancy to be more than that. For instance, specimens from a small pond near Hale Pond, West Branch of the Penobscot River, show a slightly shorter head, the scales run a little smaller, and there are sometimes fewer anal fin rays, with some other minor differences. The male sometimes becomes brilliantly red along the side of the abdomen from behind the pectoral fin to the lower base of the caudal. This color persists to some extent on individuals until fall.

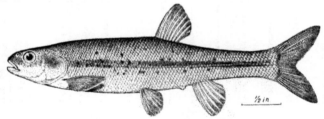

FIG. 1.—*Leuciscus carletoni* Kendall, new species.

Besides the type locality, this species has been taken in Smith Brook, which is the outlet of Haymock Lake, tributary to Eagle or Heron Lake of Allagash River and in thoroughfare between Long and Cross lakes, East Branch of the Fish River waters or Eagle Lakes of Aroostook County. It is doubtless widely distributed.

The presence of a barbel possibly should inhibit placing it in the genus *Leuciscus*. In this respect it is allied to *Semotilus* and *Couesius*. In *Couesius* the barbel is nearer the extremity of the maxillary than in this form, which in this respect is closer to *Semotilus*, to which the pharyngeal teeth would take it, being most commonly 2, 4-5, 2, which seems not to occur in a large series of *Couesius plumbeus* from Maine, or in other species. In fact, the generic distinction of *Couesius* seems to be the presence of only 4 teeth in each of the main rows. While teeth and barbel would suggest *Semotilus*, the incomplete and sometimes broken lateral line forbids that disposition of it. The barbel, however, is absent from a few of the cotypes and seems not to be present on the Southards Pond specimens.

Assuming that *Leuciscus* is the proper genus for it, which other minor characters would suggest, then it is closely allied to *Leuciscus margarita* Cope, under the subgenus *Phoxinus*. *Leuciscus margarita*, however, is proportionately a shorter and deeper fish, with slender caudal peduncle and larger scales. Our specimens have been compared with specimens of *L. margarita* from West Virginia and Lake Ontario, and barbels were found on some from the latter locality. Of 12 specimens, 4 had barbels on each side, 2 had a barbel on one side and none on the other, and 6 had no barbels at all. Barbels on our specimens can not be accounted for by age, for they are present on large and small, and of those from which they were absent one is a large individual. Its general form, and especially of the head, is that of *Couesius dissimilis*, but in the latter the mouth is much larger than in *Leuciscus carletoni*.

It would therefore seem that it must be called a *Leuciscus*, but the absence of barbels can no longer be considered a generic character by which to distinguish this group. It is very possible that this is the species reported by Cox from New Brunswick under the name of *Couesius plumbeus*, while *Couesius prosthemius* is undoubtedly *Couesius plumbeus*, as is shown by specimens sent by Dr. Cox to the United States National Museum.

Proportional measurements of Leuciscus carletoni. 10 cotypes.

Total length in millimeters.	Length, without tail.	Head in length, without tail.	Depth in length.	Eye in head.	Snout in head.	Maxillary in head.	Mandible in head.	Pharyngeal teeth.	Scales.	Dorsal rays.	Anal rays.	Longest ray of dorsal in head.	Longest ray of anal in head.	Length, pectoral in head.	Length, ventral in head.
124	116	4.29	5.52	5.40	3.85	3.50	3.50	1,5–4,2	12–68–9	8	8	1.58	1.80	1.58	1.92
123	116	4.46	5.80	5.20	3.71	3.71	2.88	2,4–5,2	13–71–9	8	8	(?)	(?)	(?)	(?)
84	69	4.31	5.30	4.57	3.55	4	2.90	2,4–5,2	13–73–9	8	8	1.39	1.39	1.23	1.88
76	63	3.93	4.84	4.57	3.20	4	3.20	2,4–5,2	12–68–8	8	8	(?)	(?)	(?)	(?)
100	84	4.20	4.94	4	4	3.81	3.33	2,4–5,2	13–66–9	8	8	1.35	1.53	1.21	1.66
85	71	4.18	4.73	4.25	3.40	4.25	2.83	2,4–5,2	12–70–8	8	8	1.66	1.81	1	1.88
87	72	4	3.60	4.50	3.60	3.60	3.27	2,4–5,2	13–70–8	8	8	1.24	1.38	1.24	1.71
73	61	4.06	4.35	3.75	3.33	3.75	3	2,4–5,1	12–72–8	8	8	1.25	1.50	1.25	1.66
73	61	4.06	4.35	3.75	3.33	3.75	3	2,4–5,1	12–71–8	8	8	1.40	1.50	1.21	1.77
74	62	4	4.76	3.82	3.87	3.82	3.10	2,4–5,1	12–69–8	8	8	1.29	1.40	1.29	1.82

Proportional measurements of Leuciscus carletoni from other localities.

[Southards Pond, August 19, 1901.]

Total length.	Head.	Depth.	Eye.	Snout.	Maxillary in head.	Mandible in head.	Teeth.	Scales.	Dorsal.	Anal.	Height of dorsal.	Height of anal.	Pectoral in head.	Ventral in head.	Sex.
111	4.12	5	4.20	3.50	3.50	3	2,5–4,2	11–73–10	8	7	1.50	1.61	1.61	2	....
115	4.34	5	4.60	3.83	3.83	3.13	2,5–4,2	12–65–10	8	7	1.64	1.52	1	2.09	....
115	4.21	4.85	4.60	3.80	5.59	3.29	2,5–4,2	14–68–10	8	8	1.53	1.69	1.64	1.93	....
109	4.09	5.62	4.40	3.66	4	3.38	2,5–4,2	12–71–11	8	8	1.57	1.57	1.57	1.83	....

[Smith Brook, September 27, 1901.]

76	3.82	5	4.25	3.77	3.40	3.40	2,5–4,2	13–68–8	8	8	1.41	1.36	1.30	1.88	Male.
70	3.86	4.83	3.75	3.75	3	3.75	2,5–4,2	66	9	8	1.36	1.50	1.36	1.87	Male.
76	3.93	4.84	4	2.66	3.20	2,5–4,2	69	8	8	1.33	1.45	1.33	1.88	Male.
75	4.13	5.16	4.20	3.33	3	2,4–4,1	68	8	8	1.36	1.50	1.50	2	Male.
46	4	4.73	4	3.60	3	1,5–4,1	12–66–8	8	6	1.64	1.80	1.64	2	Fem.
79	3.82	4.33	4.25	3.40	2.83	3.40	1,4–4,2	12–69–8	8	8	1.54	1.70	1.41	2.12	Fem.

[Cross Lake Thoroughfare, October 23, 1901.]

53	4.20	5	4.75	3.20	2.71	1,5–4,2	12–69–8	8	8	1.87	1.72	1.72	1.90	Fem.
49	4.25	5.23	4.57	3.20	3.55	2,5–4,1	13–69–8	8	8	1.45	1.88	1.60	2	Fem.
81	4.18	5.15	4.57	4	2.90	2,5–4,2	14–73–9	8	8	1.45	1.60	1.60	2	Fem.
100	4.10	4.68	4.44	3.33	3.33	2,85	2,5–4,2	14–72–8	8	8	1.66	1.53	1.25	1.81	Male.
77	4	4.42	4.57	3.20	2.90	2,5–4,2	13–69–8	8	8	1.33	1.33	1.23	1.77	Male.
75	4.20	4	4.28	3.33	3	2,4–4,2	13–73–8	8	8	1.30	1.50	1.25	1.87	Male.

Proportional measurements of Leuciscus margarita. (?)

[From Lake Ontario (Cemetery Creek near Watertown, N. Y.), for comparison with above.]

70	5	4.14	4.14	3.62	4.83	2.90	(?)	12–59–7	9	7	1.45	1.61	1.45	1.93	....
66	4.30	4.30	4.33	3.71	3.71	2.60	1,5–5,2	11–64–6	9	7	1.44	1.62	1.52	1.85	....
62.5	4.16	4.16	4.16	4.16	4.16	2.66	1,4–(?)	11–64–7	9	8	1.66	1.25	1.66	....
61	4.16	4.34	4	4	3	(?)	11–61–7	8	8	1.33	1.50	1.20	1.71	....
57	4.90	4.26	3.33	3.33	(?),5–4,1	11–58–6	8	8	....
59	4.25	4.63	4	4	3	1,5–4,1	11–59–7	9	8	....

Pimephales anuli Kendall, new species. *Blunt-nose Minnow.*

Among the fishes collected in the "thoroughfare" connecting Mud Lake and Cross Lake, October 23, 1901, were two specimens of *Pimephales*, or what were at the time thought to be *P. notatus*. But upon comparison with description and specimens of that species there were found notable differences. It was by the same methods decided that it could not be *P. promelas*, having very little in common with that species save the incomplete lateral line which at once distinguishes it from *P. notatus*. August 31, 1902, a large number of smaller individuals, which seem to be the same species, were taken in Lunkasoos Lake. Though they are apparently *Pimephales*, they bear a most striking resemblance to *Notropis*, which at first they were thought to be, rather than *Pimephales*.

These little fish in Lunkasoos Lake were fairly swarming about the shores on this and other days, and were fed upon by trout and eels. The only other minnows found in the lake were *Leuciscus neogæus* and *Couesius plumbeus*. Apparently the only other fishes than these are *Catostomus commersonii*, *Salmo sebago* (introduced)' *Salvelinus fontinalis*, and *Anguilla chrysypa*. This lake empties into the East Branch of the Penobscot through several miles of brook, though the lake in a direct line is only 1.5 miles from the river. The water is inaccessible to any fish except eels, owing to steep falls of considerable height. This lake is about 3 miles long. The shores are mainly rocky and in many places bold, but the lake is nowhere very deep. Trout here reach a large size, 5 or 6 pounds or more. From these conditions it might be expected that its fish inhabitants possibly might differ somewhat from other waters not so landlocked.

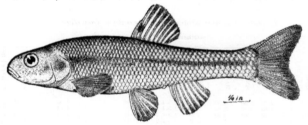

Pimephales anuli Kendall, new species.

Although this minnow has been found only in these two localities, it probably occurs in many other suitable waters. The type (No. 50830, U. S. N. M.) is the larger individual, a male fish, from Cross Lake Thoroughfare.

Named in honor of Hon. Edgar E. Ring (Latin, *anulus*), one of the commissioners of inland fishes and game of Maine.

Total length of type, 68 mm. Head 3.8 in length; depth 3.8; eye 5 in head; snout 3.33; scales 46,-13; D. i, 9; A. i, 7; teeth 4-4. General appearance of body is that of *P. notatus*, though somewhat deeper and with more arched back. The head is blunt, but with straighter profile than *P. promelas*. Lateral line very incomplete, pores upon about 16 to 20 scales; about 26 scales before the dorsal, which is inserted midway between tip of snout and base of caudal fin. Peritoneum black; intestine elongate, but not so long as in either *P. promelas* or *P. notatus* examined by the writer, about 1.5 the length of the body, not more than twice the length which is given as a generic character of *Pimephales*.

Color when fresh, light olive on back, with white sides and belly; head dark on top; an indistinct lateral stripe along axis of body, and a dark bar across base of caudal; a dark olive line from occiput splitting and passing each side of dorsal fin, reuniting behind and continuing to upper base of caudal; upper part of head, snout, and upper opercles dusky; a black spot in front of dorsal and a somewhat indistinct one similarly situated behind; other fins all pale.

The other specimen or cotype (No. 2745, U. S. F. C.) is the link that connects the above specimen with those of Lunkasoos Lake. Its total length is 55 mm. Head 4 in length; depth 3.38; eye 4.60 in head; snout 3.83; scales 45,-13; D. 8; A. 8. Similarly colored, but with no spots on dorsal fin.

The Lunkasoos Lake specimens are much smaller than the type—about the size of the cotype, but of much darker coloration. The lateral stripe is black; top of head, snout, and opercles black; scales outlined with black on back and indistinctly on sides. No spots on dorsal fin of any specimen, but on some specimens a dusky shade on front base of dorsal is distinguishable. There seem to be about 28 scales before the dorsal, which in some specimens is situated farther back than in the type, midway between pupil and base of caudal. Peritoneum black; intestine slightly more than 1.5 body without the tail. Pores on 6 to 12 or more scales. Front ray of dorsal not reaching tip of last ray when depressed; front of anal just about reaching tip of last rays.

Proportional measurements of Pimephales anuli from Lunkasoos Lake, August 31, 1902.

Total length, in millimeters.	Head.	Depth.	Eye.	Snout.	Scales.	Dorsal.	Anal.
58	3.84	4.36	4.16	3.12	45–13	i.8	i.7
48	3.72	4.55	4.40	3.66	46–13	i.8	i.7
55	3.91	4.71	4	3.42	45–13	i.8	i.7
51	3.90	4.80	4.40	3.66	46–13	i.8	i.7
56	3.75	4.71	4.16	3.57	46–13	i.8	i.7
51	3.90	4.30	4.40	3.66	45–13	i.8	i.7
51	3.81	4.66	4.40	3.57	46–13	i.8	i.7
52	3.58	4.52	4.80	3.42	46–13	i.8	i.7
52	3.50	4.20	4	3.42	46–13	i.8	i.7
50	3.81	4.66	4.90	3.57	46–13	i.8	i.7

Notropis bifrenatus (Cope). Bridled Minnow.

This little minnow, collected by Cope in the Schuylkill River and described under the name of *Hybopsis bifrenatus*, has not previously been recorded east or north of Massachusetts. In 1898 it was found in abundance by the writer in Sebago and Little Sebago lakes. It inhabits quiet, weedy coves, streams, "bogs" or "logans." It seems to attain a length of not over 2 inches. The form is very similar to that of the members of *N. heterodon* group of minnows. The lateral line is very incomplete, usually on 5 or 6 scales. Anterior rays of dorsal and anal, when depressed, extending considerably beyond last rays; anal somewhat falcate.

Color after preservation in formalin and subsequently in alcohol generally as follows: Scales on back finely dotted with dark brown, most intense on edges; dark-brown line from top of head, which is of a like color, to front of dorsal, less distinct from dorsal to upper base of caudal; a broad, shiny black stripe from snout through eye to base of caudal, where it ends in a small jet-black spot; fins all pale. In life, while in the water, the back seems of brick-red hue.

Proportional measurements of Notropis bifrenatus from Little Sebago Lake, July 27, 1898.

Total length in millimeters.	Head.	Depth.	Eye.	Snout.	Scales.	Scales before dorsal.	Dorsal rays.	Height of dorsal in head.	Anal rays.
52	4.3	4.77	3.33	4	31–9	14	8	1	7
49	4	4.44	3.33	4	31–9	12	8	1, 11	7
46.5	4.27–	4.27	3.6	4.5	32–8	12	8	1–	7
50	4.15	4.6	3.33	4	34–9	14	8	1–	7
49	4	4.4	3.33	4–	35–8	14	8	1–	7
46	4.11	4.35	3	3.6	35–9	14	8	1	7
43	4.37	5.38	3.2	4–	35–9	14	8	1	7
44	4.23	4.8	3.4	4.25	34–10	14	8	1	7
47	4.22	5.4	3	3.6	34–10	13	8	1	7
45	4.23	4.8	3.4	4.25	33–9	14	8	1–	7

Cottus gracilis (Heckel).

In a small brook tributary to Aroostook River, utilized by the State hatchery at Caribou, Me., for fish-cultural purposes, Mr. John A. Story, who is connected with the State Fish and Game Commission at that place, collected a good series of *Cottus* for the writer, who has never observed finer or larger specimens elsewhere. It seems to be locally known in the region as "rock cusk," deriving its name probably from a fancied resemblance to the "cusk" (*Lota maculosa*). In the Synopsis of the Fishes of North America, Jordan & Gilbert included all the common fresh-water sculpins under the one genus *Uranidea* of DeKay, restricting the genus *Cottus* to that group now recognized as *Myoxocephalus* in Jordan & Evermann's Fishes of North and Middle America. In the latter work, however, the fresh-water Cottidæ are comprised in two genera—one of which is *Cottus* Linnæus and the other *Uranidea* DeKay—which are distinguished one from the other only by the number of ventral rays. The genus *Cottus* is supposed to have a ventral fin formula of i, 4, while *Uranidea* has i, 3.

Out of 28 specimens otherwise essentially alike from Caribou 18 had 3 ventral rays in each ventral fin, 6 had 4 rays in each fin, and 4 had 4 rays on one side and 3 on the other. Of 15 specimens from six other localities in northern Maine, 4 had 3 rays in each ventral, 7 had 4 on each side, and 4 had 3 on one side and 4 on the other. Six specimens from Bear River, Newry, in the western part of Maine, had uniformly 3 rays in each fin.

Accordingly it would seem that the number of rays in the ventral fin will not serve to distinguish the two genera, and DeKay's *Uranidea* will have to be dropped in favor of the older name of *Cottus*. It is possible that more material will show that one or more species under *Cottus* in Fishes of North and Middle America may be identical with the form here identified as *Cottus gracilis* (Heckel).

Color after one year's preservation in alcohol: Head above, back, and sides dark gray thickly speckled with small black spots, some of which coalesce; lower parts a soiled pinkish white; soft fins all dusky, barred with pale gray on rays; spinous dorsal with jet black membrane, gray spines, margined with white which is orange in life. There are individual and sexual variations in the intensity and pattern of the colors. On some specimens the ventral and anal are wholly pale and the other fins much lighter than the above. In fact, the general shade is lighter.

Proportional measurements of Cottus gracilis, from Caribou, Me.

Total length in millimeters.	Head.	Depth.	Eye.	Snout.	Maxillary in head.	Mandible in head.	Interorbital in head.	Maxillary extends to (relative to pupil)—	Dorsal.	Anal.	Longest spine of dorsal in head.	Longest ray of dorsal in head.	Longest anal ray in head.	Pectoral in head.	Number of ventral rays.
115	3.31	5.05	5.27	3.75	2.23	2.07	5.5	Middle.	VIII, 16	13	4.83	1.81	1.93	1.03	3; 3
110	3.28	5.41	5.60	3.50	2.33	2.14	5	....do..	VII, 12	12	4.66	2.33	2	1.03	3; 3
107	3.38	5.86	6.50	3.25	2.36	2.16	4	....do..	VII, 18	12	4.83	1.85	1.83	1.04	3; 3
96	3.22	5.26	5.41	3.50	2.45	2.13	3.5	Front..	VII, 16	11	4.45	2.13	2.04	1.06	4; 4
114	3.28	6.11	5.60	3	2.33	1.20	4	Middle.	VII, 16	11	4.66	1.24	1.64	1	3; 3
110	3.37	5.35	5.10	3.37	2.25	1.92	4.5	Front..	VII, 17	12	4.50	1.92	1.80	1	3; 3
94	3.25	5.20	4.80	3	2.40	2	3.5	....do..	VI, 17	11	4.36	1.84	1.71	1.04	3; 3
92	3.84	5.13	5.11	3.53	2.56	2.09	3	....do..	VII, 17	12	3.83	2.30	1.76	1.02	3; 3
97	3.20	5	5.55	3.07	2.77	2.08	4	....do..	VII, 17	12	4.16	2.08	1.56	1.08	4; 3
101	3.23	5.09	6.50	3.46	2.26	2	4	....do..	VII, 16	12	4.83	2.16	1.85	1.08	3; 3
104	3.14	5.31	5.40	3.37	2.45	2.07	4	....do..	VII, 18	12	4.50	2.07	1.80	1	3; 3
111	3.28	5.41	5.09	3.50	2.33	2	4	Middle.	VIII, 16	11	3.50	2.16	1.64	1.03	3; 4
95	3.50	5.92	6.28	3.66	2.44	2.20	3	Front..	VIII, 16	11	3.66	2	2	.95	4; 3
118	3.34	5.10	5.27	3.22	2.07	1.93	4.5	Posterior...	VII, 16	11	4.12	1.93	1.15	.96	4; 4
103	3.40	4.72	5	3.57	2.17	1.92	3.5	Front...	VII, 16	11	4.16	1.85	1.79	.96	3; 4
93	3.45	5.06	4.40	3.14	2.20	2	3	....do...	VII, 16	11	4.40	1.76	1.76	1	3; 3
94	3.30	5.06	5.11	3.28	2.30	2.19	3	Middle.	VII, 17	12	4.18	1.84	1.78	.93	4; 4
88	3.47	1.70	1	3.50	2.75	2.10	3	Front..	VII, 17	11	4.20	1.90	1.90	.95	4; 4
85	3.50	5	5	3.63	2.50	2.22	3	....do...	VII, 17	12	3.63	1.90	1.90	.95	4; 3
102	3.40	5.86	5	3.33	2.27	2	3.5	Middle.	VIII, 16	12	4.54	2	1.85	1.04	3; 3
88	3.27	5.14	5.50	3.66	2.41	2.20	3	....do...	VII, 16	11	3.66	2	1.83	1	4; 4
89	3.47	4.86	5.25	3	2.21	1.90	3	Front..	VIII, 17	11	3.50	1.75	1.61	.95	3; 3
90	3.36	5.48	4.88	3.11	2.58	2.20	3	....do...	VII, 17	11	4.40	2	1.83	1	3; 3

THE WHITE-FISHES OF MAINE.

One species of white-fish has for many years been known to occur in certain Maine waters. Holmes mentions two species under the names of *Coregonus albus* and *Coregonus (Argyrosomus) clupeiformis*. The former the writer has decided must be the species formerly recognized as *C. labradoricus*, and the other doubtfully as *C. quadrilateralis;* but they are assigned to no particular locality. In the first report of the State Fish Commission, 1867–68, Mr. Charles G. Atkins, the commissioner, says, under the heading " White-fish (*Coregonus*)," page 25:

Of this genus we possess at least one, and probably more than one, species. They occur principally in the central, northern, and northeastern portions of the State. The species found abundantly on the St. John and its tributaries has been referred to the species *C. albus*, but we doubt whether that is correct. Whether or not our white-fish is identical with the famous white-fish of the Great Lakes, it certainly partakes of that excellence which is a characteristic of all the members of this genus. In the Fish River region, in Moosehead Lake, in Schoodic Grand, they pronounce the white-fish the best of fishes. Like nearly all the salmon family, to which they belong, they spawn in the autumn and seem to prefer running water. On the Schoodic they resort to Pocompus and Grand lakes, where the water is flowing from 3 to 5 feet deep and the bottom sandy and gravelly. In November each year small quantities of them are taken here with the spear. One night Mr. B. W. French, of Calais, set a net 30 feet long at this thoroughfare, and in the morning had a barrel of white-fish. In Moosehead Lake they sometimes take the fly. In June last we saw one taken with a fly near Mount Kineo by Artemas Libby, esq., of Augusta. It weighed 1½ pounds. Two trout weighing a pound each were taken at the same cast. They can be taken with the hook at any season of the year in deep water. Almost any bait will answer, but the best is a piece of small fish. The most of them are taken in winter. The greatest success is obtained by sinking through a hole in the ice, at the end of a line, a "cusk" thoroughly gashed with a knife. This remains there one day and tolls a great many white-fish around. They are then taken by smallest baits on small hooks. One winter many of these Moosehead Lake white-fish were sold in Augusta, and their weight was so uniformly one pound that they received the name of "pound fish," and the trouble of weighing was dispensed with by the mutual consent of seller and buyer.

The white-fish differs from most of its family in being nearly or quite destitute of teeth. Its mouth is small and tender. It has therefore none of the fierce predatory character of the trout and togue. It probably feeds mostly on small aquatic animals of various kinds, such as insects, crustaceans, and mollusks, being guiltless of the death of any of its fellow-fishes.

Several other annual reports of the State fish commission allude to these fish under the general name of " white-fish," but give no localities besides those mentioned above by Atkins and nothing further indicating more than one species.

For many years the common white-fish of Maine bore the name of *Coregonus labradoricus*, but a few years ago the well-known ichthyologist, Dr. Tarleton H. Bean, announced the identity of this species with *Coregonus clupeiformis*, or the common white-fish of the Great Lakes." Whatever changes the names may undergo, the fish remains the same for the table, unexcelled by any other fresh-water fish in Maine.

During most of the year this species (the others, too, for that matter) affects the deep water of the lakes or streams. It is essentially a lake fish, but is found throughout the year in some fresh-water streams, probably having strayed from its lacustrine home over falls which were barriers to its return. In the lakes early in the evening and throughout twilight these fishes often appear at the surface to feed upon insects, and their " rises" may be seen everywhere at some distance from the shore. The white-fish rarely, if ever, leaps from the water, and his "wake" is incon-

---

spicuous compared with that of the trout or most other fishes. At this time it will occasionally take an artificial fly, as also sometimes on cloudy days, but the most successful method of angling for it is that described above by Atkins. It may be caught in gill nets[a] if set in deep water The writer has taken white-fish in August in gill nets set at the bottom, extending from a depth of 75 to 115 feet, and from rocky to soft muddy bottom. The white-fish were about midway of the net, but this is most likely due to the part being more favorably constructed. Gill nets should be tan-colored.

This white-fish feeds upon small animals of various kinds and probably almost any kind. White-fish taken in First Debsconeag Lake August 12 and 24, 1901, contained large quantities of larvæ of a species of dipterous or mosquito-like insect.

The height of the spawning season seems to be about November 25 in the Fish River Lakes, where the fish run up the thoroughfares at night and descend before morning. They spawn in running water over gravel and where the water ranges from 1 to 3 feet deep. They also ascend streams for this purpose, but so far as

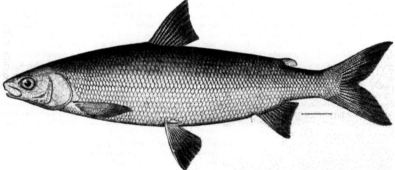

Coregonus labradoricus Richardson.

known do not spawn on shoals or shores of lakes. The nearest approach to shore spawning known to the writer is in First Debsconeag Lake, where they seek the lake end of the shallow strait connecting the lake with Debsconeag Deadwater. Atkins states that the fecundity of a 2-pound white-fish is 25,076 ova. In some Maine lakes this species attains a weight of 4 or 5 pounds, but the average is 1 to 2 pounds.

It is not known to the writer that the young of this species has been observed, except the fry at fish-hatcheries, or where they are to be found after leaving their birthplace in the thoroughfares and streams, or at what age they leave these places. It is probable that when quite young they go to deep water, where having thus escaped their enemies of the streams they become the prey of the rapacious fishes of the lake. Young individuals ranging from 4.63 to 9.5 inches long were collected in the Allagash and St. Francis waters in October, 1901, with a drag seine, along the shores of the lake. The method employed was to bait the shore about dark with fish and ruffed-grouse entrails or with corn-meal mush, and in about an hour draw the seine over the baited ground, when these fish were taken, together with hornpout,

[a] This method is unlawful in Maine except by special permit from the commissioner of inland fishes and game.

suckers, minnows, round white-fish, eels, trout, etc. In some lakes white-fish afford the principal food for trout and salmon.

This species is known to occur in Maine in the St. Croix waters—both east and west branches—Moosehead Lake, Debsconeag lakes, Allagash, St. Francis, and Fish rivers. It undoubtedly is a resident of nearly all, if not all, of the larger lakes of Maine. It is propagated to some extent by the State Fish Commission.

There is another white-fish found in Maine which is not so well known as the above, consequently no one disputes the right to its name of *Coregonus quadrilateralis*, or round white-fish, Menominee white-fish, frost-fish, shadwaiter, pilot-fish, chiven, Chateaugay shad, black-back, etc., according to the locality in which it occurs. It is found from New Brunswick westward through the Adirondacks and the Great Lakes, thence northward into Alaska. It may be distinguished from other Maine species by its more elongate, rounder body, more pointed snout, and much smaller mouth. Its habits are similar in almost every respect to the above, but it is more seldom noticed owing to its smaller size and less abundance, perhaps, and from its never being taken on a hook. It has doubtless been observed by residents of the

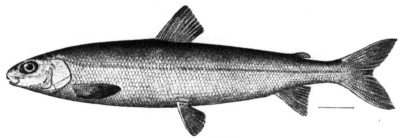

Coregonus quadrilateralis Richardson.

State and its difference from the others noticed, but it has been previously recorded from but one locality in the State—Clearwater Pond, Industry.

In 1901 the writer collected this white-fish in Umsaskis Lake, October 3, and the Cross Lake thoroughfare of Eagle lakes, Aroostook County, October 23. Late in November some were also received from Mr. John Story, who collected them with the common white-fish in Square Lake thoroughfare of the same region. It is doubtless more commonly distributed in the State than recorded observations indicate.

A NEW WHITE FISH FROM MAINE.

Supported by the opinions of such eminent ichthyologists as Dr. Jordan, Dr. Evermann, and Dr. Bean, and an abundance of material and data, the writer has no hesitation in describing a new white-fish from Maine, which will be designated *Coregonus stanleyi*. It was found in abundance upon its spawning-beds in the thoroughfare from Mud Lake to Cross Lake on the night of October 23, 1901. At one haul of a hundred-foot seine fully two barrels of these little fish were captured,[a] with them being one large specimen of *C. quadrilateralis*, several small *S. sebago*, numerous common suckers (*Catostomus commersonii*), and a few *Catostomus catostomus*.

----

a All but a few were liberated.

This spawning-bed was fine gravel covered with 1 to 2 feet of moderately flowing water. Before the haul was made the abundance of fish there was evinced by the constant "flipping" of their tails on the water surface, where it is said the spawning takes place with this as well as other species of white-fish, though the eggs sink.

There are but two instances of the introduction of non-indigenous white-fishes into Maine waters. One was *Coregonus clupeiformis*,[a] the other *Coregonus albula*, with either of which *Coregonus stanleyi* is unidentifiable. Regarding the former, in a letter dated April 1, 1901, Commissioner H. O. Stanley says:

Some twenty years ago the United States Commission sent me some white-fish eggs, I think from one of the lakes in Michigan. I hatched them at Rangeley and planted them in the upper lake—Rangeley. This winter they have been caught with hook and line in considerable numbers in Umbagog Lake, which is the fourth lake below. This is the only lake in which fishing through the ice is allowed. It is a pickerel lake. These white-fish were caught with a small live minnow. I have had some sent me twice this winter; they run in size from 1½ to 2 pounds. I presume they are in the lakes just the same and could be caught if fished for in the same way. It seems queer that they should turn up in the lower lake first, some 40 miles or more away. They are surely white-fish and none has ever been seen in Rangeley waters, to my knowledge, till this year, and I have been familiar with them all my life.

The other case was a single plant, concerning which Superintendent Charles G. Atkins, of Craig Brook Station, writes that having searched the records, as well as his own memory, he finds that he has knowledge of only one introduction of such species - namely, that of *Coregonus albula*, of which an importation of eggs was hatched at Craig Brook in the spring of 1886, and all the resulting fry, estimated at 51,000, were planted in Heart Pond at East Orland April 21 of that year.

Coregonus stanleyi Kendall, new species.

Description: Head 4.52 in length; depth 4.33; eye 4.66 in head; snout 3.81; D. 10; A. ii, 14; scales 10–82–7; gillrakers 10÷17 and 11 - 17, the longest 1.6 in eye. Body fusiform, not very deep, somewhat compressed, back gradually curving from the tip of snout to front of dorsal; head rather sharply rounded, not truncate as in *C. labradoricus*; vertical height of head from edge of branchiostegal membranes to occiput about 1.6 in length of head; maxillary reaching front of eye, 3 in head (maxillary measured from tip of snout); mandible nearly 3.5 in head; dorsal inserted in front of ventral nearer snout than base of caudal, its anterior rays extending considerably beyond tips of posterior rays when depressed, the longest 1.23 in head; pectoral 1.27; ventral 1.4, and anal 1.82 in head; anterior rays of anal not nearly reaching tip of posterior rays when depressed; caudal deeply forked, the peduncle slender, compressed, the distance from anal to first lower rudimentary rays of caudal equal to distance from adipose to upper rudimentary rays of caudal and equal to length of base of anal.

Body and head covered with white tubercles, small and dot-like on the back and belly, 1 and 2 on each scale, large and more prominent on the head and sides of body, those of the sides raised and elongate, arranged in linear series, one on each scale.

Color after preservation in formalin and subsequently in alcohol: Back, top of head, tip of snout, and around eyes, blue-black; sides and under parts yellowish, the scales margined with dusky dots. The white tubercles give the body the appearance of being striped with narrow white lines. Dorsal and caudal with blue-black rays and pale membranes; pectorals, ventrals, and anal with pale rays and slightly dusky membranes. The color when fresh was somewhat lighter, the belly and sides being more nearly white. The present color is intensification of the original shades.

---

[a] We were presented by Professor Baird, from the establishment of Frank N. Clark, Northville, Mich., 1,000,000 white-fish eggs. Owing to the extreme cold weather, long distance of transportation, and tenderness of the eggs, the percentage of loss was large, should judge about 25 per cent of the eggs hatched. They were received in February; were hatched and turned loose March 20. About 15,000 of these were put in Rangeley, the balance were turned loose in Mooselucmeguntic Lake. (Report of the Commissioner of Fisheries and Game of the State of Maine, 1881 (1882), 16.)

Type, No. 50828, U. S. N. M., a male 222 mm. (nearly 9 inches) long, from thoroughfare between Mud and Cross lakes, Maine, October 23, 1901. Cotype No. 2746, U. S. F. C.

Named for Hon. Henry O. Stanley, one of the commissioners of inland fishes and game of Maine.

The males of this species are conspicuously marked with the white tubercles, and many, though not all, of the females have them to a lesser extent. In this respect they resemble *Coregonus williamsoni*, but otherwise differ markedly. The tables of proportional measurements do not reveal a great many characteristics to distinguish this fish from its closely-related cogener, *C. labradoricus*, of the same waters. But to the eye, specimens compared may be readily distinguished by the sharper, less truncated snout, shorter appearance of the head, having more the general appearance of *C. quadrilateralis*, yet differing from this in conspicuous details, such as body less slender, shape of head, less curved profile, less compressed snout, and larger mouth, etc.

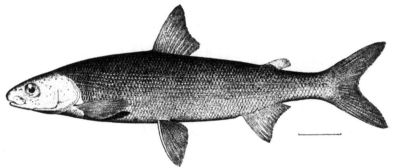

Coregonus stanleyi Kendall, new species.

Then again, it is of much smaller adult size than *C. labradoricus*, which in the same waters attains a weight of 5 pounds. In the breeding season it is more conspicuously and characteristically tuberculated, and the height of the spawning season on the same grounds is one month earlier. On the average there are more gillrakers, which are considerably longer than in the other species.

A female 230 mm. (a little over 9 inches) in length contained 3,447 eggs.

This fish abounds in the chain of Eagle Lakes, and is doubtless a conspicuous item in the menu of salmon, togue, and trout. Reports of small white-fish from other parts of the State indicate that this species may be common in other waters. It is never, or very seldom, seen, except in breeding season, but very likely could be caught with fine-meshed gill nets made of fine twine if set in deep water. It is an excellent pan fish.

Proportional measurements of Coregonus stanleyi (cotypes).

Total length in millimeters.	Head.	Depth.	Eye.	Snout.	Maxillary.	Mandible.	Scales.	Dorsal.	Anal.	Height of dorsal in head.	Length of anal in head.	Pectoral in head.	Sex.	Teeth on tongue.	Longest gillraker in eye.	Number of gillrakers.	Whole number of gillrakers on each side.
238	4.76	4.47	3.90	4.30	3.07	2.38	10-82-7	10	13	1.30	1.53	1.28	♀	0	2.50a	9+17; 9+17	26; 26
235	4.71	4.40	4.42	4.20	3.23	2.33	10-82-8	10	13	1.23	1.90	1.27	♂	0	1.90	9+17; 9+16	26; 26
217	4.77	4.66	4.40	4.82	3.14	2.44	10-90-7	11	14	1.18	1.69	1.22	♂	0	1.81	10+17; 9+17	27; 26
215	4.47	4.28	4.70	3.91	3.35	2.35	10-86-7	10	13	1.23	1.88	1.23	♀	0	1.81	10+18; 10+17	28; 27
221	5	4.75	4.50	4.22	3.16	2.12	10-82-6	10	12	1.11	1.21	1.18	♀	0	1.60	9+18; 9+17	27; 26
222	4.87	4.14	4.21	3.80	3.33	2.38	10-82-6	12	13	1.11	1.30	1.25	♂	0	2.11a	10+16; 9+17	26; 26
231	4.77	4.47	4.88	4.19	3.14	2.31	10-87-8	11	13	1.18	1.76	1.25	♂	0	1.80	10+18; 9+19	28; 28
230	4.64	4.48	4.42	4.30	3.23	2.31	10-89-8	11	13	1.23	1.61	1.35	♀	0	1.90	10+19; 11+18	29; 29

a These exceptional results are explained by the unusually large eye in these 2 specimens

Coregonus labradoricus.

[Eagle Lakes, November 23, 1901.]

Total length in millimeters.	Head.	Depth.	Eye.	Snout.	Maxillary.	Mandible.	Scales.	Dorsal.	Anal.	Height of dorsal in head.	Length of anal in head.	Pectoral in head.	Sex.	Teeth on tongue.	Longest gillraker in eye.	Number of gill-rakers.	Whole number of gillrakers on each side.
435	4.75	4.22	5.33	4	3.63	2.75	10-87-8	11	12	1.31	1.81	1.31	♀	Few .	2.14	11+15; 10+15	25; 25
420	4.93	4.97	5	3.94	3.57	2.58	10-87-8	9	11	1.29	1.70	1.22	♂	..do ..	2.14	9+13; 9+13	22; 22
4	4.69	3.61	5.40	4.05	3.24	2.61	10-83-8	11	14	1.35	1.87	1.30	♀	..do ..	2.14	9+17; 9+17	26; 26
380	4.44	3.66	4.80	4.80	3.60	2.66	10-87-8	11	12	1.20	1.71	1.24	♀	..do ..	2.30	9+15; 10+15	24; 24

[Debsconeug Lake, August 1, 1901.]

3.75	4.63	4.15	4.92	3.83	3.28	2.46	10-82-9	11	12	1.13	1.81	1.13	♀	(?)	2.37	9+16; 9+17	25; 26
3.80	4.35	4	5.57	3.90	3.39	2.60	10-85-9	11	13	1.28	2.05	1.50	♂	(?)	(?)	11+17; 11+16	28; 27
3.30	4.30	3.83	4.81	3.94	3.09	2.60	10-84-9	11	13	1.30	1.71	1.32	♀	Few .	2.70	11+17; 11+17	28; 28
3.60	4.76	4.42	5.00	3.82	3.25	2.70	10-82-8	11	13	1.14	1.62	1.30	♀	(?)	2.83	10+15; 10+16	25; 26

Comparative average measurements of Coregonus stanleyi and Coregonus labradoricus (8 specimens of each).

Species.	Head.	Depth.	Eye.	Snout.	Maxillary.	Mandible.	Scales.	Dorsal rays.	Anal rays.	Height of dorsal.	Height of anal.	Pectoral.	Longest gillraker in eye.	Gillrakers.
C. labradoricus, Cross and Debsconeug lakes	4.60	3.98	5.10	4.03	3.38	2.62	10-84-8	10.7	12.5	1.25	1.78	1.29	2.30	25.3; 25.3
C. stanleyi, Cross Lake..	4.74	4.41	4.42	4.28	3.08	2.33	10-85-7	10.6	13	1.19	1.48	1.25	1.92	27.1; 26.7

Comparative measurements of young Coregonus labradoricus and Coregonus stanleyi of the same length.

Species.	Total length in millimeters.	Length without tail.	Head.	Depth.	Eye.	Snout.	Maxillary.	Mandible.	Scales.	Dorsal.	Anal.	Height of dorsal.	Height of anal.	Pectoral.	Longest gillraker in eye.	Gillrakers.
C. stanleyi	230	205	4.61	4.18	4.42	4.20	3.23	2.83	10-80-8	11	13	1.23	1.61	1.35	1.9	29; 29
C. labradoricus ...	230	190	4.75	4.75	4.21	4.44	3.23	2.43	10-83-8	11	13	1.17	1.02	1.29	2.37	25; 25
C. stanleyi	235	198	4.71	4.10	4.42	4.20	3.23	2.33	10-82-8	10	13	1.23	1.90	1.27	1.9	26; 26
C. labradoricus ...	235	200	4.08	4.44	4.44	4.08	3.30	2.45	10-79-8	13	13	1.28	1.44	1.82	2.87	25; 23
C. stanleyi	235	210	4.77	4.47	4.88	4.19	3.14	2.31	10-87-8	11	13	1.18	1.76	1.25	1.8	29; 28
C. labradoricus ...	210	210	4.77	4.40	4.40	4.38	2.58	10-81-8	11	13	1.25	2	1.33	2	28; 26

CONTRIBUTIONS TO THE BIOLOGY OF THE GREAT LAKES.

THE PLANKTON ALGÆ OF LAKE ERIE, WITH SPECIAL REFERENCE TO THE CHLOROPHYCEÆ.

By JULIA W. SNOW,

Instructor in Botany, Smith College, Northampton, Mass.

F. C. B. 1902—24

THE PLANKTON ALGÆ OF LAKE ERIE, WITH SPECIAL REFERENCE TO THE CHLOROPHYCEÆ.

BY JULIA W. SNOW,

Instructor in Botany, Smith College, Northampton, Mass.

INTRODUCTION.

The unicellular algæ, which in themselves show most interesting characteristics in their structure and life history, come to have a double significance when considered in connection with their environment. Investigation shows the presence of an intimate connection and interdependence between them and their surroundings. While they depend upon substances in the water for nutrition, they in turn probably perform a valuable function, the same that has been proved by Bokorny '94 and Strohmeyer '97 in the case of some higher algæ, that of purifying the water, reducing the amount of bacterial growth accompanying decay, and rendering the medium fit for higher life. Their value also as a food supply to the aquatic fauna is well known. In any biological study of a body of water the algæ must therefore receive attention, and should be considered with reference to their environment rather than as independent unrelated entities.

A study of this kind should be continued for a number of years, for aside from the desirability of repeated observations, it is necessary on account of variations in the flora from year to year. Certain species may be abundant each year, but others are periodic in their appearance, being found only at intervals of three or four years; and forms, more or less polymorphic, have been known to appear almost exclusively in one condition one year and in another condition the next, so that their identity has not been known until their life history has been traced. Such variations must be due to variations in environment, so that before these phenomena can be understood the environment must be known and its influence determined.

In the natural state, the elements in environment are so numerous and so connected that to know definitely which of these produce a certain effect on an organism is impossible. This must be ascertained under artificial conditions and experimentation must be resorted to for this purpose. Under these circumstances the environment may be altered, certain of its elements may be eliminated and the effect of others studied, so that after repeated trials we may arrive at more definite knowledge of the life principles of these organisms than would be possible in the native state. When the relation to environment is definitely known, then we may go still further and, by changing this environment, exert a certain control over these

371

organisms, causing them at will to reproduce, or to assume any stage in their development which we may desire.

At present great confusion exists in the nomenclature of these lower vegetable organisms. Many, in certain stages of their development, can not be distinguished from one another, and even some polymorphic filamentous algæ have often been confused with unicellular forms, as they may assume a unicellular condition in which even a skilled observer is unable to distinguish them with certainty from true unicellular plants. The entire life history should therefore be traced, and although there may be stages in the development of different ones which can not be readily distinguished, a broader knowledge must aid in recognition.

To determine accurately the life history of a species, observation should be made from pure cultures. The need of this has been pointed out by Klebs '96, Artari '92 and Senn '99, and has often been suggested by the inaccurate work of a number of investigators, in which many species have been confused. If one start with a single cell or a small cluster of cells, all of which are known to be the same, and from these procure an unlimited supply of absolutely pure material, then one can assert with certitude that whatever developments occur they are characteristic of that species, whereas if the material be not pure one is easily misled as to the connection of different forms. But even a pure culture under one set of conditions is not sufficient. Material should be subjected to all possible conditions which might ever occur in nature, and the effect of these conditions studied in all phases of development. When this is done we may venture to classify the organism, and then many phenomena not now understood will probably be explained.

In the present study, which was continued during the summer of 1898, 1899, and 1900 at Put-in Bay, Lake Erie, and at Ann Arbor, Mich., in 1900, the work has been largely preparatory, and has been confined to comparatively few of the numerous forms present. The first summer was devoted principally to becoming acquainted with the forms found in Lake Erie, and in experimenting with culture media, that pure cultures of the different forms might be obtained and the conditions governing development be determined. It was soon found that although algæ existed side by side in the water of the lake, the conditions which determined their growth were not the same—that the favorable conditions for development must be determined for each genus, and often for each species individually. Comparatively few would live at all in the media which are so generally used for more hardy forms found in stagnant pools. After determining the favorable media for some of the most common forms in the plankton, the following summers were devoted to tracing their life history and studying such biological facts as could be determined.

As the amount of work to be done was so great, it was thought best to limit investigation to some special group, and as the *Chlorophyceæ* are more easily maintained in culture and are more varied in their development, requiring more constant observation, they were taken first. All species to which special attention is given in this paper, unless a statement to the contrary is made, were taken from the plankton, and an abundance of pure material was obtained by cultivation. Cultures were made in the ordinary Stender dishes, and parallel with these were also continued hanging drop cultures, where the development of the same individuals could be observed from day to day and no step be overlooked.

Aside from tracing the development of a number of the members of this group, a partial list is added of the algæ occurring in the plankton, though this is by no means complete, as comparatively little time was given to the determination of the Diatoms. The material was merely preserved for future examination. The determination of the Desmids was mainly left to Mr. A. J. Pieters, who has given a list of this family. (Pieters '01.) A number of the *Cyanophyceæ* also, which are common in the plankton, have not been determined. These are minute gelatinous forms occurring as flocculent masses in the water, and though the structure of the cells is constant, the form of the colony is more or less variable, depending apparently on the age of the colony and the kind and amount of nutritive substances in the water. To make an accurate list of these would require careful comparison and a more perfect knowledge of their life conditions than we now have. Undoubtedly many of them are undescribed species. Though the list of *Chlorophyceæ* here given is fuller than that of other classes, it is by no means complete. Some unrecognized forms were met with where attempts at cultivation failed, due to inappropriate culture conditions, and as a result classification could not be made.

In the examination of fresh plankton material, the more conspicuous forms were easily detected, but there were always a number of minute forms, such as *Chlorella*, *Chlorosphæra*, and *Chlamydomonas*, which easily escaped notice, or, if observed, they appeared as single green cells which could not be identified. That these might be taken account of, and not be altogether overlooked, large cultures were started from the fresh plankton, and in these cultures developed many such forms which had escaped observation in the examination of the fresh material. Some of these were isolated from all other forms of algæ and their development studied. It is believed, however, that farther study in this line will give many additional species and many interesting biological facts, for as yet but few of these larger cultures have been thoroughly examined and the species determined.

For the names of species, where a detailed study is not given, the determination is based on the simple descriptions of other authors. The list is given, however, only as a temporary guide to the forms present, for it is believed that new methods of investigation, when applied to the development of even some of the best recognized genera, will change the nomenclature considerably. Some forms which have been classed together may prove to be distinct species, and possibly others which show variation should be combined to form one species.

In the physiological work done on these forms, by far the greatest amount of attention has been given to the subject of nutrition and culture media. Temperature is of less importance, for relatively great variations do not seem to affect them. The water in the natural condition never reaches a temperature so high as to kill them, and low temperature—even freezing, at least in some cases—does not end their existence, but seems to affect them mainly in reducing their rate of increase. The degree of light, too, in which they can live would seem to vary largely, as they are often found at considerable depth, as well as at the surface. The belief of many recent investigators that algæ with chromatophores may make use of both organic and inorganic substances in their nutrition, is supported by the experiments of Artari '01 and Knörrich '01, both of whom found that the algæ used in experimentation thrived much better when organic substances were present in addition to the inorganic. Artari even found that at least certain forms could live and remain green in total darkness.

It has been the experience of the writer that great variation exists among different algæ on this point and that the kind of substance and the amount best suited to development must be determined for every alga selected for culture. While some grow more luxuriantly in a purely inorganic solution, others, among which are the unicellular blue-green algæ, seem to prefer a solution where at most but a trace of mineral matter is present. The culture medium most favorable in a large number of cases was a decoction made from the organic matter of the plankton. This seemed especially favorable if large quantities of *Anabæna flos-aquæ* were present. This observation that the organic matter of the water could be used by the algæ has suggested a possible explanation of the great increase of algæ at certain seasons, causing the "water bloom." This phenomenon has been observed by the writer but three times, but at each time it was known that an unusual amount of dead organic matter was in the water of that vicinity. At one time the matter was in the form of numerous small dead fish floating on the water; at another time a quantity of refuse had been emptied into the bay where the water bloom was noticed; and a third time large areas of the surface of the water were covered with the skins of Ephemera which are shed before the insect reaches the imago state. Such phenomena as these can be explained only experimentally, and it is along these lines of increase and source of nutrition that further investigation should be carried.

THE STRUCTURE AND LIFE HISTORY OF CERTAIN PLANKTON ALGÆ.

Chlamydomonas gracilis Snow, new species.

This species of *Chlamydomonas* (fig. 1) in its most vigorous and normal motile condition is cylindrical, ovoid or ellipsoidal in shape, rounded at the posterior end and bluntly pointed at the anterior end. Length 10.5 to 13 μ; breadth 5 to 6.5 μ. In the nonmotile condition the cells are ovoid or spherical, and often motile individuals of the same shape are noticed, with a diameter of 9 to 10.5 μ. The chloroplast and entire contents are sometimes withdrawn from the membrane either at the anterior or posterior end. When at the anterior end the two protoplasmic flagella can be seen to be continuous with the protoplasm within. The flagella are somewhat longer than the cell.

The single hollow chloroplast lines the membrane throughout, except for a very small area just at the anterior end, at which point two pulsating vacuoles can be seen. The color is a dull bluish-green, rather than a vivid green. Oil is always present. The pyrenoid is in the extreme posterior end of the cell. The pigment spot is a conspicuous dull-red disk, and is often situated as far back as midway between the two ends or even farther. The nucleus occupies a position between the center and the anterior end of the cell. After division the cells are liberated by the enveloping membrane becoming dissolved at one point, through which the new individuals escape, leaving the empty membrane behind.

This species, like most species of *Chlamydomonas*, grew and reproduced readily in a 0.2 to 0.4 per cent Knop's solution, and this culture medium was used to trace the life history of the species.

On transferring material from Knop's solution to water, individuals were formed which were taken to be the gametes, though only in one instance was indication of copulation noticed (fig. 1, 4). These were in all respects like the ordinary motile form, except that they were smaller, ovoid in shape, and had no membrane (fig. 1, 3). Though the species resembled *Chlamydomonas debaryana* Goros., it is much smaller and more cylindrical in shape than that species.

This species was found 2½ miles north of Kelley Island, in Lake Erie. It is by no means widely distributed in the water of the lake.

Chlamydomonas communis Snow, new species.

This species in the motile stage resembles closely the preceding species, but after cultivating the two forms in pure cultures side by side for over two years, and finding characteristics which are distinguishing and constant, they have been separated into two species. The size and shape of the two are almost identical, the shape being oval or ellipsoidal and pointed at the anterior end (fig. 11).

The dimensions are 10.5 to 13 μ long and 6.5 to 8 μ broad. The color is a brighter and yellower green than that of the preceding species; the pyrenoid, instead of being at the extreme posterior end of the cell is near the center, and the pigment spot is an inconspicuous elongated strip of dull red which can rarely be distinguished, except when viewed at the side. In all other respects the structure of this species resembles that of the preceding. The division is longitudinal. *Chlamydomonas communis*, though in general appearance greatly resembling *Chlamydomonas media* Klebs, is smaller, the largest cells being only about half as large as the largest of that species. The mode of division also in the two species is so different that the two could not be classified together.

This species was found in many collections taken at the western end of Lake Erie.

Chlamydomonas globosa Snow, new species.

In the natural condition in the plankton of Lake Erie this species exists abundantly, but in a form not easily recognized as a *Chlamydomonas*. In appearance it resembles *Pleurococcus regularis* Artari, consisting of one or more clusters of spherical cells, more or less separated from each other, and all imbedded and held in place by a thick, gelatinous covering. When first placed in culture the gelatinous envelope disappears, the cells become isolated and the normal appearance of a *Chlamydomonas* is assumed; but when division occurs the alga takes again the cluster form as found in the plankton.

In the motile form the cells are spherical or slightly ellipsoidal, with a diameter of 5 to 7.8 μ. No anterior beak is present. There are two flagella, as long or slightly longer than the cell, and a small inconspicuous pigment spot at the side, about half way between equator and cilia (fig. III).

The chloroplast extends to the extreme anterior end of the protoplast, and is much thickened at the posterior end, in which portion the pyrenoid lies. The pyrenoid is enveloped by a thick layer of starch. Only a single pulsating vacuole can be distinguished at the anterior end, but this is unusually large in size. Several globules of oil are present in the anterior portion of the cell. Often the cell contents are withdrawn from the membrane, either at the anterior end, the posterior end, or at all points. Gametes were not found.

After division the cells are liberated by the cell wall becoming gelatinous. In 0.2 per cent Knop's solution, where division took place normally and rapidly, the cells existed in clusters of four, which resembled in every respect some of the cell compounds found in the plankton. This species of *Chlamydomonas* was cultivated for a period of two years, and during this time no variation was noticed.

Scenedesmus bijugatus var. flexuosus Lemm.

The form under consideration is identical with *Scenedesmus bijugatus* var. *flexuosus* described by Lemmerman '99, except that in a coenobium 32 cells seem to occur more frequently than 16 (fig. IV, 1). Both numbers frequently appear in the plankton of Lake Erie, however, and the two forms are undoubtedly the same. This variety was cultivated by the author for about a year under a large number of conditions, and as some points were observed, not noted in Lemmerman's description, they are given here.

It was first thought from its general resemblance to *S. bijugatus* that it might be this species which had assumed a greater development due to unobstructed light and the inexhaustible supply of oxygen, carbon dioxide, and nutritive substances which are constantly supplied by the ever-moving water of the lake, but cultivation of the species for some months, during which many generations were traced, proved that the great number of cells was characteristic for the organism, and that when placed under the artificial conditions, where the supply of air and nutrition were not so constantly renewed as in the lake, it did not necessarily revert to the usual form of *S. bijugatus* with 8 cells. It is true that, under special conditions, where the vitality was low, it sometimes produced an 8-celled coenobium, but in the same culture where 8 cells were found coenobia of 16 or 32 cells were also found. The coenobia of 8 or 16 cells produced again coenobia of 32 cells, so that it would seem that the larger number of cells was normal, rather than abnormal.

The greatest diameter of the cells of a mature coenobium is 20.8 μ, while the shortest is 8.9 μ. A young coenobium of 32 cells measured 160 μ in length, while an older one measured 364 μ. The great length of one of these individuals strongly suggests a filamentous alga. The shape of the cells in young coenobia is cylindrical, with slightly rounded ends. In older individuals which are passing into a resting condition the ends become more rounded and the shape more ellipsoidal. In the mature resting stage the cells are spherical (fig. IV, 2).

The membrane is perfectly smooth without processes or markings of any kind. The composition of the membrane is cellulose, turning blue when treated with iodine and sulphuric acid. In the younger individuals the membrane is comparatively thin, but when the cell passes into a resting condition the membrane becomes very much thickened, is 2.5 to 3.25 μ in diameter, and two, three, or sometimes four layers are distinguishable. The thick inner layers are also of cellulose, while the outermost layer becomes to a greater or less degree cutinized. As the cells pass into a resting stage and become spherical in shape, the surface of contact between two adjoining cells becomes less and less, and finally they break away from each other and exist singly.

The chloroplast, under natural conditions, is a thin, homogeneous layer, irregularly interrupted at the center, and forming a lining to the membrane. At one side near this point a large pyrenoid is present. Under cultivation, in most media, the chloroplast assumes a granular appearance on the surface and the perforations are obscured. Later a large amount of oil is developed which is readily dissolved in absolute alcohol. As the cell passes into a resting condition this oil gradually assumes an orange color. On account of the ease with which the cells pass into a resting condition the normal condition of the chloroplast can with difficulty be maintained under cultivation.

The nucleus is small and lies near the pyrenoid, sometimes on one side, sometimes on the other. Staining with hæmatoxylin brings out the presence of several large vacuoles in the cell cavity.

In its relations to external conditions this variety seems in many ways to deviate from most other algæ. In a number of solutions, found generally to be favorable for algal culture, this variety simply passed into a resting condition. The only solution tried which really proved to be favorable was a solution of decaying *Anabæna flos-aquæ*, which occurred at times in great quantities on the surface of the lake. In this the development seemed normal. In an organic solution (decaying peas) and in 0.2 per cent Knop's solution the color became green and healthy, but no reproduction occurred, at least for many weeks. A solution from the organic material of the plankton proved favorable to reproduction, but old and young cœnobia alike soon became filled with orange-colored oil, passing into a resting condition, and remained in this condition until the nutrition of the medium was finally exhausted, or until they were transferred to a fresh and favorable solution.

Of the inorganic solutions, Sachs's, Knop's, Oelmann's and Knop's solution without calcium, Sachs's solution was the only one that was at all favorable. Here reproduction occurred readily, and the cells assumed a normal appearance, but even in this solution, after a time, the cœnobia gradually passed into a resting condition.

Staurogenia apiculata Lemm.

This species, which is very generally, though not universally, found in the plankton of Lake Erie, is undoubtedly that described by Lemmerman '98 as *Staurogenia apiculata*. His figure and measurements agree quite closely with those of the Lake Erie species, though his description leaves us in some doubt in regard to details.

In Lake Erie this alga may occur either as individual cœnobia, composed of 4 cells lying in one plane (fig. V, 4, 5), or these cœnobia may be united into large, more or less irregular rectangular plates of cells, measuring 50 to 150 μ on a side (fig. V, 1).

The 4 cells of the cœnobium are either lemon-shaped or oval, and are arranged to form a rectangle with a diamond-shaped space at the center. Of all the species of the *Cœnobiæ*, this is apparently the most constant in regard to the number of cells. In other members of this tribe the number of cells of any daughter individual depends very largely upon external conditions and the vitality of the parent cœnobium, but of the many thousand cœnobia of this species examined under widely varying conditions, only one individual showed any deviation as to number of cells. In this case division was incomplete and but 3 cells were formed instead of 4, though four pyrenoids were present.

The large plate-like structures which are found in the tow, and which also occur in cultures, arise from the daughter cœnobia remaining after liberation in the position in which they are formed (fig. V, 1), being held in place by a colorless, gelatinous substance which surrounds each individual. As each of the 4 cells gives rise to a daughter cœnobium of 4 cells, all of which lie in the same plane, a plate of 16 cells is formed, and as each of these again produces a cœnobium, a compound cœnobium of 64 cells is produced. This process continues, but the plate-like structure soon becomes more or less broken and distorted, as is seen in the material from the plankton, and the irregularity increases as reproduction continues. As the gelatinous substance which holds these together is invisible without reagents, one receives the impression that each plate-like mass is a cœnobium or individual, whereas in reality it is made up of many.

Under normal conditions of growth the cells of this species are ovoid or lemon-shaped, with the membrane projecting into a very short and almost obscure wart at one or both ends of the cell (fig. v, 4, 5). The typical shape is evidently that of a lemon which is more or less unsymmetrical with reference to its long axis. But in the cœnobia where two cells are in contact their opposed ends often become more or less flattened. The projecting portions of the membrane are then often not formed, and the cells are quite distinctly ovoid, the broader portions of the adjoining cells turned toward each other. The mature cells measure 5.2 to 8 μ long and 3.25 to 5.2 μ broad, while in young cœnobia they are 5 to 5.8 μ long and 3.25 to 4 μ broad.

The membrane is very thin and consists of cellulose as shown when treated with iodine and sulphuric acid. Surrounding the membrane and enveloping the whole cœnobium or compound cœnobium is the homogeneous gelatinous substance which unites the separate cœnobia. This is apparently excreted from the cells and is not a dissolved portion of the membrane, as the membranes of one or more preceding generations are sharply defined and lie embedded in this substance (fig. v, 1). The ways in which this responds to different stains are various. Fuchsin-iodine green neither stains the gelatinous substance nor is capable of penetrating it. It therefore leaves the cell contents uncolored. Hæmatoxylin, fuchsin, and safranin stain the cell contents, but not the gelatinous envelope; the latter, however, takes a deep color with gentian violet. The structure is best brought out by tannate vesuvine which stains it brown. With this stain single cœnobia show but a single layer of this substance, often thicker than the diameter of the cells themselves, though varying somewhat in amount. In the large compound cœnobia several layers are made visible, each successive outer layer being less dense than the adjoining inner layer. These different layers are the gelatinous envelopes developed during the different generations and retained from one generation to the next. The inner denser layer is sharply outlined from the others and is 3 to 3.5 μ thick. With the tannate vesuvine fine radiating lines are brought to view at right angles to the surface of the cell, and undoubtedly indicate a prismatic structure (fig. v, 2) such as described by Klebs '86 for *Zygnema*. The second or next outer layer shows no such striations, but is quite definitely outlined, while the third and outermost visible layer is more or less indistinct and gradually vanishes into the surrounding medium.

The chloroplast is thin, parietal, and forms a close lining to the membrane. In some young cœnobia, on the side of the cell next to the central space, there was seen to be an opening through the chromataphore, but in mature specimens no trace of an opening could be detected. Lying imbedded in the chloroplast is a single, relatively large pyrenoid surrounded by a thick layer of starch. The position of the pyrenoid in the cell is in no wise constant, as sometimes it lies nearer one end of the cell and sometimes nearer the other. Its position also in reference to the nucleus is not constant; the latter, however, occupies different positions in the cell according to the age.

In young cœnobia the nucleus invariably occupies a position near the wall adjoining the central space (fig. v, 6a), while in older individuals it moves toward the center or near to one end of the cell. The minuteness of the nucleus renders a detailed study difficult, but material stained with hæmatoxylin showed strands of protoplasm radiating from it, and in one case a nucleolus was plainly visible (fig. v, 8). In the same material also a stained network appeared throughout the cell and was undoubtedly due to the arrangement of the protoplasm and vacuoles. Small globules of oil were always present, and occasionally larger globules. This oil became darkened by osmic acid and was dissolved in 10 per cent potassium hydrate. It was not dissolved in absolute alcohol, which would show the oil to be of a fatty rather than an ethereal nature.

The new individual arises from the successive bipartition of the contents of any cell of the cœnobium. The first division is a transverse one (fig. v, 4). The second division, which occurs in each of the products of the first division, is at right angles to the first and in the same plane in the two products, so that the elements are arranged in the form of a rectangle while still within the mother membrane. The division of the different contents of the cell is not simultaneous, as the chromataphore is divided before the pyrenoid, and judging from their relative position in the cell at this time, both of these divide before the nucleus. The four daughter nuclei occupy a position near the point of contact of the four chromataphores (fig. v, 4a), while the pyrenoids lie toward the opposite end of the daughter cell (fig. v. 4). The nuclei retain this position in the cell for some time after the cells are liberated (fig. v, 6). Before liberation each cell becomes invested with a membrane.

The daughter cœnobium is set free by the membrane of the mother cell becoming ruptured from the middle of the base to the apex, so that one longitudinal half is loosened and thrown back like a lid, thus setting free the daughter individuals (fig. v, 7). Under conditions where but little gelatinous substance is present and only single cœnobia are produced, many such empty membranes are found in the surrounding medium. Where the gelatinous substance is in great abundance, and large plates of cells are formed, these remnants of membranes for two or more generations may be found clinging to the sides of the cells.

Physiology.—This species of *Staurogenia* is greatly affected by the kind of substances available for nutrition. Cultures were made in a large number of different solutions—Knop's solution, sugar solution, decoctions of mixed vegetable and animal matter found in the plankton, decoctions of pure vegetable matter, and decoctions of earth. Other solutions were also used which will be referred to later. In most cases cultures were made in a number of different concentrations of the same solution, so that amount of substance, as well as kind, was taken into consideration. For rapidity of increase the organic solutions seemed to be of far more importance to the species than the inorganic, and the solutions found most favorable to this were the decoctions of organic matter from the plankton, though a solution from decaying peas was also favorable. In fresh cultures in these solutions the plates usually consisted of sixty-four cells (fig. v, 1), one-half usually being folded back upon the other, but in the older cultures, where the material had increased greatly, the plates were small, consisting in a large number of cases of sixteen cells (fig. v, 2).

Knop's solution, which is a favorable medium for a large number of algæ, did not prove favorable to either reproduction or development in this species. In cultures of 0.4 per cent, 0.1 per cent, and 0.05 per cent of this solution, very little increase took place, the color was pale and the regularity of the cœnobia was lost, while, owing to the absence of gelatinous substance, the large masses were never formed. In 0.1 per cent and 0.05 per cent growth was more abundant than in the 0.4 per cent. The cells showed no regularity of arrangement, however, and no trace of gelatinous substance could be detected even when treated with tannate vesuvine (fig. v, 3). In the decoctions of earth development was normal, increase was rapid, and the general condition was vigorous. In a 2 per cent sugar solution the species lived but a short time. In cultures where nutrition had become exhausted the cells assumed a dull rust color, due to the presence of oil. This probably was a resting stage, though all attempts to resuscitate it after it had been dried in this condition failed; when not dry, however, the green color soon returned if fresh nutrient solution was added.

As this species did not flourish in Knop's solution the question arose as to what element in the compound was detrimental to the alga and prevented development. It was thought that it might be the large amount of calcium in Knop's solution which produced this effect, and to determine this point Knop's solution was then made without calcium, and a solution without calcium used by Oelmann in the cultivation of sphagnum was tried. Cultures were made in various concentrations of both these solutions with the following results: In a 0.4 per cent Oelmann's solution development was far more natural than under any other artifical conditions, the group of cells being much larger than in other cultures. The same appearance, though to a less extent, was found in other concentrations—0.1 per cent and 0.05 per cent of the same solution and also in 0.5 per cent and 0.25 per cent of Knop's solution without calcium. Apparently, then, calcium interferes with development, and in nature the organism probably would not find a habitat in water containing much of this element, but would seek a soft water rather than a hard. The number of clusters which were formed in these cultures without calcium was few in comparison to the number in organic solutions and the cells were a trifle smaller. The direct cause of the development being higher where the individuals are fewest was not determined in this case, but it seems to be true of the other algæ as well as of this.

Fusola viridis Snow, new genus and new species.

In the natural condition, as found in a stagnant pond in Middle Bass Island, in Lake Erie, the cells of this species were single, fusiform in shape, and sometimes slightly sigmoid (fig. vi, 3). They vary from 27 to 39 μ in length and 8.5 to 21 μ in width. A medium size is 28.5 μ long and 8 μ broad.

A distinguishing character is the gelatinous covering, usually about 6.5 to 8 *u* thick, surrounding each cell. This is excreted from the cell and is of such a consistence as to be plainly visible under the microscope, but shows no laminated structure. In mature cells no further structure is visible in the gelatinous substance; but in rather young individuals, if cells be stained with tannate vesuvine, and sometimes without staining, two portions of the ruptured mother membrane are seen lying imbedded in this gelatinous substance (fig. VI, 4).

The membrane of the cell is composed of cellulose, which shows the blue color with iodine and sulphuric acid. The gelatinous envelope remains unaltered in appearance with these reagents. The chloroplast occupies the larger portion of the cell, leaving near the center only a relatively small space, which in direct view appears as a lighter circle. In this lies the nucleus. A large pyrenoid is prominent, lying also near the center of the cell. Small globules of oil occur and become darkened by osmic acid. In old cultures the cell assumes something of a brownish color, which may represent a resting stage, but when such a brownish cell was allowed to dry it could not be induced to grow again.

The reproduction takes place by a single transverse division of the contents of the mother cell (fig. VI, 1, 2). The two parts thus formed gradually elongate in opposite directions from the point where division took place, one slipping by the other in the process (fig. VI, 3, 4) and both becoming invested with a membrane. As the gelatinous material is excreted from the cells they become gradually separated from each other, the surrounding membrane is diagonally ruptured, and the cells are set free. The division is rapidly repeated so that the appearance is as if 4, 8, or 16 cells originated at once from a single cell, but in the present study in no case were more than two cells seen to originate at one time from a parent individual.

The shape of the cells and the formation of colonies is largely controlled by the nutrient medium in which the alga grows. A large number of cultures were made in different solutions, and it was found that 0.05 per cent Knop's solution most nearly reproduced the species in the form in which it was first found. In a weak decoction of earth and in an infusion of *Anabæna flos-aquæ* the cells assumed a much longer and more slender form, while on agar mixed with 0.4 per cent Knop's solution all resemblance to the original form was lost, the cells becoming perfectly spherical, with dark contents, and a wide, gelatinous envelope. In a solution containing organic matter from the plankton, also in the decoction of earth, gelatinous masses were formed as large as a pea, while in 0.05 per cent Knop's solution the cells usually existed either singly or in smaller clusters of 4 to 16 cells. After some months these cultures appeared as a vivid green jelly, due to the great increase in the number of cells.

In some of the general characteristics this species resembles those species of *Oocystis*, where the cells lie imbedded in a gelatinous matrix, but the fusiform shape, the greater density of the gelatinous envelope, and the structure of the chloroplast would all indicate that it can not be classified with *Oocystis*.

Oocystis borgei Snow, new species.

In frequency of appearance and the form in which it occurred this species showed great variation during the three summers when observations were made. In 1898 it was not noted at all in the plankton, while in 1899 it was the most abundant of all of the *Chlorophyceæ* and appeared in large complexes of many cells grouped into twos, fours, or eights, and all imbedded in a homogeneous, transparent, gelatinous substance. In a very few cases colonies of two cells were noted. In 1900 these large gelatinous masses were never observed, but the small colonies of two or four cells, such as described by Borge ('00) occurred frequently. From the large gelatinous masses pure cultures were easily obtained. The cells measure 13 *u* long and 9 *u* broad, and the shape is ellipsoidal or slightly fusiform (fig. VII, 1, 2, 3).

The membrane is a thin layer enveloping the contents. Outside of this is a thick, gelatinous, covering which unites the cells into colonies, and varies in thickness from one-half to twice the diameter of the cell (fig. VII, 4, 5). The membrane consists of cellulose, taking a blue color with iodine and sulphuric acid. Cells in the natural condition and young cells in culture showed the membrane to be of the same thickness at all points, but some older cells in culture, though not all, showed the membrane to be somewhat thickened at the ends. This thickening, however, did not take the nature of a wart or projection, such as has been noted in other species. The outer gelatinous envelope becomes stained with hæmatoxylin and also with tannate vesuvine. With the latter it is homogeneous and shows no prismatic structure characteristic for such coverings in many similar forms of algæ. Two layers can often be detected, however, the inner one the more dense. The outer one probably belongs to the preceding generation (fig. vII, 4).

There is but a single, vivid green, homogeneous, parietal chloroplast, through which there is an opening at one side or near one end. From this opening the chloroplast gradually increases in thickness toward the opposite side, where it incloses a large pyrenoid (fig. vII, 1, 2). The central portion of the pyrenoid shows a crystalline character, and is surrounded by a thick starch envelope. The single spherical or elongated nucleus lies near the center (fig. vII, 1a). With hæmatoxylin a network throughout the cell is brought to view, and is probably due to the arrangement of protoplasm and vacuoles. Oil is found in greater or less quantity in all cells. This oil is turned brown by osmic acid and becomes dissolved in absolute alcohol.

Reproduction occurs by means of bipartition or repeated bipartition of the cell contents, so that two, four, or eight cells are formed from one individual. The first division is a transverse one (fig. vII, 3a); then, if the process continues, the next divisions, dividing the products of the first, take place at right angles to the first division and at right angles to each other (fig. vII, 3b). Each part then becomes invested with a membrane and the outer gelatinous substance begins to be excreted by the cells before they are set free from the mother cell. The enveloping mother membrane becomes much distended, probably by means of the gelatinous substance (fig. vII, 4), until finally it becomes ruptured at one point and the two, four, or eight cells are set free, leaving the remnant of the old cell wall clinging to the outer surface of the gelatinous covering (fig. vII, 3, 5). Though in young individuals there is but a single chromatophore, this very soon becomes divided into two or four, long before the division of the cells occurs, so that in a culture the great majority of cells contain four chromatophores, and it was first thought that four was the normal number. Preceding the division of the chlorophyl body occurs the division of the pyrenoid. The division of the nucleus does not occur until just before the formation of the daughter cells and long after the division of the chloroplast.

Physiology.—Though in the natural conditions the cells are usually found united into larger or smaller complexes, the aggregated form of growth is by no means necessary to existence and is characteristic only under certain conditions, for under the various environments to which the alga was submitted in artificial culture, it was found that either the isolated or aggregated condition of the cells could be produced at will. Cultures were made in various media, such as different concentrations of Knop's solution, decoctions of earth, and solutions containing animal and vegetable matter taken in the tow from the lake and stagnant ponds. Of the various solutions used the organic solutions seemed best to reproduce the aggregated form as it is found in the plankton, but even here the masses were not quite as large as those in the natural condition, although the appearance of the individual cell was perfect (fig. vII, 4). The concentration of the organic solution had a marked effect in producing the isolated or aggregated condition in the development. The exact amount of substance in solution was not determined, but it was found that in the solution which was taken as a standard the cells were all grouped in colonies, and several of these were united into compound colonies. In the same solution, but of one-half the standard concentration, and even in the above solution after the concentration had become reduced by the growth and increase of the algae, only isolated cells were formed, which were distributed throughout the liquid instead of resting on the bottom of the culture glass, as occurred in most cultures. Each individual cell was surrounded by a gelatinous covering one-half to twice the diameter of the cell in thickness, but these were not held together by a common envelope (fig. vII, 5). The vigor of the culture, however, seemed just as great as where the families were formed. It is probable that the greater amount of water present in proportion to the organic matter reduced the consistency of the gelatinous substance, and the connection between the cells was broken.

Knop's solution of different concentrations did not seem to be favorable to development, for in no case was the appearance normal. Cells were usually isolated and a great amount of oil was developed in the contents. In 1 per cent and 0.4 per cent increase was slight, while in 0.1 per cent and in 0.05 per cent, notwithstanding a very large amount of oil being present (fig. vII, 5), increase was rapid. In the decoction of earth growth was abundant and normal, except for the presence of a large amount of oil.

This species certainly resembles closely the *Oocystis lacustris* described by Chodat '97, but after cultivating the two forms, both of which occur in Lake Erie, and obtaining an abundance of pure material of both, each was found to show certain characteristic differences which separate them into different species. The most striking of these was the protruding point at the ends of the cell of *Oocystis lacustris*, while in the other species, if any thickening was noticeable at the poles, it did not project in the form of a wart, but was a gradual thickening of the membrane. Another difference, which was constant, was the longer and more slender shape of *Oocystis lacustris*.

From the figure given by Borge, 1900, of a form occurring in Sweden, it would seem that the species in question must be the same, though the dimensions are slightly smaller than those given by Borge. In recognition of Borge having first figured the species it has been called *Oocystis borgei*.

Chodatella citriformis Snow, new species.

This new species is distinguished from other species of this genus by the shape of the cell, there being a short, obtuse elongation at either end, and at the base of these are arranged the whorls of spines which characterize the genus (fig. VIII, 1, 2, 3).

The length of the cells varies from 8 to 10 μ. The spines are very delicate, often 33 to 36 μ long and but 0.5 μ broad at the base. In different individuals six, seven, eight, and nine spines were found. It was not thought by the author, however, that these represented different species, although some authors seem to distinguish different species by the number of spines. Unfortunately, large cultures of this species were not obtained; and although cells were observed through several generations in hanging drop cultures, it could not be determined whether or not the number of spines was constant in the descendants of a single individual, for as soon as an individual was confined in a hanging drop, the spines became gradually indistinct, finally disappearing, and the daughter individuals possessed either no spines or very rudimentary ones. Apparently the spines were of a gelatinous nature. In the test for cellulose with iodine and sulphuric acid they quickly disappeared when the acid was added. The reaction for cellulose was obtained in the membrane.

The chlorophyl is contained in a single parietal chloroplast, leaving the opposite side colorless. A pyrenoid lies embedded in the chloroplast.

The reproduction occurs by the cell contents becoming divided into two, four, and sometimes eight parts. Each becomes invested with a membrane, and forms a daughter individual (fig. VIII, 3). Though the actual process of liberation of the cells was not witnessed by the author, due to insufficient material, it was inferred that they were set free by the rupturing of the membrane, as membranes were found which were undoubtedly the empty mother membranes of this species.

Chodatella citriformis was found in surface tow and at a depth of about 10 meters, near North Bass Island, in Lake Erie.

Pleurococcus regularis Artari.

One of the most conspicuous and common of all the plankton algæ is a form determined by the writer as *Pleurococcus regularis* Artari (fig. IX, 1). It consists of cell complexes composed usually of 4, 8, 16, or 32 clusters of cells, more or less separated from each other and embedded in a transparent, gelatinous substance (*a*). Each cluster in turn is composed of 4, 8, 16, or 32 cells, which may be in contact with each other or may be separated from each other and held in place by the same gelatinous material. Without doubt it is these complexes which Senn '99 regards as stages in the development of *Coelastrum microporum* Naeg., and Chodat as stages of *Coelastrum sphaerium* Naeg. The striking resemblances of these complexes to *Coelastrum* was noted by the writer when the alga was first seen in the plankton. The view held by Chodat as to the identity of the two forms was then accepted and the difference in appearance was regarded simply as different phases of the same form, due to different conditions. It was noted, however, that side by side in the large plankton cultures, as well as in the fresh collections, both forms were found. The one corresponding to *Coelastrum* consisted of single, isolated coenobia composed of many closely arranged cells with very little surrounding gelatinous substance. The other form consisted of many clusters, widely separated from each other and embedded in a common gelatinous matrix. The individual cells also, as well as the coenobia, were more or less separated from each other according to age. With a view to determining definitely whether these were the same species in different stages or distinct forms, each was isolated and placed in culture under the same conditions. The difficulty attending the cultivation of the Pleurococcus-like form was very great, and it was not until after a very large number of attempts had been made that vigorous cultures were obtained from which conclusions could be drawn with any degree of certainty. All ordinary solutions, so commonly used in the cultivation of algae, failed as culture media. In 0.2 to 0.4 per cent Knop's solution the form simply assumed a yellowish color and passed into a resting condition. In Knop's solution without calcium it remained green for at least six weeks, but showed no indication of reproduction. Organic solutions of various compositions and concentrations proved more favorable to its existence, but even in these not more than three generations could usually be obtained in any culture. The only solution tried that seemed really favorable was from a quantity of decaying Anabæna flos-aquæ. The chemical composition of this solution and the degree of concentration were not determined, but in it reproduction took place rapidly, and, as far as could be determined, normally. This solution proved equally favorable for Cœlastrum. Pure cultures were then made of each of the forms and placed in conditions as nearly alike as possible.

In the course of four weeks an abundance of material of both forms was obtained. These cultures were then repeated several times, and the results agreed each time. Even externally a difference in the cultures could be detected. The Cœlastrum form showed as a very thin green covering on the bottom of the culture glass, while in the other culture a thick cloudy layer, 3 to 4 mm. deep, covered the bottom. A minute examination showed a still more marked difference, and one which was identical with the two forms from which the cultures were made. The thick layer on the bottom of the Pleurococcus culture was composed of the floating loose compound clusters embedded in jelly exactly as in the plankton, except perhaps that the arrangement of the cells was less regular (fig. IX, 2-4). Unless reproduction had just occurred all the cells were more or less separated from one another, sometimes widely so, but all were held in place by the surrounding jelly. In a fresh Cœlastrum culture, where there were many hundreds of cœnobia, but one instance was noticed where any separation of a cell from a cœnobium occurred. Each of the other cœnobia was complete and distinct, without any connection with the other cœnobia in the culture (fig. IX, 5). In old cultures the disintegration of the cœnobia was more frequent, but in no case after disintegration had occurred were the individual cells connected by a surrounding gelatinous substance. There is, as Senn (1898) has stated, a relatively thin gelatinous envelope to the cells, but after the cells have become appreciably separated from one another this envelope no longer connects them. In no case did Cœlastrum microporum form compound clusters. As a result of these experiments, it seems evident to the writer that the two forms are distinct species and can not be united.

Artari ('92), in his description of Pleurococcus regularis, says nothing about the presence of a gelatinous envelope, but from his figures it is evidently present, as the cells, though loosely arranged, are held in place by some substance not shown. The gelatinous envelope of a cluster, when taken from the plankton, as well as of those in the artificial culture, is homogeneous, but somewhat denser near the cells. When treated with tannate vesuvin the jelly is colored brown, but no prismatic radiations from the cell are shown, as in Staurogenia apiculata Lemm. The thickness of the envelope seems to vary somewhat with age, being in young individuals less in diameter and denser in consistency than in older compounds (fig. IX, 2-3). When distilled water is added the substance becomes at least partially dissolved.

The membrane is thin and consists of cellulose, turning blue when iodine and sulphuric acid are added. The chloroplast is, as Artari states, a hollow sphere closely lining the membrane, but with a circular opening at one side. In each chloroplast is a single pyrenoid. When stained with hematoxylin the single nucleus may be seen.

The new complexes arise by the division of the cell contents into 2, 4, 8, 16, and possibly 32 parts (fig. IX, 1). Each of these becomes invested with a cell membrane, and the enveloping mother membrane becomes more or less irregularly ruptured and the cell complex is set free. In this process of division the first visible step is the division of the pyrenoid, then the division of the chloroplast. At just what time the nucleus divides in reference to the division of the other parts was not determined. After the chloroplast has undergone one or more divisions the appearance is that of a cell with several chloroplasts, each with its own pyrenoid. It is a question whether mistakes have not been made at times by investigators in different species of algae in taking these portions of the divided chloroplast to be entire chloroplasts and characteristic for the species. The mother membrane, after it is cast off, remains for a time embedded in the gelatinous substance (fig. IX, 2, 3a), but in the older complexes it is no longer visible (fig. IX, 1, 4).

As the type species of *Pleurococcus, Pleurococcus vulgaris* Menegh., reproduces by means of simple vegetative division involving both contents and membrane, it would seem that this species could not rightly be called *Pleurococcus*. In respect to the mode of reproduction it agrees with *Chlorella*, but the presence of the thick, gelatinous envelope is not characteristic for that genus. The physiological characteristics also of this species vary widely from those of *Chlorella*. The correct systematic position seems to be near to *Kirchneriella*, as the chief point of distinction between the two forms is the shape of the cells, the cells of this species being spherical, while the cells of *Kirchneriella* are crescent-shaped. The formation of cell complexes is the same.

Pleurococcus aquaticus Snow, new species.

Pleurococcus aquaticus shows in its highest development the typical structure of *Pleurococcus*, consisting of clusters of cells 2, 4, 8, 16, 32, or even more in number, arranged in cubical form (fig. x, 1). These clusters arise from the repeated division of cells, alternating in three directions of space.

The diameter of the cells of a cluster varies from 4 to 7 μ. The membrane is thin and gives no reaction with iodine and sulphuric acid. The chloroplast is single, concave, thin, parietal, a vivid green in color, and has an opening on one side, which, however, is rarely distinguishable while the cells are arranged in the cluster. No pyrenoid is present. The small spherical or oblong nucleus lies near the center of the cell.

The large clusters of cells evidently do not increase indefinitely in size, for after a period, under ordinary conditions, the cells undergo a dissociation, the contact between them becomes destroyed, and the cells fall into formless masses (fig. x, 5). The individual cells may then divide and either produce again the large clusters (fig. x, 1, 4), or they may divide and remain in the isolated state in which they were (fig. x, 2). It has been noticed that if the cells are some distance apart they produce again the large clusters, but if they are closely crowded, instead of remaining united after division, they become separated and exist as many single cells, which, except for the absence of a pyrenoid, can with difficulty be distinguished from an ordinary *Chlorella vulgaris* Beyerinck. They are spherical or, before division, ellipsoidal, and the opening on one side of the chromatophore is conspicuous. At any time these cells, when separated from each other, again form large clusters. In large cultures both single cells and clusters were usually present, and in only one instance were the single cells wholly lacking. This was in a culture started for other purposes in a tube made from collodion, similar in shape to a test tube. This was filled with a 0.2 per cent Knop's solution, the algæ inserted and the tube sealed. The whole was then immersed in a 0.2 per cent Knop's solution. After a few weeks the increase had been great, but only the large masses were present. Why the alga did not undergo a dissociation in this mode of cultivation, as well as in others, was not determined. The cells were also somewhat larger than in ordinary cultures, all having a diameter of 6.5 to 7 μ (fig. x, 3).

This species was in no case found in fresh material, but was found in a large plankton culture and also in a culture taken from washings of *Chara* growing among *Scirpus americanus* Pursh. and *Sagittaria rigida* Pursh., in Squaw Bay, South Bass Island. As it was found in but a single plankton culture, it is probable that it is one of the many littoral forms which at times are found in the plankton, and that it had been carried out into the plankton by the action of the water. As it was never seen except in culture, it is difficult to say in what condition it exists in the natural state, whether as isolated cells or in large cell complexes. It is probable that the dissociated form is more usual, as the large cell complexes would be less apt to be overlooked.

Chlorococcum natans Snow, new species.

The cells are spherical or slightly ellipsoidal, the greatest diameter noticed being 13 μ. In appearance they resemble somewhat *Chlorococcum infusionum* Menegh., but are smaller in size, are of a lighter, more transparent green, and the contents, instead of being granular, are mottled, due to thicker and thinner places in the chloroplast (fig. xi, 1).

The shape of the chloroplast is a hollow sphere through which is a circular opening. On the side opposite this lies a pyrenoid with a starch envelope. The membrane is of cellulose. In the young stages there is a single nucleus, but shortly before reproduction the nucleus divides so that, for a period, the cells are multinucleate.

If the material be cultivated in a 2 per cent Knop's solution, almost all individuals produce gonidia; that is, the contents become divided into two, four, or eight portions, as if to produce zoospores, and these divisions, instead of becoming liberated as zoospores, become invested with a membrane and germinate while still within the mother membrane. These in turn may produce gonidia before they become liberated, so that two, and possibly three, generations may be included within a single cell wall (fig. XI, 2). When cultivated in organic solutions and in 0.4 per cent Knop's solution, oblong gonidia are formed, but the enveloping membrane becomes gelatinous, and the alga passes into a palmella condition (fig. XI, 3).

If the material be transferred from a nutritive solution to water, zoospores are formed. If transferred from 0.2 per cent or 0.4 per cent Knop's solution, the shape of the zoospores is cylindrical (fig. XI, 4). They measure 6.5 to 8 μ long and 2.5 to 3.25 μ broad and they move with a rapid motion. If the same material be transferred from 0.4 per cent Knop's solution to organic solution, the zoospores are oval (fig. XI, 5), 7.8 to 10 μ long and 3.2 to 6.5 μ broad, somewhat amœboid in nature, and they move with a slow, lethargic motion. Apparently they are the cells which, if they had not been transferred, would have produced gonidia. The structure in both cases is the same, there being a concave chloroplast in which is embedded a pyrenoid about equally distant between the two ends. At the anterior margin of the chloroplast is a red pigment spot. There are two cilia slightly longer than the cell, and at the base of these, two pulsating vacuoles. The zoospores become liberated by the mother membrane becoming gelatinous. In the case of the smaller cylindrical ones, as they expand, the enveloping membrane suddenly gives way at one point and the spores are liberated either in mass or singly. With the larger oval ones the process takes place more slowly. They arise by successive division of the contents of a cell (fig. XI, 7).

Although there is a difference in the size of these two kinds of zoospores, there is no indication that these represented distinct macrozoospores and microzoospores. Apparently the larger size is more of an abnormal condition of the zoospores, as in all respects the cylindrical form seemed the more natural.

The cells in the palmella condition and also the large zoospores resemble greatly *Chlamydomonas*, but the absence of a membrane in the motile form, the short period of motion, and the mode of growth of the alga in the inorganic solutions all showed a resemblance to *Chlorococcum*.

Botrydiopsis eriensis Snow, new species.

The younger stages of this species resemble that of *Botrydiopsis arhiza*, described by Borzi '95, but the later stages differ from that species. The cells are spherical, and when mature have a diameter of 18 to 21 μ. In the younger stages the chromatophores are more or less hexagonal disks, and are closely applied to the membrane, with spaces between them (fig. XII, 2, 3). In the older stages the chloroplasts are relatively smaller, more elongated, and more crowded (fig. XII, 1).

The membrane, which is thin, gives the characteristic reaction for cellulose with iodine and sulphuric acid. Within the cell no starch, pyrenoid, or oil is present, and the single small nucleus lies near the center.

The reproduction coincides in the main with that of *Botrydiopsis arhiza* Borzi, usually 16 or 32 zoospores being formed within a cell (fig. XII, 7). The successive stages of their formation were not observed, but it is probable that they arise from the repeated bipartition of the contents of the cell, as in other species of *Botrydiopsis*. In the mode of liberation of the zoospores this species deviates from *Botrydiopsis arhiza* and from other known forms of *Botrydiopsis*, where the zoospore mass, together with the inner layer of the enveloping membrane, escapes gradually through a small opening in the outer layer of the membrane. In this species the whole mass remains within the outer layer until both layers become gelatinous, and after a short period of motion within the membrane the zoospores one by one break through the membrane and escape.

The zoospores (fig. XII, 5) are 5.2 μ long, 2.5 to 3.25 μ broad, and they possess all of the characteristics of *Botrydiopsis* zoospores. They are very amœboid, changing their shape constantly as they move. Two elongated chromatophores are present, lying on opposite sides of the cells. One projects farther toward the anterior end than the other, and at the anterior extremity of this lies the pigment spot. A single flagellum is present. During the amœboid movements, when the protoplasm happens to project at the anterior end beyond the chloroplast, two contractile vacuoles may be seen, but they are discernible only under these conditions. The motion of the zoospores is of short duration, and immediately upon coming to rest they become spherical in shape, but often retain the flagellum and the pigment spot after the spherical form is assumed. The germination takes place immediately (fig. XII, 6), the cells increasing in size, and the two chloroplasts becoming divided into 4, 8, and more as the size increases. Very often in old cultures and in vigorous cultures along the sides of the culture glass at the surface of the liquid the zoospores, instead of being liberated, germinate within the mother membrane and remain united in a mass long after the surrounding membrane disappears (fig. XII, 4). Before they are mature, however, they separate and have the same appearance as cells which come from the motile spores.

Botrydiopsis oleacea Snow, new species.

This species was first found in a culture from plankton material, but many cells taken to be the same were also found in fresh collections, the distinguishing characteristic being a large brownish-red globule near the center of each cell, which as yet has not been found in other plankton algæ. The cells in the mature stage are either spherical, ovoid, or lemon-shaped, but usually become more rounded in later stages of existence (fig. XIII, 1, 2). The ovoid or lemon-shaped cells were most often noted when the cells had attained about two-thirds their natural size, the latter shape being conspicuous on account of the wart-like projection on one or both ends of the cell (fig. XIII, 3, 4). The cause of these excrescences was not determined. They occurred on some but not on all cells of the same culture. The largest cells were broadly ellipsoidal, or almost spherical, and 13.5 μ in diameter.

For some months after this form was placed in culture it was regarded as a *Chlorococcum*, so nearly did the general appearance agree with the characteristics of that genus, the only striking differences being the shape, the absence of the pyrenoid, and the presence of the large red globule near the center. On account of the very granular appearance of the cell, due to small globules of oil, the chlorophyll was thought to be in a single chloroplast, as in *Chlorococcum*, and the presence of numerous small chloroplasts characteristic for *Botrydiopsis* was not suspected until the entire development was traced. Different phases of reproduction, however, showed such a resemblance to *Botrydiopsis* that the chloroplasts were again examined. In some young cells it was evident that there were several instead of one in each, though these were not distinguishable in mature cells further than that certain areas seemed slightly darker than others. Many mature cells appeared almost black, so filled were they with the minute globules of oil. This oil became dissolved in absolute alcohol. Iodine with sulphuric acid showed the membrane to be of cellulose, and hæmatoxylin brought to view the single small nucleus a little to one side of the center.

The red globule which was always present was at first taken to be a particle of red-colored oil, but, on account of the color, the ordinary tests could not be used satisfactorily. When absolute alcohol was added the globule disappeared. Its position is apparently underneath the layer of chloroplasts. It was noticed that in the formation of the zoospores the globule did not become divided, but remained in the center throughout the whole process, though it apparently grew somewhat smaller as the process continued. When the zoospores were liberated it was cast out with them and had no further connection with the organism. This suggested the appearance described by Klebs ('96), in his work on *Protosiphon* and *Hydrodictyon*, where the cell sap did not enter into the process of zoospore formation and was cast out in the same way when these were liberated. The same phenomenon has also been noted by the author in an undescribed species of *Botrydiopsis*.

The zoospores, of which 2, 4, 8, 16, or more are formed in a cell, arise through the repeated division of the entire cell contents, except the red globule, until the final number of zoospores is reached (fig. XIII, 5, 6, 7). The zoospores are characteristic for *Botrydiopsis*, having but a single cilium and being very ameboid in their motion (fig. XIII, 8). Though their shape is constantly changing, the general form is pear-shaped, broadly rounded at their posterior end, and tapering toward the cilium. In length they vary from 5 to 7.8 μ and in breadth from 3 to 5 μ at their broadest extremity. As in the mature cells so in the zoospores the chloroplasts are obscured by oil. In the anterior end, just at the base of the flagellum, is a very refractive dull red spot, but it could not be determined whether this was the ordinary pigment spot of the zoospore or the beginning of the red globule found in the older cells. The zoospores are active and seek the light. On coming to rest they become rounded and the chloroplasts become more distinct than at any other stage of their existence. In some cases two chloroplasts are distinguishable in very young cells (fig. XIII, 9). In the liberation of the zoospores the membrane becomes to a certain extent dissolved, then the zoospores by their motion gradually expand the cell until finally the wall gives way and the spores escape. The time required for this process varies from 1 to 30 minutes.

· When the nourishment of the medium in which the organism grows becomes exhausted, the cells pass into a resting stage (fig. XIII, 10, 11), the contents assume a yellow color, and the membrane becomes thick. The red globule still remains prominent. In this condition the alga can withstand being dried, but it quickly changes to the vegetative form when fresh nutritive solution is added. It was cultivated under many different conditions, but it was constant in all.

Chlorosphæra lacustris Snow, new species.

Chlorosphæra lacustris resembles somewhat the *Chlorosphæra angulosa* Klebs, but is distinguished from it in a number of details, the principal one being the size, the largest cells of this species measuring 9 to 10.5 μ, while those of *Chlorosphæra angulosa* measure from 15 to 30 μ.

The shape of the single cell of *Chlorosphæra lacustris* is either spherical, oval, or ellipsoidal (fig. XIV, 1). As in other species of *Chlorosphæra*, a vegetative division of the cells occurs in three directions of space, involving both contents and membrane, after which the cells usually remain connected, forming complexes of two, four, eight, or more cells (fig. XIV, 2, 3). In time these complexes may fall apart and the cells become spherical, as before division.

The membrane is thin and is composed of cellulose, as the test with iodine and sulphuric acid shows. The chloroplast lies close to the membrane and is of the same shape as the membrane, though relatively much thicker. No opening through the chloroplast could with certainty be detected, though a lighter area at one side was prominent. Surrounded by the chloroplast is a pyrenoid with a starch envelope.

The zoospores are 6.5 to 9 μ long and 2.6 to 4 μ broad; they are slightly broader at the posterior end than at the anterior (fig. XIV, 4). Two cilia about as long as the cell are present; also a pigment spot. The chloroplast is concave and extends nearly to the anterior end. Two pulsating vacuoles are found just back of the cilia, and a pyrenoid is embedded in the chloroplast. Four, eight, or more zoospores are formed from a cell (fig. XIV, 7). They are liberated by the membrane becoming gelatinous at one point, through which first one or two gradually force their way. The others follow in quick succession, leaving the empty membrane behind. On coming to rest the zoospore assumes at once a spherical form (fig. XIV, 5) and develops into a mature cell. In this respect it differs from *Chlorosphæra angulosa* Klebs, as in that species the zoospores retain for some time the elongated form. They originate by the successive bipartition of the cell contents (fig. XIV, 6).

Though both forms of reproduction, the vegetative division and the production of zoospores, were found in all cultures, it seemed to reproduce mainly by means of zoospores. Knop's solution of 0.2 per cent concentration seemed to be the most favorable medium for development and was used in tracing the life history. In this solution the zoospores were formed very rapidly, and when transferred to distilled water they were produced in a much shorter time than is usually required for the production of zoospores in unicellular algæ.

Chlorosphæra parvula Snow, new species.

The present species resembles somewhat the preceding species, though, aside from being smaller, it is easily distinguished from it by the gelatinous nature of the membrane after division, causing the cells to separate slightly, though held in complexes of 2, 4, or 8 cells (fig. XV, 1, 2). The two species might also be distinguished by their zoospores, those of the preceding species being more oval or ellipsoidal than those of this species. The minute points of structure of the mature cell, however, the chloroplast, the composition of the membrane, and the contents of the cell, are the same in this species as in *Chlorosphæra lacustris*. The diameter of the full-grown cell is 7.8 to 9 μ. The zoospores (fig. XV, 3), of which four are usually formed in a cell, are oval or spherical; when oval they measure 6.5 μ long and 4.5 to 5 μ broad and the spherical ones 5 to 6 μ in diameter. They have an obliquely placed concave chloroplast, a pigment spot, two cilia about 1½ times the length of the cell, and two contractile vacuoles. The zoospores are formed in the usual manner, by successive divisions of the cell contents, and are liberated by the gradual softening of the membrane.

Mesocarpus sp.

This small species of *Mesocarpus* was found so often in the plankton that it might almost be called a plankton species. At first it could not be recognized as *Mesocarpus*, for usually the chlorophyl was greatly reduced and was collected in a very small space at the center of each cell (fig. XVI, 2), and it was only after the form was placed in culture that it was recognized as belonging to that genus (fig. XVI, 1). As the zygospores were never found, the species could not be determined with certainty. By means of cultures [a] in a large number of media it was determined that the chlorophyl collects under any circumstances that are not favorable, such as in too weak or or too strong culture media (1 per cent or 0.05 per cent Knop's solution), or in media that are not adapted to the plant, as well as in old cultures when the nutrition has become exhausted. It would appear, then, that when this form occurred in the open lake, the nutrition was not qualitatively or quantitatively favorable to its most vigorous condition, though adequate to maintain its existence and even growth.

Cœlosphærium roseum Snow, new species.

Several forms of *Cœlosphærium* are almost universally found in the plankton during the summer months. Some of them are easily recognized, while others are of doubtful name, and the difficulty of cultivation makes the classification more difficult. One of those most commonly found is the form shown in fig. XVII, 1. The colonies are 34 to 52 μ in diameter; the cells measure 3.25 to 4 μ in diameter, are spherical, pinkish or brownish in color, and are closely arranged or somewhat scattered over the surface of the gelatinous sphere. Indigo solution shows the presence of a gelatinous covering to the colony, varying in thickness according to conditions. On agar cultures this covering assumed a diameter almost equal to that of the colony.

If the cells are closely arranged, and if the focus of the microscope is on the surface of the sphere, the colony appears as a typical *Cœlosphærium* where the gelatinous sphere is homogeneous (fig. XVII, 1 *a*), but if the cells are less closely arranged and the focus is at center the gelatinous portion may be seen not to be homogeneous, but to consist of a system of dichotomous gelatinous branches radiating from a common center, bearing the cells on their terminal divisions (fig. XVII, 1 *b*), as in *Dictyosphærium*. Between these branches lies gelatinous substance continuous with that surrounding the colony. The line of demarkation between this substance and the branches is more or less distinct, according to conditions. In some cases it is hardly distinguishable, while in others it is sharply defined.

As a solution from a quantity of *Anabæna flos-aquæ* was the only one in which vigorous cultures could be obtained, the alga could not be subjected to a very great variety of conditions, and yet simply transferring from the lake water to this solution produced some change. Always in cultures the cells became more closely arranged, the branches became indistinct, and in many cases, apparently when the conditions were less favorable than in the lake, the color changed from a pink to a brown, though if the concentration of the solution were right the pink color was maintained. Apparently the distinctness of the branches is controlled somewhat by the density of the surrounding medium—the denser the medium the denser the surrounding substance, and, consequently, the less conspicuous the branches. Distilled water seems to be a solvent of this substance and the cells become detached entirely from the colony. The reproduction of this species is typical in all respects for *Cœlosphærium* as described by Naegeli. If the cells become detached from the stalks, then division occurs once in three directions of space, and afterwards only in two directions, both at right angles to the surface of the sphere. As no very small colonies were found in the plankton, however, it is probable that in nature reproduction usually occurs by the division of the entire colony into two, rather than that new colonies originate from the separated cells.

Both forms mentioned—the pink, with the visible gelatinous branches, and the brown, where the gelatinous branches are not visible—were at times found in the plankton, but their identity is evident. A number of others differing slightly from either of these were found which possibly may have been connected with these, but as neither could be made to assume the characteristics of the other artificially, their identity as yet can not be assumed. In one case all traces of the enveloping gelatinous substance were wanting, and the cells, borne on what seemed to be perfectly free gelatinous stalks, were moved about freely by the action of the water (fig. XVII, 2). Other

---

[a] These cultures were continued for some weeks by Miss Anna L. Rhodes, as well as by the writer.

forms were found with conical or oval cells, which undoubtedly were the *Gomphosphaeria lacustris* of Chodat, '98, and possibly the *Coelosphaerium naegelianum* as figured by Borge, '00.

The resemblance of the form described, as well as of these other forms, to certain species of *Gomphosphaeria*, such as *Gomphosphaeria lacustris* Chodat, is fully recognized by the author, but a study of the well-known *Gomphosphaeria aponina* Kütz, and of the well-recognized species of *Coelosphaerium* has caused the writer to place it under the genus *Coelosphaerium* rather than *Gomphosphaeria*. From a study of the true *Gomphosphaeria* each cell, instead of simply resting at the extremity of the stalk, seems to lie in a capsule of the same substance as the stalk and continuous with it, as has been figured by Schmidle ('01). The outer boundary of this capsule is sharply outlined about each individual cell, an appearance which has not been noted in any of these other forms. Further, it would seem that all *Coelosphaerium* species, although the central gelatinous sphere appears homogeneous, really have at the center the dichotomously branched framework of denser gelatinous material. If a colony of *Coelosphaerium kutzingianum* Naeg. be crushed under a cover glass, it will first divide into two, then into four, and then into eight equal parts, each becoming spherical immediately, just as would occur if left to take its normal course, whereas if it were perfectly homogeneous the mass would crush without any system. In all cases of multiplication where the colony becomes divided into two, evidently the two branches of the first dichotomous division at the center become detached from each other, and the two halves, unable to hold together by the less dense gelatinous substance, are set free.

Chroococcus purpureus Snow. new species.

The cells are spherical, or, just before division, somewhat elongated, usually arranged two by two in colonies of four or eight, all cells of which are more or less separated from each other according to age, and held in place by an enveloping gelatinous substance. The diameter of the cells is 13 μ: the membrane is thin. The color in the natural condition is grayish purple (fig. XVIII).

This species is distinguished from the *Chroococcus multicoloratus* Wood in being larger, the cells more loosely associated into colonies, and in possessing a more decided purple color. When it was first noted in the plankton it was thought it might be a *Chroococcus limneticus* Lemm., which is so abundant in the plankton and which differs from it only in color, that being a blue-green, while this is a purple. Though the two species could not be maintained in artificial culture for observation during any extended period of time, still, in an organic solution *C. purpureus* was kept in a healthy condition for a number of weeks, during which another culture of *Chroococcus limneticus*, under identical conditions, was kept for comparison. This was long enough to convince one that the two forms were not the same and that the purple did not change into the blue-green. Both this and *Chroococcus limneticus*, however, did vary their hue somewhat, according to conditions, both taking on a much darker shade of their respective colors as the concentration of the organic substances in the culture medium was increased. In old solutions both became paler, the purple form assuming something of a brown tinge and the blue-green a yellowish gray. Under no conditions, while in a healthy state, did the two algae assume the same appearance. In both species, however, when cells lost their vitality, the contents contracted, the outline of the enveloping gelatinous substance became sharply outlined, and the color became a deep blue-green (fig. XVIII, *a*). In this condition they could not perhaps be distinguished. Whole clusters of them were found in the plankton, and until this phase of these two forms was noted they were supposed to be a species of *Gloeocapsa*, but were probably only pathological stages of one of these species.

DESCRIPTIONS OF NEW SPECIES.

Chlamydomonas gracilis Snow, new species (fig. 1).

Cells cylindrical, rarely oval or spherical, 10.5 to 13 μ long, 5 to 6.5 μ broad, color a dull bluish green; cilia 2 about 1¼ times as long as the cell; pigment spot a dull red disk, often equally distant from the two ends; pyrenoid at the extreme posterior end. Gametes (?) oval in shape and somewhat smaller than the vegetative individual. Locality, plankton of Lake Erie.

Chlamydomonas communis Snow, new species (fig. II).

Shape, ovoid, cylindrical or ellipsoidal, 10.5 to 13 μ long, 6.5 to 8 μ broad; color a light yellowish green, the pyrenoid near the center; pigment spot an inconspicuous red rod; cilia 2, slightly longer than the cell; division longitudinal. Locality, plankton of Lake Erie.

Chlamydomonas globosa Snow, new species (fig. III).

Cells spherical or slightly ellipsoidal, 5.2 to 7.8 *u* in diameter; membrane smooth at anterior end; two flagella as long or slightly longer than the cell; pigment spot small and inconspicuous; chloroplast much thickened at the posterior end; pyrenoid present; a pulsating vacuole at anterior end. Gametes not found. Locality, plankton of Lake Erie.

Fusola viridis Snow, new genus and new species (fig. VI).

Cells fusiform or slightly sigmoid, 27 to 29 *u* long and 6.5 to 21 *u* broad, each cell surrounded by a thick, homogeneous, gelatinous envelope, the outer line of demarcation being prominent; color a bright green. The chloroplast occupies most of the cell except for a small spherical cavity near the center, in which lies the nucleus; a pyrenoid is present. Reproduction by means of division of the contents into two, the halves gradually assuming the shape of the mother cell, during which process the enveloping membrane becomes obliquely ruptured. Membrane of cellulose. Large masses of cells may be formed in the presence of a great amount of nutritive substance. Locality, a pond on Middle Bass Island, Lake Erie.

Chodatella citriformis Snow, new species (fig. VIII).

Cells ellipsoidal with an obtuse projection at either end; length 13 to 23 *µ*, breadth 8 to 20 *µ*; spines slender, forming whorls at the bases of the projections. Chloroplast single, parietal, lying lengthwise of the cell. Reproduction by division of the contents of the parent cell into 4 or 8, each part becoming invested with a membrane and thus forming a complete individual. Found in surface and deep tow of Lake Erie.

Pleurococcus aquaticus Snow, new species (fig. X).

Cells 4 to 7 *µ* in diameter, existing either as spherical or ellipsoidal individual cells, or as somewhat angled cells combined into large cubical or irregular masses. Membrane thin, chloroplast concave, with an opening at one side; no pyrenoid. Reproduction by division of membrane and contents alternating in three directions of space. Locality, the plankton of Lake Erie.

Chlorococcum natans Snow, new species (fig. XI).

Cells spherical or slightly elongated, not exceeding 13 *u* in diameter. Membrane of cellulose; chloroplast concave, of the shape of the cell, with a circular opening at one side; nucleus single in young individuals, but just before reproduction of the same number as the zoospores. In 2 per cent Knop's solution the organism often forms gonidia, while in weak organic solution it passes into a palmella condition in which the cells are oblong. Zoospores 6.5 to 8 *u* long, 2.5 to 3.25 *µ* broad, with two flagella, a concave chloroplast, a pyrenoid, a pigment spot, and two pulsating vacuoles. Locality, plankton of Lake Erie.

Botrydiopsis eriensis Snow, new species (fig. XII).

Cells spherical, 18 to 21 *u* in diameter; the chloroplasts in mature cells elongated and irregularly arranged, in young cells appearing as hexagonal disks closely applied to the membrane. Zoospores 2.5 *u* long, 2.5 to 3.25 *u* broad, with two chloroplasts, a single flagellum, a pigment spot, and two contracting vacuoles. Usually 16 zoospores formed in a cell; when liberated the inner layer of the mother membrane emerges with the zoospores from the outer layer. Locality, plankton of Lake Erie.

Botrydiopsis oleacea Snow, new species (fig. XIII).

Cells spherical, ellipsoidal or lemon-shaped, not exceeding 16 *µ* in diameter, containing numerous minute particles of oil which obscure the outline of the chloroplasts; near the center a large, prominent, dull-red globule; membrane of cellulose. Zoospores pear-shaped, 2, 4, 8, or 16 in number, formed from repeated bipartition of the cell contents, excepting the red globule; size of zoospores 5 to 7, 8 *µ* by 3 to 5 *µ*; character amoeboid; flagellum single at smaller end; pigment spot present; chloroplasts obscured by oil, two of them discernible in germinating cells. Zoospores liberated by the softening of the entire enveloping membrane. Under unfavorable conditions a resting stage is assumed, the membrane becomes thick, and the contents assume a yellow color. Locality, the plankton of Lake Erie.

Chlorosphæra lacustris Snow, new species (fig. XIV).

Individual cells 9 to 10.5 μ in diameter, spherical or ellipsoidal, usually in complexes of 2, 4, 8, or more, formed by vegetative division, including membrane and contents; chloroplast concave; pyrenoid present; membrane thin, of cellulose. Zoospores 6.5 to 9 μ long, and 2.6 to 4 μ broad, oval, larger at the posterior end; two cilia present, a pyrenoid, a pigment spot, and 2 contractile vacuoles; 4 to 8 zoospores formed in a cell, liberated by the softening of the membrane at one point. Locality, the plankton of Lake Erie.

Chlorosphæra parvula Snow, new species (fig. XV).

Cells usually in complexes of 4 or 8, more or less separated from each other by the partial dissolution of the membrane. Diameter of cells 7.8 to 9 μ; chromatophore concave, with a circular opening near the newest portion of the membrane. Pyrenoid present; membrane of cellulose; zoospores oval or round, 5 to 6 μ in diameter, 4 formed in each cell, liberated by the softening of the entire membrane. Locality, the plankton of Lake Erie.

Cœlosphærium roseum Snow, new species (fig. XVII).

Colony 35 to 52 μ in diameter; cells spherical, pinkish or brown, 3.25 to 4 μ in diameter, arranged more or less closely over the surface of the gelatinous center; the gelatinous center not homogeneous, but containing a system of dichotomously branched gelatinous stalks, on the ends of which are borne the cells; in the spaces between the gelatinous branches and surrounding the whole is a less dense gelatinous substance. Common in the plankton of Lake Erie.

Chroococcus purpureus Snow, new species (fig. XVIII).

Cells spherical, or just before division elongated, usually arranged 2 by 2 in colonies of 4 or 8, separated from each other and held in place by an enveloping gelatinous substance; color a grayish purple, changing to brown under unfavorable conditions. Cells when dying assume a dark blue-green, and the gelatinous envelope is sharply outlined. Common in the plankton of Lake Erie.

LIST OF PLANTS DETERMINED IN LAKE ERIE.

In the following list no account is taken of the number of individuals found, or of the relative number found during the different years, as no accurate quantitative work was done by the writer; but, had such a study been made, it is probable that interesting results would have been obtained. For instance, during the summer of 1898 *Kirchneriella obesa* (West) Schmidle was one of the most common of all the *Chlorophyceæ* found in the plankton, while in 1899 it was found but a very few times, and in 1900 only occasionally. During its absence in 1899 its place seemed to be taken by *Oocystis borgei*, which the preceding year had not been noted at all, and the next season was found only in very small quantities. An equal variation was noted in the occurrence of different forms from week to week during each year. Certain forms, such as *Anabæna flos-aquæ*, appeared in quantities for a few days and then disappeared almost altogether. The number of diatoms also varied very largely at different times. An explanation of such variations would undoubtedly involve a more accurate knowledge than we now have of the composition of the water, as well perhaps as of other elements in the environment.

In the collections of the plankton numerous fragments of filamentous algæ— *Spirogyra, Zygnema, Mesocarpus, Œdogonium,* and *Bolbochæte*—were often found, but as no stages of reproduction were present they could not be determined, and so, except in a few cases, no mention is made of them.

The species that are not starred were taken from the plankton. Those marked with one star were found in washings of stones and of plants growing in the lake. Those marked with two stars were found in Lemna Pond, South Bass Island, and those marked with three stars were found in a stagnant pond on Middle Bass Island.

List of plants determined in Lake Erie.

CONFERVOIDEÆ.

Coleochæte scutata Bréb.*
Œdogonium cryptoporum Wittr.*
Prasiola sp.
Stigeoclonium tenue Kg.
Chætophora endiviæfolia Ag.*
Aphanochæte repens A. Br.
Cladophora glomerata Kg.
    var. *subsimplex* Rabh.
Hormidium nitens Menegh.
    flaccidum (Kg.) Braun.
Bumilleria sp.

SIPHOPHYCEÆ.

Protosiphon botryoides (Kg.) Klebs.

PROTOCOCCOIDEÆ.

Volvox globator Ehrb.
Eudorina elegans Ehrb.
Pandorina morum Bory.
Synura volvox Ehrb.*
Gonium tetras A. Br.*
    pectorale Müller.*
Chlamydomonas communis Snow.
    gracilis Snow.
    globosa Snow.
Hydrodictyon utriculatum Roth.
Scenedesmus acutiformis Schröder.
    acutus Meyen.
    alternans Reinsch.
    bijugatus Kütz.
    var. *flexuosus* Lemn.
    brasiliensis Bohlin.
    caudatus Corda.
    var. *abundans* Kirch.
    var. *setosus* Kirch.
    dimorphus Kg.
    obliquus (Turpin) Ktz.
    opoliensis Richter.
    var. *carinatus* Lemn.
    quadricaudatus (Turp.) Bréb.
    var. *ecornis* Ehrb.
Cælastrum microporum Näg.
    proboscideum Bohlin.
    reticulatum (Dang.) Senn.
    sphæricum Näg.

NOTE.—A specimen was found agreeing in every respect with *Cælastrum cubicum* Näg. which produced typical coenobia of *Cælastrum proboscideum*, so that the species of *Cælastrum cubicum* is to be questioned.

Sorastrum spinulosum Kg.*
Pediastrum boryanum Menegh.
    var. *longicorne* Reinsch.
    constrictum Hass.**
    ehrenbergii A. Br.*
    pertusum Kütz.
    var. *brachylobum* A. Br.
    var. *clathratum* A. Br.
    var. *microporum* A. Br.
    rotula Ehrb.*
    sturmii Reinsch.
Staurogenia apiculata Lemn.
    quadrata (Morren) Kütz.
    rectangularis Näg.
Kirchneriella lunaris (Kirch.) Möb.
    var. *dianæ* Bohlin.

PROTOCOCCOIDEÆ.

Kirchneriella obesa (West) Schmidle.
    var. *contorta* Schmidle.
Golenkinia fenestrata Schröd.
Ophiocytium parvulum A. Br.*
    capitatum Wolle.
Characium ambiguum Herm.
    angustum A. Br.*
Polyedrium cruentum Näg.
    enorme D. By.*
    gigas Wittr.**
    lobulatum Näg.
    minimum A. Br.*
    muticum A. Br.*
    pinacidium Reinsch.
    trigonum Näg.*
    var. *tetragonum* Rabh.
Dictyosphærium ehrenbergianum Näg.
    pulchellum Wood.
Hormospora sp.
Tetraspora natans Kütz.
Schizochlamys gelatinosus A. Br.
Fusola viridis n. sp.***
Dimorphococcus lunatus A. Br.***
Porphyridium cruentum Näg.*
Botryococcus braunii Kg.
Glœocystis ampla Rabh.
Nephrocytium agardhianum Näg.
Oocystis borgei Snow.
    lacustris Chodat.
    solitaria Wittr.
Chodatella citriformis Snow.
Rhaphidium biplex Reinsch.
    braunii Näg.
    convolutum Rabh.*
    falcula A. Br.
    minutum Näg.
    polymorphum Fres.
Rhaphidium (?) *spirale* Turner.
Selenastrum acuminatum Lagerh.
    bibraianum Reinsch.
    gracile Reinsch.
Dactylococcus infusionum Näg.
Stichococcus bacillaris Näg.
Pleurococcus aquaticus Snow.
    regularis Artari.
Chlorella infusionum Beyerinck.
    vulgaris Beyerinck.
Chlorococcum infusionum Rabh.
    natans Snow.
Botrydiopsis eriensis Snow.
Chlorosphæra lacustris Snow.
    parvula Snow.
Scotinosphæra paradoxa Klebs.

CONJUGATÆ.

Mesocarpus sp.
Hyalotheca dissiliens Bréb.
Onychonema læve Nordst.
    var. *minus*.
Sphærozosma filiforme Rabh.
Closterium acerosum Ehrb.
    dianæ Ehrb.
    ehrenbergii Meneg.
    leibleinii Kg.
    lineatum Ehr.**

List of plants determined in Lake Erie—Continued.

CONJUGATÆ.

Closterium pronum Bréb.
    var. *acutum* Klebs.
    var *linea* Klebs.
Pleurotænium trabecula Näg.
Cosmarium crenatum Ralfs.
    euastroides N.
    granatum Bréb.
    kjellmanii Wille.
    meneghinii Bréb.
    var. *concinum* Rabh.
    punctulatum Bréb.
    pygmæum Archer.
    ralfsii Bréb.
    var. *typicum* Ralfs.
    reniforme Ralfs.
    tetraopthalmum Kütz.
    tinctum Ralfs.
Euastrum binale Ralfs.
    verrucosum Ehrb.
Staurastrum crenulatum Näg.**
    gracile Ralfs.
    oblongum N.
    polymorphum Bréb.
    var. *chætoceras* Schröd.
    striolatum (Näg.).
    teliferum Ralfs.

BACELLARIACEÆ.

Navicula cryptocephala Kg.*
    limosa Ag.
    longa Ralfs.*
Pinnularia major Sm.**
    radiosa Sm.*
Stauroptera parva (Ehrb.) Kirch.**
Stauroneis fenestra Sm.**
    phænicenteron Ehrb.*
Amphiprora ornata Bail.
Pleurosigma attenuatum Sm.*
Amphora ovalis Kg.*
Cymbella maculata Kg.
    rotundata H. H. C.
Encyonema prostratum Ralfs.
Cocconeis placentula Ehrb.*
Cocconema cistula Hempr.
    lanceolatum Ehrb.
Gomphonema acuminatum Ehrb.
    capitatum Ehrb.
    constrictum Ehrb.
    intricatum Kg.*
Achnanthes exilis Kg.*
Nitzschia linearis Sm.**
    sigmoidea Sm.
Campylodiscus cribrosus Sm.
Cymatopleura solea Bréb.
    elliptica Bréb.*
Surirella ovalis Menegh.
    saxonica Auersw.
    splendida Kg.
Synedra oxyrhynchos Kg.
    ulna Ehrb.
    var. *longissima* Wm. Sm.

BACELLARIACEÆ.

Fragilaria crotonensis (A. M. Edwards) Kitton.
    virescens Ralfs.
Asterionella formosa Hassal.
Tabellaria fenestrata (Lyng) Kg.
    flocculosa Kg.
Epithemia ocellata Kg.**
    turgida Kg.**
    ventricosa Kg.**
    zebra Kg.
Melosira arenaria Moore.
    granulata (Ehrb.) Ralfs.
    varians Ag.
Orthosira orichalcea Sm.
Cyclotella comta (Ehrb.) Kutz.
    dubia Hilse.
    meneghiniana Rabh.*
    striata (Kutz.) Grun.
Stephanodiscus niagara Ehr.

SCHIZOPHYCEÆ.

Rivularia radians Thur.
    var. *dura* Kirch.
    var. *minutula* Kirch.
Mastigonema ærugineum Kirch.
Alphanizomenon flos-aquæ Allman.
Anabæna flos-aquæ Kg.
    var. *circinalis* (Rabh.) Kirch.
Plectonema mirabile Thur.
Oscillatoria ærugineo-cærulea Kg.
    chalabea Martens.
    frœhlichii Kg.*
    imperator Wood.*
    natans Kg.
    subtilissima Kg.
    tenerrima Kg.
    tenuis Ag.
Lyngbya wollei Farlow.
Microcoleus anguiformis Harv.
Merismopedia elegans A. Br.
    glauca Näg.
    kützingii Näg.
    tenuissimum Lemm.
    violacea Kg.**
Cœlosphærium kützingianum Näg.*
    roseum Snow.
Clathrocystis æruginosa Henfr.
    roseo-persicina Cohn.
Gomphosphæria aponina Kg.
    var. *aurantiaca* Bleisch.
Polycystis ichthioblabe Kg.
Glœocapsa fenestralis Kg.
    punctata Näg.
Chroococcus pallidus Näg.
    limneticus Lemm.
    purpureus Snow.

PHYCOMYCETES.

Beggiatoa leptomitiformis Trevis.
    arachnoidea Rabenh.
Spirochæte plicatilis Ehrb.

LITERATURE.

ARTARI. A.
1892. Untersuchungen über Entwickelung und Systematik einiger Protococcoideen. Inaugural-diss. Basel. Moskau.
1901. Zur Ernährungsphysiologie der grünen Algen. Berichte der Deutschen botanischen Gesellschaft. Bd. XIX. No. 1.
BOKORNY. TH.
1894. Ueber die Betheiligung chlorophyllführender Pflanzen an der Selbstreinigung der Flüsse. Archiv für Hygiene. XX.
BORGE. O.
1900. Schwedisches Süsswasserplankton. Bot. Not.
BORZI. A.
1895. Studi algologici. II. Palermo. 1895.
CHODAT. R.
1896. Matériaux pour servir à l'histoire des Protococcoidees. Bull. de l'Herbier Boss.. v.
1897. Études de biologie lacustre. Bull. de l'Herbier Boss.. v. No. 5.
1898. Études de biologie lacustre. Bull. de l'Herbier Boss.. VI. No. 1.
KLEBS. G.
1896. Ueber die Organisation der Gallerte bei einigen Algen und Flagellaten. Untersuchungen aus d. botan. Institut. Tübingen. II.
1896. Die Bedingungen der Fortpflanzung bei einigen Algen und Pilzen. Jena. 1896.
KNÖRRICH. F. W.
1901. Studien über die Ernährungsbedingungen einiger für die Fischproduction wichtiger Mikroorganismen des Süsswassers. Forschungsberichte aus d. biolog. Station zu Plön. Theil 8.
LEMMERMANN. E.
1898. Beiträge zur Kenntniss der Planktonalgen. II. Botan. Centralblatt.
1898. Forschungsberichte aus d. biolog. Station zu Plön. Theil 6.
PIETERS. A. J.
1901. The plants of western Lake Erie. Bull. U. S. Fish Comm. 1901. 57-79. 10 plates.
SCHMIDLE. W.
1901. Ueber drei Algengenera. Berichte der Deutschen botan. Gesellschaft. XIX. No. 1.
SENN. G.
1899. Ueber einige coloniebildende einzellige Algen. Botanische Zeitung.
STROHMEYER. O.
1897. Die Algenflora des Hamburger Wasserwerkes: ein Beitrag zur Frage der Selbstreinigung der Flüsse. Leipzig.

EXPLANATION OF FIGURES.

PLATE I.

I. *Chlamydomonas gracilis* Snow.
    1.2. Motile cells.
    3. Gamete (?).
    4. Copulation of gametes?.
I. *Chlamydomonas communis* Snow.
    1-3. Motile cells.
III. *Chlamydomonas globosa* Snow.
    1-5. Motile cells.
IV. *Scenedesmus bijugatus* var. *flexuosus* Lemm.
    1. Cœnobium of 32 cells.
    2. Resting stage.
V. *Staurogenia apiculata* Lemu.
    1. Compound cœnobium of 64 cells.
    2. Compound cœnobium of 16 cells, showing gelatinous envelope as brought out by tannate resuvine.
    3. Mass of cells from .05 per cent Knop's solution.
    4. Single cœnobium in early stages of reproduction, taken from an organic solution. (a) Nucleus. (b) Pyrenoid.
    5. Single cœnobium.
    6. Diagram showing relative position of nuclei and pyrenoids in young cœnobia. (a) Nucleus. (b) Pyrenoid.
    7. Membrane of a cœnobium after daughter cœnobia have been liberated.
    8. Cell showing nucleus and pyrenoid. (a) Nucleus. (b) Pyrenoid.

PLATE II.

VI. *Fusola viridis* Snow.
    1-4. Different stages in process of division. (a) Ruptured membrane of mother cell.
    3. Typical cells.
VII. *Oocystis borgei* Snow.
    1. Single cell showing nucleus, a.
    2. Young cells from a culture in organic solution.
    3. Cells after division of chromatophore. Taken from organic solution.
    4. Small colony taken from organic solution.
    5. Single cell from 0.05 per cent Knop's solution, showing remnant of mother membrane, b.
VIII. *Chodatella citriformis* Snow.
    1. Mature cell seen from side.
    2. Cell seen from end.
    3. Cell showing reproduction.
IX. 1-4. *Pleurococcus regularis* Artari.
    1. Complex from the plankton.
    2-4. Clusters from a culture.
    5. *Cœlastrum microporum* Näg.

PLATE III.

X. *Pleurococcus aquaticus* Snow.
    1. Cell complex.
    2. Individual cells before formation of clusters.
    3. Complexes grown in collodion tubes.
    4. First stages in formation of complexes.
    5. Disintegration of the larger complexes.
XI. *Chlorococcum natans* Snow.
    1. Mature cell.
    2. Gonidia formed in 0.2 per cent Knop's solution.
    3 Gonidia formed in 0.4 per cent Knop's solution.
    4. Typical zoospores.
    5. Zoospores formed when material is transferred from Knop's solution to organic solution.
    6. Germinating zoospores.
    7. First stage in formation of zoospores.
XII. *Botryoliopsis eriensis* Snow.
    1. Mature cell.
    2.3. Young cells.
    4. Gonidia formed from nonliberation of zoospores.
    5. Zoospores. (Free hand.)
    6. Germinating zoospores.
    7. Zoospores before liberation.
XIII. *Botryoliopsis oleacea* Snow.
    1.2. Mature cells.
    3.4. Younger cells of different shapes.
    5.7. Different stages in the formation of the zoospores.
    8. Zoospores.
    9. Germinating zoospores.
    10-11. Resting condition.

PLATE IV.

XIV. *Chlorosphæra lacustris* Snow.
    1. Single cells.
    2.3. Complexes arising from division.
    4. Zoospores.
    5. Germinating zoospores.
    6.7. Stages in the formation of the zoospores.
XV. *Chlorosphæra parvula* Snow.
    1.2. Complexes formed by division.
    3. Zoospores.
XVI. *Mesocarpus* spec.
    1. Normal filament.
    2. Filament under unfavorable conditions.
XVII *Cœlosphærium roseum* Snow.
    1. Typical individual a. Surface view. b. Interior view.
    2. *Cœlosphærium* (?) showing free dichotomous gelatinous branches.
XVIII. *Chroococcus purpureus* Snow.
    Showing mode of growth in small clusters embedded in gelatinous substance.

I. Chlamydomonas gracilis   II. Chlamydomonas communis   III. Chlamydomonas globosa

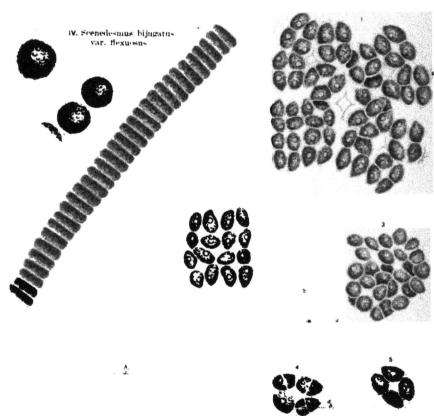

IV. Scenedesmus bijugatus
var. flexuosus

V. Staurogenia apiculata Lemm.

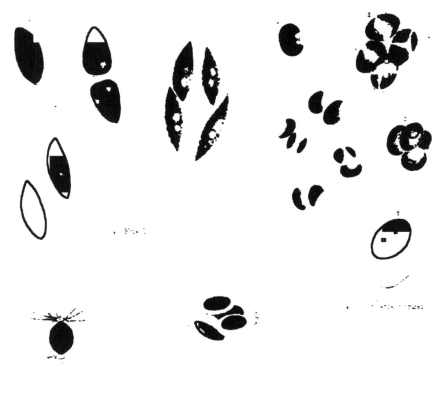

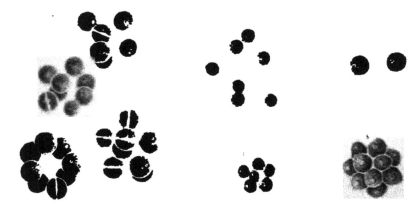

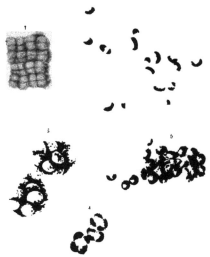

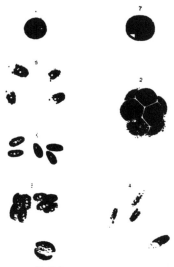

X. Pleurococcus aquaticus

XI. Chlorococcum natans

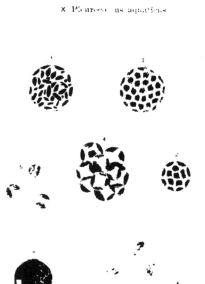

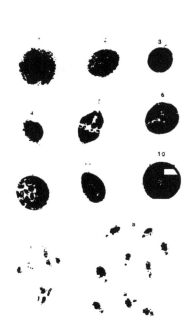

XII. Botrydiopsis eriensis

XIII. Botrydiopsis oleacea

PLATE IV

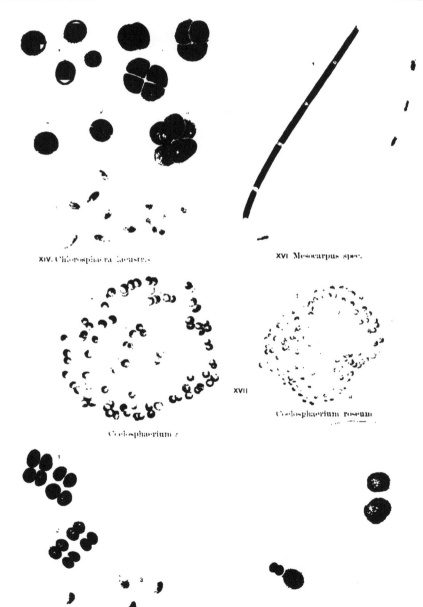

XIV. Chlorosphaera lacustris.

XVI Mesocarpus spec.

XVII

Coelosphaerium roseum

Coelosphaerium z

DESCRIPTION

OF A

NEW SPECIES OF DARTER FROM TIPPECANOE LAKE.

By WILLIAM J. MOENKHAUS,

Assistant Professor of Zoology, Indiana University.

DESCRIPTION OF A NEW SPECIES OF DARTER FROM TIPPECANOE LAKE.

BY WILLIAM J. MOENKHAUS,

Assistant Professor of Zoology, Indiana University.

During the summer of 1896, while collecting large quantities of *Percina caprodes* in Tippecanoe Lake, a single large specimen of darter was taken which could not be identified with any described species. I thought then and since, until recently, that it might be a hybrid between *Percina caprodes* and *Hadropterus aspro*, because of evident intermediate characters. After holding the specimen for six years with the hope that other specimens might be taken, I published a note in the Proceedings of the Indiana Academy for 1902 (pp. 115–116), under the title "An aberrant Etheostoma," in which I briefly described the specimen and compared it with *Percina caprodes* and *Hadropterus aspro*. Last summer the sandbar on the south side of the east end of the lake was again extensively seined, and among some 500 or 600 *Percina caprodes* 2 small specimens—probably that summer's brood—were taken which, beyond a doubt, are similar to the specimen taken six years previously in a part of the lake 3 or 4 miles distant. Among a peck of darters from a part of Tippecanoe Lake that the labels do not indicate, collected in 1898 by some students of the Indiana University Biological Station, I found 3 similar specimens, making 6 specimens of this type from different parts of the lake. There can no longer be any doubt that we have to do with a distinct species, and, so far as I can determine, the species is undescribed.

This new species is among the most beautiful and largest of the darters. It gives me the greatest pleasure to name the species for Dr. Barton Warren Evermann, ichthyologist of the U. S. Fish Commission.

Hadropterus evermanni Moenkhaus, new species.

Head 4 in length; depth 6.16; eye 3.8 in head; snout 3.95; D. xvi, 14; A. ii, 11; scales 8–79–9.

Form of body much like that of *H. aspro*, rather elongate, fusiform, somewhat compressed posteriorly, but less pointed anteriorly; mouth moderately large, maxillary reaching pupil; cleft of mouth almost horizontal, lower jaw included; eye large, about equaling snout; interorbital rather broad, flat; gill-membranes free from isthmus and separate; opercular spine and flap well developed; preopercle entire. Scales ctenoid; nape with fewer, smaller, embedded scales; median ventral line in one specimen provided with a row of closely set, slightly enlarged scales; a second specimen has 3 or 4 such scales, remaining specimens without scales; breast naked; opercle with closely set ctenoid scales slightly smaller than those on body; cheeks with fewer, still smaller, embedded ctenoid scales; lateral line complete, slightly arched over pectorals. Pectoral and ventral fins about equal in length, measuring 1.4 in head; origin of spinous dorsal one-third distance between snout and base of caudal; origins of soft dorsal and anal equally distant from snout, the distance from snout 1.56 in body length; spinous dorsal somewhat longer than soft dorsal, the latter longer than anal; these 3 fins about the same height, the order of their height in an ascending series being spinous dorsal, soft dorsal, anal; their heights being, respectively, about 2.5, 2.2, and 2.1 in head.

The color pattern suggests an intermediate type between *Percina caprodes* and *Hadropterus aspro;* side with about 19 large, distinct black blotches, which, especially along the middle region, are alternately larger and smaller, these often being the ventral ends of more or less well-developed transverse bars; dorsal side with a series of large quadrate blotches alternating and anastomosing with variously developed transverse bars; color pattern of the transverse type rather than the longitudinal characteristic of *H. aspro* and *H. macrocephalum.* In the older individual this dorsal pattern becomes more diffuse and less regular; dorsal two-thirds of opercle and upper part of cheek black; a distinct black band extending downward and another, more diffuse, forward from the eye; both dorsals and caudal fin barred; pectoral indistinctly barred; ventrals and anal plain; a black spot at base of caudal.

Table of measurements and counts of all the specimens.

No. of specimen.	Length of body.	Head in length.	Depth in length.	Eye in head.	Snout in head.	Maxillary in head.	Interorbital in head.	Pectoral in head.	Ventrals in head.	Spinous dorsal from snout.	Soft dorsal from snout.	Anal from snout.	Dorsal fin.	Anal fin.	Scales.
	mm.														
1	77.00	4.05	6.16	3.80	3.95	3.58	4.63	1.36	1.31	3.20	1.60	1.64	XVI,14	II,11	8-79-9
2	49.00	3.82	7.00	3.65	4.00	3.65	5.13	1.28	1.42	2.88	1.58	1.58	XIV,14	II,10	9-94-12
3	50.00	3.84	6.25	3.42	3.82	3.71	5.20	1.30	1.32	2.92	1.66	1.61	XV,14	II,10	9-84-11
4	55.00	3.93	6.11	3.79	4.66	4.66	4.66	1.21	1.40	3.23	1.57	1.62	XIV,13	II,11	8-82-11
5	49.00	3.92	6.30	3.90	4.17	4.17	4.17	1.56	1.39	3.06	1.58	1.58	XIV,15	II,11	9-82-12
6	51.00	3.92	....	4.30	4.33	4.23	5.20	1.30	1.44	3.18	1.58	1.59	XIV,13	II,11	8-84-11
Average		3.91	6.36	3.81	4.14	4.00	4.83	1.33	1.38	3.08	1.59	1.60	....		82

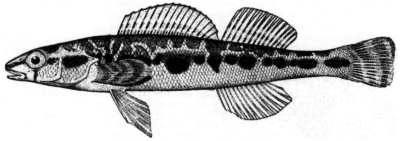

Hadropterus evermanni Moenkhaus, new species.

The species is most closely related to *Hadropterus aspro* and *Hadropterus macrocephalum.* From the former it differs most strikingly in the color pattern, especially that of the dorsal side, which is transverse in type rather than longitudinal, and in the greater number of scales, which in this species are ctenoid instead of cycloid, on the cheeks and opercles.

Type, No. 9785, Museum Indiana University; cotype, No. 9786, Museum Indiana University; cotype, No. 50834, U. S. National Museum; cotype, No. 2742, U. S. Fish Commission; cotypes have also been deposited in the British Museum of Natural History and in the Museum of Leland Stanford Junior University.

The drawing was made from the largest specimen, which is taken as the type. This has an unusual number of dorsal spines, 16, as will be seen from the table of counts.

HABITS OF SOME OF THE COMMERCIAL CAT-FISHES.

By WILLIAM CONVERSE KENDALL,

Assistant, United States Fish Commission.

HABITS OF SOME OF THE COMMERCIAL CAT-FISHES.

BY WILLIAM CONVERSE KENDALL,

Assistant, United States Fish Commission

The fresh-water cat-fishes of the United States of more or less commercial importance may be classified in a popular way as channel cats (*Ictalurus*), mud cats (*Ameiurus*), yellow cats (*Leptops*), and stone cats (*Noturus*). This arrangement is not wholly satisfactory, however, owing to the confusion of the common names, for a mud cat of one locality may be the yellow cat of another, and the yellow cat of some place be the stone cat in another locality, and so on. Then, too, there is no distinct line between channel cats and mud cats. The technical nomenclature and synonymy of these fishes are not in much better shape than the popular classification; therefore the discussion in the following pages will be more or less generic.

The cat-fishes are of such commercial value as food that there have arisen extensive and almost special fisheries for them in the South, the Mississippi Valley, and the Great Lakes region—that is to say, in the centers of their greatest abundance. Of about a dozen species appearing in the markets, probably not more than one-half are very common or merit more than passing notice. The largest are the "great forked-tail cat" of the Mississippi (*Ictalurus furcatus*) and the Great Lakes cat (*Ameiurus lacustris*). The first attains a weight of 150 pounds, the other perhaps 50 or more. Of the smaller cats the more important are the spotted cat (*Ictalurus punctatus*), Potomac channel cat (*Ameiurus catus*), and bullhead (*Ameiurus nebulosus*).

There is very little published information on the habits of any species of cat-fish, and it has been thought that it might be of use to bring together the most important published and otherwise available facts on this subject. Owing to the similarity of habits, for this purpose it is unnecessary to refer to more than those of the most common forms except in a very general way.

The cat-fishes are a hardy race, very prolific, in habits and structure comparatively safe from enemies. For these reasons wherever they occur they are usually very abundant. In late years, however, the demand for these fish has reached such dimensions that in some localities extensive inroads have been made upon their numbers and there has arisen the problem of how to repopulate the depleted waters. It has not, until recently at least, been considered necessary to resort to artificial propagation of cat-fishes, and there have been but few, if any, attempts in that direction. There are a few instances of pond culture, which will be referred to in another place.

Food qualities.—In flavor and other edible qualities the cat-fishes differ somewhat among themselves. As a rule the channel cats, especially the spotted cat (*Ictalurus punctatus* and *I. furcatus*), seem to possess more delectable qualities than the mud cats. This is possibly due to difference in habits and habitat.

Regarding *Ictalurus punctatus*, Jordan says (Bull. U. S. F. C. 1885, p. 34):

As a food-fish the channel cat is certainly better worthy of attention than any other American cat-fish. There is much less waste in the body of the channel cat than in other cat-fishes, as the latter lose more than half their weight by removal of the head, the entrails, and the skin. The flesh of the channel cat, when fresh, is very superior; it is white, crisp, and juicy, of excellent flavor, and not tough. It is much more delicate both in fiber and in flavor than that of the other cat-fishes. When well cooked, I consider it superior to that of the black bass, the wall-eye, the yellow perch, or any other percoid fishes. Among other fresh-water fishes, it is inferior only to the white-fish, the trout, and other *Salmonidæ*.

Speaking of the blue cat (*Ictalurus furcatus*), Jordan & Evermann say (American Food and Came Fishes, p. 19):

In spite of popular prejudice to the contrary, the flesh of this cat-fish is of excellent quality, firm and flaky, of very delicious flavor, nutritious in a high degree, and always commanding a fair price.

Regarding the yellow cat, which they term the mud cat, the same authors state, on page 32:

Its flesh is of fine texture and of excellent flavor, and there is really no good reason for the prejudice against it which obtains in many localities. The fact that it is a large, rather repulsive-looking fish, not too cleanly in its habits, doubtless has something to do with this.

Mr. Charles Hiesler (Bull. U. S. F. C. 1882, 76–79) has written regarding *Ameiurus nebulosus* (?):

It is one of the very best of pan fishes, and has no noticeable bones. It retains its excellence as fresh fish as long as any fish and longer than most of them. It is eaten and relished by all classes of people, and they would eat more if they could get them. It is not salted down because the demand for fresh fish exceeds the supply. Its quality for table food will ever prevent its use for any other purpose.

The great popular demand testifies to the food virtues of the cat-fishes.

Habitat.—Almost any one of the species of cat-fishes seems to be adapted to a wide range of climatic conditions, although somewhat restricted to certain immediate surroundings. *Ameiurus lacustris* is supposed to be distributed from the Saskatchewan River and the Great Lakes to Florida. *Ameiurus nebulosus* is found from Maine to Florida; but in Maine this species occurs, as a rule, only in muddy lakes and streams with plenty of vegetation, and such portions of bodies of water of other character as afford those conditions, and apparently the fish do not stray far from home. Such localities are probably the warmest ones of the region. Regarding the local habitat of the bullhead (*Ameiurus nebulosus*), Dean says (Nineteenth Annual Report State Fish Commission, New York, 1890, 302):

It is one of the hardiest of fishes, will care for itself and even thrive in the muddiest of stagnant waters. It will breed readily and will endure complacently every hardship of drought, extremes of temperature, and lack of food.

Every trait of our cat-fish bespeaks its stagnant mud-loving nature; dusky in color, sluggish, and blundering, furnished with long and tactile barbels, a shallow, slowly drained pond, furnished with an occasional deep mudhole, will suit admirably the needs of the fish. If the water does become warm in the summer, the cat-fish will survive—knowing how to survive is one of its especial virtues. In a 3-foot aquarium at college about a dozen 9-inch cat-fish were kept during very warm weather, the room temperature often in the nineties and the water changed but once a day, with but few fatal results. Should the air supply in the water fail, trust the fish to care for itself. It will come to the surface, leisurely renew the air in its swim-bladder, and even, frog-like or turtle-like, swallow air in bulk, trusting to stomach respiration. Of undoubted respiratory value, moreover, must be the scaleless, highly vascular skin, so important in the breathing economy of the frogs. Should the pond dry, and the whole pond basin be serried with mud cracks, the cat-fish will lie dormant for days, even for weeks. It has been found in a clod of mud, which served as a cocoon, until softened by the return of the water. In winter the cat-fish, like frogs, and unlike many of its neighbors, appears to hibernate. In November it becomes sluggish and refuses food, and early in December buries itself in the deepest ooze of the pond. It does not reappear till the first sharp thunderstorm in February or March. Then the fish are seen, thin and ravenous, approaching the shore so closely that their heads ripple the surface. So fearless are they in early spring in Central Park that they come in schools in shallow water and will take food almost from the hand.

The channel cats are so called owing to their apparent preference for channels of streams and clearer, cleaner water than that affected by the majority of so-called mud cats, though the native channel cat of the Potomac River, according to our present classification, is generically a mud cat (*Ameiurus*). In some Southern rivers, the St. Johns in particular, several species of cat-fish occur together with precisely the same kind of surroundings, whether muddy or sandy. The description of the method of fishing for cat-fishes in Atchafalaya River, Louisiana, given by Evermann (Report U. S. F. C. 1898, 290) indicates their habits sufficiently to warrant quoting from it under this head:

> The Atchafalaya River is in some respects a peculiar stream. It has its sources in Avoyelles and Point Coupee parishes, near where the Red River joins the Mississippi, and is at all seasons more or less connected with both of those rivers by a number of anastomosing channels and bayous. The Atchafalaya River is, in fact as well as historically, one of the mouths of the Mississippi River, and during the floods which come periodically to that region a vast amount of the surplus water of the Mississippi and Red rivers is carried to the Gulf by the Atchafalaya. * * * There are four species of commercial cat-fishes handled by the firms at Morgan City and Melville, viz: The blue cat or poisson bleu (*Ictalurus furcatus*), the yellow cat or goujon (*Leptops olivaris*), the eel cat (*Ictalurus anguilla*), and the spotted cat (*Ictalurus punctatus*). * * * All river fishing during the fall and winter is done on the bottom, while all lake fishing is at the surface. During the spring, when the country is flooded, the fish betake themselves to the woods, and the fishing is then carried on chiefly along the edges of the float roads. The old tackle, which had been previously used in rivers and lakes, is now cut up into short lengths and tied as single lines, called brush lines, to the limbs of trees in such a way as to allow the single hooks to hang about 6 inches under the water. Each fisherman ties his lines to the trees along the edges of the float roads if he can find such territory not already preempted by some one else.

The spotted cat, previously mentioned as one of the most highly esteemed channel cats, thrives equally well in pond or stream. Regarding this species Jordan says:

> The channel cat abounds in all flowing streams from western New York westward to Montana and southward to Florida and Texas. It is, perhaps, most common in Tennessee, Arkansas, and Missouri. It seems to prefer running waters, and young and old are most abundant in gravelly shoals and ripples. The other cat-fishes prefer sluggish waters and mud bottoms. I have occasionally taken the channel cat in ponds and bayous, but such localities are apparently not their preference. They rarely enter small brooks unless these are clear and gravelly. Whether they will thrive in artificial ponds we can only know from experiment.

Mr. J. C. Jones, referring to the speckled cat-fish as an artificial-pond fish, speaks of it as follows (Bull. U. S. F. C. 1884, 321):

> It is naturally a pond fish, and found only in one locality in the South, at least such is my information and observation. That locality is in Flint River, running south and emptying into the Chattahoochee some distance below Columbus, Ga. Many years ago this fish was plentiful, being found only in still water, lagoons, or ponds. The Flint River runs through the Pine Mountain. Not far south or north of the mountain these fish cease to occupy the waters and inhabit only the tributaries to the rivers,

including a space of about 50 or 75 miles. Some time since I determined to try to domesticate them, and the effort has resulted in success. * * * They love a pond of clean water and a mud bottom. All the floods that come can not wash them from their home, unless the whole of the pond is carried away. They will not go into running water if they can avoid it. Disturb them and, like a carp, they will sink in the mud and hide. They can be caught conveniently in a gill net, but with great difficulty in a seine. My pond covers 5 acres of land, the largest and best pond in western Georgia. It is a perfect mass of fish, and has been constructed only eleven months. The water is from an inch to 5 feet deep and abounds in vegetation.

Food and feeding habits.—The cat-fishes are omnivorous, subsisting upon animal or vegetable food. In a strictly wild state the food is probably to a great extent animal, but they will eat almost any kind of vegetable matter fed to them in artificial inclosures. Mr. Jones further remarked regarding his domesticated cat-fish:

The species is easily tamed or domesticated. They can be trained like pigs; increase and grow fat when well supplied with food. They subsist upon vegetation, but in the absence of it can be fed upon any kind of fruit, such as peaches, apples, persimmons, watermelons, and the like, corn, wheat, and sorghum seed. I put fifty 3 inches long in a basket and set it in my pond. I fed them well on corn shorts and dough. In the short space of six weeks they grew to be 6 and 7 inches long and trebled in weight.

Jordan (l. c.) says *Ictalurus punctatus* is an omnivorous fish, though less greedy than its larger-mouthed relatives, and that it feeds on insects, crayfishes, worms, and small fishes, and readily takes the hook.

In some localities the mud cats swarm about the mouths of sewers and other places, where they obtain refuse and offal. This garbage-eating habit is, however, not confined to the mud cats, and the channel cats occasionally indulge their tastes in that direction. Slops from the galley and refuse from the toilet rooms of the *Fish Hawk* in the St. Johns River, Florida, formed a great attraction for the two principal cat-fishes of that region (*Ameiurus catus* ? and *Ictalurus punctatus*). It is doubtful if the food, however foul, taints the flesh in any way, and this allusion to some apparently disgusting feeding habits can not consistently deter anyone who is fond of pork or chicken to forego the cat-fish solely on this account. Besides it is only occasionally and locally that these fish have access to such food.

Hiesler (l. c.) says that cat-fish appear to live on the larvæ of insects and on flies that fall into the water. "They never jump out of the water."

Writing of *Ameiurus nebulosus*, Dean (l. c.) says:

The habits of the cat-fish make it a most objectionable neighbor. * * * The stomach contents show its destructiveness to fish eggs and to young fish. * * * It will eat incessantly day and night, prowling along the bottom with barbels widely spread. It will suddenly pause, sink headforemost in the mud for some unseen prey. Nor is it fastidious in its diet, "from an angleworm to a piece of tin tomato can," it bolts them all. From the contents of miscellaneous cat-fish stomachs, however, there appears to exist a general preference for fish food. Professor Goode has already noted the attractiveness of salt mackerel or herring bait. He has, moreover, hinted incidentally that the fish will not bite when an east wind is blowing. It is in order to procure food in a lazy and strategic way that the cat-fish has been seen to sink in the mud with but barbels and dusky forehead exposed, ready to rush out and swallow the unwary prey.

In their feeding habits all species of cat-fish seem to be more or less nocturnal. They take a hook most readily from about twilight on into the night. Most set-line fishing is carried on at night. Moonlit nights, however, are more favorable than dark ones. On the St. Johns River it was noticed that the fish would begin to rise shortly after sunset, in large numbers, and the sound of their "breaks" could be heard in all directions, although a lot of garbage thrown overboard would not fail to raise more or less of them during the day. The cat-fish here were wary of a baited hook, and although freely eating of pieces of bread or meat floating at the surface, if a hook and line were attached, it would never be touched. Yet a hook baited with meat or fish and sunk would usually be satisfactorily effective, especially if "bream" (*Lepomis*) began to bite first. The presence of other more readily biting fish seemed to attract the cat-fish and render them bolder. Large cat-fish would take a small baited "bream" hook much more quickly than they would a large hook. The mud cat here bit no more greedily than the channel cat. It might be well to state in this connection that the channel cats (*Ictalurus punctatus* and *Ictalurus furcatus*) are sufficiently game fighters to give an angler not too fastidious a very satisfactory battle. These two species might justly be classed as game fishes.

In northern lakes and streams the bullhead or hornpout does not always seem to be so wily as the southern cat-fishes were usually during the daytime. Although the best time to angle for hornpout is about dusk or after dark, they are not infrequently caught in the daytime, much to the annoyance of the "still fisher" for black bass, pickerel, and other fishes. When hornpouts begin to bite, if other fish are desired, it is necessary to seek another berth. They will take live-fish or dead-fish bait or frogs with equal readiness. If, however, hornpouts are wanted, angleworms are the best bait.

Spawn-eating habits.—Dean has referred to the fish-egg-eating propensity of *Amiurus nebulosus*, and to show that this species is not alone in this ovivorous habit it may be stated that on the Potomac River a seine haul was estimated to contain about 10,000 cat-fish (*Amiurus catus* and *Amiurus nebulosus*). A large number of these fish were opened and their stomach contents examined. They were found to have been feeding almost exclusively upon herring (*Pomolobus*) eggs, to such an extent that their stomachs were distended with the eggs. Mr. Harron, at whose fishery this observation was made, told the writer that although these large hauls were not frequent, occasionally much larger ones were made. In Albemarle Sound, during one shad season, the writer frequently found cat-fish full of shad roe, but cat-fish were not abundant at this time. It is obvious, then, that cat-fishes are very destructive to the eggs of other species.

Under the heading "Salmon not injured by cat-fish," in the Bulletin of the U. S. Fish Commission for 1887, page 56, Mr. Horace Dunn makes the statement:

Word has gone out that cat-fish have been taken in Suisun Bay [California] whose stomachs were full of young fish and salmon spawn. Upon this statement the cry has been made that the cat-fish were destroying both spawn and young salmon. The facts of the case are that the cat-fish were caught in the vicinity of a salmon cannery, and that the spawn was among the fish offal thrown into the bay, and the young fish were "split-tails" and not valuable for food purposes.

The facts of the case as stated do not prove that cat-fish may not be injurious to salmon. The chances are that if they would eat salmon spawn as offal, and living "split-tails," they would eat naturally deposited spawn and young salmon of the "split-tail" size if they had access to them.

Dr. Hugh M. Smith says (Bull. U. S. F. C. 1895, p. 387):

The cat-fish have a reputation among the California fishermen of being large consumers of fry and eggs of salmon, sturgeon, shad, and other fishes. This accords with their known habits in other waters. Mr. Alexander's examination, however, of the contents of several hundred stomachs of cat-fish in California and Oregon yielded only negative results as to the presence of young fish and ova. Writing of the bullhead in Clear Lake, California, Jordan & Gilbert say that it is extremely abundant and is destructive to the spawn of other species. The scarcity of the valuable Sacramento perch in that lake, which they attribute to the carp, here as in the Sacramento River, may be partly due to the more numerous cat-fish, which feed almost exclusively on animal matter.

Breeding habits. - Probably less is actually known of the breeding habits of most of the species of cat-fishes than of their other habits, yet observations have been made upon two or more species with sufficient detail to warrant the assumption that in the main the habits of most species are essentially alike. Speaking of *Ictalurus punctatus*, Jordan says that it spawns in the spring, but that its breeding habits have not been studied. Mr. Jones says this species spawns when one year old, and twice a year - in May and in September. In the preceding spring he procured eight wild ones. After feeding them well up to this time (October 31), they had spawned in May and September and filled his pond. He says that they take care of their own young and trouble no other fish.

Ryder (Bull. U. S. F. C. 1883, p. 225) thus describes the breeding process of a pair of Potomac channel cats (*Amciurus catus*) in the aquarium at Washington:

A number of adult individuals of *Amciurus albidus* were brought from the Potomac River to the Armory building at the instance of Lieut. W. C. Babcock, U. S. N., and Colonel McDonald, and deposited in the large tank aquaria of that institution about the close of the shad-fishing season of 1883. One pair of these have since bred or spawned in confinement, and thus afforded the writer the opportunity of observing and describing some of the more interesting phases of the development of this singular and interesting family of fishes. * * * Its habits of spawning and care of the young are probably common to all the species of the genus, and are quite remarkable, as will appear from the subjoined account.

On the morning of the 13th of July, a little after 10 o'clock, we noticed a mass of whitish eggs in one of our aquaria inhabited by three adult specimens of *Amciurus albidus*, two of which were unmistakably the parents of the brood, for the reason that they did not permit the third one to approach near the mass of eggs, which one of them was watching vigilantly. One of the individuals remained constantly over the eggs, agitating the water over them with its anal, ventral, and pectoral fins. This one subsequently proved to be the male, not the female, as was at first supposed. The female, after the eggs were laid, seemed to take no further interest in them, the whole duty of renewing and forcing the water through the mass of adherent ova devolving upon the male, who was most assiduous in this duty until the young had escaped from the egg membranes. During all this time, or about a week, the male was never seen to abandon his post, nor did it seem that he much cared even afterwards to leave the scene where he had so faithfully labored to bring forth from the eggs the brood left in his charge by his apparently careless spouse. The male measured 15 inches in length, the female one-fourth inch more.

The mass of ova deposited by the female in a corner and at one end of the slate bottom of the aquarium measured about 8 inches in length and 4 inches in width, and was nowhere much over one-half to three-fourths of an inch in thickness. The ova were covered over with an adhesive, but not gelatinous, outer envelope, so that they were adherent to the bottom of the aquarium and to each other where their spherical surfaces came in contact, and consequently had intervening spaces for the free passage of water, such as would be found in a submerged pile of shot or other spherical bodies. It was evident that the male was forcing fresh water through this mass by hovering over it and vibrating the anal, ventral, and pectoral fins rapidly. There were probably 2,000 ova in the whole mass, as nearly as could be estimated. All of those left in the care of the male came out, while one-half of the mass which he had detached from the bottom of the aquarium on the third day, during some of his vigorous efforts at changing the water, were transferred to another aquarium, supplied with running water, and left to themselves. Those which were hatched by the artificial means just described did not come out as well as those under natural conditions. Nearly one-half failed to hatch, apparently because they were not agitated so as to force fresh water among them and kept clean by the attention of the male parent. * * * When first hatched, on the sixth to eighth day, the young exhibited a tendency to bank up or school together like young salmon. They also, like young salmon, tended to face or swim against the current in the aquarium, a habit common, in fact, to most young fishes recently hatched. * * *

On the fifteenth day after oviposture it was found that they would feed. While debating what we should provide for them, Mr. J. E. Brown threw some pieces of fresh liver into the aquarium, which they devoured with avidity. It was now evident that they were provided with teeth, as they would pull and tug at the fragments of liver with the most dogged perseverance and apparent ferocity. This experiment showed that the right kind of food had been supplied, and, as they have up to this time (August) been fed upon nothing else, without our losing a single one, nothing more seems to be required with which to feed them.

It is worthy of note that when pieces of liver were thrown into the aquarium the parent fishes would apparently often swallow them, with numbers of young ones eating at and hanging to the fragments. I was soon agreeably surprised to find that the parent fishes seemed to swallow only the meat, and that they invariably ejected the young fish from the mouth quite uninjured, the parent fish seeming to be able to discriminate instinctively, before deglutition occurred, between what were its proper food and what were its own young. As soon as the young began to feed they commenced to disperse through the water and all parts of the aquarium, and to manifest less desire to congregate in schools near the male, who also abated his habit of fanning the young with his fins, as was his wont during the early phases of development.

Regarding the breeding habits of *Ameiurus nebulosus*, Dean (loc. cit.) says:

In breeding habits the cat-fish still maintains its reputation for hardiness. It spawns rapidly, even when transferred to aquaria. The eggs are one-eighth inch in diameter and are adhesive, reminding one somewhat of frog spawn. The mass is deposited in shallows where the bottom is sufficiently hard to support its weight. The danger to the egg occasioned by stagnancy or muddiness of the water is carefully provided for; the male, standing guard, forces the water slowly through them. In some of the southern species, for thorough aeration, the male turns to account the operation of breathing, filling the back of the mouth often so full of eggs that the whole face and throat are distended. In the neighborhood of New York the spawning season is in the early part of April, and appears to last about a fortnight. Toward the latter part of the month the females go into deeper water. At this season (Central Park) of a dozen fish caught, ten proved to be males.

A similarity of breeding habits in *Ameiurus nebulosus* and *Ameiurus catus* is shown by comparing the observations[a] presented in a paper by Dr. H. M. Smith to the American Association for the Advancement of Science and a notice[b] of which appears in Science (February 13, 1903, 243) with the preceding record of Dr. Ryder. Dr. Smith observed:

A pair of fish from the Potomac River in the Fish Commission aquarium at Washington *made a nest on July 3,*[c] 1902, by removing in their mouths upwards of a gallon of gravel from *one end of the tank, leaving the slate bottom bare. On July 5 about 2,000 eggs, in four separate agglutinated clusters, were deposited between 10 and 11 a. m.* on the scrupulously clean bottom. *Ninety-nine per cent hatched in five days* in a mean water temperature of 77° F. The young remained on the bottom *in dense masses* until 6 days old, when they began to swim, at first rising vertically a few inches and immediately falling back. By the end of the seventh day they were swimming actively, and most of them *collected in a school* just beneath the surface, where they remained for two days, afterwards scattering. They first ate finely ground liver on the sixth, and fed ravenously after the eighth day. The fish were 4 mm. long when hatched, and grew rapidly, some being 18 mm. long on the eleventh day, and at the end of two months their average length was 50 mm. Both parents were very zealous in caring for the eggs, *keeping them agitated constantly by a gentle fanning motion of the lower fins.* The most striking

a See also Observations on the Breeding Habits of Ameiurus nebulosus. Doctor A. C. Eycleshymer. (The American Naturalist, November, 1901, 911.)

b For the complete account see Breeding Habits of the Yellow Cat-fish. Hugh M. Smith and L. G. Harron. (U. S. F. C Bull., 1902, 151-154.)

c Italics by the writer to show close similarity to Ryder's observations.

act in the care of the eggs was the sucking of the egg masses into the mouth and the blowing of them out with some force. The fanning and mouthing operations were continued with the fry until they swam freely, when the care of the young may be said to have ceased. During the first few days after hatching, the fry, banked in the corners of the tank, were at irregular intervals actively stirred by the barbels of the parents, *usually the male.* The predaceous feeding habits of the old fish gradually overcame the parental instinct; the tendency to suck the fry into their mouths continued, and the inclination to spit them out diminished, so that the number of young dwindled daily, and the 500 that had been left with their parents had completely disappeared in six weeks, although other food was liberally supplied.

In Sebago Lake, Maine, in a shallow, sandy pool, on July 6, the writer observed one cat-fish (*Ameiurus nebulosus*), sex undetermined, with a brood of young thickly clustering under it, in the manner previously described. From Dr. Smith's observations, they might have been 8 or 10 days old; from Dr. Ryder's, about 15 days of age. They were about 12 mm. long. The development doubtless would be somewhat retarded in the cooler waters of this more northern latitude.

Introduction into other waters.—Several species of cat-fish have been successfully introduced into new waters in the United States, and attempts have been made to provide some European waters with American cat fish, with uncertain results, however. A detailed account of the results of the attempts to acclimatize cat-fishes in the Pacific States may be found in the Bulletin of the U. S. Fish Commission for 1895, 379. The cat-fishes handled were *Ameiurus nebulosus, Ameiurus catus,* and *Ictalurus punctatus.* In California the cat-fishes have become very abundant and widely distributed. In the lower Columbia and Willamette rivers they are also very numerous. In 1884 ten individuals, presumably *Ameiurus catus* or *nebulosus,* were transferred from the Potomac to the Colorado River in Arizona (Bull. U. S. F. C. 1884, 212). The shipment consisted at first of 100, only 10 of which survived the journey. Their status in those waters at the present time is unknown. Some spotted cats (*Ictalurus punctatus*) have been placed in the Potomac, of which species one or two now and then make their appearance in the catches of the fishermen.

A number of years ago, at different times, small consignments of *Ameiurus nebulosus* were sent to Europe. They survived transportation very well and the last accessible records show that they continued to do well after reaching their destinations. What the ultimate results have been the writer has been unable to ascertain.

Available records of shipments of young cat-fish (*Ameiurus nebulosus*) to Europe give the following data:

Nov. 15, 1884.—One hundred were shipped to Ghent and on the 28th of November 95 were received.
July   7, 1885.—Thirty sent to Amsterdam.
June 16, 1885.—Fifty shipped, and later 49 were received in Germany.
July 18, 1885.—One hundred sent to France, and 81 were received in good condition.
June 20, 1885.—Fifty consigned to England, and 48 were received in good condition at South Kensington.

The latest information possessed by the writer regarding any of these plants is found in early bulletins of the Fish Commission. The following is quoted from the Bulletin for 1886, 197–199:

The first practical attempt in this direction was made in Belgium. Mr. Thomas Wilson, United States consul at Ghent, first suggested placing cat-fish in the Scheldt, a river which, owing to the large number of factories on its banks, does not contain many fish. It was presumed that the cat-fish would be particularly adapted to the river Scheldt, because it had been sufficiently proved in America that this fish is not much affected by the refuse from factories. After consulting with Prof. Spencer F. Baird, 100 young cat-fish arrived at Antwerp in November, 1884. By the advice of Professor Baird, these young cat-fish were not immediately placed in the river, but first in the large basins of the large aquarium. It is only after these fish have reached maturity in the aquarium and have spawned there that the young generation should be transferred to the river. This was done, and the young cat-fish received from America have provisionally been placed partly in a small pond in the Botanical Garden at Ghent, and partly in the Victoria Regia basin in the same garden. The selection of the last place we do not consider fortunate, as the temperature of the water in this basin is certainly much too high for these fish. At present there are in the Amsterdam aquarium 45 cat-fish, brought direct from New York and placed in a special basin with the hope that they will reach maturity and propagate their species. At present these fish measure from 4 to 6 inches long.

In the same bulletin, on page 138, appears the following, by Dr. Jousset de Bellesme, on the American cat-fish in the Trocadéro Aquarium of Paris:

These fish, which measured 12 cm. (about 4¾ inches) in length, were, in the beginning, owing to their small size, placed in one of the tanks for young fish in the aquarium and remained there till November, 1885, when they were put in the large basin, No. 6.

They were first fed with raw meat, but as they did not seem to take very well to this kind of food they were fed on raw fish chopped fine, which they appeared to like. As soon as they were transferred to the large basin they were fed on live fish.

The water at the disposal of the aquarium is that which comes from the Vanne, whose temperature is 15° C. (59° F.) in August and 9° C. (48.2° F.) in December. It is hardly probable that this temperature is sufficiently high for the reproduction of the cat-fish. At any rate, those which we have in our aquarium, no matter to what variety they belong, have never spawned.

When the American cat-fish were transferred to basin No. 6 they were all alive and well, although they had not grown perceptibly. Since that time none of them have died, as far as we have been able to observe, for these fish have a habit of keeping in their holes and never coming out during the day, so that they are hardly ever seen. In basin No. 1 we had some of considerable size, and in order to assure ourselves of their existence it became necessary to empty the basin and carefully search for them at the bottom between the rocks. Even then we did not always succeed in finding them. I have, therefore, reason to believe that seven cat-fish which the Acclimatization Society has given us are still in existence, and the first time the basin is emptied I will search for them again in order to make sure.

Cat-fish are preeminently a poor man's fish. They not only afford him a cheap food-fish, but become so abundant in time and there is so much demand for them that they afford a paying industry, notwithstanding their cheapness. They may be raised in artificial ponds or in ponds unsuited to other fish. They propagate rapidly and prolifically and grow fast. Therefore there can be no objection to the introduction of them into waters unsuited to other fishes or in which other fishes do not occur, provided there is no danger of escape into waters where they would prove an undesirable acquisition owing to the objectionable characters already enumerated. The past attempts to introduce them into European waters, from the records cited, would seem hardly extensive enough to prove their adaptability or unsuitability to those waters

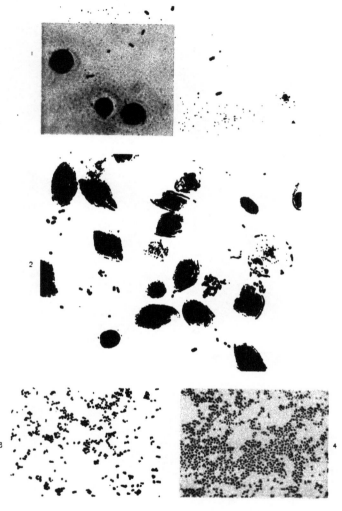

BACTERIUM TRUTLE.

   1. A smear from a characteristic local lesion in body muscles of brook trout dead after artificial inoculation. Fixed at 115° C. for 2 minutes. The organism, leucocytes and debris. Alkaline methylene blue.

   2. A smear of the heart's blood of a brook trout dead in aquarium. X 1000. The organism, erythrocytes and leucocytes. Alkaline methylene blue.

   3. From a bouillon culture 3 days old. X 1000. Washed in distilled water, fixed at 115° C. for 20 minutes. Aqueous fuchsin.

   4. From an agar culture 2 days old. X 1000. Aqueous fuchsin.

Photomicrographs by **Dr. Erwin F. Smith.**

A MORE COMPLETE DESCRIPTION OF BACTERIUM TRUTTÆ

BY M. C. MARSH,

Assistant, U. S. Fish Commission.

The organism here described was obtained from the blood of diseased brook trout at a station of the U. S. Fish Commission, at Northville, Mich., during the summer of 1901, and is the specific cause of the disease. The name and preliminary characterization appeared in *Science*.[a]

It is a pleomorphic form which assumes on nutrient agar-agar its simplest stage, that of a spherical or subspherical micrococcus, with occasional forms that are greater in one dimension. The strictly spherical forms are of an average diameter of 0.71μ, with extremes of 0.5 to 1.0μ, which are comparatively rare. Microscopically the field gives the impression of cocci, but bacillary individuals are frequent and reach a maximum length of 1.5μ. In liquid media, and in liquefying gelatin and blood serum, it has the form of a bacillus, and the microscopical field gives distinctly the impression of bacilli, while occasional spherical forms are intermingled. In bouillon the predominating rods are of a length from that of the diameter of a coccus up to a maximum of 2.35μ, and 0.48 to 0.83μ wide. The arrangement is frequently as diplobacilli. Many of the single rods show a slight constriction indicating their separation into cocci, while many give no sign whatever of such a structure. A few of the longer forms are slightly curved.

In the blood and local lesions of its host, the organism is in general somewhat larger than when growing on artificial media, and appears distinctly as a bacillus with occasional scattered cocci. They grow out infrequently into filaments of a maximum length of 6μ, but the individuals average much less, and may be not longer than the diameter of a coccus, and of a width between 0.5μ and 1.0μ, with rounded ends. When the blood or the contents of the local lesions are plated upon agar, the resulting colonies are alike and the plate contains apparently a pure culture. All the colonies are composed chiefly of cocci, which when transferred to bouillon are transformed into a culture chiefly of apparent bacilli by the next day, or when inoculated into trout reproduce the disease and are found in the blood and lesions as bacilli. This pleomorphism is one of the most interesting characters of the species, and repeated efforts failed to reduce it to a constant form. The considerable variation in morphology in a single culture can not be removed by repeated plating, and such cultures are evidently pure, notwithstanding the variety in the form of the individuals which they contain.

a *Science*, N. S., Vol. XVI, No. 409, p. 706, Oct. 31, 1902.

The organism does not form spores, and a capsule has not been demonstrated. Old cultures show no marked change in the form of the organism.

Staining reactions.—It stains readily by the ordinary aniline dyes in aqueous solution. Thionin and methylene blue give excellent results. Unstained areas are occasional, but not particularly characteristic. The reaction with Gram's stain is not of much value. It stains faintly by this method, but films stained for thirty seconds in aniline-gentian-violet will retain some stain after considerable washing in alcohol, whether the iodine solution is applied or not; so that the ordinary routine of Gram's method is of little value as indicating the applicability of Gram's stain to this organism as a differential staining property. If, however, the gentian violet is applied instantaneously and then treated with the iodine solution, it is seen that the organisms retain the stain after the washing in alcohol better than control films which have not been treated with the iodine. The iodine has at least some fixing power with the gentian violet, and the organism therefore stains by Gram's method.

Cultural characters.— Growth occurs in the ordinary nutrient media, luxuriantly at a titre neutral or +0.5 to phenolphthalein; it will not grow, or but very slightly, at +1.5; at −0.5 growth is inhibited, and at −1.0 to −1.5 scarcely occurs. The following descriptions of cultures refer to media whose titre is +0.5 to phenolphthalein unless otherwise stated. The media employed were prepared from ingredients chemically pure, or as near so as the market affords. Unless otherwise stated herein the procedures[a] recommended to the American Public Health Association by its committee of bacteriologists were followed throughout, save that Liebig's extract of beef was employed instead of fresh meat.

Bouillon: A marked growth is visible after eighteen hours, without pellicle or clouding, but with the sedimenting white growth clinging to the sides of the tube. After about five days a delicate interrupted pellicle may form, and numerous flocculæ are distributed throughout the medium, both of which sink readily and upon the slightest agitation. After ten or fifteen days a characteristic brown color makes its appearance, diffusing throughout the medium, and the sediment takes on a dirty brownish color. The color deepens with age to a dark brown.

Agar-agar: On +1.5 slants it scarcely grows. After twelve days a slight multiplication is indicated by a pale filmy streak, visible best when held in the light against a dark background, and which has not increased after several weeks. The condensation water contains a slight sediment. No color is produced. Growths on agar of this titre are not sufficient to characterize the species. On +0.5 agar moderately abundant growth occurs of a grayish-white color, which with age becomes grayish brown. On usually the third day a production of a soluble pigment becomes evident, which diffuses in the medium and does not reside in the growth itself. It is a reddish-brown shade and deepens gradually until after two or three weeks it becomes a very dark brown, and the growth itself takes on a tinge of brown. In an agar-stab culture a growth of the usual features, with nothing particularly characteristic, occurs throughout the line of puncture, and an umbilicate surface growth takes place which is nearly circular and reaches a diameter of about 5 mm. in five days. Very faint color is visible by the third day, diffusing slowly downward

a Procedures recommended for the Study of Bacteria, etc., Jour. Amer. Pub. Health Assn., vol. 23, 1898, 56.

to a depth of about 2 cm. in seventeen days, fading gradually into the pale agar. The surface growth takes on a brownish color.

Agar plate surface colonies are round, slightly convex, the outline well defined. Microscopically they are granular, and when more than two days old the deep colonies and the surface colonies near the center become grumose. The edge of young colonies is slightly erose, but usually becomes entire or rarely broadly undulate. Well-developed colonies are translucent and yellowish under the microscope by transmitted light. Plates of about 400 colonies have, after two days, surface colonies about 0.58 mm. in diameter; after five days 0.85 mm., and then increase but little. Plates of 200 colonies have 0.54 mm. surface colonies in one day; and plates of about 25 colonies have 0.83 mm. surface colonies in two days, and after about one week these reach 3 mm. and cease to increase. Plates crowded with colonies are tinged with the brown color on the second day.

Gelatine: In +1.5 gelatine there is probably a very slight multiplication, the line of puncture showing a slight growth like a nonliquefying organism. No visible surface growth occurs and no evident liquefaction. In +0.5 gelatine abundant growth and liquefaction take place. The latter at first is either crateriform or funnelform, but may finally become stratiform, reaching the walls of the tube and extending down horizontally. Occasionally the lower end of the stab liquefies faster than the portions above it and produces a terminal sac of liquefaction. Gelatine plate cultures liquefy rapidly.

Blood serum: Blood serum is liquefied; a streak culture on solid serum results in a visible growth in eighteen hours. On the second day evident liquefaction has occurred, a shallow groove without sharply defined edges having formed, the growth sedimented on the bottom and collected at the foot of the slant with the liquefied serum. After three or four days there is a marked brown color, and the slanted portion of the serum is rapidly liquefying. After about eleven days the growth becomes slow and the color very dark brown, much darker than in old agar cultures.

Potato: On ordinary acid potato no growth occurs. On potato boiled in a known quantity of distilled water, which is then titrated and neutralized to phenolphthalein and the potato boiled in it again, there is a very slight growth. It becomes visible on the third day and appears as a faint and scanty growth of white, which is not elevated above the surface of the potato. It does not increase after four or five days and never produces color.

Milk: It grows abundantly in milk and does not cause coagulation. The reaction is unchanged or becomes slightly acid. The milk is peptonized and becomes fairly clear in from one to two weeks, and pepton may be detected in the liquid.

Dunham's pepton solution: The growth resembles that in bouillon, but proceeds more slowly. The characteristic pigment begins to be evident after about three weeks. The cultures tested gave the nitroso-indol reaction on account of indol present in the pepton. The organism does not produce indol. In Dunham's pepton solution containing rosolic acid a deepening of the pink color after about two weeks indicates the production of alkali.

Temperature relations.—The exact optimum was not determined, but it is not far from the room temperature, or 20° C. In the refrigerator at a temperature between 3° and 6° C. no visible growth occurs, but the organism is not injured; 31° C. inhibits somewhat the growth and the human body temperature arrests it entirely and the organism is killed by an exposure to it of seventeen hours. The thermal death point is therefore low. For bouillon cultures it lies between 42 C. and 43° C. during an exposure of ten minutes.

Viability on media.—A culture on a sealed agar slant was still alive at the end of seven months. Upon transfer, however, it grew more slowly than ordinary cultures, and the pigment did not appear until between the sixth and tenth day. On the second transfer growth and chromogenic property were restored substantially to the normal.

Relation to free oxygen.—Agar plates in vacuo, by exhaustion with a Chapman pump and absorption by pyrogallic acid and caustic potash, show after two days very small microscopic colonies, while agar slants show a slight growth, neither of which increase after several days. No color appears. This incipient growth is probably due to incomplete absorption of oxygen at the beginning of the experiment, and the organism is probably an obligate aerobe.

Fermentation tests and products of growth.—It does not ferment the carbohydrates glucose, lactose, or saccharose. Cultures in 1 per cent glucose bouillon acquire an acidity, or an increase of acidity, of 1.2 per cent to 1.6 per cent in fifteen days, due probably to acetic acid, and the characteristic brown color is not developed. Lactose and saccharose bouillon show only a slight or no development of acidity, while the pigment production takes place much as in plain bouillon. The acidity apparently breaks up or prevents the formation of the pigment.

It reduces nitrates to nitrites and finally to ammonia. Seven-day cultures in nitrate broth contain both nitrites and ammonia. Forty-day cultures contain no nitrite, but give a strong test for ammonia. It does not produce indol, phenol, ammonia, invertin or diastatic ferments.

The characteristic pigment is produced in agar, bouillon, Dunham's pepton solution, and blood serum, but not in gelatin or upon potato. It is produced in alkaline, neutral, and acid media, and is inhibited by extremes of reaction, as the growth itself of the organism is inhibited. The pigment is soluble in alcohol[a] and colors the nutrient medium instead of the bacteria themselves, though with age the latter take on a shade of the color of the pigment. In liquid media and in crowded agar plates it colors uniformly the whole media, while in agar tubes the diffusion is slower, the part nearest the growth having the deepest color. It is produced at the room temperature. Higher temperatures inhibit the color faster than they do the growth. At 31½ C., which retards slightly the growth, the color is entirely inhibited, at least for a space of four days. In agar tubes the color appears on the third day, on blood serum after three or four days, in bouillon after two weeks, and in Dunham's pepton solution after three weeks.

Cultures do not have a marked odor.

Motility.—The organism direct from the blood or local lesions of the trout gives no sign of motility in the hanging drop, and its conduct in liquid media when recently isolated—a sedimenting growth without clouding—indicates nonmotility; but after it has remained for several months on artificial media and been repeatedly transferred a change takes place in its appearance in hanging drop and in its growth in bouillon.

[a] Dr. C. L. Alsberg, Harvard Medical School.

PLATE II

BACTERIUM TRUTTAE

PIGMENT PRODUCED IN AGAR CULTURES

THREE, SIX, AND FOURTEEN DAYS

The Brownian movement is more pronounced, and a somewhat doubtful motility is suggested. Its behavior in Stoddart's medium (water, 1,000; gelatin, 5; agar, 0.5; salt, 0.5; pepton, 1) does not give a definite answer to the question of motility, the freshly isolated culture spreading scarcely beyond the point of inoculation, while cultures long in the laboratory when planted extend beyond the original inoculation, yet do not cloud thoroughly this medium. In bouillon, however, a slight but distinct clouding of the medium is observed in such cultures. A modification of habit in the line of an approach toward motility is suggested by the conduct of the organism when newly taken from its host, as compared with that when long habituated to artificial media. It is to be remembered that it circulates with the blood of the trout which it attacks, and while an active parasite in the living trout probably has little use for the power of locomotion. In artificial media the ability of the individual to change its own position would be of value. An interesting question of variation on media in the possible acquirement of motile powers is raised.

The crucial character—the presence of flagella—has not been demonstrated. Many attempts, by various methods, to stain flagella have had negative results, and for purposes of classification their absence must be assumed. For this reason and because of the morphology in the tissues of its host, which is to be regarded as its natural habitat, the organism is placed in the genus *Bacterium* as limited by Migula.

Pathogenesis.—It is pathogenic to trout, and particularly the brook trout (*Salvelinus fontinalis*), in which the disease first appeared. It has also been isolated from the Loch Leven (*Salmo trutta levenensis*) in epidemic, and in a few cases from lake trout (*Cristivomer namaycush*). It has been found thus far only in domesticated or aquarium fish and has not been seen in wild fish from the natural waters. Healthy brook trout succumb to the disease in a few days, by direct inoculation, beneath the skin, into the peritoneal cavity, or into the orbital cavity, and after a longer time by mixing cultures with their food, the organism recoverable in all cases from the heart blood. Inoculation into the dorsal lymph sac of a frog of 1 per cent of its body weight of a bouillon culture was negative, the frog showing no effects. Trout dead of the disease may be eaten, after ordinary cooking, without ill effects. A cat has habitually eaten and thriven upon the fresh, uncooked bodies of the dead trout, and the organism is probably not pathogenic for any warm-blooded animals.

Illustrations.—The colored illustration of the pigment in +0.5 agar cultures was executed by Mr. A. H. Baldwin. The photomicrographs are due to the kindness of Dr. Erwin F. Smith, who made them in the laboratory of plant pathology of the United States Department of Agriculture.

REPORT ON COLLECTIONS OF FISHES MADE IN THE HAWAIIAN ISLANDS, WITH DESCRIPTIONS OF NEW SPECIES.

By OLIVER P. JENKINS,

Professor of Physiology, Leland Stanford Junior University.

The account here presented of fishes from the Hawaiian Islands is based mainly on a large collection made by me in the summer of 1889, with the help of Mr. George C. Price and Mr. Oscar Vaught, then students of De Pauw University. The greater part of the expenses of this expedition was borne by De Pauw University. This collection contained 140 genera and 238 species, of which 7 genera and 78 species are thought to be new to science.

The other collections which have come into my hands for study and which have also been used as material for this report are as follows: A small collection, consisting of 16 species, being the shore fishes taken by the U. S. Fish Commission steamer *Albatross* in 1891 at Honolulu, during the Hawaiian cable survey made by that vessel; a collection of 18 species obtained under the direction of Dr. David Starr Jordan by the *Albatross* in 1896, on the return of that vessel from the work of the Fur-Seal Commission; a collection made in 1898 by Dr. Thomas D. Wood, in the making of which Dr. Wood had the valuable assistance of Mr. Keleipio, at that time inspector of the fish market at Honolulu; a small but important collection sent to Stanford University in 1899 by Dr. A. B. Wood of Honolulu; a single specimen (*Ranzania makua*) sent to Stanford University in 1893 by Mr. C. B. Wilson of Honolulu; a collection made at Honolulu by Dr. Jordan and Mr. Snyder on their return from their expedition to Japan in 1900; and lastly, a small collection made by Mr. Richard C. McGregor in 1900 at various points among the islands.

These collections, together with my own, aggregate 147 genera and 254 species, of which 7 genera and 93 species were thought to be new. Besides the new species here given the list contains 62 species which are for the first time noted from the Hawaiian Islands, making in all 155 species added to the known fish fauna of this group.

In view of the fact that in the summers of 1901 and 1902 the U. S. Fish Commission, under the direction of Dr. David Starr Jordan and Dr. Barton Warren Evermann, made extensive collections of both the shore fishes and the deep-sea forms of the Hawaiian group, and under the direction of Dr. Jordan, in the summer of 1902, an extensive collection of the fishes of the Samoan group was made, thus adding very considerably to the material available for the discussion of all questions pertaining to the fish fauna of this group of islands; and, since the reports on these expeditions by these eminent specialists are soon to appear, it would seem obviously unwise and premature with the material of my collections to enter on the discussion of such questions as that of distribution. In fact, as extensive and careful collecting is necessary in other of the Pacific island groups as has been made in the Hawaiian and Samoan groups to permit one to enter with confidence on the study of the facts and laws of distribution of the Pacific fishes.

Three preliminary papers have already been published based on the collections in my hands."

Since the discovery of the Hawaiian Islands by Capt. Cook in 1778 there have been taken, at various times, small collections of fishes from the islands. The accounts of some of these have been noted in different publications. Many of these earlier-obtained species were described in the work of Cuvier & Valenciennes. A few of these descriptions are so incomplete as to render it impossible to identify any of the species in my hands with them. Quoy & Gaimard's account of the fishes in Le Voyage de l'Uranie and Bennett's accounts contain a number of descriptions of new species of Hawaiian fishes.

The fishes recorded from the Hawaiian group up to the time of the appearance of Günther's Catalogue of the Fishes in the British Museum, are fairly represented by the 54 species accredited in that work to the Hawaiian Islands, several of which are there described as new by Dr. Günther.

The most important accounts of Hawaiian fishes that have appeared since the publication of Günther's Catalogue are given below:

In the Proceedings of the California Academy of Sciences for the year 1863 Andrew Garrett described several new species from the Hawaiian Islands.

In 1873–1875 was published, in the Journal des Museum Godeffroy, Andrew Garrett's "Fische der Südsee, beschrieben und redigirt von Albert C. L. G. Günther." This splendid work contains reproductions in colors of Garrett's paintings of fishes made by him through a series of many years spent in the Hawaiian Islands, the Society Islands, and in other parts of Polynesia. This work contains records of 50 species from the Hawaiian Islands.

An account of 27 species from the Hawaiian Islands is given by Dr. Günther in the "Report on the shore fishes procured during the voyage of H. M. S. *Challenger* in the years 1873–1876."

In 1875 MM. L. Vaillant and H. E. Sauvage, in the Mag. de Zool., III, pp. 278–287, published as a preliminary report on Hawaiian fishes collected by M. Ballieu, brief descriptions of 18 species thought to be new.

"Fishes of the Hawaiian Group" by Thos. H. Streets, M. D., Bulletin U. S. Nat. Museum, No. 7: Contributions to the Natural History of the Hawaiian and Fanning islands and Lower California, pp. 56–77, 1877. This paper contains the account of 39 species.

In 1900 there appeared in the Denk. Acad. Wiss. Wien a very important paper by Dr. Steindachner giving an account of 135 species, all but 4 of which were collected by Dr. Schauinsland in the Hawaiian Islands including Laysan, a small island some 800 miles northwest from Honolulu. Dr. Schauinsland spent considerable

a Description of a new species of Ranzania from the Hawaiian Islands, by O. P. Jenkins. <Proc. Calif. Ac. Sci., second series, vol. v, 1895 (Oct. 31), pp. 779–784, with colored plate (frontispiece).

Descriptions of new species of fishes from the Hawaiian Islands belonging to the families of Labridæ and Scaridæ, by Oliver P. Jenkins. <Bull. U. S. Fish Comm. for 1899 (Aug. 30, 1900), pp. 45–65.

Descriptions of fifteen new species of fishes from the Hawaiian Islands, by Oliver P. Jenkins. <Bull. U. S. Fish Comm. for 1899 (June 8, 1901), pp. 389–404.

time on Laysan studying the flora and fauna of that island. Four of the species in this list were not taken in the Hawaiian group, and 27 were taken from Laysan only.

In the Proceedings of the Academy of Natural Sciences of Philadelphia, for 1900, Mr. Henry W. Fowler, under the caption, "Contributions to the Ichthyology of the tropical Pacific," gives an account of 101 species contained in the collections of the Academy made mainly by Dr. John K. Townsend in 1834, later by Dr. W. H. Jones, and still later, 1893, by Dr. Benjamin Sharp.

Many of the descriptions of Hawaiian species which have been made in the past have been based on alcoholic specimens in a bad state of preservation, or have been taken from dried skins. In consequence, the color has been in many cases very meagerly or erroneously described. In making my collection color notes were taken of as many living or fresh specimens as the conditions would allow. These color notes have been included in this account.

During the time of making my collection in 1889, the fishing was still largely done by native fishermen, but in recent years the Chinese and Japanese have been rapidly encroaching upon this industry. Skilled as were the native fishermen, the newer and more aggressive methods, together with the more industrious habits of the newcomers, are making common in the market fishes before only rarely or never seen by the natives. While these fishermen are adding to the known fauna by their methods, it may be said, in passing, that some of their methods are very destructive and if not regulated by opportune and wise legislation, will soon disastrously affect the fish fauna as a food supply.

The city fish market at Honolulu, the only place where fish are allowed to be sold in the city, is a large, well-appointed, and well-administered institution. Since there comes to it the catch of all kinds of fishing pursued about Oahu, and since among the native and widely diverse foreign population almost every species of fish, as well as of other marine life, finds favor as a food with some, the market proves to be an excellent resource for the student and collector.

About the only fishes which escape the fishermen are the minute forms which make their homes in the spaces of the branching corals or in the small holes in the coral rocks. A number of new species were obtained by breaking up with a hammer coral heads over a dip-net of fine mesh. Either old, dead, or living coral heads were pried off with an iron bar, and quickly lifted up over the net and broken to pieces, the contents falling into the net.

Of all situations about the island of Oahu, the submerged reef which extends from the entrance of the harbor of Honolulu to some distance past Waikiki furnishes the most prolific supply of fishes, both as to number of species and amount of the catch. This reef at low water is from a few inches to a few feet under water and extends from 1 mile to 2 or 3 miles from the shore, where the water abruptly reaches great depths. Over the surface and along the bluff of this reef may be found representatives of most of the shore fauna of the Hawaiian Islands. This reef, so favorably situated, so accessible, and so rich in material, can not fail to be of increasing interest to naturalists who may have the good fortune to devote themselves to the study of its wonderful life.

The types of all the new species have been deposited in the United States National Museum. Cotypes and series of all the species, so far as possible, have been presented to the Leland Stanford Junior University Museum and to the United States

Fish Commission. A representative collection has been presented to the British Museum and one retained by the museum of De Pauw University.

The following list gives full descriptions of all the new genera and species not given in my previous papers or that recently published by Jordan & Evermann.[a] Since, in the forthcoming report by these authors a complete list of all the species known to be recorded in the Hawaiian Islands, with a full discussion of their synonymy, is to be given, I have included only those species of which I have examples in my collections. The synonymy given is limited to that which will give the student into whose hands this paper may fall a ready reference to the most important works treating of the species here listed, or to that which seems necessary to discuss doubtful identifications. In the synonymy, localities without parentheses are type localities; those with parentheses are localities from which the species was recorded by the author cited.

Illustrations from drawings by Mr. W. S. Atkinson are given of all new species.

Family I. CARCHARIIDÆ.

1. Carcharias melanopterus Quoy & Gaimard.

Very common at Honolulu. Three were taken by me in 1889. It has been known from the Indian Ocean and Archipelago, but this is its first record from the Hawaiian Islands.

Color in life, upper part of body a very light olive, covered with pretty thickly set fine points of brown; belly nearly white; tips of all the fins inky black; whole margin of caudal black; pupil and iris very light, almost white.

Carcharias melanopterus Quoy & Gaimard, Voy. de l'Uranie, Zool., 194, pl. 43, figs. 1 and 2, 1824, Pacific Ocean; Gunther, Cat., VIII, 369, 1870, Day, Fishes of India, 715, pl. 185, fig. 3.

2. Carcharias phorcys Jordan & Evermann.

Two specimens (Nos. 245 and 546), 29 and 28 inches long, were obtained. Six examples (one of them a fœtus) were obtained at Honolulu by Jordan & Evermann in 1901.

Color in alcohol, dark gray, lighter on ventral aspect; tip of pectoral and tip of lower lobe of caudal darker; tip of dorsal and upper lobe of caudal only slightly darker than rest of fin.

Carcharias phorcys Jordan & Evermann, Bull. U. S. Fish Comm. 1902 (April 11, 1903), 163, Honolulu. (Type, No. 50812 U S. N. M.; coll. Jordan & Evermann.)

Family II. SPHYRNIDÆ.

3. Sphyrna zygæna (Linnæus).

Thirteen specimens of this shark were obtained at Honolulu, where it is very common. It is sold for food in the market.

Squalus zygæna Linnæus, Syst. Nat., ed. x, 234, 1758, Europe, America.
Zygæna malleus, Günther, Shore Fishes, Challenger, Zool., I, part VI, 59, 1880 (reefs at Honolulu).

Family III. DASYATIDÆ.

4. Dasyatis hawaiensis Jenkins, new species.

Snout 4.5 to base of tail; eye about 3.67 in interorbital space; interorbital space broader than length of snout; width of mouth 2 in interorbital; internasal space 2 in interorbital. Body somewhat pentagonal in form; length of disk 1.42 in width, the line of greatest width passing about the length of the spiracles behind them; anterior margins nearly straight; tip of snout not projecting, very obtuse; lateral margins only slightly convex; snout very broad; eye small; mouth very small, slightly undulate; teeth very small, in about 30 oblique series in the upper jaw; upper buccal flap with a fine fringe; floor

---

[a] Descriptions of new genera and species of fishes from the Hawaiian Islands, by David Starr Jordan and Barton Warren Evermann. Bull. U. S. Fish Comm. for 1902 (April 11, 1903) pp. 161–208.

PLATE I.

DASYATIS HAWAIENSIS JENKINS NEW SPECIES. TYPE.

DASYATIS SCIERA JENKINS, NEW SPECIES. TYPE.

of mouth with 5 tentacles; nostrils large, the border of the broad nasal flap with a fine fringe; interorbital space broad, more or less flattened, gill-openings of about equal length, the fifth on a level with greatest width of disk; body at i tail everywhere smooth; caudal spine broad, flattened, serrated at the sides, longer than interorbital space; tail broad and flattened anteriorly, very slender posterior to spine, its length, 1.6 the disk, a cutaneous fold above and below, the latter beginning below base of dorsal spine; pectoral rounded obtusely; ventrals broad, their width but little less than their height.

Color in alcohol, dark brown above; body, upper parts of sides, pectorals and ventrals light brown, lighter toward margins; body and fins white underneath; cutaneous folds black.

Only one specimen obtained. Length of disk 6.5 inches; length of tail 10.5 inches; width of disk 8 inches.

This description is based on the type deposited in the U. S. N. M. original No. 547, obtained by me at Honolulu.

5. Dasyatis sciera Jenkins, new species.

Tip of snout measured to orbit 3.5 in disk; interorbital 2.25. Body rhomboid, the width of disk greater than its length, the greatest width somewhat in front of the center of its length; head very broad, the anterior margins of the disk nearly straight, slightly convex, meeting at tip of snout at a very obtuse angle, the tip with a slightly projecting point; outer angle of disk slightly rounded, the lateral margins very slightly rounded; mouth small, slightly undulated, 2.2 in interorbital; about 30 teeth in upper jaw, in a very oblique series, about 24 in lower; upper buccal flap with broad fringe; 8 tentacles on floor of mouth, 4 median and 2 on each side; nasal flap with a fine fringe; nostrils large; interorbital space somewhat flattened, concave in the middle; gill-openings of about equal length, the fourth in line with the greatest width of disk; body everywhere smooth, with no indications of spines or plates; distal half of tail with small, sharp tubercles above and on sides; caudal spine had been removed before the specimen was received by me; length of tail more than twice that of disk, a cutaneous fold below only.

Color in alcohol, upper side of disk uniformly light brown; tail darker; under side of disk white.

The single specimen secured measures 13 inches to base of tail, the latter measuring 29 inches. A skate without the tail is in the collection made by the Fur-Seal Commission. It corresponds with the type in all particulars and is doubtless the same species.

Type deposited in the U. S. N. M. original No. 587, Honolulu.

Family IV. MYLIOBATIDÆ.

6. Aetobatus narinari Euphrasen.

This beautiful ray is not uncommon at Honolulu, and occasionally large examples are exposed for sale in the market. The single example obtained measured 15.5 inches to base of tail, the tail being 32.5 inches long; disk more than twice as broad as long; anterior borders slightly convex, the posterior slightly concave; outer angles pointed; origin of dorsal fin a short distance back of posterior attached margin of ventrals; ventrals nearly twice as long as broad; fontanelle on top of head gradually narrowing backward.

Color of disk, blue above, covered with numerous distinct white ocellus-like spots as large as eye; no spots on head in front of spiracles; white below; teeth in lower jaw bent, an obtuse angle projecting forward; about 5 teeth of lower jaw projecting beyond those of upper.

Raia narinari Euphrasen Vet. Ak. Nya. Handl. xi. 174, 217, Brazil.
Aetobatis narinari, Müller & Henle, Plagiostomen, 179, 1841; Jordan & Evermann, Fishes North and Mid. Amer., t. 88, 1896; Steindachner, Denk. Ak. Wiss. Wien., xx, 1889, 70, Layson.
Goniobatis meleagris Agassiz Proc. Bost. soc. Nat. Hist. 58, 1885, 88. Hawaiian Islands.

Family V. LEPTOCEPHALIDÆ.

7. Leptocephalus marginatus (Valenciennes). "Palaula."

Fairly common at Honolulu and apparently valued as food by the natives. I saw one in the market, 5 feet in length, for which $1 was asked. Four specimens were obtained, Nos. 175, 20 inches long, 2041, 21 inches; 2042, 23 inches, and 2043, 26 inches in length.

Color in life of No. 175, back to below lateral line light; under parts white; dorsal fin light brown; anal fin white anteriorly but gradually shading posteriorly to light brown; outer margin of dorsal,

caudal, and posterior of anal edged with black (in two other specimens the whole of the margin of the
caudal is black, also); anterior portion of pectoral fin with a dusky spot, posterior portion reddish.
Teeth in lower jaw strong, pointed, in a single series except in anterior portion; similar teeth in a
single series in upper jaw except an oval patch at anterior angle.

Conger marginatus Valenciennes, Voy. Bonite, Poiss., 201, pl. 9, fig. 1, 1841, Sandwich Islands; Günther, Challenger Report,
     Shore Fishes, 61, 1880 (reefs at Honolulu); Steindachner, Denks. Ak. Wiss. Wien, LXX, 1900, 514 (Laysan).
Conger noordzieki Bleeker, Act. Soc. Sci. Ind. Neerl., II, 1857, Amboina, 86; Bleeker, Atlas, IV, Mur. 26, pl. 23, fig. 2.

8. Congrellus bowersi Jenkins, new species.

Head 2.83 in length to anus, 6.16 in total; depth 2.6 in head; tip of snout to anus 1.25 in distance
from anus to tip of tail; snout 4.5 in head; eye 4 in head; cleft of mouth nearly 3 in head; teeth
small, sharp, in many series on anterior portions of both jaws and on vomer, in 2 series on posterior
portion of jaws; origin of dorsal slightly in front of gill-opening; about 47 pores in lateral line before
the anus; pectoral 3.3 in head.

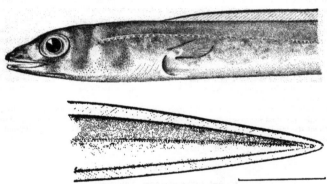

Fig. 1.—*Congrellus bowersi* Jenkins, new species. Type.

Color in life, very pale brown, white below; a narrow silvery line with golden reflections along
lateral line, a broader band below this from axil to caudal fin; membrane over branchiostegals
golden with dusky shade toward chin; cheeks and opercles each with a silvery patch; margins of
dorsal, caudal and anal black; tip of tail white.

Similar to *C. anago* (Schlegel) from Japan, with specimens of which species in the collection made
by Jordan & Snyder in Japan, in 1900, mine have been compared.

Common at Honolulu where it is used by the natives as food and is exposed for sale in the market, and where I obtained 8 specimens from 8 to 11 inches in length. The specimens are numbered
254, and 2044 to 2050.

The species is named for Hon. George M. Bowers, U. S. Commissioner of Fish and Fisheries.

Type, No. 50689, U. S. N. M. (original No. 254), Honolulu.

Family VI. OPHICHTHYIDÆ.

9. Microdonophis macgregori Jenkins, new species.

Head 5 in trunk; head and trunk 1.75 in tail; eye 2 in snout, and about equaling interorbital;
snout 5.5 in head; gape 3 in head; pectoral 3.5 in head; body cylindrical, slender; tail tapering,
ending in a blunt, horny point; head elongate, somewhat compressed; snout small, produced beyond
the mandible; eye small, nearer angle of mouth than tip of snout; mandible broad, lip of upper jaw

with a fringe of short, fleshy barbels; teeth sharp, in a single series in each jaw and on vomer; anterior nostrils with conspicuous fleshy tubes on the lower surface of snout; interorbital space convex, about equaling eye; gill-opening low, space between broad; head with numerous pores; lateral line developed throughout whole length of body and tail; skin smooth; origin of dorsal fin midway between tip of snout and gill-opening; height of dorsal about half depth of body; anal about equal to dorsal.

Color in alcohol, general color, brownish yellow, lighter below, with silvery areas; the upper half of body darker by being covered with numerous minute points of black.

This description is based on a single specimen, 10.2 inches in length, obtained by Mr. R. C. McGregor, from Lahaina, Maui, in February, 1900. (Type, No. 50721, U. S. N. M.)

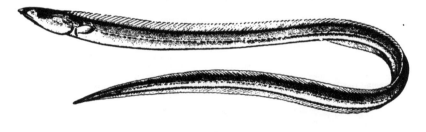

FIG. 2.—*Microdonophis macgregori* Jenkins, new species. Type.

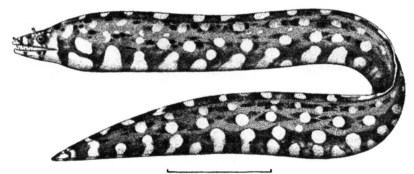

FIG. 3.—*Muræna lampra* Jenkins, new species. Type.

Family VII. MURÆNIDÆ.

10. Muræna lampra Jenkins, new species. "*Puhi-o-u.*"

Body compressed; posterior nostril with a tentacle as long as diameter of eye; head 7.2 in total length; depth slightly more than 2 in head; snout 6 in head; eye 1.65 in snout; posterior margin of eye slightly nearer angle of mouth than is anterior margin to tip of snout; vent nearer tip of snout than tip of tail by half length of head; teeth long, pointed, smaller ones interspersed in 2 series on posterior portion of upper jaw; 2 very long, pointed, depressible teeth on vomer, one behind the other, followed by a single series of smaller teeth; gill-opening a very small, narrow slit without color marking;

interorbital narrow; head with many pores; dorsal fin high, its origin considerably in advance of gill-opening, confluent with caudal and anal; anal similar to dorsal.

Color in life, very bright; ground work of light brown with conspicuous white spots, intermingled with black and brown spots; 3 longitudinal rows of white spots on body, one row on outer margin of dorsal, and about 33 spots or bars of white across the ventral aspect including anal fin; median row on body and head contains about 25 spots, each about size of eye; ventral spots in front of anus to head largest; black, as well as brown spots, small, irregularly placed, but generally following line of rows of white spots; very brilliant red on snout and jaws.

Only the type known, No. 50680, U. S. N. M. (original No. 269), a specimen 8 inches long, obtained by me from the coral rocks on the reef in front of Honolulu.

11. Muræna kauila Jenkins, new species. *"Puhi Kauila."*

Head 7.3 in length; depth 16; tail a little longer than head and trunk; snout 4 in head; eye 1.5 in snout; interorbital 2.25; mouth 2 in head.

Body elongate, compressed; tail tapering posteriorly; head elongate, pointed, sides swollen a little above and behind eyes; snout long, slightly convex in profile; mouth large; jaws arched, not completely closing, upper slightly the longer; teeth uniserial in jaws, compressed, long canines with intervening smaller teeth; 2 large depressible canines on vomer; 3 or 4 large depressible canines

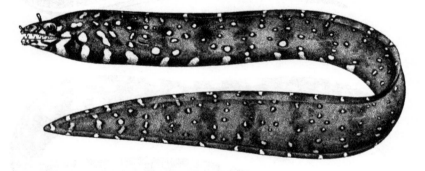

FIG. 4.—*Muræna kauila* Jenkins, new species. Ty, .

below eye, forming an inner series on each side of upper jaw; lips thin, not concealing teeth when mouth is closed; eye about midway in length of mouth; nostrils in long tubes, the posterior larger, equal to eye; interorbital space flattened; gill-opening small, 0.75 in eye; roof of mouth with a single median series of small teeth beginning below front margin of eye and running back well beyond its posterior margin; dorsal beginning nearly midway between corner of mouth and gill-opening; caudal small, rounded.

Color in life, light brown, with 2 longitudinal rows of dark brown spots about the diameter of snout gradually fading into one row on posterior portion of tail; many clear white spots as large or larger than pupil, over head, body, fins, and tail, many of the spots forming more or less distinct vertical rows over fins and dorsal portions, some confluent on throat and belly, each one surrounded by a dark brown margin; about 30 white spots crossing the ventral line; nasal tubes bright red; bright red bars on snout and lower jaw, and bright red undulations posterior to angle of mouth.

Color in alcohol, brown with the white and dark brown spots distinct; white spots edged with dark brown; bright red undulations posterior to angle of mouth fading out.

Only one specimen obtained, the type, No. 50684, U. S. N. M. (original No. 304), 13 inches long, taken by me from the coral rocks on the reef at Honolulu.

12. Gymnothorax laysanus (Steindachner).

Head 6.6 in total length; depth 2 in head; distance from tip of snout to vent shorter by one-fourth of itself than distance from vent to tip of tail; eye less than snout; gape 2 in head; gill-opening very small and inconspicuous, marked by no dark spot; anterior nostril tubular, near tip of snout above margin of mouth; posterior nostril round, inconspicuous, above and slightly forward of middle of upper margin of eye; origin of dorsal slightly in advance of gill-opening; teeth sharp and pointed; on sides of lower jaw in a single series, on sides of upper jaw in 3 series; on anterior portions of both jaws large, sharp, depressible teeth among the smaller ones; a large, sharp, depressible tooth on vomer.

Color in life, very dark brown over whole of head; body and fins marked everywhere with very many small white spots, with indistinct and irregular outlines, more irregular on anterior portions, being on these almost reticulations; no conspicuous markings at angle of mouth or at gill-openings; tip of tail white.

Eight specimens were obtained at Honolulu, ranging from 6 to 7.25 inches in length. The smaller ones show variation in color from the one above described in that the white spots are more distinct in outlines and more regularly placed. In the smallest they are almost definitely arranged in longitudinal rows of which there are 4 on the body exclusive of the fins. There is a gradation from this arrangement to the irregular arrangement seen in the one described. Found among the coral rocks.

Murzna laysana Steindachner, Anzeiger der Denks. Ak. Wiss. Wien, No. XVI, June 27, 1900, 177, Laysan Island (coll. Dr. Schauinsland, 1896–97); Ibid, Denks. Ak. Wiss. Wien, LXX, 1900, 515, pl. VI, fig. 1 (probably not fig. 2).
Lycodontis parribranchialis Fowler, Proc. Ac. Nat. Sci. Phila. 1900 (Nov. 6, 1901), 494, pl. XVIII, fig. 1, Hawaiian Islands. (Type, No. 16483, Ac. Nat. Sci. Phila.)

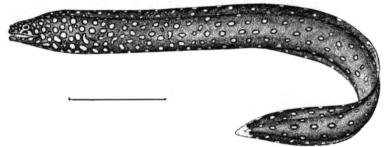

FIG. 5.—*Gymnothorax leucostictus* Jenkins, new species. Type.

13. Gymnothorax leucostictus Jenkins, new species.

Head 8 in total length; depth 0.5 in head; snout a little longer than diameter of eye, 5.5 in head; tip of snout to angle of mouth 2.5 in head; gill-opening small and inconspicuous, less than diameter of eye; body deep, compressed; tail tapering, much compressed posteriorly; head compressed; snout blunt, rounded, not projecting above the mandible; jaws even; eye small, slightly nearer corner of mouth than tip of snout; mouth horizontal; lips thin, concealing the teeth; teeth all sharp-pointed; large depressible canines in anterior parts of both jaws mingled with smaller ones, those in posterior portion of upper jaw forming wide bands, in posterior portion of lower jaw in a single series; large teeth on anterior portion of vomer followed by smaller ones in a single series; anterior nostril tubular, placed near tip of snout and above margin of lip; posterior nostril smooth, small, and well above margin of eye, somewhat in front of a vertical through center of eye; interorbital space narrow, convex; origin of dorsal fin over posterior margin of eye, fin high; anal similar, both confluent around tail.

Color in life, general color uniform dark brown; head, body, tail, and fins covered with numerous white spots, which are larger than eye on trunk, but smaller elsewhere; tip of tail white; margin of gill-opening brownish black.

This species is distinguished from *G. meleagris* (Shaw) by the more anterior insertion of the dorsal, and by the larger spots, which are fewer in number and larger on the trunk.

This description is based on 2 specimens taken from the coral reef at Honolulu, the type, 6.13 inches in length, and a cotype. These do not differ from each other in coloration.

Type, No. 50681, U. S. N. M.

14. Gymnothorax gracilicauda Jenkins, new species.

Head 9 in total length; depth 2.5 in head; tip of snout to vent 1.33 in distance from vent to tip of tail; tip of snout to angle of mouth 2.33 in head; eye 1.33 in snout; gill-opening very small, less than one-half diameter of eye, with no color marking; dorsal fin low, its origin in advance of gill-opening; body very slender and compressed; tail long and very gradually tapering to a point; teeth all long and sharp-pointed, in a single series in lower jaw, in a double series in the upper, the inner series on each side consisting of 4 longer, sharp teeth, the teeth on anterior part of each jaw and on vomer the longest; 2 teeth on the vomer; anterior nostril tubular, near tip of snout and just above margin of lip, posterior smooth and very near upper anterior margin of eye.

Color in alcohol, general color, very pale, nearly white, marked by very irregularly-formed light brown spots, arranged in about 40 ill-defined transverse bands, these lacking on ventral aspect before the vent, leaving the belly white; a very small brown spot at angle of the mouth.

The only known specimen is the type, No. 50679, U. S. N. M. (original No. 367), 8.5 inches long, obtained by me from coral rocks on the reef in front of Honolulu.

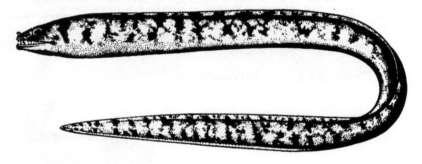

Fig. 6.—*Gymnothorax gracilicauda* Jenkins, new species. Typ

15. Gymnothorax undulatus (Lacépède).

This species seems to be the most common eel at Honolulu. It varies much in color and reaches a length of 3 feet or more. It is used as food by the natives.

Color in life (No. 132), general color drab, with many reticulations of nearly white; a yellow area on top of head, and from snout to a considerable distance behind the eyes yellow. In another specimen (No. 232) the dark ground-work was reddish-brown.

Numerous specimens, ranging from 10 to 28 inches, were taken by me at Honolulu.

Muraenophis undulata Lacépède, Hist. Nat. Poiss., v, 629, 644, 1803, South Seas.
Muraena undulata Günther, Cat., viii, 110, 1870 (Zanzibar, Cocos, East Indies, Hawaiian Islands); Streets, Bull. U. S. Nat. Mus., No. 7, 77, 1877 (Honolulu).
Thyrsoidea knupii Abbott, Proc. Ac. Nat. Sci. Phila. 1860, 477, Hawaiian Islands.
Lycodontis knupii Fowler, Proc. Ac. Nat. Sci. Phila. 1900, 494, pl. xviii, fig. 6 (Abbott's type).
Lycodontis pseudothyrsoidea Fowler, Proc. Ac. Nat. Sci. Phila. 1900, 494 (Hawaiian Islands); not of Bleeker.

16. Gymnothorax steindachneri Jordan & Evermann.

One specimen of this eel was obtained by the *Albatross* in 1896.

Gymnothorax steindachneri Jordan & Evermann, Bull. U. S. Fish Comm. for 1902 (April 11, 1903), 166, Honolulu. (Type No. 50616, U. S. N. M., coll. Jordan & Evermann.)
Muraena flavomarginata var., Steindachner, Denks. Ak. Wiss. Wien, lxx. 1900, 32, pl. vi, fig. 3 (Laysan); not of Rüppell.

PLATE II

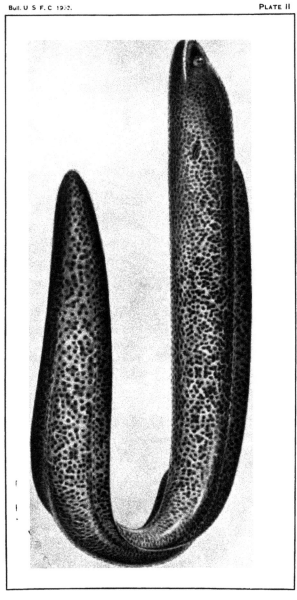

GYMNOTHORAX THALASSOPTERUS JENKINS, NEW SPECIES. TYPE.

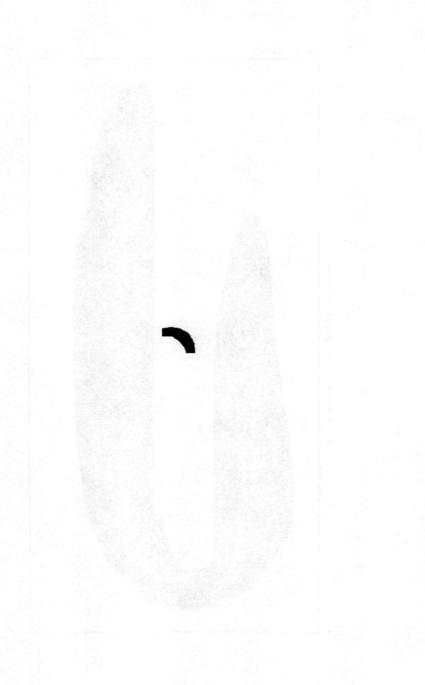

17. Gymnothorax thalassopterus Jenkins, new species.

Head 8.5 in length; depth 13.5; eye 10.5 in head; snout 5.75; interorbital 6.5; gape 2.

Body long, slender, and compressed; vent a little nearer tip of snout than tip of tail; head rather short, rather strongly compressed, broadest above; top of head strongly swollen; profile of snout nearly straight to nape, thence rather strongly elevated; interorbital space somewhat convex; mouth large, the gape long, somewhat wavy, the jaws even, closing completely; posterior edge of orbit a little nearer tip of snout than angle of mouth; teeth in jaws in a single series on each side, some of them canine-like anteriorly, those on sides compressed and directed backward, those on vomer large, fang-like, and depressible; anterior nostrils each in a short tube or papilla near tip of snout; posterior nostrils each without tube, situated just in front of line vertical from front of orbit; gill-opening a long, narrow slit, a little shorter than snout in length, its distance behind angle of mouth equaling length of gape; origin of dorsal a little in front of vertical at gill-opening; dorsal rather low, its greatest height less than snout; anal still lower, its greatest height scarcely exceeding diameter of orbit.

Color in life (No. 03548), very dark brown, nearly black, the light interspaces smoky-yellow; outer margin of vertical fins lemon yellow, below which the color is bright green, gradually losing itself in dark brown.

Color in life of No. 03375, body and fins mottled yellowish and brown, the brown forming irregular granular spots of various sizes, but all less than the pupil; fins a little darker, no pale edges; gill-opening and angle of mouth black; throat-streaks brownish and the spots on jaws smaller.

Color in life of No. 305, brown, with spots and mottlings of darker brown; black spot larger than eye about opercular opening; margins of dorsal and anal bright green; margin of caudal yellow.

Color in alcohol of the type, pale brownish, profusely covered with small, roundish or irregular darker brown spots and blotches, varying considerably in size and also in depth of color, some being

Fig. 7.—Gymnothorax leucacme Jenkins, new species. Type.

mere specks, others as large as the pupil, some pale brown, scarcely darker than the ground color, others almost black; blotches and spots often more or less coalescing, forming reticulations; dorsal and anal fins colored like the body, the edges dark; tip of tail narrowly white; head somewhat darker than body; angle of mouth somewhat dark, with a few white specks; gill-opening with a dark brown or blackish border. One specimen (No. 03722) has a narrow bluish-white line from angle of mouth over the nape. In many of the specimens the narrow white border on the tail extends some distance forward on the dorsal and anal fins.

This is one of the largest and most abundant eels found among the Hawaiian Islands. It reaches a length of 3 feet. One specimen (No. 305) 14 inches long, was secured by me at Honolulu, where numerous examples were taken by Jordan & Evermann. One example was obtained by them at Cocoanut Island, Hilo.

Type, No. 50619 (field No. 03772), U. S. N. M., 23 inches long, collected by Jordan & Evermann in 1901, at Honolulu.

18. Gymnothorax leucacme Jenkins, new species.

Head nearly 8 times in total length; depth 2.3 in head; snout 5.6 in head; anus nearer tip of snout than tip of tail; distance from tip of snout to anus, in the type, 9.5 inches; from tip of tail to anus nearly

11 inches; gill-opening narrow, longitudinal, length less than diameter of cornea; interorbital space equal to diameter of cornea; distance from posterior margin of cornea to angle of mouth about equal to distance from anterior margin to tip of snout; dorsal fin beginning at occiput, in height about 0.5 depth of body; height of anal about diameter of cornea; teeth all pointed, long, and in a single series in each jaw; 3 long sharp teeth on vomer.

Color in life, light brown, with 17 distinct wide, dark brown bands encircling body and fins, not much narrower across the dorsal, much narrower or interrupted on ventral aspect on forward part, less so on posterior part of tail; bands nearly equal to depth, on anterior portion quite equal to depth; a white spot on outer margin of dorsal on each side each dark brown band; between white spots a dark brown spot; spaces between dark bands on anal nearly white; area between eyes, angles of mouth, and borders of lower jaw, dark brown; from dark brown area between eyes to first dark brown band, yellow.

This species differs from *G. petelli* in the white spots on margin of dorsal and white areas on anal, in distinctness of bands, and their encircling the dorsal and anal.

Only one specimen known, type, No. 50682, U. S. N. M. (original No. 280), 21 inches long, obtained by me from the coral rocks at Honolulu.

19. Gymnothorax ercodes Jenkins, new species.

Head 6.6 in total length, or 3 in distance from tip of snout to vent; depth 12; snout 6.6 in head; eye 1.3 in snout; gape 2.6; tip of snout to vent 1.35 in tail; interorbital width slightly greater than eye, or nearly equal to snout. Body moderately elongate and much compressed; tail more compressed and pointed; mouth rather large, the gape reaching beyond eye a distance equal to length of snout;

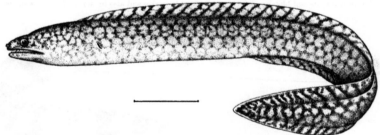

Fig. 8.—*Gymnothorax ercodes* Jenkins, new species. Type.

lower jaw scarcely the shorter, not much curved; teeth all pointed, in 2 series anteriorly and 3 series posteriorly in upper jaw; lower jaw with the teeth in 2 series anteriorly, laterally and posteriorly in a single series; 2 large, sharp-pointed, depressible teeth on anterior part of vomer, followed by a series of about 6 smaller teeth on the shaft; anterior nostril in a short tube whose length is one-fourth diameter of eye, situated near tip of snout just above the lip; posterior nostril without tube, situated above margin of eye just in front of vertical through middle of eye; gill-slit moderate, its length 1.5 in eye; origin of dorsal slightly in front of gill-opening, height of fin 3.5 in head; anal similar, but lower.

Color in alcohol, body and fins light brown on a whitish background, the brown arranged in a somewhat regular net-work, giving the appearance of rows of indistinct whitish spots surrounded by polygonal brownish interspaces, which are most distinct on tail; no white border to the fins or tip of tail, and no dark area around gill-opening.

The only specimen known is the type, No. 50843, U. S. N. M. (original No. 2354), 8.5 inches long, obtained by the *Albatross* at Honolulu in 1891.

20. Echidna leihala Jenkins, new species.

Head 7 in total length; depth 2.1 in head; tip of snout to angle of mouth 2.5 in head; eye 10 in head; interorbital 8.5; gill-opening a very small narrow slit, 3 in eye, with no distinguishing color-marking; origin of dorsal well in advance of gill-opening, 3 in head; jaws curving away from each other, closing only at tip; a few sharp, fixed teeth in anterior portions of jaws, the others all blunt; teeth in anterior portion of upper jaw sharp, in a single series; in the posterior portion a double series of blunt teeth, between which the roof of the mouth is crowded with blunt teeth, becoming as many as 6 series posteriorly; teeth in lower jaw in 2 series anteriorly, becoming blunt posteriorly and apparently in 3 series; anterior nostril tubular, near tip of snout above margin of lip; posterior nostril smooth, near the middle of the upper margin of the eye.

Color in life, uniform yellowish brown (not lighter on the belly), being distributed over the whole body in fine, granular markings; no transverse bands appearing in life, but evidence of bands, especially toward tip of tail, appearing some hours after death; snout white, angles of mouth brown, iris yellow; no other conspicuous markings.

The type measures 17 inches in total length; from vent to tip of tail, 8.4 inches; from tip of snout to vent, 8.7 inches. My collection contains 3 specimens, all from the reef in front of Honolulu. Type, No. 50844, U. S. N. M. (original No. 283), Honolulu; cotypes, No. 7783, L. S. Jr. Univ. Mus. (original No. 2368), 15.5 inches long, and No. 2752, U. S. F. C. (original No. 2369), 12 inches long.

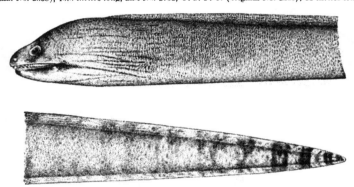

FIG. 9.—*Echidna bibola* Jenkins, new species. Type.

21. Echidna nebulosa (Ahl). *"Puhikapa."*

Color in life, ground color white, on which is a series of about 26 black spots along sides of body and tail, these spots branching into very irregular reticulations; a bright yellow spot in the center of each black spot; the black lateral spots connected under belly with broad, black bands; on the belly are also occasional large round black spots not connected with the bands; a series of spots similar in size and shape to the lateral spots extending along dorsal fin; there is a black reticulated band around the head; iris golden; inferior nasal tubes orange. This is a brilliantly marked eel in life.

The natives report that this species "goes ashore where it catches lizards" and other prey. They regard it with some fear. I obtained but one specimen (No. 292), 22 inches in length.

Muraena nebulosa Ahl, De Muraena et Ophichtho., Thunb. Dissert., III, 5, pl. 1, fig. 2, 1789, East Indies; Günther, Cat., VIII, 130, 1870.

Echidna variegata Bleeker, Atlas, IV, 90, pl. CLXVIII, fig. 2.

Echidna zonata Fowler, Proc. Ac. Nat. Sci. Phila. 1900, 459, pl. XVIII, fig. 2, Hawaiian Islands (Type, No. 16484, Ac. Nat. Sci. Phila.).

22. Echidna vincta Jenkins, new species.

Head 7.2 in total length, or 3.75 in distance from tip of snout to vent; vent about midway between tip of snout and tip of tail; depth about 2.2 in head; eye 10 in head, 1.6 in snout, or 1 in interorbital space; length of mouth 2.7 in head; body moderately elongate, compressed, the tail strongly compressed and pointed; head swollen; mouth moderate, gape reaching beyond eye a distance equal to length of snout, lower jaw shorter than upper, curved so that the mouth does not close completely; teeth bluntly conic, in a single series in front in upper jaw, in 2 series laterally; teeth on vomer bluntly conic, in a single series of 3 teeth, depressible anteriorly, in a double series of molar teeth posteriorly, about 7 teeth in each series; lower jaw with a double series of bluntly conic teeth on each side, and a median series of similar teeth.

Color in life, body crossed by 25 (by error 24 in drawing) broad, reddish brown, non-reticulating bands, the width of those at middle of body exceeding snout and eye, the bands completely encircling the body and separated by somewhat narrower light bands; tip of snout yellowish-white; the first dark band through eye broadening on interorbital space; second dark band crossing side of head and very broad on nuchal region; tip of tail narrowly white. In some of the cotypes, the dark crossbands tend to break up below and form reticulations.

This species is not rare about Honolulu among the coral rocks, where I obtained 16 specimens. It does not appear to reach a large size, the examples in hand ranging from 6 to 15 inches in length.

Type, No. 50687, U. S. N. M. (original No. 231), 13.5 inches long, obtained by the *Albatross* at Honolulu. Cotypes No. 7492, L. S. Jr. Univ. Mus. (original No. 224), 15 inches long; U. S. F. C. (original No. 282); No. 2753, Field Museum (original No. 263).

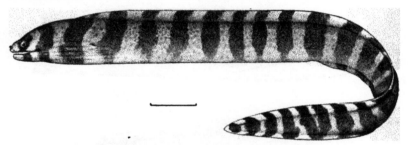

FIG. 10.—*Echidna vincta* Jenkins, new species. Type.

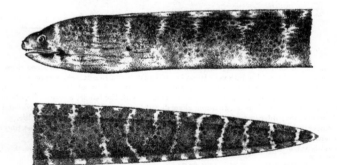

FIG. 11.—*Echidna obscura* Jenkins, new species. Type.

23. Echidna obscura Jenkins, new species.

Head 8.3 in total length; depth 17; eye 9.5 in head; snout 5.75; interorbital 5.75; gape 2.8; distance from tip of snout to vent slightly less than from vent to tip of tail; body moderately elongate, rather deep and somewhat compressed; head narrow, somewhat swollen above; mouth large, the gape extending more than an eye's diameter beyond eye; lower jaw shorter than upper and somewhat curved; eye about midway between tip of snout and angle of mouth; interorbital equals snout; origin of dorsal in front of gill-opening a distance equal to length of mouth; dorsal fin somewhat higher than anal, its height greater than length of snout; tail compressed and moderately slender; a few short conical teeth in anterior parts of each jaw; 2 series of conical teeth in each side of upper jaw; roof of mouth paved with molars, in 2 rows anteriorly, in 4 posteriorly; molars in 2 series in each side of lower jaw; gill-opening small, narrow, length less than diameter of eye; anterior nostril tubular, near tip of snout, considerably above margin of mouth; posterior nostril round and inconspicuous, near middle of upper margin of eye.

Color in alcohol, dark brownish with about 23 dark crossbands, mostly as broad as depth of body, indistinct on middle part of body, but quite distinct anteriorly and on tail; alternating with them are white ones which are narrower than eye and which extend on anal and dorsal fins, the edges of the bands jagged, the white bands widening toward the belly; extreme tip of tail brown (in the cotypes the tip is narrowly edged with white); side of lower jaw brown, angle of mouth black with white spot in front on lower jaw; gill-opening without dark border. The 2 cotypes show some variations in color. In the larger (No. 2351), 16.5 inches long, the body is more uniformly dark brown and the light crossbands are very indistinct except on tail; in the other (No. 2353), 9.5 inches in length, the white crossbands are very distinct, all completely encircling the body except 3 or 4 anterior to vent.

Three specimens obtained, from Honolulu. Type, No. 50686, U. S. N. M. (field No. 2352), 12.5 inches long. Cotypes, No. 7725, L. S. Jr. Univ. Mus. (field No. 2351), 16.5 inches long; and No. 2754, U. S. F. C. (field No. 2353), 9.5 inches long.

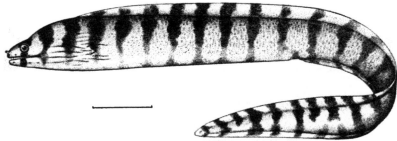

Fig. 12.—*Echidna psalion* Jenkins, new species. Type.

24. Echidna psalion Jenkins, new species.

Head 7.25 in body, or 3.4 in distance from tip of snout to vent; depth 13; snout 5.5 in head; eye slightly less than snout and slightly nearer tip of snout than angle of mouth; gape 2.5 in head; tip of snout to vent 1.2 in tail; interorbital about equal to eye; body moderately elongate and compressed posteriorly; tail slender, pointed; gill-opening very small and inconspicuous; anterior nostril tubular, about 2 in eye, near tip of snout, well above the lip; posterior nostril without tube, oval, above eye just anterior to its middle; a series of pores along upper lip and a series on each side of lower jaw; upper jaw with a single series of blunt, conic teeth in front, those on sides smaller and in a single series; roof of mouth with 2 series of large molars; vomer in front with a single series of about 3 strong, blunt, conical, depressible teeth; lower jaw with 2 series of blunt, conic teeth, the inner the larger; origin of dorsal in front of gill-opening a distance equal to one-fourth the head.

Color in alcohol, a series of 27 narrow brown bands alternating with wider light bands, the narrow bands mostly somewhat narrower than eye, the light ones mostly twice eye; a series of narrow parallel brown longitudinal lines on side of head in front of gill-opening; the anterior brown band running through eye, the second around head posterior to gape; angle of mouth brown.

Only one specimen obtained. Type, No. 50685, U. S. N. M. (original No. 2355), 13 inches long, Honolulu.

Family VIII. ELOPIDÆ.

25. Elops saurus Linnæus.

This species is very abundant; numerous examples are usually in the market, some reaching a length of 2.5 feet. It is not very highly esteemed as food.

Color in life, dorsal aspect gray with greenish and bluish reflections; belly silvery; dorsal and caudal fins dusky; pectoral, ventral, and anal yellowish brown.

Three specimens were obtained by me, 12, 13, and 11 inches in length; 3 by the *Albatross* in 1896, 11, 10, and 10 inches in length; and 1 by Jordan & Snyder in 1900, 8.5 inches in length.

Elops saurus Linnæus, Syst. Nat., ed. XII, 518, 1766, Carolina: Günther, Cat., VII, 470, 1868; Steindachner, Denks. Ak. Wiss. Wien, LXX, 1900, 513 (Honolulu): Fowler, Proc. Ac. Nat. Sci. Phila. 1900, 496 (Honolulu).

Family IX. ALBULIDÆ.

26. Albula vulpes (Linnæus).

Many specimens obtained by me and two by Dr. Wood. This is a very important food-fish in the Hawaiian Islands. Examples of 3 feet or more are often seen in the market.

Color in life, bright silvery with about 7 gray streaks parallel with the lateral line.

Esox vulpes Linnæus, Syst. Nat., ed. X, 1758, 313.
Albula conorynchus, Streets, Bull. U. S. Nat. Mus., No. 7. 76, 1877 (Honolulu).
Albula glossodonta, Steindachner, Denks. Ak. Wiss Wien. LXX. 1900, 513 (Honolulu).

Family X. CHANIDÆ.

27. Chanos chanos (Forskål). *"Awa."* The small *"Puaawa,"* the large *"Awakalamolo."*

One of the most abundant and important food-fishes at Honolulu. I obtained several specimens.

Mugil chanos Forskål, Descript. Animal, 74, 1775, Red Sea.
Chanos salmoneus Cuvier & Valenciennes, Hist. Nat. Poiss., XIX, 201, 1846, between New Caledonia and Norfolk Island;
  ' Günther, Challenger Report, Shore Fishes, 61, 1880 (reefs at Honolulu).
Chanos cyprinella Cuvier & Valenciennes, Hist. Nat. Poiss., XIX, 198, 1846, Hawaiian Islands.
Chanos chanos, Steindachner, Denks. Ak. Wiss. Wien. LXX. 1900, 514 (Honolulu).

Family XI. CLUPEIDÆ.

28. Etrumeus micropus (Schlegel). *"Makiawa."*

This species until now only known from Japan. D. 18; A. 9.

Color in life, upper part of body greenish blue with bright metallic reflections; lower two-thirds of body bright silvery; about and in front of the eyes transparent; iris white with metallic reflections; iridescent; pectoral fins olivaceous; snout translucent.

This species is at times fairly abundant and is much esteemed as food, bringing a high price. I obtained several specimens. Two are in Dr. Wood's collection.

Clupea micropus Schlegel, Fauna Japonica, Poiss., 236, pl. 107, fig. 2, 1842, Japan.
Etrumeus micropus, Bleeker, Verh. Bat. Gen., XXV, 1853, 48 (Japan).

Family XII. ENGRAULIDIDÆ.

29. Anchovia purpurea (Fowler).

This species is very abundant about the reef of Honolulu. I obtained numerous specimens. It appears to be the species described from a specimen in the Museum of the Academy of Natural Sciences of Philadelphia, by Mr. Fowler, in 1900.

Stolephorus purpureus Fowler, Proc. Ac. Nat. Sci. Phila. 1900, 497, pl. XIX, fig. 1, Hawaiian Islands. (Type, Nos. 28329 and 28330, Ac. Nat. Sci. Phila. Mus.: coll. W. H. Jones.) ·

Family XIII. SYNODONTIDÆ.

30. Trachinocephalus myops (Forster).

A specimen of this fish appears in Dr. Wood's collection. It is 7.25 inches long and is No. 2069. I did not obtain it.

Salmo myops Forster, in Schneider, Syst. Ichth., 421, 1801, St. Helena.
Saurus trachinus Schlegel, Fauna Japonica, Poiss., 231, pl. 106, fig. 2, 1842, Japan.
Synodus myops, Bleeker, Atlas Ichth., VI, 153, pl. 278, fig. 3, 1870-72.
Trachinocephalus myops, Jordan & Evermann, Fishes North & Mid. Amer., 1, 533, 1896.

31. Synodus varius (Lacépède).

Two examples of this fish were obtained (field Nos. 2070 and 2071) 5 and 5.5 inches in length. There is one 4.5 inches long in Dr. Wood's collection. It is found with *Saurida gracilis* in the coral sand on the reef, and with it exposed for sale in the market.

Salmo varius Lacépède, Hist. Nat. Poiss., v, 224, pl. 3, fig. 3, 1803, Isle de France.
Synodus varius, Steindachner, Denks. Ak. Wiss. Wien, LXX, 1900, 513 (Honolulu and Laysan).
Synodus sharpi Fowler, Proc. Ac. Nat. Sci. Phila. 1900, 497, pl. XIX, fig. 2, Hawaiian Islands. (Types, Nos. 16084 to 16086, Ac. Nat. Sci. Phila. Mus.; coll. W. H. Jones.)

32. Saurida gracilis (Quoy & Gaimard).

Color in life, upper part of body drab, lower white; upper part with many pearly white, dark, and blackish spots so intermingled as to give the colors and shadings of coral sand in which the species is found; 8 or 9 groups of black spots along the lateral line. I have 8 specimens, 5.5 to 6 inches in length. I saw none larger than 6 inches. They abound in the coral sand on the reef and elsewhere and are sold in the market.

Saurus gracilis Quoy & Gaimard, Voy. de l'Uranie, Zool., 224, 1824, Timor.
Saurida nebulosa Cuvier & Valenciennes, Hist. Nat. Poiss., XXII, 504, pl. 648, 1849, Isle de France; Streets, Bull. U. S. Nat. Mus., No. 7, 76, 1877 (Honolulu).
Saurida tumbil, Fowler, Proc. Ac. Nat. Sci. Phila. 1900, 498 (Hawaiian Islands).

Family XIV. BELONIDÆ.

33. Belone platyurus Bennett.

A number of specimens of this fish were taken by me. The longest (field No. 392) is 16 inches long. It is abundant and is brought to the market.

Belone platyurus Bennett, Proc. Comm. Zool. Soc., 1830, 168.
Belone platura, Cuvier & Valenciennes, Hist. Nat. Poiss., XVIII, 451, 1846. Gunther, Cat., VI, 237, 1866; Streets, Bull. U. S. Nat. Mus., No. 7, 75, 1877 Honolulu.
Belone carinata, Cuvier & Valenciennes, Hist. Nat. Poiss., XVIII, 437, 1846 (Sandwich Islands).
Mastacembelus platurus, Bleeker, Atlas Ichthy., VI, 50, pl. 257, fig. 1, 1872 (Singapore, Amboyna).

34. Tylosurus giganteus (Schlegel).

Two specimens were obtained by me in 1889, the larger (field No. 104), 440 mm. in length, gives: D. 23; A. 22; origin of ventrals midway between base of median caudal rays and center of eye; eye 1.5 in interorbital, 2.33 in postorbital part of head. The smaller, 195 mm. long: D. 22; A. 21; origin of ventrals midway between base of median caudal rays and front of eye.

Color in life, upper parts light green; lower parts silvery white.

Belone gigantea, Schlegel, Fauna Japonica, Poiss., 245, 1846, Japan.
Belone annulata, Cuvier & Valenciennes, Hist. Nat. Poiss., XVIII, 447, 1864, pl. 550, Gunther, Cat., VI, 240, 1866; Steindachner, Denks. Akad. Wiss. Wien, LXX, 512, 1900 (Honolulu and Samoa).
Mastacembelus annulatus, Bleeker, Atlas Ichthy., VI, 48, pl. 258, fig. 3, 1872 (East Indies).

35. Athlennes hians (Cuvier & Valenciennes).

Three specimens were obtained by me at Honolulu, the largest 26 inches in length. These are the first specimens of the species reported from the Pacific west of the American coast. My specimens do not differ from the description of *A. hians* nor in any way from a specimen of that species in the Stanford University collection from North Carolina, with which they have been compared.

D. 25; A. 27; greatest depth of body at base of ventrals 13 in length; head with beak 3.75 in length; snout 1.5 in head; eye 2 in postorbital part of head; interorbital slightly greater than eye; suborbital very narrow, about 7 in eye; pectoral a little greater than greatest depth of body, equal to distance from front of eye to margin of opercle; ventrals shorter, equal to distance from center of eye to margin of opercle, their bases midway between the bases of median caudal rays and front of arched part of upper mandible; front part of dorsal and anal falciform; longest rays of dorsal equaling distance from front of ventral to posterior margin of opercle; longest rays of anal equal pectoral, and equaling distance from front of eye to posterior margin of opercle; caudal forked, lower lobe the longer; posterior dorsal rays longer than the median ones.

Quite abundant, being brought in numbers to the market. It is esteemed as a good food-fish.

Belone hians Cuvier & Valenciennes, Hist. Nat. Poiss., XVIII, 432, 1846, Bahia; Havana.
Athlennes hians Jordan & Evermann, Fishes North and Mid. Amer., I, 718, 1896 (West Indies).

Family XV. HEMIRHAMPHIDÆ.

36. Hemiramphus brasiliensis (Linnæus).

Color in life, dark blue on top of head and body, silvery below; ventral side of beak red, tip orange, upper side dark.

Abundant at Honolulu, large numbers being brought to the market for sale. Several specimens were obtained by me. In comparing them with others from the West Indies, no other structural differences could be noted than the slightly longer pectoral fin, as may be seen from the table.

Comparative measurements of specimens of H. brasiliensis from West Indies and Hawaiian Islands.

	Hawaiian Islands.							West Indies.			
	No. 1308.	No. 186.						No. 10328.	No. 10332.	No. 11176.	No. 11175.
Length (from tip of upper jaw) ......mm..	267	247	235	238	236	179	170	185	184	203	172
Head (from tip of upper jaw) ....mm..	24	23	23	23	22	23	24	23	24	23	25
Depth ....mm..	15½	16	14	15	14	15	14	16	15	16	16
Lower jaw (from tip of upper) ......mm..	27	28	28	25	27	29	29	......	......	......	......
Pectoral ....mm..	19	20	22	19	20	21	21	17+	18+	17+	......
Ventral...........................mm..	11	11½	12	11	11	10½	11	......	......	......	......
Tip of snout to ventrals .... mm..	66	67	67	67	66	67	66	67	67	70	68
Ventral to last vertebra ....mm..	33	34	36	33	36	36	36	34	33	33	33
Eye...............................mm..	7	6	6	6	6	7	6	6	6	6	6
Interorbital.......................mm..	5½	5	5½	5½	5	6½	5½	5½	5½	5½	5½
Snout.............................mm..	8	8	7½	8	8	6½	8	8	8	8	8
Dorsal rays	15	14	14	13	14	14	15	14	14	14	14
Anal rays.............................	13	13	13	12	13	12	12	12	12	13	12
Scales in lateral line.....................	55	57	56	56	54	58	58	54	58	55	54

Esox brasiliensis Linnæus, Syst. Nat., ed. X, 1758, 314.
Hemiramphus depauperatus Lay & Bennett, Zoology Beechey's Voyage, Fishes, 66, 1839, Oahu, Hawaiian Islands; Fowler, Proc. Ac. Nat. Sci. Phila. 1900, 499 (Hawaiian Islands).
Hemiramphus pleii, Cuvier & Valenciennes, Hist. Nat. Poiss., XIX, 15, 1846 (Antilles).
Hemiramphus macrochirus, Poey, Memorias Cuba, II, 299, 1856–58 (1861) (Cuba).

37. Euleptorhamphus longirostris (Cuvier).

This is the first record of *E. longirostris* from the Hawaiian Islands. It is brought to the markets in numbers and sold for food. Seven specimens, 16 to 18 inches in length.

Color in life, dark above, a longitudinal silvery band on the side; belly white; beak black.

Hemirhamphus longirostris Cuvier, Règne Animal, ed. 2, II, 286, 1829, Indies; Cuvier & Valenciennes, Hist. Nat. Poiss., XIX, 52, 1846 (Pondicherry).

Family XVI. EXOCŒTIDÆ.

38. Evolantia microptera (Cuvier & Valenciennes).

This is the first record of the species from the Hawaiian Islands. Eight specimens, 6 to 7 inches in length, were obtained.

Color in life (field No. 229), upper parts dark blue, below silvery; dorsal bluish; pectoral somewhat dusky, especially toward tip; anal and ventrals white; no definite markings on any of the fins.

Exocœtus micropterus Cuv. & Val., XIX, 127. pl. 563, 1846; Günther, Cat., VI, 279, 1866; Bleeker, Atlas Ichthy., VI, 77, pl. 249, fig. 1, 1872.

Cypsilurus micropterus, Bleeker, Nederl. Tydsch. Dierk., III, 126, 1865 (Amboyna).

Erolontia, Snodgrass & Heller, MSS.

39. Parexocœtus brachypterus (Solander). *"Malolo."*

D. 13; A. 14; origin of dorsal just over that of anal; pectoral reaching just beyond origin of anal. Color in life, upper parts of head and body down to lateral line a brilliant indigo blue, below this silvery; dorsal with a large dark blue blotch toward tip; pectoral transparent; caudal, anal, and ventrals colorless. The Hawaiian specimens were compared with one in the Stanford University Museum from Pensacola, Fla. No differences could be detected further than the slightly shorter dorsal fin in the Florida specimen, which difference is probably only an individual variation. Very abundant, large quantities being brought to the market, where they bring a good price. Eight specimens were taken, 6 to 7 inches in length, and 2 are in Dr. Wood's collection. I observed none over 7 inches in length.

Exocœtus brachypterus Solander, in Richardson, Ichthy. China, 265, 1846, China; Günther, Cat., VI, 290, 1866 (Otaheiti and China); Streets, Bull. U. S. Nat. Mus., No. 7, 75, 1877 (Hawaiian Islands); Steindachner, Denks. Ak. Wiss. Wien, LXX, 1900, 512 (Honolulu and Laysan).

Exocœtus hillianus Gosse, Nat. Sojourn Jamaica, II, 11, pl. 1, fig. 1, 1851, Jamaica; Günther, Cat., VI, 294, 1866; Poey, Mem., II, 301.

Parexocœtus mesogaster Jordan & Evermann, Fishes North and Mid. Amer., I, 728, 1896; Fowler, Proc. Nat. Ac. Sci. Phila. 1900, 500 (Hawaiian Islands); Evermann & Marsh, Fishes of Porto Rico, Bull. U. S. Fish Com. 1900, 103.

The figure of *Exocœtus mesogaster* Bloch, XII, 17, pl. 399, can not be of this species.

40. Exocœtus volitans Linnæus.

A single specimen, 5.5 inches in length, was picked up on the beach at Honolulu, and one 6 inches in length was collected by the *Albatross* in 1891, in lat. 28° 03' N., long. 143° 10' W.

Color in life, back very dark blue, with greenish reflections, especially along the sides; belly silvery white; pectoral rays dusky, the membrane not so dark. The pectoral reaches root of caudal.

Exocœtus volitans Linnæus. Syst. Nat., ed. X, 316, 1758; Fowler, Proc. Ac. Nat. Sci. Phila. 1900, 500 (Hawaiian Islands).

Exocœtus evolans Linnæus. Syst. Nat., ed. XII, I, 521, 1766; Günther, Cat., VI, 282, 1866.

Halocypselus evolans, Jordan & Evermann, Fishes North and Mid. Amer., I, 729, 1896.

41. Cypsilurus simus (Cuvier & Valenciennes).

Dorsal profile gently and evenly convex from tip of snout to caudal peduncle; ventral profile straight from head to front of ventral fin; depth 5 in length; head 4.25 in length; snout wide, its anterior straight transverse border 1.3 in length of snout; length 3.5 in head; lower jaw not projecting; depth of head equaling distance from tip of snout to middle of preopercle; eye 3 in head; interorbital a little concave (varies from flat to considerably concave), 3.5 in head; teeth conical (varying from tricuspid in smaller specimens to bicuspid in larger ones and conical); gillrakers short, flat, mostly of equal length, 5 in interorbital, 16 on lower limb of arch; pectoral 15, reaching to base of caudal, first ray simple, second divided; ventrals 6, reaching past middle of anal; D. 13 (varying from 12 to 14); third and fourth rays longest, 2.5 in head; last ray longer than the penultimate ray, 4 in head; A. 8, shorter and lower than dorsal (rays varying from 7 to 9), second and third rays longest, 3 in head; lower lobe of caudal much longer than upper, length of tip from last vertebra about 3.2 in length of body; length of upper lobe of caudal about 1.6 in lower, 5 in length of body; least depth of caudal peduncle but slightly less than its length, 3 in head; 50 scales on lateral line, tube on each scale with numerous fine convoluted branches on lower half of scale only; 31 scales on mid-dorsal line in front of dorsal fin; 12 scales on oblique row between front of dorsal and anal fins, 8 from dorsal to lateral line.

Color in life, upper parts dark blue, back almost black; belly silvery white; generally a dusky area behind opercle at base of pectoral; ventrals and caudal transparent; pectoral varying much in coloration. In 1 specimen the first pectoral ray was white, the membranes of the next 7 rosy-brownish, the membranes of the rest transparent. Another had the pectoral fin nearly colorless, with only a small amount of dusky clouding and several small black spots on the membrane. In others the pectorals are closely covered with round or oval black spots, varying in different specimens from many to few. Others again have none. Since there are no discoverable structural differences between the specimens having spotted pectorals and those having no spots on those fins, and since the number of the spots varies so greatly on those that possess them, there are no grounds on which the specimens with spotted pectorals can be separated from those without spots. Hence the synonymy given below.

Eight specimens were collected at Honolulu in 1889 (field Nos. 166, 167, 168, 191, 192, 193, 194, and 196), and 6 specimens were collected by Jordan & Snyder at Honolulu in 1900 (field Nos. 03, 04, 05, 07, 08, and 09), and 1 specimen (field No. 2943) is in Dr. Wood's collection. These specimens are 12 to 13 inches in length. Large numbers are caught about Honolulu and brought to the market.

Measurements of nine examples of Cypsilurus simus.

	Pectoral unspotted.					Pectoral spotted.			
	No. 167.	No. 08.	No. 196.	No. 168.	No. 193.	No. 05.	No. 194.	No. 192.	No. 191.
Length ....millimeters..	250	270	215	208	261	254	247	245	242
Depth ....do....	19	19	18	19	19	18	19	19	18
Head ....do....	24	23	24	25	23	23	24	23	23
Snout ....do....	6	7	6½	6	7	6½	7	7	7
Eye ....do....	8½	8	8½	9	8½	8	8	8½	8
Interorbital ....do....	9	8	8	8	9	8	9	9	8½
Eye to ventral ....do....	46	48	45	47	47	45	45	46	47
Ventral to last vertebra ....do....	41	43	40	36	44	44	43	40	40
Length of pectoral ....do....	70	68	65	65	72	70	64	69	67
Length of ventral ....do....	25	28	26	27	29	29	28	29	31
Dorsal rays................	13	12	14	13	13	13	12	13	12
Anal rays	7	8	9	8	8	9	8	8	7
Scales before dorsal................	32	31	28	31	30	30	30	30	30
Scales between dorsal and lateral line................	8	8	8	7	8	8	8	8	8
Caudal peduncle................	8	8½	8½	8	8	8½	8½	9	8

Exocœtus simus Cuvier & Valenciennes, Hist. Nat. Poiss., XIX, 105, 1846, Sandwich Islands.

? *Exocœtus pœcilopterus* Cuvier & Valenciennes, Hist. Nat. Poiss., XIX, 112, pl. 561, 1846, New Holland: Günther, Cat., VI, 291. 1866; Bleeker, Atlas Ichthy., VI, 71, pl. 251, fig. 5, 1872.

Exocœtus alatus Solander, in Cuvier & Valenciennes, Hist. Nat. Poiss., XIX, 112, 1846, Otahiti.

? *Exocœtus spilopterus* Cuvier & Valenciennes, Hist. Nat. Poiss., XIX, 113, 1846, Caroline Islands: Günther, Cat., VI, 292, 1866; Bleeker, Atlas Ichthy., VI, 71, pl. 250, fig. 2, 1872.

Exocœtus neglectus Bleeker, Esp. Exoc. Ned. Tydsch. Dierk., III, 112, 1865; Bleeker, Atlas Ichthy., VI, 71, pl. 247, fig. 2, 1872 (Sumatra, Batjan); Steindachner, Denks. Ak. Wiss. Wien, LXX, 1900, 512 (Hawaiian Islands).

? *Exocœtus callopterus* Günther, Cat., VI, p. 292, 1866, Panama.

42. Cypsilurus bahiensis (Ranzani).

One specimen of this fish was obtained by me in 1889, 15 inches in length (field No. 195), and one by Dr. Wood (field No. 12043) 13.5 inches; and one by Jordan & Snyder (field No. 04).

Exocœtus bahiensis Ranzani, Nov. Comm. Ac. Sci. Inst. Bonon., V, 1842, 362, pl. 36, Bahia: Bleeker, Atlas Ichthy., VI, 71, pl. 249, fig. 2, under the name *E. spilonopterus*: Jordan & Evermann, Fishes North and Mid. Amer., I, 739, 1896; Steindachner, Denks. Ak. Wiss. Wien, LXX, 1900, 512 (Honolulu).

Exocœtus vermiculatus Poey, Memorias, II, 300, Cuba.

Exocœtus spilonopterus, Bleeker, Nederl. Tydsch. Dierk., III, 113, 1865 (Sumatra).

43. Cypsilurus atrisignis Jenkins, new species.

Head 4.3 in length; depth 5.5; D. 15; A. 10; P. 14; lateral line about 60; scales before the dorsal fin 34; scales between origin of dorsal and lateral line 9; body elongate, broad dorsally, narrowing ventrally, broadest just in front of base of pectorals, where it is nearly as broad as the depth; top of posterior portion of head broad, narrowing toward tip of snout, somewhat concave between the eyes; interorbital space equaling distance from posterior margin of eye to margin of opercle; eye large, its center anterior to center of head; snout less than eye, somewhat pointed, lower jaw slightly projecting; maxillary included and falling considerably short of anterior margin of eye; pectoral reaching tip of last dorsal ray; ventral reaching to one-third the base of the anal; its origin halfway between eye and base of caudal; origin of dorsal much in advance of vent, its distance from first caudal ray 1.4 times head, the longest ray, the anterior one, about .5 the head; lower lobe of caudal the longer.

Color in alcohol, dark purple above, light below; dorsal fin with black spot about 0.7 the diameter of eye between eighth and eleventh spines; caudal and ventrals colorless, unmarked; ventrals white, without spots; pectoral rays and membranes very dark purple above, the rays light below, the membranes with black spots on anterior and posterior portions.

One specimen (field No. 197) 13.5 inches in length, was taken by me in 1889 at Honolulu. Type, No. 50713, U. S. N. M.

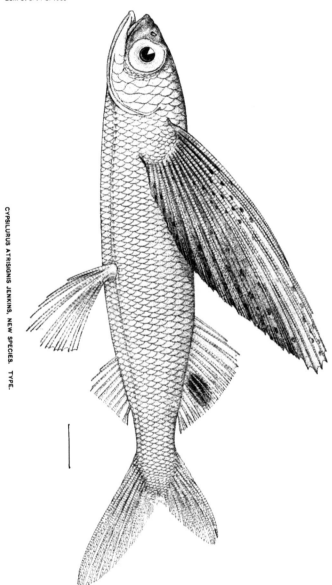

CYPSILURUS ATRISIGNIS JENKINS, NEW SPECIES. TYPE.

Family XVII. AULOSTOMIDÆ.

44. Aulostomus valentini (Bleeker). "*Nunu.*"

Color in life (field Nos. 165 and 179), dark brown, with light crossbars, about 17 in number, between eye and dorsal fin, 5 others posteriorly; also about 4 longitudinal whitish bands, much more distinct when the fish is alive in the water; base of dorsal and anal black, with the anterior portion of each red, posterior portion of each paler but also tinged with red; caudal olivaceous, a black spot on upper margin and one opposite on lower margin; posterior to these spots the margins are red; a black line extending forward from each eye, meeting along snout, this interrupted with about 5 whitish bars across snout; ventrals each with a spot at base. Although there were individual variations from the above, I did not see any of the bright yellow forms figured by Günther in Fische der Südsee.

This species is fairly abundant at Honolulu, where I caught four specimens with a dip-net off the wharf, ranging from 6.5 to 8 inches in length. One specimen (field No. 03583) was obtained by the *Albatross* in 1896. Three others, each 13.5 inches in length, are in Dr. Wood's collection. Others much longer were seen. They are very beautiful objects in the water. They are able to dart with great rapidity through the water and were seen frequently feeding on something at the surface.

Polypterichthys valentini Bleeker, Nat. Tyds. Ned. Indie, IV, 1853, 60s, Ternate.
Aulostoma sinensis Schlegel, Faun. Japon., Poiss., 320.
Aulostoma chinense, Streets, Bull. U. S. Nat. Mus., No. 7, 74, 1877 (Honolulu); Steindachner, Denks. Ak. Wiss. Wien, LXX, 1900, 502 (Honolulu, Laysan); Günther, Fische der Südsee, VII, 221, pl. 123, figs. B and C, 1881 (Hawaiian Islands).
Aulostomus chinensis, Fowler, Proc. Ac. Nat. Sci. Phila. 1900, 500 (Oahu).

Family XVIII. FISTULARIDÆ.

45. Fistularia petimba Lacépède.

This species is quite abundant at Honolulu. Thirteen examples were taken by me in 1889, and 20 were collected by the *Albatross* in 1899 at Tahiti.

Fistularia petimba Lacépède, Hist. Nat. Poiss., V. 349, 1803, New Britain, Isle of Reunion, equatorial Pacific.
Fistularia depressa Günther, Rep. Shore Fishes, *Challenger*, 69, pl. 32, fig. D, 1880, Sulu Archipelago.

46. Fistularia serrata Cuvier.

Color in life, upper parts dark drab; lower white; tips of dorsal and anal and lobes of caudal rosy with dusky shades; pectoral transparent. Fifteen specimens were taken by me at Honolulu.

Fistularia serrata Cuvier, Règne Animal, ed. 1, vol. II, 349, 1817, America; after Bloch; Streets, Bull. U. S. Nat. Mus., No. 7, 74, 1877 (Honolulu).
Fistularia petimba Jordan & Snyder, Proc. U. S. Nat. Mus., vol. XXVI, 67, 1902 (Japan).

Family XIX. ATHERINIDÆ. The Silversides.

47. Atherina insularum Jordan & Evermann.

Head 4 in length; depth 4.75; eye 3 in head; snout 4; interorbital 2.8; maxillary 2.5; mandible 2.2; D. VI-I, 11; A. 17; scales 46, 6 rows from anterior base of anal upward and forward to spinous dorsal. Body oblong, compressed; head triangular, the sides compressed, top flat; mouth large, oblique, maxillary reaching front of pupil, lower jaw included; teeth in rather broad villiform bands on jaws, vomer, and palatines; interorbital space very broad and flat; snout broad, truncate; origin of spinous dorsal slightly posterior to vertical at vent, slightly nearer tip of snout than base of caudal; longest dorsal spine about 2.4 in head, reaching nearly to vertical at front of anal; distance between spinous and soft dorsals equal to distance from tip of snout to middle of pupil; edge of soft dorsal concave, anterior rays somewhat produced, their length 1.9 in head; last dorsal ray about one-half longer than one preceding; base of soft dorsal 1.8 in head; origin of anal considerably in advance of that of soft dorsal, the fins similar, anterior rays about 1.7 in head, base of anal 1.3 in head; caudal widely forked, the lobes equal; ventral short, barely reaching vent; pectoral short, broad, and slightly falcate, its length about 1.4 in head. Scales large, thin, and deep, 19 in front of spinous dorsal, 6 rows between the dorsals and 9 on median line of caudal peduncle.

Color when fresh, clear olive green with darker edges to scales; lateral stripes steel blue above, fading into the silvery belly; fins uncolored.

Color in alcohol, olivaceous above, silvery on sides and below; scales of back and upper part of side with numerous small, round, coffee-brown specks, disposed chiefly on the edges; median line of back with a darkish stripe; middle of side with a broad silvery band, plumbeous above, especially anteriorly, more silvery below; top of head and snout with numerous dark brownish or black specks; side of head silvery, opercle somewhat dusky, sides and tip of lower jaw dusky; dorsals and caudal somewhat dusky, other fins pale; pectoral without dark tip.

This small fish is common inside the reef in shallow bays everywhere in the Hawaiian Islands. Many individuals were seen off the wharf at Lahaina on Maui. Jordan & Evermann's collections of 1901 contain 20 specimens from Kailua, from 1.5 to 3.5 inches long; 43 from Hilo, 1.5 to 2.25 inches long; and 1 from Honolulu, 2.25 inches in length. Numerous specimens were obtained by the *Albatross* at Honolulu in 1902, 1 of which is taken as the type and 3 others as cotypes.

Five specimens were obtained by me in 1889.

Atherina insularum Jordan & Evermann, Bull. U. S. Fish Comm. for 1902 (April 11, 1903), 170, Honolulu. (Type, No. 50819, U. S. N. M.; coll. *Albatross*, 1902.)

Family XX. MUGILIDÆ.

48. Mugil cephalus Linnæus. *"Amaama"; Mullet.*

This is the most highly prized food-fish about the islands, always bringing good prices in the market. Besides being caught in nets as they run in schools, they are kept in large ponds from which they are taken to supply the market. These are portions of the sea inclosed in favorable places by walls with openings through which the fish are allowed to run, but from which they are prevented from escaping. They remain here feeding until of sufficient size to market. Many of these ponds have been maintained from times previous to the discovery of the islands by Captain Cook. Other fishes accompany the mullet into the ponds and are likewise restrained with it, notably the "Awa," *Chanos chanos.* The natives have different names for different sizes of the mullet. The very small, about 1 inch or less, is "Pua"; about 6 inches is "Pua-ama-ama"; the larger ones are "Anae."

My collection contains 70 examples from 2.5 to 7.2 inches in length. I have also examined a specimen 10 inches long, collected by Mr. R. C. McGregor, and Dr. Wood's collection contains one (field No. 6114), 9 inches. The Hawaiian specimens compared with specimens in Leland Stanford Junior University Museum identified as *Mugil cephalus*, from Naples; La Paz, Mexico; Callao, Peru; and Japan, show no structural differences. The older individuals have a greater mandibulary angle, it being more acute in the small ones.

Mugil cephalus Linnæus, Syst. Nat., ed. x, 316, 1758, Europe.
Mugil dobula Günther, Cat., III, 420, 1861, Aneiteum, Australia; Fische der Südsee, VI, 214, 1877 (Hawaiian Islands); Steindachner, Denks. Ak. Wiss. Wien, LXX, 1900, 501 (Honolulu).

49. Chænomugil chaptalii (Eydoux & Souleyet).

Color in life, gray above, white below, golden spot on upper part of base of pectoral, upper portion of iris golden, the remainder white. Four specimens were obtained, the longest being 10 inches. It seems fairly common at Honolulu and is highly prized as a food-fish.

Mugil chaptalii Eydoux & Souleyet, Voyage Bonite, Zool., 1, 171, pl. 4, fig. 1, 1841, Hawaiian Islands.
Myxus (Neomyxus) edater Steindachner, Sitz. Ber. Ak. Wiss. Wien, LXXVIII, 1, 381, 1878, Kingsmill and Sandwich Islands.

Family XXI. SPHYRÆNIDÆ.

50. Sphyræna commersoni Cuvier & Valenciennes.

This species was described by me as *S. snodgrassi* in a former paper, but an examination of a larger number of specimens has led to the decision that this is the same as the East Indian species. This species reaches a large size, individuals 5 feet in length having been brought to the market.

Sphyræna commersoni Cuvier & Valenciennes, Hist. Nat. Poiss., III, 352, 1829.
Sphyræna snodgrassi Jenkins, Bull. U. S. F. C. 1899 (June 8, 1900), 388, fig. 2, Honolulu. (Type, No. 50833, U.S.N.M.)

51. Sphyræna helleri Jenkins.

This comparatively small species is fairly common about Honolulu, being frequently taken in the mullet ponds, where it preys on that fish. But one specimen (16 inches long) was taken. Apparently it is not often in the market.

Sphyræna helleri Jenkins, Bull. U. S. Fish Comm. 1899 (June 8, 1900), 387, fig. 1, Honolulu. (Type, No. 49092, U.S.N.M.)

Family XXII. POLYNEMIDÆ.

52. Polydactylus sexfilis (Cuvier & Valenciennes). "Moi."

Fairly abundant at Honolulu, where I obtained 6 specimens. Dr. Wood's collection contains one.

Polynemus sexfilis Cuvier & Valenciennes, Hist. Nat. Poiss., VII, 515, Isle de France; Günther, Shore Fishes, Challenger, 59, 1880 (Hilo, Hawaii); Steindachner, Denks. Ak. Wiss. Wien, LXX, 1900, 492 (Honolulu).
Polydactylus pfeifferi, Fowler, Proc. Ac. Nat. Sci. Phila. 1900, 501 (Hawaiian Islands).

Family XXIII. HOLOCENTRIDÆ.

53. Holotrachys lima (Cuvier & Valenciennes).

Color in life, whole body, with head and fins, bright red; iris red.

Three specimens of this beautiful fish, 4.5, 4.7, and 5.2 inches respectively, were taken at Honolulu. Dr. Wood's collection contains four, ranging from 4.2 to 5 inches, and one was taken by Jordan & Snyder 4.7 inches in length. It is common at Honolulu.

Myripristis lima Cuvier & Valenciennes, Hist. Nat. Poiss., VII, 493, 1831, Isle de France.
Myripristis (Holotrachys) lima, Günther, Fische der Südsee, 93, pl. 63, fig. A, 1873 (Mauritius, Kingsmill, Samoa, Society, and Hawaiian Islands); Steindachner, Denks. Ak. Wiss. Wien, LXX, 1900, 492 (Honolulu).

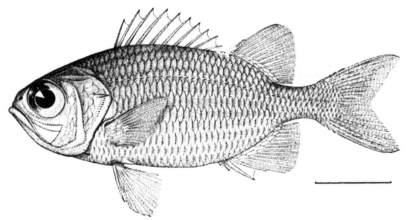

FIG. 13.—*Myripristis sealei* Jenkins, new species. Type.

54. Myripristis multiradiatus Günther.

Five examples obtained at Honolulu, 4.5 to 6 inches in length, and one is in Dr. Wood's collection, 3.75 inches in length. Dr. Günther has described this species from a single specimen, 6 inches in length, from Vavau. This seems to be the first record of the species since this description. It appears to be fairly abundant at Honolulu.

Myripristis multiradiatus Günther, Fische der Südsee, 1, 1873, 93, Vavau.

55. Myripristis sealei Jenkins, new species.

Head 3 in length; depth 2.5; eye 2.5 in head; snout 5; maxillary 1.8; interorbital 4; D. x-i, 15; A. iv, 13; P. i, 15; V. i, 7; scales 4-37-8; body elongate, deep, compressed, greatest depth at about tip of pectoral; upper and lower profiles about evenly convex; head compressed, rather elongate, a little longer than deep, its width 1.8 in its length; snout short, broad, blunt, convex, steep; upper profile of head nearly straight from above nostril to occiput; eye large, high, hardly impinging upon the upper profile of head, about equal to postocular part; mouth large, oblique, mandible slightly projecting, reaching below posterior rim of pupil; distal expanded extremity of maxillary 1.7 in eye; teeth sharp, minute, not enlarged on edges of jaws; teeth in jaws, on vomer and on palatines in bands; tongue elongate, rounded, free; suborbital rim narrow, finely serrate; lower posterior margin of maxillary smooth; lips rather thick, fleshy; nostrils close together, posterior very large, close to front rim of orbit; bones on head all finely serrate; opercle with well-developed spine; gill-opening large, filaments large; gillrakers long, fine, longest longer than longest gill-filaments; pseudobranchiae very large, free for distal half; dorsal spines slender, sharp, first 3.2, second 2, third 1.9, tenth 4.6, last 3.3 in head; anterior dorsal rays elongate, bluntly pointed, second ray 1.7, last 5.5; third anal spine large, 2.5 in head; soft anal similar to soft dorsal, second ray 1.7, last 6.4; caudal elongate, forked, the lobes pointed; pectoral small, pointed, 1.5; ventral 1.5, spine 2.25 in head; scales large, finely ctenoid; lateral line slightly convex, running down obliquely to base of caudal along upper side of caudal peduncle; four slender, sharp-pointed, graduated rays above and below.

Color in alcohol, pale brown or brownish-white, fins pale or whitish; no black or brown on edges of gill-opening or in axil of pectoral.

My collection contains 12 examples of this species, all obtained at Honolulu in 1889. They range in length from 2.2 to 5.25 inches. Type, No. 50708, U. S. N. M.

Named for Mr. Alvin Seale, curator of fishes in the Bernice Pauahi Bishop Museum at Honolulu.

56. Myripristis murdjan (Forskål).

Color in life (field No. 85), body and head pale red, first dorsal pink, outer margin orange, outer margins of second dorsal, caudal, ventral and anal fins white; immediately underneath the white margin of each fin is a bright red region, the rest of each of these fins a paler red; pectoral pale red. Seven specimens were taken, ranging from 4.25 to 6.2 inches in length; three are in Dr. Wood's collection, 5.30 to 6.25 inches in length, and one 4 inches long was taken by Jordan & Snyder. Large numbers are brought to the market.

Sciana murdjan Forskål, Descript. Animal., 48, 1775, Djidda, Red Sea.
Myripristis murdjan, Ruppell, Fische Roth. Meer, 86, pl. 23, fig. 2, 1828; Günther, Fische der Südsee, 92, plates LXI and LXII, 1873 (Hawaiian Islands), Steindachner, Denks. Ak. Wiss. Wien. LXX, 492, 1900 (Honolulu); Fowler, Proc. Ac. Nat. Sci. Phila. 1900, 501 (Hawaiian Islands).

57. Flammeo sammara (Forskål).

Four specimens taken, 7.75 to 8.75 inches in length, and one 8.5 inches long is in Dr. Wood's collection. This is the first record of the species from the Hawaiian Islands. It appears to be fairly common at Honolulu.

Sciana sammara Schneider, Syst. Ichthy., 89, 1801, Red Sea.
Holocentrum sammara, Günther, Fische der Südsee, 100, 1875 (Society and Paumotu Islands).

58. Holocentrus diadema Lacépède.

Color in life (field No. 114), body bright red, with about 9 white longitudinal stripes, somewhat diverging from the head and converging toward the tail; iris red; top of head and snout red; cheek white with a red stripe from pupil to preopercular spine; first dorsal a dark scarlet with a longitudinal white stripe broken at the sixth and seventh spines; outer tips of the first to eighth spines transparent; second dorsal red; caudal red; anal red, but membrane between longest spine and first soft ray bright scarlet; pectoral pale red; ventral with a narrow white line next and parallel to the spine; parallel to this a wider bright scarlet line, the remainder of the fin colorless.

I obtained 11 specimens of this brilliant fish, ranging from 4 to 6.5 inches in length. It is quite abundant about the reef.

Holocentrum diadema Lacépède, Hist. Nat. Poiss., IV, 335, 372, 374, pl. 32, fig. 3, 1802; Günther, Fische der Südsee, 97, 1875 (Samoa, Tahiti, Tonga,and Hawaiian Islands); Steindachner, Denks. Ak. Wiss. Wien. LXX, 1900, 492 (Honolulu, Laysan).

59. Holocentrus microstomus Günther.

Two specimens, respectively 6 and 6.2 inches in length, of this species are in Dr. Wood's collection from Honolulu.

Holocentrum microstoma, Günther, Cat., 1, 34, 1859, Amboyna; Günther, Fische der Südsee, 1, 98, pl. 64, fig. B, 1875 (Amboyna, Samoa, Tonga, Society, Kingsmill, Hervey, Paumotu, and Hawaiian Islands).
Holocentrus microstoma, Seale, Occasional Papers, Bishop Museum, 1, No. 3, 70, 1901 (Guam).

60. Holocentrus leo Cuvier & Valenciennes.

One specimen of this species was taken by Jordan & Snyder. While widely distributed throughout Polynesia, this seems to be its first record from the Hawaiian Islands.

Holocentrum leo, Cuvier & Valenciennes, Hist. Nat. Poiss., III, 204, 1829, Society Islands, Waigiou.
Holocentrum spiniferum, Günther, Fische der Südsee, I, 94, 1874.
Holocentrus spinifer, Fowler, Proc. Ac. Nat. Sci. Phila. 1899, 483 (Thornton Island).

61. Holocentrus erythræus Günther.

One specimen of this species, 8.5 inches in length, is in Dr. Wood's collection, and one, 13.5 inches long, was taken by Jordan & Snyder at Honolulu.

Holocentrum erythræum Günther, Cat., I, 32, Sea of San Christoval, 1859; Günther, Fische der Südsee, 99, pl. 63, fig. B, 1875 (Kingsmill, Society, Paumotu, and Hawaiian Islands).

62. Holocentrus diploxiphus Günther.

Fifteen specimens, from 4 to 5.5 inches long, were obtained; three are in Dr. Wood's collection, each 5 inches in length; one was obtained by the *Albatross*, 5.2 inches, in 1896, and three by Jordan & Snyder, 4.5 to 5 inches in length. It is brought to the market in large numbers.

Holocentrum diploxiphus Günther, Proc. Zool. Soc. Lond. 1871, 660, pl. 60, 2 figs., Samoa: Fowler, Proc. Ac. Nat. Sci. Phila. 1900, 501 (Hawaiian Islands; coll. J. K. Townsend).

Family XXIV. SCOMBRIDÆ.

63. Auxis thazard (Lacépède).

This important food-fish is caught by the hook and is common in the Honolulu market. It has not before been reported from the Hawaiian Islands.

Scomber thazard Lacépède, Hist. Nat. Poiss., III, 9, 1802, between 6° and 7° S. Lat., coast of Guinea.
Auxis thazard Jordan & Evermann, Fishes North and Mid. Amer., I, 867, fig. 365, pl. CXXXIII.

64. Gymnosarda pelamis (Linnæus).

This fish is abundant at Honolulu, where large numbers are brought to the market. One specimen, 17 inches in length, taken by me, has 6 narrow dark bands along the side, instead of 4, as given in descriptions. This seems to be its first record from the Hawaiian Islands.

Scomber pelamis Linnæus, Syst. Nat., ed. X, 297, 1758, open sea, locality unknown.
Thynnus pelamys, Günther, Cat., II, 364, 1860.
Gymnosarda pelamis, Jordan & Evermann, Fishes North and Mid. Amer., I, 868, 1896.

65. Gymnosarda alletterata (Rafinesque).

Common at Honolulu, although this is its first record from the Hawaiian Islands. One example (field No. 389), 15 inches in length, was taken. *G. alletterata* and the two preceding species are caught by the hook, by means of a short pole and line.

Scomber alletteratus Rafinesque, Caratteri, 46, 1810, Palermo.
Thynnus thunnina Günther, Cat., II, 364, 1860.
Gymnosarda alletterata, Jordan & Evermann, Fishes North and Mid. Amer., I, 869, 1896.

66. Acanthocybium solandri (Cuvier & Valenciennes). "*Ono.*"

I saw in the market one large example, 48 inches long exclusive of caudal. D. XXVII-13+9; A. 12—8; body covered with small scales; origin of anal just under soft dorsal; gape extending to middle of eye; pectoral 8 in body; strong keel on tail. This is the first record of this species for Honolulu.

Cybium solandri Cuvier & Valenciennes, Hist. Nat. Poiss., VIII, 192, 1831, open sea, locality unknown.

Family XXV. CARANGIDÆ.

67. Scomberoides tala (Cuvier & Valenciennes). "*Hai.*"

Color in life, upper part of body light gray, with silvery reflections, lower part silvery white; a row of about 6 very indistinct spots about as large as pupil above lateral line; a dusky blotch on first five rays of dorsal; tips of ventrals milky white; lower margin of caudal white. I obtained five specimens, 8 to 10 inches in length.   Two examples, 7.5 and 8 inches long, are in Dr. Wood's collection, and one, 6.5 inches long, was collected by the *Albatross* in 1896.   This species differs from *S. sancti-petri*, as described by Cuvier & Valenciennes, in having a simple bend in the lateral line and in having a longer pectoral.

Chorinemus tala Cuvier & Valenciennes, Hist. Nat. Poiss., VIII, 377, 1831, Malabar.
Chorinemus mondetta Cuvier & Valenciennes, Hist. Nat. Poiss., VIII, 382, Red Sea.
Chorinemus mondetta, Steindachner, Denks. Ak. Wiss. Wien, LXX, 1900, 495 (Honolulu).

68. Scomberoides sancti-petri (Cuvier & Valenciennes).

One specimen, 7.5 inches long, obtained by Jordan & Snyder in 1900.   I did not see it in 1889.

Chorinemus sancti-petri Cuvier & Valenciennes, Hist. Nat. Poiss., VIII, 379, 1831, Malabar; Streets, Bull. U. S. Nat. Mus., No. 7, 70, 1877 (Hawaiian Islands); Günther, Fische der Südsee, V, 138, 1876; Steindachner, Denks. Ak. Wiss. Wien, LXX, 1900, 495 (Honolulu).

69. Seriola sparna Jenkins, new species.

Head 3.6 in length to base of caudal; depth equal to head; eye 1.3 in snout; D. VI, 32; A. II, 20; scales 220; head conical; body fusiform; mouth somewhat below axis of body; least depth of caudal peduncle but little greater than its width at same position; eye with adipose eyelid before and behind; interorbital strongly convex, about equal to snout and slightly less than 3 in head; premaxillary

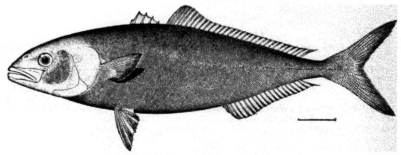

Fig. 14.—*Seriola sparna* Jenkins, new species. Type.

protractile; maxillary with supplemental bone; maxillary 2.5 in head, reaching to anterior margin of pupil, its posterior margin oblique, nearly equaling vertical diameter of eye; cheek and part of opercle scaled, remainder of head naked; teeth in villiform bands on jaws, palatines and tongue; preopercle entire; gillrakers longer than diameter of pupil, 22 on lower arm of first arch; first dorsal low, second spine longest, 6.3 in head, first less than half of second, third to fifth regularly decreasing in length, fifth shorter than first, second and third soft rays longest, equal to snout, those back of fifteenth about 2.5 in second, except last which is 1.5 in second; anal fin similar in shape to soft dorsal but shorter; first rays 1.25 in snout; 2 free anal spines; caudal deeply forked, the lobes slender, about equal, 1.3 in head (measured from last vertebra); pectoral about 2.5 in head reaching less than half way to anus; distance of origin of pectoral from origin of pectoral to snout; body entirely scaled, scales very small, about 220 in longitudinal series; lateral line slightly arched over pectoral.

Color in alcohol, upper parts pale brownish-purple with silvery reflections, lower parts silvery.

This description is based on a single specimen, type, No. 50845, U. S. N. M. (field No. 742), 9 inches in length, collected by me at Honolulu in 1889.   (σπαρνός, rare.)

70. Decapterus canonoides Jenkins, new species.

Head 4 in length; depth 5.5; eye 4 in head; snout 3; interorbital 4; D. VIII-I, 33-I; A. II-I, 28-I; scales 116, scutes 27; breadth of body three-fourths its depth; body fusiform; head conical; maxillary scarcely reaching anterior border of eye; mouth slightly oblique; center of eye slightly above axis of body; adipose eyelid well developed; teeth not evident on jaws, vomer, or tongue; small teeth on

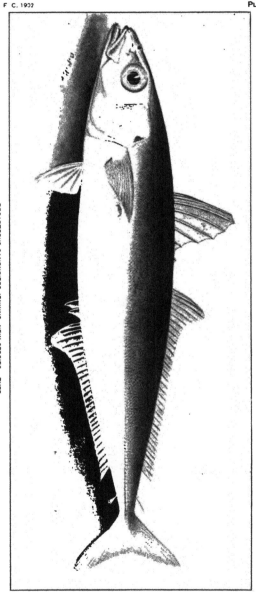

palatines; body completely scaled; cheeks, opercle and top of head to middle of interorbital space scaled, remainder of head naked; anterior portion of lateral line to twelfth soft dorsal only slightly curved, remainder straight, with about 27 armed scutes; width of one of largest scutes equaling half eye; origin of dorsal spines only slightly in front of tips of pectoral and ventral fins; spines slender; third spine the longest, 2 in head, its tip nearly reaching last spine, the last two spines very short, sinking into a groove; space between last dorsal spine and origin of soft dorsal equaling eye; origin of soft dorsal slightly in advance of that of soft anal; longest soft dorsal ray 3.3 in head; soft anal similar in form to soft dorsal, its base 1.3 the base of soft dorsal; space from each fin to detached finlet 1.7 in eye; caudal broadly forked, lobes about equal; pectoral 1.6 in head, its origin slightly in advance of origin of ventrals; ventrals 2.5 in head; vent half way from tip of snout to angle of fork of caudal.

Color in alcohol, dark bluish above, with silvery reflections, silvery below; black spot on opercle near its upper angle; dorsal fin somewhat dusky, with small black punctulations; soft anal lighter, the other fins pale.

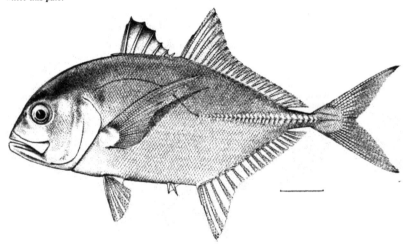

Fig. 15.—*Carangus hippoides* Jenkins, new species. Type.

This description is based on a specimen, the type, No. 50846, U. S. N. M. (field No. 2737), 9 inches in length to tip of caudal, collected by Dr. Wood at Honolulu. Two smaller examples were collected by Mr. McGregor at Kihi, Maui, each 7 inches in length. This fish is quite common at times at Honolulu and is highly prized by the natives as food.

71. Trachurops crumenophthalmus (Bloch). "*Akuli.*"

Very abundant at Honolulu, where it is highly prized as food. I obtained 7 specimens, 7.5 to 12 inches in length. Color in life, steel-blue above and silvery below.

Scomber crumenophthalmus, Bloch, Ichthy., x, 65, pl. 343, 1797, Guinea.
Caranx crumenophthalmus, Günther, Cat., II, 429, 1860, ibid., Fische der Südsee, v, 131, 1876, ibid., Rep. Challenger, Zool., I, 59, 1880 (Honolulu), Steindachner, Denks. Ak. Wiss. Wien, LXX, 1900, 495 (Honolulu).
Trachurops mauritianus, Streets, Bull. U. S. Nat. Mus., No. 7, 68, 1877 (Hawaiian Islands).
Trachurops crumenophthalmus, Jordan & Evermann, Fishes North and Mid. Amer., I, 911, 1896; Fowler, Proc. Ac. Nat. Sci. Phila. 1900, 501 (Hawaiian Islands).

72. Carangus hippoides Jenkins, new species.

Head 3.6 in length; depth 2.4; eye 4 in head; D. VIII-I-20; A. II-I, 17; armed scutes 32; profile of face rather steep, the depth of head at nape being a little greater than length of head; outline from snout to origin of second dorsal a curve somewhat more convex over the head; along the base of the soft dorsal the outline is only slightly convex; ventral outline from tip of lower jaw to origin of soft anal nearly straight, obliquely descending; base of anal similar to soft dorsal; depth of caudal peduncle less than its width; body compressed, width about 2.5 in head; interorbital equaling eye; eye mostly above axis of body, its posterior border halfway from snout to posterior border of opercle; jaws subequal, maxillary reaching to vertical through center of pupil; teeth on vomer, palatines and tongue villiform, those on jaws in a single series, conical, short, and strong; spinous dorsal with 1 procumbent spine and 8 joined spines, the third longest, somewhat more than half of head, its tip reaching tip of seventh spine, fourth spine nearly equaling third; soft dorsal and anal falcate, similar in form, the lobe about three-fourths of head, base of soft dorsal the longer, 2.5 in body to base of caudal; pectoral slender, strongly falcate, length about equaling that of soft dorsal, 1.3 in body; tip of ventrals reaching just past vent and one-half distance from origin to origin of soft dorsal; caudal deeply forked, lobes equal; anterior portion of lateral line well arched, arched portion of 55 scales reaching about to vertical from seventh anal, armed portion straight; armed scutes 32; breast naked except a very small patch of small scales in its center; cheek, postocular and upper part of opercle scaled, rest of head naked; dorsal and anal scarcely sheathed; gillrakers strong, equaling three-fourths diameter of eye, 13 developed on lower arm of first arch.

Color in alcohol, head, body and fins pale, head and body silvery; upper part of caudal peduncle dusky; no spot on opercle or on pectoral. Similar to *C. hippos*, but differing in lacking the opercular spot and the spot on pectoral, and in having a larger snout and deeper head.

This description is based upon a specimen, the type, No. 50710, U. S. N. M. (field No. 749), 9.25 inches long, taken by me at Honolulu in 1889. Another (field No. 751), 5.5 inches long, is in the collection of Dr. Wood, and another (field No. 750), 7.25 inches long, was obtained by Jordan & Snyder in 1900 at Honolulu.

73. Carangus marginatus (Gill). *"Ulua."*

D. VIII-I-20; A. II-I-17; scutes about 29.

Color in life (field No. 180), nearly white, with silvery and golden reflections; iris red; a small black spot at upper angle of opercular opening; golden areas on preopercle and opercle behind the eye; dorsal slightly dusky, lobe of soft dorsal with dusky blotch, remainder of fin yellowish; anal yellow; caudal yellow with posterior border dusky; pectoral transparent; ventral fins white. I have compared my single specimen with a specimen of *C. marginatus* from Mazatlan and they seem to be the same.

Caranx marginatus Gill, Proc. Ac. Nat. Sci. Phila. 1863 (1864), 166, Panama; Jordan & Evermann, Fishes North and Mid. Amer., I, 922, 1896.

74. Carangus latus (Agassiz). *"Ulua."*

Head 3.3 in length; depth 2.5; D. VIII-23; A. II-I-19; scutes about 38. Color in life, white, upper parts with steel-blue reflections, yellowish along the region of the scutes; lower parts silvery; no black on or behind opercle; iris yellow; first dorsal yellowish; second dorsal, lobe slightly dusky with bluish tinge; caudal slightly dusky with bluish tinge; membranes of anal spines milky white; lobe of anal slightly dusky; ventrals white.

I obtained ten specimens of this fish, ranging from 4 to 9 inches in length; and three, from 3 to 4.75 inches in length, were taken by the *Albatross* in 1896. These compared with specimens of *C. latus* from the west coast of Mexico and from Clarion Island show no differences. The native fishermen do not distinguish this species as different from *C. marginatus*, which, with it, is highly prized as a food-fish. Both are abundant.

Caranx latus Agassiz, Pisc. Bras., 105, pl. LVI-b, 1, 1829, Brazil; Jordan & Evermann, Fishes North and Mid. Amer., I, 923, 1896; Fowler, Proc. Ac. Nat. Sci. Phila. 1900, 501 (Hawaiian Islands).
Caranx hippos, Günther, Fische der Südsee, v, 131, pl. 84, 1876; Ibid., Rept. Challenger, Zool., XI, 59, 1880 (Hawaiian Islands).

75. Carangus rhabdotus Jenkins, new species.

Head 3.5 in length; depth 2.6; eye 3.75 in head; D. I-VIII-I-20; A. II-I-16; armed scutes 32. Form of body elliptical, the dorsal outline an even curve somewhat more convex than ventral outline; greatest depth of head equaling its length; center of eye slightly above axis of body; interorbital slightly greater than eye; snout somewhat shorter than eye ; maxillary with supplementary bone reaching to vertical through center of pupil, its posterior border broad, three-fourths of eye; mouth oblique; a triangular patch of strong granular teeth on vomer, a single series on palatines, small granular teeth on tongue, an outer series of enlarged teeth on upper jaw, a single series of pointed teeth on lower jaw; body completely scaled, bases of soft dorsal and anal both sheathed; cheeks and upper part of opercle scaled, remainder of head naked; lateral line strongly arched, the arched portion ending about under fifth soft dorsal ray; armed portion of lateral line straight with 32 armed scutes, the majority of which are large, the largest being 0.65 diameter of eye; dorsal with one procumbent spine; third dorsal spine longest (broken in type), 2.2 in head; soft dorsal elevated, longest rays 1.8 in head; base of soft dorsal 2.5 in body to base of caudal; soft anal of similar form, but with shorter base and with shorter elevated rays; caudal forked, lobes about equal; pectoral falcate, length about equaling head, tip of ventral reaching just past vent and halfway to origin of soft dorsal.

Color in alcohol, bright silvery, darker above, with 5 indistinct vertical dark bands nearly as wide as eye; on upper three-fourths of body, a sixth, less distinct showing on caudle peduncle; no opercular spot; fins plain except elevated portions of soft dorsal and anal, which are tipped with black.

This description is based upon the type, No. 50711, U. S. N. M., a specimen 5.5 inches in length, collected by the *Albatross* at Honolulu in 1896.

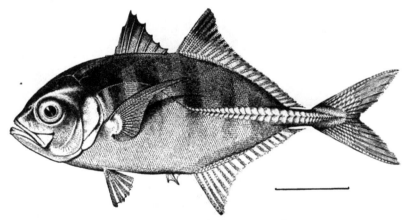

Fig. 16—*Carangus doliolus* Jenkins, new species. Type.

76. Carangus politus Jenkins, new species.   "*Maka.*"

Head 4 in length; depth 2.9; eye 4 in head; interorbital 3; maxillary 2.8; D. viii-i-24; A. ii-i, 19; armed scutes 40. Breadth of body about one-half length of head, moderately compressed; body elliptical, very symmetrically formed; dorsal and ventral profiles similar; depth of head less than length; maxillary reaching just beyond anterior border of eye; mouth oblique, lower jaw projecting; center of eye slightly above axis of body; eye with well-developed adipose eyelid; bands of rather strong teeth on vomer and palatines, fine granular ones on tongue; minute teeth in upper jaw, with a few larger ones on anterior part; teeth on lower jaw almost obsolete; body completely scaled; a well-developed scaly sheath at base of soft dorsal and anal; rows of small scales on the interradial spaces of caudal fin; scales on cheeks, upper parts of opercle and preopercle, and a patch on top of head above and behind eye, remainder of head naked; lateral line well curved anteriorly, the curved part ending under eighth dorsal ray, the armed portion straight, with 40 armed scutes, which are strongly developed on caudal peduncle, width of largest three-fourths diameter of eye; origin of spinous dorsal about on a vertical through middle of ventrals; third and fourth spines about equal, 1.9 in head, fourth reaching origin of soft dorsal; anterior soft dorsal elevated, longest rays 1.9 in head; soft anal similar in form, shorter and less high; pectoral slender, falcate, about equaling depth; origin close to margin of opercle; caudal obtusely forked, lobes equal; ventral 2.1 in head, its tip reaching posterior margin of vent and slightly past midway point from its origin to first anal spine; distance from snout to anus 2.4 in body to base of caudal.

Color in life, back light-greenish with 9 very indistinct bands; snout and lips dusky; margin of first dorsal yellow, second dorsal and caudal yellow; lower part of body silvery white; black spot on margin of opercle near upper angle, the black extending onto body.

This description is based on a specimen 8 inches in length (field No. 100) taken by me at Honolulu in 1889. Only one other specimen was obtained. It is 5 inches in length and does not differ from the type except in size. Type, No. 50709, U. S. N. M.

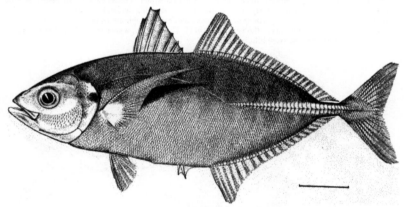

Fig. 17.—*Carangus politus* Jenkins, new species. Type.

77. Carangus affinis (Rüppell). *"Amuka."*

D. VIII-I, 23; A. II-I-20; depth 3.5 in length; head slightly shorter than depth; anterior soft dorsal and anal rays not much elevated, 2.25 in head; breast scaled; 37 scutes; lateral line moderately arched; the curved part equaling the straight part to front of caudal peduncle; teeth very small, slender, in a single series in each jaw; villiform teeth on vomer, palatines and tongue; pectoral long, somewhat curved, a little less than 3 in length; its tip reaching to front of second anal fin; caudal peduncle almost twice as wide as deep.

Color in life, upper part of body green, with bluish reflections, lower silvery white; a black spot on opercle extending on shoulder, axil of pectoral black.

This important food-fish is very abundant at Honolulu. Eight examples were obtained.

Caranx affinis Rüppell, Neue Wirbelthiere, Fische, 49, pl. XIV, fig. 1, 1835 (March, 1830), Red Sea.
Caranx (Selar) affinis, Klunzinger, Fische des Rothen Meeres, II, 19, 1879 (Red Sea); Steindachner, Denks. Ak. Wiss. Wien, LXX, 1900, 195 (Honolulu).

78. Carangoides ferdau (Forskål).

Two specimens of this fish, 9.75 and 10 inches in length, are in Dr. Wood's collection. The group of spots on the sides characteristic of this species has almost entirely disappeared in these examples.

Scomber ferdau Forskål, Descr. Anim., 55, 1775, Red Sea.
Caranx ferdau, Gunther, Fische der Sudsee, I, 131, 1876 (Red Sea, Tahiti, Seychelles, Bonin Islands); Steindachner, Denks. Ak. Wiss. Wien, LXX, 1900, 195 (Honolulu).

79. Caranx speciosus (Forskål). *"Ulua Pauu."*

Color in life (field Nos. 106 and 296, 4.5 and 5.25 inches in length), head, body and fins a bright golden yellow, the body with 11 black crossbands, narrow ones alternating with broader ones, the bands not extending on the fins. A large specimen about 20 inches in length, even when fresh, showed no trace of the bands and lacked the bright yellow color on the body. I obtained 3 specimens, and 2 are in Dr. Wood's collection, 7.75 and 8.5 inches in length, each of which shows the crossbands distinctly. The fish does not seem to be very common at Honolulu.

Scomber speciosus Forskål, Desc. Anim., p. XII. 1775, Red Sea.
Caranx speciosus, Günther, Cat., II. 444, 1860; Steindachner, Denks. Ak. Wiss. Wien, LXX. 495, 1900 (Hawaiian Islands).
Gnathanodon speciosus, Jordan & Evermann, Fishes North and Mid. Amer., I. 928, 1896.

80. Alectis ciliaris (Bloch). *"Ulua Kihikihi."*

Color in life, lead colored above, bright silvery below; first dorsal filament white, tips of the other filaments white and black; a dusky blotch on the dorsal fin. One specimen was secured, 5.5 inches in length. Two specimens in Dr. Wood's collection, each 5.5 inches in length, have the same coloration, and in addition show about 5 indistinct crossbands on the upper part of body. One specimen (field No. 1337) 2.5 inches long, was obtained by the *Albatross* in 1896. This fish seems to be rare at Honolulu, as only occasional examples are taken.

Zeus ciliaris Bloch, Ichthyolgia, VI. 27, pl. 191, 1788, East Indies.
Caranx ciliaris, Günther, Cat., II. 454, 1860; Ibid., Fische der Südsee, V. 135, pl. 89, 1876 (Hawaiian Islands).
Alectis ciliaris, Jordan & Evermann, Fishes North and Mid. Amer., I. 931, 1896.

Family XXVI. KUHLIDÆ.

81. Kuhlia malo (Cuvier & Valenciennes).

Color in life, upper parts light gray, lower part silvery. It is a good food-fish. Numerous specimens are in my collection, the largest being 8.7 inches in length. It is very common about the islands, ascending the fresh-water streams.

Dules malo Cuvier & Valenciennes, Hist. Nat. Poiss., VII. 479, 1831, Tahiti.

Family XXVII. CORYPHÆNIDÆ.

82. Coryphæna hippurus Linnæus. *"Mahihi."*

Head 4.75 to base of caudal; depth through base of pectoral, nearly equaling head; D. 56; A. 26; maxillary reaching middle of eye; dorsal beginning over anterior portion of eye, its highest portion 1.3 in head; profile nearly vertical; ventral inserted slightly behind upper ray of pectoral, under thirty-seventh dorsal ray; pectoral 1.3 in head; ventral 1.2 in head. Color in life, body bluish-gray above, silvery on belly, with golden tinges, covered with numerous small blue spots; dorsal bright blue (in alcohol the blue becomes black). I have examined several specimens in the market, some 44 inches in length. In one the maxillary reached nearly to the vertical of posterior border of eye. The example described is 37 inches in total length, or 28.5 inches to base of caudal. This is the first record of this species in the eastern portion of Polynesia. This fish is used for food in Honolulu. One specimen was obtained and several larger ones were seen in the market from time to time.

Coryphæna hippurus Linnæus, Syst. Nat., ed. X. 261, 1758, in the open sea: Günther, Cat., II. 405; Günther, Fische der Südsee, 146, 1876.
Coryphæna japonica Schlegel, Fauna Japon., Poiss., 120, pl. 64, 1842.

Family XXVIII. APOGONIDÆ.

83. Fowleria brachygrammus Jenkins, new species.

Head 2.6 in length; depth 2.6; eye 3 in head; snout 1.5 in eye=interorbital; D. VII-I, 9; A. II, 7; C. 22; P. 11; V. I, 5; scales 1½-22-4; dorsal and ventral outlines symmetrical; front of mouth on axis of body; cleft of mouth oblique; suborbital very narrow, 2.5 in snout; teeth minute, villiform, on jaws, vomer, and palatines; preopercular margin entire, a small flat spine at angle of opercle; angle of preopercle rounded; gillrakers of moderate length, only 7 well developed on lower half of first arch, 3 anterior ones rudimentary; first dorsal spine very short, second a little longer than half of the third,

third longest, 1.6 in head; median soft rays longest, equal to third spine; first anal spine very short, second equal to second dorsal spine; first soft anal rays longest, equal to longest soft dorsal rays; caudal mutilated, outline and length indeterminate; least depth of caudal peduncle 2 in head; pectoral 1.75 in head, somewhat longer than distance from tip of snout to posterior rim of orbit; ventral not quite equaling pectoral; scales very large, loosely inserted, ctenoid; anterior portion of lateral line with about 10 tubes ending just below front of second dorsal fin; the posterior portion continued to base of caudal as very rudimentary pores on the row of scales two scales below.

General coloration in life plain; pectoral light rosy; dorsal, caudal, and anal yellow; ventral dusky.

One specimen, 1.5 inches in length, taken by me in coral rocks at Honolulu. This is probably the species recorded by Streets as *Apogon auritus*. Type, No. 50899, U. S. N. M., Honolulu.

? *Apogon auritus*, Streets, Bull U. S. Nat. Mus., No. 7, 72, 1877(Honolulu); not of Cuvier & Valenciennes.

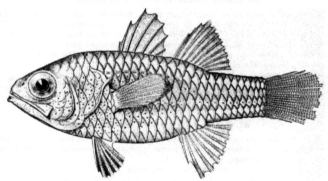

FIG. 20.—*Foa cia brachygrammus* Jenkins, new species. Type.

84. Apogon maculiferus Garrett.

Two specimens of this species, 3.75 and 4.8 inches in length, were taken by Jordan & Snyder in 1900. I did not see it in 1889. It is conspicuously marked with 6 or 7 longitudinal rows of dark dots on the sides of the body.

Apogon maculiferus Garrett, Proc. Cal. Ac. Sci., Series 1, III, 105, 1863, Hawaiian Islands; Gunther, Fische der Südsee, 1, 20, 1873 (Sandwich Islands).

85. Apogon snyderi Jordan & Evermann.

Color in life (field No. 1981), general color pale red, 2 longitudinal pearly lines on the body; iris yellow; first dorsal with dusky olivaceous anterior border; white lines along fourth, fifth, sixth, and seventh spines, olivaceous between; second dorsal with many white and some olivaceous spots; anal with a dusky line along base, red distally; caudal with base dusky, rest pale red; white spot on tip of ventral; pectoral pink.

I obtained six examples of this species ranging from 4.4 to 5 inches in length, and three 3.7 to 5.5 inches were taken by Jordan & Snyder in 1900. It is abundant at Honolulu.

Apogon snyderi Jordan & Evermann, Bull. U. S. Fish Comm. for 1902, (April 11, 1903), 180, Honolulu. Type, No. 50640, U. S. N. M., coll. Jordan & Evermann.

86. Apogon menesemus Jenkins, new species.

Head 2.75 in length; depth 2.8; eye 2.8 in head; snout 4; D. VII-I, 9; A. II, 8; C. 21; P. 13; V. I, 5; scales 3-27-6; maxillary not reaching quite to posterior margin of orbit; suborbital 3 in interorbital, which is about equal to snout; angle of preopercle rounded, somewhat produced backward, both margins serrated; teeth in jaws in villiform bands; teeth on vomer, none on palatines; gillrakers slender, 10 on lower arm of first arch; branchiostegals 6; first dorsal spine very short, third longest, 2 in head, second 2 in third, seventh 2 in second, eighth spine 2.75 in head, equal to fifth; first soft ray longest, 1.6 in head; last ray equal to eighth spine; length of caudal peduncle (from end of dorsal to base of caudal rays) 3.5, slightly tapering posteriorly; caudal notched; first anal spine very short, second 3 in head, anterior soft rays longest, 1.8 in head; lateral line complete, following curvature of back; scales finely ciliated: opercle and preopercle scaled; rest of head naked.

Color in life (field No. 303), general color pale red, finely punctate with black, brown, and deeper red: black longitudinal line from snout through eye just below edge of pupil: first dorsal with a black bar on anterior border, second dorsal and anal each with a black longitudinal band near the base, the band of the dorsal with a white band below it, that of the anal with a white band above and one below; basal part of caudal brown; a submarginal black band on dorsal and one on ventral border of caudal fin, these connected near base of fin by a transverse, crescent-shaped, black band; ventral with anterior margins black distally, olive basally; pectoral pale red. In alcohol there appears to be a narrow pale band along the lateral line.

This description is based on the type, No. 50700, U. S. N. M. (field No. 675), 5 inches in length, and eleven cotypes ranging from 5 to 6 inches in length collected by me in 1889, and five from 5 to 6 inches collected by Jordan & Snyder in 1900, all at Honolulu.

This species is quite abundant at Honolulu, where it is sold in the market.

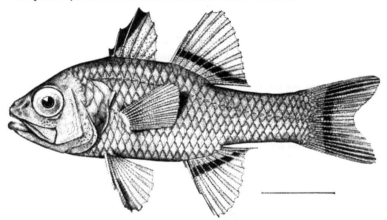

Fig. 19 —*Apogon menesemus* Jenkins, new species. Type.

Family XXIX. SERRANIDÆ.

87. Pikea aurora Jordan & Evermann.

Four specimens in Dr. Wood's collection (field Nos. 403, 682, 687, and 607), ranging from 4.5 to 7 inches in length, agree closely with Steindachner's excellent figure given in places cited below.

Pikea lunulata Steindachner, Sitzb. Ak. Wien. LXX, 1. 375, 1874, and Denks. Ak. Wiss. Wien. XLVII, 1883, pl. VI, fig. 2; not *Gryetes lunulatus* Guichenot, in Maillard, App. C. 4, Reunion, 1862, Indian Ocean.

88. Epinephelus quernus Seale.

Head 2.6 in length; depth 2.5; D. XI, 15; A. III, 9; scales 120; eye 4 in head; snout 3.25; lower jaw not projecting, snout acute; profile of head rising at angle of 45° from tip of snout to front of dorsal fin; upper limb of preopercle nearly vertical, inclined a little forward, finely serrated, the teeth enlarged at the angle; lower limb of preopercle entire; opercle with 3 small spines, the middle one nearer the lower, the latter a little farther back than the upper; suborbital narrow, 3 in snout;

interorbital 2 in snout; eye elliptical, longer diameter horizontal; posterior nostril oval; maxillary naked; mouth only a little oblique; maxillary with a narrow elongate supplemental bone; small canines in front of upper jaw; smaller depressible teeth in a band of several rows in upper jaw, widest in front, some of the anterior ones enlarged; an outer row of larger fixed teeth; teeth on vomer and palatines; dorsal fin continuous; third and fourth dorsal spines longest, 2.2 in head; first spine short, 2.5 in the third; soft dorsal higher than spinous dorsal, longest rays 2 in head; soft anal similar to soft dorsal; second and third anal spines of equal length, the second thickest; lateral line continuous.

Color in alcohol, head and body light brown, clouded with blackish-brown in irregular pattern; lips, gill-membranes, and fins black.

Two specimens of this species, 3.8 and 12 inches in length respectively, are in Dr. Wood's collection. It was not seen by me.

Epinephelus fuscoguttatus, Fowler, Proc. Ac. Nat. Sci. Phila. 1900, 522 (Honolulu); not of Forskal.
Epinephelus quernus, Seale, Occasional Papers, Bishop Museum, I. No. 4, 3, fig. 1, 1901, Honolulu. (Type No. 481, B. P. B. M.)

89. Anthias fuscipinnis Jenkins.

Three specimens, 7.5, 8, and 9.5 inches in length, respectively, are in Dr. Wood's collection.

Anthias fuscipinnis, Jenkins, Bull. U. S. Fish Comm. for 1899 (June 8, 1901), 389, fig. 3, Honolulu. (Type, No. 49695, U. S. N. M.; coll. O. P. Jenkins.

Family XXX. PRIACANTHIDÆ.

90. Priacanthus cruentatus (Lacépède). "*Aweoweo.*"

Color in life, that of head and body made up of mottlings of bright red and white; iris white, with bright-red blotches; dorsal mottled with red and white and covered on posterior portion with more or less distinct red; anal similar in color to dorsal; caudal red with rows of distinct darker red spots on membranes; pectoral pale red; ventral white with red mottlings; inside of mouth white with bright-red blotches.

Six specimens, ranging from 9.25 to 11 inches in length, were obtained; one is in Dr. Wood's collection and one in that made by Jordan & Snyder. These I have compared with examples of the West Indian species *P. cruentatus* and with specimens collected by Snodgrass & Heller at the Galapagos, and find they can not be distinguished by either color or structural differences. The young of this fish is known as the "red-fish." At various times it has occurred in immense numbers at Honolulu. It is an old belief of the natives that this phenomenon is a precursor of the death of some member of the royal family. This species is abundant at Honolulu and is an important food-fish.

Labrus cruentatus Lacépède, Hist. Nat. Poiss., III. 522, 1801, Martinique.
Priacanthus carolinus, Günther, Fische der Südsee, 17, pl. XVIII, 1873 (Otaheiti and Ralatea); Jordan & Evermann, Fishes North and Mid. Amer., III, Addenda, 2838, 1898 (Clarion Island); Jordan & McGregor, Report U. S. Fish Comm for 1898–1899, 278 (Socorro and Clarion Islands).
Priacanthus cruentatus, Jordan & Evermann, Fishes North and Mid. Amer., I, 1238, 1896.

91. Priacanthus meeki Jenkins, new species.

Head 3.4 in length; depth 2.6; eye 2.2 in head; snout 3.75; D. X, 14; A. III, 15; scales 120; body compressed, somewhat elliptical in outline; mouth very oblique, lower jaw projecting, ending in a hook; maxillary reaching well beyond anterior border of eye; interorbital somewhat more than half eye; both limbs of preopercle finely serrated, its angle terminating in a small free spine; margin of opercle entire, the flap with 2 keels ending at margin as short, blunt spines; anterior nostril small, with a raised margin; posterior nostril a long, narrow slit, one-third diameter of eye; teeth in bands on vomer, palatines, and jaws, somewhat strong, hooked; dorsal and anal fins high, longest soft dorsal ray 1.3 in head, longest soft anal somewhat shorter; caudal deeply lunate, upper lobe the longer; pectoral 1.6 in head; ventral nearly as long as head, its tip reaching slightly beyond origin of anal; head and body completely covered with small rough scales, the roughened portion of each scale forming a triangular or crescent-shaped patch on posterior portion of the scale; lateral line ascending abruptly from gill-opening, then curving gently to caudal peduncle, upon which it is straight; gillrakers 23 on lower arm of first arch, strong, longest one 3 in eye.

Color in life, uniformly red; iris bright red; inside of mouth and gillrakers bright red; tips of ventrals, soft dorsal and anal and posterior margin of caudal dusky; no spots on the fins.

Priacanthus meeki resembles *P. hamrur* somewhat, but differs from it in the much shorter head, deeper body, larger eye, higher soft dorsal and anal, and in the coloration of the dorsal and anal fins. The description is based on the type, No. 50847, U. S. N. M., 12 inches in length, and two cotypes, 4 and 4.5 inches long, in Dr. Wood's collection, all from Honolulu. The smaller examples agree with the description of the larger one, except that the spine at the angle of preopercle is more distinct.

This species is named for Dr. Seth Eugene Meek, assistant curator of zoology, Field Columbian Museum.

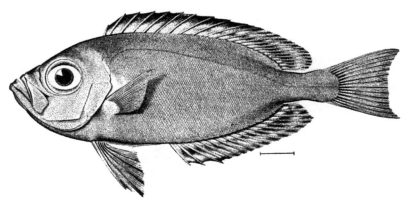

FIG. 20.—*Priacanthus meeki* Jenkins, new species. Type.

Family XXXI. LUTIANIDÆ.

92. Aphareus flavivultus Jenkins.

Color notes of the type taken when fresh (1889), overlooked when the species was originally described, are as follows: General color light-bluish; end of lower jaw, snout, face, and top of head covered by a broad band of bright yellow, the yellow band extending less broad and less distinct to origin of dorsal; dorsal fin with lower portion rosy, outer yellowish; anal yellow; caudal yellow with rosy posterior margin; ventrals yellowish with white anterior margin; pectoral rosy.

One specimen, 12.5 inches long, taken off the coast of Kona, Hawaii. The two young (Nos. 1 and 2) in the table given with the description, identified with this species, are evidently of another species, *A. furcatus* (Lacépède).

Aphareus flavivultus Jenkins, Bull. U. S. Fish Comm. 1899 (June 8, 1901), 390, fig. 4, Honolulu. (Type, No. 49691, U. S. N. M., coll. O. P. Jenkins.)

93. Aphareus furcatus (Lacépède).

The two small examples recorded by me in 1901 as *A. flavivultus* are apparently referable to this species. One large specimen was obtained at Honolulu by Dr. Wood.

Labrus furcatus Lacépède, Hist. Nat. Poiss., III, pp. 429 and 477, pl. 21, fig. 1, 1801.
Aphareus rutilans Cuvier & Valenciennes, Hist. Nat. Poiss., VI, 490, 1830, Isle de France; Gunther, Fische der Südsee, 17, 1873 (Society Islands).
Aphareus furcatus, Günther, Cat., I, 386, 1859; Jordan & Starks, Proc. U. S. N. M., XXIII, 1901, 719, pls XXVIII and XXIX (Odawara, Japan).

94. Apsilus microdon (Steindachner).

Two specimens of this species, 5.25 and 10 inches in length, are in Dr. Wood's collection. It is a common food-fish in the market in Honolulu.

Aprion microdon Steindachner, Sitz.-Ber. Ak. Wiss. Wien, LXXIV, Abt. 1, 1876, 158 (Sandwich Islands).

95. Apsilus brighami (Seale).

Head 3 in length; depth 2.8; eye 3.6 in head; snout 2.6; suborbital 2.3; interorbital 4; D. x, 11; A. III, 8; scales 7-66-15; snout wide; lips thick; lower jaw not projecting; profile rising from tip of snout to occiput at angle of about 45°, angulated at occiput, profile rising from here at a gentler slope to front of first dorsal; length of caudal peduncle 2 in head, depth of peduncle 3; interorbital flat, occipital ridges prominent upon it; 6 rows of scales on cheek; 8 rows on opercle; preopercle naked, its posterior limb almost vertical, scarcely notched below, both limbs finely serrated; pectoral a little shorter than head; longest rays 1.2 in head; dorsal fin continuous, fourth spine longest, 3.2 in head; last dorsal and anal ray elongated, equal to length of caudal peduncle, 3 in head; third anal spine longest, equal to longest dorsal spine; fine teeth in bands in each jaw, bands widest in front, lacking posteriorly on sides of lower jaw; an outer series of enlarged teeth along sides and front of each jaw; in front of each jaw a series of less enlarged innermost teeth; teeth on vomer and palatines.

Color in alcohol, pale silvery yellowish; a faint indication of 4 wide oblique crossbands on back and upper part of sides—the first on the occiput, the second through front of dorsal fin, third through middle of spinous dorsal, fourth through soft dorsal, a fifth on end of caudal peduncle.

One example of this species is in Dr. Wood's collection.

Serranus brighami Seale, Occasional Papers, Bishop Museum, vol. 1, No. 4, 7, 1901, Honolulu (type, No. 625 B. P. B. M.).

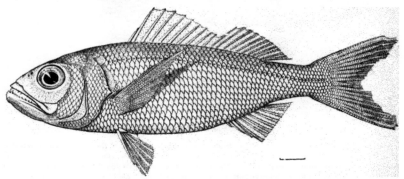

FIG. 21. —*Eteliscus marshi* Jenkins, new species. Type.

96. Aprion virescens Cuvier & Valenciennes.

Color in life, whole body and head pale blue, lighter on belly; dusky blotches between the spines in middle part of dorsal; rays of ventrals white; membranes dusky.

Two specimens of this species, 8.5 and 9.8 inches in length, were obtained by me. It is a common and valued food-fish.

Aprion virescens Cuvier & Valenciennes, Hist. Nat. Poiss., VI, 544, pl. 168, 1830, Seychelles; Günther, Fische der Südsee, 1, 16, 1873; Society and Hawaiian islands; Steindachner, Denks. Ak. Wiss. Wien, LXX, 1900, 484 (Honolulu).
Aprion (*Aprion*) *virescens*, Bleeker, Atlas, VIII, 77, pl. 293, fig. 3, 1876.

97. Etelisous marshi Jenkins, new species.

Head 3.5 in length; depth 4; eye 3.5 in head; snout 3.5; suborbital 2 in snout; interorbital equaling eye; D. x, 11; A. III, 8; scales 5-51-11; 4 canines in the front of each jaw, the lateral ones largest; an outer row of enlarged teeth in sides of jaw, within these a band of small, fixed teeth, the band widest in front of jaws; teeth on vomer and palatines; lower jaw slightly projecting; eye large, elliptical; interorbital flat; 7 vertical rows of scales on preopercle and same number on opercle; 4 oblique rows on side of occiput; rest of head naked; margin of upper limb of preopercle inclined a little forward, very slightly denticulated, lower limb entire, the angle with longer, slender points; second and third dorsal spines longest, 2.6 in head; first short, 2.5 in second, equal to last; first soft dorsal rays abruptly longer than the last spines, 1.3 in second spine; anal spines slender, third longest, equaling eighth spine; last soft ray each of dorsal and anal elongated, equaling first soft ray; caudal deeply forked; lateral line continuous.

Color in alcohol, plain pale yellowish, a little darker above (probably rosy or red in life); very faint indications of light longitudinal bands formed by a light spot on the base of each scale.

This description is based on a specimen 15 inches long, in Dr. Wood's collection from Honolulu. Type, No. 50714, U. S. N. M.

This species is named for Mr. Millard C. Marsh, of the U. S. Fish Commission.

Family XXXII. SPARIDÆ.

98. Monotaxis grandoculis Forskål. "*Mu.*"

The color markings vary in distinctness and character with the age of the fish. Color in life of No. 260, 6.75 inches in length, body light, almost white, with three broad dusky bands across back to middle of body; width of first band from head to first dorsal spine; second band, from fourth dorsal spine to past ninth; third band from in front of eleventh spine to last ray; dusky areas on dorsal, a distinct black area from last spine to fourth soft ray; posterior margin of caudal olivaceous; outer margin of anal olivaceous; anterior margin of pectoral yellow, rest of fin pale pinkish; ventrals white with rosy shadings; lips yellow; iris white, yellowish area about the eye. In another specimen, No. 323, 14 inches in length, outer margin of dorsal bright red; pectorals and ventrals red; upper lips yellow.

Four specimens, 6.75, 7.5, 9, and 14 inches in length, respectively, were obtained at Honolulu in 1889. This series shows well the variations in form and coloration remarked by Bleeker, which has led him to regard the species of authors as one. This species, while frequently seen, is not abundant at Honolulu. At the time of my visit it was not easy to obtain specimens, as from some superstition connected with it, it was in high estimation by the native fishermen, on which account when taken it was either retained to be eaten "with a friend," or an exorbitant price was asked for it.

Sciæna grandoculis Forskål, Desc. Anim., 53, 1775.
Monotaxis indica Bennett, Life of Raffles, Cat. Fish., Sumatra, 683, 1830.
Spherodon grandoculis Ruppell, Neue Wirbe. des Rothen Meeres, 113, tab. 28, fig. 2, 1835 (March, 1838). Günther, Cat., I, 465; ibid, Fische der Südsee, 67, 1873 Sandwich, Society, Friendly, Samoa, Kingsmill, Hervey, and Pelew islands).
Spherodon heterodon Günther, Cat., I, 465, Bleeker, Atlas, VIII, Taf. 299, 1876. Day, Fishes of India, I, 138, 1876 (Ceylon and Malay Arch.).
Monotaxis grandoculis, Bleeker, Atlas, VIII 105, pl. 299, fig. 1, 1876.

Family XXXIII. KYPHOSIDÆ.

99. Kyphosus elegans (Peters).

Color in life (field No. 161), a golden band on lower part of premaxillary and maxillary extending from angle of mouth horizontally back on preopercle to behind eye; a golden spot on nostril, and one behind eye; membranes of opercles golden; axil golden; longitudinal golden stripes on side between rows of scales. I have compared the 4 specimens taken at Honolulu with examples of *K. elegans* from Mazatlan and can detect no structural differences.

Pimelepterus elegans Peters, Berliner Monatsberichte K. Preuss., Ak. Wiss., 707, 1869, Mazatlan.
Kyphosus elegans, Evermann & Jenkins, Proc. U. S. N. M. 1891, 155 (Guaymas)

Family XXXIV. MULLIDÆ.

100. Mulloides samoensis Günther. "*Weke.*"

Color in life (field No. 134), general color white, with light-green shadings on the back; belly white with yellowish tinges; a bright yellow line from eye to base of caudal, wider than pupil; yellow stripes under eye; 2 faint yellow lines along side beneath the large one; first and second dorsals and caudal yellow, the other fins white; irregular rosy blotches on anterior portion of body; barbels white.

Six specimens, 7.5 to 13 inches in length, were collected by me, and 2 examples, each 11 inches in length, were taken by the *Albatross* in 1896. At times this fish is very abundant in the market, and is regarded by the natives as being "as good as the mullet."

Mulloides samoensis Günther, Fische der Südsee, 57, taf. XLIII, fig. B, 1873, Apia, Samoa.

101. Mulloides auriflamma (Forskål).   "Weke."

Color in life (field No. 203), red with yellowish border to each scale; a bright yellow band extending from each eye; margins of opercle and preopercle yellow; iris white, with red inner border; dorsal fins red toward body, yellow outwardly; caudal bright yellow; ventrals and anal reddish toward body, bright yellow outwardly; pectoral red; barbels white.

I obtained 10 examples of this species, from 7.24 to 9.62 inches in length; a 10-inch example is in Dr. Wood's collection; and 2 examples, each 7 inches long, are in the collection made by the *Albatross*. This is a common and much valued food-fish at Honolulu.

Mullus auriflamma Forskål, Descript. Anim., 30, 1775, Djidda, Arabia.
Mullus flavolineatus Lacépède, Hist. Nat. Poiss., III, 406, 1801.
Upeneus flavolineatus, Cuvier & Valenciennes, Hist. Nat. Poiss., III, 456, 1829.
Mulloides flavolineatus, Bleeker, Nat. Tyjdsch. Ned. Ind., III, 1652, 697 (Wahal); Günther, Cat., I, 403, 1859; Günther, Fische der Südsee, I, 56, 1873; Bleeker, Revision Insul. Mulloides, 15, 1874; Bleeker, Atlas, pl. 394 (Mull., pl. 4), fig. 3, 1877; Streets, Bull. U. S. Nat. Mus., No. 7, 89, 1877 (Fanning Islands).

102. Pseudupeneus chryserydros (Lacépède).

Color in life (field No. 243), dark lead color with violet and golden shadings; blue lines alternating with golden, radiating from the eye; longitudinal golden lines on the cheek; anterior portion of first dorsal golden, rest of fin violet with dusky shades; second dorsal with oblique blue lines alternating with golden; large bright orange area on upper portion of caudal peduncle; anal marked as second dorsal but less dark; caudal dusky violet; ventral rays blue, membranes golden; barbels tipped with yellow.   Another example (field No. 121) was much lighter in color, with rosy tinges; first dorsal rosy with dusky shades; the first and second spines olivaceous and with olivaceous markings along outer margin; second dorsal with oblique whitish and yellow stripes; caudal dusky with areas of olive; anal fin pale with yellow stripes; ventral rays white, membrane yellow; a large bright, light orange area on upper part of caudal peduncle extending half way down its sides; tips of barbels orange.

Four specimens of this species, 8.7 to 9.5 inches in length, were taken; three, 6.5 to 11 inches in length, are in Dr. Wood's collection, and one 9 inches long was collected by Jordan & Snyder. This fish is brought in great numbers to the market.

? *Mullus cyclostomus* Lacépède, Hist. Nat. Poiss., III, 404, pl. 14, fig. 3, 1801, Isle de France.
Upeneus chryserydros Lacépède, Hist. Nat. Poiss., III, 406, 1801; Cuvier & Valenciennes, Hist. Nat. Poiss., III, 470, 1829 (Isle de France, Sandwich Islands, Isle of Bourbon, Coromandel).
Upeneus oxycephalus Bleeker, Act. Soc. Neerl., I, 45, 1856, Manado en Macassar; Günther, Cat., I, 409, 1859.
Parupeneus chryserydros Bleeker, Revision Insul. Mulloides, 35, 1874; Bleeker, Atlas, IX, pl. 385 (Mull., pl. 3), fig. 2, 1877.
Upeneus chryserythrus Günther, Fische der Südsee, I, 60, pl. 45, fig. A, 1873, Polynesia.
Parupeneus cyclostomus, Steindachner, Denks. Ak. Wiss. Wien, LXX, 1900, 486 (Honolulu).

103. Pseudupeneus chrysonemus Jordan & Evermann.

Four specimens, 5.5 to 8 inches in length, were obtained by me, and one, 5.5 inches long, was collected by Jordan & Snyder.   Many specimens were obtained by Jordan & Evermann.

Pseudupeneus chrysonemus Jordan & Evermann, Bull. U. S. Fish. Comm. for 1902 (April 11, 1903), 186, Hilo, Hawaii Island. (Type No. 50666, U. S. N. M.; coll. Jordan & Evermann.)

104. Pseudupeneus porphyreus Jenkins, new species.

Head 3.3 in length; depth 3; eye 4 in head; snout 2; D. VIII-I, 8; A. II, 6; C. 19; P. 15; V. 1, 5; scales 2½-30-6; pectoral 1.4 in head; ventral 1.25; caudal 1.2; longest dorsal spine 1.5; longest soft dorsal ray 2; longest anal ray equal to longest soft dorsal ray; length of caudal peduncle equal to length of ventral fin, 1.25; greatest depth equal to 2.4 in head; greatest width of body at bases of pectorals, a little greater than half of head; width of middle of caudal peduncle equal to 3.5 in head; preorbital deep, 3 in head; maxillary 2.5 in head, reaching almost to vertical from posterior nostril; distance between nostrils a little less than diameter of eye; interorbital 3.5 in head; teeth in single series, present only in jaws, short, blunt, conical, rather widely separated and of unequal sizes; gillrakers 5 + 25, the uppermost and lowermost ones very short, uppermost ones of lower arm of arch longest, 2 in eye, gradually increasing in length downward; lowermost ones of upper arm about 0.6 length of uppermost of lower arm; snout blunt, almost truncate; dorsal profile of head straight, rising at angle of about 45° to nape, profile of back horizontal from here to front of second dorsal, then descending in a gentle curve to caudal fin; ventral profile of head and body almost straight from snout to middle of belly, then forming a gentle concave curve symmetrical with corresponding part of dorsal profile; mouth only very slightly oblique; posterior limb of preopercle almost vertical, lower limb horizontal, angle rounded; eye almost circular, anterior rim slightly before middle of head; center of pupil on level with opercular spine; interorbital convex; scales on snout extending a little below nostrils, several large ovate ones about nostrils; preorbital, lower part of snout, jaws, and maxillaries naked; first dorsal spine very short; third and fourth longest; third flexible at tip, not pungent, 1.8 in head; spines back of fourth regularly decreasing in length to last which is 0.3 of second; distance between last dorsal spine and first ray of second dorsal 2.5 in head; spine of second dorsal equal to sixth spine, 2.8 in head; first branched ray longest, a little less than half of head, rays gradually decreasing in length to last, which is 3.5 in head; caudal deeply forked, lobes about equal, upper of 10 rays, lower of 9; first anal spine very short, concealed within membranes about base of second, second equal to first branched ray of dorsal; others gradually decreasing in length to last, which equals last dorsal ray; pectoral pointed, upper rays longest; ventral pointed, second branched ray longest; lateral line parallel with dorsal profile, beginning above upper end of gill-slit; scales large, ctenoid, those of ventral parts

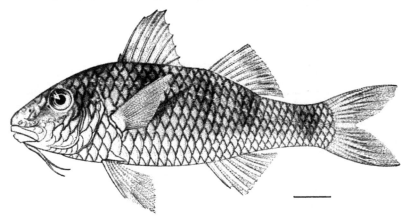

FIG. 22.—*Pseudupeneus porphyreus* Jenkins, new species. Type.

rather larger than those of back; basal half of caudal scaled; other fins naked; dorsal depressible in a groove; tubes of lateral line with numerous (13 in largest examples) radiating branches on each scale; scale before upper end of base of pectoral with large lobe overhanging base of pectoral; scale above base of ventral with prominent tapering backward prolongation; barbels slender, short, not reaching posterior edge of preopercle, 1.75 in head.

Color in life (field No. 212), head and body uniformly red, fins brighter; first dorsal tipped with white; belly lighter; iris light yellow; soft dorsal and anal with inconspicuous golden tinge on membranes.

Color in alcohol, somewhat variable, most of the specimens with pale-yellowish ground-color; all have a well-marked bright yellow or white quadrate area saddled across anterior half of dorsal surface of caudal peduncle, reaching about half way down to lateral line on side. On specimens with most color present, dorsal fins and dorsal half of head and body much clouded with dusky; a band on head from below anterior nostril to middle of front rim of eye and appearing again back of eye on side of head; a dusky blotch on opercle before angle; on free part of side of body indistinct definite yellowish lines below lateral line and 2 above it following middle row of scales.

Resembles *P. xanthospilurus* (Bleeker) in possession of yellow blotch on front half of back of caudal peduncle, differing from *P. chryserydros* in confinement of blotch to fore half of peduncle and in much shorter barbels. Differs from *P. xanthospilurus* in much shorter barbels and in shorter snout.

Variations: In most of the specimens the second spine of the dorsal is stiff and pungent. The barbels, although generally not reaching the posterior margin of preopercle, do so in some cases. In most cases the scales of the head do not extend below nostrils from top of head.

This description is based on the type (field No. 212) 10.6 inches in length, and 17 cotypes, 5 with the type collected by me in 1889, 4 by Dr. Wood, 3 by Jordan & Snyder, and 4 by the *Albatross*. These examples range from 5 to 12 inches in length. (Type, No. 50705, U. S. N. M., Honolulu; coll. O. P. Jenkins.)

105. Pseudupeneus bifasciatus (Lacépède).

One specimen, 8.5 inches in length, was taken by me, and one, 9 inches in length (field No. 1339), by the *Albatross* in 1896. This species is so much confused with *P. multifasciatus* that its range is not well known. Günther had specimens from Rarotonga, Navaii, and Solomon Islands. It was first described from the island of Bourbon, and is now certainly known to occur at the Hawaiian Islands.

Mullus bifasciatus Lacépède, Hist. Nat. Poiss., III, 404, pl. 14, fig. 2, 1801.
Upeneus bifasciatus, Cuvier & Valenciennes, Hist. Nat. Poiss., III, 468, 1829 (Isle of Bourbon); Günther, Fische der Südsee, I, 59, pl. 44, fig. A., 1873 (Rarotonga, Savaii, Solomon Islands).
Upeneus trifasciatus (in part), Günther, Cat., I, 407, 1859.
Mullus trifasciatus Lacépède, Hist. Nat. Poiss., III, 404, pl. 15, fig. 1, 1801.

106. Pseudupeneus multifasciatus (Quoy & Gaimard). "*Moano.*"

Color of fresh specimen (field No. 82), 8.5 inches long, whole body suffused with red; region over nape, eye, opercle, and to a short distance behind pectoral fin dusky; dusky band about 7 scales wide from posterior margin of first dorsal to middle of second dorsal (this band split by narrow band of ground-color at front of second dorsal); dusky band around caudal peduncle, back of middle; first dorsal red with dusky spot at tip; margin of second dorsal black; pectoral yellow; anal black with bluish blotches; ventrals dusky, suffused with red and with about 6 rather distinct light-bluish cross-bands; barbels red, with white tips; iris red.

Seventeen examples, 4.25 to 8.25 inches in length, were taken by me, 4 by Dr. Wood, 4 by Jordan & Snyder, and 2 by the *Albatross*.

This fish is very abundant at Honolulu and is a very important food-fish.

Mullus multifasciatus Quoy & Gaimard, Voy. Uranie, Poiss., 330, Atlas, pl. 59, fig. 1, 1824, Oahu.
Upeneus trifasciatus, Cuvier & Valenciennes, Hist. Nat. Poiss., III, 468, 1829 (Sandwich Islands, Caroline Islands); Streets, Bull. U. S. N. M., No. 7, 71, 1877 (Honolulu); Günther, Voyage Challenger, Shore Fishes, 59 (Honolulu); Fowler, Proc. Ac. Nat. Sci. Phila. 1900, 520 (Tahiti).
Parupeneus multifasciatus (in part) Bleeker, Revision Insul. Mulloides, 20, 1874.
Parupeneus multifasciatus, Bleeker, Atlas, IX, pl. 394 (Mull., pl. IV), fig. 4, 1877.
Upeneus velifer Smith & Swain, Proc. U. S. N. M. 1882, 130, Johnston Island. (Type, No. 28822, U. S. N. M.)

107. Pseudupeneus pleurostigma (Bennett).

Color in life, upper part of body red, lower white; first dorsal, pectoral, caudal, and ventral red; second dorsal with a dusky spot on front portion, and with about 6 bright yellow lines running across fin parallel to axis of body when the fin is extended; a dusky spot, as large as eye, on body one scale below the dorsal line and just behind the vertical from posterior margin of first dorsal.

Three specimens of this species, 6, 7.2, and 7.8 inches in length, were taken by me in 1889; four, 5 to 6 inches, by Dr. Wood; and two, 5.5 and 6.25 inches, by Jordan & Snyder.

Upeneus pleurostigma Bennett, Proc. Lond. Zool. Soc., I, 59, 1833; Günther, Fische der Südsee, I, 58, 1873 (Otaheiti, Apemana); Bleeker, Atlas, IX, pl. 393 (Mull., pl. III), fig. 3.
Upeneus brandesii Bleeker, Nat. Tijds. Ned. Ind., II, 1851, 236, Banda Neira; Günther, Cat., I, 407, 1859 (Sea of Banda Neira).
Parupeneus pleurostigma, Bleeker, Revision Insul. Mulloides, 29, 1874; Steindachner, Denks. Ak. Wiss. Wien, LXX, 1900, 481 (Laysan).

108. Upeneus arge Jordan & Evermann.

Body white, 2 brown longitudinal stripes on body above lateral line; one orange-yellow stripe from eye to base of caudal, another similar from base of pectoral to base of caudal, reddish line from eye to nostril; first dorsal transparent with dusky blotches along outer margin; second dorsal same, the dusky blotches forming 3 oblique bars on the fin; caudal fin with white bars alternating with dark bars (black and brown); barbels bright yellow.

Four examples of this species, 8, 8, 9, and 10.5 inches in length, were obtained. It is abundant at Honolulu and is highly valued as a food-fish.

Upeneus arge Jordan & Evermann, Bull. U. S. Fish Comm. 1902 (April 11, 1903), 187
Upeneoides vittatus, Streets, Bull. U. S. Nat. Mus., No. 7, 71, 1877 (Honolulu).

Family XXXV. POMACENTRIDÆ.

109. Dascyllus albisella Gill.

Eighteen examples of this species were taken by me, the majority being caught by means of a dip-net, on the reef in front of Honolulu. They range from 1.3 to 4.7 inches in length. In the smallest the white spot on the side of the body is 0.7 as broad as the head and extends from near the dorsal down the side more than 0.7 of its width, the remainder of the body being black. These specimens form a complete series, which show in the smallest the white spot relatively large and distinct, being very conspicuous; in the largest it gradually becomes less distinct and relatively smaller. In the largest the general color is gray, becoming almost white in some. A nuchal spot can not be distinguished in any of these specimens. The series also shows a gradation in length of second and last dorsal spines. One fresh specimen (field No. 259) showed each scale on the body pale blue with posterior margin black; head and fins all black; iris pale blue; white spot on side, 5 scales wide and 6 scales deep, is under bases of sixth to tenth spines.

Dascyllus albisella Gill, Proc. Ac. Nat. Sci. Phila. 1862, 149, note, Sandwich Islands; Gunther, Challenger Report, Zool., I,
    Part VI, 61, 1879 (1880), (Honolulu); Günther, Fische der Südsee, VII, 236, 1881 (Honolulu).
Dascyllus trimaculatus, Günther, Fische der Südsee, VII, 236, 1881 (Sandwich Islands); Steindachner, Denks. Ak. Wiss.
    Wien, LXX, 503, 1900 (Honolulu).
Tetradrachmum trimaculatum, Bleeker, Atlas, IX, Taf. 409, fig. 8, 1879; Fowler, Proc. Ac. Nat. Sci. Phila. 1900, 503 (Oahu).

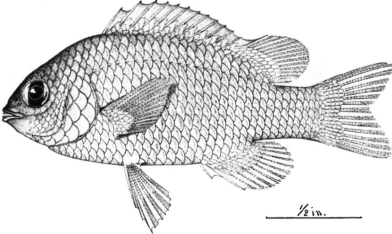

½ in.

FIG. 23.—*Chromis elaphrus* Jenkins, new species. Type.

110. Chromis elaphrus Jenkins, new species.

Head 3.4 in length; depth 2.25; eye 3 in head; snout 3.75; maxillary 3.75; interorbital 2.5; D. XII, 15; A. II, 11; P. 18; scales 2-28-8. Dorsal outline of body more convex than ventral outline; snout shorter than diameter of eye; caudal peduncle about 2 in head; eye greater than interorbital space; interorbital slightly convex; scales reaching a little below level of the single nostril; suborbital, preopercle, and opercle entire, the preopercle somewhat crenulate; the opercle ending in 2 small flat spines, the upper somewhat obscure; teeth conical, in a single series in each jaw; third dorsal spine the longest, 1.8 in head; longest dorsal rays somewhat longer than third dorsal spine; soft dorsal rounded somewhat higher than spinous portion; soft anal rounded 1.4 in head; caudal deeply emarginate, the upper lobe slightly the longer, the longest rays scarcely equaling the head; pectoral broad, 1.3 in head; ventrals nearly reaching vent; all parts of the body and head, except portion of snout anterior to nostril and tip of lower jaw, covered with scales; lateral line developed; tubes on 20 scales, reaching to within 3 scales of base of last dorsal ray, where it ceases; bases of all fins scaled.

Color in alcohol, body and head a pale brown, lighter toward the ventral region; on the lower third of body faint traces of longitudinal rows of pearly dots corresponding to the scales; fins pale and without markings, no spot on anterior spinous dorsal and none at base of pectoral.

This description is based on the type, No. 50703, U. S. N. M., 2.5 inches long, and 8 smaller cotypes taken by me in 1889. They were caught in the coral rocks in the reef in front of Honolulu.

111. Chromis ovalis (Steindachner).

Three examples, each 6 inches in length, were taken by me, and one of same length by Dr. Wood. This species appears to be the one described by Steindachner, although in his description the measurement of the body-height as 3.3 in the body-length must be an error. My description of this species was in the hands of the printer at the time Steindachner's paper appeared, and his paper did not reach me until after my paper was published.

Heliastes ovalis Steindachner, Denks. Ak. Wiss. Wien, LXX, 1900, 502. Honolulu.
Chromis velox Jenkins, Bull. U. S. Fish Comm. 1899 (June 8, 1901), 393, fig. 6, Honolulu. (Type, No. 49998, U. S. N. M.; coll. O. P. Jenkins.)

112. Pomacentrus jenkinsi Jordan & Evermann.

Numerous specimens taken by me in 1889. This small species is very common about the reef.

Eupomacentrus marginatus Jenkins, Bull. U. S. Fish Comm. for 1899 (June 8, 1901), 391, fig. 5, Honolulu. (Type, No. 49700, U. S. N. M.; coll. O. P. Jenkins), name preoccupied.
Pomacentrus jenkinsi Jordan & Evermann, Bull. U. S. Fish Comm. for 1902 (April 11, 1903), 189, Honolulu.

113. Glyphisodon abdominalis Cuvier & Valenciennes.

Color in life, bands distinct to lower two-thirds of body, yellow between the bands; white below. This fish is fairly abundant at Honolulu. Ten examples, 4.4 to 8.25 inches in length, were obtained; and two, 5.25 and 7.6 inches, are in the collection of Jordan & Snyder.

Glyphisodon abdominalis Cuvier & Valenciennes, Hist. Nat. Poiss., v, 457, 1830, Hawaiian Islands.
Glyphidodon maxillis, Streets, Bull. U. S. Nat. Mus., No. 7, 66, 1877 (Honolulu); Günther, Fische der Südsee, 229, Taf. CXXVI, 1881 (Sandwich Islands); Steindachner, Denks. Ak. Wiss. Wien, LXX, 1900, 502 (Honolulu and Laysan).
Abudefduf maxillosus, Fowler, Proc. Ac. Nat. Sci. Phila. 1900, 504 (Oahu Island).

114. Glyphisodon sordidus (Forskal).

Color in life (field No. 163, 5.8 inches), gray, with 5, not very distinct, dark crossbands; a black spot on upper part of base of pectoral; black spot on caudal peduncle next to posterior part of dorsal fin; upper part of spinous dorsal yellow.

Two adults, 5.5 and 5.8 inches long, and six from 0.8 to 1.3 inches in length, were obtained at Honolulu. Mr. McGregor obtained one 1.6 inches long at Lahaina, Maui. This species is not as frequently seen in the market as the preceding.

Chaetodon sordidus Forskål, Desc. Animal., 62, 1775, Djidda, Red Sea.
Glyphidodon sordidus, Günther, Cat., IV, 41, 1862 (China); Bleeker, Atlas, Taf. 110, fig. 5, 1877; Günther, Fische der Südsee, 231, VII, 1881 (Red Sea, east coast Africa; East Indian Archipelago; Tahiti, Raiatea, Samoa; and Bonham Island).
Abudefduf sordidus, Fowler, Proc. Ac. Nat. Sci. Phila. 1900, 504 (Oahu Island).

Family XXXVI. LABRIDÆ.

115. Lepidoplois bilunulatus (Lacépède). "A'awa."

Color in life (field No. 122), general color white with pinkish shades, many horizontal brown stripes crowded together along top of head and back; a brown stripe from angle of mouth to angle of preopercle; chin and throat white, overlaid with red spots; colors of body posteriorly gradually giving way to yellow, which becomes on the caudal fin a bright yellow; a black area on body at base of posterior part of soft dorsal, extending on caudal peduncle; iris black with red inner margin; pectoral rosy; ventrals white, with rosy and yellow shadings; a black spot nearly as large on first dorsal between first and third spine; tips of dorsal spines yellow; soft dorsal and anal bright yellow. Another specimen showed the brown lines of above red, and the caudal was orange.

Five specimens of this beautiful fish, 6.6, 7.2, 7.7, 7.8, 9.4 inches in length, respectively, were taken. It is common in the market, where it is conspicuous for its brilliant coloration.

Labrus bilunulatus Lacépède, Hist. Nat. Poiss., III, 454, 526, pl. 31, 1801.
Cossyphus bilunulatus, Cuvier & Valenciennes, Hist. Nat. Poiss., XIII, 121, 1839 (Isle de France); Günther, Fische der Südsee, VII, 240, pl. CXXX, 1881 (Mauritius, Zanzibar, Amboyna, Misol, Sandwich Islands).
Harpe bilunulata, Steindachner, Denks. Ak. Wiss. Wien, LXX, 1900, 503 (Honolulu).

116. Anampses cuvier Quoy & Gaimard.

Color in life (field No. 71, 6 inches in length), general color dark brown, becoming a bright red on the belly; a bright pearly spot on each scale half as large as pupil, making about 17 longitudinal rows; upper part of head dark with many small pearly spots, lower part blue with bright red spots; throat and breast blue with many bright spots; iris yellow; dorsal fin dark red with blue border and with 4 or 5 rows of bright pearly blue spots; anal fin bright red, with bright blue margin and bright blue base, with 3 wavy lines of blue running whole length of fin, the blue crossing in places from one line to another; caudal reddish-brown, upper and lower margins blue. One small specimen, about 2 inches in length, showed the same coloration, with the exception of having a black ocellated spot on the posterior portion of soft dorsal.

This species is common at Honolulu. I obtained eight examples, 5.25 to 10 inches in length; one 8.25 inches long was taken by the *Albatross* in 1896, and one 7 inches long by Jordan & Snyder.

Anampses cuvier Quoy & Gaimard, Voy. de l'Uranie, Zool., 276, pl. 55, fig. 1, 1824, "de l'ile Mowi."
Anampses cuvieri, Cuvier & Valenciennes, Hist. Nat. Poiss., 14, 11, 1839; Günther, Cat., IV, 136, 1862 (Hawaiian Islands); ibid., Fische der Südsee, 251, pl. 136, fig. A, 1881 (Hawaiian Islands); Fowler, Proc. Ac. Nat. Sci. Phila. 1900, 506 (Honolulu).

117. Anampses evermanni Jenkins.

This fish is not uncommon at Honolulu. Four specimens, 11 to 12 inches in length, were obtained by me; one, 10.5 inches, by the *Albatross* in 1896; and two, 10.5 and 12 inches, by the *Albatross* in 1891.

Anampses evermanni Jenkins, Bull. U. S. Fish Comm. for 1899 (August 30, 1900), 57, fig. 14, Honolulu. (Type, No. 6136 L. S. Jr. Univ. Mus.; coll. O. P. Jenkins.)

118. Stethojulis axillaris (Quoy & Gaimard).

Color in life (field No. 308, 4.4 inches long), upper parts dusky, covered with numerous minute green dots; lower parts lighter, reddish; bright orange spot on body just behind opercular flap; base of pectoral black; black ocellated spot on base of caudal; orange area behind angle of mouth; iris red.

Another specimen (field No. 126) shows, ground color olivaceous, thickly covered with bright green dots; throat and belly greenish silvery; iris bright green; base of pectoral and axil as in No. 308; 3 black spots on caudal peduncle on lateral line, the third on base of caudal fin; dorsal fin olivaceous with brownish spots, a black spot at base of last 2 rays; anal olivaceous, base green.

Fourteen specimens of this beautiful little fish, 2 to 4.25 inches in length, were obtained by me from the coral rocks on the reef in front of Honolulu, and 3 are in Dr. Wood's collection.

Julis axillaris Quoy & Gaimard, Voyage de l'Uranie, Zool., 272, 1824, Maui, Hawaiian Islands.
Stethojulis axillaris, Streets, Bull. U. S. Nat. Mus., No. 7, 65, 1877 (Honolulu); Günther, Fische der Südsee, VII, 254, taf. CXXXVI, fig. C, 1881 (Pelew, Solomon, Fiji, Navigator, Society, Hawaiian, New Hebrides, Ponape); Fowler, Proc. Ac. Nat. Sci. Phila. 1900, 508 (Hawaiian Islands).

119. Stethojulis albovittata (Kölreuter).

Color in life, upper parts green, lower lighter; a bright blue line from middle of snout to upper part of iris through iris, then upward and backward along base of dorsal fin for its whole length; another bright blue line from eye back and just below lateral line to about sixth or seventh scale; another such line from mouth just above the angle running backward just below eye over opercular flap, ending at middle of base of caudal; another bright blue line below chin curving upward and around to lower angle of preopercle (this portion of the line violet) and past it, thence to base of pectoral, which interrupts it, continuing downward and backward, ending at lower part of base of caudal; between these last 2 lines posterior to base of pectoral, a bright orange band; dorsal and caudal fins orange; anal transparent, but blue toward base; ventrals transparent; iris orange.

Twenty-one examples of this brilliantly colored and very beautiful little fish, from 4.5 to 5 inches in length, were taken. It is quite abundant about the reef.

Labrus albovittatus Kölreuter in Bonnaterre, Ichthyol., 108, fig. 399, 1798.
Stethojulis albovittata, Günther, Fische der Südsee, vii, 256, taf. cxli, fig. B, 1881 (Zanzibar, Madagascar, Hawaiian Islands). Steindachner, Denks. Ak. Wiss. Wien, lxx, 1900, 504 (Honolulu); Fowler, Proc. Ac. Nat. Sci. Phila. 1900, 508 (Hawaiian Islands).

120. Halichœres lao Jenkins. *"Lao."*

Two specimens, 3.8 and 4.25 inches in length. Description and color in reference given below.

Halichœres lao Jenkins, Bull. U. S. Fish Comm. for 1899 (August 30, 1900), 48, fig. 3, Honolulu. (Type, No. 6132, L. S. Jr. Univ. Mus.; coll. O. P. Jenkins.)

121. Halichœres iridescens Jenkins. *"Ohua Paawela."*

Three specimens, 5, 5.4, and 5.56 inches long, respectively, taken by me. Description and color given in paper referred to.

Halichœres iridescens Jenkins, Bull. U. S. Fish Comm. 1899 (August 30, 1900), 47, fig. 2, Honolulu. (Type. No. 6131, L. S. Jr. Univ. Mus.; coll. O. P. Jenkins.)

122. Macropharyngodon geoffroy (Quoy & Gaimard). *"Hinalea Akilolo."*

Five examples of this beautiful fish were taken. They measure 3.5, 4.2, 4.2, and 4.75 inches in length. This was thought to be a new species by me, but recently having access to Voy. Uranie, Zool., containing Quoy & Gaimard's description and figure based on a specimen taken at Maui, there remains no doubt that my specimens are of their species.

Julis geoffroy Quoy & Gaimard, Voy. de l'Uranie, Zool., 270, pl. 56, fig. 3, 1824, Maui; Cuvier & Valenciennes, Hist. Nat. Poiss., 13, 479, 1839 (Hawaiian Islands).
Platyglossus geoffroyii, Günther, Cat., iv, 145, 1862 (Hawaiian Islands).
Macropharyngodon aquilolo Jenkins, Bull. U. S. Fish Comm. for 1899 (August 30, 1900), 46, fig. 1, Honolulu. (Type. No. 6130, L. S. Jr. Univ. Mus.; coll. O. P. Jenkins.)

123. Julis gaimard (Quoy & Gaimard).

Color in life, bright red, with bright blue dots on the body, more numerous posteriorly; bands on head and chin green; greenish transverse area on body about region of tip of pectoral.

This fish is fairly common at Honolulu. Three specimens, 9, 10.8, and 11.5 inches in length; also one 11.7 inches long taken by the *Albatross* in 1896.

Julis gaimard Quoy & Gaimard, Voy. de l'Uranie, Zool., 265, pl. 54, fig. 1, 1824, "de l'Ile Mowi."
Coris gaimardi, Fowler, Proc. Ac. Nat. Sci. Phila. 1900, 510 (Sandwich Islands).

124. Julis pulcherrima Günther. *"Akilolo."*

Color in life (field Nos. 96 and 99, 8.6 and 10.5 inches long), head red with bright, wavy green, longitudinal stripes, 1 from snout to eye; 1 from angle of mouth to opposite base of pectoral fin; 1 from chin to base of ventral; 1 on middle of forehead; 2 from eye, the upper to base of dorsal, where it becomes a row of bright blue spots, 1 spot at base of each spine and ray; lower stripe from eye extending to opercular flap; ground color of anterior part of body reddish brown, the remainder of body through first third of caudal with dark cloudings; dorsal fin red, outer third bright red, separated from inner two-thirds by a blue line; outer margin black, inner two-thirds with many small blue spots; anal fin similar to dorsal but brighter, margin blue instead of black, with delicate shadings difficult to indicate; a bright blue spot at base of each soft ray; outer two-thirds of caudal bright yellow; pectoral rays red, membranes colorless; axil dark blue surrounded by rings of dark green; many bright spots on posterior third of body; ventral fins, outer margin dark blue, next bright red, next light blue.

This is one of the most beautiful of the brilliantly colored fishes seen at Honolulu. It is fairly common. I obtained 10 examples in 1889 from 7 to 10.5 inches in length, and 2, each 6.4 inches, were taken by the *Albatross* in 1896.

Coris pulcherrima Günther, Cat., IV, 200, 1862. Amboyna, Celebes, Tahiti, New Hebrides; Steindachner, Denks. Ak. Wiss. Wien, LXX, 1900, 507 (Honolulu).

125. Julis lepomis (Jenkins).

One specimen of this beautiful fish, 17.5 inches in length, was taken by me. Description in paper referred to below.

Coris lepomis Jenkins, Bull. U. S. Fish Comm. for 1899 (August 30, 1900), 48, fig. 4. Honolulu. (Type, No. 12141, L. S. Jr. Univ. Mus.; coll. O. P. Jenkins.)

126. Julis eydouxii Cuvier & Valenciennes.

Color in life (field No. 256, 11.3 inches long), upper parts dark brown, lower parts pink; a band running along back just below base of dorsal fin, beginning on head behind vertical from posterior border of eye, running on to caudal, wavy and blue in front, nearly white posteriorly; a yellowish-white band becoming blue posteriorly from tip of snout on middle line on top of head to origin of dorsal, then running upon dorsal as a blue anterior margin; from this band just back of tip of snout, a band (one on each side) branching and running back just along upper margin of eye and ending at a vertical from fourteenth dorsal spine, blue in front and almost white posteriorly; another band blue in front, running from mouth along lower margin of eye through opercular flap, which has a blue spot, straight back to tail; this band dividing the upper brown color from the pink below; between these bands, the brown color on body gradually changing to red on the head; pectoral and ventrals pink; dorsal dusky, almost black, outer margin bluish-white; a conspicuous yellow longitudinal band along the central portion of the fin, anterior portion with a black spot on second spine and membrane between second and third spines; caudal fin outer margin and base white, between which is a broad black band; anal black, outer margin bluish-white.

Six examples, from 6.6 to 11.3 inches in length, were obtained by me, and two, 10.75 and 12.75 inches, by the *Albatross*.

Julis eydouxii Cuvier & Valenciennes, Hist. Nat. Poiss., XIII, 455, Sandwich Islands.

127. Hemicoris baillieui (Vaillant & Sauvage).

One specimen of this fish, 9.2 inches in length, which shows well the markings shown in Dr. Steindachner's figure, is in Dr. Wood's collection.

Coris baillieui Vaillant & Sauvage, Rev. Mag. Zool., III, 1875, 285, Sandwich Islands.
Coris schauinslandii Steindachner. Anzeiger, No. XVI, Denks. Ak. Wiss. Wien, June 21 (June 27), 1900, 177, Honolulu; Steindachner, Denks. Ak. Wiss. Wien, 1900, taf. V, fig. 1, 508, Honolulu.

128. Hemicoris argenteo-striata (Steindachner).

One specimen of this species was taken by Dr. Wood. My description of this fish was published before Dr. Steindachner's paper reached me.

Coris argenteo-striatus Steindachner, Anzeiger, No. XVI, Denks. Ak. Wiss. Wien, June 21 (June 27), 1900, 176, Honolulu; Steindachner, Denks. Ak. Wiss. Wien, 1900, LXX, 507, taf. III, fig. 1, Honolulu.
Hemicoris keleipionis Jenkins, Bull. U. S. Fish Comm. for 1899 (August 30, 1900), 51, fig. 6, Honolulu. (Type, No. 6049. L. S. Jr. Univ. Mus.; coll. O. P. Jenkins.)

129. Hemicoris remedia Jenkins.

Twelve examples of this species, 5 to 6.6 inches in length, were taken by me. Dr. Steindachner has identified this fish with *Coris multicolor* (Rüppell), a species described from the Red Sea. This differs from Rüppell's species in not having the anterior dorsal spine produced, in the absence of a dark blotch on anterior dorsal, and in coloration of the head.

This very brilliant fish is fairly common at Honolulu.

Coris multicolor Steindachner, Denks. Ak. Wiss. Wien, LXX, 1900, 507, pl. V, fig. 2 (Honolulu, Laysan), not of Rüppell.
Hemicoris remedius Jenkins, Bull. U. S. Fish Comm. for 1899 (August 30, 1900), 49, fig. 5, Honolulu. (Type, No. 6138, L. S. Jr. Univ. Mus.; coll. O. P. Jenkins.)

130. Cheilio inermis (Forskål).

This fish is very common at Honolulu and varies much in color and form. **Twelve examples, 7.5** to 16 inches in length, were taken by me, and one, 10.25 inches long, was taken by the *Albatross* in 1896.

Color in life (field No. 279, 16 inches long), leaden, darker above, lighter below; margin of **each** scale faint golden; line of golden spots backward from angle of jaw, spreading into golden reticulations on opercle and preopercle; dorsal fin with golden reticulations; membranes of anal with a series of golden crossbars; membranes of caudal with faint brown spots; on the body, at a vertical from fourth and fifth dorsal spines, is an orange blotch running into a black one about the lateral line, which extends as a dark blue band about the belly.

Another example (field No. 280, 13.25 inches long) shows, general color reddish-brown; a dark longitudinal band from opercular flap to caudal; each scale with a spot, which below the lateral line is pearly; rays of dorsal brown; rays of anal greenish; reddish-brown spots on throat and chin; reticulations of the same color on sides of head, cheek, preopercles and opercles; rays of caudal greenish.

Another example (field No. 101, 9.5 inches long), body light brown, lighter on belly; each scale with a pearly spot; throat with light orange reticulations; rays of dorsal and anal light yellow.

Labrus inermis Forskål, Descript. Anim., 34, 1775, Red Sea.
Cheilio auratus Quoy & Gaimard, Voy. de l'Uranie, Zool., 274, pl. 54, fig. 2, 1824 (Maui, Hawaii).
Cheilio inermis Streets, Bull. U. S. Nat. Mus. No. 7, 65, 1877 (Honolulu); Fowler, Proc. Ac. Nat. Sci. Phila. 1900, 511 (Sandwich Islands); Steindachner, Denks. Ak. Wiss. Wien, LXX, 1900, 507 (Honolulu).

131. Thalassoma purpureum (Forskål).

Color in life (field No. 330, 10.5 inches long), general color light red, shading into orange forward and below; two rows of quadrangular blocks of color on the side, each block a bright blue shading to a bright green at center; 4 green crossbars connecting base of dorsal fin with upper row; outer margins of dorsal and anal fins bright blue; the portions of the fins next the body golden; pectoral indigo blue with the proximal region bright yellow; caudal with alternating longitudinal bands of blue and golden; no distinct markings on head; iris bright green with inner margin orange.

Two specimens of this very brightly colored fish, 5 and 10.5 inches in length, were obtained by me. This is the first record from the Hawaiian Islands.

Labrus purpureus Forskål, Descript. Animal., 27, 1775, Red Sea.
Julis trilobata Günther, Cat., IV, 187, 1862 (var. a, 188, South Africa).
Julis quadricolor Bleeker, Atlas, 1, 93, pl. 31, fig. 3, 1862 (in part, including specimen shown in fig. 3).

132. Thalassoma quadricolor (Lesson).

General color (field No. 138, 9.25 inches long), bright green; irregular, dark red, longitudinal band along upper part of body; vertical lines projecting from this at right angles; a bright red stripe from just above opercular flap to base of caudal, with vertical branches at each scale; another bright red band from near axil to base of caudal; a complex figure made by red bands on the face; a bright green band across lower part of snout; upper lip green, lower blue; chin and throat blue; cheek bright yellow; a double bright red band, somewhat reticulated, from eye obliquely downward to margin of opercle; dorsal fin, with longitudinal bands of red and green, margin blue, dark blue spot on anterior portion; anal fin with a longitudinal band next body of red, next to this a band of blue-green blotches, outer edge of band with band scalloped, next band red, scalloped; the outer band blue; caudal rays red, membranes blue, upper and lower margins green, posterior margin greenish-yellow; pectoral, base red, then line of blue, then greenish, remainder transparent except dusky blotch on tip; ventrals green. Another specimen (field No. 265, 11 inches long) recognized as different by fishermen, has stripes and bands bright red, and spaces between on both body and head green above, blue below; the markings on the fins red and green.

I obtained the two specimens just described; in addition, one, 11 inches in length, was taken by the *Albatross* in 1896, and one, 8.75 inches, by Jordan & Snyder.

Julis quadricolor Lesson, Voy. Coquille, Zool., III, 139, pl. 35, fig. 1, 1826–1830, Otaheite, Cuvier & Valenciennes, **Hist. Nat.** Poiss., XIII, 143, 1839 (Tahiti). Bleeker, Atlas, 1, 93, 1862 (in part; not the plate, which is of *T. purpureum*).
Thalassoma immaculatum Fowler, Proc. Ac. Nat. Sci. Phila. 1899, 488, pl. 18, fig. 2 Caroline Island.
Thalassoma herndti Seale, Occasional Papers, Bishop Museum, 1, Nos. 4 15, fig. 7, 1901, Honolulu. (Type. No. 681, B P. B. M., 1901.)

133. Thalassoma umbrostigma (Rüppell). *"Olali."*

Color in life (field No. 118, 4.5 inches in length), general color brown; side of body with 2 longitudinal rows of light-green oblong patches, of which the vertical length of each is greater, breadth of each less than diameter of eye; 2 bright blue stripes from upper margin of eye, the anterior meeting its fellow from other eye, the posterior not quite meeting its fellow; chin blue; blue spot on opercle; blue stripe around snout; belly blue; dorsal brown, with greenish-blue longitudinal stripe; anal similar to dorsal; caudal with alternating stripes of brown and green.

Another example (field No. 154, 7.25 inches in length), colors of body bright green and red, the red on head broken up into spots instead of in bands; blue spot on anterior dorsal.

Another example (field No. 155. 7 inches long) shows an arrangement of color much like that of No. 154, with the exception that it has bright blue where the other is green, and has no dots and bands on the head. The difference in color from No. 154 is so great that the native fishermen call this form by a different name. Specimens Nos. 156 and 157 form, however, a gradation in color pattern intermediate between Nos. 154 and 155.

Quite abundant at Honolulu. Eleven examples, 5.5 to 11.5 inches in length, were taken.

Julis umbrostigma Rüppell, Neue Wirbe., Fische, II, Taf. 3, fig. 2, 1835, Mohila and Djetta.
Julis umbrostigma Bleeker, Atlas, I, 92, Taf. 34, fig. 2, 1862; Steindachner, Denks. Ak. Wiss. Wien, LXX, 1900, 506 (Honolulu).

134. Thalassoma duperrey (Quoy & Gaimard). *"Hinalea Lauli."*

Thirty specimens of this fish, the longest 7 inches, were taken by me; 3 by Dr. Wood, 3 by the *Albatross* in 1896, and 4 by Jordan & Snyder. An examination of Quoy & Gaimard's description and figure, based on a specimen from the Hawaiian Islands, leaves no doubt of the identity of my specimens with this species. Many young of this species were taken in the coral rocks which show no evidence of the color markings of the adult, but series of sufficient completeness shows the gradual development of the adult color pattern. The following is a description of an example 2.75 inches in length.

Head 3.4 in length; depth 3.5; eye 4.75 in head; snout 3.5; interorbital about equal eye; D. VIII, 12; A. II, 11; scales 3-28-8, 20 pores before the bend. Body short, moderately compressed; dorsal and ventral outlines evenly convex; head small, longer than deep; snout moderate, pointed; mouth small, horizontal, entirely below axis of body; lips broad, the upper overhanging the lower in the closed mouth; preorbital narrow and oblique; eye small, lower edge touching axis of body; interorbital moderately broad, little convex; caudal peduncle much compressed, its least depth 2 in head; dorsal spines low, the last the longest, 3 in head; soft dorsal somewhat higher, about 2 in head; anal similar to soft dorsal, slightly higher; caudal truncate or very slightly rounded; ventrals short, reaching half-way to origin of anal; pectoral broad, reaching slightly past tips of ventrals, its length about 1.3 in head; scutes large, firm, with thin, flexible edges; scales on nape and breast but little reduced; head naked, except a patch of 6 or 7 scales on upper angle of opercle; head with numerous conspicuous pores and tubes, a series radiating from orbit on its under side, and 3 conspicuous ones on opercle; lateral line complete, beginning at upper end of gill-opening parallel with the dorsal outline to beneath fourth dorsal ray from last, where it curves downward 2 rows and continues to base of caudal; pores on upper portion mostly 3-branched, lower portion mainly single.

Color in alcohol, dark olive brown on head, back, and sides, paler below; spinous dorsal pale dusky with black on membranes between first and fourth spines; edge of dorsal pale with a narrow marginal dark line; anal dusky; caudal dusky; ventral paler; pectoral pale, somewhat dusky; the upper rays with a long dark blotch, less distinct than in most species, obsolete in some specimens; axil with a dusky spot. In some specimens the boundary between the dark of upper parts and the pale of belly is more marked.

This is perhaps the most abundant labroid at Honolulu.

Julis duperrey Quoy & Gaimard, Voy. de l'Uranie. Zool., 268, pl. 56, fig. 2, 1824, Sandwich Islands.
Thalassoma pyrrhorinctum Jenkins, Bull. U. S. Fish Comm. for 1899 (August 30, 1900), 51, fig. 7, Honolulu. (Type, No. 6138, L. S. Jr. Univ. Mus.; coll. O. P. Jenkins.)

135. Thalassoma obscurum (Günther).

Color in life, dark brown, with a purple tinge, each scale with dark-blue vertical bar, otherwise plain. Abundant at Honolulu. Nineteen specimens, 4.4 to 9.75 inches in length, were obtained.

Julis obscura Gunther, Report Shore Fishes, Challenger Zool., Part VI, 61, pl. 26, figs. A and B, 1880, Honolulu.
Julis verticalis Smith & Swain, Proc. U. S. Nat. Mus. 1882 (July 8), 135, Johnston Island.

136. Gomphosus varius Lacépède.

Color in life (field No. 214), from eye to end of produced snout dark orange; cheek and under side of head rosy; behind the eye 2 rows of dark brown spots; groundwork of body drab, belly light rosy; base of each scale with a dark brown spot, small anteriorly, gradually increasing in size posteriorly, where they become quite dark; dorsal fin dark brown with a very narrow white edge; anal dark brown, with a white edge, and with a row of golden spots along middle of fin, one on each membrane between the rays; caudal dark brown, with posterior margin yellow, then white; pectoral nearly transparent, with shades of yellowish and rosy.

Fairly common at Honolulu. Eight examples, 6.5 to 9.2 inches in length, were obtained by me in 1889; and two, 5 and 5.2 inches, were taken by Jordan & Snyder at the same place in 1900.

Gomphosus varius Lacépède, Hist. Nat. Poiss., III, 104, pl. 5, fig. 2, 1801: Steindachner, Denks. Ak. Wiss. Wien. LXX, 1900, 507 (Honolulu); Fowler, Proc. Ac. Nat. Sci. Phila. 1900, 510 (Sandwich Islands).
Gomphosus pectoralis Quoy & Gaimard, Voyage de l'Uranie, Zool., 262, 1824, "de l'île Mowi."

137. Gomphosus tricolor Quoy & Gaimard.   "*Hinalea.*"

Color in life (field No. 95, 10 inches long), a very bright dark blue over whole of body; the portions of the dorsal, caudal and anal fins projecting beyond the scales, a bright light blue; a dark violet bar on anterior part of each scale; pectoral fin with base and axil green, middle portion blue, posterior third black; green of axil extending upward on body; ventral fins blue.

Another example (field No. 242, 10.2 inches long) appeared with body a very dark green, with other markings the same as in No. 95.

Common in the Honolulu market, where its conspicuous color and odd form attract attention. Ten specimens, 6.75 to 10.6 inches in length, were taken by me; three, 6.5, 7.25 and 9.7 inches, by the *Albatross* in 1896; and four, 6.6, 6.8, 7 and 8.2 inches long, by Jordan & Snyder.

Gomphosus tricolor Quoy & Gaimard, Voy. de l'Uranie, Zool., 280, pl. 55, fig. 2, 1824, "de l'île Mowi"; Steindachner, Denks. Ak. Wiss. Wien, LXX, 1900, 506 (Honolulu); Fowler, Proc. Ac. Nat. Sci. Phila. 1900, 510 (Sandwich Islands).

138. Pseudocheilinus octotænia Jenkins.

One example, 4 inches in length, is in Dr. Wood's collection, and was taken by Jordan & Snyder. Description in paper referred to.

Pseudocheilinus octotænia Jenkins, Bull. U. S. Fish Comm. for 1899 (August 30, 1900), 64, fig. 22, Honolulu. (Type, No. 6122, L. S. Jr. Univ. Mus.; coll. Dr. Wood.

139. Cheilinus zonurus Jenkins.

Four examples, 6, 8.6, 9, and 9.5 inches in length, were obtained by me; and one, 10.25 inches, by the *Albatross* in 1896.

Cheilinus zonurus Jenkins, Bull. U. S. Fish Comm. for 1899 (August 30, 1900), 56, fig. 13, Honolulu. (Type, No. 6134, L. S. Jr. Univ. Mus.; coll. O. P. Jenkins.)

140. Cheilinus bimaculatus Cuvier & Valenciennes.

This beautiful little species shows most delicate coloring. From Dr. Bleeker's description of the East Indian species, *C. ceramensis*, it would appear that it can not be separated from the Hawaiian form.

Cheilinus bimaculatus Cuvier & Valenciennes, Hist. Nat. Poiss., XIV, 96, 1839, Sandwich Islands.
Cheilinus ceramensis, Bleeker At. p. 69, Taf. 28, fig. 4.

141. Hemipteronotus umbrilatus Jenkins.

One specimen, 4.75 inches in length, was obtained by me; and one, 7.5 inches long, by Jordan & Snyder.

Hemipteronotus umbrilatus Jenkins, Bull. U. S. Fish Comm. for 1899 (August 30, 1900), 53, fig. 10, Honolulu. (Type, No. 6135, L. S. Jr. Univ. Mus.; coll. O. P. Jenkins.)

142. Novaculichthys hemisphærium (Quoy & Gaimard.)

Two specimens, 7.5 and 8.25 inches in length, were taken by Dr. Wood, and two, 5.75 and 9.25 inches, by Jordan & Snyder.

Julia ranicorones Quoy & Gaimard, Voy. Astrol., Poiss., 704, pl. 20, fig. 1.
Novacula ranicolensis, Steindachner Denks. Ak. Wiss. Wien, LXX, 501, 1900 (Honolulu).

143. Novaculichthys woodi Jenkins.

Three specimens, 5.6, 6, and 6 inches long, are in Dr. Wood's collection. In the paper referred to below, I recognized 2 distinct species. However, an examination of a large series of fresh specimens seems to prove that both these and the one described by Mr. Seale are of one species.

Novaculichthys woodi Jenkins. Bull. U. S. Fish Comm. for 1899 (August 30, 1900), 52, fig. 8, Honolulu. (Type, No. 6029, L. S. Jr. Univ. Mus.; coll. O. P. Jenkins.
Novaculichthys entargyreus Jenkins, Bull. U. S. Fish Comm. for 1899 (August 30, 1900), 53, fig. 9. Honolulu. (Type, No. 5984, L. S. Jr. Univ. Mus.; coll. O. P. Jenkins.)
Novaculichthys tatoo Seale, Occasional Papers, Bishop Museum, vol. 1, No. 4, 5, fig. 2, 1901, Honolulu. (Type, No. 611, Bishop Museum; coll. A. Seale.

144. Iniistius pavoninus (Cuvier & Valenciennes).

General color in life, pale drab; light blue wavy lines downward and backward from eye; ventrals white; pectoral pale olivaceous; dorsal with wavy blue reticulations; a longitudinal light blue band near outer margin of anal; caudal with shade of olivaceous, posterior margin light blue; black spot on scale a short distance below the fourth spine of second portion of dorsal, just above the lateral line.

Four examples of this species, 7.5, 7.5, 9, and 9.5 inches in length, were obtained by me; and three, 5, 5.75, and 5.75 inches in length, by Jordan & Snyder. This species is quite abundant. It was recognized as distinct from *I. paro* by Cuvier & Valenciennes, who received specimens at different times from the Hawaiian Islands. Other authors have regarded it as a synonym of *I. paro*, but a study of a large number of examples and these alcoholic specimens seems to justify retaining *I. pavoninus* for the Hawaiian form. A comparison of these specimens with a specimen of *I. mundicorpus* Gill, No. 824, in the L. S. Jr. Univ. Mus., from Cape St. Lucas, seems to prove them identical in structure and color. This being true, this species becomes one of those few shore fishes which are common to the Hawaiian Islands and the Pacific coast of North America.

Xyrichthys pavoninus Cuvier & Valenciennes, Hist. Nat. Poiss., XIV, 63, 1839, Hawaiian Islands.
Novacula (*Iniistius*) *paro* Steindachner. Denks. Ak. Wiss. Wien, LXX, 1900, 505 (Honolulu); not of Cuvier & Valencienner.
Iniistius mundicorpus Gill. Proc. Ac. Nat. Sci. Phila. 1862, 145, Cape St. Lucas (coll. by John Xantus); Jordan & Evermann, Fishes North and Mid. Amer., II, 1620, 1898.

145. Iniistius leucozonus Jenkins.

Two specimens, 4.5 and 5 inches in length, were taken by me.

Iniistius leucozonus Jenkins, Bull. U. S. Fish Comm. for 1899 (August 30, 1900), 54, fig. 11, Honolulu. (Type, No. 6137, L. S. Jr. Univ. Mus.; coll. O. P. Jenkins.

146. Iniistius niger (Steindachner).

Two specimens of this fish are in Dr. Wood's collection. My description was published soon after Dr. Steindachner's paper was printed and before his paper reached me.

Novacula (*Iniistius*) *nigra* Steindachner, Anzeiger fur Denks. Ak. Wiss. Wien, 1900, No. XVI, 176 (June 27, 1900), Honolulu; Steindachner, Denks. Ak. Wiss. Wien, LXX, 1900, 505, pl. 4, Honolulu.
Iniistius verater Jenkins, Bull. U. S. Fish. Comm. for 1899 (August 30, 1900), 55, fig. 12, Honolulu. (Type, No. 5990, L. S. Jr. Univ. Mus.; coll. Dr. Wood.)

147. Cymolutes lecluse (Quoy & Gaimard).

One specimen, 5 inches in length, was taken by the *Albatross* in 1896; and five, 5, 5.5, 5.5, 6, and 6.25 inches in length, by Jordan & Snyder.

Xyrichthys lecluse Quoy & Gaimard, Voy. de l'Uranie, Zool., 284, pl. 65, fig. 1, 1824, Hawaii.
Xyrichthys microplotus Cuvier & Valenciennes, Hist. Nat. Poiss., 14, 52, 1839, Owhyhee (Hawaii). (Coll. Quoy & Gaimard.)
Cymolutes leclusii Gunther, Cat., IV, 207, 1862 (Hawaiian Islands).

Family XXXVII. SCARIDÆ.

148. Calotomus cyclurus Jenkins, new species.

Head 3.1 in length; depth 2.5; eye 5.7 in head; snout 2.2; interorbital 4.2; D. IX, 11; A. I, 11; scales 2-24-5. Body somewhat elongate, compressed; dorsal outline rising in a gently sloping, nearly straight line to origin of dorsal, from this point descending in a nearly straight line to caudal peduncle; ventral outline about evenly convex; head length a little greater than depth; snout long, bluntly conic; mouth large, horizontal, about in axis of body; lips thin, double for about two-thirds the side, lower double only a short distance; lower jaw just included; interorbital slightly convex, considerably broader than eye; caudal peduncle, least height a little less than half head; dorsal spines flexible, rather high, nearly half head; soft dorsal 2 in head; anal similar, but less high, longest ray 2.3 in head; caudal rounded, no rays produced; ventrals 1.5 in head, reaching halfway to base of third soft anal ray; pectoral broad, its tip reaching to or slightly past vertical through tip of ventral, its length 1.4 in head (in the type there are 12 rays on right side and 9 on left, which is doubtless deformed); distal border convex (on left side); origins of dorsal, pectoral, and ventral about in same vertical; scales large, firm, those on breast not reduced; those at base of dorsal hardly forming sheath; no sheath at base of anal; large scales on upper and posterior portion of opercle; 1 row of about 7 scales below and behind eye; remainder of head naked; lateral line complete, portion to bend parallel to dorsal outline, straight portion beginning below base of fourth from last soft dorsal ray; 2 or 3 supernumerary scales with tubes extending from upper portion on row just above straight portion; tubes much branched, the branching covering well the exposed portion of the scale; teeth in anterior portion of jaws distinct, pointed, imbricated, in several series; 2 posterior canines; lateral teeth in upper jaw small, distinct, in a single series; lateral teeth in lower jaw large, in a single series; 2 conical teeth within outer teeth at symphysis of upper jaw, other small teeth within outer ones at sides of upper jaw.

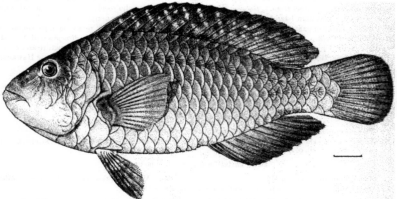

FIG. 24.—*Calotomus cyclurus* Jenkins, new species. Type.

Color in alcohol, head and body a uniform brown, with some indications of dots of lighter on some of the scales, and a wide margin of the posterior border of each scale shows paler than the base; dorsal fin brown, with faint traces of mottlings, no dark spot on anterior portion, nor darker margin; anal darker brown, with less evident mottlings and no darker margin; caudal much paler than body with upper and distal edges brown; ventral with indications of brown clouding; pectoral pale without marking, except that the base is dark brown; no marking shown on head, except that top of head and the isthmus are darker than its sides.

This description is based on a single specimen, 14 inches in length to tip of caudal, obtained at Honolulu by the *Albatross* in 1896. This species appears to be similar to *Callyodon waigiensis* (*Scarus spinidens* Cuvier & Valenciennes, Hist. Nat. Poiss., vol. 14), a small species, first described very imperfectly from the island of Waigiou. A fish from this island has been identified by Bleeker as Cuvier & Valenciennes's species, which he designates as *C. spinidens*, and of which he gives a full description and a figure. My specimen differs from Bleeker's description in the dorsal outline, in having a greater depth, a much smaller eye, and longer snout, in much longer ventral, in not having scales on the lower limb of the opercle, and in having the base of the anal dark. Dr. Bleeker had many (72) specimens and found *C. spinidens* of a limited range in distribution. (Type, No. 50849, U. S. N. M., Honolulu; coll. *Albatross*, 1896.)

149. Calotomus irradians Jenkins.

One specimen (field No. 306, 17 inches in length) of this beautiful fish was obtained by me; and one (field No. 1298, 17 inches in length) by the *Albatross* in 1896.   This species does not appear to be common, but is highly prized by the native fishermen for virtues which it is supposed to possess.

Calotomus irradians Jenkins, Bull. U. S. Fish Comm. for 1899   August 30, 1900, 58, fig. 15, Honolulu.   Type, No. 12142, L. s. Jr. Univ. Mus.; coll. O. P. Jenkins.

150. Calotomus sandvicensis Cuvier & Valenciennes.

Color in life, dusky brown, with dark mottlings; base of pectoral black; chin light brown; no other distinct markings.   This is a dull-colored fish not recognized by native fishermen as different from at least 2 or 3 other distinct species.   Cuvier & Valenciennes's description of *C. sandvicensis*, based on a specimen from the Sandwich Islands in Quoy & Gaimard's collection, is very meager, and Guichenot's redescription is not more complete, and this identification may prove incorrect.

Very common at Honolulu.   Thirteen specimens, 5 to 13 inches in length, were obtained by me.

Callyodon sandvicensis Cuvier & Valenciennes, Hist. Nat. Poiss XIV, 295, 1839, Sandwich Islands; Guichenot, Cat. Scarides, 62, 1865 Cuvier & Valenciennes's types.

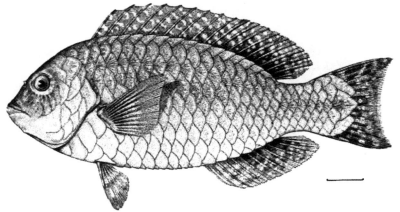

FIG. 25.—*Calotomus snyderi* Jenkins, new species   Type.

151. Calotomus snyderi Jenkins, new species.

Head 3.2 in length; depth 2.6; eye 5.6 in head; snout 2.8; preorbital 4.7; interorbital 4; D. IX, 10; A. III, 10; P. 12; scales 2-25-6.   Body short, deep, and much compressed; dorsal outline rather straight from tip of snout to nape, from which point it is gently convex to origin of dorsal, thence in a long, low curve to base of caudal peduncle; ventral outline rather evenly convex; head short, as deep as long; snout bluntly conic, lower jaw slightly included; mouth small, in axis of body; teeth in front of each jaw free, convex, incisor-like, in 2 or 3 rows; upper jaw with 2 moderately strong, recurved canines, inside and posterior to which is a row of close-set smaller teeth; side of lower jaw with overlapping series of rounded incisor-like teeth; preorbital oblique, moderately deep; eye small, high up; inter-orbital broad, low, convex; caudal peduncle 2 in head; scales large, thin, adherent, the free edges membranous; 4 scales on median line in front of dorsal; cheek with a single row of 4 scales; opercle with 2 rows of large scales, 3 scales on lower limb; lateral line complete, decurved under base of last dorsal ray, where there is usually one or more supernumerary tubes; tubes of lateral line numerously and widely branched, the branches varying from 4 or 5 to 12 or more; dorsal spines soft and flexible, the longest about equaling snout; soft dorsal somewhat elevated, the longest rays equaling distance from tip of snout to pupil; anal similar to soft dorsal; caudal somewhat lunate, the upper lobe the longer, about 1.8 in head; ventral short, reaching barely halfway to origin of anal; pectoral broad, reaching past tips of ventrals, its length 1.3 in head.

Color in alcohol, dirty yellowish-brown on head and body, marbled with light and darker; side above lateral line with a series of about 5 roundish white spots as large as pupil, and numerous smaller, irregular, less distinct white spots; side below lateral line with about 10 or 12 large, rounded, white spots and numerous small white specks and irregular markings, these especially distinct in pectoral region; head with similar white specks and markings; dorsal fin brown, with irregular paler spots; membrane between first and second dorsal spines black; soft dorsal with a large brownish-black spot at base of last 5 or 6 rays; anal similar to soft dorsal, blotches not so distinct; a series of black blotches at bases of rays, the one on last ray larger than others, covering base of last membrane; ventrals brownish, dusky at tip, a paler interspace; pectoral dusky, dark at base and in axil, pale on tip.

The only specimen known is the type, No. 50850, U. S. N. M. (field No. 1369), 10.5 inches long, obtained at Honolulu by me in 1889.

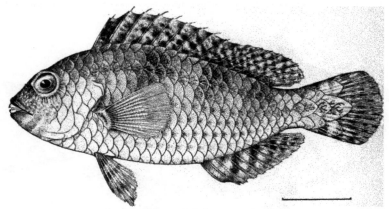

FIG. 26.—Scaridea zonarcha Jenkins, new species.   Type.

SCARIDEA Jenkins, new genus.

Scaridea Jenkins, new genus of *Scarida* (*zonarcha*).

Jaws subequal, the lower just included; dorsal spines pungent; gill-membranes broadly joined to the isthmus, not forming a fold; lateral line continuous; 1 row of scales on cheek; teeth white, distinct in anterior portion of each jaw, in more than 1 series, irregularly imbricated; lateral teeth in upper jaw small, in a single series, outer extremities free, the bases coalesced; lateral teeth in lower jaw large, in a single distinct series, crowded together at base; posterior canines present. This genus is related to *Calotomus* in the character of the teeth, but differs from that genus in having stiff, pungent spines, in which character it agrees with *S. psrisoma*. From *Scarichthys* Bleeker it differs in having pungent spines, in having the upper lip double for only a portion of its length, and in the distinct teeth. From *Callyodontichthys* Bleeker it is distinguished by the included lower jaw and the distinct teeth in upper jaw. Only 2 species known.

152. Scaridea zonarcha Jenkins, new species.

Head 3 in length; depth 2.75; eye 3.7 in head; snout 2.6; mandible 3.7; interorbital 5; preorbital 5.6; D. IX, 9; A. II, 9; scales 2-24-5, 19 with tubes before the bend. Body compressed, the dorsal outline more convex than the ventral, highest portion of dorsal outline about at third spine; head nearly as deep as long; mouth horizontal, about on line of axis of body; lips thin, the upper double for about

. half its length, the lower double only very short distance; lips covering base of teeth; eye moderate, its lower border above axis of body; interorbital somewhat less than eye, slightly concave (in alcohol); caudal peduncle 2.6 in head; dorsal spines pungent, about equal, the first being the shortest, the longest about 2.1 in head; soft dorsal slightly higher; anal similar to dorsal, but a little less high; caudal truncate, or slightly rounded; pectoral reaching 0.7 of distance to vent, its length 1.4 in head; pectoral reaching slightly beyond tips of ventrals, its length 1.6 in head; origins of dorsal, pectoral, and ventral about in same vertical line; scales large, firm, those on nape and breast not reduced; 3 scales in front of origin of dorsal; one row of 2 or 3 scales on cheek; large scales on upper posterior portion of opercle, 1 scale showing on lower limb; rest of head naked; sheath of scales at base of dorsal; none at base of anal; lateral line complete; 19 scales with tubes to the bend which occurs just below last ray of dorsal; the tubes on the portion to the bend are very much branched; jaws subequal, lower slightly included; teeth white, distinct, on anterior portions of both jaws in more than 1 series, irregularly imbricated; lateral teeth in upper jaw small, in a single series, outer extremities free, bases coalesced; lateral teeth in lower jaw large, in single series, distinct, crowded together at base; 2 developed posterior canines on one side, 1 on the other in the type, present but small in the small cotypes.

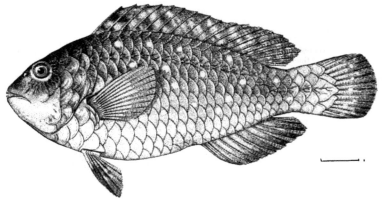

Fig. 27.—*Scaridea balia* Jenkins, new species. Type.

Color in alcohol, ground color of body and fins, except pectoral, a light brown on which are scattered numerous darker reddish-brown spots of indistinct outline about size of pupil; on the body the spots frequently correspond with the scales and show within their area small light points; on the vertical fins the spots are more distinct; on dorsal irregularly, 3 or 4 on each membrane; a conspicuous black spot on membrane between first and second spines; on anal arranged to show 4 bands running obliquely forward and downward across rays; distal half of caudal dark; ventral crossed by 3 indistinct dusky bands; pectoral transparent, base dark brown; head uniform brown.

Type, No. 50851, U. S. N. M. (field No. 2095), 5.75 inches long, obtained at Honolulu in 1889 (coll. O. P. Jenkins).

Cotypes, No. 7850, L. S. Jr. Univ. Mus. (field No. 2096), 4.5 inches long, and No. 2755, U. S. F. C. (field No. 2097), 4 inches long; collected at same time and place by O. P. Jenkins.

153. Scaridea balia Jenkins, new species.

Head 3.25 in length; depth 2.4; eye 5 in head; snout 2.9; preorbital 4; interorbital 4.3; scales 2-24-5; D. ix, 10; A. ii, 9; P. 12. Body short, deep, and compressed; dorsal outline regularly and evenly arched from tip of snout to caudal peduncle; somewhat straighter from tip of snout to origin of dorsal fin, there being no angle at last mentioned point; ventral outline evenly convex; head short and deep; snout short; mouth moderate, about horizontal, in line with axis of body, the gape reaching past vertical from nostril; lower jaw included; teeth in anterior portion of each jaw distinct, imbricated in 2 irregular series; upper jaw with 2 posterior canines directed backward; lateral teeth of upper jaw distinct, small, and in a single series; each side of lower jaw with about 5 distinct, blunt, incisor-like teeth in a single series; no teeth inside of front series of upper jaw; preorbital rather narrow; interorbital rather narrow, little convex, somewhat concave anteriorly; dorsal spines stiff and pungent, length of longest about equal to snout; soft dorsal somewhat elevated, the longest rays 2.1 in head; anal similar to soft dorsal, slightly lower; caudal rounded, 1.6 in head; ventrals short, their tips reaching scarcely half way to origin of anal; pectoral short, free edge rounded, its length a little greater than that of ventrals; scales large; lateral line continuous from upper end of gill-opening to posterior end of dorsal fin, where it curves downward 2 rows and continues to base of caudal fin; tubes of lateral line numerously and widely branched, the branches 4 to 6 or 7 in number; a few additional tubes at the bend of lateral line; 1 series of about 4 scales on cheek between which and eye are several long tubes; opercle with a series of large scales on basal portion; 4 scales on median line in front of dorsal fin.

Color in spirits, dirty rusty brown, paler below; upper part of side above lateral line with a series of about 6 roundish whitish spots larger than pupil; a similar series of about 4 spots on first row of scales below lateral line; lower part of side with 2 or more similar spots; side of head and body with a few scattered, similar, less distinct, whitish spots; dorsal fin indistinctly mottled with light and brownish; membrane between first and second spines blackish at center; anal and caudal rather uniformly plain pale brownish, without distinct markings; pectoral and ventrals lighter brownish; base of pectoral darker brown.

This species differs from *Scarida sonarcha* in greater depth of body, the much greater distance between tips of ventrals and origin of anal, the greater distinctness of the white spots on body, the less distinct mottling of the dorsal fin, and the entire absence of mottlings on anal and caudal.

This description is based on a single specimen 10 inches in length, type, No. 50852, U. S. N. M. (field No. 1985), obtained by the *Albatross* at Honolulu in 1896.

154. Scarus bennetti Cuvier & Valenciennes.

One specimen (field No. 2081) is in Dr. Wood's collection. It is 5.25 inches in length.

Scarus bennetti Cuvier & Valenciennes, Hist. Nat. Poiss., XIV, 270, 1839, Sandwich Islands.

155. Scarus brunneus Jenkins.

This species is fairly common at Honolulu. Eight examples, 5 to 9 inches in length, were taken by me; one, 8.25 inches, by Dr. Wood; and one, 7.25 inches, by the *Albatross* in 1896.

Scarus brunneus Jenkins, Bull. U. S. Fish Comm. for 1899 (August 30, 1900), 70, fig. 16, Honolulu (Type, No. 6139, L. S. Jr. Univ. Mus.; coll. O. P. Jenkins.)

156. Scarus miniatus Jenkins.

Two specimens of this large species, 5.5 and 17 inches in length, were taken by me at Honolulu; and two, 6.75 and 17.5 inches, by the *Albatross* in 1896. It does not seem to be common, but is highly esteemed by the natives as food, a high price being asked for it in the market.

Scarus miniatus Jenkins, Bull. U. S. Fish Comm. for 1899 (August 30, 1900), 62, fig. 20, Honolulu. (Type, No. 12144, L. S. Jr. Univ. Mus.; coll. O. P. Jenkins.)

157. Scarus ahula Jenkins. "Uhula"; "Pauuhumuhu."

Three examples of this species, 6, 7.5, and 8.75 inches in length, were obtained by me; and two, 4.6 and 9.5 inches, are in Dr. Wood's collection. This species does not seem to be common at Honolulu. It is not distinguished from *S. brunneus* nor from *S. paluca* by the native fishermen.

Scarus ahula Jenkins, Bull. U. S. Fish Comm. for 1899 (August 30, 1900), 64, fig. 19, Honolulu. (Type, No. 6142, L. S. Jr. Univ. Mus.; coll. O. P. Jenkins.)

158. Scarus perspicillatus Steindachner.

Color in life, each scale on anterior part of body with many blue dots, on posterior part, each scale with a vertical curved line of violet on its anterior part, the posterior part of each scale blue; head with bright blue band on violet ground; one band across forehead passing down in front of eye and bending around under it and backward a short distance; 2 short bands radiating from posterior border of eye; a band including a quadrilateral area across snout; a band or area on chin; irregular bands and dots on lower side of head and throat; opercular flap and region in front of it green, and with many green dots; dorsal pink, with outer margin blue, and a blue longitudinal bar on middle portion of posterior half; a bright blue line on body a short distance below base of dorsal parallel to it; anal colored similarly to dorsal, caudal blue; pectoral blue, upper border bright blue, lower portion dark blue.

This large scaroid is one of the most beautifully colored fishes seen at Honolulu. One specimen, 19 inches in length, was taken by me; and one, 17.5 inches, by the *Albatross* in 1896. It does not seem to be common, and brings a high price in the market.

Scarus (Scarus) perspicillatus Steindachner, Denks. Ak. Wiss. Wien. XLI. 16, Taf. IV, fig. 1, 1879, Sandwich Islands; Smith & Swain, Proc. U. S. Nat. Mus. 1882, 134 (Johnston Island).

159. Scarus paluca Jenkins.

One specimen, 7 inches in length, was obtained by me at Honolulu.

Scarus paluca Jenkins, Bull. U. S. Fish Comm. for 1899 (August 30, 1900), 60, fig. 15, Honolulu. (Type, No. 6141, L. S. Jr. Univ. Mus., field No. 297.; coll. O. P. Jenkins.)

160. Scarus gilberti Jenkins.

Five examples of this species, from 8 to 14 inches in length, were obtained by me; one by the *Albatross* in 1896; and two, 10 and 11.5 inches, by the *Albatross* in 1891. Fairly common at Honolulu and, like all scaroids, highly esteemed as food by the natives.

Scarus gilberti Jenkins, Bull. U. S. Fish Comm. for 1899 (August 30, 1900), 59, fig. 17, Honolulu. (Type, No. 6140, L. S. Jr. Univ. Mus.; coll. O. P. Jenkins.)

161. Pseudoscarus jordani Jenkins.

One specimen of this species, 35 inches in length, was obtained. This is the largest and one of the most beautifully colored of the scaroids that I have yet seen in the Hawaiian Islands. It is not common, but is highly esteemed by the natives as food, as its high price in the market shows.

Pseudoscarus jordani Jenkins, Bull. U. S. Fish Comm. for 1899 (August 30, 1900), 63, fig. 21, Honolulu. (Type, No. 12143, L. S. Jr. Univ. Mus.; coll. O. P. Jenkins.)

Family XXXVIII. CHÆTODONTIDÆ.

162. Forcipiger longirostris (Broussonet).

Color in life (field No. 284, 6.4 inches in length), upper part of snout, face, upper part of head and back to origin of dorsal, light brown; under side of lower jaw and lower side of head and as far back as origin of pectoral, light, nearly white; dorsal fin, side of body, caudal peduncle, base of caudal, and anal fin, yellow, the color growing brighter from above downward toward anal fin, where it is a very bright yellow; outer border of posterior portion of soft dorsal white; posterior two-thirds of caudal white; a jet-black spot on outer and posterior angle of anal fin; ventrals yellow; pectoral rays dusky, membranes transparent.

Fairly common at Honolulu. Four examples, 5.2, 5.7, 5.75, and 6.4 inches in length, were taken.

Chætodon longirostris Broussonet, Desc. Ichth., 1, 23, pl. 7, 1782 (Society and Sandwich islands).

Chelmo longirostris Gunther, Cat., II, 38, 1860 (Amboyna). Günther, Fische der Sudsee, I, 48, 1874 (Sandwich, Society, Paumotu, Friendly, and Kingsmill islands).

Chelmon (Forcipiger) longirostris, Steindachner, Denks. Ak. Wiss. Wien, LXX, 1900, 489 (Honolulu).

Forcipiger longirostris, Fowler, Proc. Ac. Nat. Sci. Phila. 1900, 512 (Sandwich Islands).

163. Chætodon setifer Bloch.

Life colors, very complex and conspicuous; general ground color of body white; on upper anterior portion of body 5 narrow, dusky, nearly parallel lines, running from head region upward and somewhat backward on to the dorsal fin, the most posterior of these lines from angle of opercle upward and backward to margin of dorsal; running at right angles to this posterior one and joining it are 6 similar narrow, dusky lines, which extend downward and backward, curving slightly, ending somewhat short of base of anal; in the angle formed by upper of this last group and last of first group of dusky lines, the ground color is brownish-yellow, which becomes a bright yellow toward and on the soft dorsal; in the angle of this area are 2 parallel bands of light yellow, parallel with the borders of the angle mentioned; soft dorsal bright yellow with a narrow black margin, the produced ray yellow, a large oval black spot on upper anterior portion; caudal peduncle and fin bright yellow, its outer margin ornamented with a band of 4 successive colors from within outward, light yellow, brown, orange, and white; anal fin bright yellow, with a narrow outer border of 3 colors, inner black, middle white, outer yellow; ventrals white; pectoral transparent; head white; ocular band wide as eye through eye, broader below and black, narrower above and dusky; 4 narrow orange lines across forehead from one ocular band to the other, 3 from near margin of eye, 1 above; above these on forehead a small circle of orange.

Three specimens of this beautiful fish, 6.5, 7, and 7 inches in length, were taken at Honolulu, where it is fairly common.

Chætodon setifer Bloch, Ichth., VI, taf. 426, fig. 1; Günther, Fische der Südsee, I, 36, taf. XXVI, fig. B, 1873 (Sandwich Islands); Fowler, Proc. Ac. Nat. Sci. Phila. 1900, 512 (Hawaiian Islands).

164. Chætodon multicinctus Garrett.

Color in life, body white, the white extending on dorsal and anal fins as far as the black line running lengthwise on these fins; each scale with a brown spot; the black line on dorsal and anal varying in distance from outer margins of fins, but at about one-third height of fin from margin on soft dorsal and anal; a narrow white line just outside the black line; outside of these lines the dorsal fin is yellow and the anal is white; dark line on anal fin extending on the belly as far as base of ventral as a yellow line; side of body crossed by 5 crossbands of light brown about as wide as eye; an ocular band brown below eye, black above, as wide as pupil from lower margin of opercle ending above eye in an acute angle; just above this a black spot in front of spinous dorsal; yellow at borders of upper portion of opercle and preopercle; a dark ring including brown spots around the base of caudal; a black crescent extending across caudal fin at its middle; ventrals white; pectoral transparent.

This is a very delicately colored fish. One example (field No. 318, 3.75 inches in length) was taken in 1889; and one, 3.9 inches, by the *Albatross* in 1896. Mr. Garrett states that he had found only 2 examples. These are all that have thus far been seen.

Chætodon multicinctus Garrett, Proc. Cal. Ac. Sci., III, 1863, 65 (Sandwich Islands); Günther, Fische der Südsee, II, 44, taf. XXXIV, fig. B, 1874 (Sandwich Islands). (Günther's colored plate is from Garrett's drawing.)

165. Chætodon ornatissimus Solander.

Color in life (field No. 321, 6.5 inches in length), sides of body white, the white extending on dorsal to a black line running lengthwise of fin near outer margin; breast and belly yellow; 6 bands of orange on side of body running obliquely backward, the upper 3 slightly convex toward dorsal outline, the lower 3 slightly convex toward ventral outline; head yellow, with black bands, the most posterior being vertical on opercle and joining the upper orange band; the next anterior vertical on side of head just behind eye, above joining the black band on dorsal fin; the next (ocular band) which is as narrow as pupil at eye, broadens above and below eye; a black band around mouth; upper lip yellow; chin black; spinous and soft dorsal fin yellow to the black line yellow, posterior margin of soft dorsal with a narrow black line; base of caudal white, a black band across its middle, exterior to this a yellow band, then a black crescent-shaped band, posterior margin white; base of anal white, inner portion dark brown, outer portion yellow, margin black, spinous portion yellow; ventrals bright yellow.

Two examples of this brilliantly colored fish, 6.5 and 7.25 inches in length, were obtained by me in 1889; and two, 6 and 7 inches in length, by the *Albatross* in 1896. It is not uncommon at Honolulu.

Chætodon ornatissimus Solander, in Cuvier & Valenciennes, Hist. Nat. Poiss., VII, 22, 1831; Tahiti, Günther, Cat., II, 15, 1860 (Sandwich Islands, Amboyna); Günther, Fische der Südsee, II, 38, taf. XXX, fig. B, 1874 (Sandwich Islands); Fowler, Proc. Ac. Nat. Sci. Phila. 1900, 513 (Sandwich Islands).

166. Chætodon miliaris Quoy & Gaimard.

Color and description in paper referred to below. In describing this as a species distinct from *C. miliaris* Quoy & Gaimard, I was led into error by the description and figure given by Günther in the Fische der Südsee, and by Bleeker in the Atlas. An examination of the original description of *C. miliaris* Quoy & Gaimard, which was based on a specimen from the Hawaiian Islands, shows that my examples are that species, and that the species referred to and figured by Günther in Fische der Südsee and by Bleeker is not the same. In Günther's figure and description and in that of Bleeker a blue spot is assigned to each scale on the sides of the body as far as the belly. In Quoy & Gaimard's figure and in each of a large number of fresh and preserved examples the spots are on the upper parts alone, mostly anterior, and are in vertical rows on about every third row of scales; in some, intercalary, faint vertical rows appear between the dense black conspicuous ones, making a characteristic pattern.

Common at Honolulu. I obtained nine examples, 3 to 4.8 inches in length, and examined a large number of others.

Chætodon miliaris Quoy & Gaimard, Voy. de l'Uranie, Zoöl., 380, pl. 62, fig. 6, 1824 (Sandwich Islands); Steindachner, Denks. Ak. Wiss. Wien, LXX, 1900, 489 (Honolulu, Laysan; Fowler, Proc. Ac. Nat. Sci. Phila. 1900, 512 (Sandwich Islands; not of Günther and not of Bleeker.

Chætodon mantelliger Jenkins, Bull. U. S. Fish Comm. for 1899 (June 8, 1901), 394, fig. 7, Honolulu. Type, No. 49899, U. S. N. M.; coll. O. P. Jenkins.

167. Chætodon fremblii Bennett.

Life color, body bright yellow with 7 longitudinal light blue bands on side directed slightly upward, some of these extending on posterior portion of dorsal fin, second band interrupted; a bluish-black spot in front of dorsal; a black area on caudal peduncle extending up on posterior part of soft dorsal, this area with blue border anteriorly; dorsal bright yellow, with two longitudinal bands; caudal peduncle black, base of caudal fin white, posterior to this a yellow crossband, posterior to this (the margin) white.

Fairly common at Honolulu, where 8 examples, 3.6 to 5.75 inches in length, were obtained.

Chætodon fremblii Bennett, Zoöl. Jour., IV, 42, 1828 (Sandwich Islands); Günther, Fische der Südsee, II, 39, taf. XXIX, fig. B, 1874 (Sandwich Islands); Steindachner, Denks. Ak. Wiss. Wien, LXX, 1900, 488 (Laysan).

168. Chætodon lunula (Lacépède).

Color in life (field No. 149, 5 inches long), prevailing color yellow; black ocular band broader than eye extending over head, including both eyes and reaching down on each side as far as lower edge of preopercle; front margin of this band bordered by a whitish line; immediately posterior to ocular band, not extending so far down as it, is a broad white band; space in front of ocular band to mouth yellow; tips of jaws red; a large black area on back including base of first 5 dorsal spines; a broad black band, lower portion as broad as length of caudal fin, extending from humeral region upward and backward to about origin of seventh and eighth dorsal spines, this black band bordered anteriorly and posteriorly by a bright orange yellow band; posterior and below this on the side and belly and breast, shading to olivaceous above, to bright yellow below; side with longitudinal narrow bands of orange, made by rows of orange dots, 1 on each scale; breast with orange dots; dorsal fin mostly yellow, with a narrow brown border, a yellow band running from body just in front of caudal peduncle upward on to the dorsal and along it; a long black area on middle of posterior portion of soft dorsal; caudal peduncle black, the black extending on base of caudal posterior to this, caudal yellow for nearly half its area; posterior to this a dusky crossband, then yellow, then a dusky band and most posteriorly white (the border); anal base and inner two-thirds yellow, outside of this a dusky longitudinal narrow band, then yellow, next this the brown border of the fin; ventrals bright yellow; pectoral transparent. This species varies much in the degrees of development of this pattern with age.

Another specimen, young, 1.25 inches in length, gives the following coloration: Snout red; ocular band black, next to this a broad white band followed by a broad black one; body yellow, darker above, brighter below, black spot on caudal peduncle covering whole of peduncle; white crossband on base of caudal fin, posterior to this a narrow black line, behind this dusky; black spot on soft dorsal surrounded by bright yellow line which behind becomes white.

Quite common in the market. Nine specimens, 1.3 to 6.25 inches in length, were obtained.

Pomacentrus lunula Lacépède, Hist. Nat. Poiss., IV, 507, 510, 513, 1802.

Chætodon lunula Cuvier & Valenciennes, Hist. Nat. Poiss., VII, 59, pl. 173, 1831; Günther, Fische der Südsee, II, 42, taf. XXXIII, 1874 (Tahiti, Sandwich Islands, Society Islands; Steindachner, Denks. Ak. Wiss. Wien, LXX, 1900, 489 (Honolulu).

Chætodon tau-nigrum, Fowler, Proc. Ac. Nat. sci. Phila. 1900, 513 (Sandwich Islands.

169. Chætodon sphenospilus Jenkins.

Twelve specimens were obtained. In a large number of fresh examples seen by me since the description was published, each one shows the wedge-shaped dusky area projecting from the spot on the side; in some the point of the wedge-shaped area extends nearly to the ventral outline.

Chætodon sphenospilus Jenkins, Bull. U. S. Fish Comm. for 1899 (June 8, 1901), 395, fig. 8, Honolulu. (Type, No. 49765, U. S. N. M.; coll. O. P. Jenkins.

170. Chætodon trifasciatus Park.

Color in life (field No. 183, 5.75 inches in length), general color of sides of body orange, with numerous longitudinal lines of dark brown, those above the middle line of body slightly convex toward dorsal line, those below slightly convex toward ventral outline; ocular band black, narrower than eye, with narrow bright yellow borders, the band continuous around the head above and below; in front of ocular band is a bright yellow band, continuous above and below; tips of jaws black; behind ocular band is a bright yellow band, adjoining this a narrow black band; the opercle and preopercle behind this are yellow; an orange patch on middle line of body just in front of origin of spinous dorsal; breast yellow; anterior portion of spinous dorsal orange with light green border; beginning on the body parallel with base of soft dorsal fin and including the whole of the fin are 7 bands of color; first (innermost) yellow; next black, wedge-shaped, broadest end on caudal peduncle, narrow end on posterior portion of spinous dorsal; third band bright yellow, fourth yellow, fifth dusky; sixth yellowish-green, on posterior portion round dots of orange; seventh reddish-brown (the outer); caudal base (except black of above-mentioned band) light with violet tinge, exterior to this a series of crossbands; the first about the middle, a narrow orange; next a broader black; next a narrow bright yellow; and the next, the outermost, a broader white; just above anal on body parallel to base a yellow band, exterior to this and on base of fin also a black band, then (on soft portion) next a narrow orange, then a broad, brownish-red band; last (the extreme border of soft anal) a narrow bright yellow; area covering the first 2 anal spines a bright reddish-orange; ventrals bright yellow; base of pectoral bright orange, fin transparent.

Two specimens of this most beautifully marked species, 5.6 and 5.75 inches in length, were taken by me in 1889; and one, 4.8 inches long, by the *Albatross* in 1896 at Honolulu, where it does not seem to be common. This seems to be the first record of this species from the Hawaiian Islands.

Chætodon trifasciatus Mungo Park, Trans. Linn. Soc., iii, 34, 1797, shores of Sumatra. Lacépède, Hist. Nat. Poiss., iv, 462, 494, 1802.

Chætodon vittatus, Günther, Cat., ii, 23, 1860. Günther, Fische der Südsee, ii, 41, 1874 (east coast of Africa to Paumotu group).

171. Chætodon quadrimaculatus Gray.

Color in life (field No. 293), the upper half of side of body dark brown, almost black; lower half orange, with each scale with a brown spot; 2 white areas on the upper brown portion; ocular band wide as eye, dark brown above eye, orange below, bordered on both sides through its whole length by white borders; before and behind ocular band on head orange; tip of snout brown; base of dorsal dark brown, the brown extending about half its height on soft dorsal; a white longitudinal band above this on soft dorsal, extending a distance on to the spinous dorsal; this band is bordered above by a narrow line of dark brown, above this the fin is orange, with a very narrow outer margin of dark brown, outer portion of spinous dorsal orange; caudal peduncle orange; dark-brown crossband at base of caudal fin, the fin a bright orange, with narrow posterior border of white; inner portion of anal bright orange, a curved white band, convex outwardly, extending on it from base of first spine to near posterior outer angle of soft portion; bordering this band exteriorly a narrow brown band; exterior to this both spinous and soft portions are bright orange; ventrals bright orange; pectoral transparent.

This is a very common form about the reefs at Honolulu. Nineteen examples were obtained, 3 to 5 inches in length.

Chætodon quadrimaculatus Gray, Zool. Miscell., 33, 1831–42, Sandwich Islands. Günther Cat. ii 15, 1860, Sandwich Islands (type); Günther, Fische der Südsee, ii, 38, taf. xxx, fig. A, 1874, Sandwich Islands; Steindachner, Denks. Ak. Wiss. Wien, lxx, 1900, 489 (Honolulu, Laysan); Fowler, Proc. Ac. Nat. Sci. Phila. 1900, 512 (Sandwich Islands).

172. Microcanthus strigatus (Cuvier & Valenciennes).

Sides with 6 dark brown longitudinal bands, spaces between white to gray.

This species appears only rarely at Honolulu. Three examples were obtained, 3.3, 4.1, and 4.3 inches in length.

Chætodon strigatus Cuvier & Valenciennes, Hist. Nat. Poiss., vii, 25, pl. 170, 1831, Japan. Günther, Fische der Südsee, ii, 47, 1874 (Sandwich Islands).

173. Heniochus macrolepidotus (Linnæus).

Color in life (field No. 213, 4.5 inches in length), 2 broad, black crossbands on side of body, the first including above, the first dorsal spine to base of produced spine and below the origin of ventral to origin of anal; second black band including above, most of spinous dorsal behind the produced spine, descending downward and obliquely backward, covering from lower side of caudal peduncle to tip of lobe of caudal; space in front of first black band, including head, nape, and breast, white, except a black band on forehead from one eye to the other; space between the black bands white extending on anal; just behind second black band a white band extending on soft dorsal above, below spreading from base of caudal to lower tip; the space on the soft dorsal and caudal behind this white band, yellow; anal spines black; ventrals black; pectoral transparent.

I obtained one specimen of this fish 4.5 inches in length. It seems to be rare at Honolulu, and this seems to be the first record from the Hawaiian Islands.

Chætodon macrolepidotus Linnæus, Syst. Nat., ed. x, 274, 1758, Indies.

Heniochus macrolepidotus, Cuvier & Valenciennes, Hist. Nat. Poiss., vii, 93, 1831; Günther, Fische der Südsee, 48, taf. xxxvii, 1874.

Family XXXIX. ZANCLIDÆ.

174. Zanclus canescens (Linnæus). "Kihikihi."

Color in life (field No. 78), front part of body and posterior portion of head included in a broad black band, the front margin of which is just in front of eye, down to throat, the posterior margin from origin of produced rays of dorsal down a point nearly half way from ventral to origin of anal; on this area are some nearly vertical narrow light blue streaks, one running up from origin of ventral, one running from base of pectoral, a short one behind the eye; face light blue; an orange area on each side of snout surrounded by black border; chin black; a broad space behind this broad black band extending above and below on portions of the dorsal and anal fins; front part of this space is light blue, shading into a sulphur yellow below, extending on the produced dorsal as light blue and on the spinous anal as a light blue; just back of this space a black crossband running across dorsal above and anal below; just bordering this black crossband a narrow light blue, this followed by narrow bands of yellow, next light blue, then black; caudal peduncle and base of caudal sulphur yellow, bordered posteriorly by a narrow cross-line of light blue, remainder of caudal fin black, with light blue posterior margin; anterior and posterior parts of iris golden, upper and lower dark. This complex pattern is gradually developed, consequently the young show variations. In the adult it is quite constant.

Forty-five examples of this very brilliantly colored fish, from 3.2 to 6 inches in length, were obtained by me, and 2 by the *Albatross* in 1896 at Honolulu, where it is very common.

Chætodon canescens Linnæus, Syst. Nat., ed. x, 272, 1758, Indies, after Artedi, young.

Chætodon cornutus Linnæus, Syst. Nat., ed. x, 273, 1758, Indies, after Artedi, adult.

Zanclus cornutus Cuvier & Valenciennes, Hist. Nat. Poiss., vii, 102, pl. 177, 1831, Caroline, Sandwich Islands, Tongataboo, Vanicolo, Celebes; Günther Cat., ii, 493, 1860, Amboyna, Ceram, Sandwich Islands; Jordan & Evermann, Fishes North and Mid. Amer., ii 1687, 1898, Honolulu, Steindachner, Denks. Ak. Wiss. Wien, LXX, 489, 1900, (Honolulu); Fowler, Proc. Ac. Nat. Sci. Phila 1900, 513 (Sandwich Islands).

Family XL. TEUTHIDIDÆ.

175. Teuthis achilles (Shaw).

Color in life (No. 295, 10.2 inches in length), head, body and fins, except the lines and areas to be described below, a very dark brown varying to black; the other colors very bright and look as if put on with thick paint; a narrow, bright blue band under chin; opercular flap white; 2 narrow lines of color on body at base of dorsal, the upper orange red, lower bright blue; 2 similar lines at base of anal, the outer orange red, inner bright blue; a large oval area of orange red including and in front of caudal spine; anterior two-thirds of caudal fin orange-red, followed by narrow crossband of black, this followed by blue which shades into white on margin; upper and lower margins blue, with next to them, above and below, narrow line of black; iris blue; ventral black, blue area on anterior margin.

Fifteen examples of this species, 4.2 to 10.2 inches long, were obtained by me; and one, 4.2 inches long, by Dr. Wood.

Acanthurus achilles Shaw, Gen. Zool., iv, 383, 1803; Steindachner, Denks. Ak. Wiss. Wien, LXX, 1900, 493 (Honolulu).

Teuthis achilles Fowler, Proc. Ac. Nat. Sci. Phila. 1900 513 (Sandwich Islands).

176. Teuthis olivaceus (Bloch & Schneider).

Color in life (field No. 311), body dark brown, nearly black; fins black, with show of blue; horizontal orange bar about as wide as eye and long as head, extending from upper angle of opercular opening, straight backwards; this bar with a wide border of black; chin dark blue; posterior margin of caudal between upper and lower produced rays white; an orange line along base of dorsal, a similar one along base of anal. Eleven examples, 6.5 to 10.6 inches long, were obtained by me; and one, 9.5 inches long, by Jordan & Snyder.

Acanthurus olivaceus Bloch & Schneider, Syst. Ichthy., 213, 214, 1801, Tahiti; Günther, Cat., III, 326, 1861 (Otaheiti, Feejee Islands); ibid., Fische der Südsee, IV, 113, 1875; Steindachner, Denks. Ak. Wiss. Wien, LXX, 1900, 483 (Honolulu).
Acanthurus humeralis Cuvier & Valenciennes, Hist. Nat. Poiss., X, 231, 1835, Caroline and Society Islands.

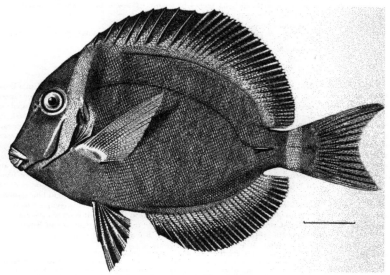

Fig. 28.—*Teuthis leucopareius* Jenkins, new species. Type.

177. Teuthis leucopareius Jenkins, new species.

Head 4 in length; depth 1.75; eye 3.25 in head; snout 1.25; interorbital 2.7; D. IV, 26; A. III, 24. Body deep, compressed, upper profile strongly convex, lower shallowly convex; head deep, compressed, upper profile steep to dorsal, lower profile nearly straight from chin to root of ventrals; jaws large, lower slightly produced; snout long, slightly concave in profile; mouth small, low; teeth broad, edges crenulate, compressed, forming a sharp cutting edge; margin of preopercle very oblique, blunt angle of which would fall below middle of eye; eye rather large, high, in posterior third of length of head; nostrils small, close together, anterior with small thin flap, circular, posterior a small slit; interorbital space broad, convex; dorsal spines strong, graduated to posterior, last 1.3 in head; first anal ray 1.25; anal spines graduated to third, longest 1.65; fourth anal ray 1.25; caudal broad, emarginate; pectoral pointed, 2.8 in body; ventral sharp-pointed, 4 in body; ventral spine 2 in head; caudal peduncle compressed, its depth 2; caudal spine moderately long, depressible in a groove, 4 in body.

Color in life of the type, a whitish band from base of first 2 dorsal spines downward behind eye, including opercle to its lower margin; also a transverse whitish band at base of caudal.

Color in alcohol, deep brown, vertical fins and ventrals darker; a light brown band running from base of first two dorsal spines, downward, behind the eye, over opercle, to its lower margin; a light brown transverse band on caudal peduncle.

The above description is based upon the type, No. 50712, U. S. N. M. (field No. 324), and 8 cotypes, all obtained by me at Honolulu.

178. Teuthis umbra Jenkins, new species.

Head 3.8 in length; depth 1.7; eye 3.75 in head; snout 1.2; interorbital 2.65; D. ix, 27; A. iii, 25. Body ovoid, greatest depth at pectoral region; head deep; upper profile nearly straight from tip of snout to interorbital space, then convex to origin of dorsal; jaws low, lower inferior; mouth small; interorbital width broad, convex; nostrils small, round, close together, anterior larger, with small, thin, fleshy flap; last dorsal spine 1.7 in head, anterior spines graduated to posterior; fifth dorsal ray 1.25; third anal spine longest, 2 in head; eighth anal ray 1.5; caudal rather broad, emarginate; pectoral broad, a trifle longer than head; ventrals sharply pointed, a trifle less than pectoral, or about equal to head; caudal peduncle compressed, 2 in head; caudal spine small, sharp, about 5 in head, depressible in a groove; scales small, finely ctenoid, very small on top of head, breast, and basal portions of vertical fins; lateral line irregular, arched at first, then sloping down to caudal spine, a good portion of it more or less straight, from below anterior portion of spinous dorsal to below middle of soft dorsal.

Color in alcohol, more or less uniform dark chocolate brown, outer portions of the fins blackish; pectoral pale olivaceous brown; ventrals blackish on outer portion.

The above description is based upon type, No. 50841, U. S. N. M. (field No. 05363), 7.5 inches long, obtained at Honolulu in 1901 by Jordan & Evermann. Only small specimens were secured by me.

179. Teuthis dussumieri (Cuvier & Valenciennes).

Color in life (field No. 140, 6.2 inches long), general color brown with very many narrow longitudinal wavy lines, olivaceous alternating with light lines; dorsal fin bright; bright light line running along its base, below which is a bright golden line; signs of indistinct longitudinal lines parallel with body on posterior part of fin; anal fin similar to dorsal but darker in color; ventrals dusky, with whitish markings; pectoral olivaceous; caudal dusky with darker spots; a ring of golden around eye.

Seven examples, 6.2 to 11.75 inches in length, were taken by me; five, 4.5 to 6.7 inches, by the *Albatross*; two, 6 and 6.25 inches, by Jordan & Snyder; and four, 4.43 to 6.86 inches, by Dr. Wood. This is a common fish at Honolulu.

Acanthurus dussumieri Cuvier & Valenciennes, Hist. Nat. Poiss., x, 201, 1835, Isle de France; Günther, Cat., iii, 335, 1861 (Mauritius), Günther, Fische der Südsee, iv, 112, pl. 72, 1873 Sandwich Islands Steindachner, Denks. Ak. Wiss. Wien, lxx, 1900, 493 (Honolulu).

180. Teuthis xanthopterus (Cuvier & Valenciennes).

A single specimen, 7.25 inches in length, was obtained.

Acanthurus xanthopterus Cuvier & Valenciennes, Hist. Nat. Poiss., x, 215, 1835, Seychelles.

181. Teuthis güntheri Jenkins, new species.

Head 4 in length; depth 1.8; eye 3.5 in head; snout 1.4; interorbital 2.7; D. ix, 26; A. iii, 24. Body ovoid, greatest depth at origin of anal; head deep, compressed, upper profile obliquely convex from tip of snout to spinous dorsal; eye rather large, high, in last third of head; nostrils small, close together, anterior the larger, rounded, with thin fleshy flap; mouth small, low, inferior; jaws blunt, slightly produced, lower inferior; interorbital space broad, elevated, convex; margin of preopercle forming an angle below anterior rim of orbit; dorsal spines slender, graduated to last, which is 1.25 in head; soft dorsal and anal not pointed behind; eighth dorsal ray 1.2 in head; third anal spine longest, 1.9; eighth anal ray 1.25; caudal long, emarginate; pectoral longer than head, 3.5 in body; ventrals sharp-pointed, 1.2 in head; ventral spine 1.9; caudal peduncle compressed, its least depth 2.2; caudal spine short, 1.5 in eye; scales small, crowded, ctenoid; very minute scales on basal portions of vertical fins; lateral line nearly concurrent with dorsal profile of back, straight from anterior dorsal spines to below middle of soft dorsal, then running down above edge of caudal spine to base of caudal.

Color in alcohol, dark brown, vertical fins darker; side plain or uniform brown, without any lines; soft dorsal and anal grayish posteriorly; both dorsals and anal with 4 broad, deep brown, longitudinal bands; caudal deep brown, apparently without spots, base of fin pale; pectoral brown on basal portion, marginal portion broadly yellowish-white.

Color in life (field No. 199), general color brown; an orange-yellow band along the back just below the base of dorsal, just above golden band a blue line; 4 golden longitudinal bands on dorsal fin, with an intercalary band which in some examples makes 5 bands; anal with 4 similar ones; pectoral yellow; yellow area through eye; yellow line over snout.

This description is based on the type, No. 50842, U. S. N. M. (field No. 199), 8.6 inches in length), and 11 cotypes ranging from 5.5 to 8 inches in length, all obtained by me at Honolulu.

Acanthurus blochi, Günther Fische der Südsee, IV. 109, LXIX, fig. B (copy of Garrett's drawing); (not of Cuvier & Valenciennes); Streets, Bull. U. S. Nat. Mus., No. 7, 68, 1877 (Honolulu).

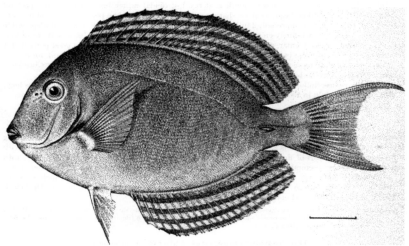

Fig. 29.—*Teuthis guntheri* Jenkins, new species. Type.

182. Teuthis matoides (Cuvier & Valenciennes).

One specimen obtained.

Acanthurus matoides Cuvier & Valenciennes, Hist. Nat. Poiss., x, 204, 1835, Oualan.
Acanthurus nigroris Cuvier & Valenciennes, Hist. Nat. Poiss., x, 208, 1835, Sandwich Islands.
Acanthurus annularis Cuvier & Valenciennes, Hist. Nat. Poiss., x, 209, 1835, Isle de France.
Teuthis annularis, Fowler, Proc. Ac. Nat. Sci. Phila. 1900, 513 (Sandwich Islands).

183. Teuthis atrimentatus Jordan & Evermann.

Color in life (field No. 141), dorsal fin with 4 distinct, longitudinal, olivaceous stripes; body dark brown, with many wavy longitudinal lines of blue; anal with 4 longitudinal olivaceous stripes; an inky black spot at posterior angle of dorsal fin (at base) and one at posterior angle of anal fin (at base); no white or golden line at base of dorsal; caudal dusky, with indistinct transverse olivaceous bars; ventrals dusky, with no white; pectoral olivaceous.

Three examples, 4.4 to 4.6 inches in length, were taken by me; two, 5.3 and 6.6 inches in length, are in Dr. Wood's collection, and sixteen, 3 to 9.5 inches long, were obtained by Jordan & Evermann.

Acanthurus lineolatus, Günther, Fische der Südsee, IV. 112, taf. LXXIII. fig. A, 1875 (Ratatea); Steindachner, Denks. Ak. Wiss. Wien, LXX, 1900, 493 (Honolulu); not of Cuvier & Valenciennes.

Teuthis atrimentatus Jordan & Evermann. Bull. U. S. Fish Comm. for 1902 (April 11, 1903), 198, Honolulu. (Type, No. 50678, U. S. N. M.; coll. Jordan & Evermann.)

184. Teuthis bipunctatus (Günther).

Color in life (field Nos. 74, 75, 4.4 and 5.5 inches in length), dark brown; face, throat and belly with many small inconspicuous spots of dark yellow.

Fairly common at Honolulu. Eleven examples, 4.4 to 7 inches in length, were taken by me, and two by Dr. Wood.

Acanthurus bipunctatus Günther, Cat., III, 331, 1861. Sea of China, Fiji Islands ; Steindachner, Denks. Ak. Wiss. Wien, LXX, 494, 1900 (Honolulu).

Acanthurus nigros, Günther, Cat., III, 332, 1861 (New Hebrides) ; ibid., Fische der Sudsee, IV, 110, 1875 (Sandwich Islands, New Hebrides, Pelew Islands, Tahiti).

185. Teuthis sandvicensis (Streets). *"Manini."*

Color in life, general ground color drab, with yellowish cloudings; belly, chin, throat, and body along base of anal white, or nearly so; side with 5 black vertical bars as wide as pupil, first on head through pupil extending slightly forward to branchiostegals; second from front of dorsal to axil of pectoral, continued by a bar originating on base of pectoral and extending obliquely backward to belly; third from about sixth dorsal spine to a point about midway between anus and anal fin; fourth from about the first soft dorsal ray to first or second soft anal ray; fifth from seventh soft dorsal ray to ninth soft anal ray; a black spot on top of caudal peduncle extending to lateral line on each side; vertical fins dusky, anal with a white margin; pectoral fins colorless; ventral fins white with dusky under surface.

One of the most common fishes about the reefs; 8 adults and a large number of young were taken. An examination of a large number of specimens of different ages shows a constancy in coloration.

Acanthurus triostegus, Günther (in part), Cat., III, 327, 1861; ibid., Fische der Sudsee, IV, 108, 1875 (Sandwich Islands; Steindachner, Denks. Ak. Wiss. Wien, LXX, 493, 1900 (Honolulu and Laysan); not of Linnæus.

Teuthis triostegus, Fowler, Proc. Ac. Nat. Sci. Phila. 1900, 513 (Hawaiian Islands).

Acanthurus triostegus sandvicensis Streets Bull. U. S. Nat. Mus., No. 7, 67, 1877 (Hawaiian Islands). (Type, No. 15898, U. S. N. M.)

186. Teuthis guttatus (Schneider).

Color in life (field No. 328), general color brown; a light band from short distance in front of dorsal, downward over opercle to its lower margin; another from about fifth dorsal spine to space between anus and anal fin; another narrower, not so light nor so distinct, from third soft ray of dorsal nearly to third soft ray of anal; dorsal fin, anal fin, posterior half of body including caudal peduncle, covered with white spots; ground color of dorsal and anal darker than that of body; a broad yellow band across base of caudal; posterior portion of caudal dark, nearly black; ventrals bright yellow with narrow dark margin; pectoral brown; chin, throat, and breast as far back as base of ventrals white; iris brown.

I obtained three examples of this species, 9, 10.75, and 11.5 inches in length; and two, 8 and 10.3 inches long, were taken by Jordan & Snyder. It seems to be rare at Honolulu.

Acanthurus guttatus Schneider, Syst. Ichth., 215, 1801. Tahiti: Günther, Cat., III, 329, 1861. Günther, Fische der Südsee, IV, 109, taf. LXIX, fig. A, 1875 (Sandwich Islands).

Harpurus guttatus, Forster, Descrip. Animal , Ed. Licht., p. 218, 1844.

Teuthis guttatus, Fowler, Proc. Ac. Nat. Sci. Phila. 1900, 513 (Hawaiian Islands).

187. Zebrasoma hypselopterum (Bleeker). *"Kihikihi."*

Color in life (field No. 189, 3.1 inches in length), 6 chestnut-brown bands across body, running somewhat obliquely backwards, the first 4 edged with light blue, between the bands and in front of the first band yellow, which color anteriorly is brighter; caudal black; second and third yellow bands extending on dorsal and below on anal, following direction of rays on fins; remaining parts of anal and caudal chestnut brown; ventrals yellow, the color on these fins being a prolongation of the yellow of the first yellow band; pectoral anteriorly orange-yellow, posteriorly transparent; lips reddish.

Seven specimens of this beautiful species, 3.1 to 9 inches in length, were obtained. It seems to be fairly common at Honolulu.

Acanthurus hypselopterus Bleeker, Nat. Tijds. Ned. Ind., VI, 313, 1854. Floris: Günther, Cat., III, 344, 1861; ibid., Fische der Sudsee, IV, 117, 1875 Feejee Islands; Steindachner, Denks. Ak. Wiss. Wien, LXX, 1900, 494 pl. IV, fig. 1, (Honolulu).

188. Scopas flavescens (Bennett).

Color in life (field No. 172, 3.2 inches in length), body, head, and fins uniformly sulphur-yellow; tips of ventral orange; lips red.

One specimen of this species, 3.2 inches long, was taken. It seems to be rare at Honolulu.

Acanthurus flavescens Bennett. Zool. Journ., IV, 40, 1828; Günther, Fische der Südsee, IV, 116, pl. 76, 1875 (Sandwich Islands); Steindachner, Denks. Ak. Wiss. Wien, LXX, 1900, 493 (Honolulu).
Acanthurus rhombeus, Günther, Cat., III, 342, 1861 (Sandwich Islands).

189. Ctenochætus strigosus (Bennett).

Color in life (field No. 205, 5.6 inches in length), dark reddish-brown, with numerous narrow, blue, longitudinal lines on body and vertical fins, those on body narrowest; pectoral brownish-orange.

Seven examples, 4.4 to 6.2 inches in length, were taken. It appears to be fairly common at Honolulu.

Acanthurus strigosus Bennett, Zoological Journal, IV, 41, 1828 (Sandwich Islands); Cuvier & Valenciennes, Hist. Nat. Poiss., X, 243, 1835; Günther, Cat., III, 342, 1861 (Sandwich Islands); ibid., Fische der Südsee, IV, 116, pl. 79, figs. B and C, 1875 (Sandwich Islands).
Acanthurus (Ctenodon) strigosus, Steindachner, Denks. Ak. Wiss. Wien, LXX, 1900, 494 (Honolulu).

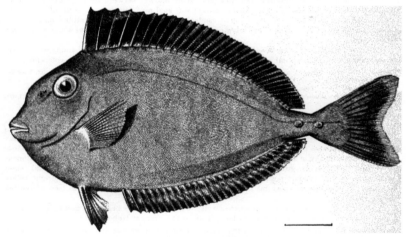

FIG. 32.—*Acanthurus incipiens* Jenkins, new species. Type.

190. Acanthurus incipiens Jenkins, new species.

Head 2.75 in length; depth 2.2; eye 3.4 in head; snout 1.4; interorbital 3; D. v, 28; A. II, 27; P. 18. Body oblong-ovate, compressed, the greatest depth under third dorsal spine; head, short, anterior profile from tip of snout to front of eye concave; a distinct prominence in front of eye at level of its lower border, from which the profile is nearly straight to origin of dorsal; ventral outline strongly convex from tip of snout to base of ventrals; dorsal and ventral outlines from origin of dorsal and ventral fins to base of caudal peduncle each in a long, low curve, most convex anteriorly; head short; snout blunt; mouth small, horizontal, below axis of body; jaws each with a single series of fine, sharp, close-set, finely serrulate, conic teeth; a short, curved groove in front of eye below nostrils, its length 2 in eye; gill-opening long and oblique, extending far anteriorly, the upper end at vertical of first dorsal spine; interorbital space not strongly convex, the median ridge low, body rough velvety; each side of caudal peduncle with 2 very low, weak, horny plates without spines, the distance between them 1.5 to 2 in eye; first dorsal spine rather strong, moderately rough, its length

1.7 in head; other dorsal spines smoother and more slender, the fifth 2 in head; soft dorsal not high, the rays of about uniform length, the longest about 2.5 in head; anal spines rather short and slender, the second equal to diameter of orbit; soft anal similar to dorsal, equally high; caudal deeply emarginate, the free edge of the 2 lobes forming a broad angle; upper lobe of caudal somewhat the longer, its length nearly equaling head; ventral spines moderate, reaching base of first anal spine, their length 2 in head; pectoral short, its length a little less than snout.

Color in alcohol, pale dusky olivaceous, lower parts palest; dorsal and anal with pale purplish brown, mottled with lighter; edge of dorsal and anal each with a narrow, blackish border tipped along the posterior portion with whitish; caudal dirty brownish, narrowly edged with white; ventrals whitish, dusky tipped; pectoral dusky at base, lighter at tip.

Only 2 examples of this species were obtained. Type, No. 50707, U. S. N. M. (field No. 382), a specimen 8.25 inches in length, obtained by me in Honolulu in 1889; cotype, No. 7726, L. S. Jr. Univ. Mus. (field No. 12048), 11 inches long, obtained by the *Albatross* in 1896, at Honolulu.

191. Acanthurus brevirostris (Cuvier & Valenciennes).

Color in life (field No. 244, 7.25 inches in length), head, body, dorsal and anal fins uniformly dark; outer margin of caudal yellow, central area light orange.

Four specimens, 7.25, 7.75, 8.25, and 8.75 inches in length, were obtained. This is the first record of the species from the Hawaiian Islands. It appears to be fairly common at Honolulu.

Naseus brevirostris, Cuvier & Valenciennes, Hist. Nat. Poiss., x, 277, pl. 291, 1835; Günther, Cat., III, 349, 1861; ibid., Fische der Südsee, IV, 121, pl. LXXIX, fig. A, 1875 (Kingsmill Island, Tahiti).

192. Acanthurus unicornis (Forskål).

Nine examples of this species, from 5 to 12 inches in length, were obtained. It is very abundant at Honolulu, where it is exposed for sale as food with all others of the *Teuthidida*.

Chætodon unicornis, Forskål, Descript. Animal., 63, 1775.
Naseus unicornis, Günther, Cat., III, 348, 1861 (Polynesia); ibid., Fische der Südsee, IV, 118, pl. 78, figs. 1 to 4, 1875 (Sandwich Islands); Streets, Bull. U. S. Nat. Mus., No. 7, 68, 1877 (Honolulu); Steindachner, Denks. Ak. Wiss. Wien, LXX, 495, 1900 (Honolulu).
Monoceros unicornis, Fowler, Proc. Ac. Nat. Sci. Phila. 1900, 513 (Hawaiian Islands).

193. Callicanthus lituratus (Forster).

Color in life (field No. 160, 8.5 inches in length), body dark drab, with darker mottlings; a bright yellow area around eye, the color extending as a curved line of bright yellow down the side of the snout to angle of mouth, where it spreads out, then extends a short distance backward; the area about the eye backward and upward to opposite base of second dorsal spine yellow, the yellow area of each eye connected by a broad band between the eyes; lips rosy; caudal spines and the area around them bright orange; dorsal inky black, soft part with white border, edged outwardly with dark; a bright blue line on body along base of dorsal; base of anal yellow, outer part orange with black edging; ventrals olive; pectoral dusky, with inner surface yellow; caudal dark drab, with posterior border white, just within which is a crossband of yellow.

Common at Honolulu. Eight specimens, 6.5 to 10 inches in length, were taken by me; one, 11 inches long, by Dr. Wood; and three, 7, 8, and 10.5 inches long, by the *Albatross* in 1896.

Acanthurus lituratus, Bloch & Schneider, Syst. Ichth. 216, 1801.
Harpurus lituratus, Forster, Descript. Animal., 218.
Naseus lituratus, Cuvier & Valenciennes, Hist. Nat. Poiss., x, 282, 1835 (Otahiti); Günther, Cat., III, 353, 1861 (Polynesia); Günther, Fische der Südsee, IV, 124, pl. 82, 1875 (Sandwich Islands); Steindachner, Denks. Ak. Wiss. Wien, LXX, 1900, 495 (Honolulu).

194. Callicanthus metoposophron Jenkins, new species.

Head 4 in length; depth 2.5; eye 3.5 in head; snout 1.9; interorbital 3; D. VI, 29; A. II, 30. Body rather oblong, greatly compressed, the greatest depth under last dorsal spine; anterior dorsal profile without protuberance of any kind, strongly and evenly convex from tip of snout to about fourth dorsal ray, thence less convex to caudal peduncle; ventral outline similar, but less convex; snout rather short; mouth small, horizontal, slightly below axis of body; teeth small, slender, close-set, and pointed, in a single row in each jaw; a short lunate groove in front of eye, its length equaling that of maxillary;

gill-opening long and oblique, the lower arm extending far forward, the upper end on a level with lower edge of orbit and directly above upper base of pectoral; interorbital space moderately broad, the 2 sides meeting at a broad, rounded angle; entire body and head finely granulated or velvety; each side of caudal peduncle with two, weak, keeled, horny plates, the distance between which is 1.4 in eye; first dorsal spine strong, roughened laterally, inserted above gill-opening; other dorsal spines slender, smooth, and pointed, the third longest, its length nearly equaling that of snout; dorsal rays slender and weak, shorter than the spines, the longest about 2.8 in head; anal spines slender and pointed, the second slightly the longer, its length equaling diameter of eye; anal similar to soft dorsal, but somewhat lower; caudal deeply lunate, the lobes not greatly produced, the upper slightly the longer; ventral spines long, rather strong, reaching base of second anal spine, their length equaling that of longest dorsal spine; pectoral of moderate length, 1.4 in head.

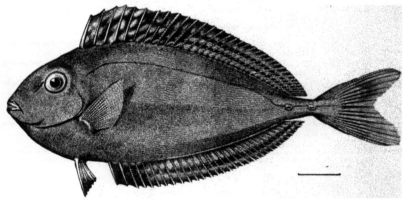

Fig. 31.—*Callicanthus metopossophron* Jenkins, new species. Type.

Color in alcohol, nearly uniform olivaceous brown; pale below; dorsal fin darker brown, crossed by 3 broad, longitudinal lighter bands; on the spinous portion the lighter and darker markings are broken up into more or less vertical bars; membrane between first and second dorsal spines with a pale or transparent area distally, edge of fin narrowly black; anal similar to dorsal, but with less distinct bands; caudal uniform dusky; pectoral dusky, paler at tip; ventral dusky.

Only 2 examples of this species were obtained by me and none has been secured by any subsequent collector. Type, No. 50706 U. S. N. M. (field No. 268), 9.25 inches long, obtained at Honolulu; cotype, No. 7727, L. S. Jr. Univ. Mus. (field No. 461), 9.5 inches long, from same place.

Family XLI. BALISTIDÆ.

195. Balistes vidua Solander. *"Humuhumu hiukole."*

Color in life, uniformly dark brown with tinge of olive; membrane of spinous dorsal olive; soft dorsal and anal white, with a narrow black border along anterior and distal margins; distal portion of caudal peduncle white, fin red, the upper and lower margins each with a narrow black line; pectoral rays bright yellow; faint violet at angles of mouth; iris yellow. The younger examples have spines on the side of the caudal peduncle, which are obsolete in the large ones.

My collection contains 5 specimens (field Nos. 020, 227, 364, 636, and 684), ranging in length from 6 to 10 inches. This species seems to be common at Honolulu.

Balistes vidua Solander in Richardson, Voy. Sulphur, Fishes, 128, pl. 59, figs. 9 and 10, Otaheiti; Gunther, Cat., VIII, 216, 1870; Streets, Bull. U. S. Nat. Mus., No. 7, 57, 1877 (Honolulu).
Melichthys vidua, Bleeker, Atlas, v, 109, pl. 217, fig. 2, 1865.

196. Pachynathus capistratus (Shaw). "*Humuhumu meemee.*"

Color in life (field Nos. 116 and 228), body uniform light brown, fins same color, plain; rosy line beginning slightly behind and below angle of mouth, extending backward and slightly downward to vertical from eye, this joined by another of same color extending under chin. In No. 116, membrane of first dorsal olivaceous, with a black blotch; scaled skin pushed back from the chin shows bright orange-yellow; outer margins of soft dorsal and anal light.

Five examples taken by me at Honolulu, where it is common. The longest is 11 inches in length.

Balistes capistratus Shaw, General Zoology, v, 417, 1804.
Balistes mitis Bennett, Proc. Comm. Zool. Soc., I, 183, 1829; Günther, Cat., VIII, 218, 1870 (Indian and Pacific oceans).
Balistes frenatus Bleeker, Atlas, v, 114, pl. 223, fig. 2, 1865.
Pachynathus capistratus, Jordan & Evermann, Fishes North and Mid. Amer., II, 1704, 1898 (Tropical Pacific).

197. Pachynathus bursa (Lacépède).

General color, light drab, with darker cloudings; a narrow distinct white line from near angle of mouth to near origin of soft anal, which returns along base of spinous anal to base of ventral spine; an olivaceous dash extending in a curve from upper part of base of pectoral upward and backward toward middle of, but not quite reaching, the first dorsal fin; another from above and through the eye downward and backward to lower part of base of pectoral; throat and belly below white line light; first dorsal olivaceous with white; second dorsal and second anal transparent, first anal black; caudal dusky; inside of mouth black.

I have 11 specimens, from 5 to 8.5 inches in length. It seems to be abundant and is sold in the market as food.

Balistes bourse Lacépède, Hist. Nat. Poiss., I, 335, 375, 1798.
Balistes bursa, Bloch & Schneider, Syst. Ichthy., 476, 1801 (Indian Ocean); Bleeker, Atlas, v, 116, pl. 223, fig. 3, 1865; Günther, Cat., VIII, 219, 1870 (Indian and Pacific oceans).
Balistapus bursa, Fowler, Proc. Ac. Nat. Sci. Phila. 1900, 514 (Hawaiian Islands).

198. Balistapus rectangulus (Bloch & Schneider). "*Humuhumu Nukunuku Apua'a.*"

Very brilliantly colored in life; ground of upper part of body and head light brown becoming lighter toward snout; 3 black bands reaching from one eye to the other, the borders and the spaces between these bands green; the most posterior green band on head passing downward and backward, where, after an abrupt bend backward, it becomes a violet line running along middle of body to a vertical from tip of third dorsal spine, where it forms an acute-angled fork, each prong a brilliant yellow line, the upper ending at about base of third from last soft dorsal ray, the lower ending at corresponding position at base of anal; within the fork are 2 other bright yellow lines parallel with the prongs of fork, forming anteriorly an acute angle on a vertical through first third of dorsal.

Twelve specimens of this very brilliant fish were obtained by me at Honolulu, the longest 9.5 inches in length. It is very abundant at Honolulu where it is conspicuous in the market on account of its colors.

Balistes rectangulus Bloch & Schneider, Syst. Ichthy., 465, 1801, Indian Ocean; Günther, Cat., VIII, 225, 1870.
Balistes cinctus Bleeker, Atlas, v, 119, pl. 228, fig. 1, 1865.
Balistapus rectangulus, Fowler, Proc. Ac. Nat. Sci. Phila. 1900, 514 (Hawaiian Islands).

199. Melichthys radula Solander. "*Humuhumu Eleele.*"

Color in life, uniformly black with slight show of bluish; axil bluish; a very distinct, conspicuous, narrow line of light blue running longitudinally on base of dorsal; a similar one on base of anal.

Apparently abundant at Honolulu, where I obtained 7 specimens, the longest being 11 inches.

Balistes radula Solander, in Richardson, Voy. H. M. S. Samarang, Fishes, 21, 1848.
Melichthys ringens Bleeker, Atlas, v, 108, pl. 220, fig. 2, 1865; not of Linnæus.
Balistes buniva Günther, Cat., VIII, 227, 1870; Streets, Bull. U. S. Nat. Mus., No. 7, 56, 1877 (Honolulu), not of Lacépède nor of Risso.
Balistes (Melanichthys) buniva, Steindachner, Denks. Ak. Wiss. Wien, LXX, 1900, 517 (Honolulu and Laysan).
Melichthys bispinosus Gilbert, Proc. U. S. Nat. Mus. 1890, 125, Clarion and Socorro islands.

Family XLII. MONACANTHIDÆ.

200. Cantherines sandwichensis (Quoy & Gaimard).

Color in life, uniformly black; dorsal, anal, and pectoral golden, the color mostly confined to the rays, the membranes being transparent; caudal rays black, membranes dusky olivaceous; dorsal spine black, membrane behind it olivaceous.

My collection contains 5 specimens, from 6 to 6.5 inches in length, and there are in the collection made by Dr. Wood 4 examples, from 6 to 10.5 inches in length. It is used as food by the natives.

Balistes sandwichensis Quoy & Gaimard, Voy. de l'Uranie, Zool., 214, 1824. Hawaiian Islands.
Monacanthus pardalis Rüppell, Neue Wirb. Fische, 57, pl. 15, fig. 3, 1835, March, 1836; Günther, Cat., VIII, 230, 1870 (Indian and Pacific oceans).
Cantherines nasutus Swainson, Nat. Hist. Fishes, etc., II, 327, 1839, substitute for *B. sandwichensis* Quoy & Gaimard.
Liomonacanthus pardalis Bleeker, Atlas Ichthy., v, 1865, 136, pl. 130, fig. 7
Cantherines carolæ Jordan & McGregor, Rept. U. S. Fish Comm. for 1898, 281, pl. 6, Socorro Island.
Cantherines sandwichensis, Fowler, Proc. Ac. Nat. Sci. Phila., 1900, 514 (Hawaiian Islands).

201. Stephanolepis spilosomus (Lay & Bennett).

Color in life, face bluish, body olivaceous; face and cheeks with dark wavy lines; sides of body covered with dark spots as large as pupil; membrane behind dorsal spine orange; soft dorsal and anal with many very narrow longitudinal yellow lines alternating with light blue; caudal very brilliant, its ground color yellow, the distal margin orange, within which is a broad transverse band of black; remainder of fin with transverse rows of black dots. My collection contains fifteen specimens from 2.36 to 5.14 inches in length, taken on the reef at Honolulu, where it is abundant in the coral; and five, 3.25 to 4 inches in length, were taken by the *Albatross* in 1896.

Monacanthus spilosoma Lay & Bennett, Zoology, Capt. Beechey's Voyage, in H. M. S. Blossom, 70, pl. 22, fig. 1, 1839, Hawaiian Islands; Günther, Cat., VIII, 233, 1870 (Sandwich Islands); Fowler, Proc. Ac. Nat. Sci. Phila, 1900, 514 (Hawaiian Islands).

202. Osbeckia scripta (Osbeck).

One specimen, a skin 23 inches in length, was obtained by Jordan & Snyder in 1900 at Honolulu. This species has been taken by Jordan & Evermann, Check-List of Fishes of North America, page 424, as the type of the new subgenus *Osbeckia*. This seems to be its first record from the Hawaiian Islands.

Balistes scriptus Osbeck, Inter. Chin., I, 144, 1757, China.
Alutera scripta, Jordan & Evermann, Fishes North and Mid. Amer., II, 1719, 1898.

Family XLIII. TETRAODONTIDÆ.

203. Tetraodon hispidus (Linnæus).

Color in life varies considerably. Upper parts (field No. 302) golden olive, lower parts white; bluish white spots over tip of head and back, becoming smaller on caudal peduncle and caudal fin; 2 bluish white concentric rings around eye; 1 distinct and 1 or 2 other not so distinct rings of white around base of pectoral fin, the white bands with olive interspaces; base of pectoral and region below black; some black blotches anterior to the lower of these; dorsal dusky yellow; pectoral bright yellow; anal orange-yellow; caudal dusky, membranes yellowish with bluish white spots.

Three specimens of this species were obtained at Honolulu, the longest being 9.7 inches. There is a great range in the distinctness of the markings and in the amount of roughness given the skin by the spines. It is offered for sale as food in the market at Honolulu, but is considered very poisonous if not cooked in a certain manner.

? *Tetraodon hispidus* Linnæus, Syst. Nat., ed. x, I, 333, 1758, India.
Tetraodon perspicillaris Rüppell, Atlas, Reise Nord Afrcn., 63, 1828, Red Sea.
Tetraodon hispidus, Günther, Cat., VIII, 297, 1870.
Tetraodon implutus, Streets, Bull. U. S. Nat. Mus., No. 7, 1877, 56 (Honolulu)
Ovoides ctelhizon Jordan & Gilbert, Proc. U. S. Nat. Mus., 1882, 631, Panama. (Type, No. 29679, U. S. N. M.)

204. Ovoides latifrons Jenkins.

One specimen, 9.4 inches in length, was obtained.

Ovoides latifrons Jenkins, Bull. U. S. Fish Comm. for 1899 (June 8, 1901), 398, fig. 10, Honolulu (Type, No. 49696, U. S. N. M. Coll. O. P. Jenkins.)

Family XLIV. TROPIDICHTHYIDÆ.

205. Tetraodon jactator (Jenkins).

Two specimens, 1.5 and 2.5 inches long, were taken on coral rocks on the reef in front of the city.

Tropidichthys jactator Jenkins, Bull. U. S. Fish Comm. for 1899 (June 8, 1901), 399, fig. 11, Honolulu. (Type, No. 49783, U. S. N. M.; coll. O. P. Jenkins.)

206. Tropidichthys oahuensis Jenkins, new species.

Head 4 in length; depth 1.9; eye 4 in head; snout 1.4; interorbital 3; D. xi; A. xi.

Body short, deep, and greatly compressed; dorsal profile rising in a nearly straight line from tip of snout to occiput, thence descending in a small regular curve to base of caudal fin; ventral outline nearly uniformly convex; interorbital space broad, nearly flat; snout long, conic; mouth small, slightly below axis of body; eyes small, high up, the supraorbital rim prominent; occiput prominent, with a distinct knob at highest point; caudal peduncle compressed, its least width about 3 in its least depth; gill-opening shorter than orbit, somewhat oblique; pectoral short, the anterior rays longest, about 2 in head, the others gradually shorter, the bases considerably exceeding diameter of orbit; origin of dorsal midway between upper end of gill-opening and base of caudal, the anterior ray longest, about 2.2 in

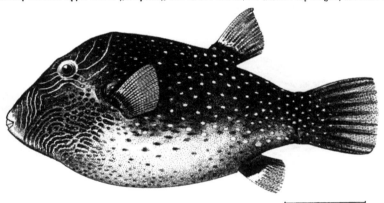

FIG. 32.—*Tropidichthys oahuensis* Jenkins, new species. Type.

head, the base of the fin about 3 in head; anal short, rounded, the anterior rays somewhat the longer and equaling the anterior rays of dorsal; caudal truncate, its length about 1.3 in head; head covered all over with small, sharp asperities, only the lips naked; belly mostly smooth, the asperities smaller and weaker; sides smooth; median line of back rough.

Color in alcohol rich brown, the side of body with numerous, very small, sky-blue spots; side of head with numerous, very narrow, whitish or bluish white lines separating dark brown or blackish lines, the latter disposed to break up into spots on the cheek; 2 narrow blue lines on cheek below eye, each curved upward at the middle; 2 or 3 similar lines backward from eye, the second one extending upward and backward to nuchal prominence, where it joins its fellow from the opposite side; small blue spots about and above the gill-opening; caudal peduncle spotted similarly to the side of body; lower part of side with irregular brownish blotches, the belly rusty brown; fins all pale, blackish at base; caudal somewhat dusky.

This description is based upon a single specimen, the type, No. 50690, U. S. N. M., field No. 326, 4.5 inches long, obtained by me at Honolulu.

207. Tropidichthys epilamprus Jenkins, new species. *"Puu Olai."*

Head 2.8 in length; depth 2.7; eye 3.6 in head; snout 1.6; interorbital 3.5; D. 9; A. 9; C. 10; P. 17. Body oblong, compressed; head long; snout pointed, its sides flattened; mouth small; teeth strong, convex, meeting in a produced point at the center; eye high up, the supraorbital rim prominent; interorbital space concave; anterior profile from tip of snout to occiput nearly straight; caudal peduncle compressed and deep, its depth 2.25 in head; gill-opening vertical, its length less than diameter of eye; nostril small, perforate, not in a projecting tube; body chiefly smooth on sides and caudal peduncle; dorsal region between eyes and dorsal fin with small, sharp prickles; a similar patch on lower part of cheek and belly; snout and interorbital region naked; lower jaw naked; posterior part of body and caudal peduncle naked; fins moderate; dorsal with the anterior rays longest, the free edge oblique, nearly straight, height of fin 2 in head; anal pointed, its length about 2.8 in head; caudal truncate, its length 1.3 in head; pectoral broad, little oblique, its length 2.6 in head.

Color in alcohol, pale brownish above, paler on sides and belly; a large blackish area on side below base of dorsal; cheek and entire body covered with small roundish brown spots; 2 dark brown lines on cheek under eye; 2 or 3 similar lines radiating backward from eye and 2 others running forward from eye; 5 narrow dark lines across head between eyes; a dark median line from tip of lower jaw to vent; side of snout with 2 vertical and 2 horizontal brown lines; fins all pale, the caudal with converging light brown lines on base.

This species is known only from the type, a specimen 3.5 inches long, collected near Kihei, Maui, by Mr. Richard C. McGregor. Type, No. 50853, U. S. N. M.

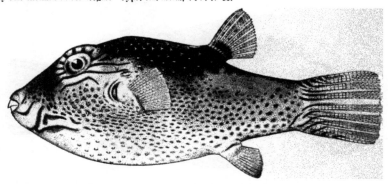

FIG. 33. *Tropidichthys psilonotus* Jenkins, new species. Type.

208. Tropidichthys bitæniatus Jenkins.

One specimen of this species, 2 inches in length, is in Dr. Wood's collection.

Eumpetrina bitæniatus Jenkins, Bull. U. S. Fish Comm. for 1899 (June 8, 1901), 100, fig. 12, Honolulu (Type, No. 49702, U. S. N. M.; coll. O. P. Jenkins.

Family XLV. MOLIDÆ.

209. Ranzania makua Jenkins.

One specimen, 2 feet in length, was sent to Stanford University from Honolulu, by Mr. C. B. Wilson in 1892. It is the type of the species and is No. 12605, L. S. Jr. Univ. Mus.

Ranzania makua Jenkins, Proc. Cal. Ac. Sci., second series, vol. v, 1895 (October 31), 779-784, with colored plate (frontispiece), Honolulu.

Family XLVI. OSTRACIDÆ.

210. Ostracion camurum Jenkins.

I obtained 4 specimens of this species, and 2 are in the collection made by Dr. Wood. They vary in length from 3.5 to 4.5 inches.

Ostracion camurum Jenkins, Bull. U. S. Fish Comm. for 1899 (June 8, 1901), 386, fig. 9, Honolulu. (Type, No. 49697, U. S. N. M.; coll. O. P. Jenkins.

211. Ostracion oahuensis Jordan & Evermann. "*Moaoauau.*"

Head 3.9 in length; depth 2.9; eye 2.9 in head; snout 1.2; preorbital 1.6; interorbital 1; D. 9; A. 9; P. 10; C. 10.   Body 4-sided; dorsal side of carapace evenly convex, its greatest width one-fourth greater than head; lateral dorsal angles not trenchant, slightly convex anteriorly, then evenly convex; snout blunt, the anterior profile ascending abruptly, then strongly convex in front of eyes; interorbital space nearly flat; cheek flat; side of body concave, its width about equal to head; ventral keel prominent, evenly convex; ventral surface nearly flat posteriorly, but little convex anteriorly, its greatest width 1.4 times length of head, its length just twice its width; gill-opening short, not exceeding two-thirds diameter of eye; least width of anterior opening of carapace 1.75 in interorbital or 1.5 times diameter of orbit, the depth nearly twice orbit; mouth small; teeth rich brown; least depth of posterior opening of carapace much less than width of anterior opening, equaling distance from lower edge of preorbital to pupil; length of caudal peduncle less than that of head, its depth 2.2 in its length; no spines anywhere; dorsal fin high, its edge obliquely rounded, its length 1.3 in head; anal similar to dorsal, the edge rounded, its length 1.2 in dorsal; caudal slightly rounded, its rays nearly equal to head; pectoral with its free edge oblique, the rays successively shorter, length of fin equal to height of dorsal.

Color in life, dark brown with blue tinges; interorbital space showing more or less golden; small whitish spots profusely covering entire dorsal surface; no spots on side of body or on face; no spots on ventral surface except a faint one of a slightly darker color than general gray color of surface; one longitudinal row of golden spots on each side of upper part of caudal peduncle from carapace to base of caudal fin; pectoral, anal and dorsal fins with transverse rows of faint spots; caudal bluish black at base, white on posterior half; a broad light or yellowish area below eye; iris golden.

Color in life, upper parts dark brown, with shades of olive; belly brown; sides of body, back, tail and caudal fin covered with bright blue spots, mingled with which are dark brown spots; face and top of head and snout with bright blue lines, between which in region of the eyes is golden; cheeks and below with close-set dark blue lines and with dots, the color of which is nearly white; base of anal blue; outer portions of dorsal, anal and pectorals transparent, with golden tinge, a very brightly colored fish.

Color in alcohol, rich brown above, sides darker, ventral surface paler, brownish about margins, dusky yellowish within; entire back with numerous small, roundish, bluish-white spots; upper half of caudal peduncle with similar but larger spots; forehead and snout dark brown; lips brownish-black; cheek dirty yellowish; sides and ventral surface wholly unspotted; base of caudal blackish, paler distally, the dark extending farthest on outer rays; other fins dusky, with some obscure brownish spots.

This species is related to *O. camurum* Jenkins, from which it differs in the smaller, more numerous spots on back, the entire absence of spots on side, the smaller size of the spots on the caudal peduncle, and the brighter yellow of the suborbital region.   Only 2 specimens known, both from Honolulu.

Type, No. 50668, U. S. N. M. (field No. 03443), a specimen 5.6 inches long, obtained by Jordan & Evermann, July 25, 1901.   Cotype, No. 7478, L. S. Jr. Univ. Mus. (field No. 2156), 5.25 inches long, collected at Honolulu in 1898 by Dr. Wood.   The species was not obtained by me.

Ostracion oahuensis Jordan & Evermann, Bull. U. S. Fish Comm. for 1902 (April 11, 1903), 200, Honolulu.

212. Ostracion lentiginosum Bloch & Schneider.

Two specimens (5 and 5.43 inches in length) obtained by me on the coral reef at Honolulu.

Ostracion lentiginosum Bloch & Schneider, Syst. Ichthy., 501, 1801, Indies.
Ostracion punctatus Steindachner, Denks. Ak. Wiss. Wien, LXX, 517, 1900 (Honolulu).

213. Lactoria galeodon Jenkins, new species.

Head 2.8 in length; depth 2; eye 2 in head; snout 4.5; interorbital 1.2; D. 9; P. 11; A. 8.   Carapace 4-sided; a pair of long, slender, slightly divergent spines, their direction slightly upward; a similar pair terminating the lateral ventral angles, horizontal and not divergent; middle of back with a strong, compressed, triangular spine, notched on posterior border, projecting slightly backward, and resembling a shark tooth; snout short, the anterior profile concave; dorsal lateral angles little convex, the ventral angles more convex; 12 plates along its edge from snout to spine; 8 plates in lateral dorsal angle, no spine at its middle; ventral surface with 11 or 12 plates in longitudinal median series, 7 in transverse series.

Color in alcohol, dirty yellowish or olivaceous above; middle of side with an oblong dark or blackish area; ventral surface yellowish.

Closely related to the East Indian species *O. diaphanus* Bloch & Schneider, from which it is readily distinguished by the entire absence of median spines on the lateral ventral keel, by the longer and straighter frontal and ventral spines, the character of dorsal spine, and the opaque carapace.

A single specimen was obtained by me in 1889. It is 1.3 inches in total length and is taken as the type, No. 50717, U. S. N. M. The specimens recorded from Laysan and Hawaii by Steindachner probably belong to this species.

Ostracion diaphanus Steindachner, Denks. Ak. Wiss. Wien, LXX. 1900, 547. Laysan and Hawaii; probably not of Lacépède.

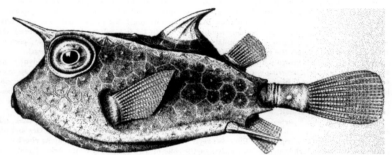

FIG. 34.—*Lactoria galeodon* Jenkins, new species. Type.

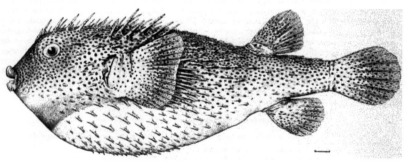

FIG. 35.—*Diodon nudifrons* Jenkins, new species. Type.

Family XLVII.   DIODONTIDÆ.

214. Diodon nudifrons Jenkins, new species.

D. 15; A. 12; forehead sloping upward from snout at angle of about 45°; interorbital 1.3 in head; spines mostly short, not longer than eye, except those back of pectoral, the longest of which are equal to length of pectoral and about 2 in head; no spines on forehead below level of upper margins of eyes; foremost spines of head few and short; head, back, sides, all the fins and the membranous sheaths of the spines closely covered with small, roundish, black spots, much smaller than pupil; below pale; a brown band from below gill-opening forward along lower angle of head to below eye, then across throat continuous with corresponding band of opposite side. Known only from one specimen, 21 inches long, obtained by me at Honolulu. Type. No. 50854, U. S. N. M.

215. Diodon hystrix Linnæus.

One large specimen, 25 inches in length, was obtained fresh, and one smaller specimen was obtained by the *Albatross* in 1896. It is not very common at Honolulu where the natives use it as food, but regard it as poisonous unless receiving certain treatment in cooking.

Diodon hystrix Linnæus, Syst. Nat., ed. X, 335, 1758. India: Günther, Cat., VIII, 306, 1870.

Family XLVIII. CIRRHITIDÆ.

216. Cheilodactylus vittatus Garrett. "*Kīkākāpu.*"

This interesting form was described by Garrett in 1868 from a single specimen which now seems to be lost. A copy of Garrett's painting is given in Günther's Fische der Südsee. The species had not been seen again until a single specimen was taken by Professor Schauinsland in 1897, and recorded by Steindachner. The specimen here recorded was collected by Dr. Rosenstern and presented to the California Academy of Sciences. It agrees fairly well with Garrett's figure reproduced in Fische der Südsee.

Head 3 in length; depth 2.6; eye 3.2; D. xvii, 29; A. iii, 7; lateral line 1.63; maxillary not reaching anterior margin of eye; no teeth on vomer or palatines; bands of villiform teeth on jaws; opercle and preopercle entire, the angle of opercle ending in a weak, flat spine; dorsal outline rising abruptly from interorbital space to base of fourth dorsal spine, to a height about three-fifths the length of head above the eye. Since the fourth dorsal spine is very high, about equaling length of head, when the fin is raised the fish has a peculiarly deformed appearance. Length of specimen, 5 inches.

The alcoholic specimen shows 4 black bars across the head, the most anterior a small black area over the anterior end of snout, the second not complete, passing over forehead along anterior margin of eye on the cheek, ending near angle of preopercle; third across head, running obliquely backward through posterior portion of eye and ending just below middle of opercle, the black at base of pectoral and in axil in line with this as if an extension of it; two broad bands on body slightly broader than eye, the first including the first 3 dorsal spines and base of fourth and running obliquely backward, ending on belly just behind origin of pectoral; the second beginning at tips of fourth, fifth, and sixth, reaching body at base of ninth spine, running along dorsal side of body covering remainder of dorsal spines except tips, coming to lie almost wholly on body below soft dorsal, just including bases of its rays, the band covering caudal peduncle except a narrow space on ventral side and extending on lower lobe of caudal; pectoral black. The parts of the body and fins not included in the bands described are white in alcohol.

Cheilodactylus vittatus Garrett, Proc. Calif. Ac. Sci. 1863, 103. Hawaiian Islands: Günther, Fische der Südsee, 73, taf. LI, fig. B, 1873 (Sandwich Islands, Garrett, one specimen) Steindachner, Denks. Ak. Wiss. Wien, LXX, 490, 1900 (Honolulu).

CIRRHITOIDEA Jenkins, new genus.

Cirrhitoidea Jenkins, new genus of *Cirrhitidæ* (*bimacula*).

No palatine teeth; teeth on vomer; jaws with narrow band of small canine-like teeth; intermaxillary denticulate; preopercle finely toothed; dorsal single of 10 spines and 12 rays; the 5 lower rays of pectoral simple, the upper of which is elongate, 1.8 in head; snout long, pointed, 3 in head.

This genus is allied to *Oxycirrhites* Bleeker, from which it differs chiefly in the shorter snout.

217. Cirrhitoidea bimacula Jenkins, new species.

Head 2.6 in length; depth 3; eye 4.6 in head; snout 3.8; maxillary 2.7; D. x, 12; A. iii, 6; scales 3-37-7. Body short, deep and compressed, the dorsal profile strongly arched from tip of snout to base of first dorsal spines; back very narrow, trenchant, ventral outline nearly straight; head rather long, pointed; snout long and pointed; mouth moderate, slightly oblique, the jaws equal; maxillary reaching pupil; jaws with small, close-set, canine-like teeth, small teeth on vomer, none on palatines; preopercle serrate; opercle ending in a long flap; fins rather large; dorsal spines slender, weak, their length equal to distance from tip of snout to middle of pupil; dorsal rays somewhat shorter; second anal spine longest, about equal to longest dorsal spine; anal spines similar to those of soft dorsal; caudal slightly rounded; ventrals rather long, reaching past vent; pectoral moderate, the middle rays longest, about 1.2 in head; scales rather large, lateral line complete, beginning at upper end of gill-opening and running a little nearer dorsal outline posteriorly; scales on nape, breast, cheek and opercle.

Color in life, about 10 red crossbands, some running into each other, whitish between; iris bright red; black spot on opercle, black spot on body extending somewhat on dorsal fin at base of eighth, ninth, and tenth soft rays, the crossbands on body extending on dorsal; pectoral, ventrals, and anal reddish; ventrals and anal with dusky tips.

Color in alcohol, dusky; body crossed by about 7 rather broad, darker, vertical bars, the first at origin of dorsal, second under middle of spinous dorsal, third under beginning of soft dorsal, last two on caudal peduncle; head dusky yellowish; a large brownish-black spot on opercle; another large, round, brownish-black spot on side above lateral line and under posterior third of soft dorsal; fins somewhat dusky, the anal darkest.

This description is based on the type, No. 50702, U. S. N. M. (field No. 275), 2 inches in length, and one cotype, both taken by me in living coral at Honolulu.

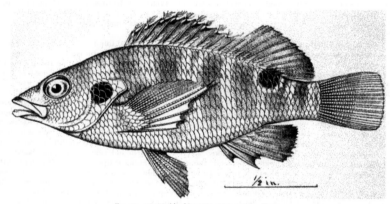

FIG. 36.—*Cirrhitoidea bimacula*, new species. Type.

218. Paracirrhites forsteri (Schneider). *"Hilupilikoa"; "Pilikoa."*

Color in life (field No. 89), spots on lips and face very bright red; on opercles, preopercles and body above the pectoral, red and black; on chin and throat, red; on belly as far as ventral fins, red and black; on upper part of posterior half of body and caudal peduncle, a row of large black patches fusing into one another; iris red; dorsal fins red, the first with dusky outer margin, the second with membranes dusky in parts; caudal red; rays of pectoral red, with red spots on its base and in its axil; ventral yellow; rays of anal yellow with dusky membranes.

Ten examples, from 4.75 to 7.25 inches in length, obtained by me, and five, 4.4 to 5.25 inches, taken by Jordan & Snyder. Common at Honolulu, appearing in numbers in the market.

Grammistes forsteri Bloch & Schneider, Syst. Ichthy., 191, 1801.

Paracirrhites forsteri Bleeker, Atlas, VIII, 143, pl. CCCXLIX, 5, 1876–77.

Cirrhites forsteri Günther, Cat., II, 71, 1860 (Mauritius, Cape Seas, India); Günther, Fische der Südsee, I, 69, taf. XLIX, fig. A, 1873 (Indian Archipelago, east coast of Africa, Red Sea, Polynesia, Hawaiian Islands); Streets, Bull. U. S. Nat. Mus., No. 7, 73, 1877 (Honolulu, Oahu); Steindachner, Denks. Ak. Wiss. Wien, LXV, 1900, 480 (Honolulu); Fowler, Proc. Ac. Nat. Sci. Phila. 1900, 502 (Honolulu).

219. Paracirrhites cinctus Günther. *"Pilikoa"; "Oopuka haihai."*

Color in life, upper part of head dark with many small blue dots, lower part with large blue and red dots; body back of fourth spine with 4 broad, bright-red crossbands, anterior one mingled with brown, the spaces between the bands white; dorsal fin red, membranes at tips of spines transparent, except the filaments, which are bright red; breast white with golden-brown spots; caudal red; anal transparent, with olive and red markings; ventral rays red, membranes white; pectoral rays with brown spots, membranes transparent.

Thirteen examples, from 3.6 to 4 inches in length, were obtained. Jordan & Snyder collected five, 3.5 to 4 inches long. This very beautiful little fish is quite abundant at Honolulu, some specimens being almost always present in the market.

Cirrhites cinctus Günther, Cat., II, 73, 1860, Hawaiian Islands, Madagascar, Isle de France; Günther, Fische der Südsee, II, 72, pl. 52, figs. A and B, 1874 (Hawaiian Islands and Mauritius); Steindachner, Denks. Ak. Wiss. Wien, LXX, 490, 1900 (Honolulu).

220. Paracirrhites arcatus (Cuvier & Valenciennes).

Color in life (field No. 92), body suffused with red, becoming brighter toward the dorsal and posterior portion; spinous dorsal bright red; lower part of soft dorsal red, membranes half-way out dusky, the outer portions colorless; caudal red, pectoral and ventral pinkish; the anal fin and anterior border of ventrals olivaceous; lips bright red; a small area behind eye bordered by an irregular line made up of 3 colors, very bright, the outer blue, middle red, inner orange; nostrils orange; 3 or 4 bright-yellow orange spots on suboperele.

Another from which the one described above can not be distinguished structurally is of much lighter general color and bears on the posterior portion of body and on caudal peduncle a white longitudinal band. The two forms are constantly found together, and are doubtless, as Günther thinks, of the same species.

Cirrhites arcatus Cuvier & Valenciennes, Hist. Nat. Poiss., III, 74, 1829, Isle de France; Günther, Fische der Südsee, II, 70, pl. 49, figs. B and C, 1874.
Cirrhites (Amblycirrhites) arcatus Steindachner, Denks. Ak. Wiss. Wien, LXX, 1900, 490 (Honolulu).

221. Cirrhites marmoratus (Lacépède).

Color in life (field No. 76), general color light, nearly white; belly white with dark cloudings over upper part of body; golden yellow vertical bars on upper lip and wavy stripes of same color on head; golden spots on the posterior part of body; spots on posterior portion of body and on vertical fins bright red; pectoral and ventral fins pale pink; iris red.

This species is caught with *C. forsteri*, and, like it, is an abundant and important food-fish. I obtained fourteen specimens, 4.8 to 8.5 inches in length; three, 5.8 to 6.6 inches in length, are in Dr. Wood's collection; and one was obtained by Jordan & Snyder.

Labrus marmoratus Lacépède, Hist. Nat. Poiss., III, 492, pl. 5, fig. 3, 1801.
Cirrhites maculatus Lacépède, V, 3, 1803; Günther, Fische der Südsee, III, 71, pl. 51, fig. A, 1874 (Red Sea, east coast Africa, Hawaiian Islands, Society Islands, Cook Islands).
Cirrhitichthys maculatus Günther, Cat., II, 74, 1860 (Hawaiian Islands, India, Polynesia, Isle de France).
Cirrhites (Cirrhitichthys) maculatus Steindachner, Denks. Ak. Wiss. Wien, LXX, 1900, 490 (Honolulu, Laysan).
Cirrhitus alternatus Gill, Proc. Ac. Nat. Sci. Phila. 1862, 122 (Hawaiian Islands).
Cirrhitus marmoratus Gill, Proc. Ac. Nat. Sci. Phila. 1862, 107 (Hawaiian Islands).

Family XLIX. CARACANTHIDÆ.

222. Caracanthus maculatus (Gray).

Head 2.6 in length; depth 1.6 to 2 in length; snout 2.25 in head; eye 4.3 in head, suborbital equals eye; D. VIII, 12. Body deep, short, compressed; profile from tip of snout to before center of eye almost vertical in largest specimens, forming a conspicuous angle before eyes with profile of head above eyes, which rises sharply to a gentle curve to front of dorsal fin. In small specimens (1.1 inches long), profile much more inclined from tip of snout to before eye, forming a continuous curve with part of profile above eye from tip of snout to front of dorsal fin; some of intermediate sizes have profile below eye inclined but forming angle with part above; greatest depth through front of dorsal fin;

mouth below mid-longitudinal line of body, so that ventral profile of head is much less convex than dorsal; back of front of body, dorsal and ventral outlines almost symmetrically converging in gentle curves to base of caudal peduncle; dorsal curvature meeting peduncle in advance of ventral curvature, so that posterior end of body is unsymmetrical and the peduncle in most specimens is bent somewhat upward; depth of caudal peduncle at base 7 in length of body; 4 in greatest depth of body; mouth short and somewhat oblique; lips rectangular, being straight in front and on sides; teeth in jaws in bands in front, villiform in upper jaw and an outer series of very slightly enlarged teeth; no teeth on vomer or palatines; posterior margin of opercle running from above downward and forward, not reaching to posterior margin of gill-opening and leaving branchiostegals exposed on side of head, 2 flat spines at its angle near upper end of gill-slit; posterior limb of preopercle with 5 flat, short, wide spines, the lower two more slender, elongate and curved upward; interopercle with a long, strong spine directed backward; preorbital with a large, flat spine directed downward and backward, lying in groove above maxillary and close to it; eye almost circular, or elliptical, with longer diameter vertical; horizontal length of space back of eye 1.8 in length from tip of snout to angle of opercle, which is a very oblique line, angle of 45° with line from middle of caudal peduncle to middle of greatest depth of body; interorbital flat, narrow, three-fifths of eye; nostrils of equal size, posterior above anterior, each with slightly raised margins forming very short tubes, anterior with elongate flap on upper margin; head and body scaleless; skin of body everywhere roughened by numerous minute warty elevations; top of head covered with small rough ossifications, specially large ones reaching from between eyes to nape; other parts of head and body, especially fore part of back, finely villous, villi in some very small and inconspicuous, in others comparatively large and prominent; lateral line beginning above upper end of gill-opening, descending with gentle convexity upward to middle of base of caudal peduncle, obsolete on peduncle in some, in others extending in straight line to end of peduncle; dorsal spines all short, the fifth longest, 3.6 in head; the first very short, second to fifth abruptly longer and of almost equal height, the next 3 regularly descending again to size of first; soft dorsal higher than spinous, middle rays longest, 2.6 in head. The degree of separation of the two parts of dorsal fin varies considerably. In some specimens the two are definitely discontinuous and in some distinctly continuous, while most of them are intermediate in this regard, so that probably the types of the species M. maculatus Gray and M. unipinna Gray are simply two extremes of the same species, since otherwise they do not differ. Caudal fin rounded, median rays 5 in length of body; base of fin covered by skin of peduncle; pectoral rays directed upward and backward at angle of 45°, middle ones longest, 6.8 in length of body.

Color in life, head and body drab, lighter below, covered with small, bright red spots; fins unmarked.

Nine specimens, 1.5 to 1.75 inches in length, were taken from coral heads on the reef in front of Honolulu. This species is quite common among the branches of coral, where they are so able to hide and fasten themselves that they are dislodged with difficulty.

Micropus maculatus Gray, Zool. Miscellany, 20, 1831 (Owaihi and Hao); Günther, Cat., II, 147, 1860 (Owaihi and Hao); Fische der Südsee, III, 86, 1871 (Otaheite and Sandwich Islands)
Micropus unipinna Gray, Zool. Miscellany, 20, 1831, Pacific; Günther, Cat., II, 147, 1860; Fische der Südsee, II, 86, 1871 (Sandwich Islands, Otaheite, Vavau, Fiji, Pelew, Maduro.
Caracanthus typicus Kröyer, Naturhist. Tidsskr., I, 261 and 267, 1844
Amphiprionichthys apistis Bleeker, Nat. Tyds. Ned. Sud., VIII, 170, 1855 (Cocos Islands); Günther, Cat., II, 144, 1860 (Kokos Islands); Kner, Sitzb. Ak. Wiss. Wien, 185, 17, pl. III, fig. 8.
Centropus staurophorus, Kner, Sitzb. Ak. Wiss. Wien, 1860, 3 (Zanzibar).
Caracanthus apistis, Bleeker, Atlas Ichthy. Ind. Neer., IX, pl. 416 (Scap., pl. VI, fig. 5, 1877.

Family L. SCORPÆNIDÆ.

223. Sebastopsis kelloggi Jenkins, new species.

Head 2.5 in length; depth 2.7; eye 3.3 in head; snout 4 in head; interorbital about half eye; D. XII–1, 9; A. III, 5; P. 19; lateral line with 23 tubes, about 28 scales in transverse series. Body moderately elongate, compressed posteriorly, greatest depth about under sixth and seventh spines; snout blunt; jaws subequal; mouth large, oblique, below axis of body; maxillary broad, reaching to posterior border of eye; eye large, its lower border above axis of body; interorbital space narrow, deeply concave without ridges; a large, broad, dermal flap on upper border of anterior nostril, a thin cirrus on anterior upper margin of eye; a conspicuous tentacle as long as half the eye diameter at posterior upper margin of eye; slender cirri along lateral line. Spines on head as follows: a short, sharp, conical nasal spine above and within anterior nostril, 3 supraocular, 1 anterior, 2 close together, posteriorly; no coronal spine; behind the last supraocular spine, a row of 3 spines in an irregular line with it, just within the last a small spine; a tubercle on the posterior border of orbit, behind which is a row of 3 spines, the last at the angle of the gill-slit; upper angle of opercle with 2 diverging flattish sharp spines (the lower on right side double in the type); end of preopercle with 4 spines, the upper and lower larger than the middle ones; preorbital with a tubercle near border of eye; suborbital ridge with 5 spines, the last abutting against the upper of the preorbital spines, below the fourth on the ridge a small spine; spinous dorsal low, longest spine equal to eye; soft dorsal higher, 2.6 in head; second anal spine longest, 2.2 in head, its tip reaching tip of third spine; soft anal 2 in head; caudal rounded, 2 in head; pectoral 1.3 in head, the middle rays longest; scales moderate; body, top of head, cheeks, opercles, and preopercles scaled; fins naked except base of caudal and proximal half of anterior surface of pectoral, which have fine scales; lateral line evident and in a nearly straight line from upper angle of gill-slit to base of caudal.

Very bright in coloration in life, but unfortunately details of color were not taken.

Color in alcohol, body and head gray, well covered with dark-brown mottlings, gathered on sides into 3 indistinctly outlined areas or broad bands; 2 indistinct brown bands from lower border of eye,

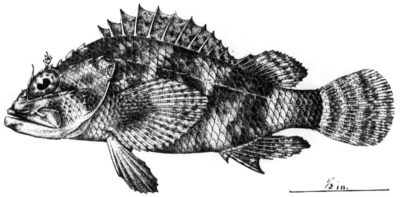

FIG. 37.—*Sebastopsis kelloggi* Jenkins, new species. Type.

1 from middle of border toward posterior end of maxillary, the other from posterior border to margin of opercle across bend of preopercle; spinous dorsal dark brown; base of soft dorsal dark brown; distal portion transparent, with few dark brown spots; base of anal dark brown; distal portion lighter with dark brown mottlings; ventrals uniformly dark brown or black; dark-brown area on proximal portion of pectoral, the remainder lighter with brown dots; caudal with dark brown across base, with 3 or 4 narrow dark-brown bands across the rays, interspersed with white.

This species is based on 2 specimens, 2 inches (the type) and 1.3 inches in length, taken by me in the coral rocks on the reef at Honolulu.

Named for Prof. Vernon Lyman Kellogg, department of entomology, Leland Stanford Junior University.

Type, No. 50694, U. S. N. M. (field No. 234).

224. Sebastapistes corallicola Jenkins, new species.

Head 2.5 in length; depth 2.75; pectoral slightly less than 3 in length; ventral 3.3; caudal equal to ventral; eye 4 in head, a little shorter than snout; D. XII, 9; A. III, 5; C. 19; P. 16; V. I, 5; scales 40 in the lateral line, 6 in series from fourth dorsal to lateral line, 14 from origin of anal to lateral line; mouth but little oblique, lower jaw projecting very slightly; maxillary 1.6 in head, projecting beyond posterior margin of eye; teeth all small and simple, in bands in upper and lower jaws, bands interrupted at front; teeth on vomer in a V-shaped patch, in bands on the palatines equal in **length** to width of vomerine patch; suborbital 1.5 in eye; a pit below anterior lower angle of eye; **anterior nostril** transversely oval, with a tentacle in the inner posterior part of rim, posterior nostril simple, circular; snout with a triangular median elevation, the apex between the anterior nostrils; **between** each anterior nostril and apex of rostral elevation is a strong short spine; 6 spines on the upper **half** of ocular rim, first at upper anterior angle, second on upper rim over center of pupil, third **over posterior** margin of the pupil, fourth on level with upper edge of pupil, fifth back of center of pupil, **sixth** on level of lower edge of pupil; sixth bifid on each side, fifth bifid on right; occipital depression **with** 2 spines at each angle, at the anterior angles one is lateral to the other, at the posterior angles one **is** caudal to the other; a strong spine at upper end of opercle; posterior to this spine and a little **above** it 2 smaller spines just before upper end of gill-slit; posterior to these a single spine at upper end **of** gill-slit; two large diverging spines on opercle; suborbital with a bony ridge without spines **except a** small one on its posterior end; preorbital with 3 spines, 2 directed downward over upper **edge of** maxillary, the other forward over edge of premaxillary; at angle of preopercle an upper small and a lower larger spine, below these on arm of preopercle are 4 decreasingly smaller spines; **supraorbital**

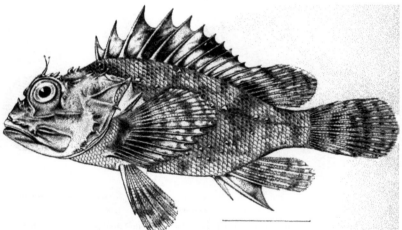

Fig. 38.—*Sebastapistes corallicola* Jenkins, new species. Type.

tentacle well developed, just back of supraorbital spine; a tentacle back of posterior vertical **spine,** lapping over edge of maxillary; a strong spine at angle of shoulder-girdle above base of pectoral, **sharp-** pointed, projecting upward and backward; a small, less prominent spine back of this one; **gillrakers** short, 5–10; interorbital space slightly concave with 2 prominent ridges diverging posteriorly **and** ending in the 2 twin spines of those at anterior angles of the occipital depression; dorsal profile **of the** body much more convex, greatest depth at fourth dorsal spine; depth of caudal peduncle equal to **eye,** 3.5 in greatest depth of body; fourth dorsal spine the longest, a little greater than half of head; **third** and fifth spines equal, but little shorter than fourth; second 1.5 in third, first 1.75 in second; **spines** back of fourth regularly decreasing in length to next to last, which is 2 in fourth; last longer, **equaling** second; soft rays abruptly longer than last dorsal spines, second, third, and fourth longest, **equaling** longest spine; last soft rays equal to next to last spine; second anal spine longest and much thicker **than** others, 1.75 in head; first slender, 2 in second; third slender, 0.8 of second; first and second soft **rays** longest, 1.6 in head; caudal slightly rounded; pectoral round, middle rays longest; ventral **rounded,** second ray longest, its spine equal to sixth dorsal spine; head and fins naked.

Characters very constant. In smaller specimens the posterior spines of orbital rim not so evident as in type and in most the humeral spine smaller. Size of supraorbital tentacle varies much, in some very small or absent, in others very large, fringed, length greater than eye, equal to second dorsal spine.

In alcohol the color varies considerably; in some, fins distinctly banded, in others, fins plain. Some lack the black blotch on posterior part of spinous dorsal, others have it present but small, others have it well developed and reaching from sixth to eleventh spine. A series of dermal flaps along lateral line, also a number of smaller ones on lower half of sides; lateral line simple, slightly convex downward posteriorly.

Color of type (field No. 236) in alcohol, head and body mottled with lighter and darker shades of brown, plain pale below; a wide pale transverse band on nape (very indistinct); spinous dorsal with dusky blotch from seventh to tenth spines on distal half of fin; other fins mottled with brown in triangular transverse bands; a dusky rim above margin of eye on eye membrane; dermal flaps white.

Color of fresh specimen (field No. 223), whitish, with brownish cloudings and many bright red spots on head, body, and fins; black blotch on dorsal fin on eighth to tenth spines, fin clouded with dark bars; dermal flaps white.

Another fresh example (field No. 206), golden brown on body and fins, with many very bright red spots; a black blotch on spinous dorsal on seventh and ninth spines longer than eye but not so deep.

Closely related to *Scorpæna* Jordan & Snyder from Japan, but differs in lacking knob at symphysis of lower jaw, in having no spine on suborbital except on its end, in having 2 spines instead of only one at each anterior angle of occipital depression, in greater length of maxillary, not reaching beyond posterior rim of orbit in *S. onaria;* in having vomerine teeth both V-shaped and not V-shaped. Otherwise very similar.

Close to *Scorpæna nuchalis* Günther, from Raratonga Island (Fische der Südsee, I, 76, 1873); differs from this species in having maxillary reaching past the posterior rim of eye; in having the third, fourth, and fifth spines largest, instead of the fourth to the seventh largest; in having the black blotch on the posterior part of the spinous dorsal (7-10 spines) instead of on the fore part.

My collection contains 3 specimens, all from Honolulu. Type, No. 50691, U. S. N. M. (field No. 236); cotype, No. 7729, L. S. Jr. Univ. Mus. (field No. 223); and cotype, No. 2756, U. S. F. C. (field No. 206).

225. Sebastapistes coniorta Jenkins, new species.

Head (to end of bony opercle) 2.5 in length; depth 2.5; eye 2.75 in head; snout 3.3; interorbital 2 in eye = suborbital; pectoral 3.6 in length; ventral 4; caudal equal to ventral; D. XII, 9; A. III, 5; C. 19; P. 15 (lower 9 simple); V. I, 5; lateral line 8-46-10. Head and body compressed, greatest width through base of pectorals, 4.25 in length of body, 1.6 in depth; dorsal profile of body very convex, greatest depth through base of third dorsal spine; ventral profile of body only gently convex; profile of snout steep, forming an angle before eye with very gently rising part of profile of head behind eye; interorbital area deeply concave, with 2 well-developed longitudinal ridges, diverging posteriorly behind, but not ending in a spine, although a spine arises just behind the posterior end of each; back of each of these are 2 occipital spines; 5 supraocular spines, a row of 5 postocular spines, the last above the upper end of gill-slit, a spine just above penultimate spine of postocular row of spines, 2 flat spines at angle of opercle, upper the larger; 6 spines on lower limb of preopercle; suborbital stay mostly small, with 2 small spines posteriorly; preorbital with 4 spines—2 directed forward, 1 posteriorly, and 1 downward and posteriorly; no dermal appendages on head, except a short wide flap on upper edge of anterior nostril; teeth in jaws in bands, widest in front, in short narrow bands on palatines, in V-shaped patch on vomer; 2 strong flat spines at humeral angle above base of pectoral, lower the larger; nasal spine at inner edge of each anterior nostril; fourth dorsal spine longest, 2 in head; first short, 2.5 in fourth; eleventh spine slightly greater than first; twelfth 3 in head, longer than tenth, slightly larger than second dorsal spine; second dorsal rays abruptly longer than last spine, first rays longest, 2.2 in head; last ray connected by membrane with caudal peduncle; caudal slightly rounded; pectoral broad with wide base, border rounded, median rays longest; ventrals rather broad, first and second rays longest, spines strong; second anal spine longest and thickest, 1.6 in head, longer than longest dorsal spines and equal to caudal fin; first spine short and slender, a little less than half length of second spine; third spine slender, equal to third dorsal spine; first soft anal rays longest, equal to second anal spine, longer than longest dorsal rays; gillrakers 4-10, the lowermost one on lower arm of arch rudimentary, all short, upper ones of lower arm of arch longest but less than half length of pupil; branchiostegals 7; lateral line beginning at last postocular spine, above upper end of gill-slit, ending at middle of base of caudal, slightly convex anteriorly, slightly concave posteriorly; scales ctenoid, entire body at posterior part of head scaled.

Color of fresh specimen (field No. 278), light olive, with dark-brown mottlings; body, head, and fins covered thickly with small brown spots; posterior margin of caudal red; fins color of body.

Color in alcohol, brown, clouded with darker dusky brown; head and spinous dorsal covered with very small round dusky spots.

This description is based on the type, 2 inches in length, and numerous cotypes taken in the coral rocks on the reef in front of Honolulu. (Type, No. 50683, U. S. N. M.)

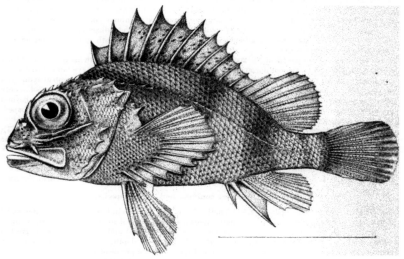

FIG. 39.— *Sebastapistes contorta* Jenkins, new species. Type.

226. Sebastapistes galactacma Jenkins, new species.

Head 2.3 in length; depth 2.6; eye 3.1 in head; snout 3.5; interorbital 2 in eye; D. xii, 9; A. iii, 5; P. 16; lateral line with 23 tubes. Head and body compressed; dorsal outline convex, its highest point at about base of fifth spine, 2 prominences in outline of head, 1 at snout and 1 at the eye, a notch between; the outline slightly concave behind the eye; from the fifth dorsal spine outline gently descending to roots of caudal rays; ventral outline but slightly convex; eye large, circular in outline, wholly above axis of body; mouth large, somewhat oblique, below axis of body; maxillary broad, reaching past pupil; suborbital about 3 in eye; interorbital area narrow, deeply concave, smooth, without ridges; spines on the head distributed as follows: on snout, nasal spines short, nearly erect, just within the anterior and in front of posterior nostrils; 3 supraocular spines, behind these and in line with them 3 occipital spines; no coronal spines; from the middle occipital spine, diverging outward and backward, a bony ridge ending in a spine; a small spine on posterior margin of orbit, behind which extends a series of 3 postocular spines, the last at angle of gill-opening; 2 strong diverging spines ending at angle of opercle, their origin near together at its anterior margin, the upper the longer; 5 spines on margin of preopercle (4 in 2 smaller specimens); suborbital stay well developed, ending in 2 spines, the last abutting on the upper preopercular spine, a small spine at its margin below eye; preorbital with 3 spines, two directed forward, one downward and backward; a short wide dermal flap at upper edge of anterior nostril, a well-developed tentacle about equal to diameter of eye at base of middle supraocular spine; no dermal appendage apparent on other portions of body; a strong spine at the humeral angle above base of pectoral; fourth dorsal spine longest, 2 in head; first the shortest, 2 in fourth, 1.5 in eleventh; soft dorsal rounded, its longest rays nearly equaling the fourth spine, membrane of last adhering its whole length to caudal peduncle; caudal very convex, its middle rays 1.4 in head; third anal spine longest, 1.6 in head, strong, curved; first spine half second; third slender, 0.8 of second; soft portion rounded, 2 in head; ventrals 1.5 in head, reaching halfway between vent and origin of anal; pectoral with rounded outline, broad, its middle rays the longest; teeth villiform, in bands, wide anteriorly, narrow posteriorly, in both jaws, in a V-shaped band on vomer and in a narrow band on palatines; gillrakers 4 · 9, short, rounded, the most anterior rudimentary; branchiostegals 7; lateral line beginning at the last postocular spine and running nearly straight to vertical of last soft dorsal where it bends slightly, ending at middle of base of caudal; head naked, body completely scaled, scales smooth, entire.

Color in alcohol, general effect a light gray, with brown mottlings; white under chin, throat, and belly; head and body covered with thickly-set, minute, bright white points, which are more numerous and more minute on head and anterior part of body, being equally distributed over the gray and

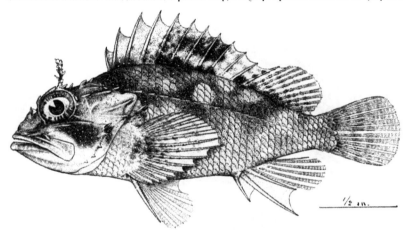

Fig. 40.—*Sebastopistes galactacme* Jenkins, new species. Type.

brown; preopercle dark brown, upper part of opercle a lighter brown; a dark brown area between base of pectoral and humeral spine; middle and upper part of body with brown mottlings; the white points showing to a greater or less extent on all the fins except the caudal; spinous dorsal white, with a blackish area at bases of fifth, sixth, seventh spines, another on distal portion of membrane between sixth and ninth spines; dusky spots on soft dorsal; caudal colorless with faint show of dusky spots near base; anal colorless except white points on the rays; ventrals white, with bright white points on rays; pectoral white, with white points and dusky spots on rays, dark area at base.

This description is based on the type, No. 50692, U. S. N. M. (field No. 2175), 2.6 inches in length, the longest in the collection, and 80 cotypes, all taken by me from the coral rocks on the reef at Honolulu.

227. Scorpænopsis cacopsis Jenkins.

Only one specimen obtained.

Scorpænopsis cacopsis Jenkins, Bull. U. S. Fish Comm. for 1899 (June 8, 1901), 401, figs. 13 and 14, Honolulu. (Type, No. 49690, U. S. N. M.; coll., O. P. Jenkins.)

Scorpænopsis cacopsis, Seale, Occasional Papers, Bishop Museum, I, part IV, fig. 5, 1901 (Honolulu).

228. Dendrochirus chloreus Jenkins, new species.

Head (to end of bony operculum) 3 in length; depth 2.5; pectoral 2.25; ventral 3; caudal 3.2;
D. XIII, 9; A. III, 5; C. 19; P. 18; V. I, 5; scales 8–38–13. Head and body much compressed, greatest
width through bases of pectorals, 5 in length, 2 in greatest depth; dorsal profile a little more convex
than ventral, greatest depth at vertical through base of fifth spine; eye oval, longest diameter horizontal, 3 in head; interorbital deeply concave, 2 in horizontal diameter of eye; profile of snout almost
straight, inclined at angle of 45°; length of snout equal to vertical diameter of eye; suborbital 5 in
head; maxillary reaching to posterior margin of pupil, 2 in head; tip of upper jaw with a toothless depression between inner ends of premaxillaries which receives a slight knob on the upper surface of the
symphysis of the lower jaw; teeth very fine, in bands in jaws, in reniform patch on vomer; no palatine
teeth; posterior nostril large, simple, round, close to rim of orbit; anterior nostrils smaller, each with
flat tentacle on posterior margin; spines on head all small, a small movable spine at outer corner of
each posterior nostril; several small spines on upper margin of orbit, unsymmetrical on the 2 sides;

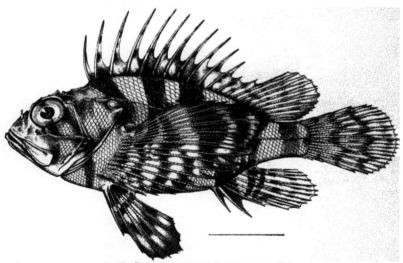

Fig. 41.—*Dendrochirus chloreus* Jenkins, new species. Type.

no occipital depression; 2 spines on occiput just back of eyes on each side, 2 others on each side of
nape; a horizontal series of temporal spines; a moderately long spine at angle of preopercle, 2 smaller
ones below it; a flat spine at angle of opercle; suborbital stay with a series of several very inconspicuous
spines reaching from below anterior margin of orbit to angle of preopercle; a small pit below anterior
lower angle of eye; opercle and preopercle scaled, other parts of head naked; branchiostegals 7;
gillrakers 6 + 10; fin rays not excessively developed; dorsal fin beginning above middle of opercle;
fourth to eighth spines longest, 1.5 in head; first spine 2.75 in head; twelfth shortest, three-fifths of first,
2.5 in fourth; thirteenth spine longer, equal to first and to eleventh; soft dorsal short; median rays
longest, but slightly shorter than longest spine; last soft ray equal to last spine; caudal slightly rounded,
median rays 1.25 in head, 3 uppermost and 3 lowermost rays short and spine-like; first anal spine
shortest and slender, 3.6 in head, a little longer than twelfth dorsal spine; second and third spines of
equal length, but second much thicker than third, each 1.5 in longest dorsal spine, a little less than half
of head; longest pectoral rays reaching to posterior end of base of anal fin; fourth to seventh rays, from
above, longest, a little less than half of length of body; uppermost ray equal to third dorsal spine;

lowermost ray shorter, four-fifths of uppermost; first ray and lower 9 rays unbranched; ventral equal to caudal, second ray longest; spine rather strong, 2 in head; base of innermost ray connected by membrane with body; scales mostly ctenoid, those of base of pectoral, belly, and head cycloid; lateral line beginning a little above end of gill-slit, running backward and downward to middle of end of caudal peduncle; slightly arched anteriorly, slightly convex toward ventral line, posteriorly.

Color in alcohol, light brown, with 6 indistinct dark brown vertical bands on body, and with transverse bands on fins. Color of a fresh specimen (field No. 301), vertical green band on head below eye; 6 olive-green vertical bands on side of body, the first three extending upon dorsal fin; the first at base of first two spines, the last on caudal peduncle; spinous dorsal olive-green with red spots; soft dorsal with oblique bands of alternating red and white; anal fin anteriorly with 2 olive bands crossing lower end of 2 bands of same color on body which extend upon anal fin, posterior part of fin with alternating red and white bands; caudal with alternating crossbands of red and white (posterior part of dorsal and of anal and caudal with alternating red and white bands); ventrals mottled with dark olive; pectoral proximally with 3 bands of dark olive, distally with spots of same color; dark spots just beneath opercular flap; another similar spot just behind the 2 preopercular spines; 2 dusky patches on breast; parts dark in alcoholic specimen were green in life.

The coloration is very similar to that of *Pterois barberi* Steindachner, but differing from that species in the number and arrangement of cephalic spines and in the absence of the conspicuous tentacle on the head.

Probably most closely related to *Pterois brachyptera* Cuvier & Valenciennes; known from Zanzibar to Samoa.

This description is based on the type, No. 50701, U. S. N. M., 5.25 in length, and 8 cotypes, 3.5 to 2.2 inches in length, all obtained by me at Honolulu. This fish is found among the coral rocks and is not common.

Family LI. CEPHALACANTHIDÆ.

229. Cephalacanthus orientalis (Cuvier & Valenciennes). "*Lolohau.*"

Color in life (field No. 255), body drab above, white below, upper part of body with darker and greenish mottlings and with many chestnut brown spots as large as pupil; dermal fringes on lower lip red; lower surface of preopercular spine red; upper surface of pectoral fin dark, with brown spots at the base and around margins, especially anterior margin; a blue spot on distal margin; 4 or 5 bright yellow spots in middle of fin about one-third the distance from the base, lighter yellow beyond; under surface dark bluish without markings; ventrals red and yellow; first spine and filament of dorsal fin black, the remainder of the spinous portion dusky, with dark brown spots; soft dorsal, membranes transparent, rays with dark brown spots; caudal similar; anal transparent, with dark spot at base of last 2 rays. In alcohol the red and yellow disappear.

One specimen of this species was secured.

Dactylopterus orientalis Cuvier & Valenciennes, Hist. Nat. Poiss., iv, 134, pl. 76, 1829, Indian Ocean; Günther, Fische der Südsee, vi, 169, 1877 (Sandwich, Society, and Paumotu Islands).

Family LII MALACANTHIDÆ.

230. Malacanthus parvipinnis Vaillant & Sauvage.

Two specimens are in Dr. Wood's collection from Honolulu, where it seems to be fairly common.

Malacanthus parvipinnis Vaillant & Sauvage, Rev. Mag. Zool. (3), iii, 1875, 283, Sandwich Islands.
Malacanthus hoedtii Günther, Fische der Südsee, v, 160, taf. xcviii, fig. B, 1876 (Tahiti, Yap, and Sandwich Islands); Steindachner, Denks. Ak. Wiss. Wien, LXX, 1900, 497 (Honolulu); not of Bleeker.

Family LIII. GOBIIDÆ.

231. Eleotris sandwichensis Vaillant & Sauvage.

Head 3.25 in length; depth 4.5; eye 6.5 in head; D. vi, 9; A. i, 8; scales 80–21; head wide, depressed, width greater than the depth, 1.5 in length; body anteriorly wide, rapidly becoming compressed posteriorly; caudal peduncle very flat; width of body at middle of second dorsal equals 0.5 width through pectorals; dorsal profile of snout convex, profile above eye concave, profile from eye to front of dorsal rising in a gentle curve; depth of caudal peduncle 2 in head; eye lateral, = bony interorbital, 2 in snout from tip of lower jaw; maxillary reaching to below center of pupil; least width of preorbital 0.6 of eye; teeth in rather wide bands in each jaw, an outer series of enlarged teeth in each; vomer and palatines toothless; entire head, except jaws, scaled; fourth and fifth dorsal spines longest, a little less than 2 in head; membrane from last spine reaching to front of second dorsal fin; sixth soft ray longest, a little greater than 2 in head; A. 1-8, length of base only 0.8 that of soft dorsal; pectoral rounded, median rays 1.5 in head; third and fourth rays of ventral longest, 2 in head; space between ventrals 0.6 of eye; branchiostegals 6; gillrakers very short.

Color in alcohol, brown, paler below; vertical fins spotted with small dusky brown spots, a dusky brown spot on upper part of base of pectoral. Large specimens 6 inches in length are plain dark brown, with the fins dusky. Small specimens, the size of the type, are much paler especially along the sides of the body and head. Several dusky brown bands radiating backward from eye; side of body with small dusky brown spots. Still smaller specimens (2 inches long) are much paler, side of body mottled with dark brown, and side and lower part of head punctate with minute dusky spots; vertical fins pale with distinct dark crossbars.

One fresh specimen was olivaceous with 6 dark bands over the body. A spine on angle of pre-operculum directed downward and forward.

I obtained examples, 1.5 to 9 inches in length, in fresh-water streams and in salt water along shore and in marshes about Honolulu, and some at Hilo. Two examples are in Dr. Wood's collection, and 7 were obtained by the *Albatross* in 1896 from Honolulu.

Eleotris sandwichensis Vaillant & Sauvage, Mag. de Zool., III, 1875, 290.
Eleotris fusca Gunther, Rept. Shore Fishes Challenger, 69, 1880, Honolulu ; Fowler, Proc. Ac. Nat. Sci. Phila., 1900, 516 (Honolulu).
Culius fuscus, Streets, Bull. U. S. Nat. Mus., No. 7, 57, 1877 (Oahu).

232. Asterropteryx semipunctatus Rüppell.

Head 3.25 in length; depth 2.6; eye 4 in head =snout; D. vi, 11; A. 10; C. 22; P. 19; V. i, 5; scales 25-7; least depth of caudal peduncle 2 in head; depth of base of pectoral 2.3; teeth in upper jaw in a band, outer row enlarged, others small, villiform; teeth of lower jaw similar to those in upper, but no enlarged teeth in back part of sides of jaw; lower pharyngeals with the lower ends triangularly expanded, having villiform teeth; third dorsal spine filamentously prolonged, reaching to base of third ray of second dorsal; second dorsal spine longest of rest, 1.6 in head; posterior dorsal rays increasing slightly in length, tenth longest, 2 in head; caudal rounded; anal similar to second dorsal, next to last ray longest, equaling second dorsal spine, 1.6 in head; median rays of pectoral longest, upper 4 unbranched; fourth ray of ventral longest; inner very slender, unbranched; dorsal profile of head and body a little more convex than the ventral; upper profile of head sloping upward at angle of about 45° from snout to front of first dorsal, gently rounded, descending in gentle curve to posterior end of base of second dorsal; upper profile of caudal peduncle straight, horizontal; mouth but little oblique, lower profile with less inclination than upper, mouth on level of lower third of pectoral fin; ventral profile of body very gently and regularly curved from base of ventrals to base of caudal fin; greatest depth of body at front of first dorsal; interorbital very narrow, width less than half diameter of pupil; entire head and body scaled except interorbital, top and sides of snout and jaws; scales all large, those of head scarcely smaller than those of body; scales below eye cycloid, the rest ctenoid; preorbital narrow, less than diameter of pupil; body somewhat compressed, width of head 1.75 in length, widest part of body very slightly narrower than head.

Color in alcohol, body faded, general color brown, about 6 unequally defined, dusky, vertical bars on side back of pectoral, a similar band over nape midway between eyes and front of first dorsal spine; traces of pale blue spots on side; dorsal and anal fins dusky, other fins pale.

Color in life (field No. 221), uniform dark with rows of minute blue dots.

Many specimens were obtained at Honolulu, where it is abundant.

Asterropteryx semipunctatus Rüppell Atlas Fisch., 138, pl. 34, fig. 4 1828, Red Sea; Klunzinger, Fische des Rothen Meeres, 484, Verh. K. K. Zool. bot. ges II, Wien, 1871
Eleotris cyanostigma Bleeker Tyds. Ned. Ind., VIII, 452, 1855, Cocos Islands.
Eleotrodes cyanostigma, Bleeker, Enum. Spec. Pisc. Arch Ind., 112, 1859.
Eleotris cyanostigma, Gunther, Cat., III 119 1861 (s.a of Booroo and Kokos Islands).
Brachyeleotris cyanostigma, Bleeker, Arch. Neer. Sci. IX, 306 1874, X 464, 1875, Streets, Bull. U. S. N. M. No. 7, 58, 1877 (Oahu).

EVIOTA Jenkins, new genus.

Eviota Jenkins, new genus of *Gobiidæ* (*epiphanes*).

Related most closely to *Oxymetopon* Bleeker, from which it is distinguished by the following characters: Body not greatly elongate, head not compressed into a keel, dorsal fins separate, neither dorsal nor anal elongate.

233. Eviota epiphanes Jenkins, new species.

Depth 4 in length; head a little greater than the depth; eye 3 in head; snout shorter than eye, about 4 in head; D. vi, 10; A. 9; scales 25–6. Body not elongate, not compressed; head not compressed and without keel; inferior pharyngeal bones not united with each other, each enlarged toward its lower end, making a triangular expansion, each armed for almost its entire length with rather long, slender, tapering villiform teeth; teeth in upper jaw in a band, widest in front; teeth villiform, short, conical, and a little curved inward; in front of side of each half of jaw in outer series and more prominently curved inward, several large canine-like teeth; teeth in lower jaw, villiform, straight, slender, tapering, in band widest in front, where there is a group of enlarged, backward-curved, canine-like teeth; vomer and palatines toothless; branchiostegals 5; pectoral 17 rays; the second to seventh (from

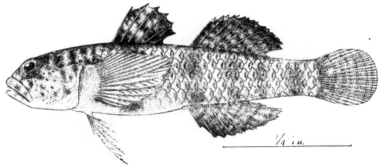

Fig. 42 —*Eviota epiphanes* Jenkins, new species. Type.

the lower edges inclusive, branched, the others simple, the seventh longest; a group of about 6 long, slender, bristle-like spines projecting from the segments of tips of the rays, and extending basally through second third terminal segments of ray, as in *Eleotris*; ventrals close together, of one short spine and 5 rays, fifth ray of each rudimentary, fourth longest, others successively shorter; first about half length of fourth; all the rays, except fifth, branched, the branches very short, lateral, on the outer side only of the rays; base of pectoral large; longest ray equal to head; ventrals shorter than pectoral; dorsal fins well separated, by a distance greater than a third of the head; second, third, and fourth dorsal spines longest, about 2 in head; first soft dorsal ray unbranched; anal and soft dorsal rays of about uniform height, about 1.6 in head; interorbital very narrow, less than pupil; eyes almost contiguous; caudal peduncle length 4.5 in body, depth 6.5 in body; a prominent anal papilla.

Color in life, general color blue; dark vertical bands imbedded in the body; caudal fin orange; the markings on head, body, and scales bright brown; belly blue, in some specimens very bright; the body translucent.

Color in alcohol, orange, marked everywhere with small, round black dots; dots on top of head and dorsal part of body grouped into areas which form spots between eyes and first dorsal spine also, scattered on sides of head; in some grouped into 2 or 3 short radiating bands from below eye and on side of body, 5 or 6 vertical bands, the first just back of axil, the last before base of caudal fin; numerous very small dots on base of pectoral; on sides of body are small crescentic groups of dots on each scale; dorsal and anal fins punctate dusky; other fins colorless.

This description is based on the type and 20 cotypes taken by me at Honolulu in 1889. The largest of these is 15 mm. in length. This is the smallest vertebrate that has up to this time been described. This minute species was obtained by breaking up heads of coral over a dip net. (Type, No. 50720, U. S. N. M.)

234. Gobius albopunctatus Cuvier & Valenciennes.

Color in life, whitish, with shades of pale brown and dark cloudings; head and body covered with many small, pearly, white spots; coloration of dorsal, caudal, and pectoral same as body; ventral milky white with some dusky marking; anal white with dusky margin; belly and throat white. When seen from above, the cloudings make 3 rather distinct broad bands across the back.

One hundred and thirty specimens were obtained at Honolulu, ½ inch to 4 inches in length. It is very abundant at Honolulu along the shores, in marshes and mouths of streams.

Gobius albopunctatus Cuvier & Valenciennes, Hist. Nat. Poiss., XII, 57, 1837. Isle de France: Günther, Cat., III, 25, 1861; Günther, Fische der Sudsee, VI, 172, pl. CX, fig. A, 1877 (Fiji and Society Islands).
Gobius nebulo-punctatus Rüppell, Neuewirb , Fische, 139, 1835.
Gobius punctillatus Rüppell, l. c., p. 138.

235. Awaous genivittatus (Cuvier & Valenciennes).

Color in life, general color pale; sides of belly with shades of pink; body with 5 or 6 more or less distinct transverse bands and mottlings of dark along back; broad black band downward and obliquely backward; dark spot on upper part of base of pectoral; caudal transversely barred; dorsal and anal with longitudinal bars. Another specimen was very pale, with about 11 crossbars, less distinct forward, more distinct posteriorly; an indistinct, dark, longitudinal band along middle of side; dorsal fin transparent, with black spots; caudal tinged with olive; a broad black band from eye downward and backward. All these colors fade more or less quickly on being taken out of water.

Seven specimens, 1.8 to 5.7 inches in length, were obtained by me; and two, 5.1 and 5.4 inches in length, by Dr. Wood, at Honolulu. It occurs in fresh waters.

Gobius genivittatus Cuvier & Valenciennes, Hist. Nat. Poiss., XII, 64, 1837. Otahaiti: Günther, Cat., III, 13, 1861 (Otaheite): Günther, Fische der Sudsee, VI, 170, taf. CX, fig. C, 1877 (Tahiti, Fiji, and Sandwich Islands).
Awaous genivittatus, Fowler, Proc. Ac. Nat. Sci. Phila. 1900, 517 (Sandwich Islands).

236. Awaous stamineus (Cuvier & Valenciennes). "Oopu."

Color in life, body yellow, but somewhat transparent; pectorals and ventrals plain; caudal with 4 bars of black; first dorsal with black dots; second dorsal with 4 longitudinal rows of black dots and reticulations; body covered with black dots and reticulations; a black spot at base of caudal; belly white; golden spot on upper branchiostegal and opercle.

Thirty-two specimens of this species, 1 to 9 inches in length, were obtained by me in fresh water at Honolulu, 4 by the *Albatross* in 1896, 2 by Dr. Wood from Honolulu, and 4 by Mr. McGregor from a small ditch at Hilo in 1900. The largest of all of these is 9 inches in length.

Gobius stamineus Eydoux & Souleyet,Voy. Bonite, Poiss., 179, pl. 5, fig. 5, 1841, Sandwich Islands; Günther, Shore Fishes, Challenger, Zool., I, part VI, 59, 1880 (fresh waters of Honolulu and Hawaii).
Awaous crassilabris, Streets, Bull. U. S. Nat. Mus., No. 7, 59, 1877 (Oahu): Fowler, Proc. Ac. Nat. Sci. Phila. 1900, 517 (Sandwich Islands).

237. Sicyopterus stimpsoni (Gill).

One specimen, 3.4 inches in length, was taken by me from fresh-water streams at Honolulu.

Sicydium stimpsoni Gill, Proc. Ac. Nat. Sci. Phila. 1860, 101, Hilo, Hawaii, in fresh water: Günther, Cat., III, 93, 1861, Günther, Fische der Sudsee, VI, 183, 1877 (Gill's description).
Sicyopterus stimpsoni, Streets, Bull. U. S. Nat. Mus., No. 7, 59, 1877 (Oahu, in fresh water).
Sicydium nigrescens Günther, Challenger, Zool., vol. I, part VI, 60, pl. XXVI, fig. C, 1880 (Hawaii, fresh water: Honolulu, fresh water).

CHLAMYDES Jenkins, new genus.

Chlamydes Jenkins, new genus of *Gobiidæ* (*laticeps*). Distinguished from the genus *Gobius* by the presence of scales on the sides of the head.

238. Chlamydes laticeps Jenkins, new species.

Head 3 in length; depth 4; width of head 0.8 of its length, depth 1.5 in its length; D. vɪ-ɪ, 9; A. ɪ, 8; C. 17; ventral fins united ɪ, 5; scales 38,-14; head depressed; ventral profile almost straight; dorsal profile rising in very gentle curve from tip of snout to nape; dorsal and ventral outlines of body straight and parallel from base of first dorsal to front of anal, from here slightly converging to base of caudal fin; height of caudal peduncle 2.25 in head; 1.6 in height of body at front of dorsal fin; mouth almost horizontal; snout flat, broad, equal to eye, 3.5 in length of head; interorbital very narrow, less than diameter of pupil; eyes inclined at angle of 45° on sides of head; snout bluntly rounded from above; top and sides of head scaled to posterior border of pupils; branchiostegals 4; teeth villiform in bands on each jaw; an outer series of enlarged teeth in the upper jaw; fourth dorsal spine longest, 2.5 in head; rays of second dorsal of nearly uniform height, slightly longer than fourth spine, 2.3 in head; caudal rounded; median rays 1.5 in head; middle rays of anal longest, 2 in head; median pectoral rays longest, 1.8 in head, the lower 14 rays normal, above these numerous fine silk-like rays; median rays

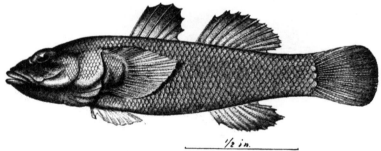

FIG. 43.—*Chlamydes laticeps* Jenkins, new species. Type.

of ventral 2 in head; basal membrane with a well-developed lobe on each side; scales ctenoid, covering body and top and sides of head; those on posterior part of body and on caudal peduncle but little enlarged; those on head smaller than body scales.

Color in alcohol, plain chestnut brown, pale below; a few darker mottlings on side of body; ventral and pectoral fins dusky brown, ventral pale.

This description is based on a single specimen, 1.5 inches long, taken by me in the coral rocks on the reef in front of Honolulu. (Type, No. 50716, U. S. N. M.)

239. Gobionellus lonchotus Jenkins, new species.

Head 3.75 in length; depth 4.75; eye 3.6 in head; snout 3; D. vɪ-ɪ, 12; A. ɪ, 13; scales 75-28; head and body much compressed; width of head 2 in its length; depth of head 1.5 in its length; width of body through middle of soft dorsal 4 in head, 0.5 width of head; height of caudal peduncle 3 in head; caudal fin long and pointed; median rays 2.5 in length of body; mouth oblique; snout blunt; profile of lower jaw roundly convex; profile of head steeply rising to before middle of eye, rounded from here to top of head; dorsal and ventral profiles gently converging to base of caudal from its greatest depth, which is just back of base of ventrals; interorbital extremely narrow, about 3 in eye; least width of preorbital equaling vertical diameter of eye; teeth in single series in the upper jaw; in bands in lower jaw with the outer series enlarged; no teeth on vomer or palatines; fourth and fifth dorsal spines longest, 1.3 in head; eleventh soft dorsal ray longest, a little less than head; twelfth soft anal ray longest, equal to longest dorsal spine; caudal fin long and pointed, median rays 2.5 in length of body; pectoral pointed, median rays longest, very slightly greater than head; ventrals equal to head; branchiostegals 4; gillrakers very small and soft, about 6 on lower arm of arch; scales apparently very small and very deciduous, only a few preserved on any of the specimens.

Color in life, pale translucent, with the markings very indistinct, these brought out by the alcohol.

Color in alcohol, brown; about 10 vertical darker brown bars on side, indistinct in older, darker examples; a transverse band of same color on nape just back of eyes; another band of dark brown running downward from middle of lower border of eye to back of angle of mouth; a large, vertically elongate, oval spot on base of pectoral; soft dorsal and caudal fins finely banded transversely with darker; do not show on larger specimens.

This description is based on the type, 4.3 inches in length, and several cotypes, all from Honolulu. They were caught in great numbers by the Chinese and sold in the market, where they are salted and eaten without cooking. (Type, No. 50808, U. S. N. M.)

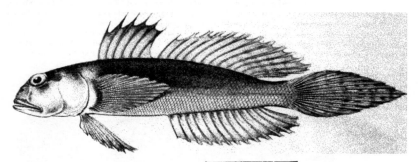

Fig. 44. *Gobionellus iomelas* Jenkins, new species. Type.

Fig. 45. *Enypnias oligolepis* Jenkins, new species. Type.

240. Enypnias oligolepis Jenkins, new species.

Head 4.5 in length; depth 5.75; eye 4 in head; snout 4.5; D. vi, 12; A. 7; V. i, 5; P. 15. Body elongate, compressed, greatest depth about under last dorsal spine; head compressed, its depth 1.25 in its length, its width 1.5 in its length; interorbital narrower than pupil; profile of snout bluntly rounded; top of head flat; dorsal outline from head to first dorsal slightly concave, from this gently curving to caudal; ventral outline curving from tip of lower jaw to ventral fin, nearly straight to origin of anal, curving to caudal; spinous dorsal well separated from soft dorsal; first to fifth spines about equal in length, about 1.3 in head; the last, shorter; soft dorsal slightly higher than spinous dorsal; anal about equal to dorsal; middle rays of pectoral the longest, 0.25 longer than head; ventral fin about equals head; caudal fin somewhat longer; teeth in lower jaw large, in a wide band anteriorly, in a single series posteriorly, inner ones slender, straight; outer enlarged, notably serial toward front of sides of jaw, which are canine-like and bent backward; teeth in upper jaw similar but with fewer canine-like teeth; vomer and palatines toothless; body apparently scaleless with the exception of a few very minute scales on the posterior portion.

Color in alcohol, plain brown, minutely punctate with black; about 12 dark brown vertical bars on sides of body, those on caudal peduncle very indistinct; generally 1 or 2 poorly defined similar bands across nape; generally several short radiating bands from lower border of eye; the brown bands on side of body much wider than pale narrow interspaces.

This description is based on the type and 12 cotypes, each about 1.2 inches long, caught by breaking up coral rocks on the reef in front of Honolulu. (Type, No. 50715, U. S. N. M.)

Family LIV. PTEROPSARIDÆ.

241. Osurus schauinslandi (Steindachner).

Two specimens obtained by Dr. Wood. My description of this fish as a new species was published before I received Dr. Steindachner's paper. It seems to be fairly common at Honolulu.

Percis schauinslandi Steindachner, Anzeiger, No. XVI, June 27, 1900, for Denks. Ak. Wiss. Wien, LXX, 1900, 496, pl. III, fig. 5, Honolulu.
Parapercis pterostigma Jenkins, Bull. U. S. Fish Comm. for 1899 (June 8, 1901), 492, fig. 15, Honolulu. (Type, No. 49701, U. S. N. M.; coll. Dr. Wood.)

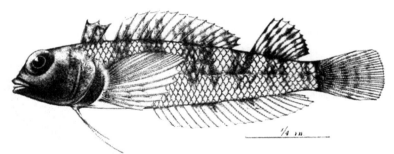

Fig. 16.—*Tripterigion atriceps* Jenkins, new species. Type.

Family LV. BLENNIIDÆ.

242. Tripterigion atriceps Jenkins, new species.

Head 3.3 in length; depth 4.6; eye 3 in head, equaling snout; D. xv, 10; A. 20; C. 19; P. 15; V. 2; scales 3-34-7; depth of caudal peduncle 3 in head, 2.5 in depth of body; length of caudal fin equal to depth of body, 4.6 in length; pectoral equal to head, 3.3 in body; ventral equal to caudal; snout acute from above and laterally; dorsal profile forming angle with that of chin of about 65°; about 45° with longitudinal axis of body; dorsal profile of snout re-entrant before lower rim of eye, thence rising steeply to straight line from tip of snout to before eye, then turning backward at much gentler inclination to foot of first dorsal fin, whose origin is situated a little before posterior border of opercle; greatest depth of body at front of second dorsal; dorsal and ventral profiles beyond this point converging; dorsal profile anteriorly gently convex; posteriorly, including profile of peduncle, slightly concave; ventral profile from ventrals to anal evenly convex; body compressed; posterior part of head wider than the body, equal to 0.6 of length of head; width of body at bases of pectorals 2 in head; back of this the width rapidly tapers to very narrow caudal peduncle, width here equal to pupil; mouth very slightly oblique; both interorbital and suborbital very narrow; interorbital but slightly wider than pupil; suborbital a little greater than interorbital; spines of first dorsal of nearly equal length, about equal to eye; anterior and middle spines of second dorsal longest, 1.75 in head; anterior rays of third dorsal again abruptly longer than last rays of second, slightly longer also than longest rays of second, 1.6 in head; secondary rays rapidly shorter to last, which is of about same length as last rays of second dorsal, slightly shorter than spines of first fin; anal fin of approximately uniform height, rays equal to spines of first dorsal; posterior border of caudal straight, upper and lower 3 rays much smaller than others, tips free from posterior edge of fin; pectoral pointed, middle rays longest, lower 6 simple, others bifid toward tip; teeth small, simple, in bands in each jaw, widest in front; an outer series in each of enlarged teeth; no canines; a V-shaped patch on vomer.

Color in alcohol, general color pale reddish brown, paler and more yellowish below; snout, sides, and ventral surface of head, gill-membranes, humeral region at base of pectoral fin, covered with close-set black dots; top of head mottled with dusky; on side of body about 8 irregular vertical dusky bands, split below with the ground color, giving them an irregular Y-shape; 2 bands on caudal peduncle, simple, apparently produced by entire splitting of a single band; caudal fin finely and irregularly crossbanded with dusky; dorsals similarly marked; other fins plain; black color of head and of black lateral bands on body not continuous coloring, but formed of numerous closely, evenly distributed, small, round, black dots. Considerable variations in color are noted. Some have no black area on head or shoulders, except small, scale-like, black spots.

Scales rather large, fine ciliated; lateral line straight, with tubes on first 19 scales, ending a little behind middle of body, here dropping to second scale row below and going farther backward for 4 more scales.

T. atriceps is related to *T. hemimelas* but differs in the number of dorsal spines and of dorsal and anal soft rays.

This description is based on the type, No. 50719, U. S. N. M., a specimen about 1 inch long, and 9 cotypes caught by me at Honolulu, by breaking up heads of coral over a dip net.

243. Salarias brevis Kner.

One specimen of this species, 4.6 inches in length, is in Dr. Wood's collection. Structurally this specimen seems to correspond to Kner's description and figure, but the grouping of the spots in my example differs from that shown in Kner's figure. The example described by Kner was from the Godeffroy Museum from Savaii, and to the time of this record was the only known specimen of the species.

Salarias brevis Kner, Sitzb. Ak. Wiss. Wien, LVIII, 1868, 334, taf. 6, fig. 18. Günther, Fische der Südsee, IV, 203, taf. 118, fig. C, 1877.

244. Salarias cypho Jenkins, new species.

Head 5 in length; depth 5; eye 4 in head; snout 4; suborbital 7.6 in head, 1 6 in eye; interorbital 3.5; D. XII, 22; A. 24; C. 17; P. 14; depth of apex of caudal peduncle 2.25 in head; width of body at bases of pectorals 1.6 in head; middle of apex of caudal peduncle equal to one-half diameter of eye; profile of top of head straight horizontal, before eyes sloping downward very sharply to mouth; lower jaw included; lower profile of head inclined gently downward from symphysis; mouth horizontal; dorsal profile almost straight and horizontal from origin of dorsal to middle of soft dorsal, inclined slightly downward from here to base of caudal fin; belly distended; profile from front of anal fin to base of caudal straight, gently inclined upward; a simple filamentous tentacle on middle of eye equal to diameter of eye; a large dermal crest on occiput, equal to snout in length, 2.4 in head; no canine teeth in either jaw, single series of comb-like teeth in margin of each jaw, upper forming a semicircle; dorsal beginning above upper end of gill-slit; anterior spines somewhat curved backward; first spine 1.75 in head; second, third, and fourth spines longest, 1.3 in head, the following regularly decreasing, last 2.75 in head; first ray of second dorsal abruptly longer than last of first dorsal, separated from it by a greater interval than that between connected spines of first dorsal; membrane between the two fins deeply notched; first soft ray very slightly shorter than first spine; rays increasing slightly in length to fifth, which is 1.3 in head, equal to longest spine; succeeding rays about equal to fifth, except posteriorly, where last short ray equals 2 in head; first anal ray hidden in membrane, weak, one-half diameter of eye, second anal ray 3 in head; succeeding rays gradually but slightly increasing in length to sixth from end, which is a little greater than half head; last anal ray a little longer than first, 3 in head; border of membrane notched between each 2 rays; tips of rays curved backward; last anal ray opposite antepenultimate dorsal ray; anal fin not connected with caudal; last dorsal ray connected by membrane with upper edge of caudal peduncle and base of upper caudal ray; posterior border of caudal rounded; pectoral pointed; tenth ray from above longest, equal to head; median caudal rays 4.3 in length; ventral of 2 simple rays; inner slightly the longer, 1.5 in head; upper nostril simple, below center of pupil and close to anterior rim of eye; anterior nostril below posterior, before middle of lower half of eye, with a fringed tentacle on upper margin.

Color in alcohol, plain dark brown above, paler brown below; dorsal fins brown with darker margins; second dorsal with numerous small, oblique, dusky brown streaks between the rays; anal brown, pale at base, becoming dusky brown at margin; tips of rays colorless; caudal fin dusky brown; pectoral brown like side; ventral pale brown; very indistinct indications of veiled dusky bands on side of body. Specimens vary much in color. Some have a ground color of pale gray, except top and front of head, which in such specimens is light brown. The fins in all pale brown; marked as in the type. On each side, however, are about 8 distinct veiled brown bands. Generally all but the last 2 are more or less distinctly split vertically into a pair of bands. In many such specimens the brown color is found on the back and the pale ground color appears as a series of 7 pale spots along the side of the dorsal fin, corresponding in position with the pale interspaces between the lateral vertical bands. There is every gradation in color between the specimens thus marked and those that are almost

FIG. 47.—*Salarias ophis* Jenkins, new species. Type.

plain. The largest specimens are always plain brown, and in general it is the smaller specimens that have the pale and banded coloration, but many of the smaller ones are almost plain brown with but slight indication of lateral bands.

This description is based on the type, No. 50897, U. S. N. M., a specimen about 4 inches in length, and 54 cotypes, all from Honolulu, where they were collected by me in 1889.

245. Salarias marmoratus (Bennett).

Five examples of this species were taken by me from holes in the rocks at low tide near Honolulu, the largest 2 inches in length.

Blennius marmoratus Bennett, Zool. Journ., IV, 35, Hawaiian Islands.
Salarias ornatus Bleeker, Nat. Tijds. Ned. Ind., VIII, 173, 1855, Cocos Island, Günther, Cat., III, 249, 1861.
Salarias marmoratus, Cuvier & Valenciennes, Hist. Nat. Poiss., XI, 305, 1836, Ceylon; Günther, Cat., III, 248, 1861 (Sandwich Islands); ibid, Fische der Südsee, 204, pl. 116, fig. B, 1877.

246. Salarias variolosus Cuvier & Valenciennes.

Color in life (field No. 277), brownish olive, lighter anteriorly, posteriorly dusky; head and face with bright red dots; upper half of anterior margin of dorsal bright red; posterior portion of dorsal dark; anal dark, almost black; caudal dark, upper margin red; pectoral olivaceous.

Five specimens of this species, 2.1 to 2.6 inches in length, were obtained at Honolulu.

Salarias variolosus Cuvier & Valenciennes, Hist. Nat. Poiss., XI, 317, 1836, Guam; Fowler, Proc. Ac. Nat. Sci. Phila. 1900, 518 (Sandwich Islands).

247. Salarias saltans Jenkins, new species.

Head 4.75 in length; depth 5.5; eye 3.6 in head; snout 5; D. vii, 20; A. i, 21; ventral 8 in length; pectoral 5.5. Body elongate, slender, compressed, width through pectorals 2 in head; profile of snout rising vertically to below eye, then bulging slightly forward, curving upward and backward around eye; profile of top of head horizontal, continuing in straight line with profile of back to posterior end of second dorsal; from here both dorsal and ventral profiles converging to base of caudal; ventral profile of head descending from mouth to posterior edge of gill-membrane; from head to foot of anal slightly convex; mouth inclined slightly upward posteriorly; eye placed in upper anterior angle of side of head, close to profile; posterior nostril simple, placed above and somewhat lateral to the anterior, before center of pupil; anterior nostril with a soft, short, branched tentacle on its upper rim; eye circular; a single, filamentous, tapering tentacle above eye over anterior half of pupil, not quite as long as diameter of eye; eyes inclined, looking laterally and upward; interorbital space very narrow, less than diameter of pupil; suborbital 6 in head; length of head behind eye contains eye 2.5 times; gill-openings large, membranes broadly united; teeth movable, forming a fine comb along margin of each jaw; well within these in back of lower jaw, a small, backward curved, canine tooth in each jaw; branchiostegals 6; dorsal fin deeply notched; dorsal spines low, first of same length as last, 2.5 in head; spines gradually increasing toward middle of fin, there longest, 4.8 in head; first soft ray a little longer than longest spine, 1.75 in head; soft rays of uniform length, from fourth to fourteenth, being 1.3 in

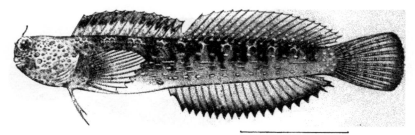

FIG. 18.—*Salarias saltans* Jenkins, new species. Type.

head; the soft dorsal being considerably more elevated than anterior dorsal, rays back of fourteenth decreasing in size, last equal to last spine; anal similar in shape and size to soft dorsal; caudal slightly rounded; median rays a little longer than head; pectoral pointed with 14 rays, median ones longest, all simple; ventral inserted before base of pectoral, below posterior ends of branchiostegals, of 2 simple rays of which the inner is the longer; caudal of 17 rays; lateral line present on anterior half of body, extending in a gentle curve from upper end of gill-slit above pectoral to middle of side of body.

Color in life (field No. 152), general color olivaceous, with brown reticulations on fins and head; a line of pearly light blue dots (longitudinal row) along body three-fourths way down; below this many spots of same color; dorsal fin with black spot on anterior part (anterior upper angle); margin brown; fin covered with many white dots; anal unmarked except a marginal band.

Color in alcohol, ground color pale brown; upper half of side crossed by 8 indistinct, wide, dusky vertical bars; general color of head pale, covered above, below, and on sides with numerous small white spots, the paler ground color appearing as a reticulation between the spots; several longitudinal series of small oval or elongate white spots on lower half of side; fins pale brownish; soft dorsal and caudal spotted with white; spinous dorsal with a black spot between first and second spine; soft anal with wide dusky border, tips of spines white; pectoral plain brown with spots toward its base.

This description is based on the type, No. 50886, U. S. N. M., 3.2 inches in length, and 2 cotypes, caught in holes in the rocks at low tide near Honolulu in 1889.

248. Salarias rutilus Jenkins, new species.

Head 4.3 in length; depth 5.3; eye 3 in head; D. xii, 20; A. i, 21; C. 15; P. 14. Profile of head and body almost identical with that of *Salarias saltans;* profile of back and head from eye to base of soft dorsal straight and horizontal; dorsal and ventral profile of body from base of soft dorsal and anal fins slightly converging to apex of caudal peduncle, whose depth is 2 in depth of body through pectoral, or 10 in length; body compressed, width decreasing posteriorly; width of apex of caudal peduncle less than diameter of pupil; profile of snout slightly receding from eye to mouth; profile before and above eye prominently bulging; eye at upper anterior angle of side of head, front of head forming an isosceles triangle; interorbital very narrow, less than diameter of pupil; suborbital 6 in head; branchiostegals ?, upper ones projecting beyond posterior end of opercle; a small, backward curved, canine tooth in back part of each side of lower jaw well within the outer teeth which form a border to each jaw of 5 small teeth; a fine, slender, simple tentacle over the eye; mouth inclined slightly upward posteriorly; chin sloping backward and downward from mouth; gill-openings large, members united, fused with isthmus; lower jaw included; dorsal spines of nearly uniform height, median one longest, 2 in head; last one considerably shorter, about 3 in head; a deep incision between first and second dorsals; first ray of second dorsal slightly shorter than longest ray of first dorsal; last soft ray of same length as last spine of first dorsal; other rays slightly elongated toward middle of fin; soft dorsal not united with the caudal; anal similar in size and shape to soft dorsal; last ray opposite penultimate dorsal ray; caudal slightly rounded, median rays nearly equal to length of head; pectoral

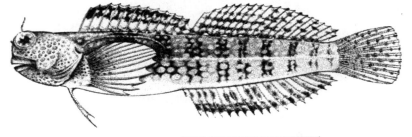

Fig. 49.—*Salarias rutilus* Jenkins, new species. Type.

bluntly pointed, median rays longest, equal to caudal, nearly equal to head; lateral line forming a gentle curve over pectoral from upper end of gill-slit to middle of side of body; ventral of 2 simple rays; inner ray a little the longer, 1.5 in head. Structure and form almost identical with that of *Salarias saltans*, the chief differences being in coloration.

Color in alcohol, ground color pale reddish brown; lower part of opercle and lower surface of head including the gill-membranes with numerous small reddish dark brown spots; spots extending on side of body, left of pectoral and on anterior surface of pectoral; side of body with 9 (in each of 4 specimens) vertical dusky bands; first just back of base of pectoral, last just before base of caudal; bands widest anteriorly, there they are wider than interspaces; posterior ones equal to interspaces, split above and below and so invaded by ground color that they have a double-H form, one H above the other; belly plain pale; a black quadrate spot between the tips of first and second dorsal spines, the part between first and second a solid black; back of the marginal half of fin, and posteriorly the whole fin with regular brown blotches forming very irregular longitudinal bands; soft dorsal similarly marked, first bands more definitely formed; on soft dorsal spots mostly on spines; caudal spotted like dorsal fins, spots forming on it vertical bands; anal fin with a dusky margin, formed of a dusky area on membranes back of tips of rays; a series of vertically elongate, oval, dark brown spots along middle of fin; ventral unmarked; on anterior part of side of body small brown spots on both black bars and interspaces.

This description is based on the type, No. 50895, U. S. N. M., 2.5 inches long and 3 cotypes, the smallest 2 inches long, caught by me in holes in the rocks at low tide near Honolulu.

249. Aspidontus brunneolus Jenkins, new species.

Head 4.3 in length; depth 5; eye 3 in head; snout 5; D. 31; A. 19; P. 14; C. 13; V. 2. Body somewhat elongate, deepest at anterior end, compressed, widest anteriorly; head compressed but slightly wider than body; profile from snout to top of head about a fourth of a circle, from this point gently rising to about middle of dorsal; profile of chin and breast convex; belly slightly bulging: from vent both dorsal and ventral outlines gradually converge to caudal peduncle; depth of caudal peduncle about 3 in depth of body; beginning slightly in front of pectoral, dorsal fin highest at about twenty-third ray, length 5 in body, 1.8 in head; last ray united by membrane along its whole length to caudal fin; caudal rounded, 1.2 in head; anal fin lower than the dorsal, middle rays the longest, about 2 in head, the last ray joined to caudal by a membrane its whole length; pectoral somewhat pointed, middle rays longest, 1.3 in head; ventrals about equal to depth of body, outer ray the longer, falling considerably short of the vent, about half its length; teeth rather long, close set, in a single series in each jaw; a backwardly curved canine tooth in the side of each jaw, the lower the larger; entire head and body scaleless; skin with numerous small warty elevations.

Color in alcohol, dark brown, fins blackish; in life uniformly black.

This description is based on the type, No. 50718, U. S. N. M., 1.25 inches long, and 18 cotypes, collected by me from the coral rocks at Honolulu.

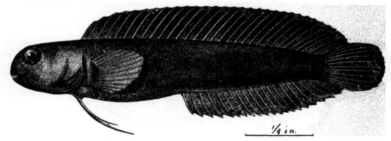

Fig. 50.—Aspidontus brunneolus Jenkins, new species. Type.

Family LVI. BROTULIDÆ.

250. Brotula marginalis Jenkins.

One example, the type, is in Dr. Wood's collection of 1898.

Brotula marginalis Jenkins, Bull. U. S. F. C. 1899 (June 8, 1901), 383 fig. 16, Honolulu (Type, No. 49894, U. S. N. M.)

Family LVII. PLEURONECTIDÆ.

251. Platophrys pantherinus (Rüppell).

Color in life of a small example (field No. 130), on the eyed side general color light, covered with numerous whitish and brownish patches, the centers of most of which contain small dots, the coloring giving the appearance of a stone covered with minute lichens; the eyeless side white. Larger specimens show the same markings, the spots being larger, and there is in the greater number a large brown spot on the lateral line two-thirds of the way back.

Twelve examples of this species, from 6 to 4 inches in length, were obtained.

Rhombus pantherinus Rüppell, Atl. Reis. Nordl. Af. Fisch., 121, pl. 31, fig. 1, 1828, Red Sea.
Rhomboidichthys pantherinus, Streets, Bull. U. S. Nat. Mus., No. 7, 57, 1877 (Honolulu); Günther, Rept. Shore Fishes, Challenger, Zool., part VI, 61, 1880 (Honolulu).
Platophrys pantherinus, Steindachner, Denks. Ak. Wiss. Wien. LXX. 1900, 511 (Honolulu).

Family LVIII. ANTENNARIDÆ.

252. Antennarius commersoni (Lacépède).

One specimen is in my collection; it was not seen fresh. In alcohol it is dark brown, almost black; side with black spot; a black spot on base of posterior rays of anal and of dorsal, a whitish spot above base of pectoral. It corresponds fairly well with the figure in Günther, Südsee, plate 103, fig. B.

Lophius commersoni Lacépède, Hist. Nat. Poiss., I. 327, 1801, South Seas.
Antennarius commersonii, Günther, Fische der Sudsee, v, 163, taf., 100 to 104, 1876 (Raiatea, Bonham, Tahiti, Sandwich, Society, Zanzibar, Huahue, Navigator Islands); Steindachner, Denks. Ak. Wiss. Wien, LXX, 1900, 497 (Laysan); Fowler, Proc. Ac. Nat. Sci. Phila. 1900, 519 (Sandwich Islands).
Chironectes niger Garrett, Proc. Cal. Ac. Sci., III, 1868, 107, Sandwich Islands.

253. Antennarius bigibbus (Lacépède).

Color in life (field No. 274, and 4 others), general color bright yellow, with brown reticulations on most examples, one yellow without the reticulations on the body; in another, the reticulations very indistinct, while on one they are very distinct; pectoral, ventral, and anal fins each with a dark brown band, also outer margin of each of these dark brown; near base of caudal 2 dark brown, almost black, crossbands; posterior margin of caudal dark brown; dorsal with a very narrow dark brown line on the outer margin; very narrow brown lines radiating from the eye.

Five examples of this small fish were obtained by me at Honolulu, by breaking heads of coral over a dip net.

Lophius bigibbus Lacépède, Hist. Nat. Poiss., I, 325, 1798.
Antennarius bigibbus Günther, Cat., III, 199, 1861 (Madagascar); Günther, Fische der Südsee, v, 165, taf. 100 v, fig. A, 1876 (Paumotu, Sandwich, Huahue Islands).

254. Antennarius rubrofuscus (Garrett).

One specimen, 5.1 inches in length, was obtained by me in Honolulu, and I saw one of about the same size in alcohol in a "curio" shop, for which the owner asked a large price.

Chironectes rubrofuscus Garrett, Proc. Cal. Ac. Sci., III, 1868, 64, Sandwich Islands.

A CATALOGUE OF THE SHORE FISHES COLLECTED BY THE STEAMER ALBATROSS ABOUT THE HAWAIIAN ISLANDS IN 1902.

BY JOHN OTTERBEIN SNYDER,

Assistant Professor of Zoology in Leland Stanford Junior University.

This paper contains a list of the species of fishes obtained by the United States Fish Commission steamer *Albatross*, Chauncey Thomas, commander, in the shore and reef work about the Hawaiian Islands during the spring and summer of 1902, under the general direction of Dr. David Starr Jordan and Dr. Barton Warren Evermann. In addition to the fishes collected from the tide pools and from the shallow water near shore, a few are included which were taken from depths of 75 fathoms or more, while some have come from near the surface far out at sea. Several species occurring in the rivers and ponds are also noted. Many of the rarer forms, which are only occasionally caught by the fishermen, were secured through the kindly interest of Mr. E. Louis Berndt, inspector of fisheries in Honolulu.

The writer takes pleasure in expressing his obligations to Dr. Charles H. Gilbert, who had immediate supervision of the zoological work of the *Albatross*, and also to Dr. Jordan, for much help and for many valuable suggestions in the preparation of this paper. Dr. Jordan's advice has been followed in all matters of nomenclature.

The following genera and species, believed to be new to science, are described:

Veternio, new genus of Leptocephalidæ.
Collybus, new genus of Bramidæ.
Carcharias insularum.
Carcharias nesiotes.
Veternio verrens.
Sphagebranchus flavicaudus.
Callechelys lutens.
Moringua hawaiiensis.
Gymnothorax nuttingi.

Gymnothorax berndti.
Gymnothorax mucifer.
Gymnothorax xanthostomus.
Gymnothorax waialuæ.
Eropterygius leucurus.
Exonautes gilberti.
Carangus cheilio.
Carangoides ajax.
Collybus drachme.

Apogon erythrinus.
Cirrhilabrus jordani.
Pseudojulis cerasina.
Hemipteronotus jenkinsi.
Chaetodon corallicola.
Holacanthus fisheri.
Stephanolepis pricei.
Antennarius nexilis.
Antennarius duescus.

CARCHARIIDÆ.

1. Carcharias melanopterus Quoy & Gaimard. Honolulu market.

The following measurements, recorded in centimeters, were taken from a female specimen. Total length 156; tip of snout to dorsal 52; to eye 12.8; to first gill-opening 30.5; to pectoral 36.2; length of gill area 7.7; height of first, second, third, and fourth gill-slits 6.3; fifth 5.6; anterior margin of pectoral 28; base of pectoral 10.8; posterior margin of pectoral 27.3; axil of pectoral to ventral 36.8; anterior margin of ventrals 12; free margin of ventrals 10; base of ventrals 10; axil of ventrals to front of anal 13.3; base of anal 8.3; anterior margin of anal 10; base of anal to caudal pit 9; base of dorsal 11; anterior margin of dorsal 19.5; free edge of dorsal 15.3; distance between dorsals 38; base of second dorsal 7.6; second dorsal to caudal pit 10; upper lobe of caudal 38; spread of caudal 35.5; lower lobe of caudal 19; width of mouth 17; preoral length of snout 9.5; girth behind pectorals 63.5; girth at front of ventrals 53.

2. Carcharias phorcys Jordan & Evermann. Honolulu; Hanalei Bay, Kauai.

3. Carcharias insularum Snyder, new species. Plate 1, fig. 1.

Head, measured to last gill-opening, 0.32 of length; tip of snout to caudal pit; depth at front of pectorals 0.163; at front of ventrals 0.16; snout 0.33 of head; interorbital width 0.5; pectoral 0.25 of length; upper lobe of caudal 0.38.

Mouth semicircular, its width equal to distance between tip of snout and posterior border of eye, distance between edge of mouth and tip of snout 1.7 times width of mouth, or a little more than distance between nostrils; upper teeth serrated from base to tips, the lower ones smooth on base, upper parts weakly serrated; teeth of upper jaw a little broader at base than they are high, the cutting edges of median ones straight; lateral teeth with edges slightly concave, concavity of outer edges deepening somewhat as they approach corners of mouth; teeth not pointing outward in either jaw, those of lower jaw much more slender than those above, the bases somewhat wider than height of teeth; cutting edges concave. There are 30 rows on each jaw, those of the two median rows minute or absent. Tip of pectoral fin acutely rounded; first dorsal broadly rounded; second dorsal slightly smaller than anal; caudal very large, under side of upper lobe with a deep notch; free edges of dorsals, pectorals, and ventrals concave; claspers of male 1.5 times as long as ventral fin is high.

In life, bluish slate color, somewhat lighter below; first dorsal broadly tipped with lighter color; second dorsal, pectorals, ventrals, and caudal with slightly darker tips. In alcohol the fins and upper parts of the body are rather indistinctly spotted with a darker shade than that of body; spots of body somewhat larger than eye, the spaces between them somewhat wider than diameters of spots; spots on fins smaller and more closely crowded.

The following measurements were taken before the specimen, a male, was preserved: Total length 213 centimeters; tip of snout to dorsal 71; to eye 17.8; to first gill-opening 40.5; to pectoral 48; length of gill area 10; height of first gill-slit 7.5; of second 8.2; of third 8.8; of fourth 8.2; of fifth 5.7; length of pectoral 39; base of pectoral 14; free edge of pectoral 37; axil of pectoral to ventral 47; anterior margin of ventral 14; free margin of ventral 12; base of ventral 12.7; axil of ventral to front of anal 17.8; base of anal 9; anterior margin of anal 12.7; anal to caudal pit 8.2; base of dorsal 21; anterior margin of dorsal 32; free edge of dorsal 23.5; first to second dorsal 47; base of second dorsal 6.3; front margin of second dorsal 9; second dorsal to caudal pit 12.7; upper lobe of caudal 59.5; lower lobe of caudal 30; spread of caudal 61; girth at front of ventrals 66; girth at front of pectorals 78.5.

Type, No. 50859, U. S. Nat. Mus., station 3845, off Diamond Head, Oahu.

Seven young were obtained from a large female of the same species taken at station 4111, between Molokai and Oahu. They measured 61 cm. in length. Color bluish, pectorals, second dorsal, anal, and lower caudal lobe broadly tipped with black; ventral surface of body and paired fins, except the terminal dark areas, yellowish; tip of first dorsal yellowish. The head measured to last gill-opening 0.35 of length; depth at front of pectorals 0.18; depth of caudal peduncle 0.18 of head; snout 0.34; interorbital width 0.48. Curve of mouth elongate instead of circular, as in adult, its width being an eye's diameter less than distance between tip of snout and anterior border of orbit. Distance between edge of mouth and tip of snout 1.1 times width of mouth. Height of dorsal 0.16 of length of head and body; length of pectoral 0.29; upper lobe of caudal 0.38. Dorsal and pectorals broadly rounded. Two of the specimens are kept as cotypes, Nos. 12788 and 12789, L. S. Jr. Univ. Mus.

This shark appears to be closely related to *Carcharias lamia* Rafinesque, of the Atlantic.

4. Carcharias nesiotes Snyder, new species. Plate 1, fig. 2.

Head, measured to last gill-opening, 0.32 of length (tip of snout to caudal pit); depth at front of pectorals 0.18; at front of ventrals 0.16; snout 0.32 of head; interorbital width 0.42; pectoral 0.27 of length; upper lobe of caudal 0.36. Mouth elliptical, not semicircular in shape, width equal to distance from tip of snout to posterior edge of orbit; width of space between tip of snout and anterior edge of mouth equal to distance between outer edges of nostrils, 0.26 of head; teeth of upper jaw strongly serrated, those near center of jaw symmetrical in shape, the width at base equal to or a little greater than height; laterally the outer edges of teeth grow concave, then notched; inner edges becoming convex, teeth pointing away from symphysis; teeth of lower jaw narrow, with wide bases, their edges smooth or very slightly serrated; symmetrical in shape on both middle and lateral parts of jaws. Pectorals pointed at tips when depressed, reaching as far back as posterior part of first dorsal, free edge concave; first dorsal bluntly pointed; second dorsal and anal equal in size, edge of anal deeply notched; edge of upper caudal lobe notched, distance from notch to tip of lobe contained 0.22 of length of lobe.

Color, bluish gray above, the fins growing darker toward the tips; ventral surface lighter.

The following measurements were made of a male taken at station 3902, off the northern coast of Molokai. Total length 224 cm.; tip of snout to dorsal 71; to eye 17.8; to gill-opening 44; to pectoral 54; length of gill area 13.5; height of first gill-slit 6.5; of second 7; of third and fourth 6.5; of fifth 5.8; anterior margin of pectorals 49; base of pectoral 14; posterior margin of pectoral 42; axil of pectoral to ventral 49.5; anterior margin of ventral 12.8; free margin of ventral 12.8; base of ventrals

10.8; axil of ventral to front of anal 19; base of anal 8.3; anterior margin of anal 12; anal to caudal pit 13.4; base of first dorsal 19.7; anterior margin of first dorsal 30.5; free edge of dorsal 26; distance between dorsals 58; base of second dorsal 7; second dorsal to caudal pit 19; upper lobe of caudal 61; spread of caudal 66; lower caudal lobe 29; width of mouth 20.5; preoral length of snout 15.

Type, No. 50860, U. S. Nat. Mus., a female about 1.48 meters long, taken at French Frigate Shoals. A smaller specimen, also a female, from Laysan Island, does not differ from the type except that it is darker in color, the under parts being quite dusky. Cotype, No. 12790, L. S. Jr. Univ. Mus.

A large and voracious shark seen everywhere about the islands. Compared with *Carcharias japonicus* of Japan, it is more robust in form, having a shorter and broader head.

5. Prionace glauca (Linnæus). Station 3801, 28° 31′ N., 141° 47′ W.

A female taken with a handline contained 47 embryos measuring 39 centimeters in length. The following measurements of the adult were made: Tip of snout to end of caudal lobe 274 cm.; to dorsal fin 110; to eye 23; to first gill-opening 55; to pectoral 65; length of gill area 18; height of first gill-slit 5; of second and third 7.5; of fourth 7; of fifth 5; length of pectoral 62; base of pectoral 23; free edge of pectoral 56; axil to ventral 77; anterior margin of ventral 17.5; free margin of ventral 20.5; base of ventral 16.5; axil of ventral to front of anal 24; base of anal 13.5; anterior margin of anal 17; anal to caudal pit 22; base of dorsal 23; anterior margin of dorsal 30.5; free edge of dorsal 28; posterior edge of first dorsal to second dorsal 63.5; base of second dorsal 13; front margin of second dorsal 13.5; posterior end of second dorsal to caudal pit 21.5; upper lobe of caudal 58.5; spread of caudal 67; lower caudal lobe 37; girth at front of ventral 76; girth at front of pectorals 91.

SPHYRNIDÆ.

6. Sphyrna zygæna (Linnæus). Station 3844, southern coast of Molokai.

SQUALIDÆ.

7. Squalus mitsukurii Jordan & Snyder. Honolulu; station 4085, off the north coast of Maui.

Head, measured to last gill-opening, 0.263 of length (snout to caudal pit); measured to first gill-opening 0.22; width of head 0.5 its length to last gill-opening; snout 0.417 of head measured to first gill-opening; interorbital space 0.417; height of first dorsal fin 0.5; second dorsal 0.29.

Teeth in both jaws similar, except that the lower ones are slightly larger than those above; placed in three closely apposed rows, pointing away from middle of jaw; outer edge with a deep notch, inner serving as cutting edge; distance between mouth and tip of snout 0.5 of length of head to first gill-opening; width of mouth 0.29; length of fold at corner of mouth equal to distance between nostrils; distance between nostril and tip of snout 0.26 of head; between nostril and middle of mouth equal to distance between nostril and tip of snout; distance between spiracles 0.44 of head; length of gill area 0.22; diameter of eye 0.21.

Length of exposed portion of first dorsal spine equals the distance from tip of spine to tip of fin; height of spine equal to length of base of fin. Second spine 0.75 as high as fin; distance between dorsals 3.66 times length of snout; pectoral, when depressed, reaching to a vertical through posterior edge of base of dorsal, the tip bluntly pointed; edges of pectoral and first dorsal concave, that of second dorsal emarginate; edge of ventrals straight; distance from anterior edge of anal opening to tip of depressed ventral 0.417 of head; upper caudal lobe 0.27 of the length; a low lateral keel on caudal peduncle.

Color, dark slaty blue above, lighter below.

Some of the specimens examined have the heads slightly narrower than examples of the same species from Japan, while others are like them in every particular.

DASYATIDÆ.

8. Dasyatis sciera Jenkins. Honolulu.

MYLIOBATIDÆ.

9. Aetobatus narinari (Euphrasen). Honolulu.

LEPTOCEPHALIDÆ.

10. Leptocephalus marginatus (Valenciennes). Honolulu; Hanalei Bay, Kauai.

11. Congrellus bowersi Jenkins. Honolulu.

VETERNIO Snyder, new genus.

Veternio Snyder, new genus of *Leptocephalidae;* type, *Veternio verrens*, new species.

Body without scales; lateral line present; tail much longer than head and trunk; head long, snout pointed; lower jaw much shorter than upper. No teeth; vomer, maxillaries, and mandibles with broad, smooth, hard areas; tongue free; nostrils not tubular; the anterior ones near tip of snout, with narrow rims; posterior ones oblong, near the eyes; gill-openings separate, with broad, lunate slits; fins well developed, dorsal inserted above base of pectorals. Color uniform.

The absence of teeth serves to distinguish *Veternio* from closely related genera.

12. Veternio verrens Snyder, new species. Plate 2, fig. 3.

Head, from tip of snout to upper edge of gill-opening, 0.64 of trunk; depth 0.42 of head; eye 0.13; snout 0.26; length of pectoral 0.36. Color plain; fins edged with black.

Head very long and pointed, upper profile sloping gently from tip of snout to occiput; interorbital space flat, its width 0.16 of head; snout slender, projecting beyond lower jaw a distance equal to 0.5 of diameter of eye; cleft of mouth somewhat oblique, extending beyond eye a distance equal to 0.3 of pupil; no teeth, the vomer, maxillaries, and mandibles with broad, flat, smooth surfaces; tongue free, tip rounded; lips thin, simple; anterior nostrils at end of snout, with low rims and posterior flaps; posterior nostrils close to upper anterior part of eye, without rims, rounded oval in shape; a pair of large mucus tubes at tip of snout, a tube immediately behind anterior nostril and one on each side of snout just above the latter; gill-openings lunate, their width about 0.15 of length of head; lateral line somewhat above middle of body anteriorly, gradually extending downward and reaching middle of body a short distance beyond the vent. Pectoral inserted just below middle of body, obtusely pointed, upper rays longest, lower border convex; dorsal inserted above middle of base of pectoral; height of dorsal fin at a point above tip of pectoral equal to vertical diameter of eye; at a point twice the length of head behind vent the length of rays equals length of snout; 65 rays between its insertion and a vertical through anal opening; anal inserted immediately behind vent, its height equal to 0.5 the length of snout.

Color in spirits brown, darker above than below; pectorals brownish, growing black toward tips; dorsal brownish, shading into black along edge; anal bordered with black, the band about half as wide as pupil and sharply defined.

A single mutilated specimen from the Honolulu market measures 270 mm. from snout to vent. The tail was severed 320 mm. behind the vent.

Type, No. 50862, U. S. Nat. Mus., Honolulu.

(*verrens*, trailing, in reference to the long tail.)

OPHICHTHYIDÆ.

13. Leiuranus semicinctus (Lay & Bennett).

Two specimens from Honolulu. One measures 275 mm., its tail equal in length to head and body; 23 black bands behind interorbital band. The other measures 177 mm., its tail the length of snout longer than head and body; 24 dark bands behind the interorbital band. An example from Ishigaki, Japan, measures 435 mm., length of tail equal to distance between vent and gill-opening; 22 bands behind interorbital; teeth uniserial.

14. Microdonophis fowleri Jordan & Evermann. Honolulu.

15. Sphagebranchus flavicaudus Snyder, new species. Plate 2, fig. 1.

Head, measured to upper edge of gill-opening, 0.055 of the length, 0.105 of trunk including head, 0.12 of tail; depth 0.26 of head; snout 0.18. Snout long, slender, and sharp, projecting beyond lower jaw, tip of latter reaching beyond eye a distance equal to diameter of pupil; eye midway between tip of snout and angle of mouth, its diameter contained 3 times in length of snout; anterior nostril with a short tube on ventral side of snout a little nearer its tip than to border of eye; posterior nostril without tube, placed below anterior margin of eye; upper lip with a fold extending from nostril to angle of mouth; teeth of jaws in a single series; a group of 4 canines at end of upper jaw, all being beyond end of lower jaw when it is closed; a few sharp teeth on anterior part of vomer; gill-openings inferior, converging, the distance between them about equal to diameter of eye; width of gill-opening 0.13 of head. No fins; tail pointed.

Color in alcohol, pale olive, the tail nearly white.

The description is from the type, No. 50883, U. S. Nat. Mus., 367 mm. long, from station 3874, between Mauai and Lanai, 21 to 28 fathoms.

Two examples from off the northeast coast of Hawaii. One from station 4055, depth 50 to 60 fathoms (cotype, 7509, L. S. Jr. Univ. Mus.), measures 245 mm. The head equals 0.06 of the length; 0.1 of head and trunk. In life it was pinkish anteriorly, the posterior third tinged with lemon yellow. The other specimen, from station 4061, depth 24 to 83 fathoms, measures 220 mm.; head 0.065 of length, 0.12 of head and trunk. In life it was light orange, fading to lemon yellow posteriorly; an indistinct, light, median, dorsal stripe extending from occiput to tip of tail; ventral surface slightly tinged with purple, the tint extending about twice the length of head beyond anal opening; side of head with two white spots, the anterior one just behind eyes, the posterior one indistinctly connected over the occiput with its fellow on opposite side.

16. Callechelys luteus Snyder, new species. Plate 3, fig. 5.

Head, measured to upper edge of gill-opening, 0.06 of the length, 0.095 of head and trunk, 0.15 of tail; snout 0.14 of head; cleft of mouth 0.29. Body extremely long and slender, tapering gradually from head to tail; depth at gill-opening 0.37 of head; width of body 0.72 of depth; gill-pouches greatly expanded, making head deeper and broader than body; snout sharp, projecting two-thirds of its length beyond lower jaw; eye midway between tip of snout and angle of mouth; tongue small, free, on sides and at tip; teeth on jaws and vomer projecting backward, movable though not depressible; 3 large canines just posterior to nostril tubes, the median one being anterior to tip of lower jaw; 2 short rows of teeth on vomer, a single row on maxillaries and on lower jaw; anterior nostril with a tube equal in length to diameter of eye, inferior in position, halfway between tip of snout and end of closed lower jaw; posterior nostrils on lip, below the eye, provided with an anterior, valve-like flap; gill-openings slit-like, inferior, distance between lower edges of openings equal to half the length of snout; width of gill-opening equal to distance from tip of snout to posterior border of eye. Dorsal inserted on occiput above angle of mouth; height at a point above gill-slit equal to distance between the tip of the snout and the posterior border of the eye, above anal opening equal to width of gill-slit; fin not reaching tip of tail; membrane thin, the rays distinctly visible. Anal inserted immediately behind vent, its height equal to half the width of gill-opening. Tip of tail sharp; no caudal fin; pectorals absent.

Color in alcohol white, rather finely blotched with brownish black, the spots not so numerous on ventral surface as elsewhere; fins colored like body. In life, the upper parts, including dorsal fin, are white mottled with black and lemon yellow; under parts white, rather sparsely mottled with black, except on throat, where the spots are numerous.

One specimen, 83 cm. long, from station 3821, near the southern coast of Molokai. Caught while swimming about the ship at night, attracted by the lights.

Type, No. 50864, U. S. Nat. Mus., southern coast of Molokai.

MORINGUIDÆ.

17. Moringua hawaiiensis Snyder, new species. Plate 3, fig. 6.

Head, measured to gill-opening, 0.065 of length; tail 0.3; depth 0.24 of head. Body cylindrical and extremely elongate, the tail tapering to a sharp point; snout pointed, its length 0.15 of head; lower jaw projecting beyond upper a distance equal to diameter of pupil; cleft of mouth extending beyond eye a distance equal to pupil; teeth on jaws and vomer sharp, long, and fang-like anteriorly; tongue adnate to floor of mouth; eye very small, the diameter equal to about 0.2 of snout; gill-opening a vertical slit, equal to 0.6 of length of snout; lateral line slightly arched above branchial chamber, discontinued about a head's length from tip of tail; number of pores 113. Pectorals present, minute, the rays easily distinguishable; the base equal to half the gill-opening, length a little less than diameter of pupil; dorsal and anal fins scarcely developed, indicated by slight ridges commencing about a head's length behind anal opening, growing larger and more distinct in region where lateral line ceases; caudal fin distinct, pointed, its length equal to width of interorbital space.

Color in alcohol pale brown, no spots or bars.

One specimen, 320 mm. long, from Honolulu reef. Type, No. 50865, U. S. Nat. Mus., Honolulu.

Closely related to *M. jaranica* of the East Indies, but differs from that species as described, in having pectoral fins with distinct rays, longer head, and longer tail.

MURÆNIDÆ.

18. Muræna kailuæ Jordan & Evermann.   Station 3881, Napili Harbor, Mani; Honolulu.

Muræna lampra Jenkins, Bull. U. S. Fish Com. 1902, 423, fig. 3.
Muræna kailia Jenkins, l. c., 424, fig. 4.

Color in life, light brown; dorsal half of body with dark-brown vertical bars, which fade out and disappear below and become indistinct on tail; everywhere with light-colored ocellated spots, elongate vertically, fused together, forming narrow vertical bands on dorsal and anal fins; crossbands on belly and chevrons on lower jaw and throat; white spots and bands clear and bright below and on throat; soiled white or light gray on upper and posterior parts; dark portions bounding the spots and bars, brownish black; snout, throat, and sides of head suffused with bright orange red; both pairs of nasal tubes bordered with same color; traces of brick red on anterior part of dorsal fin and on body

19. Gymnothorax laysanus (Steindachner).   Honolulu; Hilo.

20. Gymnothorax meleagris (Shaw).   Honolulu.

21. Gymnothorax gracilicauda Jenkins.   Station 3834, southern coast of Molokai.

One specimen, 130 mm. long, from the coral rocks.

Head, measured to gill-opening, 0.125 of length, 0.23 of tail.

Color in spirits white, tinged with brown, spotted or clouded with dark brown, the markings rather indefinitely arranged in crossbands.   The brown figures of the body extend upward on the dorsal; the fin is bordered with white; anal white.

22. Gymnothorax steindachneri Jordan & Evermann.   Honolulu.

23. Gymnothorax undulatus (Lacépède).   Honolulu: station 3824, off the southern coast of Molokai, in 222 to 498 fathoms.

24. Gymnothorax thalassopterus Jenkins.   Honolulu.

25. Gymnothorax nuttingi Snyder, new species.   Plate 4, fig. 7.

Head, measured from tip of snout to gill-opening, 0.145 of the length, 0.28 of head and trunk, 0.31 of tail; depth 0.5 of head; cleft of mouth 0.45; snout 0.19.   Brown with white spots.   Snout rounded, jaws equal, closing completely; lips very thick; teeth in a single series, firmly embedded, close-set; largest below middle of snout, growing gradually smaller posteriorly; basal halves with finely serrated edges; no median fangs; vomer with very short, blunt teeth; eye on vertical passing midway between tip of snout and corner of mouth; distance between eyes 0.74 of the snout; anterior nostril tube equal in length to half diameter of eye; posterior nostril located above and just anterior to margin of eye, its opening with a low rim; gill-opening a narrow slit equal to vertical diameter of eye, situated on a level with pupil; origin of dorsal on a vertical anterior to gill-opening a distance equal to length of snout, the membrane fleshy, though not greatly thickened; height in region of vent equal to length of snout; anal inserted just behind vent, its height near middle of tail equal to diameter of orbit; tail not slender and pointed, but rather stubby, the dorsal, caudal, and anal forming a bluntly rounded terminal fin.

Color in spirits brown, covered with white spots; those on head minute and close together, scarcely discernible on snout and end of lower jaw; on the body, larger and more elongate, growing round on tail, where their diameter is about equal to half that of pupil; gill-opening and corner of mouth brown; dorsal spotted like tail, the spots on edge of fin elongate, narrow, and close together, coalescing posteriorly to form a white border; anal spotted, with a white border.

The species is represented by a single individual, 79 cm. long, obtained in the Honolulu market. Type, No. 50899, U. S. Nat. Mus.

Of the spotted Hawaiian eels this species can only be confused with *Gymnothorax goldsboroughi*. The latter may be distinguished at a glance by the slender, pointed tail, the larger, circular spots, and dark throat-patch, its most striking characters.

Named for Prof. Charles Cleveland Nutting.

26. Gymnothorax goldsboroughi Jordan & Evermann.   Honolulu.

27. Gymnothorax leucacme Jenkins.   Honolulu.

28. Gymnothorax pictus (Ahl).   Honolulu; Puako Bay. Hawaii.

29. Gymnothorax berndti Snyder, new species.   Plate 4, fig. 8.

Head, measured to gill-opening, 0.15 of the length, 0.27 of tail; depth 0.58 of head; snout 0.21; cleft of mouth 0.5.   Profile, a gently sloping straight line between tip of snout and posterior part of interorbital space, whence it abruptly curves upward over the greatly swollen occipital region; snout slender and pointed; lower jaw projecting slightly beyond upper; mouth closing completely; teeth in a single series in each jaw, large, smooth-edged, close set, firmly embedded, the anterior ones somewhat longer than the others; median canines absent; five small teeth on the vomer; nostril tubes equal in height to diameter of pupil; posterior nostrils without rims, located above and just posterior to border of eye; orbit round; width of space between eyes equal to half the distance between tip of snout and center of pupil; gill-opening located on a level with eye, the slit equal in width to diameter of eye; origin of dorsal on a vertical midway between gill-opening and corner of mouth, fin membrane thick and fleshy; height of fin near middle of tail equal to half the length of snout; anal inserted immediately behind vent, appearing as a ridge of skin, the highest part about 0.6 diameter of eye; length of caudal equal to vertical diameter of eye.

Color gray, with fine brown reticulations over which is a coarse network of brown bands.

Color in alcohol white, tinged with brown, more clear along the upper lip, on lower jaw, and on belly; finely clouded and reticulated with brown, except on jaws and anal fin, all overlaid with a brown-colored, coarse network of rather broad bands, the meshes becoming finer on head and broken up into elongate, crooked spots on jaws; gill-opening brown; dorsal with oblique bars which connect with reticulations of body; anal blackish brown, with a broad, white border.

This description is of the type, No. 50867, U. S. Nat. Mus., an example 93 cm. long, obtained in the Honolulu market through the kindness of Mr. E. Louis Berndt. Two other specimens of about the same size were likewise obtained. One has the fine reticulations less distinct than those of the type and the bands of the coarse ones a little narrower, about equal to width of pupil; the snout measures 0.22 of the head, jaws equal. The other, a female, 78 cm. long (cotype, No. 12791, L. S. Jr. Univ. Mus.), when compared with the type has a more slender head, the occipital region being less swollen; the color is similar, except that the bands of the large reticulations are narrower and the dorsal is conspicuously, though narrowly, edged with white; there are two large, depressible fangs in the anterior median portion of the upper jaw.

G. berndti may be distinguished from all other Hawaiian eels by the broad brown reticulations on the body.

Named for Mr. E. Louis Berndt, the efficient inspector of fisheries in Honolulu.

30. Gymnothorax mucifer Snyder, new species. Plate 5, fig. 9.

Head, measured to gill-opening, 0.43 of trunk; depth 0.44 of head; snout 0.21; cleft of mouth 0.43. Snout rather slender and pointed, jaws equal, closing completely; teeth in one series, slender, lance-like with slight constrictions near base, their edges smooth; 3 depressible median canines in upper jaw, the longest (posterior) one equal in length to diameter of eye; a row of small, sharp teeth on the vomer; eye midway between tip of snout and angle of mouth; width of space between eyes contained twice in snout; gill-opening a narrow slit equal to diameter of eye, located on a level with upper lip; nostril tubes equal in length to half diameter of eye; posterior nostrils with scarcely perceptible rims, located above and just anterior to eye; origin of dorsal anterior to gill-opening a distance equal to space between tip of snout and posterior border of eye; height of fin above gill-opening equal to diameter of eye, about 1.33 times as high near middle of tail; the membrane not very fleshy; anal inserted immediately behind vent, appearing for much of the length like a thickened fold of the skin.

Color in alcohol, rich dark brown with flakes of white, which are gathered in clouds and more or less definite vertical bars; the flakes scattered rather evenly on head, scarcely perceptible on lower jaw and snout; throat and belly lighter than other parts, the white and brown being about equal; dorsal growing darker toward the edge, where it is nearly black, with white flakes like those of the body arranged in oblique bars; anal edged with white; corner of mouth dark; no spot at gill-opening.

The species is represented by a single example, type, No. 50868, U. S. Nat. Mus., from the Honolulu market. It measures 34 cm. from tip of snout to vent. The tail, which has been injured, several centimeters having been lost, is 37.5 cm. in length.

31. Gymnothorax xanthostomus Snyder, new species. Plate 5, fig. 10.

Head, measured to gill-opening, 0.125 of the length, 0.22 of tail; depth 0.68 of head; snout 0.2; cleft of mouth 0.62. Snout acutely rounded, lower jaw projecting slightly; profile from tip of snout to interorbital area convex and gently rising, that of occipital region rising abruptly, nuchal muscles well developed; diameter of eye 0.4 length of snout; width of space between eyes 0.78 of snout; mouth closing completely, the cleft extending about one-third its length beyond posterior margin of orbit; teeth of jaws in a single series close-set and firmly imbedded; those at symphysis small; lateral ones large anteriorly, growing gradually smaller posteriorly, the basal two-thirds of their edges denticulate; a median, depressible canine near tip of upper jaw; vomer without teeth; anterior nostril tube equal in length to diameter of pupil; posterior nostril with a minute rim; gill-opening oval, the diameter equal to 1.5 times that of eye, the lower margin on a level with mouth. Origin of dorsal on a vertical passing midway between corner of mouth and anterior edge of gill-opening; membrane fleshy; height of fin near vent, 0.75 of the snout; anal inserted immediately behind the vent, where it is but a low ridge of skin; much higher and less fleshy posteriorly; height near its middle portion equal to half the length of snout; caudal slightly longer than diameter of eye.

Color in life, yellowish olive on anterior third, becoming a rich brown posteriorly. Head and body covered with conspicuous, light, ocellated spots, the light part of which is clearly defined, the dark part more intense next the white, growing diffuse without; spots on head very small, 0.1 to 0.2 diameter of eye, placed from 1 to 3 times their width from each other, their centers tinged with yellow; behind the gill-opening the spots grow rapidly larger for a short distance, then very gradually increase in size to the tail, where they are nearly as large as the eye and 1 to 2 or 3 times their diameter apart; posteriorly and on the fins the spots are pure white or cream colored; opercles with a brownish black margin; mouth, within and at corners, bright lemon yellow.

The color in alcohol differs but little from that of the living specimen.

Described from the type, No. 50869, U. S. Nat. Mus., 91 cm. long, obtained in the Honolulu market. Of two other examples from the same place, one agrees closely with the type, except that the lower jaw projects beyond the upper a distance equal to the diameter of the eye; belly without spots. The other (cotype), No. 12792, L. S. Jr. Univ. Mus.) has the body very thick and robust, nuchal region greatly enlarged; head 0.134 of length, 0.23 of tail; depth 0.6 times length of head.

This species may be known from all other Hawaiian eels by the yellow mouth and the very large, dark-bordered, white spots.

32. Gymnothorax waialuæ Snyder, new species. Plate 6, fig. 11.

Head, measured to gill-opening, 0.125 in the length; depth 0.5; tail 0.55; snout 0.2 of head; cleft of mouth 0.4. Body compressed, the width in middle of trunk equal to half the depth; interorbital space slightly convex; jaws equal; cleft of mouth extending beyond eye a distance equal to longitudinal diameter of eye; width of suborbital space equal to vertical diameter of eye; gill-opening an oblique slit equal to vertical diameter of eye; teeth in jaws mostly long, sharp, and depressible, the two in anterior median part of upper jaw longest; those below eye in two series, the outer ones short and close-set; three short, sharp teeth on vomer; anterior nostril tubes near tip of snout, their height equal to diameter of eye; posterior nostrils without rims, located above and a little anterior to eyes; dorsal inserted on head anterior to gill-opening; fin highest posteriorly; its height behind middle of tail equal to longitudinal diameter of eye; anal inserted immediately behind the vent, about half as high as dorsal; caudal slightly longer than height of dorsal.

Color in alcohol, white tinged with yellow, with 20 black bands, nearly all encircling the body and extending on fins; tip of snout white, tip of tail black; the first black band covers snout except the tip between the nostrils, extends backward beyond eye, and sends a line downward to corner of mouth, where it meets a round, black blotch; chin and throat white; sides of lower jaw black; a white space between eye and corner of mouth; the second band passes over occiput, not complete below; third band incomplete, passing over back between gill-openings, a dusky prolongation passing downward behind gill-opening; other bands complete, anterior ones broader above than below, posterior ones of about equal width throughout; a narrow, dusky stripe extends forward along lower surface from vent to a point a little anterior to gill-openings.

This species closely resembles G. leucacme and may eventually prove to be the young of that form. The species differ in color and in dentition. The light spaces on the body of G. leucacme are reddish brown; on the anal fin they are white, on the dorsal reddish brown bordered with white near edge of fin. Caudal tipped with white; snout reddish brown. Teeth in a single series, those of the jaws not depressible, except 2 or 3 on anterior median portion of upper jaw.

A single specimen, 107 mm. long, from a small tide pool in the reef at Waialua Bay, Oahu. Type, No. 50870, U. S. Nat. Mus.

33. Echidna zebra (Shaw). Honolulu.

34. Echidna nebulosa (Ahl). Honolulu.

35. Echidna vincta Jenkins. Honolulu.

36. Uropterygius marmoratus (Lacépède). Many young individuals collected, as follows:

No. of speci- mens.	Length.	Station.	Locality.	Depth.
	mm.			*Fath.*
5	23 to 30	3847	South coast of Molokai	23 to 24
2	73 87	3850	.....do................................	43 46
1	110	3872	Between Maui and Lanai....	32 43
12	76 120	3876	......do................................	28 43

In life, the young are dark brown, the throat and lower jaw much lighter, almost white in some specimens. There are no dark markings as in the adult. The mucus pores on the head are white.

The following measurements are in hundredths of the length:

Length in millimeters	119	109	106	.79
Length of head........................	0.10	0.11	0.105	0.11
Length of head and trunk	.45	.47	.46	.44
Depth of body.........................	.035	.035	.04	.045

37. Uropterygius leucurus Snyder, new species. Plate 6, fig. 12.

Head, measured to gill-opening, 0.12 of the length; tail 0.54; depth 0.38 of head; snout 0.2; cleft of mouth 0.36; lower jaw shorter than upper, tip extending to base of nostril tubes; teeth of jaws in two series, outer ones small and close-set, inner ones fang-like and widely spaced; a median, depressible fang in upper jaw; a single row of sharp teeth on vomer; anterior nostrils with tubes equal in length to diameter of eye; posterior nostrils without rims, located above eyes; eye located above middle of cleft of mouth; gill-opening a horizontal slit equal to diameter of eye. The dorsal fin becomes evident at a point about half the length of head from tip of tail, it being represented anteriorly by a mere fold of the skin which extends to occiput; caudal pointed; a mere trace of an anal which joins the caudal.

Color brown, finely spotted above with white; ventrally the spots become elongate and unite, also increasing in size until on the belly the color is white with fine reticulations of brown; upper parts with figures formed by the union of elongate spots; end of snout, upper lip, lower jaw, and throat white; fin around end of tail white.

This species resembles the young of *U. marmoratus*. It differs in color, that form being neither spotted nor otherwise figured with white. The jaws of *U. marmoratus* are equal, and no dorsal fin is evident on the tail.

One specimen, 112 mm. long, was taken in 28 fathoms of water, station 3874, between Maui and Lanai. Type, No. 50871, U. S. Nat. Mus.

ALBULIDÆ.

38. Albula vulpes (Linnæus). Hanalei Bay, Kauai.

CLUPEIDÆ.

39. Etrumeus micropus (Schlegel). Honolulu.

ENGRAULIDÆ.

40. Anchovia purpurea (Fowler). Honolulu.

SYNODONTIDÆ.

41. Trachinocephalus myops (Forster). Honolulu; Hanalei Bay, Kauai.

42. Synodus varius (Lacépède). Honolulu.

43. Saurida gracilis (Quoy & Gaimard). Hanalei Bay, Kauai.

BELONIDÆ.

44. Belone platyura Bennett. Honolulu.

45. Athlennes hians (Cuvier & Valenciennes). Lahaina, Maui.

HEMIRHAMPHIDÆ.

46. Hyporhamphus pacificus (Steindachner). Laysan Island.

47. Hemiramphus depauperatus Lay & Bennett. Honolulu; station 3834, southern coast o
Molokai, depth 13 fathoms.

48. Euleptorhamphus longirostris (Cuvier). Honolulu.

EXOCŒTIDÆ.

49. Exocœtus volitans Linnæus.

An example came aboard ship at night somewhere between stations 3804 and 3805, about 24° N
by 151° W.; another at station 3808, near Oahu. These have not been compared with specimens from
the Atlantic. The ventrals are inserted midway between tip of snout and base of eighth ray of anal in
one individual; between snout and tenth ray in the other. The pectorals extend to the base of th
caudal; the first ray simple, the second divided, the third forming tip of fin. The dorsal and ana
each have 13 rays, the former inserted immediately above the latter; base of anal longer than that o
dorsal by an amount equal to space between the two posterior rays. The length of the ventrals i
0.56 of the base of anal. There are 35 scales in a series between upper edge of base of pectoral an
the caudal, 21 or 22 between occiput and dorsal fin. Head measured to end of opercular flap equal
0.25 the length; snout 0.22 of head, including opercular flap; eye 0.31; interorbital space 0.35. Th
pectorals are dusky with a light posterior margin. Dorsal, anal, and ventrals without color.

50. Parexocœtus brachypterus (Solander). Honolulu; station 3829, off southern coast of Molokai

51. Exonautes gilberti Snyder, new species. Plate 7, fig. 13.

Head, to end of opercular flap, 0.215 of length; depth 0.147; width of body at base of pectoral
0.145; depth of caudal peduncle 0.28 of head; eye 0.32; snout 0.28; interorbital space 0.38; D. 10; A. 10
scales in lateral series beginning above base of pectoral 48; between occiput and base of dorsal 32
between lateral line and dorsal 6.

As indicated by the above measurements of the body, this is one of the most slender of the flyin
fishes. Body quadrangular in section; back broader than belly, convex; some of the median scales o
back with low keels; interorbital space concave; snout a little shorter than diameter of eye; lower jav
slightly projecting beyond upper; maxillary extending to posterior border of nostril. No teeth o
tongue or roof of mouth, those on jaws scarcely perceptible; gillrakers on first arch 25, long an
slender; lateral line disappearing near end of anal fin.

The pectoral fin extends to within about an eye's diameter of base of caudal; of 18 rays, first an
second simple; second a third of its length longer than first, which is 3.81 times diameter of eye
third ray divided near tip of first; tip of fin formed by branches of fourth ray, those of fifth bein
slightly shorter; second ray of dorsal fin longest, 0.52 of head; base of fin equal to 2 times diamete
of eye; anal inserted the width of a scale posterior to dorsal, its base shorter than that of dorsal b
an amount equal to the space between 2 rays; height of first and second rays about equal to that o
fourth dorsal; ventrals inserted midway between base of caudal and a point anterior to edge o
opercle, a distance equal to diameter of pupil; extending posteriorly a little beyond the base of anal
not beyond tip of last ray when depressed. Upper lobe of caudal pointed, its length 3 times th
width of the interorbital space; lower lobe an eye's diameter longer.

Color in alcohol, dark brown above, silvery below, pectoral without spots, dusky; the free edg
with a white area as wide as pupil, proximal to which is a blackish band two-thirds as wide as the eye
upper or anterior edge of fin light, an indistinct dark area extending along the first to fourth rays
lower or posterior edge of fin along the last four rays white. Middle rays of ventrals dusky, the fir
indistinctly bordered with white. Free edge of caudal bordered with white. Dorsal dusky; anal white

In life, steel blue above, silvery below.

The species is seemingly related to *Exonautes rondeletii* (Cuvier & Valenciennes) of the Atlantic
It is represented by one individual 26.5 centimeters long (snout to end of lower caudal lobe), which
came aboard the *Albatross* at night, between stations 3799 and 3800, somewhere near 28½° N. by 140° W.

Type, No. 50872, U. S. Nat. Mus.

Named for Dr. Charles Henry Gilbert.

52. Cypsilurus simus (Cuvier & Valenciennes). Honolulu; Waimea and Hanamaula Bay, Kauai
station 3860, Napili, Molokai.

AULOSTOMIDÆ.

53. Aulostomus valentini (Bleeker). Honolulu; Laysan Island.

FISTULARIIDÆ.

54. Fistularia petimba Lacépède. Honolulu; Hilo; Hanalei Bay, Kauai; Necker Island.
Skeletons found at Necker Island, where fishes had been carried ashore by birds.

SYNGNATHIDÆ.

55. Doryichthys pleurotænia Günther. Honolulu.

ATHERINIDÆ.

56. Atherina insularum Jordan & Evermann. Laysan Island; station 3834, southern coast of Molokai; 3880 and 3870, between Maui and Molokai; station 3905 north coast of Molokai. Taken at night with the surface net.

MUGILIDÆ.

57. Mugil albula Linnæus. Waimea River, Huleia River, Hanapepe River, Kauai; Honolulu; station 3844, off southern coast of Molokai.

SPHYRÆNIDÆ.

58. Sphyræna commersonii Cuvier & Valenciennes. Honolulu.

POLYNEMIDÆ.

59. Polydactylus sexfilis (Cuvier & Valenciennes). Honolulu.

HOLOCENTRIDÆ.

60. Holotrachys lima (Cuvier & Valenciennes). Honolulu; Laysan Island.
61. Myripristis murdjan (Forskal). Laysan Island.
62. Myripristis berndti Jordan & Evermann. Honolulu.
63. Myripristis multiradiatus Günther. Laysan Island.
64. Flammeo sammara (Forskål). Honolulu; Laysan Island.
65. Holocentrus erythræus Gunther. Honolulu.
66. Holocentrus xantherythrus Jordan & Evermann. Honolulu.
67. Holocentrus diploxiphus Günther. Honolulu; Puako Bay, Hawaii; Laysan Island.
68. Holocentrus diadema Lacépède. Honolulu; Laysan Island; station 3834, southern coast of Molokai, depth 8 fathoms.

SCOMBRIDÆ.

69. Germo sibi (Schlegel). Honolulu.
70. Acanthocybium solandri (Cuvier & Valenciennes). Honolulu.

LEPIDOPIDÆ.

71. Ruvettus pretiosus Cocco. Honolulu.

XIPHIIDÆ.

72. Xiphias gladius Linnæus. Honolulu.

CARANGIDÆ.

73. Scombroides sanctipetri (Cuvier & Valenciennes). Honolulu.
74. Seriola purpurascens Schlegel. Honolulu.
75. Trachurops crumenophthalma (Bloch). Hanalei Bay, Kauai.
Said to appear in large schools. One of the most important food-fishes at Hanalei.
76. Carangus latus (Agassiz). Hanalei Bay, Kauai; Puako Bay, Hawaii.
77. Carangus affinis (Rüppell). Honolulu.

78. Carangus helvolus (Forster). Honolulu.

The locality from which Forster obtained this species is not known, but as he visited the Hawaiian Islands with Captain Cook it is possible that his specimen came from that region. An example 38 cm. long, from the Honolulu market, agrees perfectly with the description of this species.

Head, including opercular flap, equal to 0.3 the length to base of caudal fin; depth 0.4; snout 0.35 of head; lower jaw projecting somewhat beyond upper; maxillary 0.4, reaching to a vertical through anterior edge of pupil; eye 0.23, a horizontal from tip of snout passing through center of pupil; width of interorbital space 0.35. No teeth on vomer, palatines, or tongue, those of jaws in a single series. Tongue dead white in color; roof of mouth similar, becoming abruptly blue-black posteriorly, the white extending backward as a V-shaped prolongation; the membranous flap white on the part touched by tongue, black on sides; lower jaw below tongue dark, the flap white beneath tongue, dark on sides. Head with scales behind the eye and on the cheeks; a narrow, naked space on the occiput, extending backward to spinous dorsal; breast naked; plates in straight portion of lateral line 35, each plate with a keel forming a sharp ridge. D. v, 28; A. ɪ, 22.

79. Carangus cheilio Snyder, new species. Plate 8, fig. 14.

Head, measured to end of opercular flap, 0.32 of length to base of caudal; depth 0.36; depth of caudal peduncle 0.15 of the length of head; diameter of eye 0.14; width of interorbital space 0.31; length of snout 0.43; maxillary 0.37; pectoral fin 0.9; ventrals 0.41; height of first dorsal ray 0.32; anal ray 0.28; length of upper lobe of caudal 0.88; D. vɪɪɪ-ɪ, 24; A. ɪɪ, 21; scales in lateral series about 116; between lateral line and spinous dorsal, counting upward and forward, about 23; plates in straight portion of lateral line about 38. Snout pointed, anterior contour of head somewhat concave in the region of interorbital area. Lower jaw slightly shorter than upper; maxillary not reaching a vertical through anterior edge of orbit by a distance about equal to diameter of pupil; lips very thick, the width of upper near its middle equal to half diameter of eye; teeth short and blunt, in a single series on the jaws; none on vomer and palatines: a few very short teeth on tongue; gillrakers 7 : 25, the longest equal in length to diameter of iris.

Scales on occiput, interorbital area, cheek, and upper parts of opercles; other parts of head naked; body, including breast and a sheath along base of dorsal and of anal with small scales, scales of posterior part of body with minute ones along their edges; no scales on membranes of fins; lateral line a little more curved than contour of back, the straight part beginning below tenth articulated ray. Plates highest and broadest near middle of caudal peduncle, their width at that point equal to half the width of maxillary at posterior part. First dorsal spine short and closely adnate to second; second spine longest, 0.33 of length of head; rayed portion of fin with a scaled sheath along its base, the height of anterior part of which is equal to diameter of pupil; posteriorly the sheath gradually grows lower, disappearing near end of fin; anal spines thick and strong, their height a little less than diameter of pupil; base of fin with a sheath similar to that of dorsal. Caudal deeply forked, the lobes sharply pointed; pectoral falcate, sharply pointed; ventrals not reaching anal opening.

Color silvery, a little darker above than below; upper edge of opercular flap with a dark spot about half the size of pupil; axil dusky.

A single specimen, 77 cm. long, from the Honolulu market.

Type, No. 50873, U. S. Nat. Mus.

80. Carangoides ajax Snyder, new species. Plate 8, fig. 15.

Head, measured to end of opercular flap, 0.28 of the length to base of caudal; depth 0.37; depth of caudal peduncle 0.13 of the length of head; diameter of eye 0.2; width of interorbital space 0.29; length of snout 0.42; maxillary 0.41; D. 19; A. 16; plates in straight portion of lateral line about 32.

Anterior profile elevated, the contour rising abruptly to a point above posterior margin of orbit. Lower jaw slightly longer than upper; cleft of mouth almost horizontal, maxillary reaching a vertical passing through center of pupil; width of suborbital area 0.26 of length of head. Teeth villiform, in bands on jaws, vomer, palatines, and tongue. Gillrakers on lower limb of first arch 14, the longest equal in length to width of posterior part of maxillary.

Head naked, except a small area behind and below eye, where there are small, deeply imbedded scales. Body mostly naked, there being an irregularly outlined area along lateral line with small, imbedded scales; lateral line much more arched than dorsal contour, the highest point of curve just anterior to insertion of dorsal, the straight part beginning below base of twelfth dorsal ray; 3 or 4 posterior plates large, their length about half the depth of caudal peduncle; other plates growing smaller anteriorly, almost disappearing before curved portion of lateral line is reached. Spinous dorsal not present; anal spines absent; anterior rays of both fins elevated, their height about 0.4 of length of head. Caudal deeply forked, lobes of equal length, 0.85 of head; pectoral falcate, 0.95 of head; ventrals short, pointed, 0.43 of head.

Color silvery, darker above, indistinctly marbled with dusky along the back; base of pectoral colored on posterior side, upper half brownish black, lower dead white; dorsal fin with a dusky margin.

One specimen 97 cm. long, from the market at Honolulu. Type, No. 50874, U. S. Nat. Mus.

81. Caranx speciosus (Forskal). Honolulu.

82. Alectis ciliaris (Bloch). Honolulu; Hanalei Bay, Kauai.

BRAMIDÆ.

COLLYBUS Snyder, new genus.

Collybus Snyder, new genus of *Bramidæ*; type, *Collybus drachme*, new species.

Body deep, ovate, greatly compressed. Teeth in narrow bands, about 2 or 3 rows on jaws, none on vomer or palatines; 2 small fangs on each side of lower jaw near tip; teeth all small, weak, sharply pointed. Gillrakers long and slender; pseudobranchiæ large. Pyloric cæca 4; 2 of them about equal in length to diameter of pupil, the others nearly as long as stomach. Vertebræ 38. Mouth very oblique, nearly vertical. Opercle, subopercle, interopercle, and preopercle smooth. Scales short; very broad (vertically), the upper and lower edges sharply pointed; strongly ctenoid, each scale with a median, thickened, vertical ridge having a conspicuous tubercle in the center; number of scales in a lateral row between opercle and base of caudal about 50. No lateral line. Dorsal inserted on a vertical passing just behind base of ventral, rays 34, the anterior 3 or 4 without articulations; anal 30; a row of scales along base of fin. Caudal deeply forked. Ventral inserted on a vertical through posterior half of base of pectoral.

Color silvery, dusky on head and back.

The genus *Collybus* differs from *Taractes* in not having teeth on the vomer and palatine bones, in having the caudal deeply cleft, the ventrals inserted posterior to the middle of the bases of pectorals, and in not having the opercular bones denticulated.

83. Collybus drachme Snyder, new species. Plate 9, fig. 16.

Head, measured to end of opercular flap, 0.3 of length (snout to base of caudal); depth 0.6; depth of caudal peduncle 0.28 of head; eye 0.37; snout 0.21; maxillary 0.45; interorbital space 0.29; D. 34; A. 30; scales in lateral series 51; in a vertical series counting upward and backward from insertion of anal 19.

Body greatly compressed, its width at the widest part equal to length of maxillary; upper contour rather evenly curved from snout to caudal peduncle; lower contour much more convex, the base of anal not curved; mouth nearly vertical; lower jaw projecting somewhat beyond upper; posterior edge of maxillary reaching a vertical through anterior edge of pupil. Teeth small, weak, sharply pointed, in narrow bands (2 or 3 rows) on jaws; 2 larger, fang-like teeth on each side of tip of lower jaw; no teeth on vomer and palatines. Pseudobranchiæ large, the filaments equal in length to twice diameter of pupil. Gillrakers 4-10, slender, long, and sharply pointed; edges of opercle, interopercle, subopercle, and preopercle smooth; lower jaw, snout, and interorbital area naked; other parts of head, including the maxillary and the body, closely scaled. Scales strongly ctenoid, the ridges with minute tubercles; each scale with a high vertical ridge, on the middle of which is a prominent knob; the ridges of the scales are hidden by the overlapping softer parts, while the knobs project, lying in longitudinal rows. The scales are short, but very broad vertically, the upper and lower edges sharply pointed. Scales of head, at base of pectoral and along the back are much smaller than the others. No evident lateral line.

Dorsal inserted on a vertical passing behind base of pectoral a distance equal to diameter of pupil; rays 34, the anterior 3 or 4 without articulations; fin elevated anteriorly, the longest ray 0.7 of length of head; posterior rays 0.28. Anal rays except first one or two articulated; length of anterior rays 0.28 of head; caudal deeply forked; pectoral pointed, 0.35 of the length; ventrals inserted on a vertical passing through posterior half of base of pectoral.

Color bright silvery, dusky on upper part of head and along back; a silvery spot about the size of pupil at insertion of dorsal. Upper and lower rays of caudal dusky, central part yellowish white; anterior rays of dorsal dusky.

The type, No. 50875, U. S. Nat. Mus., is a specimen 0.81 mm. long, from station 4176, off Niihau, evidently near the surface. Other examples, among which are cotypes, 7737, L. S. Jr. Univ. Mus., were obtained from the stomach of a *Coryphæna* at Honolulu. Small squids and fishes were taken from the stomach of the species here described.

KUHLIIDÆ.

84. Kuhlia malo (Cuvier & Valenciennes).

Puako Bay, Hawaii; Hanalei Bay, Huleia River, Waimea River, Kauai; Laysan Island; station 3844, southern coast of Molokai.

APOGONIDÆ.

85. Apogonichthys waikiki Jordan & Evermann.

Honolulu; station 3872, between Maui and Lanai, depth 32 to 43 fathoms; station 3876, between Maui and Lanai, depth 28 to 43 fathoms. In 8 specimens, measuring 45 to 55 mm., the depth of body is contained 2.5 times in length. D. vii-i, 9; A. ii, 8.

86. Fowleria brachygramma Jenkins.

Honolulu; stations 3847 and 3849, southern coast of Molokai; stations 3872, 3873, 3875, 3876, between Maui and Lanai; in 23 to 73 fathoms.

Color in life, bright carmine, pale pink on throat and belly, snout lighter and suffused with yellow; basal half of caudal suffused with brassy; base of dorsal yellowish, otherwise the fins are of same color as the body.

87. Apogon maculiferus Garrett.

Station 3875, between Maui and Lanai, depth 34 to 65 fathoms.

88. Apogon snyderi Jordan & Evermann.

Honolulu; Hanalei Bay, Kauai; Laysan Island; station 3834, southern coast of Molokai, depth 8 fathoms.

89. Apogon menesemus Jenkins.

Honolulu; Laysan Island; station 3834, southern coast of Molokai, depth 8 fathoms.

90. Apogon erythrinus Snyder, new species. Plate 9, fig. 17.

Head, including opercular flap, 0.39 of length; depth 0.4; depth of caudal peduncle 0.17; eyes 0.37 of head; snout 0.21; maxillary 0.45; D. vi, i ; 9; A. ii, 8; P. 14; scales in lateral line 26; between lateral line and spinous dorsal 3; between lateral line and insertion of anal 7; between insertion of spinous dorsal and occiput 5. Width of body at pectorals about half the depth; caudal peduncle slender, distance between last anal ray and base of caudal 0.32 of length.

Head short, snout blunt and rounded, lower jaw included; interorbital space flat with a slight median elevation, width equal to diameter of the eye; mouth oblique, the maxillary extending to the posterior border of eye, the expanded portion with a slightly convex posterior border; both margins of preopercle weakly serrated; teeth on jaws, vomer, and palatines, the latter covering a small anterior area of bones; gillrakers on vertical limb of arch mere papillæ except a long slender one at the angle; those on horizontal limb long and slender near the angle, gradually reduced in length to near middle of limb, where they are short and rudimentary; scales weakly ctenoid; cheeks and opercles with scales; first dorsal spine short and weak; second very strong, its length equal to 0.6 of head, when depressed reaching base of second dorsal ray; remaining spines graduated in length to the last, which is about 0.3 length of second; spine of soft dorsal very slender, equal in length to fourth spine of first dorsal; longest rays 0.65 of head. Anal rays about equal in length to those of dorsal. Caudal 0.3 of the length; its margin with a deep notch. Pectorals 0.22 of the length, ventrals 0.22.

Color reddish orange, scales edged with a narrow band of a somewhat deeper hue; occiput and a spot on opercle dusky; a small black spot at origin of spinous dorsal; minute dark specks on nape, along base of dorsals, at base of caudal, on breast and on opercles; fins immaculate.

Distinguished from the other Hawaiian species by the bright reddish color, the absence of large spots or bars on the body and fins, and by having the second dorsal spine largest.

Type, No. 50876, U. S. Nat. Mus., Puako Bay, Hawaii. Length 0.36 mm. Other specimens, among which are cotypes, No. 7733, L. S. Jr. Univ. Mus., are from Honolulu, Hanalei Bay, Kauai, Laysan Island.

SERRANIDÆ.

91. Pikea aurora Jordan & Evermann. Honolulu.
92. Epinephelus quernus Seale. Honolulu.
93. Anthias fuscipinnis Jenkins. Honolulu.

PRIACANTHIDÆ.

94. Priacanthus alalaua Jordan & Evermann. Honolulu; Laysan Island.

LUTIANIDÆ.

95. Aphareus flavivultus Jenkins. Honolulu.
96. Platyinius microdon (Steindachner). Honolulu.
97. Apsilus brighami (Seale). Honolulu.
98. Aprion virescens Cuvier & Valenciennes. Honolulu.
99. Etelis evurus Jordan & Evermann. Honolulu.
One specimen 70 cm. long, measured to base of caudal. Upper lobe of caudal about 34 cm. long.

SPARIDÆ.

100. Monotaxis grandoculis (Forskål). Honolulu.

KYPHOSIDÆ.

101. Kyphosus elegans (Peters). Laysan Island.

MULLIDÆ.

102. Mulloides auriflamma (Forskål). Honolulu.
103. Mulloides flammeus Jordan & Evermann. Puako Bay, Hawaii.
104. Mulloides samoensis Günther. Honolulu.
105. Pseudupeneus chryseredros (Lacépède). Honolulu.
106. Pseudupeneus porphyreus Jenkins. Hanalei Bay, Kauai; Honolulu.
107. Pseudupeneus multifasciatus (Quoy & Gaimard). Puako Bay, Hawaii; Laysan.
108. Pseudupeneus pleurostigma (Bennett). Honolulu.
109. Upeneus arge Jordan & Evermann. Honolulu.

CIRRHITIDÆ.

110. Paracirrhites forsteri (Bloch & Schneider). Honolulu.
111. Paracirrhites cinctus (Günther). Honolulu.
112. Paracirrhites arcatus (Cuvier & Valenciennes). Honolulu.
113. Cirrhitus marmoratus (Lacépède). Honolulu; Puako Bay, Hawaii.

POMACENTRIDÆ.

114. Dascyllus albisella Gill. Honolulu; station 3868, French Frigate Shoals.
115. Pomacentrus jenkinsi Jordan & Evermann. Honolulu; Waialua Bay, Oahu; Napili Bay, Molokai; Hanalei Bay, Kauai; Puako Bay, Hawaii; Laysan Island; station 3881, between Maui and Molokai.
116. Glyphisodon abdominalis Cuvier & Valenciennes. Honolulu; Hilo; Hanalei Bay, Kauai; Puako Bay, Hawaii; Laysan Island; station 3834, southern coast of Molokai, 8 fathoms.
117. Glyphisodon sordidus (Forskål.) Honolulu; Puako Bay, Hawaii; Napili Bay, Molokai; Necker Island; Laysan Island.

LABRIDÆ.

118. Lepidaplois bilunulatus (Lacèpéde). Honolulu.

119. Anampses cuvier Quoy & Gaimard. Honolulu.

120. Anampses evermanni Jenkins. Honolulu.

121. Stethojulis axillaris (Quoy & Gaimard).

Honolulu; Puako Bay, Hawaii; Hilo; off southern coast of Molokai at stations 3829, depth 20 fathoms; 3834, depth 8 fathoms; 3837, depth 13 fathoms. Individuals of this species were frequently attracted by an electric light hung just below the surface of the water.

122. Stethojulis albovittata (Kolreuter). Honolulu.

123. Halichœres ornatissimus (Garrett). Honolulu.

The species described by Dr. Jenkins as *Halichœres iridescens* (Bull. U. S. F. C. 1899, p. 47) is no doubt identical with this. The anal has 3 spines, 12 rays. Garrett, having probably mistaken a spine for a ray, gives 2 spines, 13 rays.

124. Pseudojulis cerasina Snyder, new species.

Head, measured to end of opercular flap, 0.32 of length to base of caudal; depth 0.23; depth of caudal peduncle 0.125; length of snout 0.35 of head; eye 0.2; interorbital space 0.19; D. IX, 11; A. III, 12; pores in lateral line 28; scales in lateral series 26; between lateral line and dorsal fin 1; between lateral line and anal 7. Body notably long and slender, head conical, snout pointed; mouth very small, the cleft smaller than eye; 2 canines in each jaw, the upper pair wide apart, allowing the lower ones to fit between them; canines but little longer than the other teeth, which grow gradually smaller posteriorly; no posterior canines; preopercle not serrated. Head naked; scales of breast smaller than those of body; lateral line abruptly bent downward below ninth dorsal ray; dorsal low, the longest, eighth, spine equal in length to snout, the rays about a tenth longer; longest anal ray 0.32 of head; spines slender, the third with a short cutaneous filament. Caudal slightly rounded, 0.65 of length of head; pectoral 0.18 of head; ventral pointed, 0.5 of head.

Color in spirits (perhaps somewhat similar in life), head pale orange, upper part of opercle with a purple tint, the lower part silvery; nape purple; a narrow dorsal area, reddish orange fading ventrally to light orange; a rather indefinitely outlined, broad, pinkish stripe from opercle to base of caudal; lower part of body light orange; a reddish orange spot somewhat smaller than the pupil at lower edge of base of pectoral; each scale row with a narrow line of a darker shade than the color area on which it occurs. Fins orange, the spinous dorsal suffused with red; scaled portion of caudal reddish orange.

One specimen, the type measuring 87 mm. long, was collected at Honolulu by Mr. Berndt. No. 50877, U. S. Nat. Mus.

125. Julis pulcherrima Günther. Honolulu.

126. Julis eydouxi Cuvier & Valenciennes. Honolulu.

127. Julis flavovittata Bennett. Laysan Island.

Only one specimen; length 46 mm. Head, measured to end of opercular flap, 3.12 in length; depth 4; snout 3.3 in head; eye 4.5; D. IX, 12; A. III, 12; scales in lateral series 88; above lateral line 7; between lateral line and insertion of anal 27.

Color in alcohol white with black longitudinal stripes. A black stripe, pointed anteriorly, broken and irregular posteriorly, extends from middle of snout to end of dorsal; a second runs from tip of snout, through eye to base of caudal, where it is connected with the one on opposite side by a band passing over the caudal peduncle; a third extends from lower jaw over base of pectoral to base of caudal; a fourth passes from throat to end of anal; breast and belly dusky; dorsal black, edged with white, a few white blotches along middle of fin; caudal white at base and on margin, middle of fin with a black lunate band; anal black, bordered with white; pectorals white; ventrals dusky at base; a posterior canine tooth is present.

128. Hemicoris ballieui (Vaillant & Sauvage). Honolulu.

129. Hemicoris rosea (Vaillant & Sauvage). Honolulu.

130. Hemicoris venusta (Vaillant & Sauvage). Honolulu.

131. Cheilio inermis (Forskal). Honolulu.

132. Thalassoma purpureum (Forskal). Honolulu.

133. Thalassoma quadricolor (Lesson). Honolulu.

134. Thalassoma duperrey (Quoy & Gaimard).

Honolulu; Puako Bay, Hawaii; station 3881, between Maui and Molokai; Laysan Island.

The young have a black stripe, about as wide as the eye, extending from the snout to the upper half of the base of caudal, where its end is slightly broadened and rounded. Below and parallel with the dark stripe is a pinkish white one of about equal width. Some specimens have a narrow, short, indistinct, dusky stripe below the latter. With increasing age the stripes become indistinct and disappear, the general color grows darker, while the broad, light band behind the head appears. The caudal is at first rounded, later becoming concave, the upper and lower rays being much produced in the adult. The colors here noted are of alcoholic specimens.

135. Thalassoma umbrostigma (Rüppell). Laysan Island.

136. Thalassoma ballieui (Vaillant & Sauvage). Honolulu; Laysan Island.

137. Gomphosus varius Lacépède. Honolulu.

138. Gomphosus tricolor Quoy & Gaimard. Honolulu; Puako Bay, Hawaii.

139. Cirrhilabrus jordani Snyder, new species. Plate 10, fig. 18.

Head measured to end of opercular flap 0.36 of length; depth 0.29; depth of caudal peduncle 0.14; length of snout 0.32 of head; eye 0.25; interorbital space 0.23; D. xi, 8; A. iii, 8; scales in lateral line 16+8; between lateral line and dorsal 1; between lateral line and anterior part of anal 5.

Snout pointed; cleft of mouth equals 0.66 diameter of eye; 6 rather widely spaced canines in upper jaw; inner pair projecting forward; the others, of which the outer pair are much the longer, curve outward and backward; 2 small canines in lower jaw; no posterior canines; other teeth very small, sharp, and closely apposed; preopercle with a finely serrated margin. Cheeks and opercles with scales; bases of dorsal and anal, each with a row of long, pointed scales; 3 large scales covering the greater part of basal half of caudal. Dorsal and anal fins high. Dorsal spines slender, the longest (posterior ones) equal in length to twice diameter of eye, each with a fleshy, spine-like prolongation, a continuation of the thickened membrane surrounding the spine, extending upward and backward and acting as a support for the membrane which extends above spines. First ray equal in height to preceding spine and its thickened attachment, the following rays gradually growing shorter. Anal spines with thickened membranous attachments similar to those of dorsal, the rays longer than the spines, their length contained about 1.5 times in head; caudal rounded; ventrals sharply pointed, not greatly elongated, reaching to vent when depressed; upper rays of pectoral longest.

Color in spirits plain, a few small, white spots below base of dorsal.

Flesh color in alcohol, probably red in life; a pale purple stripe indistinctly outlined extends along body between base of dorsal and lateral line; a few small white spots scattered along the back above lateral line; three distinct, narrow, light stripes along side of abdomen. Fins plain, probably yellow in life; anal with a narrow dusky band on margin.

The specimen described is a male, type, No. 50878, U. S. Nat. Mus., from station 3876, between Maui and Lanai; depth 28 to 43 fathoms.

Other examples, females from the same locality, among them cotypes No. 7728 ichthyological collections Stanford University, have the spinous dorsal lower than the rayed portion of the fin. The thickened portions of the membrane are less developed. The anal has no dusky border.

Named for Dr. David Starr Jordan.

The following measurements of the type and cotypes are given in hundredths of the length:

	No. 1.	No. 2.	No. 3.
Length from tip of snout to base of caudal, in millimeters	65	56	50
Length of head, including opercular flap	.36	.33	.33
Length of longest (11th) dorsal spine	.16	.15	.15
Length of longest (1st) dorsal ray	.23	.2	.18
Length of longest (3d) anal spine	.13	.13	.1
Length of longest (4th) anal ray	.25	.17	.2
Length of caudal fin	.27	.26	.29
Length of pectoral fin	.22	.23	.24
Length of ventral fin	.23	.21	.2
Number of dorsal rays	XI, 8	XI, 8	XI, 9
Number of anal rays	III, 8	III, 9	III, 9
Number of scales in lateral line	16+8	15+7	16+8

This species belongs with *Cheilinoides* Bleeker, a section of the genus *Cirrhilabrus*, having short ventrals.

140. Pseudocheilinus evanidus Jordan & Evermann.

Stations 3873, depth 32 fathoms, and 3876, depth 28 to 43 fathoms, between Maui and Lanai; 4073, depth 69 to 78 fathoms, southern coast of Oahu.

The eye of this species, and probably of all the others included in the genus, is remarkably modified. The cornea is greatly thickened, the tissue being differentiated to form two lens-like structures. The iris and lens show no unusual characters. The retina has thickened areas which superficially appear as folds extending forward and backward from the point of entrance of the optic nerve.

141. Pseudocheilinus octotænia Jenkins.   Honolulu.

142. Cheilinus hexagonatus Günther.

Honolulu; station 3834, in 88 fathoms, off the southern coast of Molokai.

The example from Molokai is 203 mm. in length.   Color in life: Head on upper part and on snout and maxillary bright olive; chin, throat, and under parts of head blue green, shading to olive along a horizontal line extending backward from mouth, finally becoming brassy red on cheeks; sides of head with orange-red lines radiating from eye, fading to orange on subopercle; body orange red below and on lower half of sides, shading into greenish blue on back, each scale with a dark, brick red, semilunar mark; caudal peduncle yellowish olive, an indistinct pale band crossing it at end of dorsal and anal; dorsal greenish olive with two longitudinal orange bands, which broaden and become diffuse posteriorly; pectoral pale orange on base, becoming lilac toward the tip; ventrals reddish, with a wide, brown crossband; anal bright orange red, each ray dusky; rays of caudal greenish olive with reddish tips, the membrane pale orange.

143. Cheilinus bimaculatus Cuvier & Valenciennes.   Honolulu.

144. Novaculichthys tæniurus (Lacépède).   Honolulu.

145. Novaculichthys kallosoma (Bleeker).

A single specimen, 114 mm. long, was obtained at Honolulu.   It has been compared with a specimen of this species recently received by Stanford University from the island of Negros, Philippine Islands (coll. Dr. Bashford Dean), and they are found to agree perfectly.

146. Hemipteronotus jenkinsi Snyder, new species.   Plate 10, fig. 19.

Head, including opercular flap, 0.28 of the length measured to base of caudal; depth 0.36; depth of caudal peduncle 0.44 of head; eye 0.17; width of interorbital space 0.2; length of maxillary 0.34; D. ix, 12; A. iii, 12; lateral line 22 + 6.

Eye located 2.66 times its diameter above angle of mouth; mouth nearly horizontal, on a level with upper edge of base of pectoral, the maxillary extending to a vertical through anterior edge of orbit; lower jaw slightly longer than upper; lips with rather thin, fleshy folds.   Outer row of teeth strong, conical; those on sides of jaws posterior to canines gradually decreasing in size from before backward; canines curving outward and forward, the lower pair, which are slightly the larger, fitting between the upper ones; inner teeth short and blunt, in narrow bands.   Pseudobranchiæ present; gillrakers on the first arch 6 + 11, short and sharply pointed.   Edge of preopercle smooth.

Head naked, except for a narrow, vertical, scaled area extending downward from eye to a horizontal passing along edge of flap of upper lip; first row with 7 scales, curving upward behind eye; second row with 5, the third with 4 scales.   Scales of breast about half as large as those on sides of body; 27 series of scales between upper edge of gill-opening and caudal fin; 9 in series between insertion of anal and lateral line, 2 between dorsal and lateral line; lateral line curving upward over first 6 scales, then following the dorsal contour, approaching the back near end of dorsal, discontinued after twenty-second scale, beginning again on third scale below and passing along middle of caudal peduncle. First two dorsal spines somewhat closer together than others, but not separated from them, the membrane being continuous; height of first spine 0.39 of length of head; the second shorter; remaining spines equal in height to 0.25 of length of head; height of rays 0.35; end of soft dorsal when depressed just reaching base of caudal fin.   Anal spines small and slender, the rays equal in height to those of the dorsal; base of anal and also the tips of the rays when depressed extending farther posteriorly than corresponding parts of dorsal.   Caudal rounded, the basal fourth with scales, the length 0.65 of the head; pectoral 0.66 of the head; outer rays of ventral filamentous, just reaching vent.

Color in spirits, head plain, without spots, bars, or lines; a conspicuous black spot on back covering 2 scales above sixteenth in lateral line, its distance behind the opercular flap equal to distance between that point and tip of snout; a yellowish white spot on side of body, rather indistinctly out-

a *P. evanidus, P. octotænia,* and *P. hexatænia* have been examined.

lined, covering an area equal to width of 5 scales and height of 3 or 4, the spot partly covered by pectoral when depressed; scales of body, except on breast, belly, and part covered by the large light spot, each with a vertical pearly bar which grows wider on the ventral scales, covering over half the scale in region above base of anal; soft dorsal and anal with oblique dark bars, those of the anal not so broad as those of dorsal; caudal, pectorals, and ventrals plain.

Described from a single specimen 25 cm. long. Type, No. 50879, U. S. Nat. Mus., from Puako Bay, Hawaii. The species may readily be distinguished from other closely related Hawaiian forms by the great depth of the body and by the small dark spot, the posterior location of which is notable.

Named for Dr. Oliver Peebles Jenkins.

147. Iniistius pavoninus (Cuvier & Valenciennes). Honolulu; Puako Bay, Hawaii.

An individual measuring 105 mm. represents the bright juvenile color phase described by Dr. Jenkins as *Iniistius cozonus.*[a] Two dark lines pass over the interorbital space connecting the eyes, and 2 similar lines pass from the eye downward to lower part of opercle; a few small spots on nape and opercle behind eye; corresponding with the rows of scales, the body has longitudinal rows of small spots, darker and more distinct on crossbands; first dorsal spine with narrow black bars; first spot on dorsal very distinct, second and third ocellated; ventrals almost black, anal very dark below each vertical band of the body; caudal black at base. With increasing age, the lines of the head, the small spots of head and body, the bars of first dorsal spine, and the spots of dorsal fin, all or in part disappear; the dark bands of the body and fins grow less conspicuous. The black scale below spinous dorsal always remains distinct.

148. Cymolutes lecluse (Quoy & Gaimard). Honolulu.

SCARIDÆ.

149. Scarus ahula Jenkins. Honolulu.
150. Scarus paluca Jenkins. Honolulu.
151. Scarus perspicillatus Steindachner. Puako Bay, Hawaii.

CHÆTODONTIDÆ.

152. Forcipiger longirostris (Broussonet). Honolulu.
153. Chætodon multicinctus Garrett. Honolulu.
154. Chætodon setifer Bloch. Honolulu.
155. Chætodon ornatissimus Solander. Honolulu.
156. Chætodon miliaris Quoy & Gaimard. Honolulu; Hanalei Bay, Kauai; Laysan Island.
157. Chætodon lineolatus Cuvier & Valenciennes. Honolulu.
158. Chætodon fremblii Bennett. Honolulu; Laysan Island.
159. Chætodon lunula (Lacépède). Honolulu; Waialua Bay, Oahu; Hilo; Puako Bay, Hawaii.
160. Chætodon unimaculatus Bloch. Honolulu.
161. Chætodon quadrimaculatus Gray. Honolulu.
162. Chætodon corallicola Snyder, new species. Plate 11, fig. 20.

Head, measured to edge of opercle, 0.32 of length to base of caudal fin; depth 0.6; depth of caudal peduncle 0.32 of head; eye 0.42; snout 0.31; interorbital space 0.32. D. XIII, 21; A. III, 18; scales 4-30-12. Anterior profile between snout and dorsal almost straight, with a slight convexity over eye; snout short, its length somewhat less than diameter of eye; jaws equal; teeth fine and brush-like; pseudobranchiæ very large; gillrakers on first arch 5+14, short and pointed; scales on top of head and on snout very minute, those on cheeks, opercles, and breast larger; width of scales near middle of body about equal to diameter of pupil; those on caudal peduncle greatly reduced in size; rayed portions of dorsal and anal closely scaled, the scaled area extending forward on spinous portion of dorsal fin, decreasing in height from near tip of eighth spine to base of first, leaving the membranes of the anterior spines largely naked; lateral line curved upward and constantly approaching the back until it disappears near end of dorsal fin, not extending on caudal peduncle, with 36 pores, scales very small. Except the first the dorsal spines are high anteriorly; height of second to sixth equal to distance between tip of snout and center of eye; height of first spine about equal to diameter of pupil; length

a *Iniistius leucozonus* Jenkins, Bull. U. S. Fish Commission 1900, p. 55.

of longest rays about equal to that of longest spines; membrane deeply notched between anterior spines, the notches growing shallow posteriorly as the scales approach edge of fin; second anal spine longest, about equal to highest dorsal spine; membrane deeply notched between first and third spines, the latter closely connected with rayed portion of fin; border of anal fin extending a little farther posteriorly than that of dorsal; caudal truncate, upper rays slightly longer than lower, 0.78 of head; pectoral 0.85 of head; ventrals not quite reaching vent.

Color in alcohol, dull silvery, brownish along the back, the dark color extending downward in region of pectoral fin; scales on greater part of body with darker centers and lighter edges; posterior part of body with many dark spots about half as large as pupil; head with a vertical brownish-black band, the posterior border of which passes from insertion of dorsal through posterior border of eye, thence curving backward to origin of ventrals; width of band somewhat less than diameter of eye; interorbital area, snout, and upper lip dark brown; dorsal and anal narrowly bordered with dusky, rays tipped with white; scaled portion of dorsal with large irregular dusky spots separated by narrow light spaces, which take the form of a network; caudal light, broadly bordered with dusky; pectoral plain; ventrals blackish toward the free margins.

Described from the type, No. 50880, U. S. Nat. Mus., 59 mm. long, from station 4032.

Other specimens, among which are cotypes, No. 7732, L. S. Jr. Univ. Mus., differ little from the type. Specimens were taken off the southern coast of Oahu at station 4032, depth 27 to 29 fathoms; station 4031, depth 27 to 28 fathoms; station 4034, depth 28 fathoms.

163. Microcanthus strigatus (Cuvier & Valenciennes). Honolulu.

164. Heniochus macrolepidotus (Linnæus). Honolulu.

165. Holacanthus fisheri Snyder, new species. Plate 11, fig. 21.

Head, measured to end of opercle, 0.29 of length (to base of caudal fin); depth 0.46; depth of caudal peduncle 0.12; eye 0.34 of length of head; snout 0.4; interorbital space 0.32; D. xiv, 15; A. 17; scales in lateral series 28; in a slanting series from origin of anal toward origin of dorsal 21. Suborbital with 3 prominent spines which curve downward and backward; lower or anterior one very small, in some specimens preceded by a fourth minute spine; the upper or posterior two about equal in length to diameter of pupil; preopercle with a strong, slightly curved spine at its angle, which extends to a vertical through posterior border of opercle, excluding the flap; length of spine 0.34 of head; margin of preopercle above spine with sharp denticulations; below spine with 2 small spines, the upper of which is the larger; opercular spines not grooved; interopercle with denticulations; teeth fine, brushlike, trilobed, the lateral lobes much shorter than the median, all being sharply pointed; pseudobranchiæ large; gillrakers 5 + 15, slender, pointed.

Scales large, regular, strongly ctenoid; those on snout, chin, and interorbital space minute; dorsal, anal, and caudal densely scaled; lateral line arched over the pectoral, extending along back near base of dorsal, disappearing near end of soft dorsal fin. Dorsal spines growing gradually longer from the second to the last; the first half as long as the second; the last equal in length to distance between tip of snout and posterior border of eye; membranes of first 5 spines notched, membrane between first and second spines nearly cleft to base, the notches growing successively more shallow between the following spines; membranes between first and third spines without scales, the scaled area beginning behind third spine; fin rounded posteriorly, its edge reaching a vertical through base of caudal; third anal spine longest; membranes between spines deeply cleft, without scales except on a narrow area along the base; third spine closely attached to rayed portion of fin; posterior edge of fin pointed, extending nearly to a vertical through middle of caudal; edge of caudal truncate or slightly convex; tips of rays without scales. Ventral sharply pointed, the first ray being filamentous at tip, extending to origin of anal; pectoral when depressed extending to a vertical through vent.

Color in alcohol, light brown with a yellowish tinge; lips, interorbital area, chin and throat somewhat dusky; membranes covering suborbital and opercular spines dark; scales of body with rather indistinct dusky edgings; a round, brownish-black spot somewhat larger than eye, just above base of pectoral; dorsal narrowly edged with black, the border widening on posterior edge to form a well-marked spot; anal narrowly bordered with pearly white, posterior part of fin blackish; spine and first ray of ventral pearly white; caudal with upper and lower borders dusky, the lower part much the darker, the central area yellowish; pectorals immaculate.

Color in life, bright reddish orange, posterior two-thirds suffused with dusky; spot above pectoral brownish black; preopercular spine dark blue; dorsal and anal colored like body, the dorsal narrowly edged with black, the black spot on posterior part with an indistinct boundary; anal with a broad blackish margin narrowly edged with blue on the outside; middle of caudal lemon-yellow; pectorals orange; ventrals orange suffused with dusky near margins, spine and first ray pearly blue.

Different examples vary somewhat in intensity of color. Small specimens have a broad lemon band on the anal edged above and below with blue, the outer blue line narrowly edged with black.

Type, No. 50881, U. S. Nat. Mus., a specimen measuring 78 mm.; cotype, No. 7738, L. S. Jr. Univ. Mus.; from station 4032 off Diamond Head, Oahu; depth 27 to 29 fathoms. Other specimens are from station 3847, southern coast of Molokai; stations 3872 and 3876, between Maui and Lanai; stations 4031, 4033, and 4034, southern coast of Oahu, in 14 to 43 fathoms.

Named for Walter Kenrick Fisher.

The following measurements are recorded in hundredths of the length measured from snout to base of caudal fin:

	Station 4032.			Station 3876.	
Length (in millimeters) to base of caudal fin.....	56	58	54	51	44
Head to edge of opercle	.30	.28	.29	.29	.30
Opercular spine	.11	.125	.11	.105	.10
Fourth dorsal spine	.13	.125	.12	.13	.13
Seventh dorsal ray	.19	.18	.19	.185	.19
Third anal spine	.19	.16	.165	.16	.15
Ninth anal ray	.23	.20	.195	.22	.20
Length of caudal	.26	.25	.23	.235	.26
Length of pectoral	.23	.24	.22	.24	.25
Length of ventral including filament............	.32	.26	.27	.29	.30

ZANCLIDÆ.

166. Zanclus canescens (Linnæus). Honolulu; Puako Bay, Hawaii; Laysan Island.

TEUTHIDIDÆ.

167. Teuthis achilles (Shaw). Honolulu; Puako Bay, Hawaii.

168. Teuthis olivaceus (Bloch & Schneider). Honolulu; Puako Bay, Hawaii.

169. Teuthis umbra Jenkins. Puako Bay, Hawaii.

170. Teuthis argenteus (Quoy & Gaimard). Honolulu.

171. Teuthis atramentatus Jordan & Evermann. Laysan Island.

172. Teuthis bipunctatus (Günther). Puako Bay, Hawaii; station 3834, southern coast of Molokai, depth 8 fathoms.

173. Teuthis sandvicensis (Streets).

Teuthis elegans Garman, Deep Sea Fishes, p. 70, Plate L, fig. 2.

Honolulu; Waialua Bay, Oahu; Hanalei Bay, Kauai; Puako Bay, Hawaii; Hilo; Necker Island; Laysan Island.

A series of specimens showing all stages of growth from the larval to the adult form was secured. In the young measuring about 20 mm. the body is perfectly transparent except a broad, vertical, silvery band extending across the head and visceral region. Anterior edge of band passes obliquely downward and backward, just in front of orbit, to a point a little in advance of insertion of anal fins. The posterior boundary extends from nape, behind axil of pectoral, to insertion of anal. Between the lateral line and the pectoral is a posterior prolongation of the silvery band, about as large as eye. Nape, interorbital space, and a narrow band at base of caudal dusky; a row of dark dots along base of anal and a dusky spot on tip of caudal. The dusky, vertical bands of adult are first seen faintly outlined in the young of 31 mm. length; in others, no larger, the color pattern of the adult is perfectly developed. In the young the snout is shorter, anterior profile more rounded, and body deeper than in adult; head and body covered with long, narrow, vertically placed, scale-like plates; first dorsal spine serrated on anterior edge. Shortly after assuming the adult color, the length being about 32 mm., the serrations of the spine disappear and the plates are replaced by minute scales.

Teuthis elegans Garman is the young of this species. Living examples of the larval form are almost perfectly transparent except the silvery area and dusky spots, there being no blue or red tints.

174. Teuthis guttatus (Bloch & Schneider). Puako Bay, Hawaii.

175. Zebrasoma hypselopterum (Bleeker). Honolulu.

176. Scopas flavescens (Bennett). Honolulu; Puako Bay, Hawaii.

177. **Ctenochætus strigosus** (Bennett). Honolulu.
178. **Acanthurus brevirostris** (Cuvier & Valenciennes). Honolulu.
179. **Acanthurus unicornis** (Forskal). Honolulu; Puako Bay, Hawaii.
180. **Callicanthus lituratus** (Forster). Honolulu; Puako Bay, Hawaii.

BALISTIDÆ.

181. **Balistes vidua** Solander. Honolulu.
182. **Pachynathus capistratus** (Shaw). Honolulu.
183. **Pachynathus bursa** (Lacépède). Station 4082, off Diamond Head, Oahu; 27 to 29 fathoms.
184. **Balistapus aculeatus** (Linnæus). Honolulu.
185. **Balistapus rectangulus** (Bloch & Schneider). Honolulu; Puako Bay, Hawaii.
186. **Melichthys radula** Solander. Honolulu; Puako Bay, Hawaii; station 3824, southern coast of Molokai, depth 222 fathoms; 3814, southern coast of Molokai, depth 8 fathoms.

MONACANTHIDÆ.

187. **Cantherines sandwichensis** (Quoy & Gaimard). Honolulu; Puako Bay, Hawaii.
188. **Stephanolepis spilosomus** (Lay & Bennett).

Honolulu; station 4180, near Niihau, from stomach of *Coryphæna;* Necker Island, carried in by a bird; station 4147, near Bird Island, 26 fathoms; station 4167, near Bird Island, 18 to 20 fathoms; station 4148, near Bird Island, 26 to 33 fathoms.

Color in life: Head and belly pearly blue shading into light brassy, the color of other parts of the body; head and body with lines and spots of brownish black. Membrane of dorsal deep orange with brownish black spots, the spine bluish; dorsal and anal banded with lemon and pearly blue; caudal deep orange, narrowly bordered with lemon; a subterminal band of black; fin spotted with black. Iris brassy. Teeth orange.

189. **Stephanolepis pricei** Snyder, new species. Plate 12, fig. 22.

Head 0.33 of length measured to base of caudal fin; depth between insertion of dorsal and anal 0.38; eye 0.3 of head; interorbital space 0.3; snout 0.71; depth of caudal peduncle 0.38; D. 39; A. 36.

Snout rather pointed, upper and lower contours concave; gill-slit small and narrow, its height equal to width of base of pectoral, two-thirds diameter of eye; ventral flap notably narrow, its width equal to half diameter of eye; dorsal spine inserted above pupil, its length equal to distance between angle of mouth and upper edge of gill-opening, reaching the insertion of dorsal fin when depressed; with 6 lateral spines which project downward and slightly backward; 3 or 4 small granules in a row below the spines; anterior part of spine with prickles which point upward; length of base of dorsal about equal to length of head; height of fin equal to diameter of eye; length of base of anal equal to distance between tip of snout and posterior edge of orbit; height equal to that of dorsal; rays of dorsal and anal rough on basal halves; caudal round, the alternate rays with strong prickles; length of fin equal to length of snout; length of pectoral equal to twice the length of gill-slit; ventral spine large, length of movable part about equal to length of gill-opening, the sides with large spikes which project backward. Body and head evenly covered with prickles, those of the dorsal parts slightly coarser than the others; no enlarged spines on caudal peduncle.

Color silvery, dusky along top of head and back; membrane of dorsal spine blue-black; 3 small, round, dark spots in a line extending upward from base of pectoral; dark clouds somewhat larger than the eye extending downward at insertion of dorsal, from posterior half of dorsal, and on the caudal peduncle; a similar cloud extending upward from posterior half of base of anal.

One specimen, 65 mm. long, station 4021, vicinity of Kauai, depth 286 to 399 fathoms. Type, No. 50882, U. S. Nat. Mus.

The species is named for Dr. George Clinton Price.

TETRAODONTIDÆ.

190. **Tetraodon hispidus** Linnæus. Honolulu; Necker Island.

A specimen was picked up on the rocks on Necker Island, where it had been carried by a bird.

TROPIDICHTHYIDÆ.

191. Tropidichthys jactator Jenkins. Honolulu; Laysan Island.

As was stated by Dr. Jenkins,[a] this species is very similar to *T. punctatissimus* (Günther).[b] Its distinction rests on a difference in the color pattern, the spots being fewer and generally more widely separated than those of *T. punctatissimus*. The distended belly, an alleged distinctive character seen in the type specimen of *T. jactator*, is merely the result of its having been preserved while distended with air. In 3 specimens from Laysan Island, measuring 65, 78, and 93 mm., respectively, the spots on the sides of the head are nearly as large as those on the body; those on the upper part of the snout are about half as large; there are 7 or 8 on a line between upper part of eye and tip of snout. Those on the snout and upper part of head and nape are narrowly bordered with dark brown. On the body there are about 13 spots in a line between the dorsal and anal fins, and 6 in a vertical line near the middle of caudal peduncle. The largest example has an indefinite dark spot below the base of dorsal fin. In life the spots are light blue. Most of them are as large as the pupil and so close together that the brown ground color appears as a network.

Three examples from the reef at Honolulu measure 37 mm. The spots on the upper part of the snout and head are very small. There are 5 in a line between upper part of eye and tip of snout. They are ocellated, as are also the spots along the back to the base of dorsal. There are 7 or 8 spots in a line between anal and dorsal fins, and 4 in a vertical line near middle of caudal peduncle. The cotype collected by Dr. Jenkins in Honolulu also has large spots on the snout.

T. punctatissimus, represented by 8 specimens from Panama, has from 7 to 10 small ocellated spots in a line on upper part of snout. The spots on the back from nape to base of caudal are small and have dark margins. There are from 11 to 23 spots between anal and dorsal and from 8 to 15 on the caudal peduncle. One example has 4 short lines extending backward from the eye. A specimen from the Galapagos Islands referable to *T. punctatissimus* has the spots on the sides of the snout fused, forming vertical bands. There are 3 short bands or elongate spots radiating backward from the eye.

OSTRACIIDÆ.

192. Ostracion camurum Jenkins. Puako Bay, Hawaii.

193. Ostracion oahuensis Jordan & Evermann. Honolulu.

194. Ostracion lentiginosum Bloch & Schneider. Honolulu.

195. Lactoria galeodon Jenkins. Honolulu.

Several specimens were taken from the stomach of a *Coryphæna*.

DIODONTIDÆ.

196. Diodon hystrix Linnæus. Honolulu.

197. Diodon holocanthus Linnæus. Laysan Island.

One specimen 235 mm. long. The fins are immaculate. There are 10 or 12 small dusky spots scattered over the body. A broad, dark bar, interrupted in the middle, extends between the eyes. There is a similar bar on the nape, a spot as large as the eye above and behind the pectoral, a median brown bar on the back anterior to the dorsal, and a blotch surrounding the base of the dorsal.

198. Chilomycterus affinis (Günther). Honolulu.

SCORPÆNIDÆ.

199. Sebastopsis kelloggi Jenkins. Honolulu.

200. Sebastopsis parvipinnis Garrett. Honolulu.

201. Merinthe macrocephala (Sauvage). Honolulu.

Two specimens. D. xii, 11; A. iii, 5. Orbital tentacles barred with dusky; in one specimen they reach the base of third dorsal spine when depressed, in the other they measure 1.66 times diameter of eye, not reaching base of first dorsal spine; second anal spine large and heavy, a deep groove running lengthwise of each side.

202. Sebastapistes corallicola Jenkins. Honolulu; Hilo.

---

[a] *Tropidichthys jactator* Jenkins, Bull. U. S. Fish Com. for 1899 (June 8, 1901), 399.
[b] *Tetrodon punctatissimus* Günther, Cat. Fish. Brit. Mus. 1870, viii, 302.

203. Sebastapistes coniorta Jenkins. Honolulu.
204. Scorpænopsis catocala Jordan & Evermann. Honolulu.
205. Dendrochirus hudsoni Jordan & Evermann. Honolulu.

CEPHALACANTHIDÆ.

206. Cephalacanthus orientalis (Cuvier & Valenciennes). Honolulu; Pukoo, Molokai; Hanalei Bay, Kauai; Lahaina, Maui.

MALACANTHIDÆ.

207. Malacanthus parvipinnis Vaillant & Sauvage. Honolulu; Lahaina, Maui.

GOBIIDÆ.

208. Eleotris sandwichensis Vaillant & Sauvage. Honolulu; Waimea River, Hanalei River, Huleia River, Hanapepe River, Kauai; Anahulu River, Oahu.
209. Asterropteryx cyanostigma (Bleeker). Honolulu.
210. Quisquilius eugenius Jordan & Evermann. Laysan Island.
211. Gobius albopunctatus Cuvier & Valenciennes. Honolulu; Waialua Bay, Oahu; Puako Bay; Kealakekua Bay, Hawaii; Hilo.
212. Awaous genivittatus (Cuvier & Valenciennes). Honolulu; Waimea River, Hanapepe River, Hanalei River, Huleia River, Kauai; Anahulu River, Oahu.
213. Awaous stamineus (Eydoux & Souleyet). Anahulu River, Oahu; Waimea River, Hanalei River, Hanapepe River, Hulei River, Kauai; Wailuku River, Maui.
214. Sicyopterus stimpsoni (Gill). Lahaina and Wailuku rivers, Maui.
215. Enypnias oligolepis Jenkins. Honolulu.

In each of 2 specimens the dorsal has 6 spines, 11 rays; the anal 8 rays.

PTEROPSARIDÆ.

216. Osurus schauinslandi (Steindachner). Honolulu.

ECHENEIDIDÆ.

217. Echeneis remora Linnæus. Honolulu market; Hanalei Bay, Kauai; stations 3869, 3879, 3887, 3834, 3838, 3973. Of 14 specimens, 4 have 17 pairs of laminæ, the others having 18.

BLENNIIDÆ.

218. Salarias zebra Vaillant & Sauvage. Honolulu; Hilo; Puako Bay, Hawaii; Hanalei Bay, Kauai; station 3829, Lanai; Laysan Island; station 3881, between Maui and Lanai.
219. Salarias marmoratus (Bennett). Laysan Island.
220. Salarias gibbifrons Quoy & Gaimard. Honolulu; Hilo.
Salarias saltans Jenkins. Bull. U. S. Fish Com. 1902, 508, fig. 18.
Salarias rutilus Jenkins. l. c., 509, fig. 19.

This species, as with several others of the *Blenniida*, has a distinct type of coloration for each sex. The one described as *S. saltans*, having the soft dorsal and caudal spotted with white, is the male; the other, described as *S. rutilus*, having the dorsal and caudal narrowly barred with black, is the female; 34 males and 30 females were examined.

221. Aspidontus brunneolus Jenkins. Honolulu.

BROTULIDÆ.

222. Brotula marginalis Jenkins. Honolulu.

Two specimens 34 cm. long. The maxillary reaches beyond the eye, its length being contained 2 times in the head, including opercular flap. The dorsal, anal, and caudal merge. Dorsal + anal 229; dorsal to near middle of caudal 124. Caudal pointed posteriorly, not notched.

ANTENNARIIDÆ.

223. Antennarius commersoni (Lacépède). Honolulu.

One specimen from the coral reef. Color black, with many round spots of a deeper black, clouded with light gray.

224. Antennarius rubrofuscus (Garrett). Honolulu.

One example from the reef. This species is admitted as distinct from *A. commersoni* with considerable doubt. The specimen is light orange-red, clouded and spotted with dusky. This and the one identified as *A. commersoni* differ only in color, the one being almost an exact negative of the other.

225. Antennarius bigibbus (Lacépède). Honolulu reef. Three specimens 20 to 25 mm. in length.

226. Antennarius nexilis Snyder, new species. Plate 13, fig. 23.

This species differs from all others of the genus in having the third dorsal spine very closely attached or bound down to the back, and in having the soft dorsal extending posteriorly to base of caudal. First spine short, equal in length to longitudinal diameter of eye; the fleshy tentacle half as long as the spine, with 7 filaments. Second spine curved backward, its length equal to 1.5 times the longitudinal diameter of eye; when depressed, the tip not reaching over half way between its base and the base of third spine, no membrane connecting posterior part of spine with the head; third dorsal spine equal in length to distance between its base and tip of snout; very closely bound down throughout its length to the back, the tip with a movable joint; soft dorsal with 12 rays, the middle ones equal in height to distance between tip of snout and base of third spine; fin extending posteriorly to bases of caudal rays; anal rays 7, equal in length to the dorsal rays; edge of fin rounded, extending posteriorly as far as the dorsal. Caudal rounded posteriorly, 3.5 in the length; pectoral rays 12.

Body and fins covered with granules and prickles, the latter usually bifid or trifid, many of them having fleshy tentacles; a lateral line of pores begins on snout, passes over eye, curves downward to a level with lower margin of eye, extends backward to a point below base of second or third dorsal ray, then bending downward and backward to a point above the origin of anal, from which it runs backward to lower edge of base of caudal; another line of conspicuous pores extends from the chin downward, curving far below the mouth, then upward, joining the lateral line behind the eye; other large pores are present on the chin and head.

Color gray, with dusky spots and clouds, large and close together on the dorsal parts of body; eye with radiating dark and light elongate spots; a large, irregular, reddish orange spot on the nape; a few small spots of same color on snout and face; fins closely covered with black spots a little larger than the pupil, the membranes of the fins near their edges white; pectorals and ventrals white and almost without spots on ventral sides; inside of mouth without dark color.

The description is of the type, No. 50883, U. S. Nat. Mus., taken at Honolulu. In another example, cotype, 7735, L. S. Jr. Univ. Mus., the upper parts of the head and body are almost covered with reddish clouds, the tint more intense anteriorly. First spine 1.33 times as long as diameter of eye.

227. Antennarius duescus Snyder, new species. Plate 13, fig. 24.

Head, body, and fins, except the edges of the latter, covered with bifid and trifid prickles; small dermal filaments scattered here and there, a conspicuous one, somewhat longer than diameter of eye, above and a little behind base of pectoral; gill-opening small, circular, located far back, half way between axil of pectoral and anal opening; first dorsal spine slender and hair-like, the length equal to depth of caudal peduncle, the fleshy tip a flat, folded membrane with minute tentacles; second dorsal spine seated close to first, slender, without a membrane, its shaft covered with minute granules, the tip with a small, fleshy knob; slightly shorter than first spine, not quite reaching base of third when depressed; third spine strong, curved backward, its length equal to distance between gill-opening and anus; capable of free movement up to a vertical position, the posterior membrane fleshy; dorsal rays 12, the highest contained 3 times in base of fin; fin extending far posteriorly; the length of the free caudal peduncle equal to diameter of pupil; anal rays 7, equal in length to those of the dorsal; caudal rounded posteriorly, its length contained 3.5 times in head and body.

Color in spirits, pale brick red, the dorsal, anal, and caudal darker on the edges; rayed portion of pectorals and ventrals gray below, dusky above; head and body sparsely clouded and spotted with dusky and gray; a large, irregular crossband on chin, extending upward a little beyond mouth; a dusky cloud above pectoral; a large, gray spot, bordered with dusky, on the head between snout and pectoral; a small, ocellated gray spot below the latter, and a similar one on body midway between gill-opening and dorsal fin; caudal peduncle with a narrow, vertical, gray band bordered with dusky; mouth immaculate within; prickles white.

In life, purplish lilac throughout (the color of the algæ brought up in the trawl), save for a few pinkish spots and the tips of pectorals and ventrals, which were whitish.

Described from type, No. 50884, U. S. Nat. Mus., 40 mm. in length. A smaller one, 19 mm. long, cotype, No. 7736, L. S. Jr. Univ. Mus., differs from type only in size; in life it was light bronze colored on upper parts, yellowish bronze below, a wide pinkish crescent on upper part of opercles; station 3872, between Maui and Lanai, depth 32 to 43 fathoms.

Another, 19 mm. long, is from station 4128, vicinity of Kauai, depth 75 fathoms; body brownish black except on nape, where there is a small cloud of reddish color; fins narrowly edged with red.

The species is distinguished by the following set of characters: First and second dorsal spines with thickened fleshy tips; dorsal and anal extending far posteriorly, length of free caudal peduncle equal to diameter of pupil; gill-opening located midway between axil of pectoral and the anal opening.

STANFORD UNIVERSITY, CAL., *May 14, 1903.*

LIST OF PLATES.

PLATE 1

1 CARCHARIAS PISCIS ARUM ...11

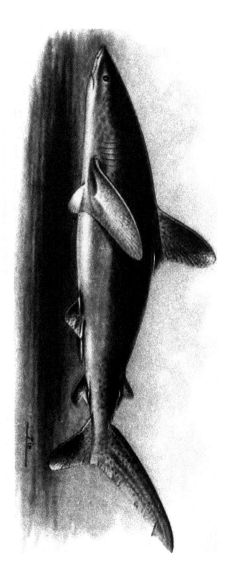

3. VETERNIO VERRENS. TYPE

4 SPHAGEBRANCHUS FLAVICAUDUS. TYPE

5 CALLECHELYS LUTEUS TYPE

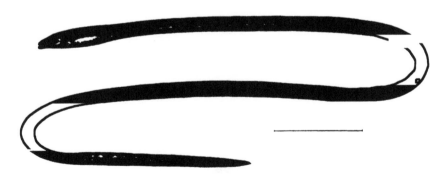

6. MORINGUA HAWAIIENSIS. TYPE.

7. GYMNOTHORAX NUTTINGI TYPE.

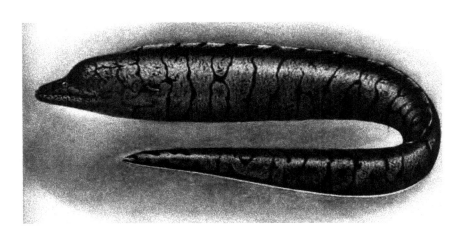

8. GYMNOTHORAX BERNDTI TYPE.

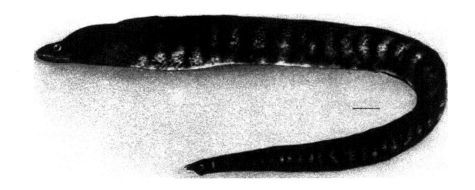

9 GYMNOTHORAX MUCHERI TYPE

10 GYMNOTHORAX XANTHOSTOMUS TYPE

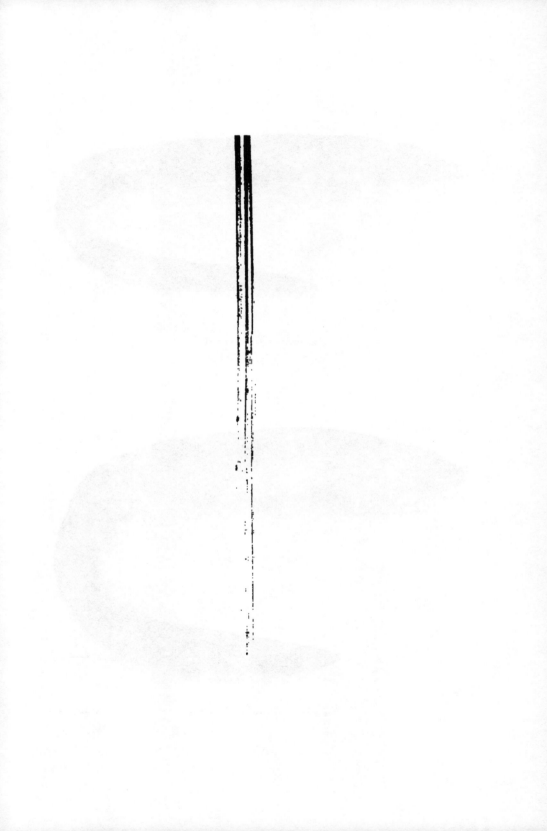

'1. GYMNOTHORAX WAIALUAE, TYPE

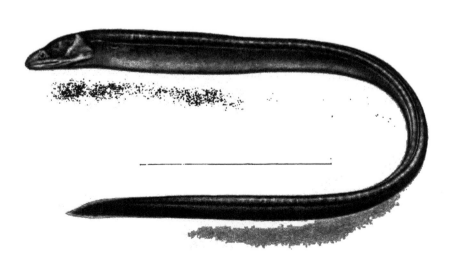

PLATE 7

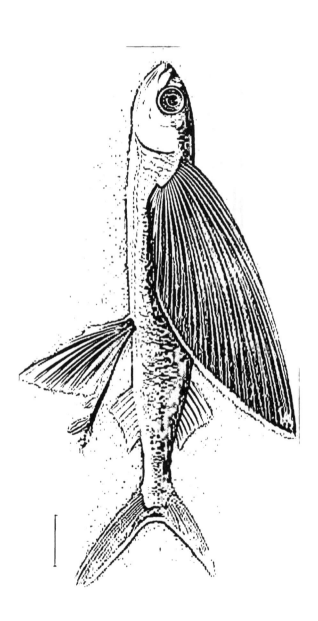

13. EXONAUTES GILBERTI. TYPE.

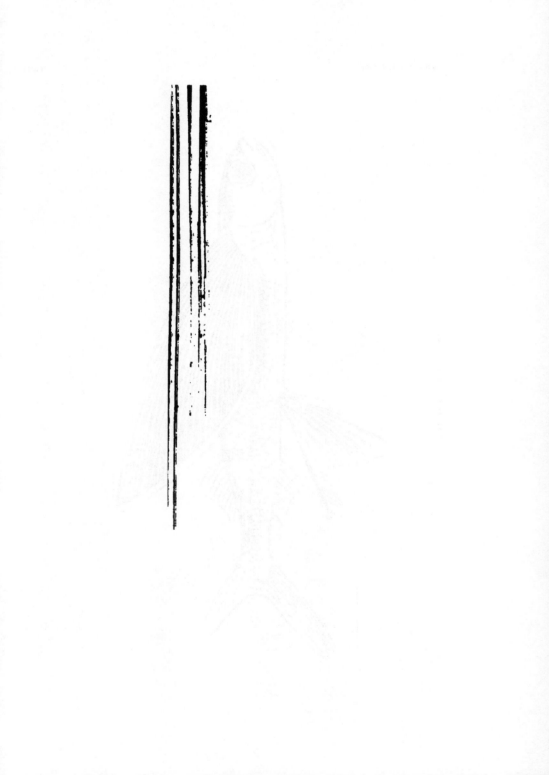

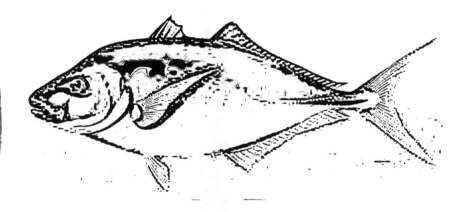

14. CARANGUS CHELIO TYPE

15. CARANGOIDES AJAX, TYPE.

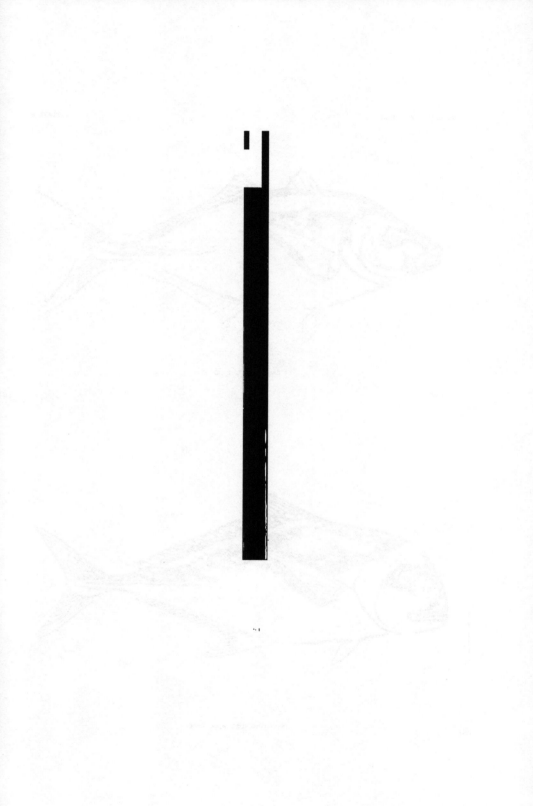

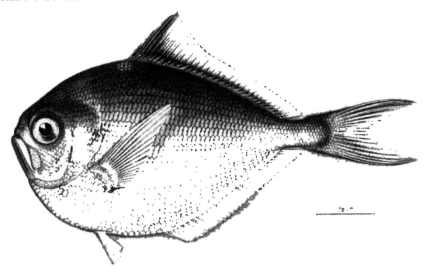

APOGON PANDIONIS TYPE.

A. HOEN & CO. L.

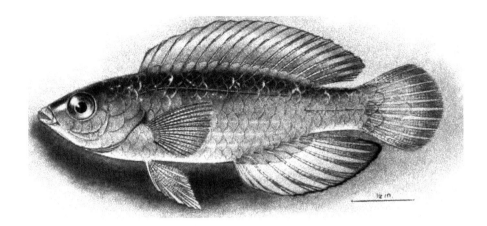

18. CIRRHILABRUS JORDANI TYPE

19. HEMIPTERONOTUS JENKINSI TYPE

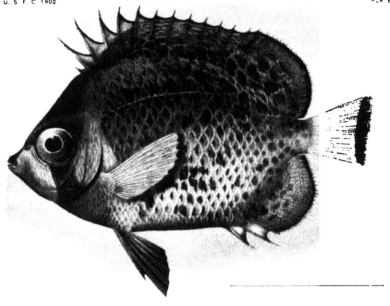

20. CHÆTODON CORALLICOLA. TYPE

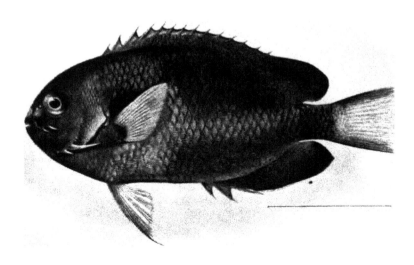

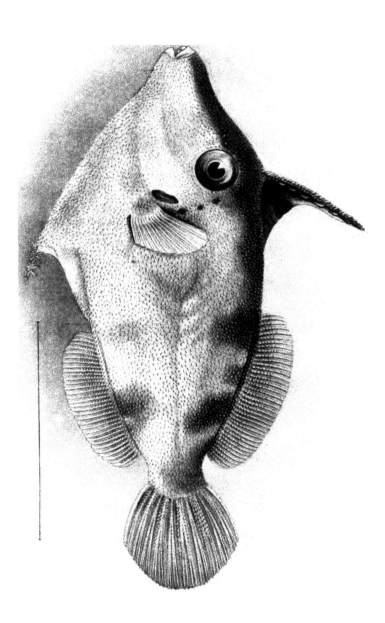

22. STEPHANOLEPIS PRICEI. TYPE.

23 ANTENNARIUS ALTIPINNIS, TYPE

24 ANTENNARIUS QUESCUS, TYPE

NOTES ON FISHES COLLECTED IN THE TORTUGAS ARCHIPELAGO.

By DAVID STARR JORDAN.

Dr. Joseph C. Thompson, United States naval surgeon, formerly stationed at Fort Jefferson, on Garden Key, one of the outlying archipelago of coral islands known as the Dry Tortugas, has sent to us a very interesting collection of small fishes from the coral reefs. Of the 21 species, 4 are new to science and 8 others are new to the waters of the United States. Series of these species are in the U. S. National Museum and U. S. Fish Commission; the others are in the collection of Stanford University.

The accompanying drawings are by Mr. Kako Morita.

Family SPHYRÆNIDÆ.

Sphyræna barracuda (Walbaum).

One specimen, very young, of the common large-scaled barracuda or picuda.

Family HOLOCENTRIDÆ.

Holocentrus siccifer (Cope).

Holocentrum sicciferum Cope, Trans. Amer. Phil. Soc., xiv, n. s., 1871 (Dec.), 465, New Providence, Bahamas.

One fine specimen, the only one known except the original type, which came from New Providence. It agrees well with Professor Cope's description.

Depth 2.75 in length; head 3; mouth small; last dorsal spine very short, scarcely visible; third anal spine moderate, 1.5 in head; scales 45; opercle with two subequal spines, little divergent; preopercular spine short; dorsal spines low, 2.33 to 2.5 in head; soft dorsal 1.75 in head; maxillary rather less than eye, 3.33 in head; head broad above, with short snout.

Color in spirits, violet-silvery above, soiled silvery on sides, silvery below; 9 lengthwise streaks on side, along the rows of scales, 4 of them dark purple, the other 5 yellowish and marked by many dark points, giving a soiled appearance, the 2 uppermost of these broader than the others; axil with a jet-black spot; head plain, dotted with black; a little black on tip of opercle; spinous dorsal with a dusky area toward the base of each membrane, a dusky cross shade toward tip; membranes of first two spines largely jet-black with a white streak below and one above, forming an ocellate black blotch, continuous with the dusky median shade of the fin; soft dorsal and all other fins pale, doubtless red in life, with no dark edgings.

The low fins, small scales, striped body, and dusky spinous dorsal with an ocellate black area in front, well distinguish this pretty species.

Family APOGONIDÆ.

Apogon sellicauda Evermann & Marsh.

Apogon sellicauda Evermann & Marsh, Fishes of Porto Rico, p. 143, fig. 40, 1900 (Dec. 29).

One very fine specimen, about 3 inches long, twice as long as the type, from Culebra Island, Porto Rico, agrees closely with the figure. The black dorsal blotch is larger, extending on the base of the fin; a faint dusky area in front of dorsal; the caudal saddle and the opercular blotch are about as in Evermann's figure. Body and fins uniform bright scarlet. Both edges of preopercle finely serrated.

Family SCIÆNIDÆ.

Eques pulcher (Steindachner).

One fine specimen, about 3 inches long. Olivaceous; three black, ribbon-like stripes along the side, the middle one reaching to end of dorsal and extending forward to eye and across forehead between eyes; faint dark streaks across middle of each interspace; a dark bar downward from eye, one behind mouth; four black cross streaks on forehead and one backward to nape; upper bands of body meeting across forehead; front dorsal black, white edged; base of soft dorsal with a black stripe anteriorly; pectoral with a large black stripe above; anal with a black crescent medially; ventrals almost entirely black; fins otherwise pale.

Family POMACENTRIDÆ.

Pomacentrus caudalis (Poey).

A small specimen about an inch long. Body deep, depth 2.2 in length. Olivaceous (bluish in life), posteriorly bright yellow; fins all yellow; no white or blue spot on last rays of anal; a large black ocellus on front of soft dorsal, a smaller one on back of tail, none on opercle nor in axil.

This little fish corresponds with Poey's *Pomacentrus caudalis* and with no other described species. It is probably a valid species, although specimens similarly colored have been considered the young of *Pomacentrus fuscus*.

Microspathodon chrysurus (Cuvier & Valenciennes).

A fine young specimen, 1.5 inches long. Color blue black, very dark; body and head with round light-blue spots about as large as pupil, rather regularly arranged; five with some fainter ones in a line from eye to tail; 5 or 6 spots on side of head; caudal fin and most of soft dorsal abruptly yellow; the other fins all blackish like the body. Dr. Thompson states that this fish is extremely active in the water, being caught only with great difficulty.

Family LABRIDÆ.

Halichœres bivittatus (Bloch).

Three young specimens. The largest is perhaps a representative of a distinct species. Lateral stripe very black, the lower one black also, although narrower, the space between them pale, and a lunate black spot longer than pupil at base of anterior soft rays. The smaller specimens are green, without dorsal ocellus; the lateral stripe is light brown; the lower almost obsolete; opercle with a small spot. Specimens like the first with the dorsal ocellus are occasionally taken, and the writer has hitherto regarded it as a deep-water variety, which it may be.

Thalassoma nitidum (Günther).

A fine young specimen, 3.5 inches long. Side with a broad violet black stripe from snout to caudal, where it divides, extending to tip of either lobe; the band is broadest medially, and there about twice diameter of eye; a paler black stripe along middle line of back and base of dorsal; a yellowish-brown area between these; dorsal with a median broad black stripe, narrowly pale above and below it, the anterior spines in a jet-black blotch; a jet-black axillary spot; belly white; caudal yellowish, except for the black streaks; fins otherwise pale; opercle black, pale edged posteriorly.

Ventrals short, 1.5 in the short pectorals. *Thalassoma nitidissimum* (Goode), said to have the ventrals still shorter, must be the same species.

Doratonotus decoris Evermann & Marsh.

Doratonotus decoris Evermann & Marsh. Fishes of Porto Rico, p. 231, pl. 29, 1900.

One small specimen, beautifully colored, nearly 1.5 inches long. Body bright green with bronze dots and markings essentially as shown in Evermann's figure of the type of *Doratonotus decoris* from Ponce, but there is no bronze streak along the side of the head and the throat and the breast are green like the rest of the body.

Three nominal species of this genus have been described, and possibly all are valid. *Doratonotus megalepis* Günther, from St. Kitts, is said to have the profile straight and the number of scales 19. The color is lost in the typical example. *Doratonotus thalassinus* Jordan & Gilbert, from near Key West, has the profile concave above eye, the snout more elongate, number of scales 20, and the long spine of the dorsal with filamentous tips. The color is a little different from that of *D. decoris*. The type of *D. decoris*, like our specimen, has the profile a little convex above eye, a little concave above snout, no filaments on the dorsal spines, the scales 26 (we count 23 on our specimen, and Evermann's figure shows no more). Probably *Doratonotus thalassinus* is the male and the others the female of the same species. Perhaps the three are distinct. Except for the number of scales, we should identify *D. decoris* with *D. megalepis*.

Family SCARIDÆ.

Cryptotomus beryllinus (Jordan & Swain).

One small specimen of this common species.

Family CHÆTODONTIDÆ.

Chætodon capistratus Linnæus.

One very small specimen, an inch long, typical of *Chætodon bricei*. It is highly probable that this is the young of *Chætodon capistratus*, from which it differs solely in the presence of a second smaller ocellus above the large one on the side. The large ocellus is, however, vertically oblong in *Chætodon bricei* and round in *Chætodon capistratus*. The two species may be identical.

Pomacanthus arcuatus (Linnæus). One small specimen.

Family GOBIIDÆ.

Gobius soporator (Cuvier & Valenciennes).

Numerous examples of this very common species, very pale, as usual in coral-reef examples.

Ctenogobius glaucofrænum (Gill).

Three specimens, the longest about 1.5 inches long.

This species is known at once by the two black spots in a vertical line at base of caudal. Color very pale olive; fins translucent; sides with 3 rows of vertical olivaceous spots; a blackish streak backward from eye to shoulder; a dusky patch on preopercle; a slight dusky shade from eye to angle of mouth; two small black spots at base of caudal.

This little species was described as from the coast of Washington. This is an error, as Dr. Eigenmann has shown. It is not found in the Pacific, while it does occur at the Dry Tortugas, from which region Eigenmann had four examples.

Ctenogobius tortugæ Jordan, new species. Plate 1, fig. 1.

Head 3.4 in length; depth 4.66; D. vi—10 or 11; A. i, 10; scales 22 or 23, 9; eye 3.66 in head, a little shorter than snout; maxillary 2.75; pectoral equal to head, caudal a little shorter; united ventrals about four-fifths head; body elliptical-fusiform, a little slenderer and a little more compressed than in *Gnatholepis thompsoni*. Head lower and more pointed, the interorbital space scarcely more than one-third the rather large eye; mouth moderately oblique, the maxillary reaching front of pupil, lower jaw very slightly projecting, upper jaw protractile; teeth small, subequal, as usual in this group; snout about as long as eye; gill-openings very slightly continued forward below; cheeks and opercles naked; nape partly naked, with a low median ridge of skin; body covered with large thin scales, which are but slightly ctenoid; dorsals rather low, the species slender, the anterior longest; caudal subtruncate, the lower rays apparently rather longest; pectorals moderate; ventrals long, reaching vent.

Color very pale yellowish, almost white; a row of small blackish spots along base of dorsal, those behind smaller and reduced to dots; a row of small spots along side; a small spot or group of dots on humeral region; a row of small specks behind it in a line; a distinct short blackish bar at base of middle rays of caudal; a jet-black spot as large as pupil behind eye, above opercle; a faint dark bar below eye; a fainter one behind preopercle; a faint bluish streak behind eye; fins all plain whitish, the ventral faintly dusky; some scales with traces of a faint whitish spot.

A single fine specimen 2.2 inches long (type, No. 8363, L. S. Jr. Univ. Mus.) was sent by Dr. J. C. Thompson from Garden Key, one of the coral islands known as the Dry Tortugas.

Gnatholepis thompsoni Jordan, new species. Plate 1, fig. 2.

Head 3.75 in length; depth 4.25; D. vi-i, 11; A. i, 10; scales 28, –9; eye 3.5 in head, equal to snout; maxillary 2.75; pectoral equal to head; caudal equal to head; ventrals a little shorter. Body fusiform, shaped as in the percoid genus *Bolcosoma*; head gibbous above eye, the snout short and decurved; interorbital space half width of the small eye; mouth low, small, horizontal, the lower jaw included; upper jaw protractile; teeth moderate, curved, subequal, apparently in a single row; preorbital broad; maxillary scarcely reaching past front of eye; cheeks and opercles each with about 4 rows of large scales; nape with smaller scales; body covered with large scales; gill-openings rather broad, extending forward anteriorly; spinous dorsal low, its outline rounded, the spines slender; anal moderate, caudal sublanceolate, the middle rays slightly produced; pectorals moderate; none of the rays silky; ventrals large, the membrane across the base well developed.

Color pale straw-yellowish, with some mottlings of darker olive, these forming a row of faint shades along back, and 6 faint quadrate spots, largest and darkest anteriorly, on lower part of side; a black blotch above axil; a narrow, sharply defined black bar, like a pen-mark, below eye, a faint dark shade across eye to humeral spot; no spot at base of caudal; a dark streak backward a short distance on base of pectoral; both dorsals with rows of very small, pale olive dots; a few dots on caudal; pectoral plain; anal pale, with a dusky shade along the edge; serrated ventrals dusky. The dark color on ventrals and anal is probably found only in the males.

A single fine specimen, about 2.25 inches long, type, No. 8364, L. S. Jr. Univ. Mus., was sent by Dr. Joseph C. Thompson, for whom the species is named. The genus *Gnatholepis*, to which it belongs, differs from *Ctenogobius* only in the scaly head. It is equivalent to the later-named genus *Hazeus*, Jordan & Snyder, based on a Japanese species.

Dr. Thompson gives the following note on this species:

"Coral-sand colored, with a vertical stripe through eye and lower cheek.

"Life color: About 10 faint dark squares down back. In these dots so arranged ∴ Below this yet a light line; another row of blotches on level of eye, extending to tail; below this another light line; at level of pectoral 6 large blotches, the darkest; the dots in all these blotches follow parallel lines; caudal faintly speckled; dorsal more so; ventral cloudy gray; pectoral color of body, which is coral-sand colored; above each pectoral a round fawn-colored spot; anal and lower half of caudal tinged with gray. (This portion of body buried in sand when at rest.) Iris yellow, over eye a dark brown lid as long as eyeball is deep; not as thick as pupil; below eye extends vertically downward: below eye this line is a mite broader, nearly as broad as pupil.

"Habit: Lies on bottom, moves very quickly from spot to spot, about 6 inches at a time. When seen by a *Pomacentrus fuscus* it was attacked by it.

"Locality: Coral sandy bottom, around coral heads, inside Bush Key. (Half a dozen seen.) Depth of water, 3 feet."

Elacatinus oceanops Jordan, new genus and species (*Gobiida*). Plate 2, fig. 3.

Head 4.25 in length; depth 4.66; D. vii-1, 12; A. i, 10; eye 3.66 in head; maxillary 3.5; pectoral 1.1; caudal 1.2; spinous dorsal 1.66; ventral disk 1.66 in head. Body fusiform, little compressed, covered with firm skin, which is entirely naked; head moderate, not rounded above, the interorbital space about equal to the small eye; mouth small, horizontal, inferior, with thick lips, resembling the mouth of *Rhinichthys*, the snout projecting beyond it for a distance equal to half the eye; each side of snout with two notching pores; teeth rather strong, sharp, somewhat close-set, apparently in two series below and somewhat unequal; maxillary extending to posterior edge of pupil; upper jaw scarcely protractile; gill-openings small, separated by a wide isthmus, the opening as wide as base of pectoral; skin smooth; lateral line indicated by a series of pores, each with three openings in a vertical line; spinous dorsal low, the spines slender, the median longest; soft dorsal and anal higher; caudal subtruncate, the median rays a little the longest; ventrals united, short, entirely free from the belly; anal papilla very small.

Color light blue; a narrow, jet-black streak from above eye along bases of dorsal and anal fins nearly to upper angle of caudal, these broader in some specimens and coalescing into a median dorsal stripe; a jet-black stripe through eye above pectoral, broadening in the axil, becoming nearly twice width of eye, and extending broad and black to tip of lower half of caudal; body and fins otherwise pale, apparently light blue in life. A color sketch by Dr. Thompson shows the dark stripes to be dark brown, in life, the dorsal stripe sky-blue, fins pinkish.

Five specimens, the largest 2 inches long, from the coral reef of Garden Key, collected by Dr. Joseph C. Thompson.

Type, No. 8365, L. S. Jr. Univ. Mus.; Cotype, No. 2757, U. S. Fish Commission.

This species is the type of a distinct genus, *Elacatinus*, allied to *Gobiosoma*, but differing in the small, inferior, minnow-like mouth and in the form of body. It is one of the handsomest of the gobies, its color suggesting that of the Matasami (eye of the sea), *Malacanthus lativittatus*.

On this species Dr. Thompson has the following notes:

"Blue-striped coral fish. Habit, clinging to coral heads; endeavors to shelter in grooves. When swimming free from one head to another, it moves a few inches—2 to 8—with great rapidity, then comes to a perfect halt and alters its course, then moves again, thus:

"The dots represent halts. Locality, coral heads, at a depth of 3 to 8 feet. Common."

Family BLENNIIDÆ.

Acteis moorei (Evermann & Marsh).

Malacoctenus moorei Evermann & Marsh, Fishes of Porto Rico, p. 309, fig. 97, 1900.

Numerous specimens, each about 2 to 2.25 inches in length. Head 3.66 in length; depth 3.66; D. xxi, 9 or 10; A. ii, 20; V. i, 3, the inner ray very small; scales 3–42–9; eye 3.33 in head; snout 3.33; maxillary 3.33; pectoral equal to head; ventral a little longer, reaching part front of anal; caudal 1.2 in head; longest dorsal spine 1.75; longest soft ray 1.25; fourth dorsal spine 2.66 in head. A single simple cirrus lower than the eye at the nape, a simple one above eye, and a smaller one at the nostril; mouth small; outer teeth in jaws rather large, with a very small band of villiform teeth behind them; teeth on vomer, none on palatines.

These specimens agree very closely with the figure and description given by Evermann & Marsh, with this exception, that the first spine of the dorsal is longer than any of the others, the second and third are progressively shorter, as usual among related species. The original type was a very small example, 1.4 inches long, and it had not the dorsal spines fully developed.

Color brown, with nine crossbands of darker brown about as wide as the interspaces, these more regular and more distinct than in Evermann and Marsh's figure. Head more or less distinctly spotted or freckled below, conspicuously so in one specimen, less so in others; in one specimen the pale interspaces are marked above by a paler spot, the dark crossbands encroaching a little on the caudal. Fins all plain, light brown, the anal with a dusky shade toward the edge, the tips of the rays slightly paler.

This species may be regarded as the type of a distinct genus, *Acteis* (α, without; κτεις, comb), distinguished from *Malacoctenus* Gill by the absence of the broad comb of filaments at the nape. This is represented by a single thread as in *Acteis moorei, lugubris*, and *culebra*, or altogether wanting as in *ocellatus, rarius*, and *macropus*, provisionally referred to the same genus.

In all these species referred to *Malacoctenus, Acteis*, and *Lepisoma*, there are three soft rays in the ventrals, the last ray being very short. Vomerine teeth are present in all we have examined, and there is a narrow band of small teeth, besides the row of larger teeth in each jaw, in all except *Malacoctenus delalandi*.

Ericteis kalisheræ Jordan, new genus and species (Blenniidæ). Plate 2, fig. 4.

Head 3.33 in length; depth 4; maxillary 2.2 in head, reaching to opposite middle of eye; pectoral 1.5 in head; ventrals short, 1.66 in head; first dorsal spine 3 in head, longest 2.2, eye longer than snout, 3.33 in head, its cirrus 1.5 in eye; lower jaw longest. D. xviii, 11, or xix, 11, A. ii, 18; V. i, 3; scales 4–49 to 52–13. Teeth rather strong; a single external series with a villiform band behind, this very narrow in lower jaw, but well developed in upper; vomer medially bare, but with a few small teeth on each side; a row of 4 or 5 strong blunt teeth on the palatines, first 4 dorsal spines much lower than the others, graduated backward, a small tuft of cirri at the nostril; a larger tuft over the eye; a comb of cirri on each side of the nape. Length, 2¼ inches.

Color dark brown, much flecked with lighter and darker brown; side with 5 dark-brown, irregular, crossbars, these growing more irregular with age and extending on vertical fins; 8 dark bars across the dorsal, 6 or 7 across anal; fins all, including ventrals closely spotted with light and dark brown, the colors forming irregular crossbars; pectoral with a black semicircle near its base parallel with the succeeding dark bars; ventrals with 5 or 6 dark bars; head mottled with dark, the markings not definite except a dark blotch, pale-edged below and behind, on opercle; throat with pale mottlings.

Two specimens, type No. 8366, Ichth. Coll., Stanford Univ.; the other retained by Dr. Thompson. On this species Dr. Thompson has the following notes:

"Toad-fish like; life color, head greenish brown, front part of lower lip much lighter; two white dots near mouth angle; body brown; six brown bands extending on dorsal below lateral line; third, fourth, fifth, and sixth bordered with whitish scales; end of caudal peduncle three light spots, middle largest, joined together; abdomen yellowish brown with white spots; under part head, 6 white dots; dorsal brown, with darker brown bands that extend upward and forward, where there cross the spines obliquely 2 white dots; caudal rows of light dots; pectoral spines dotted in rows, with characteristic ventral brown dots; anal like dorsal, only a bit darker; opercle border light, white spot under middle of preopercle.

"Habit, lives in coral crevices; quite fearless.

"Locality, inside Bush Key; depth 3 feet. November 24, 1902."

This species seems to be well distinguished from all others thus far described. It may be nearest *Lepisoma bucciferum*, but the scanty description of the latter species does not well fit it. In all species of *Lepisoma* (*Labrosomus* Gill, but not of Swainson, whose type is rather *Clinus gobio*) and *Malacoctenus* there are 3 soft rays in the ventral fins, the outer often very short. From the type of *Lepisoma* the present species apparently differs in the strong palatine teeth and larger mouth, as well as in the form of the dorsal fin. From *Malacoctenus* it certainly differs in the presence of villiform teeth behind the larger ones. Apparently *Malacoctenus* should be united with *Lepisoma*, or else additional genera must be established. The generic name *Erictis* is proposed for the present species.

These provisional genera may be defined as follows:

a. Nape with a comb of filaments on each side.
  b. Jaws each with a row of strong teeth only; no palatine teeth; dorsal fin notched; mouth small (*delalandi*) . *Malacoctenus*
  bb. Jaws each with a band of villiform teeth behind the row of strong teeth; mouth large.
    c. Palatine teeth none; spinous dorsal fin not notched (*nuchipinnis*) .... *Lepisoma*
    cc. Palatines with a few strong teeth; spinous dorsal fin notched (*haloherus*) .... *Erictis*
aa. Nape with a single filament on each side or with none at all; mouth small; jaws with a narrow band behind the
    strong teeth; palatine teeth; dorsal with the first spine longest (*moorei*) .... *Acteis*

Named, at the request of Dr Thompson, in honor of Miss Kalisher, of San Francisco.

Blennius favosus Goode & Bean.

Blennius favosus Goode & Bean, Proc. U. S. Nat Mus A, 1882, 416 Garden Key, Florida.

Numerous examples agreeing very closely with the original description, the reticulating lines on the side of the head inclosing hexagonal honeycomb-like areas being very distinct, as also the blackish spot on membrane of first and second dorsal spines. In these specimens there are very few dark spots on body or fins, the body being plain light brown, with obsolete darker clouds. The multifid cirrus above eye is characteristic.

The specimens from the Dry Tortugas noted by Garman (Bull. Iowa Lab Nat. Sci., 1896, 89) as *Blennius pilicornis*, belong to this species, which seems to differ from the Brazilian *pilicornis* by the fewer fin rays (D. xii, 18. A. ii, 20). The trifid cirrus is characteristic of *Blennius favosus*.

Certain other specimens from Dr. Thompson's collection differ in color, but seem to agree in every other respect. These may be females of the same species; they are light olive brown with 8 dark crossbars made of blackish dots; interspaces closely speckled with blackish spots and with two rows of quadrate pale areas, one near the back, the other near the belly, sides of head mottled but with no distinct honeycomb marks; a dark bar below eye; dorsal pale, with small black dots numerous on spinous dorsal, few on soft dorsal; a black spot covering upper half of first membrane, smaller, higher and less ocellated than in the true or male *favosus*; caudal nearly plain; anal nearly plain; black spots along base, a dusky shade toward tip ends of rays white; pectoral and ventral plain; two dusky bars across lower side of head behind the jaws. Dr. Thompson has the following notes:

"Life color; along side 7 shiny blue-white spots with a tendency to be rectangular, tips of pectoral and caudal orange, pupil emerald, belly from vent to ventral fins silvery blue white; on second dorsal spine a black blotch; at base of spines a dot; upper parts olive brown; head and gills plain dark brown; over entire body fine speckling of red-brown dots below and dark brown above middle line. These latter are in a row from pectoral to tail.

"Space below base of dorsal spines and top rows on body is checkered dark brown and lighter."

Family BROTULIDÆ

Ogilbia cayorum Evermann & Kendall.

One specimen, 2 inches long, apparently belonging to this species. Scales thin and embedded, not appreciable until the fish is partly dried.

Color pale olivaceous (pale fawn in life), body everywhere closely dotted with darker olive, posterior part of dorsal, anal, and outer two-thirds of caudal rendered dusky with dark points.

Several other specimens of this species were collected by Dr. Thompson and sent by him direct to the National Museum.

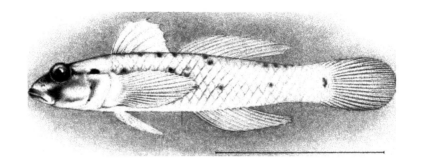

1. CTENOGOBIUS TORTUGÆ. TYPE.

2. GNATHOLEPIS THOMPSONI. TYPE.

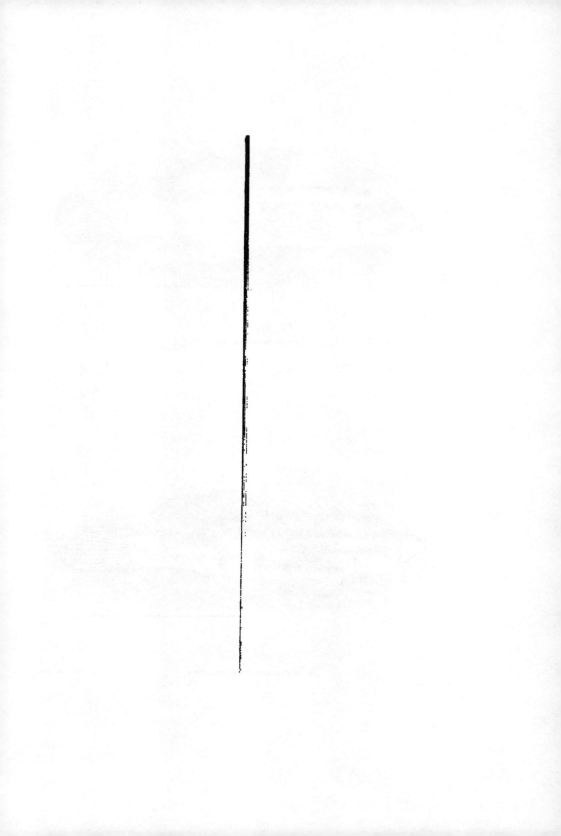

3. BLACATHUS OCEANOPS, TYPE

4. ERCTES HALISHERAE, TYPE

A. HOEN & CO., LITH.

CONTRIBUTIONS FROM THE BIOLOGICAL LABORATORY OF THE U. S. FISH
COMMISSION AT WOODS HOLE, MASSACHUSETTS.

THE ECHINODERMS OF THE WOODS HOLE REGION.

BY

HUBERT LYMAN CLARK,

Professor of Biology, Olivet College.

CONTENTS.

546

THE ECHINODERMS OF THE WOODS HOLE REGION.

By HUBERT LYMAN CLARK,
Professor of Biology, Olivet College.

As used in this report, the Woods Hole region includes that part of the New England coast easily accessible in one-day excursions by steamer from the U. S. Fish Commission station at Woods Hole, Mass. The northern point of Cape Cod is the limit in one direction, and New London, Conn., is the opposite extreme. Seaward the region would naturally extend to about the 100-fathom line, but for the purposes of this report the 50-fathom line has been taken as the limit, the reason for this being that as the Gulf Stream is approached we meet with an echinoderm fauna so totally different from that along shore that the two have little in common. This deep-water fauna, characterized by such species as *Antedon dentata, Schizaster fragilis, Astropecten americanus,* etc., is exceedingly interesting, but unfortunately too little is known about it at present to make a report in any degree satisfactory or complete. This paper, therefore, concerns itself only with the shallow-water species, all included herein having been taken in less than 40 fathoms, and, with one or two exceptions, in less than 15. No species is included which has not been actually taken within the above-mentioned limits.

The purpose of this article is not merely to aid students and collectors at Woods Hole to identify the echinoderms which they find, but also to furnish descriptions and figures of such accuracy that students of echinoderms elsewhere may be able to determine beyond question whether the species which they collect are identical with those occurring at Woods Hole. At present very few of the common littoral echinoderms of America are described in such a way as to make their positive identification possible by one not a specialist, and good figures are even more rare. Particularly is this true of the asteroids and holothurians. Thanks to Lyman and Alexander Agassiz, ophiuroids are well described, and echinoids well figured. In both these cases, however, the publications referred to are not only inaccessible to the average student, but are out of print. It therefore seems wise to gather into one report all the echinoderms of this region.

The only previous list is that of Verrill (1873b), who names 22 species, 5 asteroids, 5 ophiuroids, 4 echinoids, 7 holothurians, and 1 crinoid. The latter, however, and one of the ophiurans (*Amphiura abdita* Verrill), have never been taken within the Woods Hole limits, nor is it probable that either occurs there. Of the 5 starfishes, 2 are undoubtedly identical (*Asterias forbesi* and *arenicola*), so there are really only 19 bona fide Woods Hole species in the list. During the past four summers 5 other species (2 starfishes, 1 brittle star, and 2 holothurians) have been collected within the above-given limits, so that now 24 species are properly credited to Woods Hole, more than one-third of which are holothurians.

The number of species listed is small, but includes representatives of two quite distinct faunas, making the region one of much interest. The southern and western portions of the region mark the northern limit of several distinctively southern species (such as *Mellita pentapora, Ophiura brevispina, Thyone briareus*), while the northern and eastern portions mark the southern limit of several distinctly northern species (*Solaster endeca, Ophioglypha robusta, Cucumaria frondosa.*[a]) It is interesting to note that not fewer than 9 of our 24 species occur on the northern shores of Europe, while 4 occur in Bermuda and the West Indies. Only 10 can be called common in the Woods Hole region, and 4 of these common only in very restricted areas, so that, unless the exact locality is known, the chances are against collecting more than a half-dozen species. Moreover, the common species vary greatly in their abundance from year to year, a species easily found one summer being looked for in vain the next season. For example, the starfish *Asterias forbesi* was very abundant in Woods Hole harbor in 1895; in 1899 it was almost wanting and only small specimens occurred; in 1902 it was again abundant. Similar facts could be given regarding the common sea-urchin, *Arbacia*, and the small red starfish, *Cribrella*.

The best collecting grounds for echinoderms in this region are undoubtedly on the shoals east of Chatham and Nantucket. On the fishing ground known as Crab Ledge, off Chatham, in about 17 fathoms of water, we collected on August 22, 1902, 4 species of starfish, 3 brittle stars, 1 sea urchin, and 1 holothurian. Southeast of the Round Shoal light-ship, and about 8 miles off Sankaty Head, Nantucket, in 12 fathoms, August 13, 1902, we collected 4 species of starfish, 2 brittle stars, 3 sea urchins, and 1 holothurian. Of the 9 species at Crab Ledge, 3 were not taken at Sankaty, while of the 10 species of the latter station, 4 were not found at Crab Ledge; more than half of all the species of this region were therefore taken at these two stations, and 2 other species are known to occur there. In Vineyard Sound, including the deep water off Gay Head, 15 species have been taken, but at least 8 of these are rare and very seldom occur in the sound. At no one station have more than 6 species been taken. In Buzzards Bay, including the deep water off Cuttyhunk, only 8 species occur, and 1 of these is extremely rare.

From the point of view of human economy, the echinoderms of the Woods Hole region are of little importance. None of the 24 species is directly valuable for any purpose. The large *Cucumaria frondosa* has occasionally been used as food, and no less an authority than Dr. William Stimpson recommends it, but it is too rare in the Woods Hole region to be of any use. The starfish, *Asterias forbesi*, is of commercial significance wherever oysters are cultivated, because of its destructiveness to that

---

[a] The recorded occurrence of this species on the Florida reefs is not beyond question; but, even if correct, there is no doubt that the shoals east of Nantucket are the southern limit of the species as a littoral form.

mollusk; but this whole subject has been so thoroughly investigated and reported by Mead ('99) that there is no reason to discuss it here.   As food for fishes some of the echinoderms deserve mention, particularly *Echinarachnius parma* and *Ophiopholis aculeata*, which, in spite of their most unprepossessing appearance from a gastronomic point of view, and their firm, calcareous skeletons, seem to be important items of food with the cod and some other fishes.   The tilefish is known to feed very largely on a species of brittle star, *Amphiura*, its stomach often being completely distended with them.   The smaller holothurians are also frequently eaten by fishes, though it is at least an open question how important an article of diet they may be.

For the proper study of echinoderms, especially for purposes of identification, freshly killed material is the best.   Alcohol is the best preservative, and should be used quite strong.   Formalin is worthless, tending to make the tissues swell and become slimy, and dissolving the calcareous parts.   All echinoderms are very susceptible to Epsom salts (MgSO$_4$), and it is probably the best available narcotic.   In most cases it is not necessary to take any precautions, and the salt may be added to the sea water at once in considerable quantity.   With some holothurians, however, more care is needed, and the salt should be added to the water a little at a time.   Once thoroughly stupefied, echinoderms of all classes, except holothurians, can be killed nicely expanded by putting them into alcohol, 50 per cent or stronger.   Some holothurians, especially *Thyone briareus*, are more difficult to kill satisfactorily, and hot water or acid reagents serve the purpose better, as a rule.   Of course, when it is desired to identify the specimen, acid reagents should carefully be avoided, for the calcareous parts of all echinoderms are essential to accurate identification, and especially is this true of holothurians.   Starfishes, brittle stars, and sea urchins are much more easily identified from well-dried specimens than from the best alcoholic ones, all important specific and most generic characters being based on calcareous parts.   To prepare dry starfishes or brittle stars, they should be thoroughly narcotized, then placed in fresh water for a short time, then in strong alcohol for twenty-four hours, more or less, according to size, then dried as rapidly as convenient.   The colors of many species are readily washed out in alcohol or even in fresh water, so that specimens prepared in the manner just described rarely retain a natural color.   It is said the color may be preserved very satisfactorily by killing in 50 per cent alcohol and drying rapidly by artificial heat.   Sea urchins may be prepared in a similar manner, but care must be taken to see that the interior of the animal is thoroughly soaked in strong alcohol before drying.   Most of the specific characters of all echinoderms, except holothurians, are easily seen with a hand lens, and usually with the unaided eye, even in young and small specimens.   But for the proper identification of holothurians, a compound microscope is essential, as the minute particles of lime in the skin furnish the most important specific characters.   To examine these all that is necessary is to cut out a small piece of the thin body wall, and after soaking it in water a few minutes, clear and mount in glycerin.   If the body wall is thick, only the surface layer should be taken, and it should be treated with caustic potash (10 per cent or stronger) before being washed in water and cleared in glycerin.   The same treatment is desirable when the body wall is heavily pigmented, or when the calcareous particles are extremely numerous and crowded into more than one layer.   It is sometimes necessary to leave the piece of skin in caustic potash for several days, especially if the specimen has been in alcohol for some time.

The illustrations for the present report were made - in all but one instance, *Trochostoma œöliticum* -from specimens taken in the Woods Hole region. The figures showing the external appearance of the different species, and the denuded tests of the sea urchins, are from photographs made by Mr. J. D. Figgins. So far as possible, these photographs were from living or freshly killed specimens, and are represented natural size. It is due entirely to Mr. Figgins's patience and persistence that the results have been so generally satisfactory, and it gives me pleasure to express my sincere thanks to him. The other figures are from drawings, and are intended to show those characters which are of importance in the identification of species.

In the choice of names, I have endeavored to make as few changes as possible from those in common use, but pre-Linnæan names have been rejected, as no good reason has ever been given for introducing them into modern literature. The name of each species is followed by that of its describer; if, however, that describer did not place the species in the genus in which it is now included, his name is inclosed in parentheses. There is no good reason for appending the name of the writer who placed the species in its present genus, and the custom is bad in its effects. Popular names are given where any are in common use, but none has been coined for any of the less-known species. No attempt has been made to give a complete synonymy, but the names used in the most important papers dealing with the echinoderms of the New England coast, and in standard works on the various classes, are given under each species. A bibliography also is appended, in which the titles of such works will be found, and there is an index to all names, including synonyms.

The work upon which this report is based has been carried on exclusively in the laboratory of the United States Fish Commission at Woods Hole, in part during the summers of 1898, 1899, and 1900, but particularly in 1902. It gives me pleasure to acknowledge my indebtedness to the Commissioner, Hon. George M. Bowers, for the opportunity thus given me, and to Dr. H. M. Smith, who, as director of the Woods Hole laboratory, has afforded me every facility for carrying on my investigations. Like all other workers at the Woods Hole station, I am under special obligations to the veteran collector of the Commission, Mr. Vinal N. Edwards, who has supplied me with many important data concerning the local distribution and habits of echinoderms. Mr. George M. Gray, of the Marine Biological Laboratory, has also assisted me with information and specimens.

ECHINODERMATA.

Radially symmetrical animals, with a well-developed water-vascular system.

Key to the classes of Echinodermata of the Woods Hole region.

Body more or less flattened dorso-ventrally, with the mouth at the center of the lower surface; external skeleton well developed.
    Body with the radii extended as more or less elongated arms.
        Arms unjointed, with a longitudinal furrow on the lower side, in which are the numerous tube feet...ASTEROIDEA
        Arms made up of numerous joints, without a longitudinal furrow on the lower side....................OPHIUROIDEA
    Body hemispherical or discoidal, without extended radii ....ECHINOIDEA
Body more or less elongated, with the mouth at the anterior end, surrounded by a circle of tentacles;
        no well-developed external skeleton...HOLOTHURIOIDEA

ASTEROIDEA.

STARFISHES.

Of the 6 starfishes which occur in the Woods Hole region, no fewer than 4 belong to the large and widely distributed genus *Asterias*, the species of which are, even to the present day, greatly confused. For this reason it is impossible to avoid the use of some technical terms in the artificial key, as well as in the descriptions, although the endeavor has been made to have the descriptions as clear and free from technicalities as possible.

Following are the terms used which do not carry their own meaning:

Abactinal=aboral; opposite the mouth; the upper surface.

Actinal=oral; the lower surface.

Adambulacral spines=the spines borne on the plates which form the margin of the actinal longitudinal furrow on each arm.

Ambulacral furrow=the actinal longitudinal furrow on each arm.

Madrepore plate=the more or less prominent somewhat circular body situated abactinally in one of the interradii.

Oral plates=the skeletal plates, especially the adambulacral plates, immediately surrounding the mouth.

Papulæ=the tentacle-like outgrowths of the body wall which project between the meshes of the skeleton and contain prolongations of the body cavity.

Papular areas=spaces occupied by papulæ between the meshes of the skeleton.

Pedicellaria=the minute forcep-like modified spines, consisting of two hard jaws moved by muscles, and occurring in many starfish, especially on the abactinal surface, and at the base of the large spines.

It must be added, moreover, that no description or figures will serve to distinguish positively very young individuals of our 4 species of *Asterias*. The young of *forbesi* seem to be the first to show the specific characters, and individuals 15 mm. in diameter can generally be recognized without much trouble. The young of *vulgaris, tenera*, and *austera* resemble each other so closely when less than 10 mm. in diameter that it is practically impossible to separate them positively, but when 15 mm. in diameter they can usually be distinguished from each other on careful comparison, and by the time they have reached a diameter of 20 mm. the specific characters are generally quite marked. It must be borne in mind, however, as shown by Mead ('99), that the age of a starfish is not shown by its size, and, furthermore, that spines and pedicellariæ increase in number as the age increases. Thus some small but old specimens may have the specific characters well defined, while others much larger but younger may have them barely indicated.

Key to the Asteroids of the Woods Hole region.

Rays rough and spiny, since the meshes of the skeletal network are coarse and bear prominent spines,
singly or in groups of 3 or 4; pedicellariæ numerous; feet in 4 longitudinal rows in each
furrow.
   Adults usually over 100 mm. in diameter: papulæ numerous in small groups.
      Rays tending to be cylindrical and blunt; skeleton quite firm; spines rather few and coarse;
      pedicellariæ on adambulacral spines short and blunt; madrepore plate usually bright orange. ASTERIAS FORBESI
      Rays tending to be somewhat flattened and acuminate: skeleton open and rather soft; spines
      numerous, tending to form a noticeable median longitudinal row on the abactinal side of
      each ray: pedicellariæ on adambulacral spines numerous, long, slender, and sharp; madre-
      pore plate pale.. ASTERIAS VULGARIS
   Adults usually much less than 80 mm. in diameter; papulæ few, 1, 2, or 3 in a place.
      Rays tending to be cylindrical and tapering: skeleton moderately firm, with numerous spines;
      pedicellariæ very numerous, especially on actinal surface, forming wreaths, generally very
      noticeable, on the spines there... ASTERIAS TENERA
      Rays decidedly flattened, rather wide and blunt: skeleton coarse and very firm; spines few;
      pedicellariæ on adambulacral spines few and blunt, but a prominent series of them just
      within the edge of the ambulacral furrow... ASTERIAS AUSTERA
Rays rather smooth, since the meshes of the skeletal network are fine and bear numerous very small,
   delicate spines: no pedicellariæ; feet in 2 rows—
      With 5 or 6 rays.. CRIBRELLA SANGUINOLENTA
      With 9 to 11 rays... SOLASTER ENDECA

1. Asterias forbesi (Desor). *Common starfish.* (Pl. 1, figs. 1, 2; pl. 4, figs. 14, 15.)

   Asteracanthion forbesi Desor, 1848.
   Asterias arenicola Stimpson, 1862.
   Asteracanthion berylinus A. Agassiz, 1863.
   Asterias forbesii Verrill, 1866 et seq.

Description.—Rays normally 5, occasionally 6, rarely 4 or 7. R=75 to 130 mm., r=10 to 25 mm.,
R=4.5 to 8 r. Breadth of ray, near base, 15 to 30 mm.. R=3.5 to 5 br. Rays stout, blunt, and somewhat rounded at the tip, abactinal surface normally arched. Disk moderate or large, often highly
arched. Interbrachial arcs rather acute. Abactinal area covered with stout plates, closely soldered
together into a firm skeleton, with no constantly regular arrangement. These plates carry single
prominent spines, 1 to 2 mm. high, which are usually blunt and minutely rough or thorny at the tip,
but in young individuals may be quite sharp. About the base of these large spines are often grouped
2 to 4 smaller ones, 0.5 to 1 mm. high, also blunt. All the spines are more or less fully encircled
about midway between base and tip with a cluster of very small, blunt pedicellariæ. Scattered more
or less freely all over the abactinal surface are pedicellariæ a little less blunt and somewhat larger.
Sides of rays well rounded, but oftentimes with a well-defined longitudinal arrangement of spines,
about halfway between actinal and abactinal surfaces. Below this row is a space of very variable width
which is usually free from spines but bears a few scattered pedicellariæ. Beneath this space, and well
on to the actinal surface of the ray is a prominent longitudinal series of plates, each of which bears an
obliquely arranged pair or trio of very prominent spines (like those of the abactinal surface, and
wreathed by pedicellariæ in the same way), of which the most distal is nearest the ambulacral furrow.
This series is usually clearly defined, but in some individuals it is more or less irregular. Between it
and the adambulacral spines is a longitudinally extended space, which may be perfectly open and occupied by only a few pedicellariæ, but in many specimens is obliterated by a single series of large spines,
which are more or less square cut or even clavate at the end, and are often slightly beveled on one side.
The presence or absence of this row is not a matter of size, but it is very probably a matter of age, though
that is not yet proved. Adambulacral plates with 1 or 2 (usually 2) rather long (2 to 4 mm.), somewhat
flattened and slender spines square cut or blunt at the end, and many of them, especially near the
mouth, bearing small blunt pedicellariæ, though these are often wanting in young individuals. In
a specimen with R=24 mm. there are about 60 adambulacral plates on each side of the ambulacral furrow; in one with R=67 mm. there are about 100; and in one with R=80 mm. there are about 130.
Oral spines not peculiar. Papular areas variable in size, but usually large, with numerous papulæ,
generally 5 or more in each group. Madrepore plate of moderate size, 2 to 5 mm. in diameter, with
numerous narrow furrows, and not surrounded by any special circle of spines. Tube feet quadriserial,
crowded. Color in life very variable; the most common shades are brown, purple, orange, green, and
bronze. Spines generally light and madrepore plate usually bright orange-red. After death the
bright tints are generally lost and in preserved specimens the madrepore plate is usually yellow or
brown.

Range.—Maine to the Gulf of Mexico, rare or local north of Cape Ann; low water to 27 fathoms.

PLATE 1.

Figs. 1, 2. *Asterias forbesi* (small specimen, natural size).—1. Aboral view.　2. Oral view.
Figs. 3, 4. *Asterias vulgaris* (small specimen, natural size).—3. Aboral view.　4. Oral view.

Remarks.—This is the common starfish of the Woods Hole region. It occurs abundantly in the harbor, in Vineyard Sound, and in Buzzards Bay. It was originally described from the shoals of Nantucket, but we did not find any off Sankaty Head or Crab Ledge. It is most common near low-water mark, but occurs down to 18 or 20 fathoms. The habits have been observed and described so well by Mead ('99) that any account of them here would be superfluous. The larval stages and development of this species have been described by Agassiz ('77), but curiously enough little has been done on the embryology during the past twenty-five years. The young stars occur in great abundance on the eelgrass in the Eel Pond at Woods Hole during August. The great variety of color in the adults has yet to be explained; it is apparently not associated with age or sex, nor has its correlation with the environment been proved.

2. Asterias vulgaris Verrill. *Northern starfish.* (Pl. 1, figs. 3, 4; pl. 4, figs. 16, 17.)

Asterias rubens Gould, 1841.
Asteracanthion rubens Desor, 1848.
Asteracanthion violaceus Stimpson, 1853.
Asteracanthion rubens Stimpson, 1853.
Asteracanthion pallidus A. Agassiz, 1863. No description.
Asterias vulgaris Packard, 1863. No description.
Asterias vulgaris Verrill, 1866 et seq.
Asterias stimpsoni Verrill, 1866 (pars.).
Asterias pallida Goto, 1898.

Description.—Rays normally 5, rarely 4 or 6. R=75 to 150 mm. or even more, specimens 425 mm. in diameter being reported from Nova Scotia, r=9 to 30 mm., R=4.5 to 8.5 r. Breadth of ray near base, 15 to 30 mm., R=3.5 to 5 br. Rays more or less flattened, the sides somewhat vertical, tapering to a more or less acuminate point. Disk usually rather large, sometimes considerably arched. Inter-brachial arcs somewhat acute. Abactinal area covered by a network of narrow plates with large meshes, not forming a very firm skeleton. Almost always there is a median longitudinal series on the arm, with large papular areas (sometimes 4 or 5 mm. across) on each side. All the plates bear blunt spines 1 or 2 mm. high, usually singly but occasionally 2 or 3 together. The spines are rough or minutely thorny at the tip, and are encircled by a more or less complete wreath of pedicellariæ, which are remarkably blunt. Pedicellariæ scattered over abactinal surface much larger and quite acute. Along side of ray a very well-marked lateral series of spines, below which is a longitudinally extended area of greater or less extent quite free from spines, but with numerous pedicellariæ. The lateral series varies greatly in position, in some specimens being quite near the abactinal surface, while in others it may be scarcely visible when seen from above. It is made up of plates bearing 2 spines, side by side, so that there is apparently a single line of spines. In older specimens, however, there is often a third spine beneath the distal one of each pair, and frequently a fourth spine occurs above or beside the proximal one. Well down on the actinal surface of the ray is another series of spines, the largest and most prominent of all. These are usually over 2 mm. in length and are often 3 or 4 mm. long; they are very blunt, even square cut or clavate, in some specimens deeply so. Three spines form an oblique row on each plate, the most distal being nearest to the ambulacral furrow. Although these spines bear pedicellariæ, often in great numbers, they are not so nearly wreathed by them as are the abactinal spines. Adambulacral plates with 1 or 2 (usually 2) rather long (2 to 4 mm.) somewhat flattened slender spines, pointed, square cut or clavate at the end, most of which carry from 1 to 6 long, slender and very acute pedicellariæ (fig. 17). A few small pedicellariæ occur on the adambulacral plates within the furrow. The acute pedicellariæ on the adambulacral spines are present even in very small individuals (15 to 20 mm. in diameter). Adambulacral plates are more numerous in this species than in any of the other forms of *Asterias* occurring at Woods Hole. In a specimen with R=18 mm. there are 65 or more plates on each side of the furrow; in another with R=24 mm. there are about 75; in another with R=37 there are only about 110; in another with R=107 there are not less than 170. Oral spines not peculiar. Papular areas generally large, with 3 or more papulæ in each group. Madrepore plate of moderate size, 2 to 5 mm. in diameter, with numerous narrow furrows, and not surrounded by any special circle of spines. Tube feet quadriserial, crowded. Color in life very variable; the most common shades are yellow and purple, but cream-colored, yellow-brown, brown, orange, pink, and even bright red individuals occur. Spines generally light, and madrepore plate light yellowish.

Range.—Labrador to Cape Hatteras; but south of the Woods Hole region rarely seen in shallow water; low water to 358 fathoms.

Remarks.—Although less common than *A. forbesi*, this species occurs in abundance in many places near Woods Hole. In the harbor it is occasionally found with *A. forbesi*, and specimens have been described to me which were apparently hybrids between the two species; but I have never seen such, nor any individuals, however small, which could not readily be assigned to the proper species. Even when only 10 mm. in diameter, *forbesi* has a much stouter appearance. Young specimens of *vulgaris* were taken in August off Gay Head and Cuttyhunk, as well as at Crab Ledge, and off Sankaty Head. Most of the adults taken in the Woods Hole region are less than 180 mm. in diameter. The best specimens that we collected were taken at Crab Ledge. The habits and development of *vulgaris* do not seem to differ essentially from those of *forbesi*, though some slight differences in the early stages have been noted by Alexander Agassiz ('77). The larval stages have been well described by Field ('92), and further points have been carefully investigated by Goto ('98). As in the case of *forbesi*, the great variety of color does not seem to be associated with either age, sex, or environment. All of the earlier writers regarded this species as identical with *Asterias rubens* of Europe. Stimpson, early in 1863, first suggested in a private letter that it might prove distinct, and proposed the name *vulgaris*, but he gave no description. Later in the same year Alexander Agassiz proposed the name *pallidus*, but gave no adequate description. Still later Packard used the name *vulgaris* Stimpson, in a published list, but he also failed to give a description. Finally, in 1866, Verrill published a description under the name *vulgaris* Stimpson. Clearly, however, Stimpson's name is the barest kind of a *nomen nudum*, and Verrill is properly the describer of the species. Whether it is really distinct from *Asterias rubens*, however, has never yet been proved, and Verrill ('76) now thinks they may prove identical.

3. Asterias tenera, Stimpson (pl. 2, figs. 5–7; pl. 4, figs. 20, 21).

Asterias tenera Stimpson, 1862, p. 269.
Asterias compta Stimpson, 1862, p. 270.
? *Asteracanthion flaccida* A. Agassiz, 1863.
Leptasterias compta Verrill, 1873.
Leptasterias tenera Verrill, 1874.
Asterias (Leptasterias) compta Sladen, 1889.

Description.—Rays 5, rarely 4 or 6. R=30 to 40 mm., r=4 to 7 mm., R=6 to 7 r. Breadth of ray near base, 5 to 10 mm., R=4 to 6 br. Rays not flattened, nearly terete, slender and pointed. Disk small. Interbrachial arcs acute. Abactinal area covered by a fine network of narrow plates with rather large meshes, forming an open but fairly firm skeleton, with no clearly defined median row on the ray. All the plates carry prominent, though rather delicate, spines about 1 mm. long, more or less rough and pointed at the tip. These spines, except on the disk, are encircled at or near the base by more or less complete wreaths of blunt pedicellariæ; the contrast between disk and rays in this respect is often marked. Pedicellariæ most numerous near the tip of the ray. On sides and actinal surface of rays, the spines are somewhat larger and tend to form longitudinal series of single spines placed side by side. There are generally four such series quite clearly defined, of which the one nearest the ambulacral furrow consists of the largest spines. In all the series the spines are more or less densely wreathed with pedicellariæ which become more numerous approaching the ambulacral furrow or passing toward the tip of the ray. Adambulacral plates with 1 or 2 slender spines 2 mm. long; at middle of ray plates carrying 1 spine tend to alternate regularly with those carrying 2, but at both base and tip of ray the plates more frequently bear a single spine. Adambulacral spines with numerous pedicellariæ which often form wreaths about the spines near the middle. Similar pedicellariæ also occur on the adambulacral plates within the furrow. Adambulacral plates rather less numerous than in *vulgaris;* a specimen with R=18 mm. has about 60 plates on each side of the furrow; another with R=24 has about 68 plates, and in one with R=37 there are about 100. Oral spines not peculiar. Papular areas of moderate size, with few papulæ, usually 1 or 2, in each area. Madrepore plate small, with few, rather wide furrows, surrounded by an imperfect circle of 6 or 8 spines. Tube feet quadriserial, but not crowded. Color in life varying from purplish-pink to nearly white, the smallest ones having the least color; madrepore plate and spines nearly white.

Range.—Nova Scotia to New Jersey; in 14 to 85 fathoms.

Remarks.—This small starfish is very common off Sankaty Head, but we did not collect any specimens at Crab Ledge, nor has it been taken in Vineyard Sound. Verrill ('73) records it from south of Marthas Vineyard in 20 to 25 fathoms. A careful study of Stimpson's ('67) original descriptions of *tenera*, based on 20 specimens from Massachusetts Bay, and *compta*, based on a single individual 3 inches in diameter, has convinced me that the two forms are identical, and this belief is confirmed by a comparison of numerous specimens, undoubtedly *tenera*, taken off Sankaty Head, with 2 large specimens (72 mm. in diameter) labeled *compta* in the collection of the Fish Commission from outside the Woods

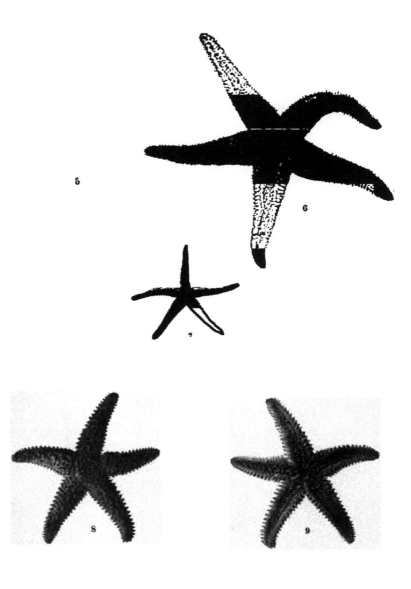

Figs. 5-7. *Asterias tenera* (natural size).—5. Large specimen; aboral view. 6. Large specimen; oral view. 7. Young; aboral view.
Figs. 8, 9. *Asterias austera* (adult, natural size).—8. Aboral view. 9. Oral view.

Hole region. The latter were probably identified by Verrill. Sladen ('89) expressed his suspicion that *tenera* and *compta* were identical, but owing to lack of material withheld his decision. Verrill ('95) thinks *tenera* is a poorly nourished, slender variety of *compta*, but the name *tenera* has precedence. The group of starfish (regarded as a genus by Verrill and a subgenus by Sladen) called *Leptasterias*, and including *tenera* and one or two other forms, does not seem to rest upon characters of sufficient importance and constancy to warrant recognition. Very little is known of the habits or development of *Asterias tenera*, though the latter is said to be without metamorphosis, the young being cared for by the mother, to whom they are attached. Whether this species is really distinct from *Asterias mülleri* of Europe remains to be proved.

4. Asterias austera Verrill. (Pl. 2, figs. 8, 9; pl. 4, figs. 18, 19.)

Asterias austera Verrill, 1895.

Description.—Rays 5. R=24 to 35 mm., r=6 to 10 mm., R=3.5 to 4 r. Breadth of ray near base 6½ to 10 mm., R=3.5 br. Rays rather short, wide, and depressed, somewhat angular from the prominent row of lateral spines. Interbrachial arcs rather acute. Abactinal area covered with rather stout plates, apparently with no regular arrangement, the median radial series being very zigzag. These plates are very thick, though the width is quite variable. They carry short, very blunt spines, considerably less than 1 mm. in length. Spines of the median radial series sometimes noticeably larger than the disk spines. Each spine is encircled at the base by a few small, blunt pedicellariæ, usually from 4 to 10 in number, but these are often wanting on the disk. Occasional isolated pedicellariæ of much larger size occur scattered here and there on the abactinal framework. Sides of ray practically vertical, the upper edge being marked by a series of 18 to 28 large stout plates, which start at the level of the disk, abactinally, and run distally with a downward slope, so that the side of the ray is almost twice as high at base as at tip. Each of these plates carries a prominent, stout, blunt spine, about 0.7 of a mm. long, encircled at base by a cluster of 8 to 15 small pedicellariæ on the upper and outer side. Between the two series of spines the lateral surface of the ray is free from projections of any kind, except occasionally a few large pedicellariæ. Between the actino-lateral series and the adambulacral plates there is apparently only a single row of plates, and these are usually confined to the basal half of the ray. They often bear stout spines nearly 1 mm. long, which are apparently opposite and not alternate with the spines of the actino-lateral series. They have no cluster of pedicellariæ at the base, but sometimes bear one or two small pedicellariæ near the tip. Adambulacral plates with one or two nearly cylindrical blunt spines, considerably over a mm. in length; they do not alternate alike even on the two sides of the same ray. These spines are mostly free from pedicellariæ, but rarely a small blunt one is present near the tip. Within the ambulacral groove, however, above the bases of the spines, are numerous small pedicellariæ. These are sometimes specially prominent on the oral plates. Adambulacral plates rather few; a specimen with R=18 has only 49 on each side of the furrow, and my largest, with R=24, has only 60. Madrepore plate of medium size, with few, wide furrows, and surrounded by a rather incomplete circle of 6 to 8 spines. Tube feet quadriserial, but not at all crowded. Color in life white, cream color, or yellowish, more or less marked abactinally by dark green, purple, or reddish. Digestive cœca often show through the abactinal surface, especially in small individuals (as in young *vulgaris* and *tenera*), adding much to their beauty.

Range.—Georges Bank and off Cape Cod; Crab Ledge, off Chatham, Mass.; 17 to 35 fathoms.

Remarks.—After some hesitation I have referred a small star-fish, of which we took 12 specimens at Crab Ledge, to Verrill's species *austera*, although, owing to the brevity of his description and the lack of figures, there is some room for doubt; but the largest specimens answer very well to his description, and as he reports his species from off Cape Cod in 33 to 35 fathoms, it is highly probable that the species here described and figured is *austera*. It is a handsome little star-fish, and seems to be common at Crab Ledge. We did not find it elsewhere. Nothing is known of its habits or development.

5. Cribrella sanguinolenta (O. F. Müller). (Pl. 3, figs. 10, 11; pl. 4, fig. 22.)

Asterias sanguinolenta O. F. Müller, 1776.
Asterias pertusa O. F. Müller, 1776.
Asterias oculata Pennant, 1777.
Asterias spongiosa Fabricius, 1780; Gould, 1841; Desor, 1848.
Linckia oculata Forbes, 1839; Stimpson, 1853.
Cribella oculata Forbes, 1841.
Linckia pertusa Stimpson, 1853.
Cribrella sanguinolenta Lütken, 1859; Verrill, 1866 et seq.
Cribrella oculata Al. and E. C. Agassiz, 1865; Sladen, 1889.

Description.—Rays normally 5, occasionally 6. The proportions vary to an extraordinary degree. R=30 to 60 mm., r=6 to 13 mm., R=2.5 to 5.5 r. Breadth of ray near base, 10 to 14 mm., R=2.4 to 5.4 br. Rays varying from nearly cylindrical to long conical, but usually with a rounded tip, and not flattened. The basal part is oftentimes swollen and then contracted close to the disk, which is of very variable size. Interbrachial arcs are sometimes rounded, but more often acute. Abactinal surface covered by a network of plates, which form a more or less firm skeleton, according to the size of the meshes, these ranging from under 0.5 mm. to over 1.5 mm. As a rule these plates have no regular distribution, but in some specimens there is, on the sides of the ray, an approach to a transverse arrangement, either in vertical or oblique rows. Upon the plates are numerous minute spines, which differ greatly in size and form in different individuals. They are sometimes very delicate and sharp, sometimes stout and blunt, and the length varies from 0.2 to 0.5 mm. These spines occur singly, or more often in groups of from 2 to 12, and as they are so small, so numerous, and of such nearly uniform length (in any one specimen) the surface of the ray does not appear spinulose in the living animal. On the actinal surface the plates tend to form longitudinal rows, especially near the base of the rays, and there is almost invariably one well-defined row just outside the adambulacral plates. The latter are not very numerous, a ray 42 mm. long having only about 65 on each side of the furrow. The armature of the adambulacral plates is made up of 10 to 12 blunt spines, of which 1 is on the side of the plate, well up in the furrow, while the remainder are on the face of the plate. Although they vary considerably in size, in different specimens, the arrangement, on the whole, is fairly constant. The furrow spine is very slender, but is sometimes quite long, over 0.5 mm. Of the others, that nearest the furrow is the largest (1.5 mm.) and usually stands more or less alone. Behind, or outside of it, are 1 or 2 a trifle smaller, and the remainder, arranged more or less irregularly in pairs, diminish rapidly in size, so that those on the outer edge of the plate are of about the same size as the spines on the adjoining plates. In young individuals the adambulacral spines tend to form a narrow, transverse series on the plate, continuous with the transverse series on the sides of the ray, but in adults such an arrangement is obscured, if not obliterated. Oral spines long, but not peculiar. Papular areas small, with usually 1 papula, sometimes 2 or 3. Madrepore plate small, with few wide furrows, the ridges between which frequently bear spines. Feet biserial. Color in life, abactinally usually bright red, with a slight orange cast, the actinal surface deep yellowish. There is, however, more or less variety. In some specimens the red is faded in spots to yellowish, or even the whole abactinal surface to orange, yellow, or cream color; in other specimens the red is deepened to purple, often of a very rich shade, while again the purple is faded to lavender or nearly white.

Range.—Greenland and Labrador to Connecticut, off New Jersey, and even Cape Hatteras; littoral only as far south as the Woods Hole region; northward it is common from low water to 220 fathoms, while southward specimens have been reported from 1,350 fathoms. Also Spitzbergen, Nova Zembla, and Iceland to Great Britain and northwestern France. Northwestern coast of Asia (Brandt), Bering Sea (Ludwig).

Remarks.—The abundance of this star-fish in Woods Hole Harbor varies greatly in different years, and although it is often common the specimens are always small. Much the finest specimens found in this region come from Crab Ledge, where the species is very common, and exhibits the greatest variety of form and color. Careful measurements were made of 142 specimens to see whether there was any tendency to diverge into two or more forms, but none appears. While some specimens have the breadth of the ray over 36 per cent of the length, others have it only 18 per cent, while the majority range about 28 per cent. The curve of variation falls off quite uniformly in both directions. Nor could I find that the diversity of either form or color was in any way correlated with age, sex, or environment. This species breeds in the early spring; the eggs are large and contain much yolk; development is abbreviated, but is still imperfectly known.

6. Solaster endeca (Retzius). (Pl. 3, figs. 12, 13; pl. 4, fig. 23.)

Asterias endeca Retzius, 1783.
Solaster endeca Forbes, 1839.

Description.—Rays usually 10 or 11, frequently 9, occasionally 7, 8, 12, or 13. R=75 to 150 mm., r=25 to 50 mm., R=3 r. Breadth of ray near base 12 to 30 mm., R=5 to 6 br. Rays arched abactinally, flattened actinally, regularly tapering, bluntly pointed. Disk large, generally highly arched; in a specimen with R=106, the disk is 35 mm. high. Interbrachial arcs rather rounded. Abactinal surface covered with a very close net-work of plates, making a firm skeleton. These plates

Figs. 10, 11. *Cribrella sanguinolenta* (natural size).—10. Slender-armed form; aboral view. 11. Stout-armed form; oral view.
Figs. 12, 13. *Solaster endeca* (small specimen, natural size).—12. Aboral view. 13. Oral view.

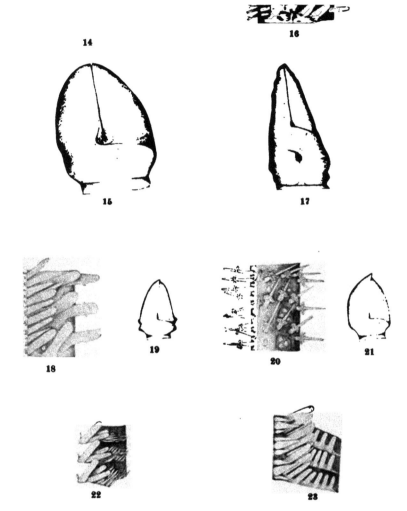

Figs. 14, 15. *Asterias forbesi.*—14. Part of oral surface of one side of ray near middle (× 2½). 15. Large pedicellaria from oral side (× 75).
Figs. 16, 17. *Asterias vulgaris.*—16. Part of oral surface of one side of ray near middle (× 2½). 17. Large pedicellaria from oral side (× 75).
Figs. 18, 19. *Asterias austera.*—18. Part of oral surface of one side of ray near middle (× 5). 19. Large pedicellaria from oral side (× 75).
Figs. 20, 21. *Asterias tenera.*—20. Part of oral surface of one side of ray near middle (× 5). 21. Large pedicellaria from oral side (× 75).
Fig. 22. *Cribrella sanguinolenta.* Part of oral surface of one side of ray (× 6).
Fig. 23. *Solaster endeca.* Part of oral surface of one side of ray (× 6).

bear low columnar elevations, each of which is crowned with a group of from 10 to 12 (more or less) delicate spines, 0.3 to 0.5 mm. in length. These spines are so short, and of such nearly uniform length, that the general impression in the living starfish is that of an almost smooth surface. In some specimens the spines radiate slightly from the top of the column while in others they are very closely erect. Along the sides of the rays, the columns tend to form longitudinal series, and 4 or 5 such series are sometimes well defined. On the actinal surface there is a well-marked series of plates forming a lateral margin to the ray. On the distal portion of the ray these plates are in contact with the adambulacral plates, but as the disk is approached, they diverge and form a margin for the actinal surface of the disk. The plates carry transverse ridge-like elevations, 0.5 mm. high (more or less), which are crowned by a group of 12 to 25 small spines, of nearly uniform length, arranged in a more or less perfectly double series. Actinal, interradial spaces, closely covered with plates, bearing low columnar or ridge-like elevations, crowned with clusters of spines somewhat larger than those of the abactinal surface. These clusters have a more or less definite arrangement, tending to form rows parallel to the adambulacral plates. The row adjoining the latter runs out about halfway to tip of ray, while the others are successively shorter, the interradial series being the shortest. Adambulacral plates with 2 series of spines, one in the furrow, and one on the face of the plate. Furrow series of 3 spines in a row parallel to the furrow; 1 (or even 2) of these spines is often absent, but when all are present, the middle one is the longest (a little over 1 mm.), while the distal one is generally the shortest. Near the mouth these furrow spines are usually more or less fused at the base. Spines on face of plate somewhat pointed, 6 to 8 in number, 1 to 2 mm. long, arranged in a single curved transverse series, concavity of curve away from mouth; largest spine nearest furrow. Oral plates very prominent, bearing along the margin a series of 14 to 18 spines, of which those at the oral end of the plate are much the largest (2 to 6 mm. long), the next 1 or 2 pairs a little smaller, and the remainder markedly smaller. On the face of each plate is a ridge on which are borne 3 to 12 spines of very variable length; the longest are nearest the oral end of the plate and may be 2 to 3 mm. long. Papulæ small, and generally single. Madrepore plate small, with numerous fine furrows. Feet biserial. Color in life, abactinally, dark red or deep rose-purple, rarely orange or dull yellowish; actinally, orange or yellowish of some shade.

Range.—Greenland and Newfoundland to Crab Ledge, off Chatham, Mass., low water to 150 fathoms; also on the coasts of Great Britain and Northern Europe; possibly circumpolar.

Remarks.—This very striking and easily recognized northern starfish just enters the Woods Hole region, as we found it not uncommon on Crab Ledge. Of the 9 specimens we took, 4 have 10 rays, and 5 only 9. Both the largest specimen (220 mm. in diameter in life) and the smallest (40 mm.) are among the 9-rayed individuals. Little is known of the habits, and nothing of the development, of this species.

OPHIUROIDEA.

BRITTLE STARS, SAND STARS, SNAKE STARS, OR SERPENT STARS.

The ophiuroids are a very large class, and the separation of nearly allied species is a task sufficient to test the skill of a specialist; but the Woods Hole region contains only 5 species, and these are so unlike each other that even a beginner can easily distinguish them. The terms which are used in describing a brittle star usually carry their own meaning, so that it is only necessary to introduce here a few of the less easily understood names:

Adoral plates.—A pair of plates on the face of each of the 5 jaws, beside or in front of the oral shield.

Oral papillæ.—Teeth-like projections along the edges of the jaws.

Oral shield.—The large plate on the face of each jaw, near the base.

Radial shields.—A pair of plates on the upper side of the disk, at the base of each arm; sometimes very prominent, again wholly covered.

Tentacle scales.—Small scales on the lower side of the arm, on each side of the under-arm plate; 1, 2, or several at the base of each tentacle.

Tooth papillæ.—Small teeth-like projections at the point of each jaw.

Key to the Ophiuroids of the Woods Hole region.

Arms simple, unbranched.
  Arm spines short, small, and more or less appressed to the arm.
    Arms nearly terete; disk covered with a fine granulation.....................................OPHIURA BREVISPINA
    Arms flattened; disk covered with scales.....................................OPHIOGLYPHA ROBUSTA
  Arm spines prominent, at a marked angle with the arm.
    Upper-arm plates surrounded by a series of small plates; arm spines 5 to 6 ....OPHIOPHOLIS ACULEATA
    Upper-arm plates not surrounded by small plates; arm spines 3.....................AMPHIPHOLIS SQUAMATA
Arms dichotomously branched ....GORGONOCEPHALUS AGASSIZII

1. Ophiura brevispina Say. (Pl. 5, figs. 28–30; pl. 7, figs. 37, 38.)

    Ophiura brevispina Say, 1825.
    Ophioderma olivaceum Ayres, 1852.
    Ophioderma serpens Lütken, 1856.
    Ophiura brevispina Lyman, 1860 and 1882.
    Ophiura olivacea Lyman, 1865; Verrill, 1873.

    Description.—Arms normally 5, occasionally 4, rarely 6, of moderate length, nearly terete, but flattened on the oral side. Diameter of disk, 10 to 15 mm. Length of arm, 40 to 60 mm.; breadth of arm at base, 2½ to 3 mm. Arm spines 7 to 8, shorter than the arm joints, approximately equal, or the lowest shortest, closely appressed to the arm. Disk more or less perfectly pentagonal, covered with granules of nearly uniform size, about 100 to 180 to a square millimeter, completely concealing the radial shields. At the base of each arm the granulation extends out on each side, so that the first 3 upper-arm plates form a narrow ridge running in toward the center of the disk. On each side of the third plate is a group of about 10 little scales. Upper-arm plates broadly in contact with each other; at the base of the arm they are nearly oblong, twice as wide as long, but as the tip is approached they become more rounded on the sides, the outer edge becoming curved, the inner markedly narrowed. Under-arm plates broadly in contact, nearly square with rounded corners, usually somewhat longer than wide. First under-arm plate much wider than long, with rounded sides; second much longer. Tentacle scales 2, of which the inner is nearly half the length of the under-arm plate, while the outer is about half as long and covers the base of the lowest spine. Oral shields oval, plainly longer than wide. Adoral plates small, lying entirely at sides of oral shields, roughly triangular, with rounded corners. All of the oral surface granulated as above, except the oral shields and adoral plates. Oral papillæ about 7 on each side of each jaw, of which the one next to the under-arm plate is small, often wanting, while the next one is the widest and largest of all. Teeth 5, blunt, the lowest the smallest. No tooth papillæ. In each interbrachial space there are 4 genital openings. Color very variable, but never very bright; some shade of green or brown is the most frequent; the disk is generally mottled or spotted, and the arms frequently banded with alternate rings of light and dark shades. More or less uniform olive green is a frequent color, uniformly brown specimens being less frequent; nearly black specimens are occasionally found.
    Range.—North Falmouth, Mass., to Bahia, Brazil, low water to 122 fathoms.
    Remarks.—This very widely distributed brittle star reaches its northern limit in Buzzards Bay. It is abundant in about 1 fathom of water, on a bottom covered with eel grass, in North Falmouth Harbor, and has been taken in similar situations in Marion, New Bedford, and Dartmouth, Mass., on the other side of the bay. It also occurs at Sag Harbor, Long Island. The habits, movements, and development have been so well described by Grave (1900) that it is unnecessary to discuss them here. Verrill ('99) considers the northern form of this species as separable from the form occurring from Florida southward, and would regard it as a variety *olivacea*.

2. Ophioglypha robusta (Ayres). (Pl. 6, figs. 31, 32; pl. 7, figs. 39, 40.)

    Ophiolepis robusta Ayres, 1851.
    Ophiura fasciculata Forbes, 1852.
    Ophiura squamosa Lütken, 1854.
    Ophioglypha robusta Lyman, 1865.
    Ophioglypha tenorii Ljungman, 1866.

    Description.—Arms 5, finely tapering. Diameter of disk, 7 to 10 mm. Length of arm, 24 to 35 mm.; breadth of arm at base 1 to 1.5 mm. Arm spines 3, rounded and acute, the upper one largest, nearly 1 mm. long; at tip of arm the lowest spine is flattened and bears 1 or 2 small hooks. Disk rounded or slightly pentagonal, covered with small, irregular scales, 4 or more to the square millimeter; in young specimens, the plates at center of disk show more or less definite arrangement. Radial shields inconspicuous, about as broad as long, barely touching without. The disk is notched in each radius

PLATE 5.

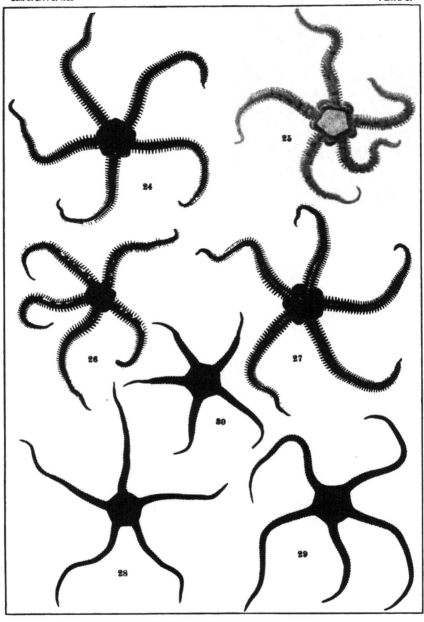

Figs. 24-26. *Ophiopholis aculeata.* Aboral view (natural size).
Fig. 27. *Ophiopholis aculeata.* Oral view (natural size).
Figs. 28, 29. *Ophiura brevispina.* Aboral view (natural size).
Fig. 30. *Ophiura brevispina.* Oral view (natural size). Shows regenerating arms.

by the upper-arm plates, and on the side of the notch are 7 or 8 short conical projections or papillæ. This series is continuous with another one of 12 small, flat papillæ along the genital slit. Just outside the disk notch are small papillæ on each side of the upper-arm plate. In young specimens all or nearly all of these papillæ may be wanting. Upper-arm plates rounded distally, more or less pointed proximally, separated by the side-arm plates on the distal part of the arm in adults, on the whole length of the arm in the young. First under-arm plates oval or triangular with rounded corners; the remainder broader than long, the distal edge slightly convex, the proximal with a little point. Side-arm plates strongly developed, completely separating the under-arm plates. Tentacle scales small and with rounded ends, 7 to 9 for the mouth tentacles, 3 or 4 for each of the first two arm tentacles, and 1 for each tentacle thereafter. Oral shields somewhat broader than long, pointed within, and broadly curved on the outer side; 1.5 mm. wide, more or less. Adoral plates long and narrow, lying along the inner sides of oral shields, and meeting within. Oral papillæ 3 or 4 on each side of each jaw, one or more of them at the point of the jaw; outermost one broadest. Teeth 5 or 6, equal, sharp and flat. No tooth papillæ. Interbrachial spaces with small scales, and 1 pair of genital slits. Color in life variable, but not bright; generally some shade of gray, more or less variegated with brown, reddish, or black; radial shields usually distinctly lighter; arms often banded with gray and greenish.

Range.—Greenland to Crab Ledge, off Chatham, Mass.; and possibly in deep water to Porto Rico. (Fish Hawk collection, 1899.) Also from the Arctic Ocean to Denmark. Alaska? Possibly circumpolar. Low water to 150 fathoms.

Remarks.—This is another of the northern echinoderms which just comes within the northeastern border of the Woods Hole region. Amidst the gravel and broken shells brought up in the dredge on Crab Ledge we found 4 small specimens of this species, the largest with the disk only 3 mm. across. The figure on plate 6 is taken from this specimen, enlarged. Apparently nothing is known of the habits or development of the species.

3. Ophiopholis aculeata (Linnæus). *Daisy Brittle-Star.* (Pl. 5, figs. 24–27; pl. 7, figs. 41, 42.)

> *Asterias aculeata* Linnæus, 1767. Müller, 1776.
> *Ophiura bellis* Fleming, 1828.
> *Ophiocoma bellis* Forbes, 1839.
> *Ophiolepis scolopendrica* Muller & Troschel, 1842.
> *Ophiocoma aculeata* Desor, 1848.
> *Ophiopholis aculeata* Gray, 1848. Verrill, 1866 and 1873. Lyman, 1882.
> *Ophiopholis scolopendrica* Stimpson, 1853.
> *Ophiopholis bellis* Lyman, 1865.

Description.—Arms, 5; rather wide and flat. Diameter of disk, 15 to 20 mm. Length of arm, 60 to 80 mm. Breadth of arm at base, without spines, 3 to 4 mm.; with spines, 6.5 to 8 mm. Arm spines, 5 or 6, of which the two middle ones are the largest (1 to 1.5 mm. long by 0.3 to 0.6 mm. wide), the upper one or two slightly, the lower two considerably, smaller, the lowest spine smallest of all; these spines are borne on a prominent vertical ridge on each side arm plate and stand at nearly a right angle to the arm. In adults these spines are very blunt, but in young individuals they are slender and acute. Beginning rather more than halfway out on the arms, the lowest spine becomes bent at the tip until finally it is little more than a sharp-pointed hook, usually with 1 or 2 minute teeth on the concave side. Disk circular, often bulging considerably between the arms, scaled, but more or less covered by a coat of very unequal granules, which are small and spheroidal near the center of the disk, and become much more prominent and spine-like on the interbrachial portions. In very few cases, however, does the granular coat cover the disk with even approximate uniformity. In the very great majority of specimens, from 6 to 36 more or less circular scales or plates are left bare, and as these are always symmetrically arranged, the disk has a very ornate appearance. In the center is 1 plate, around which are grouped 5 others, placed radially; there are then 10 sets of 1, 2, or rarely 3 somewhat smaller plates, lying in rows radiating from the center, 5 radial and 5 interradial. In extreme cases the radial shields may also be left partially bare. Some or all of the plates are surrounded by definite circles of small granules. Upper arm plates nearly elliptical, about twice as wide as long; in large specimens those near the base of the arm are broken into two pieces. Each plate is surrounded by a single series of about 12 very large, somewhat angular, flat grains; but between any two adjoining upper arm plates there is only 1 row of these grains, so that those which serve as an anterior border for one plate are also the posterior border of the next distal plate. In some specimens, as the tip of the arm is approached, the bordering grains become more numerous and nearly circular, while in other specimens they are less numerous, so that the upper arm plates come in direct contact. Under arm plates nearly rectangular, but with rounded corners and slightly concave sides; except the first 1 or 2 they are distinctly wider than long. Tentacle scales single, large, oval, more than half the length of the under arm plate. Oral shields more or less elliptical, much wider than long, the outer side often flattened; the madrepore plate is usually distinctly longer and larger than the others. Adoral plates large, more or less rounded at each end, on the inner side of the oral shields, but not meeting within. Oral papillæ, 3 or 4 on each side of each jaw, wide, flat, and thin edged, of approximately equal size. Teeth about 12, narrowest above, the broad, lower ones sometimes broken in two. No tooth papillæ. Interbrachial spaces loosely covered with plates, each of which bears 1 to 3 large granules or small blunt spines; in each space there is 1 pair of genital slits. Color extraordinarily variable; no two specimens seem to be colored just alike; shades of brown, red, yellow, purple, and green are most common; unicolor specimens are very rare, the disk being always blotched, or marked in some regular pattern, while the arms are banded or longitudinally striped; actinal surface generally light, most often yellowish.

Range.—Greenland to New Jersey, low water to 1,000 fathoms; rare or local south of Cape Cod; also Iceland and Spitzbergen; along the coasts of Great Britain and northern Europe to Ireland and the English Channel. Bering Sea (Ludwig).

Remarks.—Although this beautiful ophiuran has been known to occur in the colder waters off Gay Head and Watch Hill and in 38 fathoms even off the coast of New Jersey, it has always been regarded as a rarity south of Cape Cod. The reported cases of its occurrence in Vineyard Sound are almost certainly cases of mistaken identification. In 1894 Mr. Vinal Edwards took a number of very fine specimens on some fishing banks about 15 miles ESE. of Sankaty Head, Nantucket, in 24 fathoms. In 1902 we took hundreds of specimens off Sankaty Head in 12 fathoms and on Crab Ledge in 17. Those taken off Sankaty Head were all small, very few having a disk diameter of over 7 mm. and none over 10, while the specimens from Crab Ledge were of good size, many being over 15 mm. across the disk. Next to the remarkable variety of color, which is really beyond description, the most extraordinary thing about these brittle-stars is the way in which even large specimens secrete themselves in cavities and crannies, among rocks, shells, and barnacles. They are eagerly sought as food by codfish, and their colors and habits are doubtless protective. The development of this species has been partially described by Fewkes ('86).

4. Amphipholis squamata (Delle Chiaje). (Pl. 6, figs. 33, 34; pl. 7, figs. 43, 44.)

> *Asterias squamata* Delle Chiaje, 1828.
> *Ophiura neglecta* Johnston, 1835.
> *Ophiocoma neglecta* Forbes, 1841.
> *Ophiolepis squamata* Müller and Troschel, 1842.
> *Ophiolepis tenuis* Ayres, 1851.
> *Amphiura tenera* Lütken, 1859. Lyman, 1865.
> *Amphiura tenuis* Lyman, 1860.
> *Amphiura elegans* Norman, 1865.
> *Amphipholis lineata* Ljungman, 1871.
> *Amphipholis elegans* Lütken, 1871. Verrill, 1873 et seq.

Description.—Arms 5, slender, of moderate length, 2.5 to 4.5 times the diameter of the disk, which is from 3 to 5 mm. Length of arm 12 to 20 mm. Breadth of arm at base, without spines, 0.5 to 0.8 mm. Arm spines 3, blunt, nearly equal, the upper a little the stoutest; more slender and acute in the young. Disk nearly circular, rather flat, covered with scales of nearly uniform size, 20 or more to the square millimeter. Radial shields conspicuous, narrow (about 3 times as long as broad), in close contact with each other, barely separated at each end. Margin of disk rather sharply defined by a line where the edges of the scales on the interbrachial spaces meet the scales of the disk. Upper arm plates mostly wider than long, with the outer edge slightly curved, the inner e"ge and sides forming a common curve; all except the first 2 or 3 separated by the side arm plates. In the young, the upper arm plates are more widely separated, and are nearly pointed on the inner edge. Under arm plates about as long as broad, the outer edge nearly straight, the sides straight, or reënteringly curved, the inner edge pointed; all are separated by the side arm plates. Tentacle scales 2, quite large. Oral shields of medium size, wider than long, rounded without, but pointed within. Adoral plates rather large, meeting within. Oral papillæ 3 on each side of each jaw, the basal one very much the widest, the other 2 small and nearly equal; when pressed together the oral papillæ can completely close the mouth slit. No tooth papillæ. Teeth 5, flat and thin, the lowest the smallest. Interbrachial spaces more finely scaled than disk, with 1 pair of genital slits. Color in life quite uniformly brownish or gray,

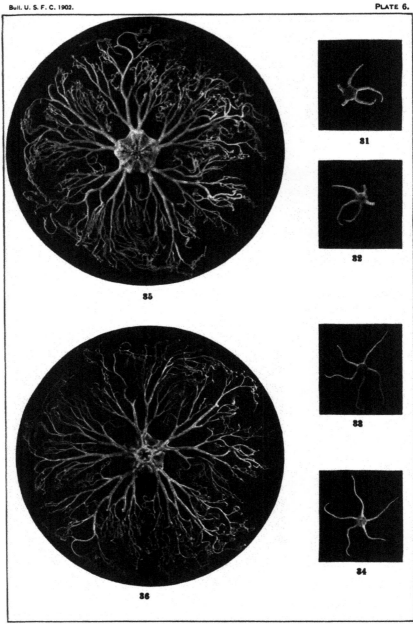

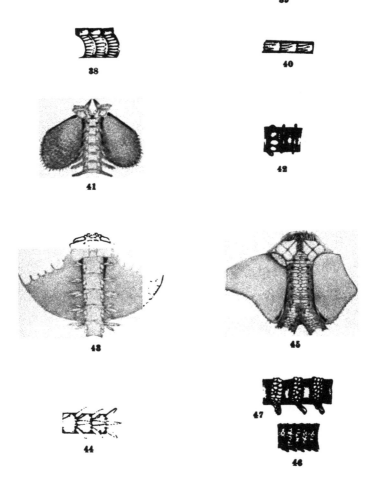

Figs. 37, 38. *Ophiura brevispina.*—37. Oral view of base of arm, with adjacent parts (× 5). 38. Side view of three arm joints, to show arm spine (× 5).

Figs. 39, 40. *Ophioglypha robusta.*—39. Oral view of base of arm, with adjacent parts (× 10). 40. Side view of three arm joints, to show arm spines (× 10).

Figs. 41, 42. *Ophiopholis aculeata.*—41. Oral view of base of arm, with adjacent parts (× 5). 42. Side view of three arm joints, to show arm spines (× 5).

Figs. 43, 44. *Amphipholis squamata.*—43. Oral view of base of arm, with adjacent parts (× 10). 44. Side view of three arm joints, to show arm spines (× 10).

Figs. 45–47. *Gorgonocephalus agassizii.*—45. Oral view of base of arm, with adjacent parts (× 2). 46. Side view of arm joints near middle of arm (× 2). 47. Side view of arm joints near tip of arm (× 10).

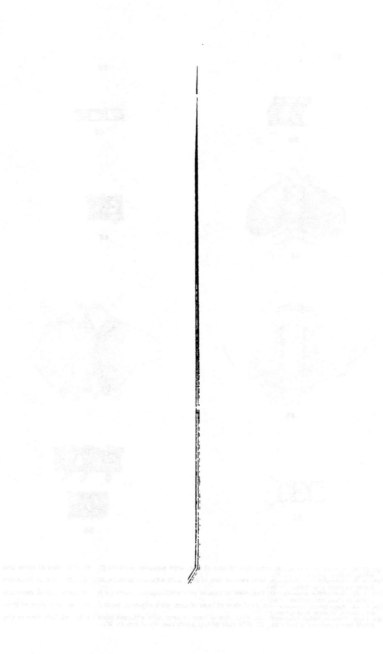

more or less finely mottled with whitish, the lower surface yellowish. Outer ends of radial shields almost always very light, giving the appearance of a white spot at the base of each arm. Very young individuals have the disk bright orange.

Range.—Arctic Ocean to New Jersey, low water to 60 fathoms. Also on the coast of Europe well into the Mediterranean. Lyman reports specimens even from near the Cape of Good Hope, 98 fathoms, and near Australia, 120 fathoms, but these probably represent species which in life would show marked differences from *squamata*.

Remarks.—This is the commonest and most widely distributed of the ophiurans in the Woods Hole region. It is common in Vineyard Sound, especially just east of Nobska, and is reported abundant at Newport. We also took it at Crab Ledge and off Sankaty Head, and specimens were brought in from Ram Island and the Eel Pond at Woods Hole. The latter were very small, and were found on the eel grass. I am indebted to Dr. H. F. Perkins for one of these, a beautifully stained and mounted specimen with disk half a millimeter in diameter. Adults occur on rocky or shelly bottom and generally are found in the interstices and cavities of the stones and shells. The breeding season is in the summer, and in July and August the adults nearly always contain eggs or young, for, as is well known, this species is viviparous. Often one of the interbrachial spaces will be found swollen and of a slightly pink shade. If opened, it will be seen to contain the bright pink eggs of a parasitic crustacean. The eggs of the brittle star are yellow, orange or reddish. The development of this species is without metamorphosis and has been studied in part by several observers in Europe and by Fewkes ('87) in this country, but much still remains to be done. Fewkes's paper contains interesting notes on the breeding and other habits of adults.

5. Gorgonocephalus agassizii (Stimpson). *"Basket Fish"*; *"Spider."* (Pl. 6, figs. 35, 36; pl. 7, figs. 45–47.)

> *Euryale scutatum* Gould, 1841 (not of Blainville).
> *Astrophyton agassizii* Stimpson, 1853.
> *Gorgonocephalus agassizii* Lyman, 1882.

Description.—Arms 5, dividing dichotomously, but unequally, 11 to 15 times, and perhaps more; branches of unequal length, and alternating, so that if the left-hand branch of the first fork is the longer, the right-hand branch will be longer at the next fork, the left hand at the third, the right hand at the fourth, and so on. Disk 40 to 80 mm. in diameter; length of arm 140 to 280 mm.; breadth of arm at base 8 to 10 mm.; at first fork 10 to 18 mm. True arm spines wanting (Lyman). Disk more or less flattened, with radial shields very prominent as raised ridges, covered with a thick skin, which is usually quite smooth, but often bears granules or low blunt spines, sometimes in large numbers, especially at the center of the disk. Margin of disk sharply defined by a band of plates connecting the outer ends of the radial shields. The latter are long and narrow and nearly meet at center of disk; although sometimes nearly smooth, they usually bear numerous knobs or thick, blunt spines, a millimeter high, more or less. Upper surface of arms beautifully curved; on the last few subdivisions there is a faint longitudinal groove; the surface may be smooth near the base and along the sides for some distance, though generally covered with coarse granules which are at first irregularly disposed, but after about the second or third fork begin to form incomplete rings about the arm. After about the sixth fork the granules are confined almost wholly to these rings, which soon become raised ridges, about 2 granules wide. Undersides of arms nearly flat, smooth, the numerous small, irregular underarm plates quite evident in dry specimens. Tentacle scales usually 3, sometimes 4, at base of arm 2 or 1; the first tentacle pore has none. These scales are spine-like, and about a millimeter long. Near the tips of the arms there are only 2 or even 1, but there they assume the form of little, toothed hooks. No oral shields or adoral plates, but the jaws are surrounded by a wide circle of calcareous plates connecting the bases of the arms. Teeth, teeth papillæ, and mouth papillæ, all alike and spiniform, about 20 in number on each jaw, and about 2 mm. long. Interbrachial spaces nearly triangular with a rounded point inward, covered with a thick, smooth, or more or less granular skin, and containing a single pair of genital openings near their outer edges. At the inner point of one of the interbrachial spaces is the single, large madrepore plate, which is wider than long. Color in life, disk and interbrachial spaces brown, of variable shade, but usually dark; radial shields often lighter; arms cream color, yellow, brownish-yellow, or reddish; some specimens have a very strong reddish cast.

Range.—Arctic Ocean and Gulf of St. Lawrence to Crab Ledge, off Chatham, Mass., and perhaps even on Nantucket Shoals. Also reported from Vadsö, Norway. Low water to 800 fathoms.

Remarks.—This remarkable animal is one of the rarest of the Woods Hole Echinoderms, and its right to a position in this list is based on the reported capture of specimens by codfishermen near Nantucket Shoals and at Crab Ledge. We failed to find it at the latter place, although trustworthy fishermen have assured Mr. Edwards and myself that it is common there, "if you get into deep enough water." It is very abundant off Race Point, Provincetown, Mass., in 35 fathoms, where, in August, 1902, I brought up 123 fine specimens in a single haul of a small dredge. At that place the basket fish breeds during the latter part of August, and these specimens were full of the ripe reproductive cells. It is called by the fishermen "spider," and the particular spot where it occurs is known locally as the "spider ground." Almost nothing is known of the habits and absolutely nothing of the development of this remarkable ophiuran. Young specimens with the arms only once forked have the disk covered with scales, and are said to resemble *Ophiopholis.* One of the specimens taken at Provincetown had 5 madrepore plates, but dissection showed that there was only 1 stone canal. The history of the discovery of the basket fish, and its original description, are very interesting, and are given very fully by Lyman ('65).

ECHINOIDEA.

SEA-URCHINS, SAND-DOLLARS, CAKE-URCHINS, SAND-CAKES, ETC

The echinoids are less represented in the Woods Hole region than any other class, for there are only 4 species known to occur, and of these 1 is extremely rare, and 1 is found only in deep, cold water. The 4 are so unlike each other that there is no danger of confusing them; but it must be borne in mind that the young are often unlike the adults, for not only do the primary spines increase in number with age, but the number and arrangement of the poriferous plates also undergoes a marked change in many cases. In the cake-urchins (clypeastroids), moreover, there is a marked change in the shape of the test, position of the anus, number and appearance of lunules, etc., as the animal increases in size. The following terms, used in the descriptions, require some explanation:

Abactinal system=the group of plates forming the apex of the test (or near it) including the genital and ocular plates, and in true urchins the anal plates also.
Ambitus=the line of largest horizontal circumference of the test.
Buccal plates=a circle of plates on the peristome around the mouth.
Branchial incisions=notches in the edge of the peristome, between the ambulacra and the interambulacra.
Coronal plates=any vertical series of plates running from the abactinal system to the peristome.
Genital plates=the five large plates terminating the interambulacra abactinally.
Imperforate tubercles=tubercles the top of which is not centrally depressed or vertically perforated.
Lunules=slit-like openings piercing the test from abactinal to actinal surface.
Miliary spines=the smallest spines of the test, usually on very insignificant tubercles.
Ocular plates=the five plates terminating the ambulacra abactinally.
Peristome=the portion of the actinal surface surrounding the mouth, covered with a membrane.
Petals=the figures formed by the poriferous zones of the ambulacra, of flat or irregular echinoids.
Poriferous zones=the vertical areas occupied by the pores through which the feet pass.
Primary spines=the large spines situated on the largest tubercles of the test.
Secondary spines=spines intermediate between primaries and miliaries.

Key to the Echinoids of the Woods Hole region.

Test nearly hemispherical, with spines of moderate or large size.
    Spines long; color, deep red, purple, or brown to nearly black .... ARBACIA PUNCTULATA
    Spines short, numerous; color, green or yellowish; spines sometimes reddish or purple-tipped,
                                               STRONGYLOCENTROTUS DRÖBACHIENSIS
Test discoidal, flat, with very numerous minute spines.
    Test without lunules .... ECHINARACHNIUS PARMA
    Test with five lunules .... MELLITA PENTAPORA

1. Arbacia punctulata (Lamarck). *Common sea-urchin.* (Pl. 7, figs. 48–52.)

Echinus punctulatus, Lamarck, 1816.
Arbacia punctulata, Gray, 1835.
Echinocidaris davisii, Agassiz, 1863; Verrill, 1866.

Description.—Diameter of test 30 to 50 mm.; height 15 to 25 mm. D=2 H. more or less. Length of longest spines 20 to 25 mm.; diameter of anal system 4 to 6 mm.; diameter of whole abactinal system 10 to 15 mm.; diameter of peristome 15 to 25 mm. Test somewhat flattened, sloping markedly toward ambitus, which is nearly circular; actinal surface flat, the peristome only slightly sunken. Branchial incisions deep, with prominent everted edges. Anal plates normally 4, occasionally 3 or 5. Ocular plates (radials) excluded from the circumanal ring in young specimens, but in old ones 1 or 2 sometimes enter it slightly. Genital plates (basals) large, the madrepore plate evidently largest. Ambulacra straight, narrow above the ambitus, but wider below. Poriferous zones narrow, with large pores, in simple pairs dorsally, then in arcs of 3, and on the very edge of the peristome polyserial. Spines few near the abactinal system, the upper half of the median interambulacral space being entirely free from them; secondary and miliary spines altogether wanting; primaries longitudinally striated, longest at ambitus, shortest near abactinal system; those above ambitus pointed or blunt; those at ambitus and below flattened and more or less rounded at tip, and often with a median longitudinal ridge there; those nearest peristome shorter and more or less spatulate. Primary tubercles smooth, imperforate, in a double series on each ambulacrum; in 4 to 8 or more series on each interambulacrum, of which the 2 middle series are smallest, with only 4 or 5 small tubercles each, while the outermost are as long as the ambulacral series. Transverse rows of interambulacral tubercles oblique. All tubercles diminish in size from ambitus upward. A specimen with D=36 mm., H=18 mm., has 13 coronal plates. Buccal plates, 5 pairs, prominent. Color in life, reddish or purplish brown of some shade' varying from a light dull reddish to almost black; tube feet, brownish red.

Range.—Nantucket Shoals and Woods Hole to west Florida and Yucatan; low water to 125 fathoms.

Remarks.—This is the common sea-urchin of the Woods Hole region. It is abundant at many places in Vineyard Sound, and is common in Hadley Harbor and, at times, around the Fish Commission wharf. We took one specimen off Sankaty Head, but none at Crab Ledge. In spite of its abundance, we know little of its habits, and its development is only partially known. Garman and Colton ('82) have published some notes on the development, and Fewkes ('81) has also contributed to our knowledge of the early stages.

2. Strongylocentrotus dröbachiensis (O. F. Müller). *Green sea-urchin.* (Pl. 9, figs. 53–57.)

Echinus dröbachiensis O. F. Muller. 1776.
Echinus neglectus Lamarck. 1816.
Echinus granularis Say, 1827.
Strongylocentrotus chlorocentrotus Brandt, 1835.
Echinus granulatus Gould 1841. Desor, 1848.
Toxopneustes dröbachiensis Agassiz, 1846.
Eurychinus granulatus Verrill, 1866.
Eurychinus dröbachiensis Verrill, 1866.
Strongylocentrotus dröbachiensis Al. Agassiz, 1872.

Description.—Diameter of test 60 to 80 mm.; height 25 to 45 mm.; D = 1.75 to 2.25 H. Length of longest spines 10 to 14 mm.; diameter of anal system 6 to 8 mm.; diameter of whole abactinal system 15 to 20 mm.; diameter of peristome 18 to 25 mm., very much larger in proportion in young specimens, sometimes 60 per cent of the diameter. Test more or less flattened, curving at first very slightly then abruptly to the circular ambitus, actinal surface flattened, the peristome sunken 4 to 6 mm. Branchial incisions rather small, and not very deep. Anal plates at first 2 or 3, but increasing in number with age, adults having 35 to 40. Ocular plates (radials), two entering circumanal ring to a marked degree. Genital plates (basals) very large, the madrepore much the largest. All the plates of the abactinal system carry miliaries, though they are few on the madrepore. Ambulacra broad at the ambitus, narrower at the peristome, though there they are much wider than interambulacra. Poriferous zones broad, with numerous small pores; pairs of pores in oblique transverse series, abruptly bent at outer end; number of pairs in each series varies somewhat with age, but is usually from 4 to 6; obliquity of series varies much with age, in very small specimens approaching the vertical—in old specimens more nearly horizontal. Spines numerous all over the test; primaries longitudinally striated, pointed but not very sharp, longest at and above ambitus, shortest around peristome; secondaries similar, but much shorter; miliaries very slender. Primary tubercles smooth and imperforate, a double series on each ambulacrum, and also on each interambulacrum. On the ambulacra the series of primaries are separated by a double series of secondaries, while 2 or 3 series of secondaries on the poriferous zones form transverse lines between the arcs of the pores. On the interambulacra there are about 8 series of secondaries, 4 between and 2 outside of each primary series. Miliary tubercles occur all over the test, on both ambulacra and interambulacra. A specimen with D = 30 mm., H = 17 mm., has 22 coronal plates, while one with D = 60 mm., H = 27 mm., has 35. Buccal plates, 5 pairs, large and bearing miliaries. Pedicellariæ numerous, long stalked. Color in life prevailingly green; the test green or greenish white, purple or purplish white, the poriferous zones markedly lighter than the rest of the test; spines green with yellow, red, or purple cast, especially in young specimens; sometimes the actinal spines are bright violet, while the abactinal may be tipped with red or violet; pedicellaria and miliary spines whitish; tubercles white; tube feet whitish or pale violet.

Range.—Circumpolar; southward in the western Atlantic to New Jersey (not in shallow water south of Cape Cod); in the Eastern Hemisphere to Great Britain and Norway; in the North Pacific from Kamchatka to Puget Sound; low water to 640 fathoms.

Remarks.—In the Woods Hole region this northern urchin is found in abundance off Sankaty Head, Nantucket; it is common at Crab Ledge; it has been taken at several points in Vineyard Sound, and it occurs in 10 to 20 fathoms off Gay Head. At Crab Ledge the specimens we took were all small, but off Sankaty Head a large number of good-sized individuals were secured. In spite of the fact that this is one of the commonest and best known of sea-urchins, no connected account of its development has ever been published, although the egg, segmentation stages, pluteus, and young are all well known.

3. Echinarachnius parma (Lamarck). *Sand Dollar.* (Pl. 10, figs. 58–62.)

> *Scutella parma.* Lamarck, 1816.
> *Echinarachnius atlanticus.* Gray 1825. Stimpson, 1853.
> *Echinarachnius parma.* Gray 1825.
> *Scutella triforia* Say, 1826.

Description.—Test greatly flattened, closely covered with minute spines, which are shortest and most uniform on the abactinal surface, longer at the margin, and longest in the interradial areas at the peristome. Ambulacra obvious abactinally as widely open, somewhat obtuse "petals," extending more than halfway to margin; actinally the ambulacra appear as furrows, widest at peristome, and when more than halfway to the margin, giving off a prominent branch on each side at an angle of about 45°; all 3 furrows run to the margin, and the main furrow may be continued abactinally. Abactinal system approximately central; genital pores 4. Anal opening abactinal in very young specimens, marginal in adults, actinal in very large or old specimens. One of the latter gives the following measurements: Longitudinal diameter, 78 mm.; transverse diameter, 78 mm.; vertical diameter, 12 mm.; diameter of abactinal system, 8 mm.; length of anterior petal, 26 mm.; length of posterior petal, 24 mm.; length of spine at margin, 1 mm.; length of spine at peristome, 3 mm.

In very young specimens the proportions are somewhat different; thus a specimen 8 mm. long is only 7 mm. broad. The greatest tranverse diameter is not always through the abactinal system, but may be considerably back of it, and the abactinal system may be considerably in front of the center of test. Thus in a specimen 12 mm. in longitudinal diameter the center of the abactinal system is only 5 mm. from the anterior edge of the test, and the greatest tranverse diameter is through a point only 4 mm. from the posterior edge. The relative width of the petals varies greatly; it may be anywhere from 37 to 50 per cent of the length. Color in life, dull brownish-red, varying from flesh-red in very young specimens to a deep reddish-brown in adults; interambulacra distinctly lighter than the ambulacra, which are quite red. When placed in fresh water or alcohol, or even when simply dried, the color changes to a bright, though dark green, which afterwards, in dry specimens, becomes a dull brown. The bare, bleached tests are of course white.

Range.—Labrador to New Jersey; also both sides of the Pacific Ocean, from Vancouver to Japan, and (according to Agassiz) India, Australia, and the Red Sea. Low water to 888 fathoms.

Remarks.—This curious species is very common on sandy bottoms in Vineyard Sound and on the Nantucket Shoals. The finest specimens were taken near the Great Round Shoal Lightship, Nantucket, in 12 fathoms, many of them being 3 inches or more in diameter. The sand dollar is said to be an important article of food for flounders and codfish. It lives more or less buried in the sand, moving about very slowly, chiefly by means of the spines. The development has been studied and partially described by Fewkes ('86).

Figs. 53–57. *Strongylocentrotus dröbachiensis* (natural size).—53. Aboral view of test with spines. 54. Oral view of test with spines. 55. Side view of test without spines. 56. Aboral view of test without spines. 57. Oral view of test without spines.

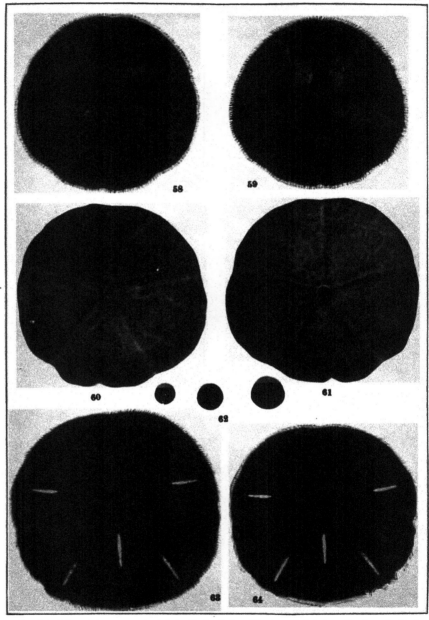

Figs. 56–62. *Echinarachnius parma* (natural size).— 56. Aboral view with spines. 59. Oral view with spines. 60. Aboral view without spines. 61. Oral view without spines. 62. Young specimens without spines, to show form of test and tion of anal o nin .

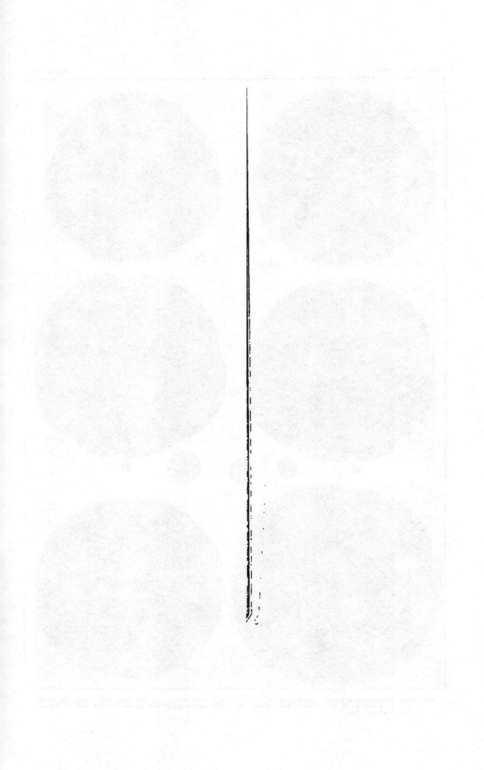

4. Mellita pentapora (Gmelin). *Key-hole Urchin.* (Pl. 10, figs. 63-64.)

Echinus pentaporus Gmelin, 1788.
Scutella quinquefora Lamarck, 1816.
Encope pentapora L. Agassiz, 1841.
Mellita quinquefora L. Agassiz, 1847.
Mellita testudinaria Gray, 1857.
Mellita pentapora Lutken, 1864. Verrill, 1873, et seq.
Mellita testudinata Alex. Agassiz, 1872.

Description. Test very flat, often wider than long when fully grown, truncated posteriorly, covered with very short, delicate spines, which are longest actinally, especially near the margin in the interradii, around the peristome or bordering the lunules. There are 5 of the latter, 1 in the posterior interradius and 1 in each radius, except the anterior one. Ambulacra obvious abactinally as bluntly rounded, nearly closed "petals," not quite reaching the lunules, the posterior pair longer; actinally the ambulacra appear as shallow furrows, the anterior running to the margin, the others ending at the lunules. Abactinal system eccentric, lying anterior to the center while the mouth is directly below; genital pores 4. Anal opening actinal; in adults, at the proximal end of the interradial lunule.

A good-sized specimen gives the following measurements: Longitudinal diameter 110 mm.; transverse diameter 115 mm.; vertical diameter 10 mm.; diameter of abactinal system 9 mm.; length of anterior petal 32 mm.; length of posterior petal 39 mm.; length of interradial lunule 22 mm.; width of interradial lunule 3 mm.; distance from margin 23 mm.; length of postero-radial lunule 24 mm.; width of postero-radial lunule 2 mm.; distance from margin 3 mm.

The difference between the longitudinal and transverse diameters is not always as marked as in the specimen given; moreover, very young specimens have the lunules only partly formed or wanting. In a specimen 3 mm. in diameter there are no lunules, but the position of the interradial one is indicated by a slight actinal depression. A specimen 12 mm. in diameter has the interradial lunule fully formed, the deepening of this actinal depression having continued until the abactinal surface was pierced; the radial lunules are arising as notches in the edges of the test. These marginal notches deepen, and finally the outer sides grow together, thus inclosing the lunule. Color in life, brownish-yellow; in alcohol, rather greenish.

Range.—Nantucket to Brazil, in shallow water; rare and local, north of Cape Hatteras.

Remarks.—This species is admitted to the list of the Woods Hole echinoderms on very scanty evidence. Verrill ('73 b) records it from Nantucket on Agassiz's authority, and dead specimens (bare tests) are occasionally taken in Vineyard Sound. Where these come from is a question yet to be answered. Mr. Gray tells me one was taken in the sound in the summer of 1901. Dr. Caswell Grave (1902) has published a brief and partial account of the larva of this species, and I am indebted to him for some of the details of the above description of the adult.

HOLOTHURIOIDEA.

HOLOTHURIANS. SEA-CUCUMBERS.

The holothurians make up more than one-third of the Woods Hole echinoderms, but although the number of species is considerable, in number of individuals the asteroids and echinoids far outrank them. None of the nine species are sufficiently common and generally distributed to be noticed by an inexperienced observer, and it is the exception rather than the rule to find a holothurian in the dredge anywhere in the Woods Hole region. There are, however, three species (*Thyone briareus* and the two *Synaptas*), which one who knows where and how to look can always obtain, and these we may fairly call common. The remaining six are of uncertain occurrence, and while two of them may be expected to occur at the proper locality, the others are distinctly rare and of uncertain occurrence. I have never seen any of these four living. The various species can be distinguished from each other with comparative ease, and few of the terms used in the descriptions will need any explanation to anyone at all familiar with echinoderm anatomy. The names applied to the calcareous particles in the skin are purely arbitrary, but are quite generally used, and the figures given will prevent any misunderstanding. It must be borne in mind, however, that as holothurians increase in age, the calcareous parts undergo considerable change, either becoming larger or smaller, more irregular or less so, and fewer or more abundant. The ambulacral appendages, especially the feet, are frequently much more numerous in adults than in young, and pigment is always more abundant with age. But the following key ought to enable anyone to distinguish even very young individuals of the species given.

Key to the Holothurians of the Woods Hole region.

Ambulacral appendages in the form of feet (pedicels) present, at least on the ambulacra.

  Pedicels chiefly or wholly confined to ambulacra; some scattered ones may be present on the back.

    Size large, up to 300 mm.; color some shade of brown .... CUCUMARIA FRONDOSA.

    Size small, less than 60 mm.; color white or whitish.. CUCUMARIA PULCHERRIMA.

  Pedicels scattered over the whole body, though the ventral ambulacrum may be distinctly defined.

    Size large, 75 to 225 mm.; body not noticeably attenuated posteriorly; color very dark .... THYONE BRIAREUS.

    Size small, less than 75 mm.; body much attenuated posteriorly; color brown..................... THYONE SCABRA.

    Size small, less than 60 mm.; body not attenuated posteriorly; color white or whitish .... THYONE UNISEMITA.

Ambulacral appendages in the form of pedicels wanting.

  Tentacles 15; posterior end of body tail-like.

    Caudal appendage long; no reddish deposits in the skin .... CAUDINA ARENATA.

    Caudal appendage short, abrupt; reddish deposits in the skin............................ TROCHOSTOMA OÖLITICUM.

  Tentacles 12; no tail-like appendage.

    Color white or yellowish; radial pieces of calcareous ring pierced for passage of nerves .... SYNAPTA INHÆRENS.

    Color red or pinkish; radial pieces of calcareous ring simply notched............. .... SYNAPTA ROSEOLA

1. Cucumaria frondosa (Gunnerus). *Sea-cucumber.* (Pl. 11, figs. 65, 66; pl. 12, figs. 76–80.)

Holothuria frondosa Gunnerus, 1770.
Cladodactyla pentactes Gould, 1841.
Cucumaria frondosa Forbes, 1841.
Botryodactyla grandis Ayres, 1851.
Botryodactyla affinis Ayres, 1851.
Pentacta frondosa Stimpson, 1853.

Description.—Length in life, normally extended, 250 to 300 mm.; may extend to 600 mm. or more; diameter of body 90 to 100 mm., or much less when considerably extended. In life ventral surface much considerably, flattened, with sides curving upward; anterior end truncated; posterior end bluntly rounded. When disturbed, the body contracts to such an extent that it becomes ovoid or ellipsoidal, or almost spherical, and museum specimens usually show more or less of such contraction. Tentacles 10 (sometimes 9 or 11), of approximately equal size and much branched; rather short and stout. Pedicels rather large, and forming a broad series on each ambulacrum, while slightly smaller and less perfect ones are scattered over the dorsal interambulacra; all lack the usual terminal, perforated calcareous plate. Calcareous deposits consist of irregular, usually smooth, perforated plates (fig. 78), the size, number, and distribution of which vary greatly, though they are apparently most abundant in the young. The largest plates (fig. 80) are near the cloacal opening, though they do not form so-called "anal teeth." At the base of the pedicels and tentacles the plates become more irregular (fig. 79), and often bear minute projections, or more or less prominent ridges. Calcareous ring very slender for so large an animal, and more or less imperfectly developed, perhaps according to age, being more perfect in smaller specimens; radial pieces somewhat wider than interradial, with a very wide and deep notch in the posterior margin; the interradial pieces not notched posteriorly. Stone canal single, of moderate size, and provided with 1 to 6 madrepore plates (fig. 77). Polian vessel usually single, very long, 100 mm. more or less. Color in life deep reddish- or purplish-brown, darkest on the dorsal side and much lighter below, sometimes nearly white; pedicels often with a strong roseate tinge.

Range.—Greenland to Nantucket; also Iceland and Spitzbergen to Norway and the south coast of England; low water to 200 fathoms. The reported occurrence of this species on the Florida Reef is almost certainly a case of mistaken identification, and the records of its occurrence on the coast of Alaska and in the North Pacific are very probably based on *Cucumaria japonica* Semper, which seems to be quite a distinct species.

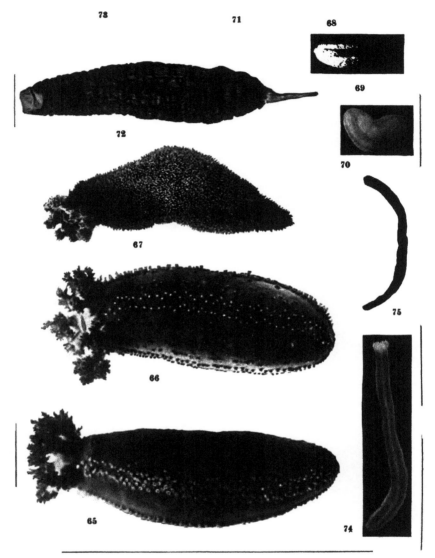

Figs. 65, 66. *Cucumaria frondosa* (one-half natural size).—65. Side view. 66. Dorsal view.
Fig. 67. *Thyone briareus*. Dorsal view (natural size).
Figs. 68, 69. *Thyone unisemita* (natural size).—68. Side view. 69. Ventral view.
Fig. 70. *Cucumaria pulcherrima*. Side view (natural size).
Fig. 71. *Thyone scabra*. Side view of a young specimen (natural size).
Fig. 72. *Trochostoma ooliticum*. Side view of contracted alcoholic specimen (natural size).
Fig. 73. *Caudina arenata*. Side view of contracted alcoholic specimen (natural size).
Fig. 74. *Synapta inhaerens*. Dorsal view (natural size).
Fig. 75. *Synapta roseola*. Dorsal view (natural size).

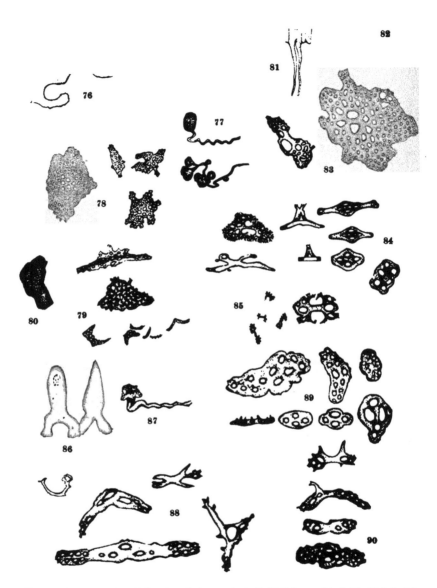

Figs. 76–80. *Cucumaria frondosa*.—76. Two pieces of calcareous ring (× 5). 77. Madrepore plates (× 10). 78. Perforated plates from skin (× 156). 79. Perforated plates from near base of tentacles (× 156). 80. Perforated plates from near cloacal opening (× 45).
Figs. 81–85. *Cucumaria pulcherrima*.—81. Two pieces of calcareous ring (× 5). 82. Madrepore plate (× 10). 83. Calcareous plates from cloaca (× 156). 84. Tables from body wall (× 156). 85. Particles from tentacles (× 156).
Figs. 86–90. *Thyone unisemita*.—86. Two pieces of calcareous ring (× 5). 87. Madrepore plate (× 10). 88. Supporting rods from tentacles (× 156). 89. Perforated plates from body wall (× 156). 90. Supporting rods from pedicels (× 156).

Remarks.—This large and noticeable, one might even say handsome, holothurian, so abundant on the coast of Maine, just enters the Woods Hole region as a resident of the shoals east of Nantucket. In October, 1894, Mr. Edwards took 6 specimens in 23 to 25 fathoms, some 15 miles east-southeast of Sankaty Head, Nantucket, and in August, 1902, we took 2 good specimens in 12 fathoms, about 8 miles off Sankaty Head. The latter were ready to breed, the reproductive glands being fully ripe. This sea-cucumber feeds upon small particles of organic matter picked up by the tentacles. It seems to prefer generally rocky or shelly bottoms. The larva is red, and passes through a metamorphosis, but the development is not fully known. Dr. William Stimpson reports this species as not only edible, but, when boiled, "as palatable as lobster."

2. Cucumaria pulcherrima (Ayres). (Pl. 11, fig. 70; pl. 12, figs. 81–85.)

Pentamera pulcherrima Ayres, 1854. Verrill, 1873b.
Thyone pulcherrima Semper, 1868.
Cucumaria pulcherrima Lampert, 1885.

Description.—Length 50 mm. or less; diameter 20 mm. or less. Body ovate, the two ends strongly upcurved (at least in preserved specimens), so that the ventral ambulacra are much longer than the dorsal. Tentacles 10, the 2 ventral much smaller than the others. Pedicels numerous, confined entirely to ambulacra. Calcareous deposits chiefly in the form of tables (fig. 84), which are very densely crowded together, so that the skin is quite hard. These tables, when simplest and most symmetrical, have a disk perforated with 4 holes and a small spire, usually made up of 2 rods with few teeth at the apex; but very frequently the disk is irregular in shape, and has 6, 8, 10, or more holes; in the pedicels the disks of the tables are elongated until near the tip of the foot they are simply supporting rods, usually having the ends perforated; terminal plates present; the tentacles contain very few supporting rods (fig. 85), and they are chiefly small and of very irregular shape. Cloacal opening surrounded by 5 tufts of pedicels, 3 or 4 in each, which are almost rigid with their crowded deposits; just within the cloacal opening is a ring of crowded calcareous plates (fig. 83), but so far as one can judge from preserved material there are no true "anal teeth." Close to the cloacal opening the calcareous deposits become very much crowded and increase in size, thus coming to resemble large irregular perforated plates, often with scarcely a trace of the spire left. Calcareous ring (fig. 81) well developed, quite high, the radial pieces with very long and slender posterior prolongations; stone canal single, terminating in a large madrepore plate (fig. 82). Polian vessel single, small. Color white or whitish.

Range.—Vineyard Sound to Fort Johnson, South Carolina; low water to 5 fathoms.

Remarks. Perhaps no one of our holothurians is less often seen alive than this one, nor is there any of whose habits less is known. The original specimen was taken in shallow water, on the coast of South Carolina, buried 2 inches in the sand. Later the species was found at Fort Macon, North Carolina, and in Vineyard Sound. Prudden and Russell dredged specimens "off Holmes Hole" (Vineyard Haven) in 4 to 5 fathoms, so Verrill ('73b) reports, but he gives no date nor any other facts. Large numbers are frequently washed up on the beach near Nobska Light, on the north shore of the Sound, after long-continued or hard easterly storms; but in spite of very thorough dredging and trawling in all depths from the shore outward, across to the Vineyard, Mr. Edwards tells me he knows of no specimens having been found, so that the habitat and habits of this species are still an enigma, and naturally the life history is unknown. During the winter of 1903 there were a number of specimens washed up on the bathing beach at Woods Hole, on the eastern side of Buzzards Bay. Mr. Gray, who very kindly sent me the specimens, says that there had been a long-continued period of heavy westerly winds. In these specimens the reproductive glands were very well developed, of a bright orange yellow, and their condition would seem to indicate that breeding occurs in the late winter or early spring.

3. Thyone briareus (Lesueur). *Common Thyone.* (Pl. 11, fig. 67; pl. 13, figs. 95–102.)

Holothuria briareus Lesueur, 1824.
Sclerodactyla briareus Ayres, 1851.
Anaperus bryareus Pourtalès, 1851.
Thyone briareus Selenka, 1867.

Description.— Length up to 225 mm., according to the state of contraction, a fair-sized specimen in normal condition being 85 to 100 mm. in length and 25 to 30 mm. in diameter. Posterior end of the body rather abruptly tapering and pointed when normally extended, but not at all attenuated; often more or less blunt and rounded according to amount of contraction. Cloacal opening terminal, surrounded by 5 groups of slender papillæ; just within the cloaca is a well-developed calcareous ring (figs. 101-102) with 5 prominent radial projections, which appear from the outside as teeth, and are usually referred to as "anal teeth." Tentacles 10, the 2 ventral much smaller than the others, which are capable of considerable extension and are much branched. Pedicels rather small, very numerous all over the body, occasionally divided by very narrow lines into 5 broad, apparently radial, bands; many of the pedicels of the dorsal side taper to a point, and are thus more or less papilliform. Calcareous deposits in adults, wanting in most parts of the body wall, but present at each end of the body, and in the pedicels and tentacles; more numerous in young than in old individuals. These deposits are in the form of tables (fig. 97), plates, and rods; tables with a more or less square disk, perforated by about 8 holes, and a spire made up of 4 rather short rods, with one cross bar, ending in single teeth. Such tables are confined almost wholly to the two extremities of the body. In the pedicels, which are provided with a large terminal plate (fig. 98), the disks of the tables are elongated and curved to form supporting rods, and as such they often lack spires (fig. 99). In the tentacles the calcareous supporting rods (fig. 100) are so numerous as to make the trunk and principal branches almost rigid, and occur even to the tips of the smaller branches; around the base of the tentacles are some scattered plates perforated by six or more holes. In very young individuals, specimens an inch long or less, the body wall is often crowded with tables. Calcareous ring (fig. 95) well developed, rather stout, the radial pieces with moderately long, slender prolongations. Stone canal single, with a large madrepore plate (fig. 96). Polian vessels usually 1 or 2. Color in life dull brown or black, the pedicels lighter, often quite reddish, the disks frequently yellow.

Range.—Vineyard Sound to Texas, low water to 10 fathoms.

Remarks.—This is undoubtedly the best known, to American students, of the Woods Hole holothurians, as it is the form commonly used for laboratory work. Hadley Harbor furnishes most of this material, *Thyone* being abundant there, but it is also to be found at Waquoit and near Cuttyhunk. Verrill ('73b) reports it from Buzzards Bay and Vineyard Sound. Usually it lies buried in soft mud in shallow water, either the posterior end alone or both ends above the surface. The currents of water repeatedly and continuously driven from the cloacal opening are often quite apparent, especially in very shoal water. The food consists of the fine organic particles gathered by the tentacles. Although this species apparently breeds in the summer, nothing is known of its development. In the winter of 1903 a number of very small specimens were washed up on the Buzzards Bay bathing beach at Woods Hole, after a period of heavy westerly winds, in company with specimens of *Cucumaria pulcherrima.* Mr. Gray kindly sent them to me for examination.

4. Thyone scabra Verrill. (Pl. 11, fig. 71; pl. 13, figs. 91-94.)

Thyone scabra Verrill, 1873a.

Description.—Length up to 90 mm. (Théel), usually much less; Verrill says, "Length, in alcohol, about 2 inches." All the specimens which I have seen from the Woods Hole region were less than 50 mm., with a diameter of about 8 mm. Posterior third of the body quite attenuate to a rather sharp point. Cloacal opening terminal, with 5 sets of small papillæ, but, so far as could be determined from preserved material, without the so-called "anal teeth." Tentacles 10, branched from the base, the 2 ventral smaller than the rest, which are not very large. Pedicels rather long and slender, somewhat rigid, quite numerous, and irregularly arranged, provided with terminal plates. Calcareous deposits very abundant, especially in pedicels, and tending to form a rather brittle, thin, and very rough layer over the whole body surface; deposits consist chiefly of tables (fig. 93) with more or less irregular disk, pierced by 6 or more holes (Verrill says 20 to 24), and a more or less prominent and rather solid spire, made up of 2 or 3 stout vertical rods, connected by 1 or 2 crossbars and terminating in a number of teeth; in pedicels, disks of tables elongated, more or less bowed, with 4 holes, and ends expanded and perforated; in tentacles, numerous supporting rods (fig. 94), which are usually more or less perforated, especially at ends. Calcareous ring (fig. 91) well developed, radial pieces with a remarkably deep notch and long, slender prolongations posteriorly. Stone canal single with a large madrepore plate (fig. 92). Polian vessel usually single. Color in life not recorded; alcoholic specimens are brown or yellowish-brown, pedicels lighter.

Range.—Georges Bank and Bay of Fundy to Vineyard Sound, Narragansett Bay, and perhaps even to Delaware (Théel); 10 to 640 fathoms.

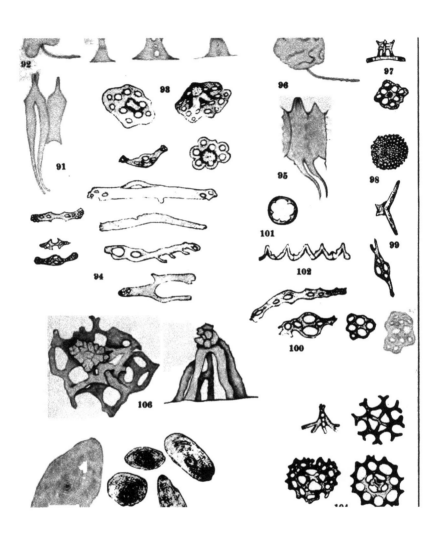

Remarks.—This is another of those holothurians of which we know altogether too little. It was first dredged on Georges Banks in 1872, and Verrill afterwards found it in the Bay of Fundy. Théel (1886) reports specimens from off the coast of Delaware; they were remarkably large, and he says "the anus possesses five calcareous teeth." Verrill does not mention "anal teeth" nor calcareous ring. The latter, Théel says, resembles that of *Thyone fusus.* I have never seen a living specimen of this species, but in the fall of 1899, Dr. H. C. Bumpus sent me some alcoholic specimens which I understood were dredged in Narragansett Bay, and I have also had 5 specimens dredged by the *Fish Hawk* in Vineyard Sound in 1901. None of these specimens were much over 40 mm. in length. They answer well to Verrill's description, except that the disks of the tables rarely contain 20 holes, while he says 20–24. Théel says that in his specimens the holes were sometimes as many as 20. As for the "anal teeth" which he mentions, they may have been due to the age of his specimen, which was twice the size of any of mine. It is by no means certain that *scabra* is really distinct from the European *fusus*, but the matter cannot be decided without more material. Naturally we know nothing of its habits or development.

5. Thyone unisemita (Stimpson). (Pl. 11, figs. 68, 69; pl. 12, figs. 86–90.)

Cucumaria fusiformis Desor, 1818 (non Forbes).
Anapeus unisemita Stimpson, 1851.
Stereoderma unisemita Ayres, 1851; Verrill, 1873b; Théel, 1886.
Thyone unisemita Ludwig, 1892.

Description.—Length 75 mm. (Verrill) or less; diameter, when normally extended, about one-third of the length or less. Body tapering to both ends, which are more or less curved upward. Cloacal opening, terminal, not provided with "teeth," and with no specially prominent papillæ about it. Tentacles 10, 2 ventral smaller than the others, which are rather long and slender, stalked, and not profusely branched at tip. Pedicels short, quite uniformly distributed over the dorsal surface and on the sides; ventrally there is a distinct double row in the midventral radius, and on each side a narrow strip of skin, wholly free from pedicels; it is from this peculiar arrangement that the species has received its name. Calcareous deposits, perforated plates (fig. 89), mostly regular, and with 4 holes, but often larger, irregular, and with more holes; plates usually smooth, frequently with knobs in the pedicels, which lack terminal plates; supporting rods (fig. 90) rather few, broad, flat, with narrow holes, sometimes with knobs, and occasionally projecting spines; in the tentacles, supporting rods (fig. 88) large, perforated, and rather numerous; near the tips of the branches they are much curved. Calcareous ring (fig. 86) well developed, wide, radial and interradial pieces of nearly equal size, but the former with a wider notch in the posterior margin, though there are no posterior prolongations. Stone canal single, with a small and poorly developed madrepore plate (fig. 87). Polian vessel single. Color in life white or yellowish-white, tentacles orange-yellow.

Range.—Grand Bank, Newfoundland, to Narragansett Bay; 17 to 22 fathoms, probably more.

Remarks.—Another uncommon holothurian, of which we collected half a dozen specimens in August, 1902, at Crab Ledge, on sandy and gravelly bottom. All of these were small, from 8 to 25 mm. in length, but several had the ovaries full of apparently mature eggs. Stimpson's specimens from Grand Bank and Massachusetts Bay were about 2 inches in length, while a specimen dredged by Packard, south of Marthas Vineyard, was about 3 inches long. Specimens kept alive at the laboratory were very sluggish, and extended the tentacles very little; the color of the latter is in striking contrast to that of the body. Besides the localities already mentioned, this species has been taken on Nantucket Shoals, off Gay Head, and in Narragansett Bay.

6. Caudina arenata (Gould). (Plate 11, fig. 73; plate 13, figs. 103, 104.)

Chirodota arenata Gould, 1841.
Caudina arenata Stimpson, 1853.

Description.—Length 100 to 175 mm., with a diameter about one-eighth as great; the posterior third of the animal constitutes what we may call the caudal portion, and this has a diameter of only a few millimeters. Integument translucent and smooth, or finely granular. Cloacal opening terminal, surrounded by 5 very small papillæ. Tentacles 15, equal, each with four short, finger-like digits. Calcareous deposits (fig. 104) in the form of tables, with smooth, flat, nearly circular or oval disks, each with a large central hole, and a more or less regular peripheral series of 8 to 12 holes; the central hole appears from above like 4, as the 4 legs of the spire cross it; there are often a few small holes outside the peripheral circle; the spire is made up of 4 rods, which are united close to their apex, and by a crosspiece near the middle. Calcareous ring (fig. 103) well developed, of moderate width, the radial pieces with not very long, stout posterior prolongations, deeply but narrowly separated. One stone canal, with a single, terminal madrepore plate. Polian vessel single. Color in life, pale to deep flesh-red, pink, or even purplish.

Range.—Pointe du Chene, New Brunswick, to Cuttyhunk; low water to 18½ fathoms.

Remarks.—This and the following species are the rarest of the Woods Hole holothurians. It is admitted to this list on the strength of Verrill's ('73b) statement that Professor Webster took it at Woods Hole, and on the existence of 3 small specimens in the collection of the U. S. National Museum, labeled "Off Cuttyhunk, 18½ fathoms." I have never seen *Caudina* alive, but it is said to be abundant at Revere Beach, Mass., at certain seasons. Strangely enough, however, trawling and dredging offshore in that region failed entirely to bring up specimens. Gerould's ('96) admirable paper on this species leaves nothing to be said as to habits or anatomy. Nothing whatever is known of the development. Théel's variety, *armata*, was taken in 898 fathoms, in latitude 35° 44′ 40″, and in 1,242 fathoms, in latitude 41° 24′ 45″.

7. Trochostoma oöliticum (Pourtalès). (Plate 11, fig. 72; plate 13, figs. 105–108.)

Chirodota oöliticum Pourtalès, 1851.
Molpadia borealis Sars, 1861.
Molpadia oölitica Selenka, 1867. Verrill, 1873b.
Trochostoma oöliticum Danielssen and Koren, 1878.
Trochostoma thomsonii Danielssen and Koren, 1878.
Trochostoma boreale Danielssen and Koren, 1879.

Description.—Length 125 to 150 mm., with a diameter about one-sixth as great; the caudal portion of the body is only about one-eighth of the total length. Integument rather thin and usually quite smooth. Cloacal opening terminal, with minute surrounding papillae. Tentacles 15, each usually with 2 (sometimes possibly more) digits. Deposits in the skin of 2 very distinct kinds, irregular tables (figs. 106, 107), and reddish or brown discoidal or ellipsoidal bodies (fig. 108). Tables sometimes wanting, apparently most frequent in young specimens and becoming less frequent with age; they are quite irregular in form; disk pierced by holes which vary greatly in number and size, and spire also variable in size and form. "Brown bodies" vary greatly in size and shade of color, for they may be mere grains or nearly as long as the diameter of a table disk, and the shade ranges from brownish-yellow to a very deep reddish-brown; these "brown bodies" may be rather scattered or more or less crowded. As a rule, the fewer the tables the more the "brown bodies," and vice versa. In typical *oöliticum*, there are no tables, and the integument is literally packed with "brown bodies." Calcareous ring (fig. 105) very stout, the radial pieces with very prominent posterior prolongations. Stone canal one, with a single madrepore plate. Polian vessel one. Color of alcoholic specimens very variable, according to the abundance of "brown bodies;" where they are very small and very few the color is dull gray, and the head and tail are always that color or lighter; where the "brown bodies" are more numerous they form brown patches on the surface; if still more abundant, the animal appears yellowish or reddish brown with gray spots of greater or less size; and finally, in typical *oöliticum*, the color appears uniformly deep brown or even almost black. Verrill (73b) says of a living specimen, "uniform flesh color."

Range.—Banks of Newfoundland to south of Marthas Vineyard, and in the Arctic Ocean north of Norway and Siberia; reported from Florida Reef also; 18 to 600 fathoms, but usually over 50.

Remarks.—This distinctly northern form is admitted to the list of Woods Hole echinoderms solely on the record of one small specimen taken by Professor Packard and reported by Verrill ('73b). Even the locality of this specimen is in doubt, for on page 715 Verrill says, "Off Block Island, 29 fathoms, sandy mud," while on page 510 he says "15 miles east of No Mans Land," and Block Island is 30 miles west of No Mans Land. The genus *Trochostoma* is probably more imperfectly known and its legitimate species less well defined than any other genus of holothurians. After the examination of a fairly large series from the collection of the U. S. National Museum, I am convinced that the differences which were supposed to separate *oöliticum* and *boreale* are unimportant and that Sars's name is really a synonym of Pourtalès's. I can not agree with Ludwig (1900), however, that *arcticum* v. Marenzeller is identical with *boreale*, for specimens of the former are easily separable from the latter by several good characteristics. As to Verrill's *turgidum*, if the characters given are constant, it is also a good species.

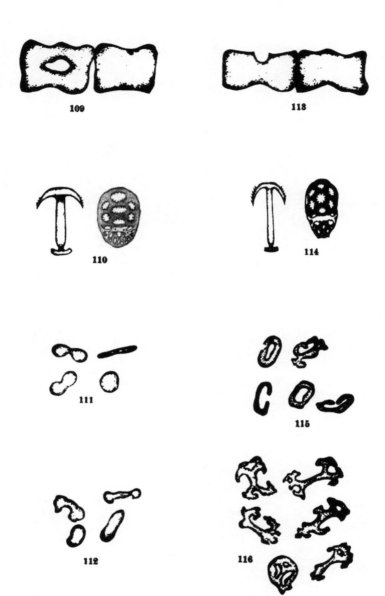

Figs. 109-112. *Synapta inhærens.*—109. Two pieces of calcareous ring (× 45). 110. Anchor and plate (× 156). 111. Particles from longitudinal muscles (× 450). 112. Particles from tentacles (× 450).
Figs. 113-116. *Synapta roseola.*—113. Two pieces of calcareous ring (× 45). 114. Anchor and plate (× 156). 115. Particles from longitudinal muscles (× 450). 116. Particles from tentacles (× 450).

8. Synapta inhærens (O. F. Müller). *Common Synapta.* (Pl. 11, fig. 74; pl. 14, figs. 109–112.)

Holothuria inhærens O. F. Müller, 1788.
Synapta inhærens Duben and Koren, 1846.
Synapta tenuis Ayres, 1851.
Synapta girardii Pourtalès, 1851.
Synapta pellucida Ayres, 1852.
Synapta ayresi Selenka, 1867.
Synapta gracilis Selenka, 1867.
Leptosynapta tenuis Verrill, 1867.
Leptosynapta girardii Verrill, 1873b.

Description.—Length 100 to 180 mm., more or less; diameter 5 to 10 mm. Body slender, very extensile, cylindrical. Integument thin, more or less translucent, sometimes minutely rough. Cloacal opening terminal. Tentacles 12, with 3 to 7 pairs of digits pinnately arranged, and 10 to 20 or more very small, sensory cups on the inner surface near the base. Calcareous deposits minute discoidal or irregularly rounded bodies (fig. 111) in the external layers of the longitudinal muscles; similar but more irregular bodies (fig. 112) at the base of the tentacles, besides curved knobbed rods in the digits; and anchors and plates (fig. 110) everywhere in the body wall, smallest anteriorly and largest posteriorly; anchors with minute teeth on the outer side of flukes; plates with 7 principal holes, with toothed margins, and 3 large and several small holes with smooth margins at posterior end. Calcareous ring (fig. 109) well developed, with no projections either anteriorly or posteriorly; the radial pieces pierced for passage of radial nerves. Stone canal and Polian vessel single. Color in life white, with or without a more or less pronounced yellow tinge; particles of red pigment frequently lie scattered in the skin, sometimes in sufficient quantity to give a pink cast to the whole animal.

Range.—Massachusetts Bay to South Carolina; also from the Arctic Ocean to the Mediterranean Sea in the Old World; and from Sitka, Alaska, to Pacific Grove, Cal.; probably circumpolar; above low water to 116 fathoms.

Remarks.—This is the most uniformly distributed, and perhaps the most common, Woods Hole holothurian, occurring in abundance along the shores of Buzzards Bay, and also about the islands of Uncatena, Nonamesset, and Naushon. It is less common along the Sound shore, but is found near the entrance to the Eel Pond. Although usually preferring a clean sand bottom, it often occurs in soft mud, even though very black, and is common in some very gravelly spots. It is often found above low-water mark. An account of this species and the next has already been published (Clark '99), so no record of the habits need be given here. Very little is known of the development, beyond the fact that segmentation is total and equal.

9. Synapta roseola (Verrill). (Pl. 11, fig. 75; pl. 14, figs. 113–116.)

Leptosynapta roseola Verrill, 1873b.
Synapta roseola Théel, 1886.

Description.—Length 100 mm., rarely more, usually much less. Body very slender. Integument translucent, very thin, soft, and delicate. Cloacal opening terminal. Tentacles 12, with 2 or 3 (rarely 4) pairs of digits pinnately arranged, and 7 to 15 sensory cups on the inner surface near base. Calcareous deposits, C or doughnut-shaped bodies (fig. 115) in the longitudinal muscles; branched, curved, and perforated rods and plates (fig. 116) in the tentacles; and everywhere in the body wall anchors and plates (fig. 114) similar to those of the preceding species, sometimes more slender and delicate, but not always. Calcareous ring (fig. 113) rather narrow, radial pieces merely notched for passage of radial nerves. Stone canal and Polian vessel single. Color in life, rosy red, varying from very pale to quite deep, rarely reddish-yellow; due to numerous pigment granules in the thin integument.

Range.—Provincetown, Mass., to New Haven Conn.; also very abundant at the Bermuda Islands, where it is the commonest holothurian; near low-water mark, above and below.

Remarks.—At Woods Hole I found this species only on the southeastern side of Buzzards Bay, from the breakwater southwestward, but Verrill ('73b) records it from Naushon. It seems to prefer gravelly and stony beaches, and often occurs under stones above low-water mark. Its anatomical characters and habits, so far as known, have already been recorded (Clark '99). Nothing is known of its development, though Dr. Wesley R. Coe, of Yale University, assures me that artificial fertilization is easily accomplished, and segmentation is total and equal.

OLIVET COLLEGE, MICHIGAN, *October, 1903.*

LIST OF PUBLICATIONS REFERRING TO THE ECHINODERMS OF THE WOODS HOLE REGION, OP TO WHICH REFERENCE IS MADE IN THE PRECEDING PAGES.

'63. Agassiz, A. On the Embryology of Echinoderms. Proc. Amer. Academy.
'64. —— On the Embryology of Echinoderms. Memoirs Amer. Academy, vol. IX.
'72-'74. —— Revision of the Echini. Ill. Cat. Mus. Comp. Zool., No. 7.

The classical work on the Echini, and absolutely indispensable to every student of the group.

'77. —— North American Starfishes. Mem. Mus. Comp. Zool., vol. v., No. 1.

Discusses and figures the hard parts of several Woods Hole species in addition to a number from other regions.

'65. Agassiz, A. and E. C. Seaside Studies in Natural History. Marine Animals of Massachusetts Bay. Boston.

An interesting popular account of many echinoderms.

'41. Agassiz, L. Monographe de Scutellidæ.
'46. —— Catalogue raissoné * * * des Echinoderms. Ann. des Sci. Nat., III ser., tome VI.
'47. —— Ditto. Tome VII.
1901. Arnold, Augusta Foote. The Sea Beach at Ebb Tide. Century Co., New York City.

An admirable attempt to provide a popular guide to marine botany and invertebrate zoology. Most of the illustrations are excellent, but the nomenclature of the echinoderms is antiquated and misleading in some respects.

'51-'54. Ayres, W. O.--Notices of Holothuriæ and other Echinoderms. Proc. Bost. Soc. Nat. Hist., vol. IV.

Some of the earliest but most interesting accounts of many of the Woods Hole species.

'35. Brandt, J. F.—Prodromus Descriptionis Animalium ab H. Mertensio observatorum. Petropoli.
'99. Clark, H. L.—The Synaptas of the New England coast. Bull. U. S. Fish Commission.

Deals with the synonymy, anatomy, and physiology.

1901. —— Synopses of North American Invertebrates. The Holothurioidea. American Naturalist, vol. XXV, No. 414.
'78. Danielssen and Koren.—Echinodermer fra den Norske Nordhavs Expedition. Nyt Mag. for Naturvid., vol. XXIV.
'79. —— Ditto, vol. XXV.
'28. Delle Chiaje, S.—Memorie * * * degli Animali senza Vertebre, etc. Naples.
'48. Desor, E.—Echinoderms of Nantucket Shoals. Proc. Bost. Soc. Nat. Hist., vol. III.
'46. Duben and Koren.—Ofversigt af Skandinaviens Echinodermer. K. Vet. Akad. Handl. Stockholm.
1780. Fabricius, O.—Fauna Groenlandica. Hafniæ et Lipsiæ.
'28. Fleming, John.—A History of British Animals. Edinburg.
'81. Fewkes, J. W.—On the Development of the Pluteus of Arbacia. Mem. Peabody Acad. of Sci. No. 6.
'86. ——. Preliminary Observations on the Development of Ophiopholis and Echinarachnius. Bull. Mus. Comp. Zool., vol XII, No. 4.
'87. ——. On the Development of the Calcareous Plates of Amphiura. Bull. Mus. Comp. Zool., vol. XIII, No. 4.
'88. ——. On the Development of the Calcareous Plates of Asterias. Bull. Mus. Comp. Zool., vol. XVII, No. 1.
'92. Field, G. W.—The Larva of Asterias vulgaris. Quar. Jour. of Mic. Sci., Nov., 1892.
'39. Forbes, Ed.—On the Asteriadæ of the Irish Sea. Mem. Wernerian Soc. Edinburgh, T. VIII, p 1.
'41. ——. A History of British Starfishes, etc., London.

A classic work most interestingly written; treats of several species which occur at Woods Hole.

'52. —— Monograph of the Echinodermata of the British Tertiaries. Palæontological Society. London.

'88. Ganong, W. F.—The Echinodermata of New Brunswick. Bull. Nat. Hist. Soc., New Brunswick, No. VII.

A most interesting and useful list of 28 species, though the illustrations are poor.

'82. Garman and Colton.—Some notes on the Development of Arbacia punctulata. Studies Biol. Lab. Johns Hopkins Univ., vol. II.

'96. Gerould, J. H.—The Anatomy and Histology of Caudina arenata Gould. Proc. Bost. Soc. Nat. Hist., vol. XXVII.

The most important paper yet published dealing with the anatomy of one of our echinoderms.

1788. Gmelin, J. F.—Linnaei Systema Naturae. Editio XIII. Lipsiae.

'98. Goto, S.—The Metamorphosis of Asterias pallida, etc. Jour. Coll. Sci., Imp. Univ. Tokio, vol. X, pt. 3.

'41. Gould, A. A.—Report on the Invertebrata of Massachusetts. Cambridge, Mass.

1900. Grave, C.—Ophiura brevispina. Mem. Nat. Acad. Sci., Baltimore.

An important contribution to the life history of one of the Woods Hole brittle-stars.

1902. —— Some points in the Structure and Development of Mellita testudinata. Johns Hopkins Univ. Circulars, No. 157.

'25. Gray, J. E.—On Echinarachnius parma. Ann. Phil., p. 6.

'35. —— On Arbacia punctulata. Proc. Zool. Soc., London, p. 58.

'48. —— List of British Animals, etc. Part I, Centroniae. London.

'51. —— On Mellita testudinaria. Proc. Zool. Soc., London, p. 36.

1770. Gunnerus, J. E.—Beschreibung dreier norwegischer Seewürmer. Abh. d. Kgl. Schwed. Akad. d. Wiss. Deutsche Ausgabe. Leipzig.

'35. Johnston, G.—Illustrations in British Zoology. London's Magazine Nat. Hist., vol. VIII.

'81. Kingsley, J. S.—Contributions to the Anatomy of the Holothurians. Mem. Peabody Acad. Sci., vol. I, No. 5.

1901. —— Preliminary Catalogue of Marine Invertebrata of Casco Bay. Proc. Portland Soc. Nat. Hist. vol. II.

Gives 29 species of echinoderms, of which 24 have been taken in less than 50 fathoms, and 13 occur in the Woods Hole region.

'16. Lamarck, J. B. P. A. de.—Histoire Naturelle des Animaux sans Vertèbres. T. III. Paris.

'85. Lampert, K.—Die Seewalzen. Wiesbaden.

An important monograph, but not nearly so useful as Théel's Challenger Report, 1886.

'24. Lesueur, C. A.—Description of several new species of Holothuria. Jour. Acad. Nat. Sci. Phila, vol. IV, pt. I.

1767. Linnaeus, C.—Systema Naturae. Editio duodecima. Holmiae.

'66. Ljungman, A. V.—Ophiuroidea viventia huc usque cognita. Öfvers. Kongl. Vet.-Akad. Förh.

'71. —— Same journal.

'92. Ludwig, H.—Die Seewalzen. In Bronn's Thierreich, Band II, Abt. 3, Buch 1.

The best monograph on holothurians ever published, but not dealing with the systematic identification of forms below genera.

1900. —— Arktische und subarktische Holothurien. In Fauna Arctica, Band I, Lieferung 1.

'54. Lutken, C. F.—Oversigt over Gronlandshavets Ophiurer. Vidensk. Medd. November.

'56. —— Oversigt over Vestindiske Ophiurer. Nat. For. Videns. Med. January and February.

'59. —— Additamenta ad Historiam Ophiuridarum, pt. III.

'64. —— Kritiske Bemærkninger over forskjellige Sostjerner, etc. Vidensk. Medd. Kjöbenhavn.

'71. —— Same journal.

'60. Lyman, T.—New Ophiurans, etc. Proc. Bost. Soc. Nat. Hist., vol. VII.

'65. —— Ophiuridæ and Astrophytidæ. Ill. Cat. Mus. Comp. Zool., No. 1.

One of the classic works, indispensable to all systematic students.

'82. —— Report on the Ophiuroidea. Challenger Reports, vol. v., pt. 14.

Another invaluable work, bringing the systematic history of ophiurans down to date and containing much valuable anatomical matter also.

'99. Mead, A. D.—The Natural History of the Starfish. Bull. U. S. Fish Commission, vol. XIX.

An unusually important and very interesting paper.

1776. Muller, O. F.—Zoologiae Danicae Prodromus. Hafniae.

'42. Muller and Troschel.—System der Asteriden. Braunschweig.

'65. Norman, A. M.—On the Genera and Species of British Echinodermata. Ann. Mag. Nat. Hist. (3) vol. XV.

'63. Packard, A. S., jr.—A list of animals dredged near Caribou Island. Canadian Naturalist, vol. VIII.

1777. Pennant, T.—The British Zoology, vol. IV. London.
'51.   Pourtalés, L. F.—On the Holothuriæ of the Atlantic Coast of the United States. Proc. Amer.
       Ass. Adv. Sci. Washington.
1783. Retzius, A. J.—Asteriæ Genus. Vetensk. Acad. Nya Handlingar, IV.
'61.   Sars, M.—Oversigt af Norges Echinodermer. Christiania.
'25.   Say, T.—On the Species of the Linnæan Genus Asterias, etc. Jour. Acad. Nat. Sci. Phila.,
       vol. V., pt. 1.
'26.   ——— Same journal.
'27.   ——— Same journal.
'67.   Selenka, E.—Beiträge zur Anatomie und Systematik der Holothurien. Zeit. f. wiss. Zool.,
       Bd. XVII.
'68.   Semper, C.—Reisen im Archipel der Philippinen. Bd. I. Holothurien. Leipzig.
       A standard work, specially remarkable for the unusually good illustrations, the colored plates being particu-
       larly attractive.
'89.   Sladen, W. P.—Report on the Asteroidea. Challenger Reports, vols. XXX and XXXI.
       A most important monograph, the only serious defect of which is the absence of a bibliography.
'51.   Stimpson, W.—Descriptions of new Holothurians. Proc. Bost. Soc. Nat. Hist., vol. IV.
'52.   ——— New Ophiurans. Proc. Bost. Soc. Nat. Hist., vol. IV.
'53.   ——— Marine Invertebrata of Grand Manan. Smithsonian Contributions, VI.
'62.   ——— On New Genera and Species of Starfishes, etc. Proc. Bost. Soc. Nat. Hist., vol. VIII.
'86a. Théel, H.—Report on the Holothurioidea, pt. II. Challenger Reports, vol. XIV.
       The one absolutely indispensable work to every worker on holothurians; can not be praised too highly.
'86b. ——— Report on the Holothurioidea * * * of the "Blake." Bull. Mus. Comp. Zool.,
       Cambridge, vol. XIII, No. 1.
1901. Tower, W. L.—An abnormal Clypeastroid Echinoid. Zool. Anzeiger. No. 640.
'66.   Verrill, A. E.—On the Polyps and Echinoderms of New England. Proc. Bost. Soc. Nat.
       Hist., vol. X.
       An interesting and important paper.
'67-'71. ——— Notes on Radiata. Trans. Conn. Acad., vol. I, pt. 2.
'73a. ——— Results of Recent Dredging Expeditions on the Coast of New England. Am. Jour.
       Sci., 3d ser., vol. X, No. 26.
'73b. ——— Report on the Invertebrate Animals of Vineyard Sound. U. S. Fish Commission
       Report. Washington.
'74.   ——— Explorations of Casco Bay * * * in 1873. Proc. Am. Ass. Adv. Sci., Portland
       meeting.
'76.   ——— Note on some of the Starfishes of the New England Coast. Am. Jour. Sci., 3d ser.,
       vol. XI.
'95.   ——— Distribution of the Echinoderms of Northeastern America. Am. Jour. Sci., 3d ser.,
       vol. XLIX, No. 290.
'99.   ——— North American Ophiuroidea. Trans. Conn. Acad., vol. X, pt. 2.
       Although these 8 titles are far from completing the list of Verrill's valuable papers on the echinoderms of
       northeastern America, they comprise those reports which bear most directly on the Woods Hole fauna.
       Those dated '66, '73b, '76 '95, and '99 are the most useful, '73b being extremely helpful to all students of
       marine zoology at Woods Hole.
1670. Winthrop, J.—Concerning * * * a very curiously contrived Fish. Philosophical Trans-
       actions. IV. London.
       A most interesting account of Gorgonocephalus. (See Lyman, '60.)

INDEX.

575

LIST OF FISHES DREDGED BY THE STEAMER ALBATROSS OFF THE COAST OF JAPAN IN THE SUMMER OF 1900, WITH DESCRIPTIONS OF NEW SPECIES AND A REVIEW OF THE JAPANESE MACROURIDÆ.

BY DAVID STARR JORDAN AND EDWIN CHAPIN STARKS.

In the early summer of 1900 the steamer *Albatross*, Jefferson F. Moser, U. S. N., commanding, returning from the South Seas, spent a few weeks in dredging along the coast of Japan. Dredge hauls were made at moderate depths in Owari Bay, Totomi Bay off Owai Point, Suruga Bay off Ose Point, Sagami Bay off Sune Point, Manazuru Point, and Enoshima, in Matsushima Bay off Nagane Point and Doumiki Point, and off the Island of Kinkwazan. In general the bottom was soft and the depths from 20 to 60 fathoms. One hundred and eleven species of fishes were secured, about 58 of these being new to science at the time of discovery. In June of the same year Messrs. Jordan and Snyder visited Japan, and in various papers in the Proceedings of the United States National Museum the collection made by the *Albatross* has been treated, in connection with the very extensive collections secured by them through other methods. The type specimens of the different new species in the collection made by the *Albatross* are all deposited in the U. S. National Museum. The figures used in the original descriptions of the new species from the *Albatross* collection are reproduced in the present paper, through the courtesy of the U. S. National Museum. The new plates are the work of Mr. William S. Atkinson and Mr. Sekko Shimada. The account of the Macrouridæ was prepared by Dr. David S. Jordan and Dr. Charles H. Gilbert.

MYXINIDÆ.

Myxine garmani Jordan & Snyder. Type.

1. **Myxine garmani** Jordan & Snyder.

Myxine garmani Jordan & Snyder, Proc. U. S. Nat. Mus., XXIII, 1901, 731; off Misaki, Japan.

Station 3757, off Suno Point, Sagami Bay, 50 to 41 fathoms.

SQUALIDÆ.

2. Centroscyllium ritteri Jordan & Fowler.

Centroscyllium ritteri Jordan & Fowler, Proc. U. S. Nat. Mus., XXVI, 1903, p. 635; Totomi Bay.

Station 3730, off Owai Point, Totomi Bay, 34 to 37 fathoms.

Centroscyllium ritteri Jordan & Fowler.

RAJIDÆ.

3. Raja tengu Jordan & Fowler.

Raja tengu Jordan & Fowler, Proc. U. S. Nat. Mus., XXVI, 1903, p. 654; Matsushima Bay.

Station 3770, off Nagane Point, Matsushima Bay, 42 to 45 fathoms.

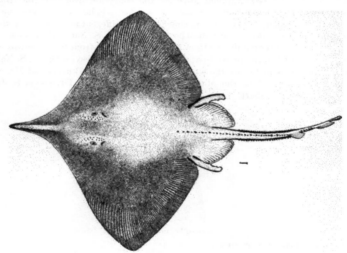

Raja tengu Jordan & Fowler.

PTEROTHRISSIDÆ.

4. Pterothrissus gissu Hilgendorf.

Station 3709, Suruga Bay, 173 to 260 fathoms; station 3715, Suruga Bay, 64 to 65 fathoms; station 3721, Suruga Bay, 207 to 250 fathoms; station 3770, Matsushima Bay, 42 to 45 fathoms; station 3772, Matsushima Bay, 79 fathoms; station 3773, Matsushima Bay, 78 fathoms.

The young are very silvery, with the silvery mucous partitions in the cavernous suborbital region very distinct.

ALEPOCEPHALIDÆ.

5. Xenodermichthys nodulosus Günther.

One fine specimen 21.5 cm. long, from station 3697, Sagami Bay. It agrees well with Günther's description and excellent figure.

Head 6 in length; depth 7; D. 32; A. 31. Color entirely black. Lateral line well developed, with scale-like structures; the rest of the body naked, covered with fine longitudinal wrinkles; luminous nodules all black.

AULOPIDÆ.

6. Aulopus japonicus Schlegel.

Specimens from stations 3708, 65 to 125 fathoms; 3713 in 500 to 600 fathoms; 3714 in 48 to 60 fathoms, and 3720, 63 fathoms, all in Suruga Bay, and from station 3730, Totomi Bay, 37 fathoms.

7. Chlorophthalmus albatrossis Jordan & Starks, new species. (Pl. 1, fig. 1.)

Type, No. 51446, U. S. Nat. Mus., from station 3698, Sagami Bay, in 153 fathoms. Cotypes, No. 8394, Stanford University.

Head 3 to 23; depth 5 to 12; D. 1, 10; A. 8; scales 53–16; eye 2.5 in head; snout 3.75; maxillary 2.5; interorbital space 4.3 in eye.

Body cylindrical, depressed anteriorly; eyes very large, close together above, with the range largely vertical; mouth terminal, very oblique, maxillary reaching to opposite front of pupil; lower jaw projecting; teeth small, in narrow bands on jaws, in very narrow straight bands on palatines, and in two small, widely separated patches on vomer; body covered with firm scales, those on breast much smaller; cheeks and opercles scaly; no lateral line; dorsal fin inserted well in advance of ventrals, the ventral base nearly under middle of dorsal; dorsal fin rather short and high; anal fin small; ventrals large, inserted well forward; pectoral long, 1.1 in head; adipose fin well developed; caudal widely forked. Color olivaceous, with numerous irregular dark cross-shades on back, these extending upward and backward; scales of anterior parts below silvery, with numerous black dots; axil black; inner rays of ventrals jet black; some black shading at base of caudal, fins otherwise plain.

One specimen 17 cm. long, and two smaller ones, from station 3698, Sagami Bay, 153 fathoms; six still smaller specimens from station 3717, Suruga Bay, 65 to 125 fathoms.

CHAULIODONTIDÆ.

8. Chauliodus emmelas Jordan & Starks, new species. (Pl. 1, fig. 2.)

One specimen 20.8 cm. long, from station 3697, in Sagami Bay, 120 to 265 fathoms; type No. 51464, U. S. Nat. Mus.

Close to *Chauliodus sloani*, the body more slender and the color entirely jet black.

Head 7.5 in length; depth 8; D. 6; A. 10; scales about 60; eye 5 in head; snout 4; longest fang 2; ventral fins unusually long, 5.5 in body; dorsal filament 2.6; pectoral 1.4 in head; barbel at chin pale.

General appearance of *Chauliodus sloani*. Luminous spots similar, 19 in a series from isthmus to ventrals; distance from pectoral to ventral, 4.2 in body; lateral fangs of lower jaw larger than those of upper. Color entirely jet-black, the fins a little paler. In the accompanying plate the artist has restored the squamation, lost in the specimen, from Garman's plates.

GONOSTOMIDÆ.

9. Neostoma gracile (Günther).

Gonostoma gracile Günther, Ann. & Mag. Nat. Hist. 1878, 187. Günther, Deep Sea Fishes, Challenger, p. 174, pl. XLV. fig. c, 1887; south of Japan.

One specimen, 2.5 inches long, from station 3712, Suruga Bay (surface); also two others, very small and in poor condition.

Head 5.5 in length; depth 9.5; D. about 10; A. 26; eye 6 in head; maxillary 1.1. Body very elongate, considerably more slender than in *Cyclothone microdon*; teeth stronger, numerous, slender, curved; short canines in each jaw; lower jaw much projecting; body apparently scaleless; anal fin beginning near middle of body, two-fifths of a head's length in advance of dorsal, the fin 3 times as long as dorsal fin; first ray of dorsal over eighth of anal; ventrals short, 1.6 in head, not quite reaching anal; pectoral long and narrow, 1.5 in head; no adipose fin.

Color black; lower side of head and fins abruptly pale; a row of photophores from isthmus to base of pectoral; a double row from between lower pectoral rays along each side of body just above base of anal and along lower side of caudal peduncle to base of caudal fin; a row from behind pectoral base to ventral; two on side above end of ventral; two above anal; two behind shoulder girdle above pectoral; one below eye.

This species has the anal fin more advanced than any other referred to *Gonostoma* or to *Cyclothone*. This trait may be regarded as of generic value. As it appears also, although to a less degree, in *Cyclothone bathyphila*, the type of the nominal genus *Neostoma* of Vaillant, we may provisionally adopt the genus *Neostoma* for these two species, which have the anal inserted well in front of the dorsal.

MYCTOPHIDÆ.

10. Neoscopelus alcocki Jordan & Starks, new species. (Pl. 2, figs. 1 and 2.)

One specimen, 19 cm. long, from station 3709, in Suruga Bay, in 173 to 260 fathoms; type, No. 51477, U. S. Nat. Mus.

Head 3 in length; depth 4; D. 13; A. 12; scales 4-33-4; eye 5 in head; snout 3.5; maxillary 2.

Body rather robust, subfusiform, head rather pointed in profile, broad and somewhat depressed above; mouth large, oblique, maxillary extending to below posterior margin of orbit, not dilated behind, posterior border truncate; teeth small, in villiform bands; eye moderate, cheek broad, not oblique in position; scales large, entire, firm, roughened on the surface, nearly all fallen in specimen examined; lateral line well developed; luminous spots, large, in about 6 rows on breast, about 14 in a lengthwise series from isthmus to ventrals, then a median and two lateral rows, to opposite front of anal, 10 spots in outer row, the posterior one smaller; an oblong circle of 10 small photophores about the vent; a row of 15 small photophores, continuous with inner lateral row before vent, from opposite vent to base of caudal, most of the median members of this series double; there is also an inner series of minute white dots along base of anal rays; a median row of small photophores behind anal below caudal peduncle. Dorsal rather large, inserted before ventral, its longest rays about half head; longest anal ray 2.4 in head; caudal well forked; pectoral long, 1.1 in head; ventral long, 1.75; gillrakers long and slender, 3 + 12 in number.

Color pale or brownish above, belly black; a dusky shade at base of caudal and pectoral; inside of mouth black; luminous spots pale, with a dark ring.

This species is very close to *Neoscopelus macrolepidotus* of the Atlantic. The sole important difference apparently is in the arrangement of the photophores on the posterior part of the body. In the figures (Nos. 108 and 109) given by Goode & Bean, the arrangement is quite unlike that seen in the Japanese fish. In the plate, the two lateral rows of spots found on the abdomen are represented as continuous to the base of caudal. In the Japanese fish the outer row is not continued behind the front of anal. The inner lateral series is continued, the spots becoming smaller. There is a ring of little spots about the vent, and a series of little dots along base of anal.

The species abundant about Hawaii, called *Neoscopelus macrolepidotus* by Gilbert & Cramer, seems to be the same as the Japanese fish.

11. Diaphus watasei Jordan & Starks, new species.

Head 3.8 in length; depth 5.2; eye 4 in head; snout 5.5; maxillary 1.4; D. 11, 13; A. 11, 13; scales 36.

Body moderately elongate and compressed, more slender than in *Diaphus* (*Æthoprora*) *effulgens*, the nearest related species, head more pointed; eye small; snout very short, truncate at tip; cheeks very oblique; scales caducous, all fallen in type; those along lateral line preserved in one specimen, and considerably enlarged; a lunate luminous gland in front of eye, extending backward to a point a little behind front of pupil; a short luminous tract above this; all these coalescent over snout in a large pale area like the headlight of an engine, the front part distinctly paler than that before the eye; luminous spots distinct, 4 + 4 + 1 + 4 + 4 + 6 + 1 + 5 + 4; the last 4 in a curved row at base of caudal, 5 before this, then 1 posterolateral spot before which are 6 in a curved row continuous with it; then 4 mediolateral spots in a V-shaped series, 4 more, 1 of them out of line, then a single spot above base of ventral; 4 spots before ventral and 4 before pectoral. One single spot shows the characteristic division, or theta-form (θ), which suggested the name *Diaphus*. It is probable that in life more spots had this form, and that there is no real difference between *Æthoprora* and *Diaphus*. Dorsal fin high, its last ray about over first of anal, its longest ray 1.2 in head; anal fin moderate; pectoral short and broken, about 3 in head; ventrals 1.6 in head; caudal broken.

Color dusky, with luminous spots, the one before eye very bright; lower jaw with 3 dark cross shades.

Of this species we have one specimen, the type, No. 51443, U. S. Nat. Mus., in good condition, dredged by the *Albatross*, at station 3698, off Atami, in Sagami Bay, in 153 fathoms, and 4 smaller examples, No. 8393, Stanford University, collected off Misaki, and presented by Professor Sho Watase, of the Imperial University.

In scales, fins, and luminous spots this species agrees with *Diaphus effulgens* of the Atlantic, but the latter species has the head notably shorter, deeper, and more blunt. The species is still nearer *Diaphus engraulis*, from which it apparently differs in the number and arrangement of the postero-anal and posterolateral photophores. The former diverge backward from the mid-ventral line, forming a continuous series with the latter. They are 6 or 7 in number instead of 5.

Diaphus watasei Jordan & Starks, new species.

STERNOPTYCHIDÆ.

12. Polyipnus stereope Jordan & Starks, new species. (Pl. 2, fig. 3.)

Head with projecting lower jaw 3 in length; depth 1.6; dorsal 13; anal 15; eye 2.25 in head; maxillary 1.33; pectoral 1.2; ventral 3.

This species differs from *P. spinosus* Günther (its nearest relative), in the character of the nuchal process. The upper of the 3 spines into which the process is divided is long, sharp, and straight; it extends at an oblique angle upward and backward; the distance from its tip vertically down to the outline of the back equals half diameter of pupil. This spine in *P. spinosus*, according to Dr. Günther's plate, extends but little above this outline; second spine shorter, and evidently variable in length. In the type it is at least three-fourths as long as the upper spine; in the 2 cotypes it is reduced to a very small inconspicuous spinule. Lower spine as long as, or not noticeably shorter than upper spine, curved downward and only slightly backward.

Mr. C. Tate Regan has kindly examined the type of *P. spinosus* and describes the nuchal process as follows: "The suprascapular spine has a few denticulations above and below. The 2 below, near its base, are somewhat enlarged, and might be called accessory spines, but the lower does not nearly approach in length the main spine."

A drawing accompanying Mr. Regan's letter shows the lower spine straight, short, and pointing nearly directly backward. As compared with Günther's plate, the spine on preopercular angle, and the dark crossbar which extends downward from just in front of dorsal, are longer in this species.

The species differs from *P. nuttingi* Gilbert, MS., from Hawaii, in the character of the nuchal process, in being brighter silvery without dark specks scattered over the silver ground, in having the crossbar much smaller, and in having the series of photophores on lower side of caudal peduncle continuous with the supraanal series.

Three specimens taken in Sagami Bay, at station 3698. The type is the largest, 65 mm. in length. No. 51451, U. S. Nat. Mus. A cotype is No. 8392, Stanford University.

SYNAPHOBRANCHIDÆ.

13. Synaphobranchus jenkinsi Jordan & Snyder.

Synaphobranchus jenkinsi Jordan & Snyder, Proc. U. S. Nat. Mus. XXIII, 1901, 845; Sagami Bay.

One specimen, No. 49727, U. S. Nat. Mus., from station 3696, off Enoshima, Sagami Bay, in 110 to 175 fathoms.

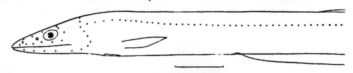

Synaphobranchus jenkinsi Jordan & Snyder

14. Synaphobranchus affinis Günther.

Synaphobranchus affinis, Jordan & Snyder, Proc. U. S. Nat. Mus. XXIII, 1901, 844.

Specimens from station 3780, Totomi Bay, off Owai Point, in 37 fathoms, and from station 3697, Sagami Bay, off Manazuru Point, in 120 to 265 fathoms.

LEPTOCEPHALIDÆ.

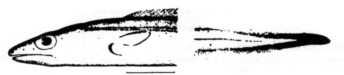

Congrellus megastomus (Günther).

15. Congrellus megastomus (Günther).

Congrellus megastomus, Jordan & Snyder, Proc. U. S. Nat. Mus. XXIII, 1901, 854.

From station 3730 in Totomi Bay, off Owai Point, in 34 fathoms.

OPHICHTHYIDÆ.

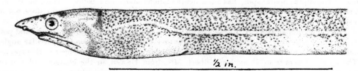

½ in.

Sphagebranchus moseri Jordan & Snyder. Type.

16. Sphagebranchus moseri Jordan & Snyder.

Sphagebranchus moseri Jordan & Snyder, Proc. U. S. Nat. Mus., XXIII, 1901, p. 864; Suruga Bay.

One example, No. 49728, U. S. Nat. Mus. from station 3700, Suruga Bay, in 100 fathoms.

MACRORHAMPHOSIDÆ.

17. Macrorhamphosus sagifue Jordan & Starks.

Macrorhamphosus sagifue Jordan & Starks, Proc. U. S. Nat. Mus., XXVI, 1903, p. 69; Suruga Bay, Totomi Bay, Sagami Bay.

Many specimens, from station 3707, off Ose Point, Suruga Bay, in 68 to 70 fathoms; station 3730, off Owai Point, Totomi Bay, in 34 fathoms; station 3715, off Ose Point, Suruga Bay, in 64 to 65 fathoms; station 3716, off Ose Point, Suruga Bay, in 65 to 125 fathoms; station 3717, off Ose Point, Suruga Bay, in 65 to 125 fathoms; station 3741, Suruga Bay, in 71 fathoms, and station 3763, off Sune Point, Sagami Bay, in 49 fathoms.

Macrorhamphosus sagifue Jordan and Starks.

HIPPOCAMPIDÆ.

18. Hippocampus sindonis Jordan & Snyder.

Hippocampus sindonis Jordan & Snyder, Proc. U. S. Nat. Mus., XXIV, 1902, p. 17; Totomi Bay.

From station 3727, Totomi Bay, example No. 47930, U. S. Nat. Mus.

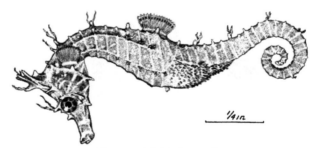

Hippocampus sindonis Jordan & Snyder.

MONOCENTRIDÆ.

19. Monocentris japonicus (Houttuyn).

Monocentris japonicus, Jordan & Fowler, Proc. U. S. Nat. Mus., XXVI, 1903, p. 19; Suruga Bay, Owari Bay.

Examples from station 3742, Suruga Bay, and station 3762, Owari Bay, in 42 fathoms.

TRACHICHTHYIDÆ.

20. Hoplostethus mediterraneus Cuvier & Valenciennes.

Hoplostethus mediterraneus, Jordan & Fowler, Proc. U. S. Nat. Mus., XXVI, 1903, p. 7; Sagami Bay.

From stations 3895 and 3897, Sagami Bay, in 120 to 265 fathoms.

21. Gephyroberyx japonicus (Döderlein).

Gephyroberyx japonicus, Jordan & Fowler, Proc. U. S. Nat. Mus., xxvi, 1903, p. 6; Suruga Bay.

Station 3716, off Ose Point, Suruga Bay, 65 to 125 fathoms.

22. Paratrachichthys prosthemius Jordan & Fowler.

Paratrachichthys prosthemius Jordan & Fowler, Proc. U. S. Nat. Mus., xxvi, 1903, p. 9; Totomi Bay.

Station 3730, Owai Point, Totomi Bay, in 34 fathoms.   One specimen, No. 50575, U. S. Nat. Mus.

Paratrachichthys prosthemius Jordan and Fowler. Type.

ZEIDÆ.

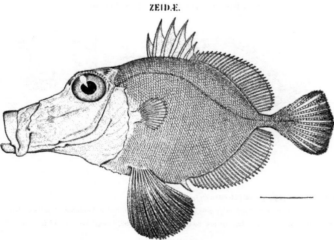

Zen itea (Jordan & Fowler).

23. Zen itea (Jordan & Fowler).

Callopsis itea Jordan & Fowler, Proc. U. S. Nat. Mus., xxv, 1903, p. 519; Suruga Bay.

Station 3738, Suruga Bay, in 167 fathoms; No. 50582, U. S. Nat. Mus.

24. Zeus japonicus Schlegel.

Station 3713, off Ose Point, Suruga Bay, in 50 to 60 fathoms.

CARANGIDÆ.

25. Carangus equula (Schlegel).

One specimen, from station 3738, Suruga Bay, 167 fathoms.

26. Seriola purpurascens Schlegel.

Many young taken at the surface in Sagami Bay.

APOGONIDÆ.

27. Apogon lineatus Schlegel.

Apogon lineatus, Jordan & Snyder, Proc. U. S. Nat. Mus., XXIII, 1901, p. 888; Owari Bay.

Station 3722, Owari Bay, in 3 to 9 fathoms.

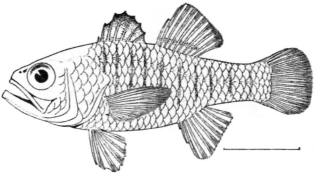

Apogon lineatus Schlegel.

28. Synagrops japonica (Steindachner & Döderlein).

Melanostoma japonicum, Jordan & Fowler, Proc. U. S. Nat. Mus., XXV, 1901, p. 573; Suruga Bay.

Station 3707, off Ose Point, Suruga Bay, in 68 to 70 fathoms.

SERRANIDÆ.

29. Labracopsis japonica (Steindachner & Döderlein).

Station 3713, off Ose Point, Suruga Bay, in 500 to 600 fathoms; one young specimen.

30. Chelidoperca hirundinacea (Schlegel). One specimen, station 3715, Suruga Bay.

30(b). Epinephelus, sp.

A very small example, blackish, the caudal white, from station 3708, off Ose Point, Suruga Bay, in 65 fathoms.

31. Pseudanthias japonicus (Steindachner & Döderlein).

Station 3707, off Ose Point, Suruga Bay, in 68 to 70 fathoms; station 3708, off Ose Point, Suruga Bay, in 65 to 125 fathoms; station 3716, off Ose Point, Suruga Bay, in 65 to 125 fathoms; station 3741, Suruga Bay, in 71 fathoms; station 3755, off Sune Point, Sagami Bay; stations 3764 and 3765, off Sune Point, Sagami Bay.

In life orange yellow with broad cross shades of orange red, one of these very bright scarlet from spinous dorsal to vent, half wider than eye, not extending on fins; the other at base of caudal; traces of both of these showing in spirits as darker shading after the colors have faded; shoulders with deep scarlet shading; space between bands deep orange yellow; cheek orange with crimson wash above; belly and fins pale; first dorsal and caudal shaded with pink, ventrals with yellowish; nose orange; top of head dull crimson, becoming red at nape.

ANTIGONIDÆ.

32. Antigonia rubescens (Günther).

Antigonia rubescens, Jordan & Fowler, Proc. U. S. Nat. Mus., XXV, 1903, p. 523; Suruga Bay, Totomi Bay.

Station 3707, off Ose Point, Suruga Bay, in 63 to 75 fathoms; station 3713, off Ose Point, **Suruga Bay, 500 to 600 fathoms; station 3715, off Ose Point, Suruga Bay, 64 to 65 fathoms; station 3717, off Ose Point, Suruga Bay, 65 to 125 fathoms; station 3729, off Owai Point, Totomi Bay, 37 fathoms, and station 3734, off Owai Point, Totomi Bay, 48 to 36 fathoms.

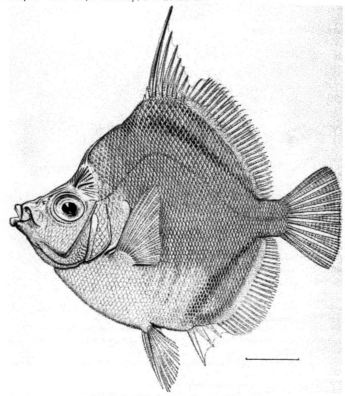

Antigonia rubescens (Günther).

LABRIDÆ.

33. Pseudolabrus japonicus (Houttuyn).

Pseudolabrus japonicus, Jordan & Snyder, Proc. U. S. Nat. Mus., XXIV, 1902, p. 625.

Station 3729, off Owai Point, Totomi Bay, in 37 fathoms; station 3730, off Owai Point, Totomi Bay, 34 fathoms; station 3734, off Owai Point, Totomi Bay; station 3761, Sagami Bay, 44 fathoms.
Color much more red than in specimens from shallow water.

CHÆTODONTIDÆ.

34. Chætodon nippon Steindachner & Döderlein.

Chætodon nippon. Jordan & Fowler, Proc. U. S. Nat. Mus., xxv, 1903, p. 537; Totomi Bay.

Station 3730, off Owai Point, Totomi Bay, in 37 fathoms.

TROPIDICHTHYIDÆ.

35. Tropidichthys rivulatus (Schlegel).

Eumycterias rivulatus, Jordan & Snyder, Proc. U. S. Nat. Mus., xxiv, 1902, p. 255; Totomi Bay.

Station 3729, off Owai Point, Totomi Bay, in 37 fathoms.

MONACANTHIDÆ.

36. Stephanolepis cirrhifer (Schlegel).

Stephanolepis cirrhifer, Jordan & Fowler, Proc. U. S. Nat. Mus., xxv, 1902, p. 264.

Station 3730, off Owai Point, Totomi Bay, in 34 fathoms.

OSTRACIIDÆ.

37. Aracana aculeata (Houttuyn).

Aracana aculeata, Jordan & Fowler, Proc. U. S. Nat. Mus., xxv, 1902, 284, Sagami Bay, Suruga Bay, Owari Bay.

Station 3707, off Ose Point, Suruga Bay, 68 to 70 fathoms; station 3754, off Sune Point, Sagami Bay, 50 fathoms; station 3762, Owari Bay, in 42 fathoms.

SCORPÆNIDÆ.

38. Sebastolobus macrochir (Günther).

Sebastolobus macrochir, Jordan & Starks, Proc. U. S. Nat. Mus., xxvii, 1904, p. 94; Sagami Bay.

Station 3697, off Manazuru Point, Sagami Bay, in 205 to 120 fathoms.

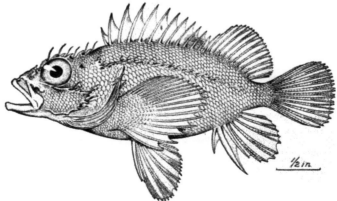

Thysanichthys crossotus Jordan & Starks.

39. Thysanichthys crossotus Jordan & Starks.

Thysanichthys crossotus Jordan & Starks, Proc. U. S. Nat. Mus., xxvii, 1904, 123; Suruga Bay.

Station 3720, off Ose Point, Suruga Bay, 63 fathoms.

40. Setarches albescens (Steindachner & Döderlein).

Setarches albescens, Jordan & Starks, Proc. U. S. Nat. Mus., XXVII, 1904, p. 138; Totomi Bay.

Station 3729, off Owai Point, Totomi Bay, in 37 fathoms.

41. Helicolenus dactylopterus (De la Roche).

Helicolenus dactylopterus, Jordan & Starks, Proc. U. S. Nat. Mus., XXVII, 1904, p. 126; Sagami Bay, Suruga Bay.

Station 3698, off Manazuru Point, Sagami Bay, in 153 fathoms; station 3717, off Ose Point, Suruga Bay, 65 to 125 fathoms; station 3719, off Ose Point, Suruga Bay, in 70 to 100 fathoms.

42. Lythrichthys eulabes Jordan & Starks.

Lythrichthys eulabes Jordan & Starks, Proc. U. S. Nat. Mus., XXVII, 1904, p. 140; Suruga Bay.

Station 3708, off Ose Point, Suruga Bay, 65 to 125 fathoms.

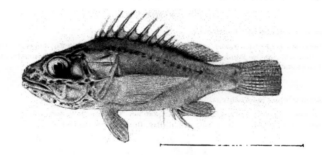

Lythrichthys eulabes Jordan & Starks.

43. Sebastiscus marmoratus (Cuvier & Valenciennes).

Sebastiscus marmoratus, Jordan & Starks, Proc. U. S. Nat. Mus., XXVII, 1904, p. 124; Suruga Bay.

Station 3718, off Ose Point, Suruga Bay, in 65 fathoms.

44. Sebastiscus albofasciatus (Lacépède).

Sebastiscus albofasciatus, Jordan & Starks, Proc. U. S. Nat. Mus., XXVII, 1903, p. 126; Suruga Bay, Totomi Bay.

Station 3707, off Ose Point, Suruga Bay, 68 to 70 fathoms; station 3708, off Ose Point, Suruga Bay, 65 to 125 fathoms; station 3714, off Ose Point, Suruga Bay, 48 to 60 fathoms; station 3715, off Ose Point, Suruga Bay, 64 to 65 fathoms; station 3720, off Ose Point, Suruga Bay, in 63 fathoms; station 3730, off Owai Point, Totomi Bay, in 34 fathoms; station 3734, off Owai Point, Totomi Bay, in 37 fathoms.

45. Erisphex potti (Steindachner).

Erisphex potti, Jordan & Starks, Proc. U. S. Nat. Mus., XXVII, 1904, p. 170; Matsushima Bay.

Station 3771, Matsushima Bay, in 51 fathoms.

46. Scorpæna izensis Jordan & Starks.

Scorpæna izensis Jordan & Starks, Proc. U. S. Nat. Mus., xxvii, 1904, p. 134; Suruga Bay, Totomi Bay.

Station 3708, off Ose Point, Suruga Bay, 65 to 125 fathoms; station 3713, off Ose Point, Suruga Bay, 500 to 600 fathoms; station 3715, off Ose Point, Suruga Bay, 64 to 65 fathoms; station 3717, off Ose Point, Suruga Bay, 65 to 125 fathoms; station 3720, off Ose Point, Suruga Bay, in 63 fathoms; station 3729, off Owai Point, Totomi Bay, in 37 fathoms.

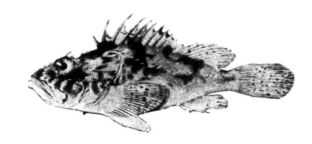

Scorpæna izensis Jordan & Starks.

Ocosia vespa Jordan & Starks.

47. Ocosia vespa Jordan & Starks.

Ocosia vespa Jordan & Starks, Proc. U. S. Nat. Mus., xxvii, 1904, p. 162; Sagami Bay, Owari Bay.

Station 3757, off Sune Point, Sagami Bay, in 50 fathoms; station 3762, Owari Bay, in 42 fathoms; station 3763, off Sune Point, Sagami Bay, in 44 fathoms.

COTTIDÆ.

48. Stlengis osensis Jordan & Starks.

Stlengis osensis Jordan & Starks, Proc. U. S. Nat. Mus., XXVII, 1904, p. 236; Suruga Bay.

Station 3738, off Ose Point, Suruga Bay, in 167 fathoms.

Stlengis osensis Jordan & Starks.

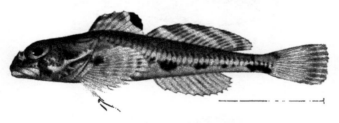

Schmidtina misakia Jordan & Starks.

49. Schmidtina misakia Jordan & Starks, new generic name, *Schmidtia* being preoccupied.

Schmidtia misakia Jordan & Starks, Proc. U. S. Nat. Mus., XXVII, 1904, p. 237; Sagami Bay.

Station 3698, off Manazuru Point, Sagami Bay, in 153 fathoms.

Daruma sagamia Jordan & Starks.

50. Daruma sagamia Jordan & Starks.

Daruma sagamia Jordan & Starks, Proc. U. S. Nat. Mus., XXVII, 1904, p. 241; Sagami Bay, Totomi Bay, Owari Bay.

Stations 3754, 3762, and 3763, off Sune Point, Sagami Bay, in 42 to 52 fathoms; station 3729, off Owai Point, Totomi Bay, in 37 fathoms; also from Owari Bay.

51. Ricuzenius pinetorum Jordan & Starks.

Ricuzenius pinetorum Jordan & Starks, Proc. U. S. Nat. Mus., xxvii, 1904, p. 243: Matsushima Bay

Station 3773, off Kinkwazan Island, in Matsushima Bay, in 78 fathoms.

Ricuzenius pinetorum Jordan & Starks.

52. Pseudoblennius totomius Jordan & Starks.

Pseudoblennius totomius Jordan & Starks, Proc. U. S. Nat. Mus., xxvii, 1904, p. 315: Sagami Bay.

Station 3759, off Sune Point, Sagami Bay.

Pseudoblennius totomius Jordan & Starks.

53. Hemitripterus villosus (Pallas).

Hemitripterus villosus, Jordan & Starks, Proc. U. S. Nat. Mus., xxvii, 1904, p. 326: Matsushima Bay.

Station 3771, Matsushima Bay, in 51 fathoms. *Hemitripterus sinensis*, Sauvage, Nouv. Archiv. Mus., 1873, p. 53, from China, seems to be identical with this species, which differs but slightly, if at all, from H. *americanus* of the Atlantic. *Hemitripterus nipponicus* Ishikawa (Proc. Nat. Hist. Tokyo, I, p. 14, pl. vii, fig. 2, 1904; Hakodate) seems to be the young of the same species.

spine, two-thirds length of eye, preceded by a sharp ridge and extending posterior to all other parts of head; opercle with a small, short, sharp spine, preceded by a low, sharp ridge; a small spine over front of eye, a high spine over posterior part of eye; a very high sharp occipital spine, two-thirds diameter of eye; no spines above muzzle; lower jaw with many barbels, the outermost a long branched brush over two-thirds length of head, and reaching to within a diameter of pupil of the vent; breast with soft skin; fins moderate; spinous dorsal beginning between second and third spines of dorsal series of plates; tips of spines when depressed reaching to base of second dorsal ray; anal projecting posteriorly beyond soft dorsal, its origin opposite that of the latter. Tip of pectoral reaching fifteenth spine of upper series of plates; ventrals reaching just past middle of vent; caudal concave, its length 2.75 in head.

Color brown, probably red in life, with no black spots or marblings except a spot behind eye and a few dark edgings on ridges of head; pectoral black, pale-edged below; spinous dorsal black; soft dorsal, caudal, and ventral mottled; a dusky shade below last rays of soft dorsal; outer barbels of mouth black on distal half.

One specimen, the type No. 51428, U. S. Nat. Mus., from station 3698, off Manazuru Point, Sagami Bay, in 153 fathoms.

TRIGLIDÆ.

60. Lepidotrigla guntheri Hilgendorf.

Lepidotrigla guntheri Hilgendorf, Ges. Naturf. Freunde, 1879, 106; Tokyo.
Lepidotrigla longispinis Steindachner, Fische Japans, IV, 1882, taf. IV, fig. 1, 1887; Tokyo.

Station 3708, 65 to 125 fathoms; station 3713, 500 to 600 fathoms; station 3714, 48 to 60 fathoms; station 3715, 64 to 65 fathoms; and station 3717, 65 to 125 fathoms, all off Ose Point, Suruga Bay; and station 3727 in Totomi Bay.

Head 3 in length; depth 4.16; D. viii-15; A. 15; scales 55; eye 3.5 in head. Head moderately decurved, the muzzle emarginate, the lateral prominences moderate, weakly toothed, at tip a preopercular spine; a low postocular spine with a cross furrow behind it, a low spine behind this, on each side of the vertex; nuchal spine long and sharp, the inner edge of the bone strongly serrated; second dorsal spine very strong, 1.33 in head, reaching far past the other spines when fin is depressed, and well past front of soft dorsal; pectoral 1.16 in head; longest detached pectoral ray reaching to tip of ventrals. Adult examples a foot in length have the head smoother, spines on vertex obsolete, spinules on snout longer, interorbital space less concave, second dorsal spine higher, pectoral fin longer.

Color brown, with 3 brown cross-shades, one under each dorsal and one at base of caudal, these fainter with age; young with a blackish bar at tip of caudal; pectoral black within; back mottled; no black dorsal spot, but sometimes a dusky cross-shade on dorsal; no sharp line on side bounding the pale color of belly.

The long dorsal spine and the broad, serrated nuchal process are especially characteristic of this species.

Five young specimens with the body and caudal fin banded and with the spines on the head rougher, the dorsal spines lower, we regard as the young of *Lepidotrigla guntheri*. Station 3708, off Ose Point, Suruga Bay, in 65 to 125 fathoms, two examples; station 3717, off Ose Point, Suruga Bay, in 65 to 125 fathoms, two examples; station 3727, Totomi Bay, one example. Head short and deep, snout steep, little produced, emarginate at tip, its lobes slightly serrate; eyebrow elevated, coarsely toothed, posterior serra a strong spine; a sharp spine on each side of vertex; a sharp parietal spine; nuchal spine strong, broad, with inner edge serrate; humeral and opercular spines strong; dorsal spines rather long, strongly serrated, second 1.66 in head; pectoral short, 1.2 in head (young).

Head gray, marbled with blackish and with numerous black specks; pectoral closely spotted with black, so that the fin appears blackish; a blackish cross-shade on body below spinous dorsal; another below soft dorsal, these extending on fins; another at base of caudal; caudal with a round black blotch near tip.

The longest of these is 4 inches long, smallest 1.5 inches. We have also a large example nearly a foot in length from the Yokohama market.

61. Lepidotrigla abyssalis Jordan & Starks, new species.

Head 3 in length; depth 4.2; D. viii-15; A. 15; scales 56; eye 3.16 in head; snout 2.5; maxillary 2.75; interorbital space 4; first dorsal spine 2.1; second dorsal spine 1.75; pectoral 1.1; ventral 1.25; caudal 1.4.

Head rather high, about as in *L. guntheri*. Snout almost truncate at tip; a sharp spine slightly projects at each angle; interorbital rather deeply concave; a short narrow cross-furrow above posterior margin of eye on supraorbital region, but not extending across top of head medially; slight indications of ridges on parietal region; nuchal spines and ridges but little developed; a slight ridge running back from upper posterior orbital rim; humeral spine moderately strong; second dorsal spine longest, though scarcely the strongest; when fin is depressed its tip reaches to front of soft dorsal, or barely past the tips of the other spines, while the third, fourth, and fifth are coterminal; first spine reaching only a little past base of last; upper detached pectoral ray reaching tip of ventral, which reaches base of second anal ray; pectoral reaching to opposite base of sixth anal ray.

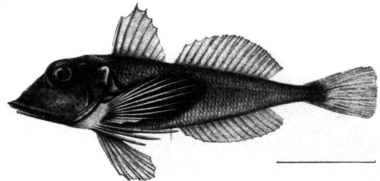

Lepidotrigla abyssalis Jordan & Starks, new species. Type.

Color mottled red; pectoral bluish black; other fins without markings; no traces of a spot on spinous dorsal.

This species differs from *L. guntheri* in having more slender dorsal spines, the second dorsal spine not enlarged, top of head without spines. These comparisons are made between fishes of about the same size. In large specimens of *L. guntheri* the top of the head is as smooth as in this species, while in very small specimens the second dorsal spine is not much enlarged.

The type and only specimen was taken in Suruga Bay, station 3713, in from 50 to 60 fathoms. It is 93 mm. in length, and is No. 51440, U. S. Nat. Mus.

62. Lepidotrigla microptera Günther.

D. ix-17; A. 17; head very smooth, the anterior profile straight; snout deeply emarginate, the prominences on each side short and sharp, each ending in a spine; preocular spine almost obsolete; no spines on top of head; nuchal spine blunt and inconspicuous; opercular spine small; humeral spine long and sharp; second dorsal spine longest, 1.5 in head; pectoral as long as head.

Color, brown; belly silvery, a pale or golden line marking the boundary above; a large, black ocellus on last dorsal spines; pectoral black.

Station 3715, off Ose Point, Suruga Bay, in 64 to 65 fathoms; station 3770, in Matsushima Bay, in 42 to 45 fathoms.

Color in life of specimen from station 3715, scarlet red above, abruptly white below. First dorsal red with a large black spot on its posterior rays; second dorsal brick red, caudal scarlet, becoming crimson behind, ventrals and anal pale, pectoral blackish purple above, paler below.

We refer these specimens to *Lepidotrigla microptera* Günther (Ann. & Mag. Nat. Hist. 1873, 241) = *Lepidotrigla strauchii* Steindachner, Ich. Beitr., v, p. 166, 1876, Hakodate.   The latter is without much doubt the same, although a shore fish, without any of the brilliant coloration of the specimens in hand. The only difference we can find is that these specimens were red, while shore examples are olive-green.

Lepidotrigla microptera may be known by the black spot on the posterior part of dorsal, the dorsal and anal slightly longer than in *L. guntheri* and *L. abyssalis*, and the short detached pectoral rays which do not reach to tip of ventral.

We have large examples of *Lepidotrigla microptera* from the Tokyo market corresponding to *Lepidotrigla strauchi* Steindachner.   In these the sharp line separating the silvery of the belly from the darker color of the back is still more apparent than in our species.   In specimens a foot long there is no trace of black on the dorsal fin, but in the young of 4 inches the black ocellus is still evident.   In these large specimens each lobe of the snout ends in a large spine and several smaller serræ.   This is the commonest species of the genus in Japan, extending its range well to the northward.   We have numerous specimens from Aomori, Hakodate, Tsuruga, Matsushima, and Hiroshima.

The remaining Japanese species of *Lepidotrigla* is *L. alata* (Houttuyn) = *L. burgeri* (Schlegel). Of this we have many specimens from the shores of southern Japan, but none taken in the dredge. In this species the snout has two long diverging processes.

Lepidotrigla japonica (Bleeker).

63. Lepidotrigla japonica (Bleeker).

Prionotus japonicus Bleeker, Nieuw Nalez, Japan, 1857, 75, taf. x, f. 1
Lepidotrigla japonica Steindachner, Fische Japans, IV, 261, 1887 Oshima, Kagoshima
"*Lepidotrigla serridens* Hilgendorf, Naturf. Freunde, 1879, p. 107, Tokyo. ' Schnabel in der Mitte nicht ausgeschnitten und jederseits mit einem etwas stärkeren Zähne versehen."

For purposes of comparison we present a figure of this rare species from one of two specimens taken in a net at Misaki.   The pectoral is longer than in other species of *Lepidotrigla*.

GOBIIDÆ.

64. Coryphopterus pflaumi (Bleeker).

Ctenogobius pflaumi, Jordan & Snyder, Proc. U. S. Nat. Mus., XXIV, 1902, p. 61, Owari Bay.

Station 3722, Owari Bay, in 3 to 9 fathoms.   The type of *Ctenogobius* (*C. fasciatus*) is said to have the tongue emarginate.   The name *Coryphopterus* may then be used for species of this type.

65. Suruga fundicola Jordan & Snyder.

Suruga fundicola Jordan & Snyder, Proc. U. S. Nat. Mus., XXIV, 1902, p. 96; Sagami Bay, Suruga Bay, Owari Bay, Matsushima Bay.

Station 3698, off Manazuru Point, Sagami Bay, in 153 fathoms; station 3708, off Ose Point, Suruga Bay, in 65 to 125 fathoms; station 3714, 48 to 60 fathoms, station 3715, 64 to 65 fathoms, station 3716, 65 to 125 fathoms, and station 3719 in 70 to 100 fathoms, all off Ose Point, Suruga Bay; station 3722, Owari Bay, in 3 to 9 fathoms; station 3723, Owari Bay, in 5 to 15 fathoms; station 3741, in Suruga Bay, 71 fathoms; station 3775, in Matsushima Bay.

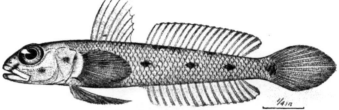

Suruga fundicola Jordan & Snyder.

66. Chæturichthys hexanemus (Bleeker).

Chæturichthys hexanemus, Jordan & Snyder, Proc. U. S. Nat. Mus., XXIV, 1902, p. 106; Owari Bay, Matsushima Bay.

Station 3722, Owari Bay, in 3 to 9 fathoms; station 3723, Owari Bay, 5 to 15 fathoms; station 3724, Owari Bay, 19 to 20 fathoms; station 3768, Matsushima Bay.

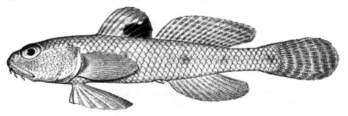

Chæturichthys sciistius Jordan & Snyder.

67. Chæturichthys sciistius Jordan & Snyder.

Chæturichthys sciistius Jordan & Snyder, Proc. U. S. Nat. Mus., XXIV, 1902, p. 107; Suruga Bay, Owari Bay, Matsushima Bay.

Station 3715, off Ose Point, Suruga Bay, in 64 to 65 fathoms; station 3722, 3 to 9 fathoms; station 3723, 5 to 15 fathoms, and station 3724, 19 to 20 fathoms, all in Owari Bay; station 3767, Matsushima Bay, and station 3768, Matsushima Bay.

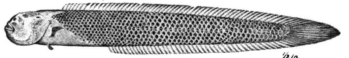

Trypauchen wakæ Jordan & Snyder.

68. Trypauchen wakæ Jordan & Snyder.

Trypauchen wakæ Jordan & Snyder, Proc. U. S. Nat. Mus., XXIV, 1902, p. 127; Owari Bay, Wakanoura.

Station 3722, 3 to 9 fathoms, and station 3723, 5 to 15 fathoms, both in Owari Bay.

CALLIONYMIDÆ.

69. Callionymus lunatus Schlegel.

Callionymus lunatus, Jordan & Fowler, Proc. U. S. Nat. Mus., xxv, 1903, p. 949; Owari Bay.

Station 3722, 3 to 9 fathoms, and station 3723, 5 to 15 fathoms, Owari Bay.

70. Callionymus flagris Jordan & Fowler.

Callionymus flagris Jordan & Fowler, Proc. U. S. Nat. Mus., xxv, 1903, p. 952; Owari Bay, Matsushima Bay.

Stations 3722, Owari Bay, 3 to 9 fathoms, and 3723, 5 to 15 fathoms, Owari Bay; station 3777, Matsushima Bay.

Callionymus flagris Jordan & Fowler.

Draconetta xenica Jordan & Fowler.

DRACONETTIDÆ.

71. Draconetta xenica Jordan & Fowler.

Draconetta xenica Jordan & Fowler, Proc. U. S. Nat. Mus., xxv, 1902, p. 939; Suruga Bay.

One specimen from station 3700, Suruga Bay, in 100 fathoms. No. 50816, U. S. Nat. Mus.

CHAMPSODONTIDÆ.

72. Champsodon vorax Günther.

Champsodon vorax, Jordan & Snyder, Proc. U. S. Nat. Mus., XXIV, 1902, p. 481; Suruga Bay.

Station 3713, 45 to 48 fathoms, and station 3714, 48 to 60 fathoms, both off Ose Point, Suruga Bay.

PTEROPSARIDÆ.

73. Neopercis sexfasciata (Schlegel).

Neopercis sexfasciata, Jordan & Snyder, Proc. U. S. Nat. Mus., XXIV, 1902, p. 467; Totomi Bay, Sagami Bay, Suruga Bay.

Station 3730, off Owari Point, Totomi Bay, 34 fathoms; station 3707, off Ose Point, Suruga Bay, 68 to 70 fathoms; station 3727, in Totomi Bay; station 3763, off Sune Point, Sagami Bay, 49 fathoms. This species frequently bears small parasitic isopods.

When fresh the specimens dredged in Suruga Bay were orange red, with black bars and markings, which on head and belly shaded into yellow. Caudal bright yellow above, orange red below, with three black cross bars on upper half; lower fins orange; pectoral golden.

74. Neopercis aurantiaca (Döderlein).

Neopercis aurantiaca, Jordan & Snyder, Proc. U. S. Nat. Mus., XXIV, 1902, p. 468; Suruga Bay, Sagami Bay, Owari Bay.

Station 3708, off Ose Point, Suruga Bay, 65 to 125 fathoms; also in Sagami and Owari bays.

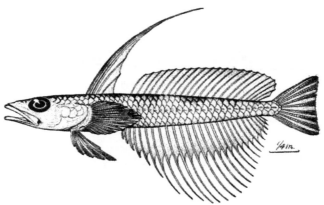

Pteropsaron evolans Jordan & Snyder.

75. Pteropsaron evolans Jordan & Snyder.

Pteropsaron evolans Jordan & Snyder, Proc. U. S. Nat. Mus., XXIV, 1902, p. 470; Sagami Bay, Owari Bay.

Station 3757, off Sune Point, Sagami Bay, 50 fathoms; station 3762, Owari Bay, 42 fathoms; station 3763, off Sune Point, Sagami Bay, 49 fathoms, No. 50008.

Also dredged in Owari Bay, in 14 to 20 fathoms.

76. Osopsaron verecundum (Jordan & Snyder).

Pteropsaron verecundum Jordan & Snyder, Proc. U. S. Nat. Mus., XXIV, 1902, p. 472; Suruga Bay.

One specimen, No. 50009, dredged at station 3716, off Ose Point, Suruga Bay, 65 to 125 fathoms.

This species seems to belong to a different genus from the preceding, distinguished by the low fins and scaly cheeks. It resembles *Acanthaphritis*, described by Günther from the Ki Islands, but is distinguished from the latter by the smooth scales, a character in which it agrees with *Pteropsaron*. For this genus we suggest the name of *Osopsaron*.

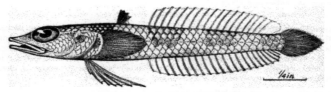

Osopsaron verecundum (Jordan & Snyder).

BLENNIIDÆ.

77. Eulophias tanneri Smith.

Eulophias tanneri Smith, Bull. U. S. Fish Comm., 1901, p. 91; Suruga Bay.

No. 49798, U. S. Nat. Mus., from station 3715, off Ose Point, Suruga Bay, 64 to 65 fathoms.

Eulophias tanneri Smith.

ZOARCIDÆ.

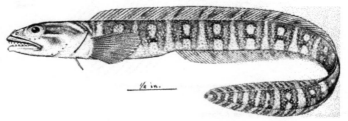

Lycenchelys poecilimon Jordan & Fowler.

78. Lycenchelys pœcilimon Jordan & Fowler.

Lycenchelys poecilimon Jordan & Fowler, Proc. U. S. Nat. Mus., XXV, 1902, p. 748; Matsushima Bay.

Station 3768, Matsushima Bay; station 3769, Matsushima Bay.

79. Bothrocara zesta Jordan & Fowler.

Bothrocara zesta Jordan & Fowler, Proc. U. S. Nat. Mus., xxv, 1903, p. 749; Sagami Bay.

Station 3086, Sagami Bay, 110 to 175 fathoms.

Bothrocara zesta Jordan & Fowler.

BROTULIDÆ.

80. Porogadus guntheri Jordan & Fowler.

Porogadus guntheri Jordan & Fowler, Proc. U. S. Nat. Mus., xxv, 1903, p. 762; Sagami Bay.

Station 3086, Sagami Bay, 110 to 175 fathoms.

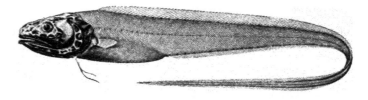

Porogadus guntheri Jordan & Fowler.

81. Watasea sivicola Jordan & Snyder. (Pl. 3, fig. 3.)

Watasea sivicola Jordan & Snyder, Proc. U. S. Nat. Mus., xxiii, 1901, p. 765; pl. xxxvii; op. cit., xxv, 1903, p. 759; Misaki.

Station 3738, Suruga Bay, 167 fathoms (one specimen ; station 3695, Sagami Bay (two specimens).

These specimens are very close to the type of *Watasea sivicola* taken at Misaki. They differ in the shorter head, smaller mouth, greater distance from the snout to the vent, and in having three or more black blotches on the dorsal fin besides dark shading on the body. None of the three specimens is as large as the type and the latter is faded through long exposure to the sunlight in the museum room of the Imperial University. We have little doubt that the present specimens represent the young of *Watasea sivicola*.

Watasea macrops (Günther) from Fiji has much in common with *Watasea sivicola*.

GADIDÆ.

82. Boreogadus saida (Lepechin).

Stations 3768 and 3769 in Matsushima Bay, off Kinkwazan Island; many young specimens.

83. Physiculus japonicus Hilgendorf.

Station 3708, off Ose Point, Suruga Bay, 65 to 125 fathoms; one small specimen.

ATELEOPODIDÆ.

84. Ateleopus japonicus Bleeker.

Station 3717, off Ose Point, Suruga Bay, 65 to 125 fathoms.

MACROURIDÆ.[a]

By David Starr Jordan and Charles Henry Gilbert.

Body elongate, tapering into a very long compressed tail, which ends in a point; scales moderate, often keeled or spinous, sometimes smooth. Suborbital bones enlarged, usually cavernous. Teeth villiform or cardiform, in bands or single series, on jaws only; tip of lower jaw usually with a barbel; premaxillary protractile. Dorsal fins 2, the first short and high, its second ray usually stiff and spine-like, the others branched; the second dorsal very long, usually of very low feeble rays, continued to end of tail; anal fin similar to second dorsal, but usually much higher; no caudal fin; ventrals small, subjugular, each usually of about 8 rays. Branchiostegals 6 or 7. Lateral line present. Gills 3.5 or 4, usually but not always with a slit behind fourth. Gillrakers small; gill-membranes free or narrowly united to isthmus, usually more or less connected; pseudobranchiæ wanting or rudimentary; pyloric cæca numerous; air-bladder present. Hypercoracoid usually without foramen.

Genera 18; species about 50, chiefly of the northern seas, all in deep water. They differ from the cod-fishes chiefly in the elongate and degenerate condition of the posterior part of the body. Dr. Gill succinctly defines the group as "Gadoidea with an elongated tail tapering backward and destitute of a caudal fin, postpectoral anus, enlarged suborbital bones, inferior mouth, subbrachial ventrals, a distinct anterior dorsal, and a long second dorsal and anal converging on end of tail."

I. BATHYGADINÆ. First branchial arch free, without fold of membrane across it; mouth large; second dorsal well developed.
  a. Gills 4, a slit behind the fourth; no elevated anterior lobe to anal fin.
   b. Coracoid foramen entirely within the hypercoracoid, as in blennoid fishes.
    c. Barbel well developed..GADOMUS
    cc. Barbel none..REGANIA
   b. Coracoid foramen between the hypercoracoid and hypocoracoid; skull papery; gill membranes jet black.
                                                                        MELANOBRANCHUS
II. MACROURINÆ. First branchial arch with a fold of membrane across its terminal portion; gills 4; a slit
      behind the fourth; foramen, so far as known, between the hypercoracoid and hypo-
      coracoid; chin with a barbel, which is rarely minute or absent.
   c. Teeth not all in villiform bands; those of the lower jaw in a single series; mouth rather large, more
      or less lateral.
    d. Upper jaw without a villiform band behind the anterior teeth; inner teeth, if present, chiefly in
       one series.
     e. Dorsal fins widely separated, interspace greater than base of first; dorsal spine serrate, scales
        nearly smooth, the ridges not spiniferous...DOLLOA
    dd. Upper jaw with a villiform band behind outer series of enlarged teeth; scales cycloid, smoothish.
     f. Dorsal spine serrate; dorsal fins not widely separate...............................CHALINURA
     ff. Dorsal spine smooth; dorsal fins well separated; pectoral fin elongate...........................ABYSSICOLA
   cc. Teeth in villiform bands above and below, the outer like the rest and scarcely enlarged or separated;
       lower band sometimes reduced laterally to a single series.
     g. Mouth wide, with considerable lateral cleft.
      h. Dorsal spine finely barbed; bones of skull rather firm, dorsals moderately separated ...CORYPHÆNOIDES
      hh. Dorsal spine entirely smooth; bones of skull very thin and papery; dorsals well separated;
         barbel small or absent...HYMENOCEPHALUS
     gg. Mouth small, inferior, with little lateral cleft; suborbital ridge usually prominent.
      i. Scales spinous, very rough.
       j. Scales distinct, regularly imbricated.
        k. Ventral rays 7 to 10.
         l. Dorsal spine serrate; snout short.....................................MACROURUS
         ll. Dorsal spine entire; snout produced, sturgeon-like ....CELORHYNCHUS
        kk. Ventral rays 13 to 15............... .. .. NEZUMIA
       jj. Scales indistinct, scarcely imbricated, the whole body rough-villous; dorsal spine smooth.
                                                                        TRACHONURUS

[a] In the account of this family all the Macrouridæ known from Japan are treated.

GADOMUS Regan.

Gadomus Regan, Ann. Mag. Nat. Hist., 1903, p. 459 (*longifilis*).

Head large, fleshy, without prominent ridges, spiny armature or external depressions; nape elevated, hump-like. Snout broad, obtuse, not produced; mouth terminal, very large, with small villiform teeth or none; suborbital ridge very low, not joined to the angle of the preoperculum. Maxillary entirely received within a groove under the prefrontal and suborbital bones, its tips narrowed and blade-like; premaxillaries protractile downward, separated anteriorly, rib-shaped, compressed vertically, very broad and without true teeth; provided posteriorly with a short flange, which is received under the maxillary; mandible received within intermaxillary bones, without true teeth, but with minute asperities, similar to those in the upper jaw; vomer and palatines toothless. Barbel well developed. No pseudobranchiæ. Gillrakers numerous, moderate, lanceolate, with minute denticulations along their inner edge. Branchiostegal membrane free from isthmus, deeply cleft. Branchiostegals 7, very stiff. Gill-opening very wide; gills 4, a slit behind last gill; anterior gill-arch free. Operculum with a blunt, spine-like prominence at its angle. A round foramen, as usual in fishes, entirely within the hypercoracoid. Ventrals below pectorals, many-rayed, anterior rays produced; dorsal consisting for the most part of branched rays, higher than anal, the first dorsal low, without differentiated spine. Scales cycloid, unarmed; lateral line strongly arched over pectoral. Deep seas.

This genus, with *Bathygadus*, differs from *Macrourus* and its allies in the structure of both the first and last gill arches. It is, perhaps, the most primitive of the family, and as such is nearest allied to the *Gadidæ*.

According to Regan, this genus differs from all other *Macrouridæ* in the presence of a foramen within the hypercoracoid bone as in ordinary fishes, not between the coracoids as in gadoid fishes, and from *Bathygadus* in having a slit behind the last gill, as usual in *Macrouridæ*. The mental barbel is well developed. In the related genus *Melanobranchus* Regan, the foramen is between the coracoids; the hypercoracoid being imperforate, there is no barbel, and a slit is present behind the last gill.

(*Gadus*, cod; ὦμος, shoulder.)

85. Gadomus colletti Jordan & Gilbert, new species.

Type 332 mm. long; from station 3721, Suruga Bay, 207 to 250 fathoms. No. 50930, U. S. Nat. Mus. First dorsal II, 10; ventrals, 9; pectorals, 21. Gillrakers 4 or 5 · 19. Branchiostegals, 7. Head 5.4 in total length; depth 7.33.

This species is related to *Gadomus multifilis, longifilis*, and *melanopterus*, the head narrow and comparatively firm, mucous canals not excessively developed, mental barbel very long. Interorbital width much less than orbital diameter or length of snout and contained 6 times in length of head. Horizontal diameter of eye a trifle less. Length of snout 3.57 in head; snout not blunt at tip. Vertical width of suborbital beneath middle of orbit 5.6 in head. Occipital crest long, the distance from its posterior end to dorsal contained 3.8 times in its distance from tip to snout. Mouth moderately oblique, the mandible everywhere included, maxillary not reaching vertical from hinder edge of orbit, its length half that of head. Teeth excessively minute, crowded, forming a wide band in premaxillaries, a much narrower band in mandibles. The premaxillary band increases in width laterally to end of second third of its length, its width there equaling one-third the orbital diameter. Individual teeth are scarcely to be made out, and constitute a fine shagreen-like surface. Barbel very long, two-thirds the length of head. Preopercle rather narrowly rounded, width at angle slightly increased, about two-fifths orbital diameter. Opercle firm, without evident ridges or spines.

The gill-membranes form a rather wide free fold across the isthmus, to which they are not joined. No trace of pseudobranchiæ can be detected. Gills 4 in number, the large slit behind the fourth arch equaling orbital diameter. Gillrakers very slender and comparatively short, the longest one-third the orbital diameter. A very deep pit marks the usual pseudobranchial area, more developed than in related species. Hypercoracoid with a well-marked foramen, as in blennioid fishes.

Second dorsal spine, second pectoral and outer ventral ray enlarged and greatly elongated; dorsal ray 3.66 times in total length; pectoral ray 3 times; ventral ray 5.5 times. Base of pectoral fin in advance of insertions of first dorsal and ventral which are vertically opposite; first dorsal spine represented by a small nodule, concealed at base of second; succeeding rays forked in their distal third, the longest articulated ray three-fifths length of head; no interval between dorsals; longest ray of the second dorsal two-fifths length of head; origin of anal fin vertically below tenth ray of second dorsal base of outer ventral ray midway between origin of anal and front of eye; vent more anteriorly placed than in related species, its distance from front of anal equaling two-thirds orbital diameter. Scales unarmed, with very fine concentric striae.   The lateral line runs posteriorly a little below middle of sides, and rises anteriorly by a gently concave curve, differing strikingly in this from most other species of the genus, in which the lateral line describes a strong convex curve above the pectorals. Scales covering top and sides of head, and present in a series along mandibular ramus; absent on opercular and gular membranes.   About 9 to 11 scales in a series between the straight portion of lateral line and base of dorsal.

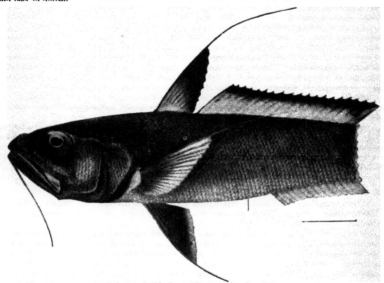

Gadomus colletti Jordan & Gilbert, new species. Type.

Color in spirits, light brownish on back and sides, silvery on cheeks and under side of head, and corresponding portions of trunk; breast and under side of head coarsely specked with brown. Upper portion of opercular membrane blackish. Snout dusky. First dorsal fin blackish throughout, including filamentous ray. Second dorsal with a black margin, fin otherwise light in anterior portion, bright blue posteriorly. Anal whitish, black near end of tail. Pectorals and ventrals blackish or dusky, the filaments and tips of rays whitish; lower lip dusky, but anterior mouth parts, including membranes, otherwise whitish; roof and posterior parts of mouth blackish; gill cavity black, save a wide margin on inner side of opercular and gill membranes, which becomes abruptly whitish and contrasts strongly; peritoneum jet black.

One specimen obtained by the *Albatross* in Suruga Bay.

(Named for Prof. Robert Collett, of the University of Christiana.)

REGANIA Jordan.

Regania Jordan, new genus (*nipponica*).

This genus differs from *Gadomus* in the absence of the mental barbel. The hypercoracoid is perforate as in *Gadomus*. There is also, as in *Gadomus* and *Melanobranchus*, a slit behind the last gill. (Named for C. Tate Regan, of the British Museum.)

86. Regania nipponica Jordan & Gilbert, new species.

Type No. 50931, U. S. Nat. Mus. 590 mm. (tail slightly injured), from station 3721, Suruga Bay, depth 207 to 250 fathoms.

First dorsal II. 10; ventrals 9; pectorals 17; gillrakers 5 - 16; branchiostegals 7. Head 5.12 in total length; depth 7.66.

Head intermediate in width and texture between *Gadomus multifilis* and *Bathygadus cottoides*, somewhat nearer former in appearance, but the barbel wholly lacking, as in latter; form elongate, head tapering regularly to a rather sharp snout; mucous canals large, covered by thin membrane, which is supported by thin, long septa; interorbital width exceeds length of snout, much exceeds diameter of large eye, and is contained 3.43 times in length of head; length of snout 3.71; longitudinal diameter of eye 4.2, the vertical diameter five-sevenths the horizontal; nostrils unusually large, anterior vertically elliptical, but little more than half the height of posterior, which is about one-fourth diameter of orbit; occipital crest short, the difference from its posterior end to origin of dorsal contained 2.25 times in its distance from tip of snout. Posterior margin of orbit a trifle in advance of middle of length of head.

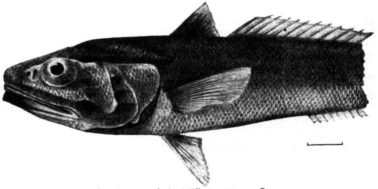

Regania nipponica Jordan & Gilbert, new species. Type.

Mouth comparatively little oblique, maxillary reaching a little beyond vertical from posterior edge of orbit, its length 1.87 in that of head. Mandible wholly included; a slight symphyseal knob: no barbel. Teeth fine, crowded, but individually visible without aid of lens, growing much smaller toward angle of mouth, distinctly arrow-shaped, as usual in the allies of *Bathygadus*; premaxillary band fully twice the width of mandibular band, its greatest width about one-seventh the orbital diameter; preopercle widened at angle, its greatest width about equaling vertical height of suborbital below middle of eye, about one-eighth the length of head; opercle firm, without evident ridges or spines. The gill membranes are injured anteriorly on middle line of throat, so the width of the free fold can not be determined; 4 full gills present, length of fourth gill slit half the orbital diameter. Gillrakers long and slender, the longest slightly more than half the orbital diameter. No trace of pseudobranchiae present. Suprabranchial very deep; origin of dorsal behind insertion of pectorals, which are vertically over ventrals; first dorsal and pectoral rays injured at tip, but the second dorsal spine and upper pectoral ray are not thickened, and were probably not greatly produced; rays of first dorsal forked; a very short interval between dorsal fins; rays of second dorsal much higher than anal rays; anal origin vertically below eleventh ray of second dorsal; vent in advance of anal a distance equaling one-third the orbital diameter; base of outer ventral ray midway between vent and front of pupil; outer ventral ray slightly produced, reaching two-thirds the distance to vent.

Scales large, somewhat more closely adherent than usual, still present over considerable areas of body, covering entire head, including mandible and gular region, but not the branchiostegals; scales unarmed, finely striated; lateral line runs posteriorly considerably below middle of sides, and rises anteriorly in a low convex curve above pectorals; chord of curve equals length of head; 9 or 10 scales in an oblique line upward and backward from posterior (submedian) portion of lateral line to base of dorsal, 21 scales upward and backward from first anal ray to base of dorsal.

Color of head and body nearly uniform light grayish, with some silvery luster, breast and belly not darker, opercular and gill membranes not black on outer surfaces; symphyseal portion of mandible, and corresponding portions of lower lip, dusky; inside of mouth and the gill cavity purplish black; posterior margins of opercular and gill membranes irregularly lighter or whitish; fins dusky, dorsal and anal blue-black posteriorly.

Known only from the type.

MELANOBRANCHUS Regan.

Melanobranchus Regan, Ann. & Mag. Nat. Hist., 1903, p. 459 (*melanobranchus*).

This genus is closely allied to *Bathygadus*, differing in the presence of a slit behind the last gill and in the absence of barbel. As in *Bathygadus*, and macrourid fishes generally, the coracoid foramen lies between the hypercoracoid and hypocoracoid instead of within the substance of the first-named bone. Skull extremely cavernous. Scales weak, caducous. Dorsal fin feeble, the second spine not serrate. Gill membranes black. Deep seas.

(μέλας, black; βραγχός, gill.)

87. Melanobranchus antrodes Jordan & Gilbert, new species. (Pl. 4, fig. 1.)

Type, 265 mm. long (tail slightly injured), from station 3696, Sagami Bay, 501–749 fathoms: No. 50932, U. S. Nat. Mus.

First dorsal II, 8; ventrals 9; pectoral 14; gillrakers 6 + 20; branchiostegals 7. Head 4.66 in total length. Depth 6.5.

Very closely related to *Melanobranchus boweri*, from vicinity of Hawaiian Islands, differing in the lighter color of anterior parts, the slightly firmer consistency of the bones of head, differing proportions of opercle and preopercle, greater development of upper opercular ridge, and the somewhat smaller scales.

Head very wide, with wide mucous canals and fragile crests: membranes covering the canals stronger than usual in this genus, and intact in all the specimens; interorbital width much longer than snout or eye, one-third length of head; longitudinal diameter of orbit one-third longer than vertical diameter, two-ninths the length of head; snout 3.4 in head, its length equaling its width opposite anterior nostrils; posterior border of orbit in middle of length of head; mouth terminal, oblique, mandible everywhere included, maxillary everywhere reaching vertical from hinder margin of orbit, its length contained 1.86 times in head; no trace of mandibular barbel; teeth minute, equal, with narrowly arrow-shaped tips, in a broad premaxillary, and a narrow mandibular band; preopercle rather narrowly rounded, greatest width of its posterior expanded portion, at angle, equaling one-eighth length of head; vertical width of suborbital below middle of orbit 5.33 in head; distance from hinder margin of orbit to preopercular angle equals less than half length of head; anterior margin of nape is slightly nearer tip of snout than front of dorsal; exposed portion of opercle much less in proportion to opercular width than in *Melanobranchus boweri*. Opercle with two diverging ridges ending in weak spines. Above upper ridge is a third much lower ridge, which ends in from one to three very weak spines, nearly as long as the one beneath them.

Gill-membranes moderately joined, free from isthmus. Four full gills, with a narrow slit behind fourth arch; outer gillrakers very long and slender, two-thirds the diameter of orbit (shorter in an older specimen). A few unmistakable free pseudobranchial filaments are present, these most abundant in the largest specimen. Coracoids thin and papery, the foramen lying between the hypercoracoid and the hypocoracoid.

Origin of dorsal is a little in advance of pectorals. All the rays of the vertical fins seem to be slender, unbranched. The dorsal rays are injured in the type, but in two cotypes the second ray is filamentous, reaching base of eleventh or twelfth ray of second dorsal, and contained 1.71 times in head. Dorsals immediately contiguous, rays of second dorsal much higher than anal rays. Upper pectoral ray filamentous, reaching as far as base of seventh anal ray, its length four-fifths that of head; it is probable that this ray was longer in life; outer ventral ray also very slender and filamentous, reaching the tenth anal ray and equaling length of head.

Scales very small, thin, caducuous, fallen in most of our specimens. As usual in the genus, they are unarmed, marked with extremely fine concentric striæ. The lateral line runs posteriorly along middle of sides, rising anteriorly in a wide, low arch, the chord of which nearly equals length of head; entire head covered with scales (except gill and gular membranes); smaller than in *M. bowersi*, be·ng little larger than the scales on sides.

Light brownish on head and body, breast and belly little darker except in young specimens, where the belly is blue-black and the breast brown; head light brown like body; opercles blackish in young. Rows of very small brown spots on top and sides of head, and along the rami of mandibles in some specimens; mouth and gill cavities and peritoneum black; fins dusky.

(ἀντροδής, full of cavities, from the spongy head.)

Station 3696, Sagami Bay, 501 to 749 fathoms; station 3711, Suruga Bay, 500 to 677 fathoms; station 3736, Suruga Bay, 480 to 599 fathoms.

DOLLOA Jordan.

Moseleya Goode & Bean, Oceanic Ichthyology, p. 417, 1896 (*longifilis*) (name preoccupied).
Dolloa Jordan, American Naturalist, XXXIV, 1900, p. 897 (*longifilis*).

Mouth rather large, upper teeth in one or two series; dorsal fins well separated, spine weakly serrate, scales feebly ridged, nearly or quite smooth; otherwise essentially as in *Chalinura*.

(Named for Louis Dollo of the Museum of Brussels.)

88. Dolloa longifilis Günther.

Coryphænoides longifilis Günther, Ann. & Mag. Nat. Hist., XXV, p. 439, 1877; south of Tokyo.
Macrurus longifil': Günther, Deep Sea Fishes of the Challenger, p. 151, pl. XXV, 1887; Coast of Japan, south of Tokyo, in 565 fathoms. One specimen 28 inches long.
Moseleya longifilis Goode & Bean, Oceanic ichthyology, p. 417, 1896, after Günther.

This species is known from the description and figure published by Günther.

(*Longus*, long; *filum*, thread.)

ABYSSICOLA Goode & Bean.

Abyssicola Goode & Bean, Oceanic Ichthyology, 1896, 417 (*macrochir*).

Upper teeth in villiform bands; lower in one series. Snout produced, four-angled; interorbital space flat and wide. Mouth wide, lateral. Pectoral fin very long, its base in line with front of dorsal and base of ventral. Dorsal fins well separated; dorsal spine smooth; scales smooth; barbel small.

Coast of Japan.

(*Abyssicola*, living in the abyss.)

89. Abyssicola macrochir Günther.

Macrurus macrochir Günther, Ann. Mag. Nat. Hist., 1877, XX, p. 435; Hyalonema ground; off Enoshima, in Sagami Bay. Günther, Deep Sea Fishes, Challenger, XXII, 1887, p. 148, pl. XXIX, fig. B, Enoshima.
Abyssicola macrochir Goode & Bean, Oceanic Ichthyology, 1895, p. 417, fig. 348; after Günther.

Of this species, well figured and described by Dr. Günther, three large specimens were dredged in Sagami Bay, by the *Albatross*, near the original locality.

(μακρός, long; χείρ, hand, from the very long pectorals, which are broadened at tip and not filamentous.)

CHALINURA Goode & Bean.

Chalinura Goode & Bean, Bull. Mus. Comp. Zoöl., x, No. 5,198, 1883 (*simula*).
Chalinurus Günther, Challenger Report, XXII, 124, 144, 1887; change in spelling.

Scales cycloid, fluted longitudinally, with slight, radiating striæ; snout long, broad, truncate, not much produced; mouth lateral, subterminal, very large; head without prominent ridges, except the subocular ones and those on snout; suborbital ridge not reaching angle of preopercle; teeth in upper jaw in villiform band, with an outer series much enlarged, those of lower jaw uniserial, large; no teeth on vomer or palatines; small pseudobranchiæ present; gillrakers spiny, strong, depressible, in double series on anterior arch; ventrals below pectorals; chin with a barbel; dorsal spine serrate; soft dorsal much lower than anal; species numerous.

This genus is allied to *Macrourus*, differing in dentition.

(χάλινος, a strap or thong; οὐρά, tail.)

90. Chalinura liocephala (Günther).

Macrurus liocephala Günther, Deep Sea Fishes, Challenger, p. 145, pl. xxxviii, fig. a, 1887; near Yokahama in 1,875 fathoms; middle Pacific in 2,050 fathoms.

This species is known to us solely from Dr. Günther's figure and description.

(λεῖος, smooth; κεφαλή, head.)

CORYPHÆNOIDES Gunner.

Coryphænoides Gunner, Trondhj. Selsk. Skrift., iii, 50, 1765 (*emp. steis*).
Branchiostegus Rafinesque, Analyse de la Nature, 1810, 86 (substitute for *Coryphænoides*).

Snout short, obtuse, high, obliquely truncated, soft to the touch, except its bony center; mouth broad, terminal, its cleft lateral; head without prominent ridges, membrane bones of side of head rather soft, but not papery; teeth villiform in both jaws, those in outer series of upper jaws somewhat enlarged. Scales spinous, second or elongate dorsal ray finely serrated in front, the serræ sometimes scarcely appreciable. Lower jaw with a barbel at tip. Deep sea. Close to *Macrourus*, differing in the larger terminal mouth.

(*Coryphæna*; εῖδος, resemblance.)

a. Dorsal spine filamentous at tip.
   b. Basal half of dorsal spine smooth; ventrals filamentous.
      c. Scales small, 10 to 12 series above lateral line; pectorals not filamentous........................ ....*altipinnis*, 91
      cc. Scales moderate, about 7 series above lateral line; dorsal spine almost entire, its spinules scarcely
         appreciable; pectoral filamentous; head 5.3 in length; scales each with about 13 to 17 ridges......................*awæ*, 92
   bb. Basal half of dorsal spine with about 6 spinules; head 5.75 in length; scales each with about 12 ridges...*marginatus*, 93
   aa. Dorsal spine not filamentous at tip.
      d. Scales large, about 5 series above lateral line, dorsal spine with its spinules evenly developed;
         head 5.5 in length..*garmani*, 94
      dd. Scales small, about 10 above lateral line; dorsal spine finely serrulate, except at base and tip;
         ventral not filamentous ....*misakius*, 95

91. Coryphænoides altipinnis Günther.

Coryphænoides altipinnis Günther, Ann. & Mag. Nat. Hist. 1877, xx, p. 439; south of Tokyo in 1,879 fathoms; off Japan in 565 fathoms.
Macrurus altipinnis Günther, Deep Sea Fishes, Challenger, 1887, p. 138, pl. xxxix, fig. A, three specimens.

This well-marked species is known to us from Dr. Günther's description and figure.

(*Altus*, high; *pinna*, fin.)

92. Coryphænoides awæ Jordan & Gilbert, new species.

Type No. 8547, L. S. Jr. Univ. Coll., 620 mm. long, from off Nanaura in Awa at the entrance of the bay of Tokyo; presented by the Imperial University.

First dorsal ii, 9; ventrals 7 and 8 respectively; pectorals 21. Head 5.3 in total length, equaling depth. Seven scales in a vertical series between lateral line and median line between dorsals.

Head and body high and compressed, greatest depth of head at posterior end of occipital crest equaling length of head anterior to upper end of preopercle; snout short and gibbous, its outline everywhere convex, its extreme tip formed by a small protruding tubercle; supraocular region depressed, longitudinal profile strongly concave, nearly flat transversely; head everywhere firm, mucus canals comparatively little developed, sides of head nearly vertical, suborbital ridge evident but very low, ending on middle of cheek; posterior nostril very large, anterior small, roundish.

Eye 4.33 in head; interorbital width 5; length of snout 3.4; posterior border of orbit in middle of length of head; preopercle not produced at angle; posterior margin vertical and convex.

Mouth but little overpassed by snout, the axial projection of which does not exceed one-half diameter of pupil; mouth nearly horizontal, comparatively short, the maxillary not passing vertical from middle of orbit; width of mouth at angle slightly exceeding length of cleft; mandibular barbel short, four-fifths diameter of pupil; teeth slender, in moderate villiform or cardiform bands in each jaw, bands tapering laterally, but not to a single series; outer teeth not at all enlarged. Gill-opening wide, membranes anteriorly united with isthmus, apparently without free fold along posterior margin. Anterior gill-slit contracted to little more than a pore just below angle of arch, its length less than half diameter of pupil. Posterior gill-slit half diameter of eye. Pseudobranchiæ wanting.

First dorsal inserted well behind pectorals and ventrals, which are in the same vertical; second spine strong, mostly smooth, with a very few distant and very weak spinules distally. These are scarcely perceptible even under the lens. Its tip is produced into a slender filament, which projects beyond first branched ray for half the length of latter; length of spine very slightly exceeds length of head; second dorsal very low and indistinct anteriorly, where the short rays are entirely disconnected and lie concealed beneath the scales; they were detected as far forward as a point distant from the first dorsal half the length of base of latter; by dissection, they could probably be traced still farther forward; anal origin below origin of second dorsal, as above described; anal opening immediately in front of first ray; ventrals wide, with rounded posterior margin, outer ray slender and filamentous, reaching vent, its length half that of head; pectorals broad, reaching vertical from the anal origin, their length three-fifths that of head.

Scales on body large, with numerous very minute spines arranged in 13 to 17 subparallel series; scales on opercles similar, but on the top and sides of head greatly reduced in size, the spines longer and with less definite arrangement. The naked area includes the gill-membranes, a narrow strip along lower portion of interopercle, all of mandible except a narrow band of scales along middle of its proximal portion, and a narrow strip along lower margin of suborbital.

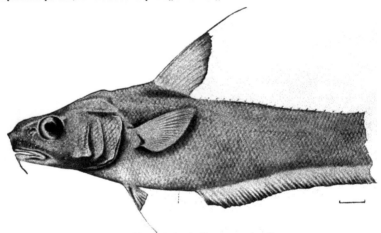

Coryphaenoides acra Jordan & Gilbert, new species. Type.

Color a uniform light grayish, probably from long exposure of specimen to light; anal with black margin, and, together with the dorsal, becoming wholly black posteriorly. The pectorals, the ventrals, and opercular membrane seem to have been bluish in life; buccal cavity light, gill cavity dark purplish.

Known only from the type specimen.

C. acra is closely related to *C. marginatus* Steindachner and Döderlein, but the important differences given in the following table prevent us from regarding the two as identical.

C. acra.	*C. marginatus.*
Head 5.3 in total length.	Head ca. 5.75.
Depth 5.3.	Depth ca. 7.5.
Eye 4.33 in head.	Eye 3.6 in head.
Snout 3.4.	Snout 3.83.
Width of head 1.83.	Width of head 2.25.
Scales with 13 to 17 minute spinous ridges.	Scales with about 12 spinous rows.
Vent distant from snout 1.6 times length of head.	Vent distant from snout twice length of head.
Basal portion of second dorsal spine smooth.	Basal portion of second dorsal spine with about 6 spinelets.
Naked area of head including lips, all but median strip of mandible, lower half of interopercle, and lower margin of suborbital.	Naked area of head confined to lips and lower margin of mandible.

93. Coryphænoides marginatus Steindachner & Döderlein.

Coryphænoides marginatus Steindachner & Döderlein, Fische Japans, IV, 284, 1887: Tokyo.

This species, evidently close to *Coryphænoides asra*, is known to us only from the original description.

(*Marginatus*, margined.)

94. Coryphænoides garmani Jordan & Gilbert, new species.

Type No. 50633, U. S. N. M., 292 mm. long, taken at station 3695, Sagami Bay, Cotype No. 8548, L. S. Jr. Univ. Museum.

First dorsal II, 10; ventrals 8; pectorals 20 or 21; 5 scales between lateral line and origin of second dorsal. Head 5.5 in total length (tail a little injured); depth 6.5.

Head and body compressed, interorbital area transversely and longitudinally convex, not depressed. Cheeks vertical; suborbital area slightly tumid, without definite ridge. Snout short, depressed, its antero-lateral profiles, seen from above, meeting to form a definite, slightly obtuse angle at tip; lower outline of snout very oblique, its length four-sevenths the ocular diameter, axial projection of snout one-third ocular diameter. Length of snout 3.8 in head; longitudinal diameter of orbit 3.25; interorbital width 3.6. Middle of length of head midway between posterior edge of orbit and pupil. Mouth

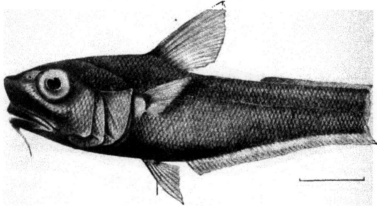

Coryphænoides garmani Jordan & Gilbert, new species.

with moderate lateral cleft, the maxillary reaching a point behind pupil, but in advance of posterior edge of orbit, its length 2.25 in head. Premaxillary teeth minute, in a broad band, outer series slightly enlarged. Mandibular teeth like inner premaxillary teeth, in a much narrower band, but not tapering laterally to a single series; all the teeth arrow-shaped. Length of barbel 3.66 times in head; preopercular angles little produced, posterior margin oblique, slightly incurved. Gill-membranes joined anteriorly and adnate to isthmus, their posterior margin narrowly free. Outer gill-cleft rather wide, contained 4.33 in length of head. Second dorsal spine slender, weak, not filamentous, its length 1.33 in head; spinules very small and numerous, appressed, equally developed throughout length of spine, except at base; soft rays of first dorsal forked; second dorsal very low, its rays distinct, the interval between the dorsal 2 to 2.5 times base of first; origin of anal under last rays of first dorsal, its distance from base of middle ventral rays equaling length of snout. The vent occupies a scaleless depression between and behind bases of inner ventral rays. The outer ventral ray develops a short filament at tip which reaches base of fifth anal ray, its length 2.3 in head. Pectoral equal to length of snout and eye. Scales rather large, adherent, those on middle of sides with 30 to 40 short, triangular, appressed prickles arranged quincuncially, somewhat reduced in size on snout, occiput, and nape, the spines not essentially modified; the scaly areas include top and sides of head, entire snout, suborbital region, and mandibles; no naked pit on breast, or between bases of ventral fins, in advance of anus.

Color, light greyish or brownish above, silvery on middle of sides; mandibles, branchial, and gular membranes, breast, and belly, and an area along anterior portion of anal black or blue-black; gular membrane coarsely vermiculated with brown; upper lip black; mouth cavity light, except the extreme posterior part of roof, which is black; gill cavity black; lower portion of shoulder girdle and isthmus silvery; hyoid arch light, the narrow posterior margin of gill and opercular membranes white; ventral portion of lining of abdominal cavity bright silvery; axil of pectorals black; fins dusky.

This species is closely allied to *Coryphænoides ctenomelas*, from the Hawaiian Islands, differing in the much smaller outer premaxillary teeth, which are scarcely enlarged in this species, and in spination of scales, as well as in many small details.

(Named for Samuel Garman.)

From station 3895, Sagami Bay, 110 to 259 fathoms; station 3697, Sagami Bay, 120 to 265 fathoms; station 3698, Sagami Bay, 153 fathoms; station 3737, Suruga Bay, 161 to 167 fathoms.

95. Coryphænoides misakius Jordan & Gilbert, new species.

Type 340 mm. long, from Misaki; collectors, Jordan & Snyder: No. 8107 L. S. Jr. Univ.

First dorsal II, 11 (II, 10 to 11, 12); ventrals 8; pectorals 21 (19 to 21); 10 scales in a vertical series between lateral line and origin of second dorsal; head 5 in total length (tail slightly injured); depth 6.5.

Coryphænoides misakius Jordan & Gilbert, new species.

In form and general appearance, including the character of the scales and lateral line, and the presence of a ventral pit, this species strongly resembles *Malacocephalus lævis* and *hawaiiensis*, to which it may have real affinity, notwithstanding the pluriserial dentition and the serrated dorsal spine.

Head compressed, with subvertical cheeks; crests very thin and papery, but membranes thick; top of head convex in all directions; snout short and high, its lower anterior profile very oblique, the distance between its tip and the premaxillaries one-fourth length of head. Width of the very convex interorbital space is 3.43 in the head; longitudinal diameter of orbit 2.77; vertical diameter 3.5; length of snout 3.25; greatest width of snout 2.75. The middle of length of head falls slightly behind posterior edge of pupil.

Mouth rather wide, with lateral cleft, but little overpassed laterally by the suborbitals, which have a very low ridge. The maxillary reaches a vertical from a point between posterior margin of orbit and pupil, its length 2.33 in head. The snout projects axially beyond the premaxillaries for a distance equaling two-thirds diameter of orbit; premaxillary teeth in a very narrow band, not more than three or four teeth wide, tapering laterally to about 2 series; outer series slightly enlarged anteriorly; mandibular teeth in a very narrow band or irregular double series, those of inner series enlarged about as much as outer premaxillary teeth. Barbel short, about one-fourth vertical diameter of eye. Preopercular angle much produced, posterior preopercular edge very oblique, incurved.

The branchiostegal membranes form a wide free fold across isthmus. Outer gill-slit wide, about two-thirds of orbital diameter. Pseudobranchiæ present, very short. Origins of first dorsal, pectorals and ventrals in the same vertical; second dorsal spine finely serrulate throughout, except for a very short distance at base and tip; tip not filamentous and fails to reach origin of second dorsal; length of spine contained 1.4 in the head; interval between dorsals contains length of base of first dorsal 2 to 2.33 times; second dorsal is very low and inconspicuous throughout; anal rather high, its origin under last rays of first dorsal; vent well forward, between basal portions of inner ventral rays, separated by a band of scales from a round, scaleless depression, which lies between middle of bases of ventral fins; distance from anal origin to base of outer ventral ray contained 3.25 times in head; outer ventral ray not produced, 2.6 in head; pectoral 1.5. Scale small, very rough and adherent, higher than wide, covered with numerous rather long, thick-set spines arranged quincuncially; spines on head much shorter; scales covering margins of shoulder girdle perfectly smooth; entire snout, suborbitals, and mandibles scaled, the latter with several series; lateral line very conspicuous, rising anteriorly in a weak, convex curve.

General color dark brownish, breast, belly, and lower side of head blue-black; opercles and posterior part of cheeks dusky; fins all dusky, a black line along base of anterior portion of anal fin; basal portion of pectorals and axil black. Mouth whitish or yellowish anteriorly, posterior portion of roof and the branchial arches black; lining of anterior outer portion of gill cavity whitish, posterior portion of gill membranes black; lining of shoulder girdle dark brown. Peritoneum dusky.

The specimens, nine in number, are all from Sagami Bay, near Misaki, taken on long lines by Kuma Aoki; two dredged by the *Albatross* at station 3695, Sagami Bay.

(*Misaki*, red point, a headland at the mouth of the bay of Sagami, famous for zoological work.)

HYMENOCEPHALUS Giglioli.

Hymenocephalus Giglioli, Pelagos, Genoa, 228, 1884 (*italicus*).
Mystaconurus Günther, Deep Sea Fishes, Challenger, 1887, p. 124 (*longibarbis*).

This genus is closely allied to *Coryphænoides*, differing in the smooth dorsal spine and membranaceous skull. First dorsal broad, placed far forward over base of pectoral; second dorsal and anal origins nearly opposite, and separated by a considerable space from the vertical from end of first dorsal; vent far from ventrals. Head large, naked, soft, and cavernous; snout abrupt, perpendicular, or parabolic; mouth lateral, wide. Eye very large, orbital margin forming part of profile of head. Barbel long. Pectoral rather narrow (10 to 16 rays). Scales thin, deciduous, with fine, short spines. Under parts in advance of ventral wholly or partly naked. Small fishes, remarkable for the papery structure of the bones of the head. (ὑμήν, membrane; κεφαλή, head.)

a. Barbel small and slender, but evident.
  b. Head 5.33 in length; depth 7.25; sides of isthmus striate .... *striatissimus*, 96
  bb. Head 6 in length; depth 7.5; dorsal spine slender, not filamentous .... *papyraceus*, 97
aa. Barbel obsolete; head 5.4 in length; depth 8; dorsal spine slender .... *lethonemus*, 98

96. Hymenocephalus striatissimus Jordan & Gilbert, new species.

Type 108 mm. long, from station 3738, Suruga Bay; depth, 167 fathoms. No. 50934, U. S. Nat. Mus.

This species seems most nearly related to *H. longibarbis* Günther, from the vicinity of the Fiji Islands, having a well-developed barbel, only 8 ventral rays, and 8 soft dorsal rays, as in that species. It differs from *longibarbis* in its shorter, higher head and blunter snout, in its much wider interorbital space, wider dentate preoperculum, and shorter barbel. The pair of lens-like spots on mid-ventral line and the minute striation of sides of isthmus are not described in *brevibarbis*, but are probably present as in all other species of the genus.

First dorsal II, 8 to II, 10; ventral 8; head 5.33 in total; depth 7.25; branchiostegals 7, the fourth much widened toward base.

In form most closely resembling *H. antræus*, from the Hawaiian Islands. Head large, subquadrate in cross section, the short, bluntly rounded snout not ending in a sharp point and not protruding beyond the premaxillaries; orbital rims greatly expanded, the median rostral ridge not greatly projecting above general level of interorbital space. As in other species the crests are all thin and papery and the membranes roofing the canals extremely delicate and easily ruptured. Eye large, circular as in *antræus*, its diameter two-fifths the length of head; interorbital width one-third of head; length of snout 4.1. The middle of length of head lies immediately in front of hinder margin of pupil.

Mouth wide, oblique, the maxillary reaching a vertical which intersects the orbit nearer to hinder margin of eye than to pupil, its length 1.8 in head. Teeth minute, all similar, in narrow bands in both jaws; barbel slender, similarly developed in the type and the two cotypes, its length half diameter of orbit. Preopercle of moderate width, the latter equaling one-third of orbital diameter; preopercular margin crinate, the margin above angle oblique and gently incurved. Gill membranes forming a free fold across the isthmus, to which they are not joined; outer gill-slits wide, the outer arch free from its angle forward for a distance equaling about one-third the length of head. No pseudobranchiæ can be detected, nor is there any trace of the deep suprabranchial pit so conspicuous in *Bathygadus*. Gill-rakers similar throughout the genus, those on inner arches short, compressed, movable, strongly spinous on inner margins, 17 or 18 on second arch. The length of the fins can not be given, as they are injured in all our specimens. The origin of the second dorsal fin can not be made out with certainty, but appears to be distant from the first dorsal 1.5 to 2 times the base of latter. The vertical from origin of anal intersects the back at a distance behind first dorsal equal to half its base length.

Color light brownish, darker along anterior portion of back, silvery on lower part of sides. A silvery streak along upper half of suborbital, a black streak occupying lower half and strongly contrasting. Corner of mouth whitish, premaxillaries black, mandible largely black but with a whitish

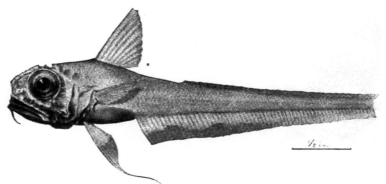

Hymenocephalus striatissimus Jordan & Gilbert, new species.

line following the lip; snout translucent; inside of mouth shining with metallic luster; posterior portion of buccal cavity black; gill cavity largely whitish, gill membranes and a blotch on opercle black. As in other species, the breast and belly are purplish, sides of isthmus and an area extending thence to above and behind the ventral fins silvery, finely cross lined with black or purple, this pattern scarcely to be made out without the aid of a lens. In the present species this striated area extends along sides of abdomen two-thirds of distance to vent.

This species is distinguished from all others by the extension of the striated area across the isthmus and anterior part of the breast, and by an extension downward in front of each of the ventral fins, forming a pair of conspicuous silvery spots, on which the fine lines are more irregularly disposed. In this species alone, the base of each ventral is wholly surrounded by the striated area. The gular membrane is crossed by two series of lines, one coarser and parallel, the other extremely fine and branching. The gill membranes are whitish on their outer face, the rays black. Base of ventrals black and very conspicuous. A lens-like body on middle of breast, and one in advance of the anus, each surrounded by a black ring, which in the posterior one includes the anal opening.

Three specimens secured, all from Suruga Bay, dredged by the *Albatross* at station 3738 in 167 fathoms, and station 3721 in 207 to 250 fathoms. A cotype is No. 8549, L. S., Jr. U.

(*Striatissimus*, most striated).

97. Hymenocephalus papyraceus Jordan & Gilbert, new species.

Type 147 mm. long, from station 3697, in Sagami Bay, depth 120 to 265 fathoms; No. 50935, U. S. Nat. Mus.

Differing from *H. lethonemus* and from all the Hawaiian species in having a distinct mandibular barbel. It also differs strikingly in the much more robust head and body.

First dorsal II, 9; ventral 11; head 6 in total length; depth 7.5; branchiostegals, 7.

Body heavy anteriorly, tapering very rapidly behind first dorsal; head very deep, with a high, median crest, which extends well above upper line of orbits; cheeks vertical, suborbitals hardly at all swollen; crests on head very high, thin, and papery, bridged over by excessively delicate membrane; interorbital width 2.6 in head; longitudinal diameter of orbit 4; length of snout 3.4; middle of length of head between hinder edge of orbit and pupil; snout massive, wide, and bluntly rounded, with a very short projecting point at tip, which extends beyond premaxillaries a distance less than half diameter of pupil.

Mouth oblique, nearly terminal, wide, the maxillary reaching a vertical from hinder margin of pupil; its length 1.9 in that of head; teeth minute, equal, in narrow bands in both jaws, the mandibular band the narrower; barbel short, but evident, less than half diameter of pupil; preopercle

Hymenocephalus papyraceus Jordan & Gilbert, new species.   Type.

greatly widened; its width an angle two-ninths that of head; its margin coarsely crenate or dentate; gill membranes widely joined anteriorly free from isthmus; first dorsal spine very slender and smooth, apparently not filamentous; the interval between dorsals equals twice the length of base of first; origin of anal is behind vertical from middle of interspace, immediately preceded by anal opening; outer ventral ray barely reaches origin of anal; its length 1.6 in head. No scales are preserved. From the size of the scale pouches there must have been two or three scales between the middle of sides and origin of second dorsal.

Color brownish anteriorly on dorsal region, posteriorly becoming light gray, very coarsely specked with black; corner of mouth whitish, mouth otherwise black; buccal cavity shining silvery, with a black cross streak behind premaxillaries and a longitudinal streak in front of tongue; posterior part of buccal cavity black; lining of gill cover anteriorly whitish, posteriorly black; gular membrane and sides of isthmus, including area above and behind base of ventrals, silvery, finely striated with purplish black, as in other species; breast and abdomen purplish black. A minute lens-like body on middle of breast and one immediately before the vent, the two connected by a black line along median line of peritoneum.

One specimen obtained in Sagami Bay.

(παπύρος, paper.)

98. Hymenocephalus lethonemus Jordan & Gilbert, new species.

Type 142 mm. long, from station 3697, in Sagami Bay, Japan, depth 120 to 265 fathoms; No. 50936, U. S. Nat. Mus. Closely allied to *H. antræus* and *H. striatulus* from vicinity of the Hawaiian Islands, but well distinguished from other species by proportions and number of ventral rays. First dorsal II, 10 or II, 11; ventrals 11. Head 5.4 in total length; depth 8. Branchiostegals 7.

Body very slender, tapering to a very long whip-like tail. Head subquadrate in cross section, interorbital area nearly flat, sides of head vertical; ridges on top and sides of head thin and papery, the large mucous canals roofed over by delicate transparent membrane; orbital rim projects above and behind as a thin membrane-like expansion; median crest on snout and interorbital protrudes but little above general level; snout terminates anteriorly in a sharply projecting point, similar to that in *H. striatulus*, but slenderer and a little longer, projecting axially beyond the mouth for a distance slightly less than half orbital diameter; interorbital width 3.33 times in head; longitudinal diameter of orbit, 3; length of snout, 3.5; middle of length of head midway between hinder margin of orbit and pupil; posterior line of occiput midway between origin of dorsal fin and anterior edge of nasal fossa; mouth large, its width nearly equal to greatest width of head, the maxillary reaching vertical from hinder margin of orbit, its length 1.87 in that of head; teeth minute, of uniform size, in very narrow bands in each jaw, the bands interrupted mesially; no trace of mandibular barbel.

Hymenocephalus lethonemus Jordan & Gilbert, new species.

Preopercle dilated to form a wide membranous expansion at its angle, the greatest width of which is two-fifths the diameter of orbit, the margin crenulate; above the angle the preopercular margin is straight and slightly oblique; opercle very thin and flexible and strengthened by two diverging ridges, the one directed downward and backward terminating in a sharp concealed spine; gill membranes united anteriorly, forming a wholly free fold across isthmus; vertical limb of outer gill arch wholly adnate, the horizontal limb free from the angle forward for a distance equaling one-third the length of head; inner series of gillrakers compressed, movable, their inner margins strongly spinous, 18 on horizontal limb of second arch; second dorsal spine slender and weak, entirely smooth, its tip sometimes a little produced beyond rays. The interval between the dorsals is 1.9 times base of first. A vertical line from origin of anal intersects this interspace slightly in advance of its middle; the anus immediately precedes anal fin; outer ventral ray filamentous in the type reaching base of tenth anal ray, its length equaling that of head; pectorals slender, 1.83 in head.

The scales are largely lost in our specimens; one from the middle of the breast is cycloid, unarmed; 3 or 4 above and behind insertion of pectoral fin bear a few weak spines, with difficulty to be detected. There were apparently 3 rows of scales between the middle of the sides and origin of second dorsal fin.

As in other species of the genus, there are two small lens-like bodies on the mid-ventral line, one immediately in front of anal opening, the other on middle of breast.

Color light-grayish or brownish above, silvery on lower half of sides; sides of body and tail marked by rather coarse brown specks, which form a definite line along middle of sides. A strip along bases of dorsal and anal fins is devoid of specks, a sharp line between this and the spotted area often marked by a series of coarser dots; dots on sides of tail arranged partly in oblique lines which seem to correspond with those separating the myotomes; a dark vertebral line between dorsals and a dark spot at base of each ray of second dorsal and anal. The first dorsal occupies a conspicuous colorless area of the back which is margined with dusky; a dark blotch behind occiput; tip of snout faintly dusky. Upper jaw black in its anterior two-thirds, the lower jaw in its anterior third; posterior portions of both jaws white. Gular membrane silvery, with a fine network of dark lines, most of which are short and transverse, arranged in a right and a left series, those of each series united by a lengthwise commissure. Lateral portions of isthmus, concealed by the gill flap, bright silvery, crossed by fine parallel hair lines of brownish purple. This striated area extends backward above and behind base of ventrals; breast and belly purplish-black. The roof of mouth shines with a silvery luster. Mouth and gill cavity light in color, region about entrance to gullet blackish. Peritoneum silvery, overlaid with brown. Fins translucent, unmarked.

One specimen was found in the Tokyo Market by Jordan & Snyder; the others were dredged by the *Albatross* at station 3697, Sagami Bay, 120 to 265 fathoms, and station 3707, Sagami Bay, 63 to 75 fathoms.

(λυθός, forgetting: ρῆνα, thread.)

9. MACROURUS Bloch.

Macrourus Bloch, Ichth., v, 152, 1787 (*rupestris-berglax*).
Macruroplus Bleeker, Versl. Med. Akad. Wet. Amsterd., VIII, 1874, 369 (*serratus*).
Macrurus Günther, Cat., IV, 392, 1862; corrected spelling.

Snout broadly conical, high, projecting beyond mouth; mouth moderate, its cleft horizontal; U-shaped, entirely inferior; teeth in both jaws in villiform bands, those of outer series not enlarged, head with roughened bony ridges, one of which, on the suborbital and preorbital, is more or less prominent; eyes very large; scales imbricate, very rough, keeled. Dorsal spine long, serrated on the anterior edge. Deep water fishes. (μακρός, long; ούρά, tail, hence correctly written *Macrurus*, but *Macrourus* is the original name as given by Bloch.)

a. Snout conical, overhanging the mouth, much longer than eye; scales each with a compact mass of spinules;
　6 scales between lateral line and spinous dorsal; dorsal spine and first ventral ray filamentous............*nasutus,* 99
aa. Snout short, little longer than eye, not much overhanging the mouth; 7 or 8 scales above lateral line; each scale with 5 ridges; dorsal spine and first ventral ray produced..............................*asper,* 100

99. Macrourus nasutus (Günther).

Coryphaenoides nasutus Günther, Ann. & Mag. Nat. Hist., XX, 1877, p. 440.
Macrurus nasutus Günther, Deep Sea Fishes, Challenger, 1887, p. 132, pl. XXX, fig. B; south of Tokyo in 565 fathoms; off Enoshima in 345 fathoms.

Short snout with a terminal and a pair of lateral spinous tubercles; mandibular rami are completely invested in scales; interorbital width slightly exceeding two-thirds the vertical diameter of orbit, which is contained 4.33 to 4.5 times in length of head. The first dorsal contains 10 rays, the second filamentous and well serrate, the pectoral 21 to 23, and the ventrals 9 (not 10, as stated by Günther), the first ray filamentous. Scales everywhere covered with closely packed long, slender spines, posterior spines well overlapping margin of scale; while not in longitudinal rows, these spines are definitely arranged in quincunx order, as can be readily seen on viewing them from different angles.

The alcoholic specimens are all very light olive-gray, fins blue or blue black; lips black; mouth, gill-cavity, and peritoneum, black; opercular membrane more or less black.

Fifteen specimens, each about 15 inches long, were dredged by the *Albatross* in Sagami Bay, near Enoshima, at station 3886, 501 to 749 fathoms; station 3889, 400 to 726 fathoms; station 3711, 500 to 677 fathoms.

Another, presented by Kuma Aoki, came from off the shore of Izu.

(*Nasutus*, long nosed.)

100. Macrourus asper (Günther).

Coryphænoides asper Günther, Ann. & Mag. Nat. Hist., 1877, xx, p. 440; south of Japan in 1,875 fathoms.
Macrurus asper Günther, Deep Sea Fishes, Challenger, p. 137, pl. xxxvi. fig. A. 1887.

This species is known from the description and figure published by Günther.
(Asper, rough.)

10. CŒLORHYNCHUS Giorna.

Cœlorhynchus Giorna, Mém. Ac. Sci. Turin, xvi, 178, 1803 ("Cœlorhynche la ville").
Krohnius Cocco, Lettera al Sig. Augusto Krohn, Pesci del Mare di Messina, 1, 1844 (filamentosus; larva).
Paramacrurus Bleeker, Versl. Med. Ak. Wetensk. Amsterd. 1874, 103 (australis).
Oxymacrurus Bleeker, Versl. Med. Ak. Wetensk. Amsterd. 1874, 103 (japonicus).

This genus agrees with *Macrourus* in all essential respects, except that the small mouth is wholly below the long-pointed, sturgeon-like snout. Dorsal spine smooth in typical species, those with serrate spine having been separated under the preoccupied generic name *Cœlocephalus* Gilbert & Cramer, which is replaced by *Malacocephalus* Berg. Species numerous.

(κοῖλος, hollow; ρὕγχος, snout.)

a. Scales smaller, 5 to 6 series between dorsal spine and lateral line.
  b. Scales each with 3 to 5 strong radiating ridges; snout nearly twice as long as eye, which is 4.5 in head; interspace between dorsals one-third more than base of first...................................*japonicus*.
  bb. Scales each with 2 to 5 strong parallel ridges; snout nearly twice eye; interspace between dorsals equal to base of first dorsal....................................*parallelus*.
  bbb. Scales each with many (16 to 19) nearly parallel ridges, which are rather weak; snout short; interspace between dorsals longer than base of first....................................*kishinouyei*.
aa. Scales large, 3 to 4 between lateral line and first dorsal; snout short and broad, suggesting a duck's bill; scales with 6 to 9 weakly divergent ridges.
  b. Interspace between dorsals 1.4 times base of first; snout 2.4 in head; eye 1.5 in snout....................................*anatirostris*.
  bb. Interspace between dorsals much less; snout longer; eye smaller than in preceding....................................*tokiensis*.

101. Cœlorhynchus japonicus (Schlegel). *Hige.*

Macrurus japonicus Schlegel, Fauna Japonica, 1846, p. 256, pl. cxii. fig. 2. Nagasaki; Günther, Deep-sea Fishes, Challenger, 1887, p. 127, pl. xxix, fig. c, Enoshima, in 345 fathoms; Steindachner & Döderlein, Fische Japans, iv, 283, Tokyo.

Dorsal ii, 9 or ii, 10; ventrals 7; pectoral 18; 5 or 6 scales in a series from lateral line to base of first dorsal. Snout long, narrow, subtrihedral, tapering to a very acute point the lateral margins with very slight curve, or none, continuing in a direct line the infraorbital ridge; width of snout opposite anterior margin of orbit is contained 1.4 to 1.6 in its length; its height at front of mouth is contained 1.75 to 2 times in its length; mouth comparatively large, length of maxillary equaling diameter of eye, reaching a vertical from posterior margin of pupil; teeth rather coarser than usual in this genus, a few of the outer teeth sometimes slightly longer than the others, all typically with arrow-shaped tips; inner pair of occipital ridges converge gently backward; outer pair decurved behind eye; 7 or 8 tubercles represent the gillrakers of outer arch. The gill membranes have a well-marked free margin behind, on median line.

Second dorsal spine smooth throughout, its height contained 1.2 in that of snout. Interspace between dorsals is four-fifths the base of first; origin of second dorsal vertically above second or third anal ray; largest pectoral rays slightly exceed length of postorbital part of head; scales on back and sides marked with 3 to 5 strongly radiating, subequal, spinous ridges, posterior spines projecting beyond margin of scale; scales on breast similar, but smaller and more crowded; on upper surface of snout the scales have usually only the median crest developed, but the median series of 12 scales attached to the subjacent bony ridge contain each 6 or 7 strongly radiating ridges; on top and sides of head scales with a single median crest predominate, but the temporal region and the upper part of the head cheeks and opercles contain also many scales with 3 to 5 ridges; lower surface of head, except gular and branchiostegal membranes, densely covered with smaller scales, bearing each a median spinous crest; nail-like process at tip of snout longitudinally rugose and spinous; no depression or naked pit in front of ventral fins.

Color is dark brownish, lighter in snout, fins all blackish, including ventral filament; lining of buccal cavity blue-black; gill cavity and peritoneum brownish black.

Seven specimens, 33 to 43 cm. in length, were obtained at Misaki. A stuffed skin was obtained at Nagasaki. The species is not infrequent in the markets of southern Japan.

This species, with others in the family, is known as *hige*, a word meaning moustache, perhaps from the large interorbital ridge. Possibly the word *grenadier* has the same origin in its relation to these fishes.

102. Cœlorhynchus kishinouyei Jordan & Snyder.

Cœlorhynchus kishinouyei Jordan & Snyder, Proc. U. S. Nat. Mus. xxiii, 1901, 376, pl. xx, Misaki. Coll. K. Otaki. No. 49396, U. S. Nat. Mus.

This species is known from the original type now in the U. S. National Museum, obtained by Professor Otaki at Misaki. No second example has been seen.

(Named for Dr. Kamakichi Kishinouye, head of the Imperial Fisheries Bureau of Japan.)

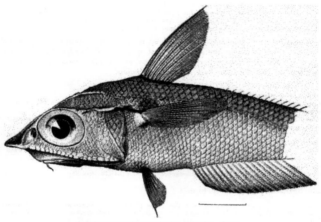

Cœlorhynchus kishinouyei Jordan & Snyder.   Type.

103. Cœlorhynchus parallelus (Günther).

Macrurus parallelus Günther, Ann. & Mag. Nat. Hist. 1877, vol. xx, p. 439, Enoshima, in 345 fathoms (south of Japan); Günther, Deep Sea Fishes, Challenger, 1887, p. 126, New Zealand, Kermadec Islands, Enoshima (two species included, pl. xxix, fig. A, representing a distinct species, *C. kermadecus*).

A specimen 245 mm. long from Misaki was presented by the Imperial University.

Scutes on head, between prominent ridges and scales on back in front of dorsal fin, are furnished each with a prominent, high, thin median crest, which bears 1 to 3 spinous points, the posterior the longer; no lateral ridges on these scales and no approach to the radiating or stellate ridges described and figured by Günther (Deep-sea Fishes, Challenger, p. 126, pl. xxix, Fig. A) from specimens from the Kermadec Islands. The scales of back and sides bear the prominent median ridge and 1 to 3 pairs of smaller parallel lateral ridges, as described by Günther. Each of these ridges consists of overlapping spines increasing in length posteriorly, the last of the median series alone projecting noticeably beyond margin of scale. The scales on breast resemble those on head and bear no lateral series of spines. Entire lower side of head, except gular and branchiostegal membranes, thickly covered with small irregular nonimbricated scales, each bearing one or several short spines; this spinous area includes lower portion of preopercular lobe and mandible. The inner pair of ridges on hind head converge backward to end of occiput, then gently diverge; branchiostegal membranes without free posterior fold where they cross the isthmus; first dorsal fin inserted over axil of pectorals; distance from inner base of ventrals to anal opening 1.4 the distance between ventrals and margin of gill-membrane on median line; interval between dorsals equals base of anterior fin; first dorsal contains 10 rays, ventral 7, pectoral 17; 5 series of scales above lateral line, and 5 cross series between dorsal fins.

The typical locality for *C. parallelus* is Japan. In view of the differences in spination of scales between these and specimens reported on by Günther from the Kermadec Islands and New Zealand, it may well prove that the latter belong to a distinct species. In the figure of a specimen from the Kermadec Islands (Deep-sea Fishes of the Challenger, pl. xxix, fig. A) the interval between the dorsals is represented as twice the length of the base of the anterior fin and as crossed by 10 transverse series of scales. If correctly shown there would seem to be no doubt of the specific distinctness of this form from the Japanese type. It may be named *Colorhynchus kermadecus* (new specific name). *C. parallelus* has been recorded from the Gulf of Manaar by Alcock (Ann. & Mag. Nat. Hist., 1889, p. 391) on the basis of "two young specimens, in bad preservation, believed to be this species." Ample verification of this record is to be desired.

104. Cœlorhynchus anatirostris Jordan & Gilbert, new species.

Type, 40 cm. long, from Misaki, No. 8550, L. S. Jr. U.

First dorsal ii. 9; ventrals, 7; pectorals, 18. Scales, 3.5 to 4 in a series between lateral line and middle of dorsal base. Snout comparatively short, wide and depressed, shaped like a duck's bill, diameter of orbit two-thirds its length. Its lateral outlines more strongly curved than suborbital ridge along sides of head. Least interorbital width nearly three-fourths diameter of eye; length of snout (without the terminal spine, which is broken) is 2.4 times length of head. Barbel short, about half length of eye. Mouth small, its width at angle of gape contained 2.4 times in width of head on

Cœlorhynchus anatirostris Jordan & Gilbert, new species.

same line, length of maxillary 4 in length of head. The mouth occupies space between the verticals from anterior nostril and posterior edge of pupil; teeth minute, not arrow-shaped, equal in size, forming rather wide bands; gill-membranes form a wide, free fold posteriorly on median line; spinous ridges on head are strongly marked; inner pair converge toward middle of their length, then gently diverge. At beginning of last third of their length a large scale with single elevated spinous crest occupies middle of space between them; this scale markedly different from those around it, but similar to those surmounting the ridges.

First dorsal fin inserted a little behind axil of pectorals; second spine smooth throughout, its length equaling that of head behind front of pupil, its tip reaching third ray of second dorsal. The interspace between the dorsals is 1.4 base of first and slightly exceeds diameter of eye; origin of second dorsal is over the third anal ray, very low and inconspicuous; longest pectoral ray equals length of snout and slightly exceeds other ventral ray; scales large, there being but 3 or 3.5 in series between lateral line and middle of first dorsal; those along middle of back and sides have 6 to 9 spinous ridges which diverge weakly, but obviously; ridges subequal, each consisting of very oblique, overlapping spines, the last spines projecting beyond margin of scales; scales on breast similar, but smaller, with shorter spines; those on top and sides of head similar to those on body, but anteriorly on snout their size is greatly decreased and the radiating ridges reduced to 2 or 3; lower side of head is wholly naked, except 3 or 4 very small scales in a series below preopercular angle. No depressed area or naked pit on breast.

Color light brownish above, under parts lighter, fins blackish; buccal, branchial, and abdominal cavities lined with black, scarcely apparent externally.

Only one specimen seen; it was taken on long lines at Misaki by Kuma Aoki and presented to us by Professor Mitsukuri.

C. anatirostris is closely related to *C. tokiensis* (Steindachner & Döderlein), but differs in its shorter snout, much larger eye, shorter maxillary, sharper keels on occiput, and much wider interspace between dorsals. There is but one instead of two series of enlarged scales along the lateral ridge of the head, and the median series on the snout are not square.

105. Cœlorhynchus tokiensis (Steindachner & Döderlein).

Macrurus tokiensis, Steindachner & Döderlein, Fische Japans, IV, 283, 1887; Tokyo.

This species is known to us from the original description only.

NEZUMIA Jordan.

Nezumia Jordan, new genus (condylura).

This genus is close to *Macrourus*, differing from that in the many-rayed ventral fins, the number of rays being 13 to 15 instead of 7 to 10, as in all other *Macrouridæ*.

(*Nezumi*, a rat, in Japanese.)

106. Nezumia condylura Jordan & Gilbert, new species.   (Pl. 4, fig. 2.)

Type 195 mm. long, from station 3721, Suruga Bay, Japan, depth 207 to 250 fathoms, No. 50937, U. S. Nat. Mus. Cotypes, No. 8551, L. S. Jr. University.

First dorsal ii, 11 (or ii, 10); ventrals 14 or 15, rarely 13; pectorals 21 to 23. Scales in a series between lateral line and anterior portion of second dorsal, 10 or 11. Head 6.3 in total length (the tip of tail broken); depth 7.4.

The profile is strongly angulated at origin of first dorsal, base of fin very oblique; head compressed, its sides vertical, its width less than two-thirds its greatest depth. Upper profile evenly and gently convex from tip of snout to origin of dorsal, without depression above the orbits. No conspicuous ridges on top of head; snout short, depressed, its tip about on level with middle of eye, its lower profile descending very obliquely to front of premaxillaries, from which it is separated by a distance equaling width of mouth, and a very little less than length of snout. The snout terminates in a median and a pair of lateral tubercles bearing rosettes of short spines; a strong ridge extends from its tip alongside of snout and suborbital to below posterior part of eye, dividing a scaly upper portion from the naked under side of head; eye large, subcircular, its horizontal diameter half longer than width of convex interorbital space, one-third length of head. Length of snout 3.5 in head.

Mouth is narrow, its width scarcely two-thirds that of head opposite angle of mouth, greatly overpassed by snout both anteriorly and laterally; length of maxillary one-third that of head, its tip reaching a vertical from behind middle of pupil; mandibular teeth small, cardiform, none of them enlarged, anteriorly in a wide band which tapers to a point near angle of mouth; premaxillary teeth similar, in a narrower band, outer series not enlarged. Mandibular barbel robust, more than half length of mandible, three-fifths diameter of orbit. Preopercular margin nearly vertical, evenly rounded at angle, the angle not at all produced backward. Branchiostegal membranes narrowly joined to isthmus, with a slight free border posteriorly. Branchiostegal rays 7. Pseudobranchiæ present, covered by the lining membrane of opercles; anterior gill slit much contracted, its width slightly less than half diameter of orbit; posterior gill slit two-thirds the anterior; base of pectorals and ventrals and origin of first dorsal approximately in same vertical; second dorsal spine slender, produced beyond soft rays but not filamentous, furnished with 10 to 14 long slender retrorse barbs, evenly spaced from near base to tip; length of spine about equal to that of head; space between dorsals equals length of base of anterior dorsal, which equals diameter of eye; second dorsal low throughout; a series of smooth cycloid scales along its base anteriorly on each side; first anal ray under or slightly behind last ray of first dorsal; pectorals five-eighths length of head; outer ventral ray filamentous, reaching base of the seventh anal ray; distance from base of ventrals to first anal ray equals length of snout and half eye; vent about equidistant between ventral base and anal and passed by all the ventral rays.

Scales all small, those on back and sides of body mostly with 7 or 8 nearly parallel series of recumbent spines, the posterior in each series projecting beyond margin of scale; series inclined a little obliquely downward, those on successive scales often falling in same straight line; on the nape the scales are smaller and furnished with much longer spines less distinctly arranged in series; on the snout and interorbital region the spines are thicker and stand erect or nearly so; on the breast the spines are suberect and arranged in more numerous series on each scale.

Color light brown, under side of head and abdomen purplish black, this color extending to above base of pectorals and backward to include the first 10 or 12 anal rays; the mouth, the branchial and abdominal cavities lined with black membrane. Basal half of first dorsal whitish, distal half blackish. Ventrals blackish.

The known specimens were dredged by the *Albatross* in Sagami and Suruga bays. Stations 3695, 110 to 250 fathoms; 3697, 120 to 265 fathoms, both in Sagami Bay; station 3721, Suruga Bay, 207 to 250 fathoms.

(*Condylura*, the star-nosed mole.)

TRACHONURUS Günther.

Trachonurus Günther, Challenger Report, Deep-Sea Fishes, XXII, 124, 1887 (*villosus*).

Scales not imbricated, separated by furrows, and densely covered with sharp spinules, so that the animal seems villous to the touch; dorsal spine smooth; dorsal much lower than anal; teeth in both jaws in villiform bands; snout obtuse, the mouth subinferior; suborbital ridge little developed. This genus is distinguished from *Celorhynchus* by the indistinct squamation. ($\tau\rho\alpha\chi\acute{v}\varsigma$, rough; $o\mathring{v}\rho\acute{\alpha}$, tail.)

107. Trachonurus villosus (Günther).

Coryphaenoides villosus Günther, Ann. & Mag. Nat. Hist. 1877, XX, p. 441; south of Tokyo in 345 fathoms; south of the Philippines in 500 fathoms.

Macrurus villosus Günther, Deep Sea Fishes of Challenger, 1887, p. 142, pl. XXXVI, f. B.

This species is known to us from Günther's account only.

(*Villosus*, hairy.)

PLEURONECTIDÆ.

108. Atheresthes evermanni Jordan & Starks, new species. (Pl. 5, fig 1.)

Head 3.3 in length; depth 3; D. 114; A. 94; scales 109; upper eye 4.75 in head; snout from upper eye 4; maxillary 1.9; pectoral of eyed side 2.1; of blind side 3.25; upper lobe of caudal 1.75.

Profile of snout on same curve with that behind eye; very slightly depressed above eye; eyes scarcely reaching to upper profile, the lower one the more anterior; interorbital appearing rather flat and moderately broad, the bone, however, narrow and convex, its width less than half diameter of pupil; nostrils close together, the posterior of eyed side in a broad short tube, anterior in a narrower, longer tube; anterior nostril of blind side with a long flap nearly a third as long as upper eye, broadening toward its tip and becoming conspicuously opaque white; snout with many pores scattered among the irregularly placed scales; mouth reaching to, or very slightly past, the vertical from posterior margin of lower eye; teeth long and slender and with lance-shaped points, in a single row on lower jaw, their length unequal; a double row of smaller teeth on side of upper jaw, the outer row the smaller; they grow larger anteriorly, become curved inward, fang-like and some of them depressible; gillrakers rather slender, the longest a trifle less than half length of eye, their number 3+10; scales very finely ctenoid, the spinules short, fine, and numerous, only seen upon careful examination with a lens; many scales have only a few irregular spinules; others are entirely without them, appearing as if they had been rubbed off; head and body everywhere with numerous, small, cycloid supplementary scales crowded in; scales of blind side all cycloid; snout, mandible, maxillary, and interorbital with numerous small cycloid scales, those on latter extending out on eyeball to edge of iris; all fins rather closely covered with fine scales; lateral line slightly bending upward from opposite tip of pectoral. Pectoral of eyed side longer and more pointed than that of blind side; first ray of dorsal inserted above anterior margin of pupil; ventral short, scarcely reaching to front of anal. Caudal shallowly concave on posterior outline.

Color uniformly dark brown without markings.

This species differs from *Atheresthes stomias*, of the Alaskan fauna, in having only a single row of teeth on lower jaw, and the upper eye not reaching the upper profile. The scales are more strongly ctenoid and the anterior nostril bears a long flap.

The type and sole specimen is 270 cm. in length; it is from station 3772 in Matsushima Bay, and is numbered 51490, U. S. Nat. Mus.

CLEISTHENES Jordan & Starks, new genus

Cleisthenes Jordan & Starks, new genus of *Pleuronectidæ (pinctorum).*

This genus is closely allied to *Hippoglossoides*, differing in having cycloid scales everywhere in the young, and an increased number of gillrakers. The adult has a single row of ctenoid scales along anterior base of dorsal and anal, a few on snout on ridge behind interorbital space, and on opercle. The dorsal begins at the orbital rim slightly on the blind side. Eyes and color on right side.

(*Cleisthenes*, the effeminate, an Athenian noted by Aristophanes.)

109. Cleisthenes pinetorum Jordan & Starks, new species.

Head 3.66 in length; depth 2.6; D. 76; A. 56; scales 80; upper eye 4.6 in head; snout from upper eye 4.6; pectoral of eyed side 2; of blind side 2.5; ventral 3; caudal 1.4.

Dorsal outline of anterior part of body and head an even concave curve to near tip of snout, broken only by protruding upper eye. Upper eye cutting into profile, and ranging nearly vertically upward, about two-fifths of it being visible from the blind side. Tip of snout blunt and rounded; mouth rather strongly curved; maxillary reaching scarcely to middle of lower eye, not covered along middle of its length by the prefrontal as in *Hippoglossoides hamiltoni;* teeth small, acute, in a single series in each jaw, scarcely enlarged anteriorly; nostrils moderate, the anterior in a short tube which does not reach to edge of preorbital; preorbital with a blunt spine on anterior edge; eyes about equal in size, separated by a flat interspace, covered with cycloid scales; gillrakers slender, equal to half the eye in length, 8 to 10 above and 24 to 27 below the angle.

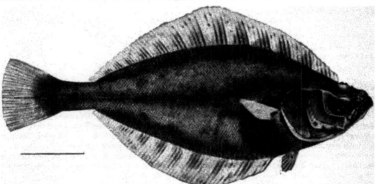

Cleisthenes pinetorum Jordan & Starks, new species.

Dorsal fin beginning slightly on blind side at edge of orbit opposite posterior margin of pupil; anal preceded by a strong spine; ventrals not reaching to anal (reaching to base of second anal ray in young); caudal evenly rounded behind; scales everywhere cycloid and with concentric rings in specimens 4 or 5 inches long.

A specimen 10 inches long has cycloid scales except a single row of ctenoid scales along base of dorsal and anal anteriorly, a few in front and behind the interorbital space, and some on opercles. The type (8.5 inches long) has only an occasional ctenoid scale along base of dorsal and anal, and the ctenoid scales on head are very sparse. A specimen 7 inches long has only a few ctenoid scales remaining on head behind interorbital space.

Color everywhere dark brown, dorsal and anal a little lighter at base of rays; membrane of caudal darker than the rays making longitudinal streaks; dorsal, anal, and caudal of blind side dark toward tips of rays.

The above description is from the type 22 cm. in length; the following is from specimens 11 cm. in length (the size of the specimen here figured):

Head, 3.5 in length; depth, 2.66; length of upper eye, 3.75 in head; snout from upper eye, 4.66; pectoral of eyed side, 2.1; pectoral of blind side, 3; ventral, 3; length of middle dorsal rays, 2.16; middle anal rays, 2.16; caudal, 1.4; interorbital space narrower and more concave than in adult; color light, with fine points scattered over a flesh-colored ground, these following more especially the outline of scales; indefinite dark brown spots scattered over body; one always present on lateral line at beginning of posterior third, one or two usually placed irregularly along lateral line anteriorly; a few following dorsal and ventral outlines of body; fins light, mottled with grayish brown.

All of our numerous specimens were dredged off Kinkwazan Island, Matsushima Bay, at stations 3769 and 3770. The type is 22 cm. in length, No. 51403, U. S. Nat. Mus.; cotypes are No. 8391, Stanford University.

110. Xystrias grigorjewi (Herzenstein).

Hippoglossus grigorjewi Herzenstein. Bull. Ac. Imp. sci., Petersb., 1890, p. 134.
Veraeper otakii Jordan & Snyder, Proc. U. S. Nat. Mus., XXIII, 1901, p. 378; Tokyo.

A large specimen, taken at station 3699, Suruga Bay, in 400 to 726 fathoms.

This species differs from others referred to *Veraeper* in having long, slender gillrakers, the scales all cycloid on blind side, and those of eyed side more finely ctenoid; spinules more even and numerous. It may be made the type of a distinct group or genus, called *Xystrias*, from the long gillrakers.

111. Pleuronichthys cornutus (Schlegel).

Two specimens taken in Suruga Bay, station 3708, off Ose Point, in 65 to 125 fathoms, and station 2356. A larva with the eyes still symmetrical perhaps belongs to this species.

112. Limanda herzensteini Jordan & Snyder.

Pleuronectes japonicus Herzenstein. not of Houttuyn. *Limanda herzensteini* Jordan & Snyder Proc. U. S. Nat. Mus., XXIII, 1901, p. 746, after Herzenstein.

One specimen dredged in Matsushima Bay, station 3768, 25 cm. in entire length, differs from Herzenstein's description in having 84 series of scales (88 pores). Others collected by Jordan & Snyder at Tsuruga, Matsushima, Hakodate, and Aomori have from 80 to 85 scales.

ALÆOPS Jordan & Starks, new genus.

Alæops Jordan & Starks, new genus of *Pleuronectidæ (plinthus).*

Allied to *Pœcilopsetta,* but with large ctenoid scales which are somewhat caducous. Eyes and color on the right side. Lateral line simple, with a broad flat-topped arch in front; mouth moderate; teeth small, in bands. Gillrakers short and sharp.

113. Alæops plinthus Jordan & Starks, new species. (Pl. 5, fig. 2.)

Head 4 in length; depth 2.4; D. 61; A. 53; scales 65 (pores); upper eye 3 in head; snout from upper eye, 4.16; maxillary 3.16; pectoral of eyed side 2.25; ventral median; caudal rays 1.

Anterior body outline strongly arched above; orbital rim of upper eye protruding beyond rest of profile; snout a little produced, blunt; anterior nostril in rather broad, short tube, which does not reach to edge of preorbital; maxillary curved, reaching to below or very slightly past anterior rim of pupil of lower eye; teeth small, in a very narrow band on eyed side, growing wider anteriorly, somewhat smaller on premaxillary. On blind side the teeth on both jaws are in moderately wide bands. Eyes equal in size, the lower slightly more anterior, separated by a narrow naked ridge; vertical limb of premaxillary short; gillrakers short and rather sharp, the longest one-half to one-third diameter of pupil, 5 – 10 in number; caudal peduncle very wide and flat, its length one-third of its width; scales large, rather finely but very evidently ctenoid on eyed side, cycloid on blind side; head on eyed side, anterior to posterior rim of pupil above, and posterior end of mandible below, without scales; lateral line turning abruptly upward at a sharp angle two-thirds the head's length behind head, and forming a conspicuous flat-topped arch, as high as half length of head; dorsal beginning slightly on blind side, a little behind middle of eye, length of first ray contained 1.1 in upper eye, longest rays near posterior end of fin, the longest 2.25 in head; pectorals equal in size; caudal broad and pointed behind; no lateral angles, the sides broadly rounded from tips of the long median rays to lateral edges of fin base.

Color pinkish slaty-brown, usually mottled with black; 2 inconspicuous semiocellated spots, one near dorsal and one near anal base a head's length anterior to base of caudal; less conspicuous dark irregular spots along side above anal and below dorsal, one below arch of lateral line; a black spot on outer rays of caudal; all fins except ventral and pectoral of blind side irregularly spotted and mottled with black. The membrane has drawn away from the scales in our specimen leaving them light at base.

The type is 155 mm. in entire length, taken at station 3708 in Suruga Bay. It is numbered 51406, U. S. Nat. Mus. Cotypes were taken at the same station, and at stations 3715, 3716, and 3717 in Suruga Bay, and at station 3725 in Owari Bay. Cotype No. 8389, Stanford University.

DEXISTES Jordan & Starks. new genus.

Dexistes Jordan & Starks, new genus of Pleuronectidæ (rikuzenius).

This genus differs from *Pseudopleuronectes* in having large scales, and large eyes narrowly separated by a high, sharp, naked interorbital ridge. Eyes and color on the right side.

114. Dexistes rikuzenius Jordan & Starks, new species. (Pl. 6, fig. 1.)

Head 3.85 in length; depth 2.75; D. 73; A. 59; scales 64 (pores); upper eye 3.1 in head; snout from upper eye 4.83; maxillary of eyed side 3.85; of blind side 3; pectoral of eyed side 2; of blind side 3; ventral 3.1; highest dorsal rays 2.5; median caudal rays 1.5.

Body moderately narrow; anterior dorsal curve slightly broken by the raised orbital rim; snout blunt; lower jaw projecting, and with a knob developed at symphysis below tip; eyes large, upper slightly the larger and placed farther back; narrowly separated by a high, sharp, naked ridge; mouth much larger on blind side; maxillary of eyed side reaching to opposite anterior edge of pupil; teeth blunt and not very even or closely set, in one moderately straight row except on blind side of lower jaw, where 3 or 4 are irregularly placed inside the row; gillrakers short and triangular, 7 on lower limb of arch, 1 developed and 2 rudimentary ones on upper limb; scales large and ctenoid on eyed side, cycloid on blind side; spinules on scales very slender, sharp, and numerous; a few scales on anterior part of interorbital where it widens on snout; upper eye with a patch of ctenoid scales, each with 2 or 3 spinules; a row of small scales running out on each fin ray; lateral line without an arch, a branch of it curves down behind eyes and around lower edge of lower eye; dorsal beginning above middle of eye; pectoral of eyed side longer and more pointed than that of blind side; ventrals equal in length, the last rays the longest; median caudal rays produced, upper edge obliquely truncate, lower slightly concave.

Color brown, with a few irregular inconspicuous dark brown spots, one on lateral line at beginning of its posterior two-fifths, one near base of caudal, one below middle of lateral line, one near top of pectoral; small ones show little color except a few brown spots, the one on lateral line the most conspicuous.

The type from which this description is taken is 22 cm. in length, and was taken at station 3774, in Matsushima Bay, in 84 fathoms. Two small cotypes were taken at station 3717, off Ose Point, Suruga Bay, in 65 to 125 fathoms.

The type is No. 51423, U. S. Nat. Mus. A cotype is No. 8388, Stanford University.

115. Araias ariommus Jordan & Starks, new genus and species. (Pl. 6, fig. 2.)

Head 3.8 in length; depth 2.6; D. 71 to 74; A. 57 to 60; scale 60; upper eye 2.8 in head; snout from upper eye 4.33; maxillary 3.75; pectoral of eyed side 1.87; of blind side 2.75; caudal 1.16.

Rim of upper orbit very slightly protruding above rest of upper profile; eyes separated by a narrow sharp ridge; anterior rim of lower scarcely or very slightly anterior to that of upper, posterior rim anterior to that of upper (to a greater degree in the type than in cotype); mouth very small, considerably larger on blind side, the maxillary reaching to just below anterior edge of orbit; teeth blunt, set in a single, very irregular row, those of lower jaw projecting around on eyed side farther than those of premaxillary; gillrakers short and triangular, 3–7 on first arch; dorsal beginning above middle of upper eye; pectoral of eyed side a little longer and not so bluntly rounded as that of blind side; caudal doubly truncate, median rays the longer; lateral line not arched, gradually curved up anteriorly; scales cycloid, with occasionally a ctenoid scale with long irregular spinules (as the spinules are easily broken, leaving no trace, it appears probable that the scales may have all been ctenoid); a few small scales running out on fin rays.

Color light pinkish brown, without definite markings; dorsal, anal, and caudal with very faint wavy cross marks.

Two specimens taken in Matsushima Bay, at stations 3770 and 3773. The type is the larger, and is 13 cm. in length. It is No. 51417, U. S. Nat. Mus. The other from station 3773 is No. 8386, Stanford University.

The new genus *Araias* is technically near *Pleuronectes*, differing in its thin, scarcely ctenoid scales and in its fragile body.

116. Clidoderma asperr'mum (Schlegel).

A larval flounder, 33 mm. in length, which we take to be the young of this species, since it agrees in fin rays, was collected in Matsushima Bay, station 3770. It is covered on both sides with small spinules of about equal size, those of blind side finer and more sparse than on eyed side; no enlarged plates, but groups of 4 or 5 spinules scattered over the eyed side probably represent them; body outline rounder and anterior curves more convex. Pectoral very short, broader than long, covered on base by a sheath of spinules, beyond which the short rays project fanlike.

Color, light cream everywhere with blended brown spots; irregular in size, position, and intensity of color.

VERÆQUA Jordan & Starks, new genus

Veræqua Jordan & Starks, new genus of *Pleuronectidæ* (*achne*).

Allied to *Microstomus* and *Limanda*.

Body rather elongate, covered with very fine cycloid scales; lateral line with a small arch in front, without accessory dorsal branch; mouth small and with about 7 large blunt teeth in a single row on blind side; eyes close together, separated by a high naked ridge which is continued backward; gillrakers very small, not numerous; no anal spine; caudal rounded; eyes and color on right side.

117. Veræqua achne Jordan & Starks, new species.   (Pl. 7, fig. 1.)

Head 4.33 in length; depth 2.87; D. 85; A. 69; scales 135; upper eye 3.16 in head; snout from upper eye 4; pectoral 2; ventral 4; highest dorsal rays 2; caudal 1.1.

Form rather slender, the outlines forming low even curves; anterior upper outline of head unbroken and continuous with body curve; mouth very small, the maxillary reaching a little past front of lower eye but scarcely to edge of pupil; 7 large and very blunt teeth, set in a single row on blind side only; eyes narrowly separated by a high naked ridge, the lower the more anterior; interorbital ridge continued backward and upward along lower margin of upper eye, forming a high, conspicuous, smooth ridge; a slight angle on lower edge where it turns upward, but no tubercles developed; nostrils close together, in short broad tubes, anterior reaching to edge of preorbital; gill-slit stopping at upper edge of pectoral; gillrakers very small, 8 on lower limb of arch. Scales very fine, everywhere cycloid; very small nonimbricated scales present on dorsal and anal nearly to tips of rays except on the brown streak behind each ray; caudal thickly covered with similar scales; scales on pectoral rays only; on base of ventral only on both rays and membrane; small imbedded scales on snout; lateral line perfectly straight and horizontal to tip of pectoral where it turns up and forms a low but conspicuous arch, the cord of its curve 3 times its height. Dorsal beginning slightly on blind side above middle of eye; low anteriorly, gradually growing higher to beginning of its last third or fourth where it reaches its greatest height; pectorals rounded, that of eyed side, in our specimen, very slightly longer than that of blind side; ventral short and rather broad, the second ray longest, making the fin pointed; caudal broadly rounded.

Color slaty brown, mottled with darker brown blended into the ground color; a brown streak behind and partly on each dorsal and anal ray; caudal uniform dark brown; pectoral with dark brown membrane.

A single specimen, the type, dredged at station 3772, Matsushima Bay, in 79 fathoms. It is 18 cm. in length, and is numbered 51447 U. S. Nat. Mus.

($\check{\alpha}\chi\nu\eta$, a whiff of foam.)

118. Microstomus kitaharæ Jordan & Starks, new species.   (Pl. 7, fig. 2.)

Head 4.25 to 4.5 in length; depth 3.5 to 3.75; D. 91 to 96; A. 75 to 83; scales 87 to 96 (pores); eye 2.83 to 3.16 in head; snout from upper eye 4.33 to 4.75; maxillary 3.75 to 4; pectoral of eyed side 1.83 to 2.33, of blind side 2.25 to 3; ventral 3.5; caudal 1.25.

Anterior upper profile evenly convex; the upper eye protruding above it; lower eye much in advance of upper, the eyes separated by a very narrow ridge; maxillary short, rather strongly curved, reaching to below anterior edge of pupil of lower eye; teeth rather blunt, in a single row, forming a continuous even cutting edge; a small bony knob developed below tip of mandible; anterior nostril of eyed side in a short broad tube; gillrakers very short, 8 on lower limb of arch; scales everywhere cycloid, the snout, maxillary, and mandible naked; lateral line conspicuous, curving up just behind tip of pectoral above upper end of gill opening, but not at all arched; dorsal beginning above posterior margin of pupil of upper eye; the longest dorsal and anal rays are at beginning of posterior fourth of body length; pectoral narrow, pointed, variable in length, the upper edge of its base distant one diameter of pupil from upper end of gill slit; ventrals reaching just to front of anal; caudal rounded or double truncate, the middle rays projecting beyond outer rays a distance slightly greater than half eye.

Color uniform brown, pectoral and caudal growing black toward tips of rays; no color on blind side except black toward end of caudal.

The type is 18 cm. in length, taken with several cotypes at station 3770, Matsushima Bay, in 42 to 45 fathoms. Other cotypes were taken near the same locality at stations 3769, 3771 (in 61 fathoms), and 3772 (in 79 fathoms); at station 3717, off Ose Point, Suruga Bay, in 65 to 125 fathoms, and station 3699, Suruga Bay, in 400 to 726 fathoms; others were collected by Jordan and Snyder in the market at Tokyo, several of which were deposited as cotypes in the Imperial University at Tokyo. Dried salted specimens were obtained in the market of Tsuruga.

The type is No. 51418, U. S. Nat. Mus. Cotypes are Nos. 8390, 8995, 8996, Stanford University.

The species is named for Mr. T. Kitahara, author of a paper on the *Scombridæ* of Japan. In the same journal of the Fisheries Bureau, 1897, the present species is figured by Mr. Otaki as *Pleuronectes cynoglossus*.

119. Pseudorhombus pentophthalmus Günther.

Head 3.33 in length without caudal; depth 2; D. 71; A. 52; pores in lateral line 68; upper eye 5 in head; snout 3.87; maxillary 2; pectoral (eyed side) 1.75; blind side 2.4; caudal 1.33.

Body broad and thin, ventral and dorsal outlines evenly curved; snout blunt, obliquely truncated, separated from anterior profile by a notch; eyes separated by a narrow sharp ridge which is continuous backward and upward above cheek; anterior edge of eyes about even, posterior edge of upper one a little more posterior than that of lower; mouth much curved, the maxillary reaching to posterior edge of lower eye; teeth sharp and curved, set in a single row on each jaw, some of them very slightly arrow-shaped at tips; on blind side teeth on premaxillary grow smaller backward and disappear opposite the middle of length of maxillary; gillrakers moderately slender and long, the longest slightly exceeding half diameter of eye, 6 - 16 in number. Dorsal beginning slightly toward blind side a little in front of anterior edge of upper eye, the first ray at notch separating the snout, anterior rays somewhat produced beyond the membrane; pectoral of eyed side longer than that of blind side; ventrals similar in size and position; caudal with the middle rays produced and with no lateral angles, the sides being broadly rounded. Scales ctenoid on eyed side, spinules short, sharp, and numerous; cycloid on blind side; scales on all fin-rays rather large, even, and ctenoid on eyed side; lateral line strongly arched anteriorly, a branch from above gill-opening running to dorsal profile above posterior edge of eye, opposite eighth ray of dorsal. The color has nearly all bleached in alcohol. There is a dark brown spot narrowly ringed with white just above arch of lateral line, and another two-thirds of the head's length behind it midway between lateral line and outline of back; traces of two similar spots on lower part of side midway between lateral line and ventral outline and slightly behind those above.

We identify this species with *Pseudorhombus pentophthalmus* described from China and recorded by Steindachner from Kobe. It has much in common with *Pseudorhombus russelli* described from Canton by Gray. It seems to differ in color and also in the larger size of the mouth. *Pseudorhombus arsius* from the Ganges, as described by Bleeker, is different from our species, and the names given by Richardson to Chinese drawings (*Platessa relafracta, Platessa balteata*) are wholly unidentifiable. *Pleuronectes chinensis* of Lacépède is probably unrecognizable. *Pleuronectes chrysopterus* of Schneider seems to be *Pseudorhombus oligolepis*.

A specimen 13 cm. in length was taken by the *Albatross* at Hakodate in 1896, having been overlooked in our earlier studies. A larger one in poor condition was taken in the Bay of Yokohama. The latter is so badly preserved that only its anterior half remains.

The genus *Pseudorhombus* is distinguished from *Paralichthys* by the presence of an accessory lateral line from the nape running forward and upward to the dorsal fin. The body is deeper than in *Paralichthys*, the mouth is smaller and the teeth are weaker. *Rhombiscus*, based by Jordan and Snyder on *Pseudorhombus cinnamomeus*, is identical with *Pseudorhombus*.

120. Engyprosopon iijimæ Jordan & Starks, new species. (Pl. 8, fig. 1.)

Head 4 in length; depth 2.33 to 2.5; D. 80 to 89; A. 69 to 72; scales 50 to 53; eye 3 in head; maxillary 3.5; pectoral of eyed side 1.2; of blind side 3; ventral 2.25; caudal equal to head. Anterior profile evenly curved, the orbits not reaching to its edge; eyes separated by a narrow sharp ridge, the lower the more anterior; mouth small, the maxillary very much curved and reaching to a little past front of orbit; teeth small and set in a single row; six very short gillrakers on lower arch of first gill. Scales finely ctenoid, the spinules on the scales slender and very numerous; blind side with cycloid scales; lateral line with a very abrupt, short, high curve, its height contained 1.83 in its chord, which is half length of head, its beginning opposite the terminal third of pectoral. Dorsal beginning in advance of eye; pectoral of eyed side long and slender, of blind side less than half as long; ventral with 6 rays, that of blind side not prolonged, its base beginning behind front of ventral of eyed side and its tip reaching further past front of anal; caudal rounded behind, its outer edges broadly rounded, scarcely angulated.

Color light brown, spotted with dark brown, ocellated spots, 3 above and 3 below lateral line, the anterior upper spot in advance of that below; 5 spots with edges more blended along body near base of dorsal, 4 similar ones along body near base of anal, these involving base of fins: one on opercle just above gill-opening; pectoral of eyed side dark brown.

Two small specimens taken in from 45 to 60 fathoms, in Suruga Bay, stations 3713 and 3714; the former, the type, 65 mm. in length, is numbered 51461, U. S. Nat. Mus.; the other is No. 8387, Stanford University.

The species differs somewhat from the type of *Engyprosopon*, but it is doubtless referable to the same genus. It is named for Dr. S. Iijima, professor of zoology in the Imperial University of Tokyo.

121. Scæops grandisquama (Schlegel). (Pl. 8, fig. 2.)

Two small specimens taken: one at station 3762, Owari Bay, in 42 fathoms, the other at station 3754, off Sune Point, Sagami Bay, in 50 fathoms.

This is a common shore fish of southern Japan.

This species is the type of a distinct genus, *Scæops* Jordan & Starks, allied to *Platophrys*, but distinguished by the large deciduous scales. In the more closely related genus *Engyprosopon*, the scales are firm, the gillrakers long and slender, and the teeth in two rows. In *Scæops*, the gillrakers are very short and triangular and the teeth in one row. *Platophrys psetturus* Bleeker, from the East Indies, is also a species of *Scæops*.

LOPHIIDÆ.

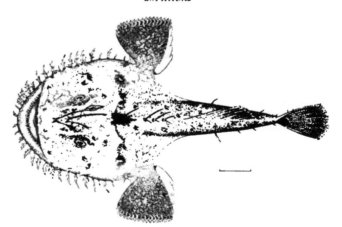

Lophius litulon Jordan.

122. Lophius litulon (Jordan).

Lophiomus litulon Jordan, Proc. U. S. Nat. Mus., XXIV, 1902, 364; Matsushima Bay.

Station 3773, Matsushima Bay, in 78 fathoms.

This species is a true *Lophius*, as Mr. C. T. Regan has already indicated, having 28 vertebræ.

SOLEIDÆ.

123. Usinosita japonica (Temminck & Schlegel).

A very young specimen taken in Suruga Bay, station 3700.

ANTENNARIIDÆ.

124. Chaunax fimbriatus Hilgendorf.

Chaunax fimbriatus Jordan, Proc. U. S. Nat. Mus., XXIV, 1902, 377; Suruga Bay.

Station 3717, off Ose Point, Suruga Bay, 65 to 125 fathoms; station 3741, Suruga Bay, in 71 fathoms.

OGCOCEPHALIDÆ.

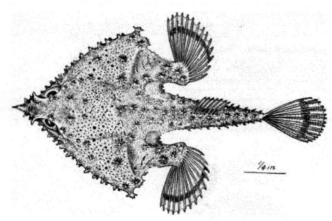

Malthopsis tiarella Jordan.

125. Malthopsis tiarella Jordan.

Malthopsis tiarella Jordan, Proc. U. S. Nat. Mus., XXIV, 1902, p. 379.

Station 3719, off Ose Point, Suruga Bay, in 70 to 100 fathoms.

LIST OF PLATES.

630

PLATE 1.

FIG. 1. CHLOROPHTHALMUS ALBATROSSIS JORDAN & STARKS.

FIG. 2. CHAULIODUS EMMELAS JORDAN & STARKS.

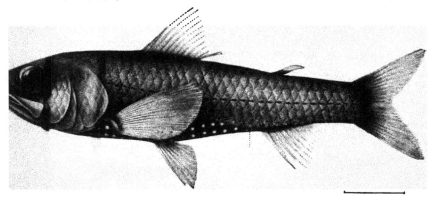

FIGS. 1, 2. NEOSCOPELUS ALCOCKI JORDAN & STARKS.

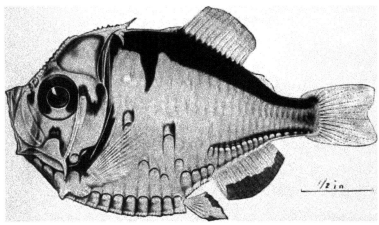

FIG. 3. POLYIPNUS STEREOPE JORDAN & STARKS.

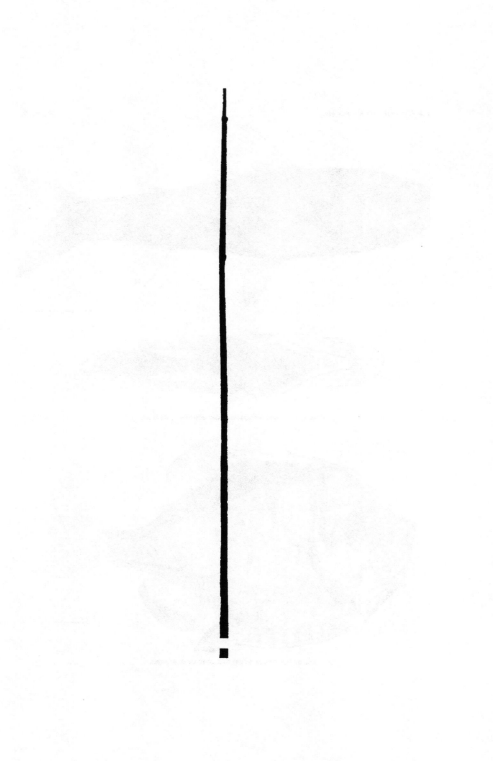

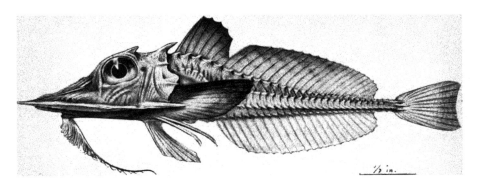

FIGS. 1, 2. PERISTEDION AMISCUS JORDAN & STARKS.

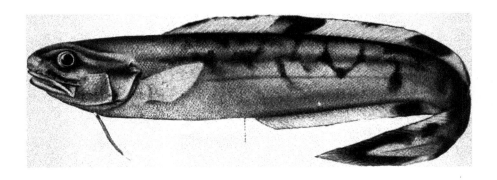

FIG. 3. WATASEA SIVICOLA JORDAN & SNYDER.

FIG. 1. MELANOBRANCHUS ANTRODES JORDAN & GILBERT.

FIG. 2. NEZUMIA CONDYLURA JORDAN & GILBERT.

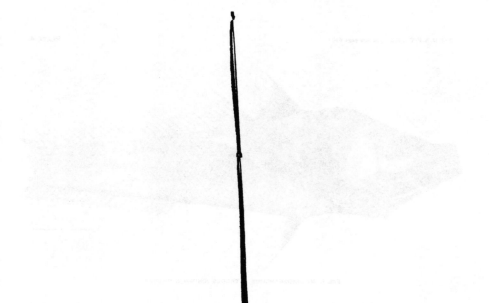

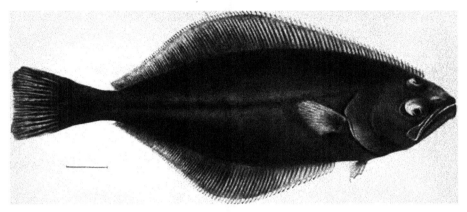

FIG. 1. ATHERESTHES EVERMANNI JORDAN & STARKS.

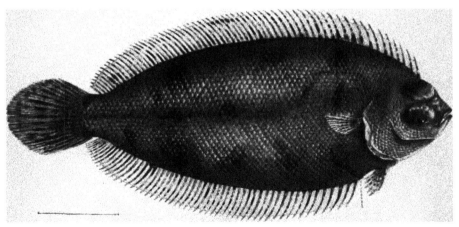

FIG. 2. ALÆOPS PLINTHUS JORDAN & STARKS.

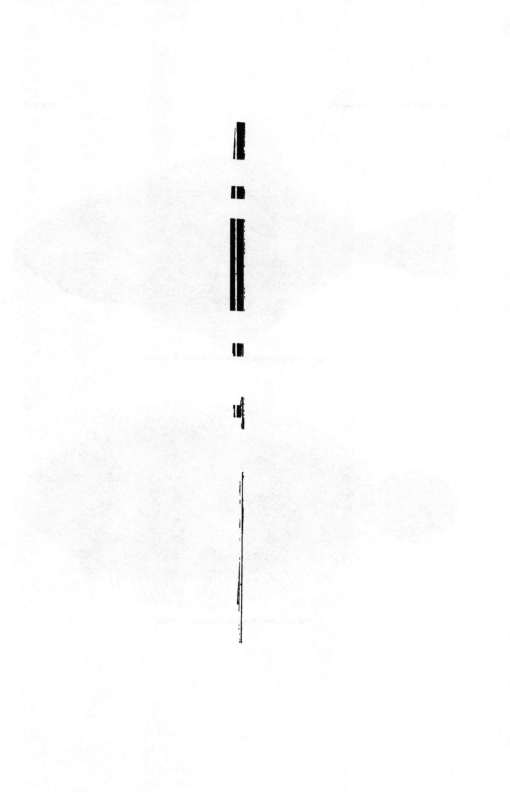

PLATE 6.

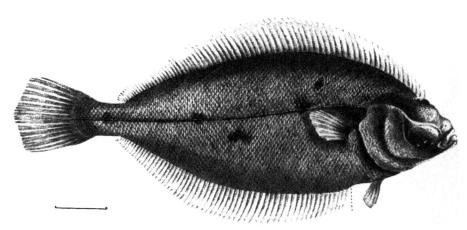

FIG. 1. DEXISTES RIKUZENIUS JORDAN & STARKS.

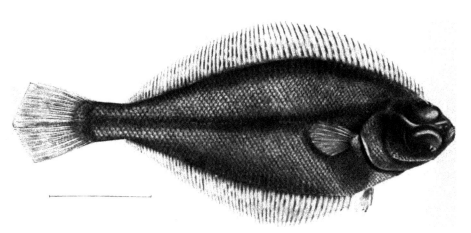

FIG. 2. ARAIAS ARIOMMUS JORDAN & STARKS.

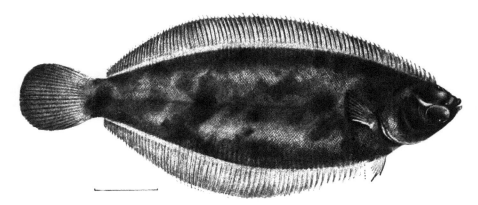

FIG. 1. VERÆGUA ACHNE JORDAN & STARKS.

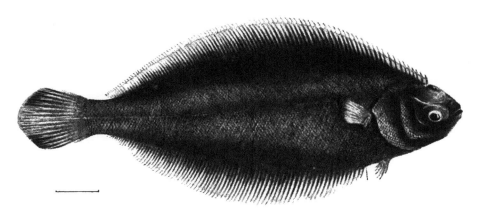

FIG. 2. MICROSTOMUS KITAHARÆ JORDAN & STARKS.

FIG. 1. ENGYPROSOPON IIJIMÆ JORDAN & STARKS.

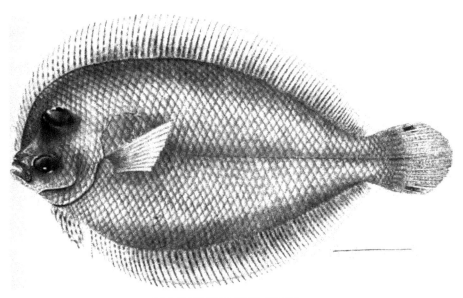

FIG. 2. SCÆOPS GRANDISGUAMA SCHLEGEL.

O

Lightning Source UK Ltd.
Milton Keynes UK
UKHW011126020119
334816UK00016B/1680/P

9 781527 650473